PE EXAM PREP

CHEMICAL ENGINEERING
PE LICENSE REVIEW

Third Edition

Dilip K. Das, MS, PE Chem. Eng. & Rajaram K. Prabhudesai, PhD, PE Chem. Eng.

BERN.
$$(Z_2 - Z_1)\frac{g}{g_c} + \frac{u_2^2 - u_1^2}{2g_c} + \frac{P_2 - P_1}{\rho} + h_L = Q + \dot{W}_s$$

FLOW IN PIPES

$$\Delta P = \left(\sum K + \frac{4Lf}{D}\right)\frac{u^2 \rho}{2g_c}$$

EQUIV. LENGTH = $\sum L_{e_i}$ + PIPE LENGTH

$$h_L = \frac{4 L_e f u^2}{2g_c}$$

$$\left(\frac{144}{62.4}\right) = 2.31$$

$$h_L = \frac{2.31 \Delta P}{S} = \frac{144 \Delta P}{\rho}$$

HEAT OF FORMATION
$$\Delta H_{RXN} = \overbrace{\Delta H_f^{PROD} - \Delta H_f^{REAC}}$$

$$\Delta H_{RXN} = \underbrace{\Delta H_c^{REAC} - \Delta H_c^{PROD}}_{\text{HEAT OF COMBUSTION}}$$

$$\Delta P^{TUBE} = \frac{2fLu^2 \rho}{g_c D} \qquad u = \frac{4\dot{Q}}{\pi D^2}$$

$$= \frac{32 f L \dot{Q}^2 \rho}{g_c (\pi)^2 D^5} \qquad \Delta P_2 = \Delta P_1 \left(\frac{D_2}{D_1}\right)^5$$

RXN ENGR.

$$F_{A0} = C_{A0} V_0 = \frac{n_{A0} V_0}{V}$$

$$\left(\frac{mol}{time}\right) = \left(\frac{mol}{vol}\right)\left(\frac{vol}{time}\right)$$

$$C_A = \frac{n_A}{V} = C_{A0}(1 - X_A)$$

$$C_B = C_{A0} X_A$$

CSTR
MOL FLOW IN: $F_{A0}(1 - X_{A0})$
MOL FLOW OUT: $F_A = F_{A0}(1 - X_A)$
MOL. DISAPP. OF A: $-n_A = -r_A V$
DESIGN EQN: $F_{A0} - F_{A0}(1 - X_A) = -r_A V$

$$\tau = \frac{1}{S} = \frac{V}{V_0} = \frac{V C_{A0}}{F_{A0}} = \frac{C_{A0} X_A}{-r_A}$$

KAPLAN) AEC EDUCATION

This publication is designed to provide accurate and authoritative information in regard to the subject matter covered. It is sold with the understanding that the publisher is not engaged in rendering legal, accounting, or other professional service. If legal advice or other expert assistance is required, the services of a competent professional person should be sought.

President: Roy Lipner
Vice President & General Manager: David Dufresne
Vice President of Product Development and Publishing: Evan M. Butterfield
Editorial Project Manager: Laurie McGuire
Director of Production: Daniel Frey
Creative Director: Lucy Jenkins

Copyright 2008 by Dearborn Financial Publishing, Inc.®

Published by Kaplan AEC Education
30 South Wacker Drive
Chicago, IL 60606-7481
(312) 836-4400
www.kaplanaecengineering.com

All rights reserved. The text of this publication, or any part thereof, may not be reproduced in any manner whatsoever without written permission in writing from the publisher.

Printed in the United States of America.

08 09 10 10 9 8 7 6 5 4 3 2 1

CONTENTS

Introduction ix

CHAPTER 1

Units and Dimensions 1
DIMENSIONS 1
SYSTEMS OF UNITS 2
UNITS FOR COMMON CHEMICAL QUANTITIES 6
CONVERSION OF UNITS 12
DIMENSIONAL FORMULAS 16
DIMENSIONAL AND DIMENSIONLESS EQUATIONS 16
DIMENSIONAL ANALYSIS 18
RECOMMENDED REFERENCES 22

CHAPTER 2

Material Balances 23
MATERIAL PROPERTIES (DEFINITIONS AND RELATIONSHIPS) 24
MASS BALANCES 28
PHASE BEHAVIOR OF SUBSTANCES 37
IDEAL GASES 40
REAL GASES (NON-IDEAL) 45
MATERIAL BALANCE CALCULATION 54
FUELS AND COMBUSTION 67
RECOMMENDED REFERENCES 76

CHAPTER 3

Energy Balance and Thermodynamics 77
THERMODYNAMIC SYSTEM 78
CONSERVATION OF ENERGY OR FIRST LAW OF THERMODYNAMICS 81
ENERGY BALANCES IN PROCESSES INVOLVING CHEMICAL REACTIONS 92
REVERSIBLE AND IRREVERSIBLE PROCESSES 103
THERMODYNAMIC PROPERTIES OF MATTER 110
POWER CYCLES AND REFRIGERATION 121
RECOMMENDED REFERENCES 129

CHAPTER 4

Fluid Mechanics 131
VISCOSITY 132
FLUID STATICS 132
FLUID MEASUREMENTS 138
PERMANENT PRESSURE LOSS 143

FLOW OF FLUIDS IN PIPES 147
PUMP CALCULATIONS 168
CONTROL VALVES 187
PARALLEL AND BRANCHED SYSTEMS 192
REPLACEMENT OF ONE PIPELINE WITH n PARALLEL PIPELINES OF EQUAL EQUIVALENT LENGTHS AS THE ORIGINAL PIPE 192
FLOW OF COMPRESSIBLE FLUIDS 193
TWO-PHASE (VAPOR/LIQUID) FLOW CONSIDERATION 201
RECOMMENDED REFERENCES 207

CHAPTER 5 — Heat Transfer 209

THERMAL CONDUCTION 210
THERMAL CONVECTION 213
THERMAL RADIATION 214
PRESSURE DROP IN EXCHANGER 251
THE EFFECTIVENESS-NTU METHOD 260
REBOILERS 262
SUBLIMATION 263
CRYSTALLIZATION 263
EXTENDED SURFACE HEAT EXCHANGER 264
HEATING AND COOLING OF LIQUID BATCHES 268
NONMETALLIC HEAT EXCHANGERS 269
HEAT PUMP 271
RECOMMENDED REFERENCES 276

CHAPTER 6 — Evaporation 279

EVAPORATION 279
RECOMMENDED REFERENCES 290

CHAPTER 7 — Filtration 291

FILTRATION EQUIPMENT 291
FILTRATION CALCULATIONS 293
RECOMMENDED REFERENCES 306

CHAPTER 8 — Membrane Separation 307

WHAT IS A MEMBRANE? 307
REVERSE OSMOSIS (RO) 308
ULTRAFILTRATION (UF) 314
ELECTRODIALYSIS (ED) 314
PERVAPORATION 314
NOMENCLATURE 315
RECOMMENDED REFERENCES 315

CHAPTER 9

Mass Transfer Fundamentals 317

MOLECULAR DIFFUSION 318

MOLECULAR DIFFUSION IN LIQUIDS 323

MOLECULAR DIFFUSION IN SOLIDS 329

CONVECTIVE MASS TRANSFER 333

TURBULENT MASS TRANSFER 336

DIMENSIONLESS NUMBERS IN MASS TRANSFER 338

INTERPHASE MASS TRANSFER 341

MASS TRANSFER IN PACKED BEDS 343

PHASE EQUILIBRIA 346

VAPOR-LIQUID EQUILIBRIA 346

ENTHALPY-COMPOSITION DIAGRAMS (BINARY SYSTEM) 365

RECOMMENDED REFERENCES 367

CHAPTER 10

Distillation 369

DISTILLATION METHODS 370

PLATE-TO-PLATE CALCULATIONS 379

BINARY SYSTEMS: McCABE-THIELE METHOD 380

FEED-PLATE LOCATION 384

COLD REFLUX 388

SIDE-STREAM WITHDRAWAL 391

USE OF OPEN STEAM 393

GRAPHICAL STAGES AT LOW CONCENTRATIONS 395

SEPARATION OF BINARY AZEOTROPIC MIXTURES 409

SHORTCUT METHODS 413

DESIGN CALCULATIONS FOR PACKED TOWERS 419

SIZING OF TRAY AND PACKED TOWERS 426

RECOMMENDED REFERENCES 432

CHAPTER 11

Absorption 433

EQUILIBRIUM SOLUBILITY CURVES 433

RECOMMENDED REFERENCES 457

CHAPTER 12

Liquid-Liquid Extraction and Leaching 459

LIQUID-LIQUID EXTRACTION 459

CONJUGATE PHASES 464

EXTRACTION EQUIPMENT 472

LEACHING 476

RECOMMENDED REFERENCES 491

CHAPTER 13 Adsorption 493

ADSORBENTS AND THEIR PHYSICAL PROPERTIES 493
TYPES OF ADSORPTION 494
EQUILIBRIUM RELATIONSHIPS IN ADSORPTION 495
ADSORPTION FROM LIQUIDS 497
STAGEWISE OPERATIONS 497
RECOMMENDED REFERENCES 510

CHAPTER 14 Psychrometry, Humidification, and Drying 511

PSYCHROMETRY AND HUMIDIFICATION 511
COOLING TOWERS 520
DRYING 525
RECOMMENDED REFERENCES 533

CHAPTER 15 Chemical Reaction Engineering 535

CHEMICAL REACTION ENGINEERING 536
INTERPRETATION OF KINETIC DATA AND THE CONSTANTS OF THE RATE EQUATION 550
REACTOR DESIGN 566
MASS AND ENERGY BALANCES 567
PRODUCT DISTRIBUTION AND TEMPERATURE 579
STOICHIOMETRIC TABLES 593
OPTIMAL TEMPERATURE PROGRESSION 597
RECOMMENDED REFERENCES 612

CHAPTER 16 Process Control 613

CONTROL SYSTEMS 613
DEFINITIONS 615
BLOCK DIAGRAMS 616
LAPLACE TRANSFORMS IN CONTROL SYSTEM ANALYSIS 617
CONTROL ACTIONS 621
PROPORTIONAL BAND 622
DEVELOPMENT OF PROCESS MODELS 623
FIRST-ORDER SYSTEMS 625
SECOND-ORDER SYSTEMS 628
CONCEPT OF STABILITY AND STABILITY CRITERIA 634
RECOMMENDED REFERENCES 637

CHAPTER 17 Corrosion and Materials of Construction 639

CORROSION TYPES AND CORROSION CONSIDERATION OF MATERIALS 639
MATERIALS OF CONSTRUCTION AND THEIR BEHAVIOR 645
LOW-TEMPERATURE APPLICATION OF METALS 652

Contents **vii**

MAXIMUM ALLOWABLE STRESS IN TENSION VERSUS TEMPERATURE 653
COST CONSIDERATIONS 653
RECOMMENDED REFERENCES 654

CHAPTER 18

Equipment Design 655

EQUIPMENT SPECIFICATION SHEETS 656
PRESSURE VESSELS 658
OPERATING VOLUME AND SURGE VOLUME 660
PIPING DESIGN 661
PROCESS DESIGN 665
SIZING AN AGITATOR 667
MASS TRANSFER COLUMNS 674
EQUIPMENT TESTING AND ANALYSIS 680
DESIGN OPTIMIZATION 682
PROCESS FLOW SHEET DEVELOPMENT 683
OPERATING MANUAL 684
RECOMMENDED REFERENCES 694

CHAPTER 19

Engineering Economics 697

TIME VALUE OF MONEY 697
CHEMICAL ENGINEERING PLANT COST INDEX (CEPCI) 706
RECOMMENDED REFERENCES 720

CHAPTER 20

Plant Safety and Environmental Consideration 723

TOXICOLOGY AND INDUSTRIAL HYGIENE 724
FIRE AND EXPLOSION ISSUES 727
HAZARD AND OPERABILITY STUDIES (HAZOP) 758
ENVIRONMENTAL CONSIDERATIONS 761
FEDERAL POLLUTION PREVENTION ACT OF 1990 762
OZONE, FRIEND AND FOE 765
ABATEMENT OF AIR POLLUTION 767
ABATEMENT OF WATER POLLUTION 768
TREATMENT OF CONTAMINATED SOIL 769
PERSONAL PROTECTIVE EQUIPMENT (PPE) 769
NOISE MANAGEMENT 770
SOLID WASTE MANAGEMENT 773
RECOMMENDED REFERENCES 781

Index 783

Introduction

OUTLINE

HOW TO USE THIS BOOK IX

BECOMING A PROFESSIONAL ENGINEER X
Education ■ Fundamentals of Engineering (FE/EIT) Exam ■ Experience ■ Professional Engineer Exam

CHEMICAL ENGINEERING PROFESSIONAL ENGINEER EXAM X
Examination Development ■ Examination Structure ■ Exam Dates ■ Exam Procedure ■ Preparing for and Taking the Exam ■ Exam Day Preparations ■ What to Take to the Exam ■ Examination Scoring and Results

ACKNOWLEDGMENTS XV

HOW TO USE THIS BOOK

Chemical Engineering PE License Review and its companion texts form a three-step approach to preparing for the Principles and Practice of Engineering (PE) exam:

- *Chemical Engineering PE License Review* contains the conceptual review of chemical engineering topics for the exam, including key terms, equations, analytical methods and reference data. Because it does not contain problems and solutions, the book can be brought into the open-book PE exam as one of your references.

- *Chemical Engineering PE Problems & Solutions* provides problems for you to solve in order to test your understanding of concepts and techniques. Ideally, you should solve these problems after completing your conceptual review. Then, compare your solution to the detailed solutions provided, to get a sense of how well you have mastered the content and what topics you may want to review further.

- *Chemical Engineering PE Sample Exam* provides complete morning and afternoon exam sections so that you can simulate the experience of taking the PE test within its actual time constraints and with questions that match the test format. Take the sample exam after you're satisfied with your review of concepts and problem-solving techniques, to test your readiness for the real exam.

BECOMING A PROFESSIONAL ENGINEER

To achieve registration as a professional engineer there are four distinct steps: (1) education, (2) the Fundamentals of Engineering/Engineer-In-Training (FE/EIT) exam, (3) professional experience, and (4) the professional engineer (PE) exam, more formally known as the Principles and Practice of Engineering Exam. These steps are described in the following sections.

Education

The obvious appropriate education is a B.S. degree in Chemical engineering from an accredited college or university. This is not an absolute requirement. Alternative, but less acceptable, education is a B.S. degree in something other than Chemical engineering, or a degree from a non-accredited institution, or four years of education but no degree.

Fundamentals of Engineering (FE/EIT) Exam

Most people are required to take and pass this eight-hour multiple-choice examination. Different states call it by different names (Fundamentals of Engineering, E.I.T., or Intern Engineer), but the exam is the same in all states. It is prepared and graded by the National Council of Examiners for Engineering and Surveying (NCEES). Review materials for this exam are found in other Kaplan AEC Education books such as *Fundamentals of Engineering: FE/EIT Exam Preparation*.

Experience

Typically one must have four years of acceptable experience before being permitted to take the Professional Engineer exam (California requires only two years). Both the length and character of the experience will be examined. It may, of course, take more than four years to acquire four years of acceptable experience.

Professional Engineer Exam

The second national exam is called Principles and Practice of Engineering by NCEES, but just about everyone else calls it the Professional Engineer or P.E. exam. All states, plus Guam, the District of Columbia, and Puerto Rico, use the same NCEES exam.

CHEMICAL ENGINEERING PROFESSIONAL ENGINEER EXAM

The reason for passing laws regulating the practice of engineering is to protect the public from incompetent practitioners. Most states require engineers working on projects involving public safety to be registered, or to work under the supervision of a registered professional engineer. In addition, many private companies encourage or require engineers in their employ to pursue registration as a matter of professional development. Engineers in private practice, who wish to consult or serve as expert witnesses, typically also must be registered. There is no national registration law;

registration is based on individual state laws and is administered by boards of registration in each of the states. You can find a list of contact information for and links to the various state boards of registration at the Kaplan AEC Web site: *www.kaplanaecengineering.com*. This list also shows the exam registration deadline for each state.

Examination Development

Initially the states wrote their own examinations, but beginning in 1966 the NCEES took over the task for some of the states. Now the NCEES exams are used by all states. This greatly eases the ability of an engineer to move from one state to another and achieve registration in the new state.

The development of the engineering exams is the responsibility of the NCEES Committee on Examinations for Professional Engineers. The committee is composed of people from industry, consulting, and education, plus consultants and subject matter experts. The starting point for the exam is a task analysis survey, which NCEES does at roughly 5- to 10-year intervals. People in industry, consulting, and education are surveyed to determine what Chemical engineers do and what knowledge is needed. From this NCEES develops what it calls a "matrix of knowledge" that forms the basis for the exam structure described in the next section.

The actual exam questions are prepared by the NCEES committee members, subject matter experts, and other volunteers. All people participating must hold professional registration. Using workshop meetings and correspondence by mail, the questions are written and circulated for review. Although based on an understanding of engineering fundamentals, the problems require the application of practical professional judgment and insight.

Examination Structure

The exam consists of 80 multiple-choice questions covering the following areas of chemical engineering (relative exam weight for each topic is shown in parentheses):

- Mass/energy balances and thermodynamics (24%)
- Fluids (17%)
- Heat transfer (16%)
- Mass transfer (13%)
- Kinetics (11%)
- Plant design and operation (19%)

For more information on the topics and subtopics covered, visit the NCEES Web site at *www.ncees.org*.

The exam is given over two four-hour sessions, with 40 questions in each session. All questions are multiple choice, with four answer choices.

Exam Dates

NCEES prepares Professional Engineer exams for use on a Friday in April and October of each year. Some state boards administer the exam twice a year in their state, whereas others offer the exam once a year. The scheduled exam dates for the next ten years can be found on the NCEES Web site (*www.ncees.org/exams/schedules/*).

People seeking to take a particular exam must apply to their state board several months in advance.

Exam Procedure

Before the morning four-hour session begins, the proctors pass out an exam booklet and solutions pamphlet to each examinee.

The solution pamphlet contains grid sheets on right-hand pages. Only the work on these grid sheets will be graded. The left-hand pages are blank and are to be used for scratch paper. The scratchwork will not be considered in the scoring.

If you finish more than 30 minutes early, you may turn in the booklets and leave. In the last 30 minutes, however, you must remain to the end to ensure a quiet environment for all those still working and the orderly collection of materials.

The afternoon session will begin following a one-hour lunch break. The afternoon exam booklet will be distributed along with an answer sheet.

Preparing for and Taking the Exam

Give yourself time to prepare for the exam in a calm and unhurried way. Many candidates like to begin several months before the actual exam. Target a number of hours per day or week that you will study, and reserve blocks of time for doing so. Creating a review schedule on a topic-by-topic basis is a good idea. Remember to allow time for both reviewing concepts and solving practice problems.

In addition to review work that you do on your own, you may want to join a study group or take a review course. A group study environment might help you stay committed to a study plan and schedule. Group members can create additional practice problems for one another and share tips and tricks.

You may want to prioritize the time you spend reviewing specific topics according to their relative weight on the exam, as identified by NCEES, or by your areas of relative strength and weakness.

People familiar with the psychology of exam taking have several suggestions for people as they prepare to take an exam.

1. Exam taking involves really, two skills. One is the skill of illustrating knowledge that you know. The other is the skill of exam taking. The first may be enhanced by a systematic review of the technical material. Exam-taking skills, on the other hand, may be improved by practice with similar problems presented in the exam format.

2. Since there is no deduction for guessing on the multiple choice problems, answers should be given for all of them. Even when one is going to guess, a logical approach is to attempt to first eliminate one or two of the four alternatives. If this can be done, the chance of selecting a correct answer obviously improves from 1 in 4 to 1 in 3 or 1 in 2.

3. Plan ahead with a strategy. Which is your strongest area? Can you expect to see several problems in this area? What about your second strongest area?

4. Plan ahead with a time allocation. Compute how much time you will allow for each of the subject areas in the exam. You might allocate a little less time per problem for those areas in which you are most proficient, leaving a little more time in subjects that are difficult for you. Your time plan should include

a reserve block for especially difficult problems, for checking your scoring sheet, and to make last-minute guesses on problems you did not work. Your strategy might also include time allotments for two passes through the exam—the first to work all problems for which answers are obvious to you, and the second to return to the more complex, time-consuming problems and the ones at which you might need to guess. A time plan gives you the confidence of being in control and keeps you from making the serious mistake of misallocation of time in the exam.

5. Read all four multiple-choice answers before making a selection. An answer in a multiple-choice question is sometimes a plausible decoy—not the best answer.

6. Do not change an answer unless you are absolutely certain you have made a mistake. Your first reaction is likely to be correct.

7. Do not sit next to a friend, a window, or other potential distractions.

Exam Day Preparations

The exam day will be a stressful and tiring one. This will be no day to have unpleasant surprises. For this reason we suggest that an advance visit be made to the examination site. Try to determine such items as

1. How much time should I allow for travel to the exam on that day? Plan to arrive about 15 minutes early. That way you will have ample time, but not too much time. Arriving too early, and mingling with others who also are anxious, will increase your anxiety and nervousness.

2. Where will I park?

3. How does the exam site look? Will I have ample workspace? Where will I stack my reference materials? Will it be overly bright (sunglasses), cold (sweater), or noisy (earplugs)? Would a cushion make the chair more comfortable?

4. Where are the drinking fountain, lavatory facilities, pay phone?

5. What about food? Should I take something along for energy in the exam? A bag lunch during the break probably makes sense.

What to Take to the Exam

The NCEES guidelines say you may bring only the following reference materials and aids into the examination room for your personal use:

1. Handbooks and textbooks, including the applicable design standards.

2. Bound reference materials, provided the materials remain bound during the entire examination. The NCEES defines "bound" as books or materials fastened securely in their covers by fasteners that penetrate all papers. Examples are ring binders, spiral binders and notebooks, plastic snap binders, brads, screw posts, and so on.

3. A battery-operated, silent, nonprinting, noncommunicating calculator from the NCEES list of approved calculators. For the most current list, see the NCEES Web site (*www.ncees.org*). You also need to determine whether or not your

state permits preprogrammed calculators. Bring extra batteries for your calculator just in case; many people feel that bringing a second calculator is also a very good idea.

At one time NCEES had a rule barring "review publications directed principally toward sample questions and their solutions" in the exam room. This set the stage for restricting some kinds of publications from the exam. *State boards may adopt the NCEES guidelines, or adopt either more or less restrictive rules.* Thus an important step in preparing for the exam is to know what will—and will not—be permitted. We suggest that if possible you obtain a written copy of your state's policy for the specific exam you will be taking. Occasionally there has been confusion at individual examination sites, so a copy of the exact applicable policy will not only allow you to carefully and correctly prepare your materials, but will also ensure that the exam proctors will allow all proper materials that you bring to the exam.

As a general rule we recommend that you plan well in advance what books and materials you want to take to the exam. Then they should be obtained promptly so you use the same materials in your review that you will have in the exam.

License Review Books

The review books you use to prepare for the exam are good choices to bring to the exam itself. After weeks or months of studying, you will be very familiar with their organization and content, so you'll be able to quickly locate the material you want to reference during the exam. Keep in mind the caveat just discussed—some state boards will not permit you to bring in review books that consist largely of sample questions and answers.

Textbooks

If you still have your university textbooks, they are the ones you should use in the exam, unless they are too out of date. To a great extent the books will be like old friends with familiar notation.

Bound Reference Materials

The NCEES guidelines suggest that you can take any reference materials you wish, so long as you prepare them properly. You could, for example, prepare several volumes of bound reference materials, with each volume intended to cover a particular category of problem. Maybe the most efficient way to use this book would be to cut it up and insert portions of it in your individually prepared bound materials. Use tabs so that specific material can be located quickly. If you do a careful and systematic review of civil engineering, and prepare a lot of well-organized materials, you just may find that you are so well prepared that you will not have left anything of value at home.

Other Items

In addition to the reference materials just mentioned, you should consider bringing the following to the exam:

- *Clock*—You must have a time plan and a clock or wristwatch.

- *Exam assignment paperwork*—Take along the letter assigning you to the exam at the specified location. To prove you are the correct person, also bring something with your name and picture.

- *Items suggested by advance visit*—If you visit the exam site, you probably will discover an item or two that you need to add to your list.

- *Clothes*—Plan to wear comfortable clothes. You probably will do better if you are slightly cool.

- *Box for everything*—You need to be able to carry all your materials to the exam and have them conveniently organized at your side. Probably a cardboard box is the answer.

Examination Scoring and Results

The questions are machine-scored by scanning. The answers sheets are checked for errors by computer. Marking two answers to a question, for example, will be detected and no credit will be given.

Your state board will notify you whether you have passed or failed roughly three months after the exam. Candidates who do not pass the exam the first time may take it again. If you do not pass you will receive a report listing the percentages of questions you answered correctly for each topic area. This information can help focus the review efforts of candidates who need to retake the exam.

The PE exam is challenging, but analysis of previous pass rates shows that the majority of candidates do pass it the first time. By reviewing appropriate concepts and practicing with exam-style problems, you can be in that majority. Good luck!

ACKNOWLEDGMENTS

It is our privilege to acknowledge individuals who assisted in the preparation of this book, particularly those who reviewed the contents of the previous edition—Thomas R. Marrero, PhD, PE, University of Missouri, Columbia; David L. Silverstein, PhD, PE, University of Kentucky; Chenming Zhang, PhD, Virginia Tech; Lloyd Hile, PhD, PE, California State University, Long Beach; and Mr. Bryon Brandt—for their dedicated review, insightful comments, and valuable constructive suggestions.

We also appreciate the help of Laurie McGuire, Karen Goodfriend, Merrill Peterson, and all employees of Kaplan Professional Publishing, the unseen heroes, who worked behind the scenes to improve the quality of this book.

Finally and most importantly, our sincere appreciation and thanks are due to Mrs. Malancha Das and Mrs. Vimal Prabhudesai for their patience, forbearance, and encouragement during the completion of this book.

CHAPTER 1

Units and Dimensions

OUTLINE

DIMENSIONS 1

SYSTEMS OF UNITS 2
Centimeter-Gram-Second (cgs) System ■ Metric or mks System ■ British Engineering System ■ Foot-Pound-Second (fps) System ■ American Engineering System (USCS)

UNITS FOR COMMON CHEMICAL QUANTITIES 6
Molar Units ■ Temperature Units ■ Pressure Units ■ Thermal Units ■ Work, Energy, and Power

CONVERSION OF UNITS 12

DIMENSIONAL FORMULAS 16

DIMENSIONAL AND DIMENSIONLESS EQUATIONS 16

DIMENSIONAL ANALYSIS 18
Raleigh's Method ■ Buckingham's π Method ■ Dynamic Similarity

RECOMMENDED REFERENCES 22

With most nations, including the United States and United Kingdom, adopting the International System of units (SI units) as the standard, there has been a progressively greater emphasis on the use of SI units. However, the cgs, fps, and metric systems are still in use and will be around for some time. Physicochemical data are reported mostly in cgs units and engineers do their calculations in either mks or fps units. Therefore, besides knowing the customary fps (engineering), mks, and cgs units, candidates for the P.E. examination should familiarize themselves with the SI units. This chapter provides a short review of the different systems of units and their conversions into one another.

DIMENSIONS

A physical quantity consists of two parts: (1) a unit, which indicates the dimension and gives its standard of measurement and (2) a quantity, which gives the numerical value of the units. For example, the distance between two points has the dimension of length and can be expressed as 3 feet. Here the foot is the unit of the length dimension, and 3 is the numerical value of the corresponding number of length units. Other examples of dimensions are mass, time, temperature, and electric charge.

Physical properties of a system are interrelated by mechanical and physical laws. Such relationships allow certain dimensions to be regarded as basic or fundamental and others as derived. Fundamental units are independently defined, whereas derived units are expressed in terms of fundamental units. The choice of number and type of fundamental units is based on convenience and is purely arbitrary. Therefore, the dimensions chosen as basic vary from one system to another, but it is usual to treat length L, time t, temperature T, and mass M as basic. In some systems, force F or both mass M and force F are used as the basic dimensions.

The units of force, mass, and acceleration are related through Newton's law, which can be written as

$$F_o = kma \tag{1.1}$$

where
F_o = resultant force acting on the body
m = mass of body, which is independent of its position or velocity
a = acceleration of body in the direction of resultant force
k = a constant that depends only on the units chosen for mass, length, time, and force.

Various systems of units have been devised for the quantities in Newton's law equation. Those of importance in chemical engineering practice are described next.

SYSTEMS OF UNITS

In this section we look at various systems of units that chemical engineers should understand: cgs and mks, British, fps, American, and SI.

Centimeter-Gram-Second (cgs) System

The basic dimensions in this system are length L, mass M, and time t. The units and nomenclature are as follows:

Dimension	Dimension Symbol	Unit
Length	L	Centimeter (cm)
Mass	M	Gram (g)
Time	t	Second (s)

In this system the constant k in Newton's law equation is arbitrarily set equal to unity. It is a pure numerical constant. The unit of force in this system is the dyne (symbol dyn), defined as the force, which imparts an acceleration of 1 cm/s² to a mass of 1g. Thus

$$1 \text{ dyne (symbol dyn)} = 1 \text{ (g)}(1 \text{ cm/s}^2)$$

The dimensions of force are then given by

$$\text{Force } F = MLt^{-2}$$

Most scientific data reported in the literature are in the cgs system.

Metric or mks System

This system is based on a length unit of meter, hence it is called the metric system. The unit of mass in this system is 1 kilogram. The cgs system is similar to the metric system but the magnitudes of quantities of the base units are smaller compared to the metric system.

In the mks or metric engineering system of units, force is taken as a fourth fundamental dimension besides mass, length, and time. Because of this the kilogram force (symbol kg_f) is defined as that force that imparts to 1 kg mass an acceleration of $g = 9.80665$ m/s². Here also Newton's law dimensional proportionality factor $g_c = 9.80665$ kg · m/kg_f · s² is introduced as follows:

Newton's Law
$$F = \frac{1}{g_c} ma$$

where

$$kg_f = \frac{1}{g_c} \times 1\,kg \times 9.80665 \frac{m}{s^2}$$

Then

$$g_c = 9.80665 \, \frac{kg \cdot m}{kg_f \cdot s^2}$$

Thus the kilogram force is equal to 9.80665 N (See page 5 for the definition of N).

British Engineering System

In this system, the foot and second are the units of length and time, but the third fundamental unit is the pound-force. The pound-force is defined as the force that gives an acceleration of 32.174 ft/s² to a mass of one pound. This is a fixed quantity and is not the same as the pound-weight, which is the force exerted by the gravitational field on a mass of one pound. The pound-weight is a variable quantity because of the variation of the gravitational field strength from place to place relative to the center of the earth. The unit of mass in the British system is the slug. This is the mass to which an acceleration of 1 ft/s² is given by 1 pound-force.

$$1 \text{ pound-force (symbol } lb_f) = (1 \text{ slug})(1 \text{ ft/s}^2)$$

or

$$1 \text{ slug} = \frac{1 \text{ pound-force}}{1 \text{ ft/s}^2}$$

Foot-Pound-Second (fps) System

The basic dimensions and units of this system are

Dimension	Dimension Symbol	Unit
Length	L	Foot (ft)
Mass	M	Pound (lb_m)
Time	t	Second (s)

The poundal, which is the unit of force in this system, is the force that gives an acceleration of 1 ft/s² to a mass of 1 lb. Thus

$$1 \text{ poundal} = (1 \text{ lb-mass})(1 \text{ ft/s}^2)$$

This implies that the constant in the Newton's law equation is set equal to one.

A comparison of the British and fps systems shows that 1 slug = 32.174 lb-mass and

$$1 \text{ lb-force} = 32.174 \text{ poundals}$$

Confusion between 1 lb-mass and 1 lb-force can be avoided by writing 1 lb-mass as 1 lb_m or simply 1 lb and 1 pound-force as 1 lb_f.

American Engineering System(AES)/US Customary System (USCS)

This system includes four basically defined units: the pound-mass, foot, second, and the pound-force. A proportionality factor g_c between the force and the mass is defined by

$$\text{Force (lb}_f) = \frac{[\text{mass (lb)}][\text{acceleration (ft/s}^2)]}{g_c}$$

For the purposes of the gravitational force unit, Newton's law proportionality factor g_c is taken as

$$g_c = 32.174 \text{ lb·ft /s}^2 \cdot \text{lb}_f$$

It should be emphasized that g_c is not a numerical constant and is a magnitude having its own dimensions (M)(L)/(lb force)(s²). Its numerical value is constant anywhere in the universe,

The poundal unit is very rarely used. As it is more common to think of the unit of force as the earth's attraction on 1 lb-mass, an entirely different set of units has become customary. If the force is required in pound "weight," it is customary to use a numerical constant g_c (no units) to divide the righthand side of Equation 1.1 so that the lb-force will be unity. Substituting $k = 1/g_c$ and $a = g$, Equation 1.1 now becomes

$$F = \frac{mg}{g_c} \tag{1.2}$$

In Equation 1.2, the numerical ratio g/g_c is practically 1 lb$_f$/lb and equals 9.81 N/kg in SI units on earth. Actually g has a value of 32.174 ft/s² only at sea level and at a latitude of 45° on earth.

Keep in mind two facts when using the American Engineering System. First, although lb is used to name both lb-mass and lb-force, they are entirely different quantities with different dimensions. Second, the quantity g_c is not acceleration. The value of g_c is so chosen that it equals the average value of acceleration of gravity at sea level and latitude 45° on earth.

SI Units (Système International)

The basic dimensions and units of this system are as follows:

Dimension	Dimension Symbol	Unit
Length	L	Meter (m)
Mass	M	Kilogram (kg)
Time	t	Second (s)

The other base fundamental units in this system are temperature, amount of substance (defined as mole), electric current, and luminous intensity. Two other supplementary units, plane angle and solid angle, are also defined. These nine units are adequate to describe physical and chemical properties in the SI system.

The unit of force in this system is the Newton, defined as the force that gives an acceleration of 1 m/s² to a mass of 1 kg. This definition is based on Newton's second law of motion $F = \frac{ma}{g_c}$ where $1/g_c$ is the proportionality factor. In the SI system, the Newton is defined so as to make g_c numerically equal to 1. Thus

$$1 \text{ N} = (1 \text{ kg})(1 \text{ m/s}^2)/g_c$$

and

$$g_c = 1 \text{ (kg)(m)/(N)(s)}^2$$

The units of g_c may be viewed in two ways. If Newton is considered as an independent unit, the constant g_c has the units given previously. However, if it is treated as an abbreviation for the composite unit (kg)(m)/s², then g_c is considered unitless. Equation 1.2 then becomes

$$F = ma \tag{1.2a}$$

Key multiples and submultiples of selected SI units are given in Table 1.1.

Good practice is to use a multiple or submultiple in such a way that the numerical value of the quantity lies between 0.1 and 1000.

There are four unit systems commonly used in chemical engineering. The values of g_c corresponding to these four systems are listed in Table 1.2 for quick reference.

Table 1.1 Prefixes and symbols for the multiples and submultiples of SI units

Multiple	SI Prefix	Symbol	Submultiple	SI Prefix	Symbol
10^{18}	exa	E	10^{-1}	deci	d
10^{15}	peta	P	10^{-2}	centi	c
10^{12}	tera	T	10^{-3}	milli	m
10^{9}	giga	G	10^{-6}	micro	μ
10^{6}	mega	M	10^{-9}	nano	n
10^{3}	kilo	k	10^{-12}	pico	p
10^{2}	hecto	h	10^{-15}	femto	f
10	deka	da	10^{-18}	atto	a

Table 1.2 Conversion factors g_c for the different unit systems

Fundamental Quantity	SI (International)	System (cgs)	AES/English Engineering	Metric Engineering (mks)
Mass M	Kilogram, kg	Gram, g	Pound-mass, lb	Kilogram mass, kg
Length L	Meter, m	Centimeter, cm	Foot, ft	Meter, m
Time t	Second, s	Second, s	Second, s	Second, s
Force F	Newton, N	Dyne, dyn	Pound-force, lb_f	Kilogram force, kg_f
g_c	$1\ kg \cdot m/N \cdot s^2$	$1\ g \cdot cm/dyn \cdot s^2$	$32.174\ lb \cdot ft/lb_f \cdot s^2$	$9.80665\ kg \cdot m/kg_f \cdot s^2$

Note: g_c is not the gravitational constant.

Example 1.1

Calculate the force exerted by 10 lb_m in terms of the following units:

(a) Lb-force (lb_f) (Engineering units)

(b) Dynes (cgs units)

(c) Newtons (SI units)

Solution

(a) $$F = \frac{mg}{g_c} = \frac{(10\ lb_m)(32.174\ \frac{ft}{s^2})}{32.174\ \frac{lb_m \cdot ft}{lb_f \cdot s^2}} = 10\ \text{lb-force}\ (lb_f)$$

(b) $$F = mg = \frac{(10\ lb_m)}{} \left| \frac{453.6\ g}{lb_m} \right| \frac{981\ cm}{s^2}$$

$$= \frac{4.45 \times 10^6}{} \left| \frac{g \cdot cm}{s^2} \right| = 4.45 \times 10^6\ \text{dyn}$$

(c) $$F = mg = \frac{(10\ lb_m)}{} \left| \frac{kg}{2.2046\ lb_m} \right| \frac{9.81\ m}{s^2} = 44.5\ N$$

UNITS FOR COMMON CHEMICAL QUANTITIES

A large number of variables or quantities defined in terms of the fundamental units are commonly used in chemical engineering problems. These secondary quantities are used because of their convenience. In this section we will review molar, temperature, pressure, thermal, work, energy, and power units.

Molar Units

In problems involving chemical reactions, it is often convenient to work in molar units. A mole of a substance equals the molecular mass expressed in given units.

Molecular mass is also commonly called molecular weight, although the term weight implies force. The mole is expressed in different units as follows:

1 kilogram-mole = molecular weight in kilograms
1 pound-mole = molecular weight in pounds
1 gmol = molecular weight expressed in grams
1 kmol = 1000 mol. (formerly 1000 g mol)

A mole consists of a certain number of molecules, atoms, electrons, or other types of particles. The mole (formerly referred to as the gram mole. SI symbol is mol) is defined by the International Committee on Weights and Measures as the amount of a substance that contains as many elementary entities as the number of atoms contained in exactly 12 grams (0.012 kg) of carbon-12. It's symbol in SI units is mol. One atomic mass unit (amu) is one-twelfth of the mass of one ^{12}C atom. This mass equals 1.661×10^{-24} g. Thus one atom of ^{12}C has 12 atomic units (amu) in it. Atomic weights of other elements are based on an arbitrary scale that assigns a mass of exactly 12 to ^{12}C. On the basis of the atomic weight scale, carbon is 12.01, hydrogen is 1.008, barium is 137.34, and so on.

The number of atoms in 12 grams (1gm-atom) of C-12 is 6.0221367×10^{23} (*Avogadro's number*). Each mole has a mass of MM amu where MM = molecular mass or molecular weight. Thus the mass of one g-atom of C-12 is obtained in the following manner:

$$m = 6.0221367 \times 10^{23} \left(\frac{\text{atoms}}{\text{g mol}}\right)\left(\frac{12 \text{ amu}}{\text{atom}}\right)\left(\frac{1.661 \times 10^{-24} \text{ g}}{\text{amu}}\right) = 12 \text{ g/g mol}$$

In the case of a carbon atom, gmol is really g-atom.

1 mol of water contains 6.0221367×10^{23} molecules. Therefore mass m of 1 mol of water is given by

$$m = 6.0221367 \times 10^{23} \frac{\text{molecules}}{\text{g mol}} \left(\frac{18 \text{ amu}}{\text{molecule}}\right)\left(\frac{1.661 \times 10^{-24} \text{ g}}{\text{amu}}\right) = 18 \text{ g/g mol}$$

One g mol has 6.0221367×10^{23} molecules.

The pound-mole (lb mol) has $6.0221367 \times 10^{23} \times (453.6)$ molecules.

The molecular weight (relative molar mass) of a compound is calculated by adding the atomic weights (masses) of the elementary atoms constituting the compound. Atomic weights are obtained from the table of atomic weights. For example, a water molecule has 3 atoms in it, 2 of hydrogen and 1 of oxygen and hence its molecular weight is 2 + 16 = 18. The atomic weight of hydrogen is 1 and 16 is the atomic weight of oxygen.

Temperature Units

Four temperature scales are in use. These are Celsius, Fahrenheit, Kelvin, and Rankine. Of these, Celsius (Centigrade) and Fahrenheit are relative scales as their zero points were arbitrarily fixed. On the other hand, the absolute scales Kelvin and Rankine have their zero points at the lowest possible temperature that can exist. Absolute zero on Kelvin scale is −273.15°C and that on Rankine scale, −459.67°F.

In practical engineering calculations, these temperatures are often rounded to −273°C and −460°F respectively.

You should know how to convert one type of temperature to another. The relations between K and °C and between °R and °C are given by the following equations

$$T_K = T_{°C} + 273.15$$

and

$$T_{°R} = T_{°F} + 459.67$$
$$T_{°R} = 1.8 T_K$$

The relation between Fahrenheit temperature and corresponding Celsius temperature is

$$T_{°C} = (T_{°F} - 32)/1.8$$

and

$$T_{°F} = 1.8 T_{°C} + 32$$

The temperature difference between the boiling point of water at atmospheric pressure (100°C or 212°F) and its freezing point (0°C or 32°F) at the same pressure is 100°C on the Celsius scale and 180°F on the Fahrenheit scale. From this it follows that 1°C temperature differencez is equal to 1.8°F temperature difference. This is the factor used in the preceding equation.

Example 1.2

The boiling point of water at atmospheric pressure is 100°C. What is its boiling point in (a) K, (b) °F, and (c) °R?

Solution

(a) Boiling point of water in K = (100 + 273) = 373 K

(b) Boiling point in °F = (100°C) $\frac{1.8°F}{1°C}$ + 32°F = 212°F

(c) Boiling point in °R = (212 + 460)°F $\frac{1°R}{1°F}$ = 672°R

Example 1.3

The heat capacity of sulfur is given by the equation

$$C_p = 3.63 + 0.640 T$$

where C_p is in cal/(gmol · K) and T is in K. What will be the expression for C_p if it is to be in units of cal/(gmol · °F) and T is in °F.

Solution

The units of K in the denominator of the heat capacity are that of temperature difference Δ°C and the units of T in K. To convert the units of C_p, first substitute T in terms of °F and then modify the units of the resulting equation.

$$T = (T_R/1.8) = (t_F + 460)/1.8 \text{ and } \Delta K = \Delta°C; \Delta°C = 1.8 \, \Delta°F$$

Substitution in the equation for C_p gives

$$C_p = \left\{3.63 + 0.64\left[\frac{t_F + 460}{1.8}\right]\right\}\frac{\text{cal}}{\text{g mol}\cdot\text{K}}$$

$$= \{167.19 + 0.356t_F\}\frac{\text{cal}}{\text{g mol}\cdot(1.8\,°F)}$$

$$= (92.88 + 0.198t_F)\frac{\text{cal}}{\text{g mol}\cdot°F}$$

where
t_F = Temperature in °F.

Pressure Units

Pressure is defined as force per unit area. It can be expressed by absolute or relative scale. The zero point of an absolute pressure scale is a perfect vacuum. The relative and absolute pressures are related by the following expression:

Absolute pressure = ±Gauge pressure + Barometric pressure.

This equation has to be used with consistent units.

Absolute pressure is always positive. Gauge pressure is positive when it is above barometric pressure and negative when it is below barometric pressure.

The unit of pressure in the SI system is the Newton per square meter, termed *pascal* (Pa). The pascal is a very small unit and is related to the *bar* as follows:

$$1\text{ Pa} = 1\text{ N/m}^2$$
$$1\text{ Bar} = 1 \times 10^5\text{ Pa}$$

A convenient unit for pressure that is used in all the systems is the standard atmosphere. This is defined by

$$1\text{ standard atm} = 1.01325 \times 10^5\text{ Pa} = 1.01325 \times 10^5\text{ N/m}^2$$

In the fps system, the unit of pressure is lb-force/in.² or psi (psig if the pressure is relative to reference pressure, usually atmospheric) or lb-force per square foot. Pressure is also measured as the height of a column of fluid under the influence of gravity, such as mm of Hg or feet of water.

Do not confuse standard atmosphere with atmospheric pressure. The latter is variable. Standard atmosphere is defined as the pressure (in a standard gravitational field) equivalent to 1 atm or 760 mm Hg at 0°C.

The values of standard atmosphere expressed in various units are

Atmosphere	1.000
psia	14.696 (usually rounded to 14.7)
ft of water	33.91
in. Hg	29.92
mm Hg	760
kPa (Pa)	101.3 (1.013×10^5).

Vacuum is a method to express pressure as a quantity measured below atmospheric pressure or some other reference pressure. Thus if atmospheric pressure is 29.8 inches of mercury, and the vacuum gauge indicates a vacuum of 26 in. Hg, the actual pressure will be 29.8 − 26 = 2.8 in. Hg. For a smaller vacuum such as 2 in. Hg, the term *draft* is used.

Example 1.4

The pressure gauge on a pipeline in a process plant reads 58 psi. At the same time the barometer reads 752.3 mm Hg. What is the absolute pressure in the pipeline in psia?

Solution

The pressure gauge is reading the relative pressure in psig and not the absolute pressure in psia. Absolute pressure in psia is the sum of gauge pressure in psig plus the atmospheric pressure (barometric) expressed in the same absolute units (in this case psia).

$$\text{Atmospheric pressure} = \frac{752.3 \text{ mm Hg}}{760 \text{ mm Hg}} \cdot 14.7 \text{ psia} = 14.55 \text{ psia}$$

The absolute pressure in the pipeline = 58 psig + 14.55 = 72.55 psia.

Example 1.5

The density of atmospheric air decreases as the altitude increases. When the pressure is 400 mm Hg, how many kilopascals is it?

Solution

The pressure in kilopascals is calculated as follows:

$$\frac{400 \text{ mm Hg}}{760 \text{ mm Hg}} \cdot \frac{1.013 \times 10^5 \text{ N/m}^2}{1} \cdot \frac{1 \text{ kN}}{1000 \text{ N}} = 53.32 \frac{\text{kN}}{\text{m}^2} = 53.32 \text{ kPa}$$

Thermal Units

For convenience, thermal energy is expressed in terms of mass, temperature, and a proportionality constant, which is the heat capacity (at constant pressure or at constant volume) of the material. Thermal energy is defined as the amount of heat required to raise the temperature of a body by one degree. Because the heat capacity varies from material to material, the heat quantities are expressed in terms of the specific heat of water at 15 °C (288 K). Thus specific heat is the ratio of the heat capacity of a substance at a given temperature to the heat capacity of water at a reference temperature usually taken as 15 °C. Although heat capacity and specific heat are two distinct quantities (one with units and the other a numerical ratio), common practice is to use them interchangeably.

In the cgs system, one calorie (commonly called thermochemical cal) is the amount of heat that will raise the temperature of one gram of water through one Celsius degree at atmospheric pressure. A kg-calorie is 1000 times a gram-calorie.

The thermochemical calorie and kilocalorie are redefined in terms of the joule as

1 calorie (thermochemical) = 4.184 J
1 kilocalorie (thermochemical) = 4184 J

Other calorie units are also defined. They are

1 calorie (mean) = 4.1900 J
1 calorie (International steam tables) = 4.1868 J
1 calorie (15°C) = 4.1858 J
1 calorie (20°C) = 4.1819 J

Corresponding kilogram-calorie units are also in use.

In the British and fps systems, the *British thermal unit* (commonly termed thermochemical Btu) is the quantity of heat required to raise the temperature of one pound mass of water through one Fahrenheit degree (from 60°F to 61°F) and the pound-calorie or Celsius heat unit is the heat quantity required to raise one pound of water (at 60°F) through one Celsius degree.

$$1 \text{ Btu (thermochemical)} = 1054.35 \text{ J}$$

Other units in this system are

$$1 \text{ Btu (international steam tables)} = 1055.04 \text{ J}$$
$$1 \text{ Btu (Mean)} = 1055.87 \text{ J}$$
$$1 \text{ Btu (59°F)} = 1059.67 \text{ J}$$
$$1 \text{ Btu (60°F)} = 1054.65 \text{ J}$$

The heat capacity values for water in various units are

cgs: cal/g·°C
fps: Btu/lb·°F
SI: 4186.8 J/kg·K (International steam table calorie)

Note that $1 \text{ Btu}/(\text{lb}_m \cdot °F) = 1 \text{ cal}/(g \cdot °C)$.

Example 1.6

In Example 1.3, obtain the expression for C_p if it is to be in units of Btu/(lb mol·°R) and T is in °R.

Solution

In this case $T = T_R/1.8$, therefore the expression for C_p is

$$C_p = \left\{ 3.63 + 0.64 \left[\frac{T_R}{1.8} \right] \right\} \frac{\text{cal}}{\text{g mol} \cdot \text{K}}$$

$$= \{3.63 + 0.356 T_R\} \frac{\text{cal}}{\text{g mol} \cdot \Delta \text{K}}$$

$$= \{3.63 + 0.356 T_R\} \frac{\text{cal} \frac{\text{Btu}}{252 \text{ cal}}}{\frac{\text{g mol}}{453.6 \text{ g mol/lb mol}} \quad 1.8 \Delta °R}$$

$$= \{3.63 + 0.356 T_R\} \frac{\text{Btu}}{\text{lb mol} \cdot °R}$$

Work, Energy, and Power

In SI units, work and energy are measured in joules (J),

$$1 \text{ J} = 1 \text{ N} \cdot \text{m} = 1 \text{ kg} \cdot \text{m}^2/\text{s}^2$$

Power is measured in watts; watts are defined as joules per second (J/s).

The unit for work and mechanical energy in the fps system is the foot-pound force (ft·lb$_f$). An empirical unit, the horsepower (hp), for measuring power is defined as

$$1 \text{ hp} = 550 \text{ ft} \cdot \text{lb}_f/\text{s}$$
$$1 \text{ Btu/s} = 778.2 \text{ ft} \cdot \text{lb}_f/\text{s}$$

Example 1.7

Two hundred kg of water is flowing through a pipe at a velocity of 3 m/s. What is the kinetic energy of this water in (a) Joules (b) ft·lb_f and (c) Btu?

Solution

(a) $\text{KE} = \dfrac{(1/2) \times 200\,\text{kg}}{} \times \left(\dfrac{3\,\text{m}}{\text{s}}\right)^2 = 900\,\dfrac{\text{kg} \cdot \text{m}^2}{\text{s}^2} = 900\,\text{J}$

(b) $\text{KE} = (1/2)\,200\,\text{kg} \times \dfrac{2.2046\,\text{lb}_m}{\text{kg}} \times \left(\dfrac{3\,\text{m}}{\text{s}} \times \dfrac{3.281\,\text{ft}}{\text{m}}\right)^2 \bigg/ 32.174\,\dfrac{\text{lb}_m \cdot \text{ft}}{\text{lb}_f \cdot \text{s}^2}$

$= 663.9\,\text{ft} \cdot \text{lb}_f$

(c) $\text{KE} = \dfrac{663.29\,\text{ft} \cdot \text{lb}_f}{778.2\,\text{ft} \cdot \text{lb}_f/\text{Btu}} = 0.8523\,\text{Btu}$

CONVERSION OF UNITS

We can convert units from one system to another by expressing the quantities in terms of the fundamental units (length, mass, time, temperature) and using the appropriate basic conversion factors as listed in Table 1.3.

Table 1.3 Basic conversion factors

Length:	1 meter = 3.281 ft	1 foot = 12 in
	1 mile = 5280 ft	1 foot = 0.3048 m
	1 mile = 1.609 km	1 foot = 30.48 cm
	1 inch = 2.54 cm	1 micron = 10^{-6} meter
	1 inch = 25.4 mm	1 Angstrom = 10^{-10} meter
Mass:	1 kg = 2.2046 lb_m	1 lb_m = 453.6 g
	1 U.S. ton = 2000 lb_m	1 lb_m = 0.4536 kg
	1 metric ton = 1000 kg	1 ton (British) = 2240 lb_m
	1 slug = 32.174 lb_m	1 gram = 15.43 grains
Time:	1 second = (1/3600) h	1 second = (1/60) minute
	1 hour = 60 minutes	1 hour = 3600 s
Force:	1 lb_f = 32.174 poundals	1 Newton = 10^5 dyn
	1 poundal = 0.138 Newton	1 lb_f = 4.448 Newton
Pressure:	1 Pa = 1 N/m²	1 atm = 1.013 Bar
	1 Bar = 1 × 10^5 Pa	1 atm = 1.03 kg/cm²
	1 atm = 33.91 ft of water	1 atm = 760 mm Hg
	1 atm = 14.696 psia	1 psi = 2.036 in. Hg at 0°C
	1 atm = 29.921 in. Hg at 0°C	1 psi = 2.311 ft of water at 70°F
Temperature:	$T_K = t_C + 273.15$	$T_R = t_F + 459.58$
	$t_C = (t_F - 32)/1.8$	$t_F = 1.8 t_C + 32$
Temperature difference:	$\Delta°C = \Delta K = 1.8\,\Delta°F$	$\Delta°F = \Delta°R$
Thermal units:	1 cal = 4.184 J	1 kcal = 4184 J
	1 Btu = 0.252 kcal	1 Btu = 778.2 ft · lb_f
	1 Btu = 252 cal	1 Chu = 1.8 Btu
	1 Joule = 10^7 ergs	1 erg = 1 dyn · cm

Examples 1.8 and 1.9 illustrate the conversion of units from one system to another.

Example 1.8

Calculate the conversion factors to convert (a) newtons to pound-force, and (b) atmospheres to pounds per square inch (psi).

Solution

(a) By definition, $1\ N = 1\ kg \cdot m/s^2$; so

$$1\ N = \frac{(1\ kg)(2.2046\ lb/kg)(1\ m)(3.281\ ft/m)}{s^2}$$

$$= 7.233\ lb \cdot ft/s^2.$$

Now, by definition, $1\ lb \cdot ft/s^2 = (1/32.17)\ lb_f$. Therefore,

$$1\ N = 7.233\ lb \cdot ft/s^2 = 7.233 \frac{1}{32.17}\ lb_f = 0.2248\ lb_f$$

(b) A standard atmosphere is defined by

$$1\ atm = 1.01325 \times 10^5 (kg \cdot m/s^2)/m^2$$

$$= \frac{1.01325 \times 10^5 [(2.2046\ lb)(3.281\ ft)/s^2]}{(3.281\ ft)^2}$$

$$= 6.8083 \times 10^4 (lb \cdot ft/s^2)/ft^2$$

$$= \frac{6.8083 \times 10^4 (lb \cdot ft/s^2)/ft^2}{g_c (lb \cdot ft/s^2 \cdot lb_f)}$$

$$= \frac{6.8083 \times 10^4 (lb_f/32.17)}{144\ in^2}$$

$$= 14.696\ lb_f/in.^2 \doteq 14.7\ psi$$

Conversion factors can be derived from the basic units and dimensions as shown in Example 1.8. However, using a table of derived conversion factors is more convenient. Some of the commonly required conversion factors are listed in Tables 1.4 and 1.5. For a more detailed discussion of units and dimensions, see McCabe and Smith, and Himmelblau. Additional conversion factors are given by Perry and Green.

As a simple example based on Table 1.5, to convert 2.5 cP into pascal-seconds, we note that $1\ cP = 0.001\ Pa \cdot s$. Hence, $2.5\ cP = 0.0025\ Pa \cdot s = 2.5\ mPa \cdot s$. Example 1.9 provides another illustration.

Table 1.4 Conversion factors

Multiply	By	To Obtain
Atm std	14.696	psia
Atm std	760	mm Hg
Atm std	29.92	In. Hg
Atm std	33.91	ft of water
Atm std	1.013×10^5	Pa
Atm std	1.013	Bar
Atm std	1.032	kg/cm^2
Atm std	101.3	kPa
Bar	1×10^5	Pa
Bar	1.0197	kg/cm^2
Bar	100	kN/m^2
Bar	14.507	psi
Btu	1054.4	Joule (J)
Btu	2.928×10^{-4}	kWh
Btu	778.2	$ft \cdot lb_f$
Btu	0.252	kcal
Btu/h	3.93×10^{-4}	hp
Btu/h	0.293	watt (W)
Btu/h	0.22	$ft \cdot lb_f/s$
$Btu/h \cdot ft^2 \cdot °F$	4.883	$kcal/h \cdot m^2 \cdot °C$
$Btu/h \cdot ft^2 \cdot °F$	5.674×10^{-3}	$kW/m^2 \cdot K$ or $°C$
$Btu/h \cdot ft^2 \cdot °F$	5.674	$J/(s \cdot m^2 \cdot K$ or $°C)$
$Btu/(h \cdot ft^2 \cdot °F/ft)$	1.49	$kcal/h \cdot °C \cdot m$
$Btu/(h \cdot ft^2 \cdot °F/ft)$	1.73	$J/s \cdot m \cdot K$ or $°C$
$Btu/lb \cdot °F$	1	$kcal/kg \cdot °C$
cP	1×10^{-3}	$N \cdot s/m^2$ or $Pa \cdot s$
cP	2.42	$lb/h \cdot ft$
cP	0.000672	$lb/s \cdot ft$
cP	3.6	$kg/h \cdot m$
cP	1	$mPa \cdot s$
ft	0.3048	m
ft^2	0.092937	m^2
ft^3	7.48	gal (U.S.)
$ft \cdot lb_f$	1.285×10^{-3}	Btu
$ft \cdot lb_f$	3.766×10^{-7}	kWh
$ft \cdot lb_f$	0.32	Cal
$ft \cdot lb_f$	1.36	Joule (J)
$ft \cdot lb_f/s$	1.818×10^{-3}	Hp
gal	3.785	Liters (L)
hp	33,000	$ft \cdot lb_f/min$
hp	42.4	Btu/min
hp	745.7	Watt (W)
$hp \cdot h$	2545.1	Btu
$hp \cdot h$	2.68×10^6	Joule (J)
joule (J)	9.478×10^{-4}	Btu
joule (J)	0.74	$ft \cdot lb_f$

Table 1.4 Conversion factors (*Continued*)

Multiply	By	To Obtain
joule (J)	1	N · m
joule (J)/s	1	Watt (W)
J/(s · m² · K or °C)	1	W/m² · K or °C
kcal	3.9685	Btu
kcal	1.56×10^{-3}	hp · h
kcal	4.184×10^3	Joule (J)
kN/m²	0.295	in Hg
kPa	0.145	Psi
kW	1.34	hp
kW	737.6	ft · lb$_f$/s
kWh	3413	Btu
kWh	3.6×10^6	Joule (J)
kip	1000	lb$_f$
m³	35.3147	ft³
m³	U.S. gallon	264.17
mile	1.609	km
micron	1×10^{-6}	meter
poundal	0.14	Newton
Pound-cal	1.8	Btu
psi	6.895	kN/m²
watt	3.41	Btu/h
watt	1	Joule/s
W/m²	0.317	Btu/h · ft²
W/m² · K	0.1761	Btu/h · ft² · °F
W/(m² · K/m)	0.58	Btu/(h · ft² · °F/ft)

Table 1.5 Units of absolute viscosity

Centipoise or 0.01 g/cm · s or 0.01 Poise	Slug/ft · s or lb$_f$ · s/ft²	lb/ft · s or Poundal · s/ft²	Pa · s or Kg/m · s
1*	2.09×10^{-3}	0.000672	0.001
47,900	1	32.2 or g	47.9
1487	(1/g) or 0.0311	1	1.487
1000	0.02088	0.672	1

Units of Kinematic Viscosity**		
Centistokes	**ft²/s**	**m²/s**
1	1.075×10^{-5}	1×10^{-6}
92,900	1.0	0.0929
1×10^6	10.7643	1

*Another conversion factor: 1 centipoise = 2.42 lb/ft · h.
**ν = kinematic viscosity = absolute viscosity/density = μ/ρ.

Example 1.9

For heating of air in turbulent flow through a tube, an experimental value of 64.3 W/m² · K was obtained for the heat transfer coefficient h. What is the value of the heat transfer coefficient in Btu/h . ft² · °F?

Solution

The conversion factor to convert from W/m² · K to Btu/h · ft² · °F is 0.1761 from the table. Therefore,

$$64.3 \text{ W/m}^2 \cdot \text{K} = 64.3 \frac{\text{W/m}^2 \cdot \text{K}}{} \times \frac{0.1761 \text{ Btu/h} \cdot \text{ft}^2 \cdot °F}{\text{W/m}^2 \cdot \text{K}} = 11.32 \frac{\text{Btu}}{\text{h} \cdot \text{ft}^2 \cdot °F}$$

DIMENSIONAL FORMULAS

The dimensional formula of a quantity in derived units shows how the fundamental units in which the quantity is measured are related. As an example, consider the derived quantity, the density of a liquid, which is measured in lb_m/ft^3. Here, the fundamental unit M represents the lb_m and L represents the length in ft. The dimensional formula of density is therefore

$$[\rho] = M/L^3$$

where symbol "$[\rho]$" is used to indicate that the righthand side of the equation is the dimensional formula of the derived quantity ρ. Likewise the dimensional formula of acceleration will be

$$[a] = L/t^2$$

because acceleration is defined as velocity per unit time (dimension t) and velocity is defined as the distance (dimension L) per unit time (dimension t).

DIMENSIONAL AND DIMENSIONLESS EQUATIONS

An equation in which all terms have the same units or dimensions is called a *dimensionally homogenous equation*. These units can be basic, or derived. Equations derived mathematically using basic physical laws do have all terms of the same dimensions and are homogenous. Hence they can be used with any system of units (basic and derived) provided the same units are used for all the terms in the equation. Units used in this way are called *consistent units*. No conversion factors are required when consistent units are used.

Example 1.10

The Hagen-Poiseuille equation for laminar flow of a liquid through a pipe of circular crosssection is as follows:

$$\Delta P = \frac{32.174 L u \mu}{g_c D^2}$$

where
 ΔP = pressure drop, lb_f/ft^2
 L = Length of tube, ft
 u = linear velocity, ft/s
 D = inside diameter of pipe, ft
 μ = viscosity, $lb_m/ft \cdot s$
 g_c = 32.174 $lb_m \cdot ft/lb_f \cdot s^2$

Is this equation dimensionally homogeneous?

Solution

This equation has been derived theoretically using principles of physics, so it should be dimensionally homogenous. We can check this conclusion by substituting the dimensions on both sides of the equation. From the data, it is apparent that four fundamental units are used, namely M, L, t, and F. The dimensions of the quantities are

ΔP	pressure drop	F/L^2
L	Length of pipe	L
u	velocity	L/t
D	diameter of pipe	L
g_c	proportionality constant	ML/Ft^2
μ	viscosity	M/Lt

Dimensions of lefthand side of equation are F/L^2.

Dimensions of righthand side are $\dfrac{L(L/t)(M/Lt)}{(ML/Ft^2)L^2} = \dfrac{F}{L^2}$.

Because all the terms in the equation have the same dimensions, the equation is homogenous.

Equations derived by empirical methods without using dimensional consistency contain terms of varying dimensions. These equations are not dimensionally homogenous. Different units of the same type such as inches and feet may appear in the same equation. For example, the equations for the rate of heat loss from the surfaces to atmosphere by free convection are of this type. One such equation for long vertical pipes is

$$h_c = 0.4 \left(\frac{\Delta t}{d_o} \right)^{0.25}$$

This equation is not dimensionally homogenous. The units of heat transfer coefficient are $Btu/h \cdot ft^2 \cdot °F$ while Δt is in $°F$ and d_o is in inches. If the temperature difference is to be expressed in $°C$, the numerical coefficient needs to be changed to $0.4(1/1.8)^{0.25} = 0.3453$.

Another example of dimensionally non-homogenous equation is the correlation for the heat transfer coefficient for condensation on horizontal surfaces.

$$\bar{h} = 0.725 \left(\frac{k_f^3 \rho^2 \lambda g}{\mu_f D_o \Delta t_f} \right)^{0.25}$$

In this equation also, if the units are changed, the value of the constant has to be changed.

DIMENSIONAL ANALYSIS

Many important physical process problems such as those occurring in fluid dynamics, heat transfer, and mass transfer cannot be solved by rigorous theoretical or mathematical methods. Many a time analysis will yield differential equations, which cannot be integrated easily. In such cases empirical relations could be found by experimentally studying the effects of different variables involved. However, if the number of variables is large (>3), the experimental work required is tremendous and time consuming. Moreover, it is very difficult to correlate such a large amount of data into an equation, which can be used for practical calculations.

The dimensional analysis method allows overcoming the difficulties of formal mathematical development and extensive experimental study. This method is based on the principle that if a theoretical relationship exists among the variables involved in the physical process, the equation expressing that relationship must be dimensionally homogeneous. It is possible to group many variables into a smaller number of dimensionless groups. The final equation consists of these dimensionless groups, thereby reducing the number of independent variables. Experimental work, though considerably reduced, is required to establish the quantitative aspect of the relationship between the groups.

There are four methods to carry out dimensional analysis. They are

1. Raleigh's algebraic method
2. Buckingham's π method
3. Differential equation method
4. Geometric, kinematic, and dynamic similarities that indicate the significance of dimensionless groups.

A detailed discussion of these methods and their application is beyond the scope of the P.E. examination and this book. However, a few examples will be given to demonstrate the application of dimensional analysis.

Raleigh's Method

This is an algebraic method. Example 1.11 illustrates its application.

Example 1.11

Consider the problem of finding pressure drop due to friction for a fluid flowing through a circular pipe of internal diameter D and of length L. The variables involved and their dimensions are given in the following table. Fundamental dimensions used are M, L, F, and t (American Engineering System). Table 1.6 lists the variables and their dimensions for this example.

Table 1.6 Variables and their dimensions for example 1.11

Variable Name	Symbol	and Units	Dimensions
Pressure drop	$-dP_f$	lb_f/ft^2	F/L^2
Pipe internal diameter	D	ft	L
Pipe length	dL	ft	L
Pipe roughness	ε	ft	L
Fluid velocity in pipe	u	(ft/s)	L/t
Fluid absolute viscosity	μ	$(lb_m)/ft \cdot s$	$M/L \cdot t$
Dimensional constant g_c	g_c	$lb_m \cdot ft/lb_f \cdot s^2$	ML/Ft^2
Fluid density	ρ	lb_m/ft^3	M/L^3

Solution

The equation in terms of the variables is

$$-dP_f = KD^a (dL)^b \varepsilon^c u^d \rho^e \mu^h g_c^j$$

where K is a constant

Substituting the dimensions of the variables listed in Table 1.6, the following dimensional equation is obtained:

$$\frac{F}{L^2} = (L^a)(L^b)(L^c)\left(\frac{L}{t}\right)^d \left(\frac{M}{L^3}\right)^e \left(\frac{M}{Lt}\right)^h \left(\frac{ML}{Ft^2}\right)^j$$

Applying the condition of dimensional homogeneity, the following set of equations is obtained by equating the exponents of dimension.

Exponent of F: $1 = -j$
Exponent of M: $0 = e + h + j$
Exponent of L: $-2 = a + b + c + d - 3e - h + j$
Exponent of t: $0 = -d - h - 2j$

There are four equations and seven constants to determine. Therefore we can determine four of the seven constants in terms of the remaining three. Let b, c, and h be these three constants. Then values of a, d, e, and f are

$$a = -b - c - h$$
$$d = 2 - h$$
$$e = 1 - h$$
$$j = -1$$

Substituting these values in the pressure drop equation, the following relation is obtained:

$$-dp_f = KD^{-b-c-h}(dL)^b \varepsilon^c u^{2-h} \rho^{1-h} \mu^h g_c^{-1}$$

or

$$\frac{g_c(-dp_f)}{\rho u^2} = K\left(\frac{dL}{D}\right)^b \left(\frac{Du\rho}{\mu}\right)^{-h} \left(\frac{\varepsilon}{D}\right)^c$$

The eight variables are now correlated by four dimensional groups. Since dP is proportional to the length of pipe, b can be set equal to 1.

$$\frac{g_c(-dp_f)}{\rho u^2} \text{ is called Euler number.}$$

$$\frac{Du\rho}{\mu} \text{ is called Reynolds number.}$$

Buckingham's π Method

Buckingham's theorem states that the functional relationship between q variables describing a physical process can be expressed in terms of $q - u$ dimensionless groups (often called π groups) where u is the number of fundamental dimensions used to describe the q quantities. [Actually u is the number of variables that will not form a dimensional group. However, in most cases u is equal to the number of fundamental units.] An example to illustrate the use of the π theorem is given next.

Example 1.12

Let us reconsider Example 1.11 with the following changes. We use three fundamental dimensions, M, L, and t. From Example 1.11, we omit the variable ε thus reducing the number of quantities to six as the proportionality factor g_c is not required. Table 1.7 lists the variables and their dimensions for this case.

Table 1.7 List of variables and their dimensions for pipe flow problem of example 1.13

Variable Name	Symbol	and Units	Dimensions
Pressure drop	$-dP_m$	$lb_m/ft \cdot s^2$	M/Lt^2
Pipe internal diameter	D	ft	L
Pipe length	dL	ft	L
Fluid velocity in pipe	u	(ft/s)	L/t
Fluid absolute viscocity	μ	$(lb_m)/ft \cdot s$	$M/L \cdot t$
Fluid density	ρ	lb_m/ft^3	M/L^3

Solution

There are six variables, so $q = 6$.
Fundamental dimensions are $u = 3$. They are mass M, Length L, and Time t.
By Buckingham's Theorem, the number of dimensionless groups or π's = $q - u = 6 - 3 = 3$. Thus,

$$\pi_1 = f(\pi_2, \pi_3) \tag{1.3}$$

It is now necessary to select a core group of variables numbering $u = 3$. These will appear in each π group and among them contain all the basic or fundamental dimensions. Among core numbers no two variables selected can have the same dimensions. Thus, both D and L cannot be selected as they have the same dimension of length. In addition we are interested in obtaining an expression for pressure drop, therefore it is preferable to keep the variable ΔP separate and independent from other groups. When these criteria are applied, four variables D, u, ρ, and μ remain. We select D, u, and ρ. Then the three dimensionless groups are

$$\pi_1 = D^a u^b \rho^c \Delta p^1 \tag{1.3a}$$

$$\pi_2 = D^d u^e \rho^f L^1 \tag{1.3b}$$

$$\pi_3 = D^g u^h \rho^i \mu^1 \tag{1.3c}$$

The variable exponents must be evaluated so that the group is dimensionless. Consider the first group (1.3a) first. Substituting the dimensions in the equation one obtains

$$M^0 L^0 t^0 = 1 = L^a \left(\frac{L}{t}\right)^b \left(\frac{M}{L^3}\right)^c \left(\frac{M}{Lt^2}\right)$$

Next by equating exponents of M, L, and t on both sides in turn, the following set of three simultaneous equations is obtained:

(L) $0 = a + b - 3c - 1$
(M) $0 = c + 1$
(t) $0 = -b - 2$

Solving these equations, $b = -2$, $c = -1$, and $a = 0$. Insertion of these values in the equation for the π_1 group, the expression for π_1 is

$$\pi_1 = \frac{\Delta p}{u^2 \rho} = N_{Eu}$$

Following the same procedure, $\pi_2 = \frac{L}{D}$ and $\pi_3 = \frac{Du\rho}{\mu} = N_{Re}$.

Combining the three groups together, we obtain

$$\frac{\Delta P}{\rho u^2} = f\left(\frac{Du\rho}{\mu}, \frac{L}{D}\right)$$

If we compare the results of Examples 1.11 and 1.12, we can see how important it is to include all the variables affecting a process. Since we omitted the variable ε (roughness factor) from this analysis, the effect of roughness factor is completely missed. Dimensional analysis will give correct description of a process as long as all the variables are included in the dimensional analysis.

Sometimes a process analysis yields a differential equation, which cannot be integrated. However, it can be used to obtain the dimensionless groups, which can be used to correlate the experimental data. Example 1.13 illustrates the method.

Example 1.13

The x component of the Navier-Stokes equation (see Geankoplis) for momentum transfer is

$$u_x \frac{\delta u_x}{\delta x} = g_x - \frac{1}{\rho} \frac{\delta p}{\delta x} + \frac{\mu}{\rho} \frac{\delta^2 u_x}{\delta x^2}$$

Use this equation to obtain the dimensional groups and explain their significance.

Solution

Dimensional equality for this equation can be written with the use of two characteristic factors u and L as follows:

$$\left[\frac{u^2}{L}\right] = [g] - \left[\frac{p}{\rho L}\right] + \left[\frac{\mu u}{\rho L^2}\right]$$

Each term in the preceding equation has the dimensions of L/t^2. In the equation the lefthand side represents the inertia force. The terms on the righthand side represent gravity force, pressure force, and viscous force respectively. By dividing each of the terms of the equation by the inertia force on the left, the following dimensional groups or their reciprocals are obtained:

$$\frac{[u^2/L]}{[g]} = \frac{\text{inertia force}}{\text{gravity force}} = \frac{u^2}{gL} = N_{Fr} \quad \text{(Froude number)}$$

$$\frac{[p/\rho L]}{[u^2/L]} = \frac{\text{pressure force}}{\text{inertia force}} = \frac{p}{\rho u^2} = N_{Eu} \quad \text{(Euler number)}$$

$$\frac{[u^2/L]}{[\mu u/\rho L^2]} = \frac{\text{inertia force}}{\text{viscous force}} = \frac{Lu\rho}{\mu} = N_{Re} \quad \text{(Reynolds number)}$$

The differential equation method not only gives the dimensionless groups but also gives physical meaning to them.

Dynamic Similarity

Two geometrically similar systems are dynamically similar if the parameters representing ratios of forces pertinent to the process situation in both the systems are equal. Thus the Reynolds', Froudie, and Euler numbers must be equal for the two systems.

The dynamic similarity is very important in obtaining data for scale up, for example, in the scale up of chemical process equipment. Experimentation with full-scale prototype is not only expensive but also difficult and therefore experiments are done with small-scale models.

RECOMMENDED REFERENCES

References reflect latest editions of the books cited. Recent older editions are equally appropriate for readers who do not have access to new editions of the books.

1. McCabe and Smith, *Unit Operations of Chemical Engineering,* 6th ed., McGraw-Hill, 2000.
2. Geankoplis, *Transport Processes and Separation Process Principles,* 4th ed., Prentice-Hall (Pearson Education Inc), 2003.
3. Himmelblau, *Basic Principles and Calculations in Chemical Engineering,* 6th ed. Prentice-Hall, 2004.
4. Perry and Green, *Chemical Engineers' Handbook,* Platinum ed., McGraw-Hill, 1999.

CHAPTER 2

Material Balances

OUTLINE

MATERIAL PROPERTIES (DEFINITIONS AND RELATIONSHIPS) 24
Density and Specific Gravity ■ Mole and Weight Fractions ■ Concentrations

MASS BALANCES 28
Definition of a System ■ The General Accounting Principle ■ Mass Balance Equation ■ Basis of Calculations ■ Chemical Equation and Stoichiometry

PHASE BEHAVIOR OF SUBSTANCES 37
Vapor Pressure of Liquids ■ Vapor Pressure of Solids

IDEAL GASES 40
The Ideal Gas Law ■ Ideal Gas Mixtures and Partial Pressure

REAL GASES (NON-IDEAL) 45
Reduced Parameters ■ Compressibility Factor ■ Equation of State ■ Virial Equations of State ■ Gaseous Mixtures

MATERIAL BALANCE CALCULATION 54
Solving Mass Balance Problems not Involving Chemical Reactions ■ Material Balances Involving Multiple Subsystems ■ Process Variants: Recycle, Bypass, and Purge Calculations ■ Material Balances of Processes Involving Chemical Reaction

FUELS AND COMBUSTION 67
Heating Values of Fuels ■ Analysis of Fuels ■ Calculation of Heating Values of Fuels ■ Terminology in Combustion

RECOMMENDED REFERENCES 76

Industrial process analysis involves application of the principles of chemistry and physics and requires understanding of the properties of materials, behavior of gases, liquids, and solids under different conditions of temperature and pressure, and mass and energy balances. Mass and energy balances are a very important aspect of process engineering and the starting points for the solution of chemical engineering process problems. Accounting of both materials and energy is necessary not only for economic reasons but also to identify the waste material streams and to find ways for their proper disposal to avoid adverse effects on the environment. No wonder in the P.E. examination a great deal of importance is attached to this aspect of chemical engineering and about 24 percent of the exam is assigned to mass, energy balances, and thermodynamics. In this chapter, we will review the following topics: stoichiometry, laws of conservation of mass, properties of materials and their phase behavior, process variants such as bypass, recycle, and purge, and combustion processes.

MATERIAL PROPERTIES (DEFINITIONS AND RELATIONSHIPS)

In carrying out mass and energy balances, many derived quantities defined in terms of the fundamental units are used because of their convenience. This section briefly recapitulates several such quantities. A few more will be defined when energy balance is reviewed later in this chapter.

Density and Specific Gravity

Density of a substance is the ratio of its mass per unit volume, ρ = mass/volume, expressed as, for example, kg/m^3 or lb/ft^3. It has a numerical value and units. Densities of solids and liquids do not change very much with pressure. Liquid densities do change with temperature and if a mixture of liquids with both temperature and composition.

Specific gravity of a substance is the ratio of the density of a substance to the density of a reference substance. Thus, specific gravity, sp. gr. = ρ/ρ_r, where ρ = density of a substance and ρ_r = the density of the reference substance in consistent units. The reference substance for liquids and solids is water at 4°C (ρ_r = 1000 kg/m^3 or 62.4 lb/ft^3). In the SI system, the density of water at 4°C is very close to 1.0000 g/cm^3 and density (in g/cm^3) and specific gravity have almost the same numerical value.

Other specific gravity conventions are in use. For example, in the petroleum industry the API gravity scale is used and is defined as $°API = \frac{141.5}{sp\,gr(at\,60°F)} - 131.5$ and sp gr(at 60°F) = $\frac{141.5}{°API+131.5}$.

Other specific gravity scales include Baumè, Brix, and Twadell. Relationships among the various systems of density are given by Perry and Green.

Example 2.1

Specific gravity of normal propyl bromide is 1.353. What is its density in
(a) g/cm^3 (b) kg/m^3 (c) lb$_m$/ft^3?

Solution

(a) Neither temperature nor reference substance are given. So assume water as the reference substance and same temperatures. Then

$$\text{Density of nPB} = \frac{1.353 \frac{g\,nPB}{cm^3}}{1.00 \frac{g\,H_2O}{cm^3}} \left| \frac{1\,g\,H_2O}{cm^3} \right. = 1.353 \frac{g\,nPB}{cm^3}$$

(b) $$\text{Density of nPB} = \frac{1.353 \frac{kg\,nPB}{m^3}}{1.00 \frac{kg\,H_2O}{m^3}} \left| \frac{1.00 \times 10^3\,kg\,H_2O}{m^3} \right. = 1353 \frac{kg\,nPB}{m^3}$$

(c) $$\text{Density of nPB} = \frac{1.353 \frac{lb_m\,nPB}{ft^3}}{1 \frac{lb_m\,H_2O}{ft^3}} \left| 62.4 \frac{lb_m\,H_2O}{ft^3} \right. = 84.43 \frac{lb_m\,nPB}{ft^3}$$

Material Properties   25

Example 2.2

Specific gravity of a petroleum liquid is 0.67 60°F/60°F. What is its °API?
Solve (a) from Table 1.11 in Perry and Green, (b) by calculation using the equation relating sp gr to °API.

Solution

(a) From Table 1.12, reading across at 0.67, °API = 79.69

(b) $°API = \dfrac{141.5}{spgr} - 131.5 = \dfrac{141.5}{0.67} - 131.5 = 79.694$

Example 2.3

1000 gal of gas oil (sp gr = 0.887) is mixed with 4000 gal of fuel oil of sp gr = 0.966. Assuming there is no volume change on mixing, calculate the following:

(a) the volume of the mixture in m^3

(b) the density of the oil mixture in kg/m^3

Solution

(a) Since volumes are additive (no volume change on mixing) we can calculate volumes of the two oils as follows:

$$\text{Volume of gas oil} = \dfrac{1000 \text{ gal}}{7.48 \text{ gal/ft}^3 \mid 35.3 \text{ ft}^3/m^3} = 3.8 \text{ m}^3$$

$$\text{Volume of fuel oil} = \dfrac{4000 \text{ gal}}{7.48 \text{ gal/ft}^3 \mid 35.3 \text{ ft}^3/m^3} = 15.15 \text{ m}^3$$

Total volume = 3.8 + 15.15 = 18.95 m^3

(b) There is no volume change.

Vol fraction of gas oil = 1000/(1000 + 4000) = 0.2
Vol fraction of fuel oil = 4000/(1000 + 4000) = 0.8
Sp gr of mixture = 0.2 × 0.887 + 0.8 × 0.966 = 0.9502

$$\text{Density of mixture} = \dfrac{\frac{0.9502 \text{ kg mixture}}{m^3}}{1.00 \frac{\text{kg H}_2\text{O}}{m^3}} \cdot \dfrac{1.00 \times 10^3 \text{kg H}_2\text{O}}{m^3} = 950.2 \text{ kg/m}^3$$

Gas Densities and Specific Gravities

Density of a gas is expressed in g/cm^3 or lb/ft^3. If not otherwise specified, the volume is at the standard conditions of temperature and pressure (0°C and 1 atm). *Specific gravity* of a gas is defined as the ratio of the density of the gas at a given temperature and pressure to that of air or any other specified reference gas, usually at the same temperature and pressure but sometimes other conditions for the reference gas may be specified. When using gas-specific gravity data from the literature, therefore, one must be careful to ascertain the specified conditions of temperature and pressure for both the gas of interest and the reference gas.

Specific volume of any substance is the reciprocal of density and is given by $\hat{V} = 1/\rho$.

We will revert to specific gravity of a gas later in this chapter when we will consider the phase behavior of substances.

Mole and Weight Fractions

Compositions of mixtures of gases, liquids, and solids can be expressed in mole fractions or mass or weight fractions. Thus if A is a component of a mixture,

$$\text{Mass fraction } A = \frac{\text{mass of component } A}{\text{Total mass of all components in the mixture}}$$

$$\text{Mass \% } A = \text{mass fraction} \times 100$$

and $\text{Mol fraction } x_A \text{ of } A = \dfrac{\text{Moles of component } A}{\text{Total moles of all components in the mixture}}$

where
x_A = mol fraction of component A.
Mol % = mole fraction × 100.

Mol fraction in vapor phase is usually denoted by y, and in liquid it is denoted by x.

Concentrations

Concentration is defined as the amount of some solute per fixed amount of solvent or solution in a mixture. There are various ways of expressing the concentration of a component in solution of two or more components.

$$\text{Volume concentration of } A = V_A/V_M$$

where
V_A = Volume of component A
V_M = Volume of mixture

V_A and V_M are at the temperature and pressure of the mixture

$$\text{Mass concentration of a species in a mixture} = (\text{Mass of species } i)/\text{Volume of mixture}$$

$$\text{Average molecular weight of a gas mixture } M_{av} = \sum x_i M_i$$

where x_i and M_i are the mol fraction and molecular weight of component i

Molar concentration of a species i in a mixture = (Moles of species i)/Volume of mixture

Molarity = M = (g mol of species i in a mixture)/1 liter of solution

Normality = N = (gram-equivalent weight of species i)/1 liter of solution

Molality = m = g mol of species i/kg solvent

ppm or parts per million is a method of expressing concentrations of extremely dilute solutions. This is reviewed in a later chapter.

Flow Rates
Flow rate is the amount of material that passes a given reference point in a system per unit of time. Mass flow rate and volumetric flow rate are the most important methods of expressing flow rates.

Example 2.4

A solution of sodium bichromate ($Na_2Cr_2O_7$) in water contains 511.6 g of the bichromate per liter. The density of this solution is 1.279 g/cm^3 at 15°C. Calculate the following:

(a) Composition in weight percent

(b) Composition in mol percent

(c) Composition in atomic percent

(d) Volumetric composition of water

(e) Molality.

Solution

Basis: One liter of solution as basis of calculation.

$$\text{Weight of 1 liter of solution} = \frac{1000 \text{ cm}^3}{\text{liter}} \bigg| \frac{1.279 \text{ g}}{\text{cm}^3} = 1279 \text{ g}$$

Mass of water in 1 liter of solution = 1279 − 511.6 = 767.4 g

Molecular weight of $Na_2Cr_2O_7$ = 2 × 23 + 2 × 52 + 7 × 16 = 262

Mols of $Na_2Cr_2O_7$ = 511.6/262 = 1.953 g mol

Mols of H_2O = 767.4/18 = 42.633 g mol

Total mols = 1.953 + 42.633 = 44.586 g mol

(a) mass (weight) percent $Na_2Cr_2O_7 = \frac{511.6}{1279 \text{ g}} \times 100 = 40\%$

mass (weight) percent H_2O = 100 − 40 = 60 %

(b) mol% of sodium bichromate = $\frac{1.953}{44.586} \times 100 = 4.38$

mol % of water = 100 − 4.38 = 95.62

(c) Atomic composition

   g atoms of Na = 2 × 1.953 = 3.90
   g atoms of Cr = 2 × 1.953 = 3.90
   g atoms of O = 7 × 1.953 = 13.65
   g atoms of H = 2 × 42.63 = 85.26
   g atoms of O = 1 × 42.63 = 42.63
   Total g atoms = 3.9 + 3.9 + 13.65 + 85.26 + 42.63 = 149.34
   Atomic % Na = (3.9/149.34) × 100 = 2.61
   Atomic % Cr = (3.9/149.34) × 100 = 2.61
   Atomic % O = [(13.65 + 42.63)/149.34] × 100 = 37.69
   Atomic % H = (85.26/149.34) × 100 = 57.09

(d) Density of water = 1 g/cm^3

$$\text{Volume of water in 1 liter of solution} = \frac{767.4 \text{g}}{1 \text{g/cm}^3} = 767.4 \text{ cc in one liter}$$

Therefore vol % water = (767.4 cc/1000 cc) × 100 = 76.74

(e) Solute per 767.4 g of water = 1.953 g mol

$$\text{Molality} = (1.953 \text{ g mol}) \times (1000 \text{ g } H_2O/767.4 \text{ g } H_2O)$$
$$= 2.55 \text{ g mol } Na_2Cr_2O_7/1000 \text{ g } H_2O$$

Example 2.5

A liquid mixture of benzene, toluene, and ethyl benzene has the following composition by weight. What is the average molecular weight of the mixture? Calculate also the component mol fractions.

$$\text{Benzene } (C_6H_6) \text{ 50\% by wt}$$
$$\text{Toluene } (C_7H_8) \text{ 30\% by wt}$$
$$\text{Ethyl benzene } (C_8H_{10}) \text{ 20 \% by wt}$$

Solution

Component	Mol. wt	kg	kg mol	Mol fraction
Benzene	78	50	0.641	0.5545
Toluene	92	30	0.326	0.2820
Ethyl benzene	106	20	0.189	0.1635
Total		100	1.156	1.000

Basis: 100 kg of mixture

Average molecular weight = 100 kg/1.156 kg mol = 86.51 kg/kg mol

Calculated mol fractions are given in the last column of preceding table.

MASS BALANCES

To make an accounting of mass or any other quantity such as atoms of element for a certain interval of time, it is necessary to define a system in which a process or processes are occurring and for which the balance is to be made, and specify the boundaries of that system. Boundaries of the system can be real or imaginary, and stationary or movable. Another aspect of mass balance is the basis on which the calculations are to be made. In addition, when the process involves chemical reaction or reactions, concepts of chemical equation and stoichiometry are basic to write and balance the reaction equations and establish stoichiometric relationship between the reactants and products. In the following sections, we review properties of materials, mass balance concepts, the basis of calculation, and chemical equation and stoichiometry.

Definition of a System

A system is defined as any arbitrary portion of the universe selected or set apart for study or analysis. It may be a specified volume in space or a given quantity of matter confined in a bounded space. It may contain a substance or group of substances undergoing a process. Everything outside the boundary of the system is called its surroundings. For practical reasons it is the immediate surroundings that are important. Figure 2.1 illustrates the concept of a system and its surroundings.

The system is enveloped by the boundary and confined within. Outside the boundary of the system are the surroundings. There are some input streams and some output streams. In addition, there might be exchange of energy in the form of heat or work (not shown in the figure) across the boundaries.

An open, also called flow, system is one in which material transfer takes place across the boundaries of the system. A closed system is one in which there is no mass transfer across the boundaries of the system during the time period chosen for accounting. For example, in a batch reactor there would be no flow of material

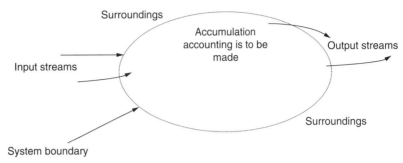

Figure 2.1 Concept of a system

into the reactor during the time interval of reaction and no product withdrawal either until the reaction process is complete. If attention is focused on the reaction time only, the system can be considered as a closed system because there is no input or output during the time of reaction.

The state of a system is described by a number of properties. The system properties depending on its mass are called *extensive* properties. *Intensive* properties of a system are independent of its mass. For example, density and temperature are intensive properties. Volume, however, is an extensive property as it depends on the extent or amount of mass of the system.

The General Accounting Principle

If a system is undergoing a process, physical, chemical or nuclear, changes in its mass, energy, or any other measurable quantity can be accounted for a given interval of time using the accounting principle, which can be expressed by the following general equation:

$$Q_E - Q_B = \Sigma Q_I - \Sigma Q_O + \Sigma Q_P \tag{2.1}$$

where
  Q = the physical quantity chosen for accounting
  Q_E = the quantity at the end of the accounting period
  Q_B = the quantity at the beginning of the accounting period
  Q_I = the quantity that enters the system during the accounting period
  Q_O = the quantity that leaves the system during the accounting period
  Q_P = the quantity created or destroyed during the accounting period.

In ordinary processes involving chemical (not nuclear) reactions, $\Sigma Q_P = 0$. If the quantity of matter in the system does not change with time, $Q_E - Q_B = 0$ and the accounting equation reduces to

$$\Sigma Q_I - \Sigma Q_O = 0$$

In this case, input equals output. The system is said to be operating under *steady-state condition*.

Mass Balance Equation

One of the most important applications of the general accounting principle in process analysis is the mass balance of a system over a chosen accounting period.

The mass balance equation in its most general form for a given system can be directly written from the accounting principle as follows:

$$M_E - M_B = \sum M_I - \sum M_O + \sum M_p - \sum M_C \quad (2.2)$$

where
- M_E = mass of the system at the end of the accounting period
- M_B = mass of the system at the beginning of the accounting period
- $M_E - M_B$ = accumulation in the system during the accounting period
- $\sum M_I$ = sum of all masses that entered the system during the accounting period
  = **input**
- $\sum M_O$ = sum of all masses that left the system during the accounting period
  = **output**
- $\sum M_p$ = mass changes resulting from physical or chemical transformations
- $\sum M_C$ = mass changes connected with atomic transmutations and relativistic effects

The term $\sum M_C$ is zero in ordinary chemical and physical transformations.

[*Note:* The equations as presented for mass balance here and later for energy balance are purposefully presented in general form for a finite difference period. From this general equation, one derives relations for steady-state flow and batch as well as unsteady state processes by applying relevant constraints.]

The preceding Equation 2.2 can also be written in non-symbolic form as

$$\begin{Bmatrix} \text{mass} \\ \text{accumulation} \\ \text{within the system} \end{Bmatrix} = \begin{Bmatrix} \text{mass input} \\ \text{through system} \\ \text{boundaries} \end{Bmatrix} - \begin{Bmatrix} \text{mass output} \\ \text{through system} \\ \text{boundaries} \end{Bmatrix} + \begin{Bmatrix} \text{mass generation} \\ \text{within the system} \end{Bmatrix} - \begin{Bmatrix} \text{mass consumption} \\ \text{within the system} \end{Bmatrix}$$

$$(2.3)$$

When applying the mass balance equation to a given system under study, the following points should be kept in mind:

1. If the system undergoes only physical changes, $\sum M_p = 0$ and $\sum M_C = 0$.

$$M_E - M_B = \sum M_I - \sum M_O \quad (2.2a)$$
$$\text{or} \quad \text{accumulation} = \text{input} - \text{output} \quad (2.3a)$$

2. If the system undergoes a chemical transformation and no atomic transmutation, $\sum M_C = 0$ and $\sum M_p$ may be zero or non-zero depending on the system chosen. For the system as a whole $\sum M_p = 0$ (law of conservation of mass applies). In this case Equations 2.2 and 2.3 reduce to

$$M_E - M_B = \sum M_I - \sum M_O \quad (2.2b)$$
$$\text{accumulation} = \text{input} - \text{output} \quad (2.3b)$$

If there is no accumulation within the system and no generation or consumption, mass balance equation reduces to

$$\sum M_I = \sum M_O \quad (2.2c)$$
$$\text{Input} = \text{output} \quad (2.3c)$$

This is the most common form of the mass balance equation in carrying out mass balances for physical and chemical processes encountered in process analysis.

3. If one of the component species is chosen as a system, $\sum M_p$ is not equal to zero.

4. If a mole balance is considered, moles may not be conserved.

5. If there is no flow in and out of the system, and $\Sigma M_C = 0$, Equations 2.2 and 2.3 reduce for conservation of one species of matter within a closed system to

$$M_E - M_B = \Sigma M_p \qquad (2.2d)$$

accumulation = (generation − consumption) of the species considered. (2.3d)

The mass balance may be a total material balance or on each species in the process. Processes that do not involve chemical reactions are typically evaporation, distillation, liquid-liquid extraction, absorption, and so on. These processes involve only material and energy balances.

Basis of Calculations

A basis is the reference point chosen to solve a particular problem. A proper selection of the basis makes it easier to solve the problem. A convenient quantity of material is specified as the basis. It may be hourly flow rate, 1 or 100 kg or 1 or 100 moles. As an example, consider the problem of Example 2.5. In this example the composition is given in wt %. So it is convenient to use 100 kg as basis rather than 100 kg mol. If the composition were given in mol %, the more convenient basis would be 1.00 kg mol, as will be clear from Example 2.5a.

Example 2.5a

In Example 2.5 assume the composition is in mol% instead of wt%. Calculate the average molecular weight of the mixture.

Solution

Here the more convenient basis will be 1 kg mol as is shown in the solution in the following table.

Component	Mol. wt	kg mol	kg
Benzene	78	0.50	39.0
Toluene	92	0.30	27.6
Ethyl benzene	106	0.20	21.2
Total		1.00	87.8

Basis: 1.00 kg mol of liquid mixture

Therefore average molecular weight is 87.8 kg/kg mol.

When a material balance is to be made for a continuous plant, hourly mass flow rate will be a more relevant and convenient basis rather than, say, annual production as equipment sizing will be done on the basis of hourly rates. On the other hand in a batch reactor operation, the material quantity of one batch would be a convenient basis. The basis should be clearly stated at the beginning of the solution.

Chemical Equation and Stoichiometry

Chemical equation represents the reaction that takes place between the reacting compounds to produce product compounds. It provides information as to how the moles of reacting compounds combine and in what proportion. Conservation laws of mass and energy are obeyed by chemical reactions. In addition, chemical reactions take place according to the following laws.

Law of Definite Proportion

This states that a chemical compound always contains the same elements and these elements will be in the same proportion in each and every molecule of the substance.

Law of Multiple Proportion

If two or more elements combine to form more than one compound, they will combine in weight multiples that are in the ratio of whole numbers.

Law of Reciprocal Proportion

The weights of two or more chemical compounds, which react separately with identical weights of a third, are simple multiples of the weights that react with each other.

It should be noted that in an ordinary chemical reaction, atoms remain undivided although this is not true in the case of nuclear reactions. Elements are conserved in a chemical reaction.

Stoichiometry is concerned with the combining of elements and compounds. When a balanced chemical equation is written, the numerical coefficient of each compound is called the *stoichiometric coefficient*.

Stoichiometric ratios are the ratios obtained from the numerical coefficients by dividing one coefficient by the other. These molar ratios permit us to calculate moles of one compound as related to another compound in the chemical equation. One should remember that the stoichiometric ratios are ratios of moles and not mass.

A chemical equation does not tell how fast the reaction will go or the extent of the reaction. It only indicates stoichiometric amounts of reactants required and products produced from the reaction if it takes place according to the equation written. Some examples will illustrate the preceding points.

Example 2.6

If 100 kg of octane is completely burned with stoichiometric quantity of oxygen,

(a) what is the stoichiometric ratio of O_2 to octane?

(b) how many kg mol of CO_2 will be produced?

Solution

Chemical equation for the combustion: $C_8H_{18} + 12\frac{1}{2}O_2 = 8CO_2 + 9H_2O$

```
1 C8H18  ─────►┌──────────────┐────► 8CO2
               │  Combustion  │
               │   chamber    │────► 9H2O
12½ O2 ────────►└──────────────┘
```

(a) Stoichiometric ratio of O_2 to octane = 12.5/1.0 = 12.5

(b) Basis 100 kg of octane. Molecular weight of octane = 96 + 18 = 114 kg/kg mol
kg of CO_2 produced =

$$\frac{100 \text{ kg } C_8H_{18}}{} \left| \frac{1 \text{ kg mol } C_8H_{18}}{114 \text{ kg } C_8H_{18}} \right| \frac{8 \text{ kg mol } CO_2}{1 \text{ kg mol } C_8H_{18}} \left| \frac{44.0 \text{ kg } CO_2}{1 \text{ kg mol } CO_2} \right. = 308.8 \text{ kg } CO_2$$

(It is recommended that in solving stoichiometric or material balance problems, four steps be followed as given below.)

1. First a simple box diagram of the process should be sketched. This diagram should show clearly the input and output streams with all pertinent data such as amount, composition, temperature, and pressure. Such a diagram helps to understand the process and to select a convenient basis for calculation. Mass balances are covered in more detail in a later section.

2. Write the chemical equations, if any.

3. A basis of calculation should be selected from the process information. In most cases, the process information will help to select a convenient basis for calculations.

4. Finally, prepare the material balance.

Example 2.7

Analysis of a certain limestone is $CaCO_3$ 93.10%, $MgCO_3$ 4.8%, and insoluble matter 2.1%. Assume both limestone and magnesium carbonate decompose completely.

(a) How many pounds of calcium oxide can be produced from 6 tons of this limestone?

(b) How many pounds of CO_2 can be obtained per lb of limestone?

(c) How many pounds of limestone are needed to make 5 tons of lime? (Lime is a mixture of CaO, MgO, and insoluble matter.)

Solution

(a) First draw a sketch of the process to understand what is happening.

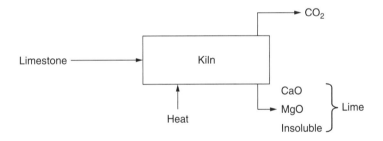

This is a stoichiometry problem. First write the equations for the decomposition of both the carbonates.

The equations are $\qquad CaCO_3 \rightarrow CaO + CO_2$

$\qquad\qquad\qquad\qquad MgCO_3 \rightarrow MgO + CO_2$

Molecular weights

	$CaCO_3$	$MgCO_3$	CaO	MgO	CO_2
Mol wt	100.1	84.32	56.08	40.32	44.0

(Basic: 100 lb of limestone)
Since the composition of limestone is given in weight percent, pounds = percent.
Prepare a table of calculations.

Limestone products

Component	lb = Percent	Lb mol	Solid Component	lb	CO_2 (lb)
$CaCO_3$	93.1	0.9301	CaO	52.16	40.924
$MgCO_3$	4.8	0.0569	MgO	2.30	2.504
Insoluble	2.1		Insoluble	2.10	
Total	100.00	0.987		56.56	43.43

The product quantities are calculated from the equations for the two reactions as given below.

Amount of CaO produced

$$= \frac{0.9301 \text{ lb mol } CaCO_3}{} \left| \frac{1 \text{ lb mol CaO}}{1 \text{ lb mol } CaCO_3} \right| \frac{56.08 \text{ lb CaO}}{1 \text{ lb mol CaO}} = 52.16 \text{ lb}$$

A similar calculation is made for the decomposition of $MgCO_3$.

(a) CaO produced $= \dfrac{52.16 \text{ lb}}{100 \text{ lb limestone}} \left| \dfrac{2000 \text{ lb}}{\text{ton}} \right| \dfrac{6 \text{ ton}}{} = 6259.2 \text{ lb}$

(b) Total CO_2 given off = 43.43 lb per 100 lb limestone.

Therefore, CO_2 produced per lb limestone = 0.434 lb

(c) Limestone required $= \dfrac{100 \text{ lb limestone}}{56.56 \text{ lb lime}} \left| \dfrac{2000 \text{ lb}}{1 \text{ ton}} \right| \dfrac{5 \text{ ton}}{} = 17680 \text{ lb}$

In the preceding examples, it is assumed that the reaction is 100% complete. In practice this is far from the truth. Reactants are almost always used in excess of stoichiometric amounts to make the reaction take place or to increase the consumption of one of the components. The excess material exits the reactor with products or sometimes separately, and can be used again. Even if excess amounts are used, the reaction may not be complete or side reactions may also occur so that products are not pure and need further treatment for purification. The following additional terms need to be defined to deal with incomplete reactions.

Limiting reactant is the reactant in a mixture of two or more reactants that disappears first, or would disappear first if the reaction were to proceed to completion. The reactant may or may not actually react completely. A simple method to determine the limiting reactant is to calculate the mol ratios of the reactants in the feed and stoichiometric mole ratios in the chemical equation.

For example, if 1 mole of octane is combusted with 15 moles of oxygen instead of the 12 ½ moles that are stoichiometrically required, the mole ratio of oxygen to octane in feed is 15, whereas according to the equation, the theoretical mole ratio is 12.5. Whether the combustion of octane is complete or not, the limiting component is octane because it would disappear first if the reaction were complete. If there are more than two reactants, we can still use the mole ratio comparison method to determine the limiting reactant. An example will illustrate the method.

Example 2.8

A reaction takes place according to the following equation:

$$A + 3B + 2C = \text{Products}$$

The reactant amounts used are A = 1.1 moles, B = 3.2 moles, and C = 2.4 moles. Which reactant is the limiting one?

Solution

Choose B as the reference reactant. Calculate the mole ratios in feed and in the equation.

Compared	Ratio in Feed		Ratio in Chemical Equation
B/A	3.2/1.1 = 2.91	<	3/1 = 3
C/A	2.4/1.1 = 2.18	>	2/1 = 2

Since B/A in feed < B/A in the equation, B is the limiting reactant with respect to A.

Also C/A in feed > C/A in the equation, so A is the limiting reactant relative to C. Hence we have

B < A < C. Therefore B is the limiting reactant overall.

Excess reactant is the reactant that is present in excess of its stoichiometric amount that is required to react with the limiting reactant. The percent excess of the excess reactant is expressed by the equation

$$\% \text{ excess reactant} = \frac{\text{moles of excess reactant in excess}}{\text{moles required to react with limiting reactant}}$$

For example, if we use 15 mol of oxygen to burn 1 mol of octane instead of 12.5 mol of oxygen required stoichiometrically, percent excess oxygen = $\frac{15-12.5}{12.5} \times 100 = 20\%$. Excess and required amounts are calculated on the entire amount of the limiting reactant.

The terms used in reference to chemical reactions are conversion, selectivity, and yield.

Conversion is the fraction of the feed or some key reactant in the feed that is converted into products.

Fractional conversion of a component is denoted by X and is given by

$$X = \frac{\text{moles of feed or moles of a reactant that react}}{\text{moles of feed or moles of a reactant in feed}}$$

% conversion is defined as follows:

$$\% \text{ conversion} = \frac{\text{moles of feed or moles of a reactant that react}}{\text{moles of feed or moles of a reactant in feed}} \times 100$$

Selectivity is the ratio of moles of a usually desired product produced to the moles of another usually undesirable product in a multi-reaction process. For example, A converts to B and C according to the following equation:

In this case if B is the desired product and C the undesirable, the overall selectivity of B with respect to C is

$$S_o = \frac{n_B}{n_C}$$

where n_B and n_C are the moles of B and C produced respectively.

Point selectivity is the ratio of the rates of the two reactions provided they are of the same order.

The *degree of completion* of a reaction is defined relative to the limiting reactant. It is usually the fraction or percent of the limiting reactant that is converted into products.

Yield for a single reactant and product is defined as the ratio of mass or moles of a product to the mass or moles of a key reactant. Percent yield is yield times hundred.

A reactant whose conversion into a desired product is of particular interest is chosen as a basis for yield calculation and is termed the key component. It may be the limiting reactant.

Example 2.9

In the manufacture of synthesis gas for NH_3 production, desulfurized natural gas is mixed with steam and passed to a reforming furnace as a first step. The nickel catalyst in the reforming furnace is kept at 1300°F–1500°F by external heat. The feed to the furnace is 22.22 mol % methane with the rest steam. The exit gas from the furnace analyzes CH_4 1.1 mol %, H_2 43.87 mol %, CO 14.62 mol %, and H_2O 40.41%. The reforming reaction takes place according to the following equation:

$$CH_4(g) + H_2O(g) \rightarrow CO(g) + 3H_2(g)$$

For this exercise, side reaction may be ignored. If total product gases are 127.2 mol/h per 20 mol/h of methane in feed, determine

(a) the limiting reactant

(b) the percentage of excess reactant

(c) the degree of completion (fraction)

(d) the percent conversion

(e) the yield based on H_2 production.

Solution

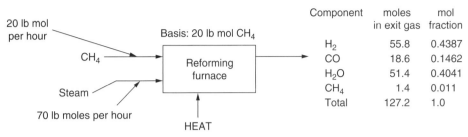

(Basis: 20 lb mol of methane feed)

Steam in feed = 20(100 − 22.22)/22.22 = 70 mol/h

(a) molar ratio of steam to methane in feed = $\dfrac{70}{20} = 3.5$

Methane is in smaller quantity; hence methane is the limiting reactant.

(b) Since steam fed to the reformer is more than stoichiometrically required to combine with methane, steam is the reactant in excess.

(c) Mol of CH_4 reacted = (mol of hydrogen produced)/3 = 55.8/3 = 18.6 moles.
The degree of completion = 18.6/20 = 0.93 (fraction)

(d) Percent conversion of methane = (18.6/20) × 100 = 93%

(e) Hydrogen produced = 55.8 lb moles

Theoretical hydrogen that should be produced = 60 lb mol

Yield = [55.8/20] = 2.79 lb mol H_2/lb mol of CH_4

Now that we have reviewed the basics of mass balance, we review the phase behavior of pure substances, as many of the mass balance problems will have to take into consideration the state and equilibrium condition of the system. In a given state or condition, a system possesses a set of unique properties such as pressure, density, temperature, and others. A change in the state will cause change in at least one of the properties. In equilibrium state of a system, there is not only absence of change but there is no tendency to change spontaneously. At a given temperature and pressure a pure compound can have a well-defined homogeneous state such as gas, liquid, or solid. This state is called the *phase*. At certain values of pressure and temperature it is possible for two or more phases to coexist, for example, ice and liquid phases of water. In the section to follow, we briefly review the properties of *pure* and *ideal* substances such as vapor pressure, their *PVT* behavior and processes of vaporization, condensation, and crystallization, saturation and partial saturation and observe how the changes in phase affect the mass balances. A short review of the properties of real substances will also be made.

PHASE BEHAVIOR OF SUBSTANCES

Vapor Pressure of Liquids

The pressure exerted by a pure component vapor in equilibrium with its liquid at a given temperature is termed its *vapor pressure*. If the gas phase consists of more than one component, the total pressure exerted on the liquid surface is the sum of the pressures exerted by the various components. The pressure contribution by a component to the total pressure is termed its *partial pressure*.

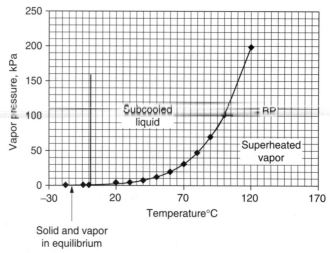

Figure 2.2 Vapor pressure curve for water

At saturation, the component partial pressure equals component vapor pressure. A gas or vapor is called saturated if it is about to condense its first drop of liquid, whereas a liquid is called saturated if it is about to start vaporizing. The temperature at which a vapor is saturated is called the *dew point* or *saturation temperature*. The temperature at which the liquid just begins to vaporize is called its *bubble point*. In between the dew point and the bubble point two phases, liquid and vapor, coexist. Any vapor or gas existing at a temperature greater than the saturation temperature at a given pressure is termed *superheated vapor*. In this state, the partial pressure of the vapor is less than its equilibrium vapor pressure. The difference between the temperature of superheated vapor and its saturation temperature is called its *superheat*. For example, from steam tables (SI units), steam at 250°C and 1500 kPa (saturation temperature = 198.32°C) has superheat of (250 − 198.32) = 51.68°C. Another term, *quality* of a vapor, refers to a wet vapor, which consists of saturated vapor and saturated liquid in equilibrium. *Quality* is the fraction of vapor in the mixture. Thus if w_e is the mass of entrained liquid, and w_v that of vapor, then quality of vapor is given by

$$x = \frac{w_v}{w_v + w_e} \quad \text{where } x \text{ is the quality of vapor.} \tag{2.4}$$

If the temperature of the liquid is less than the saturation temperature at the same pressure it is called subcooled liquid. The vapor pressure curve for water is shown in Figure 2.2. The region to the right is superheated vapor. To the left is the supercooled liquid zone.

Example 2.10

(Use of Steam Tables)

At 120°C and P = 198.53 kPa, the specific volume of a wet vapor is 0.625 m³/kg. What is its quality (fraction of vapor)?

Solution

Refer to saturated steam and water (steam table) in SI units in Perry and Green, and get the specific volumes of both the vapor and liquid at 120°C and at a pressure of 198.53 kPa. When this is done the following values are obtained:

$$\hat{V}_l = 0.0010603 \text{ m}^3/\text{kg} \quad \hat{V}_g = 0.8919 \text{ m}^3/\text{kg}$$

If x is the quality of vapor, $\hat{V}_{mixture} = 0.625 = x\hat{V}_g + (1-x)\hat{V}_l$.

Substitute the specific volumes of vapor and liquid in the equation to calculate x, the quality.

$$0.625 = (1-x)(0.0010603) + (x)(0.8919)$$

Solving the equation, $x = 0.7$

If the saturation temperature lies between two table values, double interpolation (vertical and horizontal) will be required.

The *boiling point* of a pure liquid at a given pressure is defined as the temperature at which its equilibrium vapor pressure equals the total pressure on its surface. The temperature at which a liquid boils under a total pressure of 1 atm is called its *normal boiling point*.

The *vapor pressure* of a pure substance is a unique property of that substance and increases with temperature. Two frequently used relations for the vapor pressure are as follows:

Two-constant equation: $\quad \ln p = A + B/T \quad$ (2.5)

where

p = vapor pressure
T = absolute temperature, K and
A, B = constants for the substance

The most widely used equation is that of Antoine.

Antoine equation: $\quad \ln p = A - [B/(T + C)] \quad$ (2.6)

where

$A, B,$ and C are constants and T is in degrees K.

In many cases, a polynomial fits the experimental vapor pressure data very well. Therefore polynomial correlations are also in use.

Example 2.11

The Antoine constants for ethyl bromide are $A = 15.9338$, $B = 2511.68$, and $C = -41.44$.

Calculate vapor pressure of ethyl bromide at $-10°C$ if the values of constants are based on natural logarithm.

Solution

Use Equation 2.6.

$$\ln p = A - [B/(T + C)]$$

Here $\qquad T = -10 + 273.15 = 263.15$ K

By directly substituting the constants and $T = 263.15$ K into the equation

$$\ln p = 15.9338 - [2511.68/(263.15 - 41.44)] = 4.6051$$

Then $\qquad\qquad\qquad p = e^{4.6051} = 100$ mm Hg

Vapor Pressure of Solids

The process of solids directly vaporizing into vapor phase without melting into liquid is termed sublimation. Ice sublimes below 0°C, whereas iodine crystals sublime at room temperature. Sublimation takes place when the partial pressure of the vapor in contact with the solid is less than the equilibrium vapor pressure of the solid.

IDEAL GASES

In this section we review the behavior of ideal and real gases and their relationships. We will also review PVT behavior of real gases in terms of equations of state, and the processes of vaporization and condensation, and then proceed to review the methods of solving material balance problems involving these processes.

The Ideal Gas Law

A gas is termed an ideal gas when the space or volume occupied by it is very large so that the intramolecular forces are negligible and the volume of the molecules of the gas is very small compared to the volume occupied by the gas. Ideal gas is a conceptual gas that meets the following conditions:

1. The gas obeys the following equation of state:

$$PV = nRT \qquad (2.7)$$

P = absolute pressure of the gas
V = total volume occupied by the gas
n = number of moles of the gas
R = ideal gas constant in proper units
T = absolute temperature of the gas

2. The internal energy (to be defined in Chapter 3) is independent of both the pressure and volume and is a function of the temperature alone.

For practical purposes, actual gases obey the ideal gas law only at low pressures and high temperatures within reasonable accuracy. It has been found to apply to all gases within certain pressure and temperature ranges, the accuracy increasing as the pressure decreases and temperature increases.

Sometimes the ideal gas law is written in terms of specific volume per mole of gas as

$$P\hat{V} = RT \qquad (2.7a)$$

Ideal Gases 41

Table 2.1 Standard conditions for the ideal gas

System of Units	T	P	$\hat{V}$
SI	273.15 K	101.35 kPa	22.415 m³/kg mol
Scientific	0.0° C	760 mm Hg	22.415 liters/g mol
American engineering	32.0° F	14.696 psia	379.48 ft³/lb mol
Natural gas industry	60° F	1 atm	359.05 ft³/lb mol

Table 2.2 Values of ideal gas law constant

Value of R	Units
1.987	cal/(g mol · K)
1.987	Btu/(lb mol · °R)
10.731	(psia · ft³/(lb mol · °R))
8.314	kPa · m³/kg mol · K
82.06	cm³ · atm/g mol · K
0.08206	L · atm/g mol · K
21.848	(in Hg)(ft³)/(lb mol · °R)
0.7302	ft³ · atm/lb mol · °R

where $\hat{V}$ = specific volume of the gas per mole. Equation 2.7 can be applied to a pure component or a mixture of two or more components.

Equation 2.7 represents the *equation of state* for an ideal gas.
Two other relations for ideal gases are expressed as follows:

Boyle's Law: PV = Constant at constant temperature
Charle's Law P/T = Constant at constant volume

As reviewed briefly in Chapter 1, some conditions of temperature and pressure are used as standard conditions. Table 2.1 lists the standard conditions for an ideal gas in different systems of units.

The values of ideal gas law constant R for different systems of units are listed in Table 2.2.

Example 2.12

(Use of Standard Condition)

Calculate the volume of 50 kg of NO_2 gas in cubic meters at standard conditions.

Solution

Basis: 50 kg of NO_2

Volume of 50 kg NO_2 =

$$\frac{50 \text{ kg } NO_2}{} \bigg| \frac{1 \text{ kg mol } NO_2}{46 \text{ kg } NO_2} \bigg| \frac{22.42 \text{ m}^3 NO_2}{1 \text{ kg mol } NO_2} = 24.37 \text{ m}^3 \text{ } NO_2 \text{ at std conditions}$$

Example 2.13

Calculate, from the ideal gas law, the value of universal gas law constant R if the units are to be pressure in psia, $\hat{V}$ in ft³, and T in °R.

Solution

The standard conditions are

$$P = 14.696 \text{ psia}$$
$$T = 492°R$$
$$\hat{V} = 359.05 \text{ ft}^3/\text{lb mol}$$

For 1 mol of gas,

$$R = \frac{P\hat{V}}{T} = \frac{14.696 \text{ psia}}{492°R} \left| \frac{359.05 \text{ ft}^3}{1 \text{ lb mol}} \right. = 10.73 \frac{\text{ft}^3 \cdot \text{psia}}{\text{lb mol} \cdot °R}$$

In many processes a gas goes from one initial state to other. In this case final and initial states can be related by the equation

$$\frac{P_2 V_2}{P_1 V_1} = \frac{n_2 R T_2}{n_1 R T_1} \quad \text{or} \quad \left(\frac{P_2}{P_1}\right)\left(\frac{V_2}{V_1}\right) = \left(\frac{n_2}{n_1}\right)\left(\frac{T_2}{T_1}\right) \tag{2.8}$$

and if $n_2 = n_1$,

$$\left(\frac{P_2}{P_1}\right)\left(\frac{V_2}{V_1}\right) = \frac{T_2}{T_1} \tag{2.8a}$$

In the preceding relations, the subscripts 2 and 1 refer to final and initial states. In all these equations, both the temperature and pressure are in absolute units and one must use the same units for both states of the gas. In Equations 2.8 and 2.8a, the ideal gas law, R is eliminated in taking the ratio of final to initial state.

It should be particularly noted that in the preceding equations the temperature is in K or °R, and P is the absolute (not the gage) pressure in appropriate units. The values of gas constant R will depend on the units chosen for P, V, and T. They can be calculated as in Example 2.13 or simply read from the table of R values in handbooks or textbooks. Some examples will illustrate the use of the ideal gas law.

Example 2.14

Calculate (a) the volume occupied by 20 kg of O_2 at 25°C and 750 mm Hg pressure. (b) What is its density in SI units at these conditions? (c) What is its specific gravity compared to air at 68°F and 1 atm?

(a) At the conditions of the problem, both O_2 and air can be considered ideal. We will take mol wts of oxygen and air as 32 and 29 respectively.

Volume of O_2: From Table 2.1 of standard conditions, 1 kg mol occupies 22.415 m³/kg mol at std. conditions.

$$\text{Mols of } O_2 = 20/32 = 0.625 \text{ kg mol}$$
$$T = 273.15 + 25 = 298.15 \text{ K} \quad P = 750 \text{ mm Hg}$$
$$P = 750 \text{ mm Hg}$$

$$V = \frac{0.625 \text{ kg mol}}{} \left| \frac{22.415 \text{ m}^3}{1 \text{ kg mol}} \right| \frac{760 \text{ mm Hg}}{750 \text{ mm Hg}} \left| \frac{298.15 \text{ K}}{273.15 \text{ K}} \right. = 15.5 \text{ m}^3$$

(b) Density of $O_2 = \dfrac{20 \text{ kg}}{15.5 \text{ m}^3} = 1.29 \text{ kg/m}^3$

(c) Basis: kg mol of air

Temperature of air = 68°F = (68 − 32)/1.8 = 20°C = 273.15 + 20 = 293.15 K

$$V_{air} = \dfrac{1 \text{ kg mol air}}{1 \text{ kg mol air}} \bigg| \dfrac{22.415 \text{ m}^3}{} \bigg| \dfrac{293.15 \text{ K}}{273.15 \text{ K}} \bigg| \dfrac{1 \text{ atm}}{1 \text{ atm}} = 24.056 \text{ m}^3$$

1 kg mol air = 29 kg

density of air (reference substance) = 29/24.056 = 1.206 kg/m³

Therefore specific gravity of O_2 relative to that of air = $\dfrac{1.29 \text{ kg/m}^3 O_2}{1.206 \text{ kg/m}^3 \text{ air}} = 1.07$.

Ideal Gas Mixtures and Partial Pressure

Frequently one has to deal with mixtures of gases instead of a single gas. For such calculations, the partial pressure p_i of a component i in a gaseous mixture is defined as the pressure that the component i would exert if it occupied the volume of the mixture alone at the same temperature.

Thus by definition,

$$p_i V_m = n_i R T_m \tag{2.9}$$

where
V_m = volume of gaseous mixture at temperature T_m
T_m = temperature of the gaseous mixture.

The ideal gas law gives for the total mixture,

$$PV_m = n_m R T_m \tag{2.9a}$$

The subscript m refers to the total mixture. By dividing Equation 2.9 by Equation 2.9a, one gets

$$\dfrac{p_i V_m}{PV_m} = \dfrac{n_i R T_m}{n_m R T_m}$$

which results in the relation,

$$\dfrac{p_i}{P} = \dfrac{n_i}{n_m} \tag{2.9b}$$

Since $\dfrac{n_i}{n_m} = y_i$, $\qquad p_i = y_i P \tag{2.9c}$

where y_i = mol fraction of component i in the mixture.

Dalton's law states that the total pressure exerted by a gaseous mixture at a given temperature is the sum of the component partial pressures. Thus

$$P = p_A + p_B + p_C + \cdots \tag{2.9d}$$

$p_A = y_i P$ and similar relations for other components.

Amagat's law states that the total volume of an ideal gaseous mixture equals the sum of the pure component volumes at the same temperature and pressure. Thus

$$V = V_A + V_B + V_C + \cdots \quad (2.10)$$

$V_A = y_A V$ and similar equations for other pure components.

Using the ideal gas law,

$$PV_A = n_A RT_m, \quad PV_B = n_B RT_m, \quad PV_C = n_C RT_m \quad (2.10a)$$

By addition,

$$P[V_A + V_B + V_C + \cdots] = RT_m[n_A + n_B + n_C + \cdots] = nRT_m = PV_m$$

Therefore,

$$\frac{V_A}{V_m} = \frac{\frac{n_A RT_m}{P}}{\frac{nRT_m}{P}} = \frac{n_A}{n} = y_i \quad \therefore V_A = y_i V_m$$

Example 2.15

A natural gas has the following composition by volume: methane 94%, ethane 3.1%, and nitrogen 2.9%. The gas is at 80°F and 50 psia. Calculate

(a) Partial pressure of nitrogen

(b) Pure component volume of nitrogen per 100 ft³ of gas mixture

(c) Density of gas in lb_m/ft^3.

Solution

Basis: 1 lb mol of gas

(a) Volume % = mol % for ideal gases in a mixture.

$$\text{partial pressure of N}_2 = (0.029) \times 50 \text{ psia} = 1.45 \text{ psia}$$

(b) Pure component volume of nitrogen = $y_i V_m$ where V_m is volume of mixture
$$= 0.029 \times 100 = 2.9 \text{ ft}^3$$

(c) Volume of mixture by ideal gas law

$$V_m = \frac{359.05 \text{ ft}^3}{1 \text{ lb mol}} \left| \frac{(460+80)°R}{(460+32)°R} \right| \frac{14.7 \text{ psia}}{50 \text{ psia}} = 115.9 \text{ ft}^3/\text{lb mol}$$

Mol wt. of gas = $0.94 \times 16 + 0.031 \times 30 + 0.029 \times 28 = 16.782$

$$\therefore \text{density } \rho = \frac{16.782 \text{ lb}}{\text{lb mol}} \left| \frac{1}{115.9 \text{ ft}^3/\text{lb mol}} \right. = 0.1448 \frac{\text{lb}}{\text{ft}^3}$$

Example 2.16

How many kg mol of O_2 will occupy 1000 m³ at 350 K and 110 kPa?

Solution

$$P = 110 \text{ kPa} \quad V = 1000 \text{ m}^3 \quad T = 350 \text{ K} \quad R = 8.314 \text{ kPa} \cdot \text{m}^3/\text{kg mol} \cdot \text{K}$$

$$V = 1000 \text{ m}^3 \text{ at 350 K and 110 kPa}$$

$$PV = nRT$$

$$\text{kg mol of oxygen} = n = \frac{PV}{RT} = \frac{110 \times 1000}{8.314 \times 350} = 37.8 \text{ kg mol}$$

We can also solve this problem by calculating volume of the gas at standard conditions and knowing that at standard conditions, 1 kg mol occupies 22.41 m³.

Since n remains unchanged,

$$\frac{P_1 V_1}{T_1} = \frac{P_2 V_2}{T_2}$$

where subscript 1 refers to standard conditions.

$$V_1 = \frac{P_2 V_2 T_1}{T_2 P_1} = \frac{110 \times 1000 \times 273.15}{350 \times 101.3} = 847.46 \text{ m}^3$$

$$\text{Number of kg mol of } O_2 = \frac{847.46 \text{ m}^3}{22.41 \text{ m}^3/1 \text{ kg mol}} = 37.8 \text{ kg mol}$$

Notice that in this calculation, you don't have to look for the value of R, the gas law constant, as it is eliminated in division.

REAL GASES (NON-IDEAL)

Real gases do not obey the ideal gas law over all the conditions of temperature and pressure. The $P\hat{V}$ products of real gases deviate from those calculated on the basis of ideality as the pressure increases. This necessitates some way to calculate the PVT properties of real gases. There are three different ways one can obtain, or rather predict, the PVT properties of a real gas.

These methods include (1) using *compressibility* or z *charts*, (2) using an *equation of state*, and (3) estimating. The critical state for gas-liquid transition is a set of values of P, V, and T at which the properties of gas and liquid become identical. This is the highest temperature at which liquid and gas exist in equilibrium in case of a pure component. It was found that at critical point all substances should behave the same way since at this point all substances have the same state of molecular dispersion. This observation is called the *Law of corresponding states*. It postulates that all substances should behave alike in critical state.

Temperature, pressure, and volume at critical point are denoted by T_C, P_C, and V_C respectively. The experimental values of these critical constants for various compounds can be found in relevant handbooks or textbooks. If they are not available, they can be calculated by using methods described in the book *Properties of Gases and Liquids* (Reid, Prausnitz and Polling).

If a compound is in a state above the critical point, it is called a supercritical fluid. Supercritical fluids are used in applications such as decaffeination of coffee, production of vanilla extract, and destruction of undesirable toxic organic compounds.

Reduced Parameters

These are dimensionless quantities that are the ratios of temperature, pressure, and specific volume of a substance to respective values of critical constants of the same substance. These are termed reduced temperature, reduced pressure, and reduced volume as follows:

$$\text{Reduced temperature} \quad T_r = \frac{T}{T_C}$$

$$\text{Reduced pressure} \quad P_r = \frac{P}{P_C} \tag{2.11}$$

$$\text{Reduced volume} \quad V_C = \frac{V}{V_C}$$

These are extensively used in correlations of *PVT* data for all substances.

In the case of hydrogen and helium only, pseudocritical constants are used. These are obtained as follows:

$$\begin{aligned} T'_C &= T_C + 8\,\text{K} \\ P'_C &= P_C + 8\,\text{atm} \end{aligned} \tag{2.11a}$$

where T'_C and P'_C are the pseudocritical temperature and pressure.

Compressibility Factor

PVT behavior can be represented by defining a dimensionless, empirical parameter called *compressibility factor z* as

$$P\hat{V} = zRT \quad \text{where} \quad z = \phi(T, P) \tag{2.12}$$

If pure substances exist at the same reduced conditions of temperature and pressure, they are in corresponding state. Pure gases in corresponding state have the same compressibility factors at the same reduced conditions of temperature and pressure. According to this principle, the deviations of properties of different pure fluids would show the same departure from the properties of these fluids in their ideal gaseous state at the same reduced conditions of temperature and pressure. Based on this principle, generalized compressibility charts giving values of z as functions of P_r, and T_r have been prepared by Nelson and Obert.

The compressibility at the critical point z_c is given by

$$z_c = \frac{P_c \hat{V}_c}{RT_c} \quad \text{or} \quad z = \phi(T_r, P_r) \tag{2.13}$$

which can be shown to be equal to 0.375 by the substitution of the van der Waals constants (to be discussed in a later section) for P_c, T_c, and $\hat{V}_c$.

Although the value of z_c is shown to be 0.375 as above, the actual values of z_c lie between 0.22 and 0.33. Hence, a third parameter z_c is introduced to represent z values as

$$z = \phi(T_R, P_r, z_c) \qquad (2.14)$$

Equation of State

For real fluids, the ideal-gas law does not hold over the entire range of temperatures and pressures. Hence, other equations of state have been proposed. These are of the form $\phi(P, V, T) = 0$. Some PVT equations that are cubic in volume are listed in Table 2.3.

Substituting values of the van der Waals constants (Table 2.3) into the van der Waals equation of state and using the definitions of the reduced variables, the following relation is obtained:

$$\left(P_r + \frac{3}{V_r^3}\right)\left(V_r - \frac{1}{3}\right) = \frac{8}{3}T_r \qquad (2.15)$$

This shows that the fluids obeying van der Waals' equation will have the same reduced volume $V_r = \hat{V}/\hat{V}_c$ at given values of T_r and P_r. The fluids are said to be in the corresponding states when they are at the same fraction of the critical-state values.

Table 2.3 Equations of state

van der Waals	Redlich-Kwong
$\left(P + \dfrac{a}{\hat{V}^2}\right)(\hat{V} - b) = RT$	$P = \dfrac{RT}{\hat{V} - b} - \dfrac{a}{T^{1/2}\hat{V}(\hat{V} + b)}$
$a = \dfrac{27R^2T_c^2}{64P_c} = \dfrac{9}{8}RT_c\hat{V}_c$	$a = \dfrac{0.4278R^2T_c^{2.5}}{P_c}$
$b = \dfrac{RT_c}{8P_c} = \dfrac{\hat{V}_c}{3}$	$b = 0.0867\dfrac{RT_c}{P_c}$

Peng-Robinson	Soave-Redlich-Kwong
$P = \dfrac{RT}{\hat{V} - b} - \dfrac{a\alpha}{\hat{V}(\hat{V} + b) + b(\hat{V} - b)}$	$P = \dfrac{RT}{\hat{V} - b} - \dfrac{a'\lambda}{\hat{V}(\hat{V} + b)}$
$a = 0.45724\left(\dfrac{R^2T_C^2}{P_C}\right)$	$a' = \dfrac{0.42748R^2T_C^2}{P_C}$
$b = 0.07780\left(\dfrac{RT_C}{P_C}\right)$	$b = \dfrac{0.08664\,RT_C}{P_C}$
$\alpha = \left[1 + k\left(1 - T_r^{1/2}\right)\right]^2$	$\lambda = \left[1 + \kappa\left(1 - T_r^{1/2}\right)\right]^2$
$\kappa = 0.37454 + 1.54226\omega - 0.26992\omega^2$	$\kappa = (0.480 + 1.574\omega - 0.176\omega^2)$

where b is excluded volume per mole, and ω = Pitzer acentric factor.

Virial Equations of State

By a consideration of the intermolecular forces and using the methods of statistical mechanics, the $P\hat{V}T$ relationship of gases is presented by a power series (virial equation of state)

$$z = \frac{P\hat{V}}{RT} = 1 + \frac{B}{\hat{V}} + \frac{C}{\hat{V}^2} + \cdots \qquad (2.16)$$

where B is the second virial coefficient and C is the third virial coefficient and so on. Another form of the virial equation is

$$z = \frac{P\hat{V}}{RT} = 1 + B'P + C'P^2 + D'P^3 + \cdots \qquad (2.17)$$

The constants B, C, etc in the virial equations are known as virial coefficients. These are functions of temperature only.

For engineering applications, the virial equations are truncated, i.e., few terms are retained and others are ignored because in many practical cases a finite number of terms suffices to represent experimental data very well without sacrificing much accuracy. Two forms of truncated equations are

$$z = \frac{P\hat{V}}{RT} = 1 + \frac{BP}{RT} \qquad (2.18)$$

and

$$z = \frac{P\hat{V}}{RT} = 1 + \frac{B}{\hat{V}} \qquad (2.19)$$

These are applicable up to a pressure of 15 atm. Above this pressure and up to 50 atm, an equation truncated to 3 terms is used and is given by

$$z = \frac{P\hat{V}}{RT} = 1 + \frac{B}{\hat{V}} + \frac{C}{\hat{V}^2} \qquad (2.20)$$

Other popular generalized correlations for compressibility factor z and second virial coefficient B are those of Pitzer and coworkers. This correlation takes the form

$$z = z^0 + \omega z^1 \qquad (2.21)$$

where
- ω = Pitzer acentric factor
- z^0 = compressibility
- z^1 = a correction to the compressibility for non-simple fluids. If $\omega = 0$, then the correlation reduces to z^0 and becomes identical with z.

The acentric factor is given by the relation

$$\omega = -1 - \log(P_r^{sat})_{T_r = 0.7} \qquad (2.22)$$

where P_r^{sat} is the value of the reduced vapor pressure at $T_r = 0.7$.

Lee and Kestler developed tables of z^0 and z^1 as a function of P_r and T_r (Smith et. al.).

Real Gases

When applying these equations, one should keep in mind the range of validity over which they are valid and not extrapolate outside their range.

The Benedict-Rubin-Webb equation is based on power series in $\hat{V}$ and is given by

$$P = \frac{RT}{\hat{V}} + \frac{B_0 RT - A_0 - C_0/T^2}{\hat{V}^2} + \frac{bRT - a}{\hat{V}^3} + \frac{a\alpha}{\hat{V}^6} + \frac{c}{\hat{V}^3 T^2}\left(1 + \frac{\gamma}{\hat{V}^2}\right)\exp\left(\frac{-\gamma}{\hat{V}^2}\right) \quad (2.23)$$

This equation has 8 constants and is very tedious to use in hand calculations. However, it and its modifications are widely used as it can be applied to both gases and liquids. It is best suited to computer or spreadsheet calculations.

The Beattie-Bridgeman Equation of State is a five-constant equation. It is

$$P\hat{V}^2 = RT\left[\hat{V} + B_o\left(1 - \frac{b}{\hat{V}}\right)\right]\left(1 - \frac{c}{\hat{V}T^3}\right) - A_o\left(1 - \frac{a}{\hat{V}}\right) \quad (2.24)$$

This equation fits experimental data very well (within 0.15%) for densities less than 5 moles per liter and should not be extrapolated to the critical range.

Example 2.17

Calculate the specific volume (ft³/lb mol) of SO_2 gas at 250 psia and 310°F using (a) ideal gas law, (b) van der Waals' equation of state, (c) tabular data, and (d) generalized compressibility chart.

The following data are available for SO_2:

$$a = 1730 \text{ atm} \cdot \text{ft}^6/\text{lb mol}^2 \quad b = 0.909 \text{ ft}^3/\text{lb mol}$$

$$T_c = 775.3°R \quad P_c = 1143.7 \text{ psia} \quad \hat{V}_c = 1.956 \text{ ft}^3/\text{lb mol} \quad z_c = 0.29$$

Solution

$R = 10.73$ psia · ft³/lb mol · °R, $T = 460 + 310 = 770°$ R, $P = 250$ psia.

(a) Volume by Ideal Gas Law

$$\hat{V} = \frac{RT}{P}$$

$$= \frac{10.73(\text{psia} \cdot \text{ft}^3/\text{lb mol} \cdot °R)(770°R)}{250 \text{ psia}}$$

$$= 33.05 \text{ ft}^3/\text{lb mol}$$

(b) Volume by van der Waals' Equation of State

$$a = 1730 \text{ atm} \cdot \text{ft}^6/\text{lb mol}^2 \quad b = 0.909 \text{ ft}^3/\text{lb mol}$$

$$P = \frac{250}{14.7} = 17 \text{ atm} \quad R = 0.7302 \frac{\text{atm} \cdot \text{ft}^3}{\text{lb mol} \cdot °R} \quad T = 770°R$$

Then substituting in the van der Waals equation gives

$$\left(17 + \frac{1730}{\hat{V}^2}\right)(\hat{V} - 0.909) = 0.7302(770)$$

Table 2.4

$\hat{V}$	LHS	LHS −RHS
33.05	597.30	35.04
30.00	550.47	−11.50
31.00	565.72	3.717

This is a cubic equation in $\hat{V}$, and it can be solved by iterative procedure or plotting a graph of values of the left side of the above equation at various values of $\hat{V}$. The iterative trials can be started with the value of $\hat{V}$ found by using the ideal gas law. The calculations are shown in Table 2.4.

By interpolation

$$\hat{V} = 30 + \frac{0-(-11.5)}{3.717-(-11.5)}(31-30)$$

$$= 30 + \frac{11.5}{15.217}(1)$$

$$= 30.76 \text{ ft}^3/\text{lb mol}$$

When $\hat{V}$ = 30.76, LHS −RHS = 0.2. This is very near to 0; therefore, $\hat{V}$ = 30.76 ft³/lb mol is a satisfactory value.

(The cubic equation in $\hat{V}$ can be solved by Newton's method. A computer program comes on a disc provided with the book by Himmelblau.)

(c) *Volume by the Use of Tabular Data.* The following pertinent data are reproduced from the table of the thermodynamic data for SO_2.

Temperature interpolation gives volumes at 310°F (see Table 2.5).

By pressure interpolation, volume at 250 psia and 310°F is

$$V \text{ at 250 psia} = \frac{1}{2}(0.600+0.385) = 0.4925 \text{ ft}^3/\text{lb}$$

$$V = 0.4925(64) = 31.52 \text{ ft}^3/\text{lb mol}$$

(d) Volume by Generalized Compressibility Factor Chart

$$T_r = \frac{T}{T_c} = \frac{770}{775.3} = 0.99$$

$$P_r = \frac{P}{P_c} = \frac{250}{1143.7} = 0.22$$

Table 2.5 Thermodynamic data for SO_2

t, °F	200 psia $\hat{V}$ ft³/lb	300 psia $\hat{V}$ ft³/lb
300	0.590	0.378
320	0.610	0.392
310	0.600	0.385

From the compressibility chart, $z = 0.92$. Therefore

$$\hat{V} = \frac{zRT}{P} = \frac{0.92(10.73)(770)}{250} = 30.4 \text{ ft}^3/\text{lb mol}$$

Example 2.18a

What pressure will be developed when 1 lb mol of carbon dioxide is stored in a cylinder of 3.0 ft³ volume at 150° F? Use the generalized chart. Critical constants for CO_2 are

$$T_c = 547.4° \text{R} \quad P_c = 1071.6 \text{ psia} \quad \hat{V}_c = 1.54 \text{ ft}^3/\text{lb mol}$$

Solution

$$T = 460 + 150 = 610° \text{R}$$

$$V_r = \frac{\hat{V}}{\hat{V}_c} = \frac{3}{1.54} = 1.95$$

$$T_r = \frac{610}{547.4} = 1.114$$

Since P_r is not known, it is not possible to enter the z chart, and a trial-and-error procedure is necessary. An equation relating P_r and z is required. Obtain this as follows:

$$P = \frac{zRT}{\hat{V}} = \frac{z(10.73)(610)}{3} = 2181.8(Z)$$

Now
$$P = P_c P_r = 1071.6 P_r$$

$$z = \frac{P}{2181.8} = \frac{1071.6}{2181.8} P_r = 0.49115 P_r$$

or
$$P_r = z(2.036).$$

A value of z is assumed and T_r is determined from z chart using calculated P_r. Procedure is repeated till T_r calculated equals $T_r = 1.114$. Table 2.6 summarizes the trials.

The last trial gives $T_r = 1.11$ which is very close to 1.1114. Therefore, $P = P_r P_c = 1.283(1071.6) = 1375$ psia.

Table 2.6

Z	P_r	T_r (from z Chart)
1	2.036	2.5
0.7	1.425	1.21
0.63	1.283	1.11

Example 2.18b

What pressure is generated when 1 k mol of CO_2 is stored in 0.125 m³ at 60° C? Calculations should be made by using generalized z tables by Lee-Kestler.

Solution

$T = 273.15 + 60 = 333.15$ K P to be determined, Bar $V = 0.125$ m³

$$1 \text{ k mol} = 1000 \text{ g mol}$$

$$R = 82.06 \frac{cm^3 \cdot atm}{g\, mol \cdot K} = 82.06 \frac{\frac{cm^3}{g\, mol}}{\frac{10^6 \frac{cm^3}{m^3}}{1000 \frac{g\, mol}{kg\, mol}}} \cdot \frac{1\, atm \cdot (1.0130. \frac{Bar}{atm})}{K} = 0.08314 \frac{Bar \cdot m^3}{kg\, mol \cdot K}$$

$$P = \frac{zRT}{V} = \frac{z(0.08314)(333.15)}{0.125} = 221.58 z$$

Also $P/P_C = P_r \therefore P = P_C P_r$. Since P_r is not known, trial and error procedure is needed.

Use Lee-Kessler table (you will find it in the thermodynamics book by Smith et al.) for z^0 values. (Ignore acentric factor correction z^1.)

$T_C = 304.2$ K $P_C = 73.83$ Bar $T_r = 333.15/304.2 = 1.095$

$$z = \frac{73.83 P_r}{221.5} = 0.3333 P_r \quad \text{and} \quad P_r = \frac{z}{0.3333}$$

Start with $z = 1$ then $P_r = 3.0$.
From Lee-Kessler z table, by interpolation

$$z^0 = 0.477 \text{ at } T_r = 1.1 \quad \text{and} \quad z^0 = 0.4604 \text{ at } T_r = 1.05$$

$$z^0 = \frac{1.095 - 1.05}{1.1 - 1.05}(0.477 - 0.4604) + 0.4604 = 0.4753$$

Then $P_r = 0.4753/0.3333 = 1.426$.
Thus the value of P_r appears to lie between 1.2 and 1.5.
For second trial, use $P_r = 1.426$. By interpolation, at $T_r = 1.095$ (calculation not shown here)

P_r	1.2	1.5	1.426
$T_r = 1.095$	0.583	0.4435	

$$z^0 = \frac{1.426 - 1.2}{1.5 - 1.2}(0.4435 - 0.583) + 0.583 = 0.4779$$

then $\quad P_r = 0.4779/0.3333 = 1.434$

This value is close to 1.426, so take the average of the two values for the next trial. The average value is $(1.434 + 1.426)/2 = 1.43$ for P_r.

$$z^0 = \frac{1.43 - 1.2}{1.5 - 1.2}(0.4435 - 0.583) + 0.583 = 0.47605 \quad P_r = 1.43$$

Therefore $P_r = 1.43$ and $P = P_c P_r = 73.83 \times 1.43 = 105.6$ Bar

Gaseous Mixtures

For gaseous mixtures, Kay's method is used to calculate pseudocritical constants. Thus

$$P'_C = P_{C_A} y_A + P_{C_B} y_B + \cdots$$
$$T'_{C_A} = T_{C_A} y_A + T_{C_B} y_B + \cdots \quad (2.25)$$

Example 2.19

A natural gas has the following composition (in mol %):

$$CH_4 \ 35\%, \ C_2H_4 \ 15\%, \ N_2 \ 50\%$$

at 75 atm and 100° C. What is the volume of the gas per g mol?

Solution

Basis: 1 g mol
The following data can be collected from handbooks or textbooks.

Component	T_c (K)	P_C (atm)
CH_4	191	45.8
C_2H_4	283	50.9
N_2	126	33.5

$$R = 82.06 \ \frac{cm^3 \cdot atm}{g\,mol \cdot K} \quad P = 75 \ atm \quad T = 373 \ K$$

We should first calculate the pseudocritical constants.

$$P'_C = 45.8 \times 0.35 + 50.9 \times 0.15 + 33.5 \times 0.5 = 40.42 \ atm$$
$$T'_C = 191 \times 0.35 + 283 \times 0.15 + 126 \times 0.5 = 172.3 \ K$$

Pseudocritical reduced pressure and temperature are then calculated as follows:

$$P'_r = \frac{P}{P'_C} = \frac{75}{40.42} = 1.86 \quad \text{and} \quad T'_r = \frac{373}{172.3} = 2.17$$

With the two parameters of temperature and pressure, the value of z = compressibility is obtained from the generalized z chart. $z = 0.97$

$$\hat{V} = \frac{znRT}{P} = \frac{0.97 \times 1 \times 82.06 \times 373}{75} = 395.9 \ cm^3 \ \text{at 75 atm and 100°C}$$

Using ideal gas law,

$$\hat{V} = \frac{nRT}{P} = \frac{1 \times 82.06 \times 373}{75} = 408.1 \ cm^3/g \ mol$$

The value $\hat{V}$ obtained by using ideal gas law is $\frac{408.1 - 395.9}{395.9} \times 100 = 12.2\%$ higher.

(Reading values of z from charts given in textbooks or handbooks has its pitfalls, especially in the range of $P_r < 2$. In this region it is preferable to obtain z values by interpolation from tables of Hougen et al. or those of Lee-Kestler.)

MATERIAL BALANCE CALCULATION

So far we have reviewed the elementary tools of material balances, namely, properties of substances, particularly gases and liquids, mass balance, and the conservation of mass and atoms in ordinary processes, and stoichiometry of chemical reactions. We also reviewed the concept of a system and it's boundary and surroundings. We now turn our attention to the solution of more involved material balance problems. We will cover the steady state flow problems with process variants such as recycle, bypass, and purge, and deal with unsteady state mass balance problems rather briefly.

Let us review some terms associated with a system chosen for material balance. *Accumulation* refers to either increase or decrease in mass or moles within the system over a chosen period of time. *Inputs and outputs* of material imply transfer across the system boundaries into or out of the system. A *steady state* implies that the variables within the system do not change with time and hence accumulation in the system is zero. On the other hand in *unsteady state* problems, system variables do change with time.

Material balances can be made on an individual piece of equipment, a group of equipment, or for an entire plant. These material balances may or may not involve chemical reactions. In solving these problems it should be remembered that the number of unknown variables is equal to the number of independent equations that are needed to solve the problem.

In solving these problems, especially in an exam hall, it is advisable to follow certain steps.

1. *The first step is careful reading* of the problem to know what is available and what is to be accomplished as the end result of calculations.

2. *The second step is to draw a simple sketch* of the process describing the system and its inputs and outputs in consistent units. All applicable data should be on this diagram.

3. Third, if the problem involves chemical reaction or reactions, they should be written and balanced.

4. *Then select a basis for calculation* so that you deal with and arrive at meaningful numbers. In almost all cases, the specific amount of one of the streams in the process is chosen as the basis of calculation.

5. *Last step is to make the material balance.*

The difference between the number of variables whose values are not known and the number of independent equations that are known is termed the *number of degrees of freedom*. If the number of degrees of freedom is zero, then the problem is properly specified and is solvable. The problem of material balance is then to identify the variables whose values are unknown and then write down just an equal number of independent equations to get their values.

Solving Mass Balance Problems not Involving Chemical Reactions

We shall first consider some problems that do not involve chemical reaction. Separation operations invariably involve calculations of material and energy balances. Hence we will solve some problems in the area of unit operations.

Example 2.20

(Evaporation Problem)

A given single-effect vacuum evaporator is used to concentrate an extract containing 1.2% solids to a solution containing 15% solids. Vapor from the evaporator is condensed in a condenser. Condensed water is used elsewhere. The fresh feed to the evaporator is 5000 kg per hour. Calculate the amounts of concentrated extract produced per hour and the water condensed.

Solution

Here there is only one piece of equipment; therefore, the evaporator is our system.

On careful reading (in this problem this is minimal), we know that feed to the evaporator and its concentration of solids is known. Also the product concentration is known. However, total product and water evaporated are not known. Exhibit 1 is a sketch of the problem showing the quantities involved.

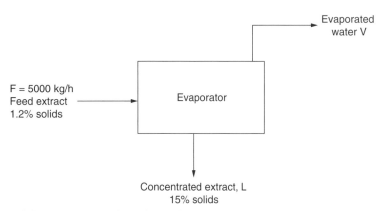

Exhibit 1 Process flow sketch for Example 2.20

In this process no chemical reactions are involved. Step 3 is to be skipped. In addition, it is a continuous process; therefore, there is no accumulation in the evaporator (system) and the mass balance reduces to input = output or $\Sigma M_I = \Sigma M_O$.

Next select a basis of calculation. The feed stream and it's solid concentration are given, and it is to be split into concentrated extract and evaporated water, so we select the feed stream as our basis of calculation.

The next step is to make material balance. For this we write as many independent equations as there are unknown variables. First material balance equation will relate to output streams and we write

$$5000 = L + V$$

This gives one equation and two unknowns, L and V. We need one more equation. Hence we make a component balance to get the second equation.

$$5000 \times (1.2/100) = V(0.0) + L(15/100)$$

Now we have two unknowns and two independent equations. Therefore the problem is properly specified and a unique solution is obtained by solving the two equations simultaneously. Note that the second equation assumes solids concentration of evaporated water to be zero, which is not mentioned in the problem. In a well-designed evaporator, entrained solids will be certainly negligible.

Solving the two equations simultaneously, we get

$$L = 400 \text{ kg/h and } V = 4600 \text{ kg/h}$$

Example 2.21

(Crystallizer Problem)

A crystallizer is fed continuously with a solution of Na_2SO_4 saturated at 40°C. The solution is cooled to 5°C in the crystallizer. Water evaporation from the crystallizer may be neglected. If one metric ton of the decahydrate is to be produced per hour, what should be the feed rate of the sulfate solution per hour?

$$\text{Solubility of } Na_2SO_4 \text{ at } 40°C = 32.6\% \; Na_2SO_4$$
$$\text{Solubility of } Na_2SO_4 \text{ at } 5°C = 5.75\% \; Na_2SO_4$$

What is the % yield?

Solution

This process does not involve chemical reaction.

System is crystallizer. Material balance required.

$$\text{Basis: 1000 kg/hr feed}$$

This a flow process, $\therefore \Sigma M_I = \Sigma M_O$ or no accumulation

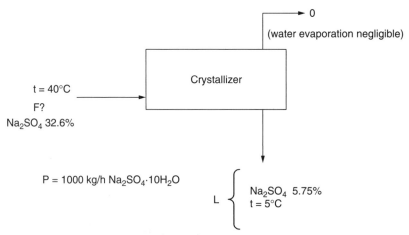

Exhibit 2 Process sketch for Example 2.21

Other data: Molecular weight of $Na_2SO_4 \cdot 10H_2O = 322$
Molecular weight of $Na_2SO_4 = 142$

$$\% \; Na_2SO_4 \text{ in crystals} = \frac{142}{142+180} \times 100 = 44.1\%$$

If $Na_2SO_4 \cdot 10H_2O$ product is P, material balance on Na_2SO_4 is given by

$$0.326 \times 1000 = 0.0575(1000 - P) + 0.441 \times P.$$
$$P = 700.1 \text{ kg/h produced from 1000 kg/h solution}$$
$$\text{Hence feed rate of solution} = (1000/700.1)(1000) = \mathbf{1428.4 \text{ kg/h}}$$

Weight of $Na_2SO_4 \cdot 10H_2O$ in 1000 kg/h original solution $= 326(322/142) = 740$ kg/h

$$\text{Then } \% \text{ yield} = \frac{700.1}{740} \times 100 = \mathbf{94.61\%}$$

Example 2.22

(Distillation Column)

An operation in a plant requires that they treat 200 kg mol/h of a mixture of heptane and ethyl benzene containing 0.45 mol % of heptane at atm pressure to produce a distillate containing 98 mol % heptane and the bottoms containing 1 mol % heptane. How many mol/h of distillate, and mol/h of bottoms will be produced. What is the recovery of heptane?

Solution

The process of distillation does not involve chemical reaction. The sketch of the process is shown in Exhibit 3.

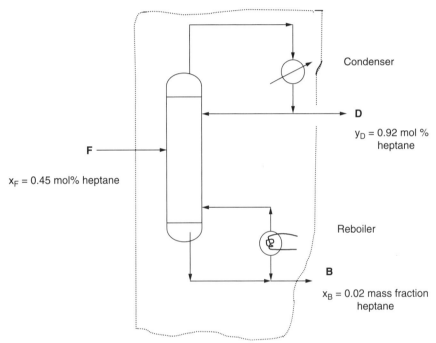

Exhibit 3 Process sketch for Example 2.22

System: distillation column as shown in Exhibit 3.
This is a continuous flow process. $\therefore \sum M_I = \sum M_O$ or no accumulation.

$$\text{Basis: 200 kg mol/h of feed}$$

Material balance equations: In terms of moles
Total mass balance

$$F = B + D$$

Heptane mass balance

$$0.45F = 0.01B + 0.98D$$

Solving the equations

$$D = 90.72 \text{ kg mol/h, and } B = 109.28 \text{ kg mol/h}$$
$$\text{Heptane in distillate} = 90.72 (0.98) = 88.9 \text{ kg mol/h}$$
$$\text{Heptane in feed} = 200(0.45) = 90.0 \text{ kg mol/h}$$

Therefore % heptane recovery = (88.9/90)(100) = 98.78%.

Tie Component

In many processes, a substance (component) may appear in only one inlet stream, go through the process unchanged, and appear only in one outlet stream.

Such a component, called the *tie substance* or *component,* can be used as a reference substance to relate the total mass flows of the streams in which it appears. Examples of tie substances are nitrogen in air used for combustion, and the inert solids in solid-liquid extraction. The following simple example will illustrate the tie component concept.

Example 2.23

Wet sand containing 4% moisture is dried in a rotary dryer continuously with a countercurrent flow of hot air to a final moisture content of 0.1% on wet basis. In one test, the analysis of hot air showed that 406.7 kg/h of water vapor are evaporated per hour. How much wet sand in metric tons/h is the dryer handling?

Solution

The process does not involve chemical reactions. Since the process is continuous, there is no accumulation. Because both inlet and outlet streams are unknown, we select 100 kg/h moisture-free, dry sand as our basis. This is a tie component. With this understanding draw a simple sketch of the process first.

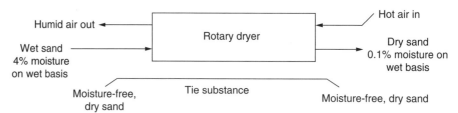

Exhibit 4 Sketch for Example 2.23

$$\text{Basis: 100 kg/h of moisture-free, dry sand}$$

Moisture in wet sand per 100 kg/h of moisture-free, dry sand

$$= \frac{4}{100-4} = 4.167 \text{ kg/100 kg/h moisture free sand}$$

Moisture in wet sand per 100 kg/h of moisture-free, dry sand

$$= \frac{0.1}{100-0.1} = 0.1 \text{ kg/100 kg moisture free sand}$$

$\therefore$ moisture evaporated per 100 kg/h of moisture free sand $= 4.167 - 0.1 = 4.067$ kg/h

Moisture-free sand fed to the dryer

$$= \frac{406.7}{4.067} \times 100 = 10000 \text{ kg/h moisture-free sand}$$

$$\text{Wet sand rate} = \frac{100}{100-4} \times 10000 = 10417 \text{ kg/h} = 10.417 \text{ metric tons/h}$$

Material Balances Involving Multiple Subsystems

In developing material and energy balances for a process plant, a chemical engineer is always faced with connected multiple subsystems as a plant is made up of several operations to treat raw materials in order to get the finished products. Principles involved in these calculations are the same as in making the balances of single systems. Most often, one makes overall material balance for the process first, ignoring intermediate units and their connections. A subsystem may not have to be equipment units but can be *junctions of streams* such as mixing points of two or more streams or a junction where a stream splits into two or more streams. Process plant design calculations are presently made with the use of modern computers and programs developed for the purpose such as Aspen, Process, and a host of other smaller programs very suitable for personal computers. However, a practicing engineer has to be conversant with the modalities of these calculations. In the following example, we review calculations of a system having a few subsystems.

Example 2.24

A two-column system for the separation of a mixture of 40 mol % benzene, 40 mol % toluene, and the rest xylene is depicted in Exhibit 5. Available data on various streams is given in the diagram. What is the flow rate of stream F_2? What is its composition?

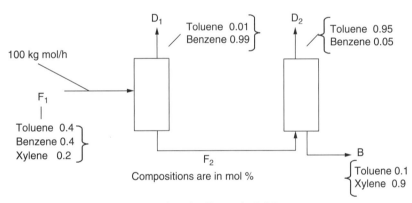

Exhibit 5 Process sketch and data for Example 2.24

Solution

This is a steady state process with no reaction. Data given is shown in the diagram.

$$\text{Basis: Feed 100 mol/h}$$

First take overall balance and write down the following equations:

$$\text{Overall balance: } F_1 = B + D_1 + D_2$$

$$\text{Toluene balance: } 0.1B + 0.01D_1 + 0.95D_2 = 40$$

$$\text{Benzene balance: } 0.0B + 0.99D_1 + 0.05D_2 = 40$$

Solving these equations simultaneously, we get

$$B = 22.62 \text{ mol}, D_1 = 38.42 \text{ mol, and } D_2 = 39.36 \text{ mol}$$

Now take a balance on the first tower only, which is a subsystem. Stream F_2 now becomes an outlet stream.

$$\text{Total mass balance: } F_1 = F_2 + D_1 = 100$$

$$\text{Toluene balance: } 0.4F_1 = x_{tol}F_2 + 0.01D_1 = 40$$

$$\text{Benzene balance: } 0.4F_1 = x_{benz}F_2 + 0.99D_1 = 40$$

There are three equations and three unknowns F_2, x_{tol}, and x_{benz}. Therefore we have a unique solution. F_1 is the feed (known) and D_1 is calculated from overall balance. By solving these equations, $F_2 = 61.68$ mol/h, and the composition of the stream is as follows:

Composition of stream in Example 2.24

Component	Mol/h	Mol %
Benzene	2.06	0.0334
Toluene	39.62	0.6423
Xylene	20.00	0.3243
Total	61.68	1.0000

Process Variants: Recycle, Bypass, and Purge Calculations

Recycle

Recycling or returning material (or energy) from a stream that leaves a process back to the process is a common practice in process plants. A schematic of recycle operation is shown in Figure 2.3.

Recycling may be done for several reasons such as to increase yield, to enrich a product, to conserve heat, or just to improve operations. For example, when a reactant is used in excess of stoichiometric proportion to completely react or maximize the consumption of the limiting reactant, the excess reactant is recycled to avoid waste. When conversion of reactants to products is limited either due to equilibrium conditions or due to very slow rate of reaction as the product concentration builds up, recycling is used to decrease the concentration of the product in order to enhance the product yield. In purely physical processes recycling is also employed. In fractionating columns part of the distillate is returned to the

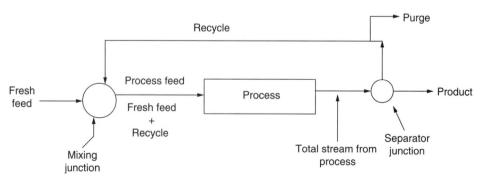

Figure 2.3 Sketch of a process with recycle stream

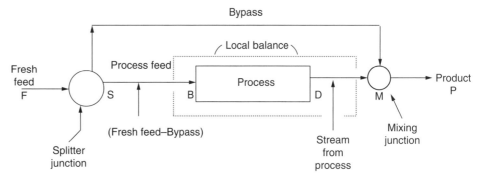

Figure 2.4 Process with bypass

column as reflux to enrich the product. In some drying operations, air humidity in the dryer is controlled by recirculating a part of the wet air exiting the dryer. Another example is the recirculation of the absorption liquid in a packed column to provide sufficient liquid rate to wet the packing better.

Recycling is a steady state operation and involves no accumulation or depletion inside the process or in the recycled stream. The same methods and steps of solving material balance problems as discussed in previous sections can be used to solve material balance problems when a recycle stream or streams are involved in the process.

Bypass Stream

The concept of a bypass stream is shown in Figure 2.4.

A bypass stream is one that skips one or more steps in the process and directly goes to another downstream stage. This involves division of a feed stream into two streams, one going to the process unit next while the other goes to some other point after. Bypass is often used to get closer control of the operation. For example, the composition of the final exit stream from a process unit is controlled by mixing the exit stream and the bypass stream. A bypass stream is an internal stream, meaning it would not be included in the overall balance of the process unit. For example in Figure 2.4, mass balance between points F and P will not include the bypass stream. A local balance will have to be made. Balances about mixing and splitting points will also be required.

Purge Stream

One difficulty with recycle stream is that inevitably inerts or impurities build up in the recycle stream. To prevent the build up a *bleed-off* or *purge stream* is used (Figure 2.5).

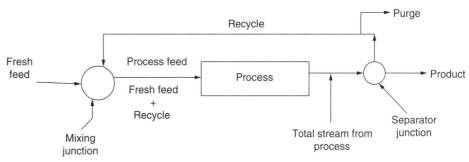

Figure 2.5 Sketch of a process with recycle and purge streams

Example 2.25

A liquid mixture containing 45 mol % benzene and 55 mol % toluene is distilled in a fractionating tower at a feed rate of 120 kg mol/h at 101.3 kPa pressure abs.

The distillate is to contain 90 mol % benzene. Bottoms are not to contain more than 4 mol % of benzene. The reflux ratio is fixed at 4:1. What is the vapor rate in kg/h to the condenser?

Solution

This is a steady state flow process without a chemical reaction. It involves reflux as recycle to the tower. There is no accumulation and the mass balance reduces to

$$\text{Input} = \text{output}$$

Draw a sketch of the distillation process and note the known and unknown variables. This is shown in Exhibit 6. The compositions are in mol %.

Basis: 120 kg mol/h feed to the column

Overall balance: $F = B + D = 120$
Benzene balance: $0.45F = 0.04B + 0.9D$

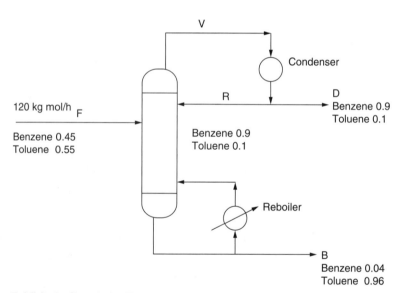

Exhibit 6 Sketch for Example 2.25

Solving the two equations, $D = \mathbf{57.2}$ kg mol/h, and $B = \mathbf{62.8}$ kg mol/h.
Now take a balance around condenser to get V.

$$V = R + D = 4D + D = 5D \text{ since } R/D = 4$$
$$V = 5(57.2) = 286 \text{ kg mol/h}$$

Compositions of R, D, and V are the same.

Molecular weight of distillate $= 0.9(78) + 0.1(92) = 79.4$
Flow of vapor to condenser $= 286(79.4) = \mathbf{22708}$ kg/h

Example 2.26

In the production of crystals of KNO_3, 1000 kg/h of a feed solution containing 25 wt% KNO_3 salt is charged to an evaporator where some water is evaporated at 422 K to produce a 50 wt% KNO_3 solution. The concentrated solution is then fed to a crystallizer from which crystals containing 96% KNO_3 are discharged as product. The mother liquor is a saturated solution and contains 37.5 wt% KNO_3. The mother liquor is recycled to the evaporator along with the fresh feed. What is the amount of recycled mother liquor and the product crystals in kg/h respectively?

Solution

This is a steady state process. Also there is no chemical reaction. Therefore there is no accumulation in the process. The mass balance is Output = Input.

Exhibit 7 gives the process flow diagram.

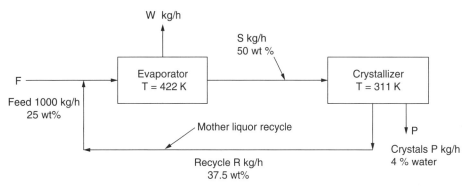

Exhibit 7 Process sketch for Example 2.26

Basis: 1000 kg/h

Overall material balance: $F = W + P$
Balance on nitrate: $0.25(1000) = W(0) + 0.96(P)$
Solving $P = 260.4$ kg/h and $W = 1000 - 260.4 = 739.6$ kg/h

To calculate the recycle stream, we can make the material balance on either the evaporator or the crystallizer. Balance on crystallizer gives

$$S = P + R = R + 260.4$$
$$0.5\,S = 0.96\,P + 0.375\,R$$

Solving these equations, $R = 958.3$ kg/h and $S = 958.3 + 260.4 = 1218.7$ kg/h

Material Balances of Processes Involving Chemical Reaction

In processes that involve chemical reactions the general mass balance Equation 2.2 or it's non-symbolic form Equation 2.3 still apply. The ΣM_C term drops out as we are considering only ordinary processes not involving nuclear reactions. Generation and consumption terms will have to be taken into account. When making balances of individual components, moles may not be conserved. However, balance on elements does not require the use of generation and consumption terms. In problems involving chemical reactions it is more convenient to work with molar units. The limiting reactant should be selected. A term frequently used in recycle operations involving chemical reactions is *conversion per pass* or *per pass conversion*, which is defined as the percentage of the limiting reactant that is converted in the combined feed to the reactor.

We will review the combustion processes separately, although they involve chemical reactions. In many of the problems involving chemical reactions, recycle and purge will have to be dealt with. Some examples will now illustrate the methods to follow in solving mass balance problems involving chemical reactions.

Example 2.27

Sulfuric acid is manufactured industrially by contact process by first burning molten sulfur to SO_2 with 100% excess air. Combustion gas is then dried in a tower and converted to SO_3 in a four-stage converter charged with V_2O_5 catalyst. The burner achieves only 90% conversion of S to SO_2. In the converter, conversion to SO_3 is 97% complete. Calculate the amount of air required in kg/100 kg of S burned and the compositions of gas streams exiting the burner and the converter.

Solution

This is a steady state process and there is no accumulation. Hence **Input = output.**
Draw a simple sketch of the process and put all relevant data on it.

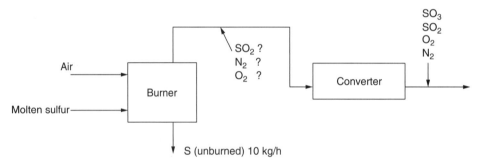

Exhibit 8 Process sketch for Example 2.27

Write down equations

$$\text{In burner } S + O_2 = SO_2$$
$$\text{In converter } SO_2 + \tfrac{1}{2}O_2 = SO_3$$

Molecular wts S = 32, O2 = 32, SO_2 = 64, and SO3 = 80

Basis: 100 kg of molten sulfur

Make a calculation
Burner exit gas

Sulfur converted = 90 kg/h = 90/32 = 2.8125 kg mol (kg atom)
SO_2 formed = 2.8125 kg mol
Total sulfur = (100/32) = 3.125 kg mol
O_2 used = 2(3.125) = 6.25 kg mol
N_2 in air fed to the dryer = (79/21)(6.25) = 23.51 kg mol
O_2 used up for 90% conversion of S = 2.8125 kg mol
O_2 − 2.8125 = 3.4375 kg mol

Burner exit gas amount and composition

Component	No of moles	mol %
SO_2	2.8125	9.45
O_2	3.4375	11.55
N_2	23.51	79.00
Total	29.76	100.00

Mass balance on converter

SO_2 converted to SO_3 = 0.97(2.8125) = 2.73 kg mol
SO_2 unconverted = (2.8125 − 2.73) = 0.0825 kg mol
O_2 consumed in converter = 0.5(2.73) = 1.365 kg mol
O_2 in exit stream = 3.4375 − 1.365 = 2.0275 kg mol
SO_3 formed in converter = 2.73 kg mol

Converter inlet and exit gas compositions

Feed to Converter			Converter Exit Gas		
Component	No of moles	mol %	Component	No of moles	mol %
SO_2	2.8125	9.45	SO_2	0.0825	0.29
O_2	3.4375	11.55	O_2	2.0275	7.15
N_2	23.51	80.68	N_2	23.51	82.93
			SO_3	2.73	9.63
Total	29.135	100.00	Total	28.35	100.00

The following problem illustrates the procedure to follow in making material balances involving both recycle and purge.

Example 2.28

Synthetic ammonia is produced by feeding 1:3 nitrogen-hydrogen mixture to a catalytic converter. A per pass conversion to ammonia is 25%. The ammonia formed is separated by condensation, and the unconverted gases are recycled to the reactor. The fresh feed nitrogen-hydrogen mixture contains 0.25 parts (mainly argon) impurities per 100 parts of N_2-H_2 mixture by volume. If the tolerable limit of impurities in the feed gas to the converter is 5 parts per 100 parts of N_2-H_2 mixture by volume, what are the amounts of recycle and purge?

Solution

This is a continuous process. Therefore Input = Output.
First draw a sketch of the process.

Basis: 100 kg mol/h of inert free fresh feed of N_2-H_2

Reaction $N_2 + 3H_2 = 2NH_3$

In this problem the inerts can serve as a tie substance. Secondly, the ratio of N_2 to H_2 is the same in the feed and recycle as both are fed in stoichiometric proportion.

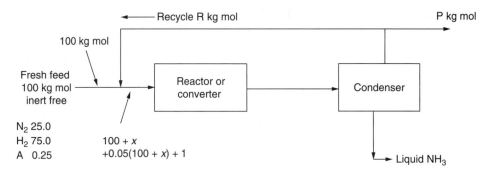

Exhibit 9 Process sketch for Example 2.28

Therefore, the mixture of nitrogen and hydrogen can be treated as one component. It is also assumed that ammonia is completely condensed in the condenser and the recycle is free of ammonia. This is to simplify calculations.

Per pass conversion = 0.25 based on the mixed or combined feed
Moles of NH_3 leaving the converter = $0.25(100 + R)/2$
Moles of $N_2 + H_2$ leaving the reactor = $0.75(100 + R)$

Moles of NH_3 formed = $\dfrac{100+R}{2}$ kg mol in the converter

Moles of argon in feed = 0.25 kg mol
Moles of argon in recycle gas = $0.05(100 + R)$ kg mol
Total moles of argon in mixed feed to reactor
  = $0.05(100 + R) + 0.25$ kg mol
Moles of argon per mole of N_2-H_2 mixture leaving the condenser

$$= \frac{0.05}{0.75} = 0.067 \text{ kg mol}$$

Moles of argon purged = 0.067 P

When steady state operation is established argon purged equals the argon in the fresh feed.
 Hence $0.0667P = 0.25$ and $P = 3.75$ kg mol inert free.
 Now take a balance around the purge junction.

$$0.75(100 + R) = R + P$$

Substituting for P and solving for R gives $R = 285$ kg mol. (Recycle)

$$NH_3 \text{ formed} = \frac{0.25(100+285)}{2} = 48.1 \text{ kg mol}$$

A summary of the material balance: Based on 100 kg mol feed (inert free)

Fresh feed of $N_2 + H_2$ = 100 kg mol (inert free)
Recycle of $N_2 + H_2$ = 285 kg mol
Purged $N_2 + H_2$ = 3.75 kg mol
NH_3 formed = 48.1 kg mol
Argon = 0.25 kg mol (in = out)
Recycle ratio = 285/100 = 2.85
Purge ratio = 3.75/285 = 0.01316

FUELS AND COMBUSTION

Combustion of fuels has been used for the generation of heat and power in process industries. Fuels are used to a large extent to supply thermal energy. In the combustion process, heat is liberated by burning the fuel with an oxidant, particularly oxygen. However, all oxidation processes are not categorized as combustion processes. In general, if the reactants involved are hydrogen, carbon, and sulfur, and the products are carbon dioxide, water, and sulfur dioxide, the process is termed combustion. Complete combustion is said to occur if the fuel burns completely to products, CO_2, SO_2, and H_2O. If one of the products is CO, then it is called *partial combustion*. Combustion of S to SO_2 is considered complete combustion even though it can be oxidized further to SO_3 in the presence of a catalyst. Combustion involves the release of heat due to an exothermic reaction. This fact must be taken into consideration while making material balances of combustion processes.

Solid Fuels
Coal is a naturally found solid fuel. Coals are ranked according to their formative stage as anthracite, bituminous, semianthracite, subbituminous, and lignite. Other solid fuels are coke, peat, wood, and sawdust.

Liquid Fuels
Crude oil is a natural liquid fuel. From it are obtained gasoline, diesel oil, alcohols, kerosene, and fuel oils.

Gaseous Fuels
Natural gas is an important gaseous fuel. Coke oven gas, blast furnace gas, liquefied petroleum gas (LPG), and refinery gases are other examples of gaseous fuels.

Heating Values of Fuels

The *heating value* or *calorific value* of a fuel is it's standard heat of combustion (heat of reaction numerically, but of opposite sign). Two types of heating values are defined. *Total heating value*, also termed *higher* or *gross heating value* (denoted as HHV), is the heat evolved per unit of the fuel under constant pressure and at a temperature of 25°C with all the water in the fuel initially present and that produced in the combustion being condensed to the liquid state at 25°C. *Net heating value*, also termed *lower heating value* (denoted by LHV), is defined in a similar manner but the final state of the water is taken as vapor at 25°C. Calorific values at constant volume are also defined in similar manner. The difference between two corresponding heating values (at constant pressure and at constant volume) is very small. The net heating value LHV is smaller than the gross heating value by an amount equal to the heat of condensation of water vapor. The heat of condensation of water at 25°C is 583.5 kcal/kg (2442.5 kJ/kg). An approximate formula relating LHV and HHV is as follows:

$$\text{LHV} = \text{HHV} - 52.5\text{H} \tag{2.26}$$

where LHV and HHV are in kcal/kg and H = weight percentage of hydrogen in coal, including hydrogen of moisture associated with coal as well as water of hydration of minerals in coal. If LHV, HHV, and λ_w are expressed in kJ/kg, then Equation 2.26 becomes

$$\text{LHV} = \text{HHV} - (18/2 = 9)\, \lambda_w \text{ (in kJ/kg)} \cdot \text{H} \tag{2.27}$$

and likewise in Btu units, it becomes

$$LHV = HHV - (18/2 = 9) \lambda_w \text{ (in Btu)} \cdot H \quad (2.28)$$
$$\lambda_w = 583.5 \text{ kcal/kg} = 2442.5 \text{ kJ/kg} = 1050.3 \text{ Btu/lb}_m$$

Analysis of Fuels

In order to make material balance calculations in combustion problems, it is necessary to have analysis of the fuel. Analyses of solid, liquid, and gaseous fuels are reported in different ways.

Coals

Two different types of analysis are used to express the compositions of coal. An *ultimate analysis* involves the determination of all major chemical elements in the coal, for example, *carbon, hydrogen, sulfur,* etc. On the other hand, in *proximate analysis* only four arbitrarily defined groups of constituents lumped together are determined. These groups are *fixed carbon, volatile matter, ash,* and *moisture.* The proximate analysis of coal is carried out according to standardized procedures. The ultimate analysis of coal is carried out for making material and energy balances.

The rank of coal is determined from it's fuel ratio. *Fuel ratio* of a coal is defined as the ratio of it's percentage of fixed carbon to that of it's volatile matter. The highest rank coal, anthracite, has a fuel ratio of between 10 and 60 while the lowest rank coal, bituminous, has a fuel ratio between $1/2$ and 3.

If HHV and proximate analysis of a coal are known, the following Calderwood equation allows us to calculate it's carbon content:

$$\text{Mass \% of carbon} = 5.88 + 0.00512(B - 40.5S)$$
$$\pm 0.0053 \left[80 - 100 \left(\frac{\text{volatile matter}}{\text{fixed carbon}} \right) \right]^{1.55} \quad (2.29)$$

where
 B = HHV in Btu/lb and
 S = mass % sulfur.

In the preceding equation, if (volatile matter/fixed carbon)(100) > 80, then the coefficient 0.0053 is negative, otherwise it is positive.

Liquid Fuels

Petroleum fuels derived from crude oil are mixtures of hydrocarbons including paraffins, napthelenes, olefins, and aromatics. Natural petroleums are rich in paraffins and napthelenes. In cracked products, however, larger quantities of olefins and aromatics may be present. Besides hydrocarbons, sulfur, oxygen, and nitrogen compounds are also constituents of petroleum fuels. The U.O.P. (Universal Oil Products Company) characterization factor of petroleum fuels is given by

$$K = \frac{T_B^{1/3}}{\text{sp gr}} \quad (2.30)$$

where
 K = U.O.P. characterization factor
 T = average boiling point, °R at 1 atmosphere
 sp gr = specific gravity of the petroleum fraction at 60° F.

A chart giving hydrogen content of a liquid fuel as a function of characterization factor for materials of constant boiling points is available (Hougen & Watson).

The elemental content of liquid fuels is also reported as ultimate analysis, as in the case of solid fuels. The hydrogen content can also be calculated by the following formula:

$$H = 26 - 15s$$

where
 H = percent hydrogen and
 s = specific gravity at 15°C {60°F/60°F}.

Gaseous Fuels

Fuel gases contain complex mixtures of saturated and unsaturated hydrocarbons. It is very difficult to analyze each component in these complex mixtures. So the carbon and hydrogen contents of saturated paraffins in the fuel are reported as C_nH_{n+2}, a hypothetical compound. Similarly, the unsaturated hydrocarbons are reported as C_aH_b. Carbon monoxide and hydrogen are separately determined.

Calculation of Heating Values of Fuels

The heating values of coals may be approximately determined by Dulong's formula.

$$\text{HHV} = 14540C + 60958\left(H - \frac{O}{8}\right) + 4050S \quad \text{Btu/lb} \tag{2.31}$$

or

$$\text{HHV} = 33950C + 144200\left(H - \frac{O}{8}\right) + 9400S \quad \text{kJ/kg} \tag{2.32}$$

C, H_a, S = weight fractions of carbon, available hydrogen, and sulfur respectively. and $H - \frac{O}{8}$ = mass fraction of net hydrogen = total hydrogen $- (1/8)$(oxygen).

The heating value of a petroleum liquid can be calculated by the formula

$$\text{HHV} = 22820 - 3780s^2 \tag{2.33}$$

Other relations are given by the following equations:

$$\begin{aligned}&\text{HHV} = 18650 + 40(\text{API gr.} - 10) \text{ Btu/lb for fuel oil} \\ &\text{HHV} = 18440 + 40(\text{API gr.} - 10) \text{ Btu/lb for kerosene} \\ &\text{HHV} = 18320 + 40(\text{API gr.} - 10) \text{ Btu/lb for gasoline}\end{aligned} \tag{2.34}$$

Terminology in Combustion

Knowledge of certain terms used in combustion of fuels (solid, liquid, and gaseous) is necessary before dealing with combustion problems. Therefore we will review these now.

Orsat Analysis   The composition of a gaseous fuel on dry basis (water free) can be established by the use of an Orsat analyzer. In Orsat analysis, volumes of products of combustion of the fuel are measured over and in equilibrium with water. Therefore each component is saturated with water vapor. Data from Orsat analysis are always on dry basis. Also SO_2 is reported along with CO_2.

Flue Gas or Stack Gas Flue gas or stack gas is the gaseous product of combustion of fuels. It primarily contains carbon dioxide, water vapor, nitrogen, and excess oxygen. It also includes sulfur dioxide. Since it contains water vapor it is also called wet gas.

Theoretical Air or Theoretical Oxygen This is the amount of air, which contains stoichiometrically required oxygen for complete combustion of the fuel.

Excess Air or Excess Oxygen Excess air is the amount of air that contains an amount of oxygen, which is in excess of the theoretically required oxygen for the complete combustion of the fuel. The calculated amount of air does not depend on how much fuel is actually burned but depends on the actual quantity available for combustion even if only partial combustion takes place. This is a similar calculation as that for excess reactant. For usual calculations, air composition is assumed to be 79 mol % N_2 and 21 mol % O_2.

The following few examples will illustrate the methods of solving problems involving combustion of fuels.

Example 2.29

In an experiment, 1 kg mol of essentially pure n-butane was burned with excess air. It was found that the combustion was incomplete and only 3.6 kg mol of CO_2 and 0.4 kg mol of CO were produced during the period of experiment. Flue gases were free of butane. Assume air is dry. What percentage of excess air was used? What was the degree of completion of combustion? Orsat analysis of flue gases was as follows: CO_2 9.49%, CO 1.05%, O_2 5.67%, and N_2 83.79% on volume basis.

Solution

We can treat this problem essentially as steady state as there is no accumulation. Draw a sketch first and place all data on the diagram.

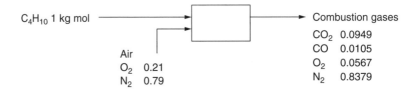

Combustion gases

Basis: 1 kg mol of butane

Equations

$C_4H_{10} + 6\frac{1}{2}O_2 = 4CO_2 + 5H_2O$ complete combustion
$C_4H_{10} + 4\frac{1}{2}O_2 = 4CO + 5H_2O$ incomplete combustion
C_4H_{10} required to produce 3.6 kg mol of CO_2 = (3.6/4)(1) = 0.9 kg mol
C_4H_{10} required to produce 0.4 kg mol of CO = (0.4/4)(1) = 0.1 kg mol

Since only 0.9 kg mol reacts to CO_2, the degree of completion is
= (0.9/1)(100) = 90%.
O_2 used to produce 3.6 kg mol of CO_2 = (3.6/4)(6.5) = 5.85 kg mol
O_2 used to produce 0.4 kg mol of CO = (0.4/4)(4.5) = 0.45 kg mol

Total O_2 used to burn 1 kg mol of C_4H_{10} = 5.85 + 0.45 = 6.3 kg mol
Oxygen in combustion gases = (5.67/9.49)(3.6) = 2.15 kg mol
Total O_2 fed for combustion = 6.3 + 2.15 = 8.45 kg mol

For complete combustion to CO_2, O_2 required is 6.5 kg mol.

$$\text{Therefore } \% \text{ excess oxygen } or \text{ air} = \frac{8.45 - 6.5}{6.5} \times 100 = 30\%.$$

Example 2.30

Natural gas of the following composition is burned with 20% excess air in a flow process. The analysis of the gas mixture is as follows:

	CH_4	C_2H_6	C_3H_8	C_4H_{10}	C_5H_{12}	CO_2	N_2
mol %	96.91	1.33	0.19	0.05	0.02	0.82	0.68

Calculate the products of combustion and the Orsat analysis. Assume both gas and air to be dry enough so that water vapor content in the gas and air can be ignored in the calculations. Assume also complete combustion of carbon and hydrogen.

Solution

This is a steady state process. Accumulation is zero and *Input = Output*. Draw a sketch of the process and place all pertinent information on the diagram.

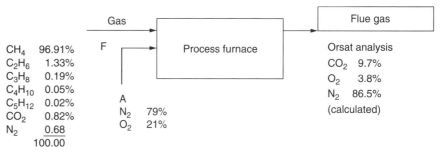

Exhibit 10 Process flow diagram for Example 2.30

Basis: 100 kg mol of natural gas mixture
To get products of combustion, prepare a table. In this problem, it is convenient to work with molar concentrations.
Reactions

$$CH_4 + 2O_2 = CO_2 + 2H_2O$$
$$C_2H_6 + 3\tfrac{1}{2}O_2 = 2CO_2 + 3 H_2O$$
$$C_3H_8 + 5O_2 = 3CO_2 + 4 H_2O$$
$$C_4H_{10} + 6\tfrac{1}{2}O_2 = 4CO_2 + 5 H_2O$$
$$C_5H_{12} + 8O_2 = 5CO_2 + 6H_2O$$

Prepare a table of products of combustions.

Table 2.7 Calculation of products of combustion

Component	Mol Fraction	kg mol	Required O_2	Air	Air	Flue Gas CO_2, N_2, O_2	H_2O Total
CH_4	0.9691	96.91	193.82			⎡ 96.91	193.82
C_2H_6	0.0133	1.33	4.66			⎢ 2.66	3.99
C_3H_8	0.0019	0.19	0.95			⎢ 0.57	0.76
C_4H_{10}	0.0005	0.05	0.33		CO_2 ⎨ 0.20	0.25
C_5H_{12}	0.0002	0.02	0.16			⎢ 0.10	0.20
CO_2	0.0082	0.82				⎣ 0.82	
N_2	0.0068	0.68		902.5	903.18	903.18	
O_2				239.9	39.98	39.98	
H_2O							
Total	1.000	100.22	199.92	1152.23	943.16	1044.42	199.02

Orsat analysis: This analysis is based on moisture-free basis.

Orsat analysis of flue gas

Total CO_2 = 101.26 kg mol, nitrogen = 903.18 kg mol, and oxygen = 39.98 kg mol

$$\% \ CO_2 = \frac{101.26}{1044.42} \times 100 = 9.7\%$$

$$\% \ O_2 = \frac{39.98}{1044.42} \times 100 = 3.8\%$$

$$\% \ N_2 = \frac{903.18}{1044.42} \times 100 = 86.5\%$$

Example 2.31

The ultimate analysis of a coal is as follows:

Carbon = 75.3%, hydrogen = 3.6%, sulfur = 2.0%,
Oxygen = 7.4%, nitrogen = 1.2%, and ash = 10.5%

This coal was burned in a furnace at the rate 250 kg per hour. The dry refuse removed from the furnace shows a combustible content of 20.1%. Orsat analysis of the flue gas is as follows:

$CO_2 + SO_2$ = 11.5%, Oxygen 7.9%, CO = 1.0%

Moisture in air = 0.005 kg/kg dry air
Moisture in feed coal = 3.06 kg/100 kg dry (moisture free) coal
What is the percent excess air the process is using?

Solution

This is again a steady state problem. Input = Output. Draw a sketch of the process.

H_2O = 0.005 kg/kg air
H_2O = 0.008 kg mol/kg mol air
H = 2(0.008) = 0.016 kg mol/kg mol air
O = 0.008 kg mol/kg mol air

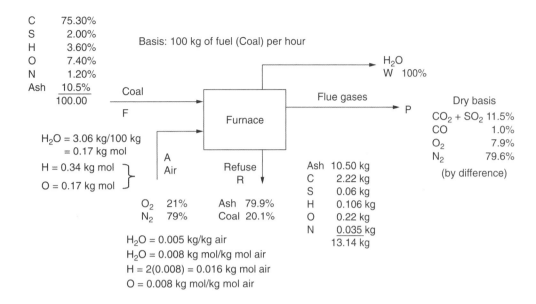

Total refuse:

Ash is a tie substance. Using this we obtain R = Total refuse.
Ash = 10.5 kg, $R = (10.5)(100/79.9) = $ **13.14** kg

Assume refuse leaving the furnace is dry.

Unburned coal in refuse = 13.14 – 10.5 = 2.64 kg

Assume composition of unburned coal same as that of fresh coal.

Carbon in refuse = $[75.3/(100 - 10.5)] \times 2.64 = 2.22$ kg

The other quantities are calculated in the same manner and the following table is prepared. In this case we write elemental balances in terms of atom moles. Quantities of elements in refuse are shown in the table.

Refuses Analysis

Component	Mass %	kg	kg mol (atom)
C	84.13	2.22	0.1850
S	2.24	0.06	0.0019
H	4.02	0.106	0.1060
O	8.27	0.22	0.0138
N	1.34	0.035	0.0025
	100.00	2.64	0.3092

Now write equations for the four components C +S, H, O, and N.
Carbon and sulfur balance

$$\frac{75.3}{12} + \frac{2}{32} = P(0.115 + 0.01) + 0.185 + 0.002$$

$$P = \mathbf{49.2} \text{ kg mol(atom)} = \text{Flue gas}$$

Nitrogen balance

$$\frac{1.2}{14} + 2A(0.796) = 2P(0.79) + 0.0025$$

Using the value of P obtained by carbon and sulfur balance, A = **49.53** kg mol Air.
Hydrogen balance

$$\frac{3.6}{1.008} + 0.34 + 0.016A = 2W + 0.106 \quad W = \mathbf{2.3} \text{ kg mol}$$

Check with oxygen balance.

$$\frac{7.4}{16} + 0.17 + 0.008A + 2(0.21A) = W + 2(0.115P) + 0.01P + 0.0138 + 2(0.079P)$$

Lefthand side of the equation equals **21.83**. Righthand side equals **21.9**. This is good agreement. The difference is only [(21.9-21.83)/21.83](100) = 0.32%.

Component	Equation	kg	kg mol	Required Oxygen
C	$C + O_2 = CO_2$	75.3 (C = 12)	6.275	6.275
S	$S + O_2 = SO_2$	2.00 (S = 32)	0.0625	0.0625
H	$H_2 + \tfrac{1}{2}O_2 = H_2O$	3.60 (H_2 =2.016)	1.79	0.895
O	—	7.4 (O_2 = 32)	(0.2313)	−0.2313
N	—	— (N_2 = 28)	—	
Ash	—	—	—	
				7.0012

Now we calculate the percent excess air.

Oxygen in the air is (49.53)(0.21) = 10.4013 kg mol

$$\text{Percent excess air} = \frac{10.4013 - 7.0012}{7.0012} \times 100 = 48.56\%$$

Example 2.32

Make a rough estimate of excess air used if Orsat analysis of flue gas from furnace burning coal is as follows: $CO_2 + SO_2$ = 17.4%, CO = 0.3%, O_2 = 3.5% and the rest nitrogen.

This is the only data available. State any assumptions you make.

Solution

The analysis of the coal is not available. Therefore, contribution of CO_2, O_2, and N_2 of coal to flue gases cannot be assessed. In this case, we will assume these to be negligible so that we can assume that contribution of nitrogen comes only from air. Now nitrogen in the air passes through the furnace as is and can be taken as a tie substance and used to calculate the amount of oxygen fed to the furnace. Also, flue gases contain no CO; therefore we can also assume complete combustion. Knowing the difference between incoming oxygen and that going out with flue gases, we can calculate % excess air.

Basis of calculation: 100 kg mol

Calculation of oxygen in flue gases:

Component	Mol %	Flu Gas kg mol	O_2 in Flue Gas kg mol
$CO_2 + SO_2$	17.4	17.4	17.4
CO	0.3	0.3	0.15
O_2	3.5	3.5	3.5
N_2	78.8	78.8	0.00
Total	100.00	100.00	21.05

Total O_2 = 21.05 kg mol

Total air fed to the furnace = $\dfrac{21.05}{0.21}$ = 100.24 kg mol

Oxygen used is 21.05 – 3.5 = 17.55 kg mol

Calculation based on O_2 used,

Percent excess air = $\dfrac{3.5}{17.55} \times 100 = 19.94$ say 20%

Calculation based on air,

Air used = 17.55/0.21 = 83.57 kg mol

Percent excess air = $\dfrac{(100.24 - 83.57)}{83.57} \times 100 = 19.94$ or 20%

Example 2.33

A coal having the following ultimate analysis is fed to a furnace and burned with 25% excess air. Coal has 3 lb moisture per 100 lb of coal. Assume air to be dry. Calculate air-to-fuel ratio, and Orsat analysis of flue gas.

Ultimate analysis of coal:

C 78.3%, S 1.2%, H = 5.3%
O 7.6%, N 1.4%, Ash 6.2%

Solution

This is again a flow process with chemical reaction. The calculation is straightforward.

Draw a sketch as usual.

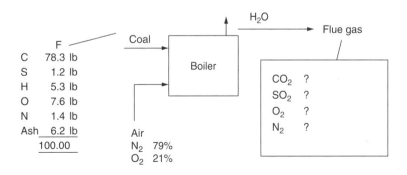

Prepare a table of combustion products and calculate oxygen used.

Basis: 100 lb of coal

Component	lb	lb mol	O_2 required lb mol	O_2, N_2 & Air in lb mol	CO_2 or SO_2 lb mol	Component in Flue Gas (dry) lb mol	H_2O lb mol
C	78.3	6.525	6.525		6.525	6.525	—
S	1.2	0.0375	0.0375		0.0375	0.0375	—
H	5.3	2.65	1.325			—	2.65
O	7.6	0.2375	(0.2375)	9.56		1.91	—
N	1.4	0.05		35.98		36.03	—
Ash	6.2						—
H_2O	3.0	0.17					0.17
Total	103		7.65	45.54		44.50	2.82

Brackets around a number in O_2 column indicates credit or negative value.

$$\text{Oxygen required} = 7.65 \text{ lb mol (see table)}$$
$$\text{Oxygen actually used} = 1.25(7.65) = 9.56 \text{ lb mol}$$
$$\text{Air used} = (1/0.21)(9.56) = 45.54 \text{ lb mol}$$
$$\text{Nitrogen} = 45.54 - 9.56 = 35.98 \text{ lb mol}$$
$$\text{Air fuel ratio} = (45.54 \times 29.0)/100 = \mathbf{13.2}$$

Flue gas analysis

$$\% \; CO_2 + SO_2 = [(6.525 + 0.0375)/44.5](100) = \mathbf{14.75\%}$$
$$O_2 \text{ in flue gas} = O_2 \text{ actually used} + O_2 \text{ in coal} - O_2 \text{ consumed}$$
$$= 9.56 + 0.2375 - (1.325 + 0.0375 + 6.525) = \mathbf{1.91} \text{ lb mol}$$
$$\% \; O_2 = (1.91/44.5)(100) = \mathbf{4.3\%}$$
$$N_2 = 100 - 4.3 - 14.75 = \mathbf{80.95\%}$$

RECOMMENDED REFERENCES

1. Himmelblau. *Basic Principles and Calculations in Chemical Engineering*, 6th ed., Prentice-Hall, 2004.
2. Hougen et al. *Chemical Process Principles Part 1*, John Wiley & Sons, 1954.
3. Henley et al. *Material and Energy Balance Computations*, John Wiley & Sons, 1969.
4. Reid, Prausnitz, and Sherwood, *Properties of Gases and Liquids*, McGraw-Hill, 1977.
5. Smith et al, *Introduction to Chemical Engineering Thermodynamics*, 6th ed., McGraw-Hill, 2001.
6. Lee-Kessler, *AIChE J*, vol 21, pp. 510–527. (1975) (Refer to preceding reference 5.)

CHAPTER 3

Energy Balance and Thermodynamics

OUTLINE

THERMODYNAMIC SYSTEM 78
Thermodynamic Properties ■ Definition of Energy Terms

CONSERVATION OF ENERGY OR FIRST LAW OF THERMODYNAMICS 81
Energy Balances for Closed Systems ■ Energy Balances for Open Systems (Without Chemical Reactions) ■ Sensible Heat and Heat Capacity ■ Tables of Enthalpy Values ■ Graphical Presentation of Enthalpy Data ■ Enthalpy Changes During Phase Transitions ■ Heat of Vaporization ■ Heat of Fusion

ENERGY BALANCES IN PROCESSES INVOLVING CHEMICAL REACTIONS 92
Laws of Thermochemistry ■ Heat of Reaction ■ Heat of Combustion ■ Energy Balance with Accounting for Heat of Reaction ■ Adiabatic Flame Temperature ■ Heats of Mixing and Enthalpy of a Solution

REVERSIBLE AND IRREVERSIBLE PROCESSES 103
Entropy and Second Law of Thermodynamics ■ Second Law of Thermodynamics

THERMODYNAMIC PROPERTIES OF MATTER 110
Residual Properties ■ PVT Relationships—Equations of State for Ideal Gases ■ Thermodynamic Property Relations for an Ideal Gas ■ Calculation of Entropy Changes for Ideal Gas ■ Third Law of Thermodynamics ■ Solution Thermodynamics ■ Fugacity and Fugacity Coefficients ■ Fugacities in Ideal Solutions and Standard States ■ Isothermal Effect of Pressure on $\hat{G}, \hat{S}, \hat{U}$

POWER CYCLES AND REFRIGERATION 121
Refrigeration

RECOMMENDED REFERENCES 129

In this chapter we review energy balance and thermodynamics, two other important topics on the PE examination. Of the 24 percent of the exam allocated to mass/energy balances, approximately 13 percent is allotted to these two topics. In this chapter we will cover energy balance (first law of thermodynamics); second and third laws of thermodynamics; sensible heat; latent heat; heats of reaction; heat of solution; estimation and correlation of physical properties; and applications requiring combination of sensible heat calculations, latent heats, heat of reaction, and heat of solution.

THERMODYNAMIC SYSTEM

We have already covered the concept of a system and its surroundings in Chapter 2. To recall briefly, a system is a portion of the universe set apart for study. It may be a specified volume in space or a given quantity of matter. Anything outside this system is termed its *surroundings*. In practice, the surroundings refer to the immediate surroundings of the system. A system with constant mass is called a *closed system*; without a constant mass a system is called an *open system*. Boundaries of a system can be real or imaginary, stationary or movable.

Thermodynamic Properties

Any system characteristic that is observable and/or measurable is called the *system property*, e.g., temperature, pressure, mass, area, volume, surface tension. A thermodynamic *state* of the system is a particular condition of the system, which is characterized and fixed by certain properties of the system. Properties are distinguished as either average or point and fundamental or derived. Temperature and pressure are fundamental point properties; mass and volume are not point properties. An *extensive property* of a system is its property that can be obtained by summing up the properties of its parts. Intensive properties of a system do not depend on the extent (mass) of the system, e.g., temperature, pressure, and specific volume.

Definition of Energy Terms

Energy exists in different forms. An understanding of energy terms is required to postulate the energy balance or the first law of thermodynamics.

Kinetic Energy, KE
All energy associated with the macroscopic motion of the mass of a system is called *kinetic energy*. Kinetic energy can be expressed as

$$\mathrm{KE} = \phi(M, u) = Mu^2/2g_c \qquad (3.1)$$

where M and u are the mass and velocity of the system, respectively.

Potential Energy, PE
Potential energy is the energy associated with a system macroscopically due to the system interactions with the gravitational, electric, and magnetic fields that exist in the surroundings of the system. Potential energy is calculated by the relation

$$\mathrm{PE} = mgh \qquad (3.1a)$$

where h is the distance from the reference point.

Heat Energy, Q
Heat Q is the energy transferred across the boundaries of a system under the influence of a temperature difference during the accounting period; it is not associated with the mass transferred across the boundaries of the system. We know that when there is a temperature difference between a body (system) and its surroundings, the heat is transferred across the boundary of the system at a certain rate. Q is the symbol used to denote the total heat transferred in a given time interval but not the heat transfer rate. Heat is a form of energy in transition and cannot be stored or created.

Heat is considered positive when received by or transferred to the system and negative when it leaves the system. It can be transferred by three mechanisms: conduction, convection, and radiation. Heat depends on the path it takes and is not a state function.

Work, W

Work W is defined as the force moving through a distance or

$$W = (\text{force})(\text{distance}) \tag{3.2}$$

Work is considered positive when it is transferred to the system and negative when transferred from the system to the surroundings. This sign convention is opposite to that used previously.

The definition is extended to include all forms of work, such as mechanical, electrical, and so on. Joule established the mechanical equivalence of work and heat. Work is a form of energy, which is transferred across the boundaries of a system during an accounting period and is not associated with either the mass transfer across the boundary or the temperature difference between the surroundings and the system. Like heat, it is a path and not a state function.

If fluid pressure is the only force acting on a system, the work is given by

$$\delta W = -PA\, d\left(\frac{V}{A}\right) \tag{3.3}$$

where P = pressure, V = volume, and A = area.

Since A is constant,

$$\delta W = -P\, dV \tag{3.3a}$$

and

$$W = -\int_{V_1}^{V_2} P\, dV \tag{3.3b}$$

Example 3.1

An ideal gas is enclosed in a cylinder by a frictionless piston. Initially the gas is at 330 K and 250 kPa pressure and its volume is 0.2 m³. When the latch on the piston is released, the gas slowly expands to a volume of 0.35 m³. Calculate the work done by the gas on the piston if the expansion takes place by two different paths:

Path 1: Expansion at constant pressure P = 250 kPa
Path 2: Expansion at constant temperature = 330 K.

Solution

We assume that the piston is frictionless and the process is ideal. This means the process is reversible. (Reversible and irreversible processes will be discussed later.)

System: the ideal gas enclosed in the cylinder

Draw a sketch of the process to get a clear idea of the process steps.

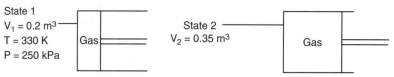

Exhibit 1 Sketch for Example 3.1

Path 1: Constant Pressure path. The gas expands and does work on the piston. This is work done by the system on the surroundings. By new convention of signs, this work has a negative value. The mechanical work done by the system "gas" for path A is then given by

$$W = -\int_{V_1}^{V_2} P\,dV \quad \text{by Equation 3.3b}$$

Since this path is of constant pressure, by integration, we get

$$W = -\int_{V_1}^{V_2} P\,dV = -P(V_2 - V_1)$$

$$= \frac{-250 \times 10^3 \text{ Pa}}{} \left| \frac{1 \text{ N/m}^2}{1 \text{ Pa}} \right| \frac{0.15 \text{m}^3}{} \left| \frac{1 \text{ J/m}}{1 \text{ N}} \right. = -37.5 \text{ kJ}$$

Path 2: Constant temperature path. In this expansion, the pressure is not constant and it is necessary to use the ideal gas law relationship to substitute P in terms of V so that we can integrate with respect to V. From the ideal gas law, $PV = nRT$, and $P = nRT/V$. Using this relation

$$W = -\int_{V_1}^{V_2} \frac{nRT}{V}\,dV = -nRT \ln\left(\frac{V_2}{V_1}\right)$$

First we must calculate n.

$$n = \frac{PV}{RT} = \frac{250 \times 0.2}{8.314 \times 330} = 0.0182241 \text{ kg mol}$$

$$W = -\frac{0.0182 \times 8.314 \times 330}{} \ln \frac{0.35}{0.2} = -27.94 \text{ kJ}$$

Thus we see that the constant pressure and the constant temperature paths yield two different values for work to reach the final state.

The value or amount of work W done during a given interval of time depends on the initial state of the system, the path taken, and the final state of the system. Types of work done on or by a system are mechanical work, shaft work, electrical work under voltage potential, and others.

Internal Energy U

Internal U or *intrinsic energy* is the energy associated with a system because of component molecules having kinetic energies from *translational, rotational,* and *vibrational* motions and having *potential energy* from intermolecular forces. The molecular level (microscopic) kinetic and potential energies are grouped together as the internal energy of the system. It is denoted by U. For a closed (i.e., constant-mass system) it is a unique-state property of the system. It is a macroscopic measure of molecular, atomic, and subatomic energies. It cannot be directly measured and; therefore, it needs to be calculated with the use of other variables such as pressure, volume, temperature, and composition. The calculation of internal energy will be reviewed later in this chapter.

For a pure component, the common practice is to use the two variables T for temperature and $\hat{V}$ for specific volume to describe the internal energy. That is to say, U is a function of T and $\hat{V}$.

Enthalpy Function H

The sum $U + PV$, which appears very often in energy balances, is given the symbol H, and is called the *enthalpy* of the system. Since U, P, and V are system state properties, enthalpy H is also a state property. Therefore, it is an exact differential. It is expressed as a function of two variables, T and P.

Both internal energy and enthalpy have no absolute values and only changes in either U or H can be calculated using some reference state.

State or Point Functions

Enthalpy, internal energy, pressure, temperature, and density are called point or state functions because their values depend only on the state of the material and not the path taken to reach that state.

CONSERVATION OF ENERGY OR FIRST LAW OF THERMODYNAMICS

The first law of thermodynamics or the law of conservation of energy in the most general form can be written for a given accounting period as

$$(U + PE + KE)_E - (U + PE + KE)_B = \Sigma(H + PE + KE)_t + Q + W - \Sigma E_c \quad (3.4)$$

where
- $(U + PE + KE)_E$ = sum of internal, potential, and kinetic energies associated with system at end of accounting period
- $(U + PE + KE)_B$ = sum of internal, potential, and kinetic energies associated with system at beginning of accounting period
- Q = heat transferred across boundaries of system during accounting period
- W = All work that crosses boundaries of system during accounting period
- $\Sigma(H + PE + KE)_t$ = energy transfer associated with mass transfer
- ΣE_c = energy converted to mass within system boundaries due to atomic transmutations ($\Sigma E_c = 0$ during ordinary physical and chemical changes. We can ignore this term in non-nuclear processes)

We can write the energy balance in words as follows (we will set aside the $\sum E_c$ term here after):

$$\begin{Bmatrix} \text{Accumulation of} \\ \text{energy within the} \\ \text{system} \end{Bmatrix} = \begin{Bmatrix} \text{transfer of energy} \\ \text{into system through} \\ \text{system boundary} \end{Bmatrix} - \begin{Bmatrix} \text{transfer of energy} \\ \text{out of system through} \\ \text{system boundary} \end{Bmatrix}$$

$$+ \begin{Bmatrix} \text{energy generation} \\ \text{within system} \end{Bmatrix} - \begin{Bmatrix} \text{energy consumption} \\ \text{within system} \end{Bmatrix} \quad (3.4a)$$

The energy consumption and generation terms will not occur in the processes we are considering in this review.

Energy Balances for Closed Systems

If we consider a closed system and apply equation 3.4 to it the following result is obtained

$$(U + PE + KE)_E - (U + PE + KE)_B = Q + W$$

or

$$\Delta U + \Delta PE + \Delta KE = Q + W$$

where ΔU, ΔPE, and ΔKE are the changes in U, PE, and KE of the system during the time interval considered.

Since there is no mass transfer across the boundaries of the system and $\sum E_c = 0$, if PE and KE are negligible, the equation further reduces to

$$U_E - U_B = Q + W \quad \text{or} \quad \Delta U = Q + W$$

ΔU, W, and Q are quantities integrated over the time interval considered. These quantities are given by

$$\Delta U = \int_{U_{t_1}}^{U_{t_2}} dU \qquad Q = \int_{t_1}^{t_2} \dot{Q} \, dt \qquad W = \int_{t_1}^{t_2} \dot{W} \, dt$$

Energy Balances for Open Systems (Without Chemical Reactions)

In open systems mass is transferred across the system boundaries and the energy transfer is associated with it. In open systems there is no accumulation in the system; therefore the general Equation 3.4 for energy balance reduces to

$$0 = \sum (H + PE + KE)_t + Q + W \quad (3.4b)$$

where the term $\sum (H + PE + KE)_t$ includes the energy transferred with mass entering the system and leaving the system. In certain flow processes, either Q or W may be absent, giving further simplification of the general equation.

When the quantities U, H, PE, KE, Q, and W are used on a unit basis, they are written with a caret; for example, enthalpy per unit mass will be denoted by $\hat{H}$. (In some books, instead of using a caret, a horizontal bar is used as in $\bar{H}$). With this notation, the various terms in the general balance equation assume the form

The accumulation term: $M_E(\hat{U} + \hat{PE} + \hat{KE})_E - M_B(\hat{U} + \hat{PE} + \hat{KE})_B$

Energy transfer into the system with mass flow in: $M_I(\hat{U} + \hat{PE} + \hat{KE})_I$

Energy transfer out of the system with mass flow out: $M_O(\hat{U} + \hat{PE} + \hat{KE})_O$

Net heat transfer into the system: Q

Net transfer of work (mechanical or electrical): W.

There is other work done on or by the system when a flowing fluid enters or leaves the system. This work is given by

$$P_I \hat{V}_I M_I - P_o \hat{V}_o M_o$$

where $\hat{V}$ is specific volume of the fluid.

The general energy balance equation in this notation becomes

$$M_E(\hat{U} + \hat{PE} + \hat{KE})_E - M_B(\hat{U} + \hat{PE} + \hat{KE})_B = M_I(\hat{U} + \hat{PE} + \hat{KE})_I \\ - M_O(\hat{U} + \hat{PE} + \hat{KE})_O + Q \\ + W + P_I \hat{V}_I M_I - P_o \hat{V}_o M_o \quad \textbf{(3.4c)}$$

Adding $P_I \hat{V}_I M_I$ to $\hat{U}_I$ and $P_o \hat{V}_o M_o$ to $\hat{U}_o$, we get two enthalpy functions as follows:

$$\hat{H}_I = \hat{U}_I + P_I \hat{V}_I M_I, \quad \text{and} \quad \hat{H}_O = \hat{U}_o + P_o \hat{V}_o M_o$$

Equation 3.4c can now be written with multiple input and output streams in terms of enthalpies as follows:

$$M_E(\hat{U} + \hat{PE} + \hat{KE})_E - M_B(\hat{U} + \hat{PE} + \hat{KE})_B \\ = M_I(\hat{H} + \hat{PE} + \hat{KE})_I - M_o(\hat{H} + \hat{PE} + \hat{KE})_o + Q + W \quad \textbf{(3.4d)}$$

Sensible Heat and Heat Capacity

Transfer of heat to a system in absence of phase change, chemical reactions, and change in composition causes the temperature of the system to rise (or decrease if heat is removed from the system). This is called sensible heat transfer to or from the system. When the system is a homogeneous substance of constant composition, according to phase rule, fixing of two intensive properties determines its state. Thus, molar or specific internal energy of a substance can be related as a function of two other state variables. These are arbitrarily selected as temperature and molar or specific volume. Thus

$$U = U(T, V). \quad \textbf{(3.5)}$$

$$\text{By differentiating, } dU = \left(\frac{\delta U}{\delta T}\right)_V dT + \left(\frac{\delta U}{\delta V}\right)_T dV \quad \textbf{(3.5a)}$$

By definition, constant volume heat capacity is given by

$$C_V = \left(\frac{\delta U}{\delta T}\right)_V \quad \textbf{(3.5b)}$$

For a constant volume process in a closed system, $dV = 0$ and Equation 3.5a becomes

$$dU = C_V dT \quad \text{at constant volume.} \tag{3.5c}$$

Similarly, the constant pressure heat capacity is defined as

$$C_P = \left(\frac{\delta H}{\delta T}\right)_P \tag{3.6}$$

The enthalpy change for a constant pressure closed system process is then given by

$$dH = C_P dT \quad \text{(constant P)} \tag{3.6a}$$

whence

$$\Delta H = \int_{T_1}^{T_2} C_P dT \tag{3.6b}$$

For ideal gases the internal energy is independent of volume or pressure. Molar specific heat at constant pressure and at constant volume for ideal gases are related by

$$C_P = C_V + R \tag{3.6c}$$

The enthalpy of a substance in a single phase can be calculated using its heat capacity, which is a function of temperature. Enthalpy difference due to sensible heat between two temperature levels at a given constant pressure is given by

$$\Delta H = \int_{T_1}^{T_2} C_P dT \tag{3.7}$$

For ideal gas mixtures, molar heat capacity is given by

$$C_{Pav} = \sum_{i=1}^{n} y_i C_{P_i} \tag{3.8}$$

Heat capacity data are usually given in terms of a polynomial as follows:

$$C_P = a + bT + cT^2 + \cdots \tag{3.9}$$

Heat capacity data for various compounds are available in handbooks and textbooks.

One can obtain enthalpy change per unit mole or mass by integration of the heat capacity equation between two desired temperatures.

Mean Heat Capacities of Gases

A mean or average heat capacity at constant pressure is given by the following equation:

$$C_{Pm} = \frac{\int_{T_1}^{T_2} C_P dT}{(T_2 - T_1)} \tag{3.10}$$

Example 3.2

Heat capacity of nitrogen is given by the equation

$$C_p = 6.895 + 0.7624 \times 10^{-3} T - 0.7009 \times 10^{-7} T^2$$

where T is in °F and C_p in Btu/lb mol · °F.

If nitrogen is heated from 77°F to 550°F, what is the enthalpy change?

Solution

$$\Delta H = \int_{T_1}^{T_2} (6.985 + 0.7624 \times 10^{-3} T - 0.7009 \times 10^{-7} T^2) \, dT \text{ in °F}$$

$$\Delta H = \int_{77}^{550} (6.985 + 0.7624 \times 10^{-3} T - 0.7009 \times 10^{-7} T^2) dT$$

$$= \left[6.985 T + 0.7624 \times 10^{-3} \frac{T^2}{2} - \frac{0.7009 \times 10^{-7} T^3}{3} \right]_{77}^{550}$$

$$= \left[6.985(550 - 77) + 0.7624 \times 10^{-3} \frac{550^2 - 77^2}{2} - \frac{0.7009 \times 10^{-7} (550^3 - 77^3)}{3} \right]$$

$$= +3413.1 \text{ Btu/lb mol nitrogen}$$

For gas mixtures, we can add the individual heat capacity coefficients multiplied by their mol fractions for the corresponding temperature power terms and integrate the resulting combined equation to get ΔH or we can integrate individual equations and multiply the resulting ΔH values by respective mol fractions of the components and add them together. In any case, it is easier to do these calculations using a spreadsheet program. If ΔH is determined relative to some reference state, then ΔH is the enthalpy relative to that reference state.

Sensible Heat Changes in Liquids

Heat capacities at constant pressure and at constant volume of a liquid are nearly equal. In general, heat capacities of liquids increase with temperature. For liquids the sensible heat change is given by

$$dQ = dH = mC_p dT \tag{3.11}$$

where m = mass of the liquid.

Tables of Enthalpy Values

The other sources of enthalpy values are enthalpy tables available in handbooks. In some tables, enthalpy values for gas phase only are given. Other tables, such as steam tables, include the enthalpy values of liquid and vapor in single phase as well as the values for phase change. The enthalpy values are tabulated with T and P as variables. If the tables give values at close range intervals, linear interpolation is reasonably accurate.

Example 3.3

(Use of Steam Tables)

Steam is cooled from 610 K and 6 Bar to 520 K and 3.6 Bar. What is the change in specific enthalpy in kJ/kg?

Solution

Use Tables 302-303 (Perry's handbook) to get relevant data for enthalpies.

P Bar	5		10	
T (K)	kJ/kg		kJ/kg	
600	H	3120	H	3109
700	H	3328	H	3322

P Bar	1		5	
T (K)	kJ/kg		kJ/kg	
500	H	2929	H	2912.4
600	H	3129	H	3120

First interpolate between temperatures.

At 610 K and 5 Bar

$$\hat{H} = \frac{610-600}{700-600} \times (3328-3120) + 3120 = 3140 \text{ kJ/kg}$$

At 610 K and 10 Bar

$$\hat{H} = \frac{610-600}{700-600} \times (3322-3109) + 3109 = 3130.3 \text{ kJ/kg}$$

Now interpolate between pressure to get value at 610 K and 6 Bar.

$$\hat{H} = \frac{6-5}{10-5} \times (3130.3) + 3140 = 3138.1 \text{ kJ/kg}$$

In a similar manner, double interpolation for the second state of steam gives

$$\hat{H} = 2963.8 \text{ kJ/kg}$$

Therefore, $\Delta \hat{H} = 2963.8 - 3138.1 = -174.3$ kJ/kg.

Graphical Presentation of Enthalpy Data

Enthalpy charts for many substances are available. A pressure-enthalpy diagram with V and T as parameters is called a P-H diagram. For a given state the enthalpy values can be directly obtained from these charts.

Enthalpy Changes During Phase Transitions

Enthalpy changes in a single phase are changes in sensible heat. Besides sensible heat characterized by heat capacity, another important consideration in enthalpy calculations is *latent heat*. During transition from one phase to the other, a substance goes through a large enthalpy change. Melting, fusion, vaporization and condensation, and sublimation are processes in which a phase change occurs.

Enthalpy change during melting is given the name *heat of fusion* and enthalpy during vaporization is termed *heat of vaporization*. During a phase transition the temperature remains constant. *Heat of condensation* is due to a vapor condensing into a liquid and is negative heat of vaporization.

Heat of Vaporization

An important equation relating molar heat of vaporization, vapor pressure, and temperature is the Clapeyron equation

$$\frac{dp}{dT} = \frac{\Delta \hat{H}_v}{T(\hat{V}_g - \hat{V}_l)} \tag{3.12}$$

where R = ideal gas law constant, ΔH_v = heat of vaporization, p = vapor pressure of liquid, and T = absolute temperature.

With the assumptions that vapor behaves like an ideal gas ($\hat{V}_g = RT/p$) and liquid specific volume is negligible in comparison with that of vapor ($\hat{V}_g \gg \hat{V}_l$), the preceding equation can be written as

$$\frac{dp}{p} = \frac{\Delta \hat{H}_v dT}{RT^2} \tag{3.13}$$

If it is further assumed that $\Delta \hat{H}_v$ is constant over a small temperature range, integration of Equation 3.13 yields the Clausius-Clapeyron equation

$$\log \frac{p_1}{p_2} = \frac{\Delta \hat{H}_v}{2.303R} \left(\frac{1}{T_2} - \frac{1}{T_1} \right) \tag{3.14}$$

Other equations useful in estimation of heat of vaporization λ_v follow.

Chen's Equation
Chen's Equation is as follows:

$$\Delta \hat{H}_v = \frac{T_b [0.0331(T_b/T_c) + 0.0297 \log_{10} p_c - 0.0327]}{1.07 - T_b/T_c} \tag{3.15}$$

where T_b = normal boiling point of the liquid in K and p_c = critical pressure in atmospheres.

$\Delta \hat{H}_v$ calculated by Chen's equation is in kJ/gmol. It gives values within 2 percent.

Watson's Empirical Equation
This equation correlating heat of vaporization and temperature is

$$\frac{\lambda_2}{\lambda_1} = \left(\frac{1 - T_{r2}}{1 - T_{r1}} \right)^{0.38} \tag{3.16}$$

where
λ_2 = Heat of vaporization cal/gmol at temperature T_2
λ_1 = Heat of vaporization cal/gmol at temperature T_1
T_{r2} = Reduced temperature at T_2 and
T_{r1} = Reduced temperature at T_1

Riedel Equation

Equation 3.16 requires one reference value λ_1. Heats of vaporization at normal boiling point are available for many liquids and therefore normal boiling point is taken as T_1. If value of λ_v at the normal boiling point is not available, it can be estimated by Riedel's equation, which is as follows:

$$\frac{\lambda_v}{RT_b} = \frac{1.092(\ln p_c - 5.6182)}{0.930 - T_{rb}} \quad (3.17)$$

where

λ_v = latent heat of vaporization, kJ/kg mol
p_c = critical pressure, kPa
T_{rb} = reduced temperature at normal boiling point

Trouten's Rule A rough estimate of latent heat of vaporization for pure liquids at their normal boiling point can be made with the use of Trouton's rule.

$$\frac{\Delta \hat{H}_{vn}}{RT_{bn}} \sim 10 \quad (3.18)$$

where

$\Delta \hat{H}_{vn}$ = latent heat of vaporization
T_{bn} = absolute temperature of the normal boiling point

The units of $\Delta \hat{H}_{vn}$, R, and T_{bn} need to be chosen so that $\frac{\Delta \hat{H}_{vn}}{RT_{bn}}$ is dimensionless.

Example 3.4

Calculate the heat of vaporization of ethyl alcohol at normal boiling point by (a) Riedel's method and (b) latent heat of vaporization at 320 K by Watson's method. Other required data are

$$T_b = 351.5 \; K, \quad T_c = 516.2 \; K, \quad p_c = 6137 \; kPa, \quad \Delta H_v = 38744 \; kg/kg \; mol$$

Solution

(a) Riedel's equation is $\dfrac{\lambda_v}{RT_b} = \dfrac{1.092(\ln p_c - 5.6182)}{0.930 - T_{rb}}$

$$T_{rb} = (351.5/516.2) = 0.681$$

$$\lambda_v = \frac{1.092(\ln p_c - 5.6182)}{0.930 - T_{rb}} RT_b$$

$$\lambda_v = \frac{1.092(\ln 6137 - 5.6182)}{0.930 - 0.681}(8.3145 \times 351.5)$$

$$= 39782 \; kJ/kg \; mol$$

(b) Watson's equation is $\dfrac{\lambda_2}{\lambda_1} = \left(\dfrac{1 - T_{r2}}{1 - T_{r1}}\right)^{0.38}$

$\lambda_1 = 38744 \; kJ/kgmol$, $T_{r2} = \left[\dfrac{320}{516.2}\right] = 0.620$, $T_{r1} = 0.681$ calculated in part (a)

Therefore,

$$\lambda_2 = \lambda_1 \left(\frac{1-0.62}{1-0.681} \right)^{0.38} = 38744 \left(\frac{1-0.62}{1-0.681} \right)^{0.38} = 41408 \text{ kJ/kg mol}$$

Heat of Fusion

When a crystalline solid at its melting point goes into a liquid state at the same temperature, heat is absorbed and the enthalpy increase occurs. The heat absorbed results in an increase of internal energy through a rearrangement of the atoms. For most elements, the ratio of $\frac{\lambda_f}{T_f}$ varies from 2 to 3, for most inorganic compounds from 5 to 7, and most organic compounds from 9 to 11.

Example 3.5a

(Energy Balance, Closed System)

One pound of a mixture of steam and water at 160 psia is contained in a rigid vessel. Heat is added to the vessel until the contents reach the condition of 550 psia and 600°F. Calculate the amount of heat added.

Solution

Energy balance with the contents of the vessel as the system gives

$$(U + PE + KE)_E - (U + PE + KE)_B = \Sigma(H + PE + KE)_t + Q + W$$

1. No mass flow in or out
2. No work done either on the system or by the system
3. Negligible potential energy and kinetic energy changes

Then, for the conditions of the example, the equation reduces to

$$\hat{U}_E - \hat{U}_B = \hat{Q} \quad \text{on unit basis.}$$

Steam tables do not contain the values of the internal energy, and therefore $\hat{U}_E - \hat{U}_B$ is to be found from the enthalpies and $\Delta(P\hat{V})$ since $\Delta\hat{H} = \Delta\hat{U} + \Delta(P\hat{V})$. The values of $\hat{H}, \hat{V}_l$, and $\hat{V}_g$ at the two conditions (from steam tables) are

Initial Condition	Final Condition
160 psia, $t = 363.5°F$	550 psia, 600°F
$\hat{V}_l = 0.01815$ ft³/lb	$\hat{H}_E = 1294.3$ Btu/lb
$\hat{V}_g = 2.834$ ft³/lb	$\hat{V}_g = 1.0431$ ft³/lb
$\hat{h}_l = 335.93$ Btu/lb	Final condition superheated; therefore no liquid present
$\hat{H}_g = 1195.1$ Btu/lb	

Let x be the quality of the steam at beginning. Now since the vessel is rigid, initial volume equals final volume

$$\hat{V}_1 = 2.834x + 0.01815(1-x) = 1.0431$$

from which
$$x = 0.364$$
$$\hat{H}_B = 1195.1(0.364) + 335.93(1 - 0.364) = 648.7 \text{ Btu/lb}$$
$$\text{Heat added} = \Delta\hat{U} = \Delta\hat{H} - \Delta(P\hat{V})$$
$$= (1294.3 - 648.7) - \frac{(550-160)(1.0431)(144)}{778}$$
$$= 570.3 \text{ Btu/lb}$$

Example 3.5b

A cylinder (see Exhibit 2) containing 1 gmol of water at its normal boiling point (373.15 K) is heated until all the liquid is converted to saturated vapor. The cylinder is fitted with a frictionless piston, which just resists a pressure of one atm.
Find

(a) from SI steam tables the heat of vaporization of water at its normal boiling point in kJ/gmol

(b) work done by the expanding vapor, and

(c) the internal energy change of water per gmol.

Solution

From SI steam tables from Perry's handbook, get the following data for water and water vapor at 373.15 K and 1 atm.

T K	Volume m³/kg Liquid	Volume m³/kg Vapor	Enthalpy, kJ/kg Liquid	Enthalpy, kJ/kg Vapor
373.15	1.044×10⁻³	1679	419.1	2676

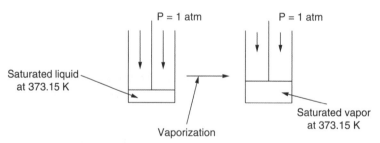

Exhibit 2 Process sketch for Example 3.5b

(a) Latent heat or heat of vaporization of water = 2676 − 419.1
$$= 2256.9 \text{ kJ/kg}$$

$$\text{Heat of vaporization} = \frac{2256.9 \text{ kJ}}{\text{kg}} \left| \frac{18.01 \text{ g}}{1000 \text{ g/kg}} \right| \frac{18.01 \text{ g}}{\text{g mol}} = 40.65 \frac{\text{kJ}}{\text{g mol}}$$

(b) Work done during expansion: This is constant pressure process.

$$W = -\int_{V_1}^{V_2} P dV = -P\int_{V_1}^{V_2} dV = -P(V_2 - V_1)$$

Initial volume of water is very small. It can be ignored in comparison with that of saturated vapor. Then

$$W = -P(V_2 - V_1) = PV_2$$

Volume per gmol $= 1.679 \dfrac{m^3}{kg} \left| \dfrac{1}{1000 \frac{g}{kg}} \right| \dfrac{18.01 \ g}{g \ mol} = 0.03024 \dfrac{m^3}{g \ mol}$

$$W = -PV_2 = -1 \times 0.03024 \times 10^5 = -3024 \ Nm = -3.024 \ kJ/g\,mol$$

The negative sign indicates the work is done by the vapor on the piston (note the new convention contrary to previous convention).

(c) Internal energy change: Basis: 1 gmol

Energy balance for this closed system reduces to

$$\Delta U = Q + W$$

From part a, $Q = 40.65$ kJ/gmol and from part b, $W = -3.024$ kJ/gmol
Therefore, $\Delta U = +40.65 - 3.024 = 37.626$ kJ/gmol
Thus the internal energy of the system has increased.

Example 3.6

(Energy Balance, Flow Process)

150 kg/h dry air is continuously compressed from 100 kPa and 270 K to 1000 kPa and 300 K. The exit velocity of the air from the compressor is 60 m/s. What is the power required for the compressor in kW? Enthalpy data for air may be obtained from Table 3.212 in Perry's handbook.

Solution

This is a flow process. No accumulation. Draw a sketch of the process first,

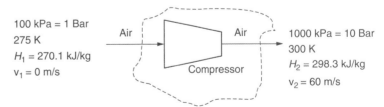

100 kPa = 1 Bar
275 K
$H_1 = 270.1$ kJ/kg
$v_1 = 0$ m/s

1000 kPa = 10 Bar
300 K
$H_2 = 298.3$ kJ/kg
$v_2 = 60$ m/s

Exhibit 3 Sketch for Example 3.6 (Basis: 150 kg of air/h)

To solve this problem we require values of three variables: enthalpies at inlet and outlet, and velocity of air at the inlet. Since at the inlet the velocity will be small, we assume it is zero. We obtain the enthalpy values from Table 3.212 from Perry's handbook, by interpolation if required. $\hat{H}_1 = 270.1$ kJ/kg, $\hat{H}_2 = 298.3$ kJ/kg.

Energy balance: steady state flow process system

$$(U + PE + KE)_E - (U + PE + KE)_B = \Sigma(H + PE + KE)_t + Q + W$$

(1) mass in = mass out therefore $M_I = M_O$

(2) Accumulation = 0 thus lefthand side = 0

(3) PE change negligible $\Delta PE = 0$

(4) $Q = 0$ since there is no heating or cooling

(5) $v_1 = 0$ assumed

Under these conditions the equation reduces to

$$W = -(\hat{H}_1 - \hat{H}_2 + \hat{KE}_E - \hat{KE}_B)m = \Delta H + \Delta KE$$

$$\Delta H = 150(298.3 - 270.1) = 4230 \text{ kJ/kg}$$

$$\Delta KE = \frac{1}{2}(m)\left(v_2^2 - v_1^2\right) = 0.5(150)(60^2 - 0)$$

$$= 0.5 \left| \frac{150 \text{ kg}}{} \right| \frac{(60\text{m})^2}{\text{s}^2} \left| \frac{1 \text{ kJ}}{\frac{1000 \text{ kg} \cdot \text{m}^2}{\text{s}^2}} \right. = 270 \text{ kJ}$$

Then $\quad W = 4230 + 270 = 4500 \text{ kJ}$

The positive sign according to new convention indicates that the work is done on the air.

$$\text{Now convert work to power, kW} = \frac{4500 \text{ kJ}}{1\text{h}} \left| \frac{1 \text{kW}}{\frac{1\text{kJ}}{\text{s}}} \right| \frac{1\text{h}}{3600 \text{ s}} = 1.25 \text{ kW}$$

ENERGY BALANCES IN PROCESSES INVOLVING CHEMICAL REACTIONS

So far we have reviewed the enthalpy changes in physical processes and how they are to be incorporated in energy balances. In processes involving chemical reactions we now have to include the heat effects that take place when the process is accompanied by chemical reaction. A chemical engineer must know how much heat is evolved or absorbed by the process due to a chemical reaction in order to design proper facilities for heating or cooling the equipment in which the reaction is carried out. This section reviews the enthalpy changes that take place because of reactions and their inclusion in the energy balance.

To begin with, we must recall some basic information about thermochemistry of chemical reactions.

Laws of Thermochemistry

In *exothermic reaction*, heat is evolved. This is because the energy required to hold the products of reaction together is less than the energy required to hold the reactants together. In *endothermic reaction*, heat is absorbed because the reverse is true.

At a given temperature and pressure, *heat of formation* of a compound from its elements is equal to the heat required to decompose the compound into its elements. The total change in enthalpy of a system depends only on the temperature, pressure, state of aggregation, and state of combination at the beginning and at the end of the reaction and is independent of the number of intermediate chemical reactions involved. This principle is also known as the *Law of Hess*. It is used to calculate the heats of formation of compounds knowing the heats of intermediate reactions. See Example 3.7.

ΔH_f^o is the enthalpy change called the *standard heat of formation* of a constituent component in a reaction. It is the enthalpy of formation of 1 mole of a compound from its constituent elements in the standard state. The standard state is defined as 25°C and 1 atm. In this state, the elements are assumed to have zero heat of formation. For a single component A, when there is no pressure effect and no phase change is involved, the heat of formation at any other temperature with reference to the standard state (normally 25°C) is given by

$$\Delta \hat{H}_A = \Delta H_{fA}^o + \int_{T_{ref}}^{T} C_{PA} dT \qquad (3.19)$$

For several components in a mixture,

$$\Delta H_m = \sum_{i=1}^{n} n_i \Delta H_{fi}^o + \sum_{i=1}^{n} \int_{T_{ref}}^{T} n_i C_{Pi} dT \qquad (3.19a)$$

where
 $i = i$th component
 n_i = number of moles of component species i in a mixture
 n = total number of species in the mixture.

If phase transition occurs, enthalpy of the phase change or the latent heat must be added to the righthand side of Equation 3.19 to yield total enthalpy of the component A.

$$\Delta \hat{H}_A = \Delta \hat{H}_{fA}^o + \left(\hat{H}_{TP} - \hat{H}_{ref}^o \right) \qquad (3.19b)$$

The standard heat of reaction is negative for exothermic reactions since heat leaves the system and is positive in endothermic reactions as heat is absorbed by the system. Standard heats of formation are available in handbooks and textbooks. They can also be calculated by group contribution method.

Heat of Reaction

Heat of chemical reaction is the change in enthalpy of the reaction system at constant pressure. This is a function of the nature of reactants and products involved and also their physical states. *Standard heat of reaction* (ΔH_R) is the change of enthalpy when the reaction takes place at 1 atm pressure in such a manner that it starts and ends with all involved materials at a constant temperature of 25°C.

Standard heat of reaction (ΔH_R) can be calculated from the heats of formation of reactants and the products using the following relation:

$$\Delta H_{reaction} = \sum \Delta H_{f(products)} - \sum \Delta H_{f(reactants)} \quad \text{at } 25°C \qquad (3.20)$$

The following relation can also be used to calculate it from the heats of combustion of the reactants and products:

$$\Delta H_R = \sum \Delta H_{c(reactants)} - \sum \Delta H_{c(products)} \quad \text{at } 25°C. \qquad (3.21)$$

Heat of reaction at any temperature can be expressed as

$$\Delta H_T = \Delta H_{T_o} + \int_{T_o}^{T} \Delta Cp\, dT \qquad (3.22)$$

where ΔH_T is the heat of reaction at T_o and ΔCp is the difference in the sum of the molal heat capacities of the products and the sum of the molal heat capacities of the reactants, each multiplied by the stoichiometric coefficient n_i or

$$\Delta C_P = \sum (n_i C_{Pi})_{\text{products}} - \sum (n_i C_{Pi})_{\text{reactants}} \qquad (3.23)$$

where C_{Pi}'s, the molal heat capacities are known as functions of temperature. When the mean molal heat capacities can be used for the reactants and products, the heat of reaction at other temperature is given by

$$\Delta H_T = \Delta H_{T_o} + \left[\sum (n_i C_{Pi})_{\text{products}} - \sum (n_i C_{Pi})_{\text{reactants}} \right](T - T_o) \qquad (3.24)$$

If $C_P^o = a + \beta T + \gamma T^2 + \square$ for each component taking part in the reaction is known, and if the reaction is of the following type:

$$n_A A + n_B B + \ldots \rightarrow n_c C + n_d D + \ldots \qquad (3.25)$$

then

$$\Delta H_T = I_H + \Delta \alpha T + \frac{\Delta \beta T^2}{2} + \frac{\Delta \gamma T^3}{3} + \cdots \qquad (3.26)$$

Heat of Combustion

Standard heat of combustion, ΔH_c, is the enthalpy change resulting from the combustion of a substance in its normal state at 25°C and atmospheric pressure, with the combustion beginning and ending at 25°C. Generally, gaseous CO_2 and liquid water are the products of combustion. It is possible to calculate the heat of formation of a compound from its heat of combustion provided the heats of formation of other substances that participate in the reaction are available. The equations of heat of combustion are to be algebraically manipulated so that the compound whose heat of formation is required falls on the right hand side of the equation.

Example 3.7

Calculate the heat of formation of ethane from the following data:

$$C_2H_6(g) + 3\tfrac{1}{2} O_2(g) = 2CO_2(g) + 3H_2O(l) \quad \Delta H_1 = -372.82 \text{ kcal/gmol} \qquad (a)$$
$$C(b) + O_2(g) = CO_2(g) \quad \Delta H_2 = -94.05 \text{ kcal/gmol} \qquad (b)$$
$$3H_2(g) + 1\tfrac{1}{2} O_2(g) = 3H_2O(l) \quad \Delta H_3 = -204.95 \text{ kcal/gmol} \qquad (c)$$

Solution

Adding 2(Equation (b)) + Equation (c) − Equation (a) and then transposing C_2H_6(g) to the right hand side of equation gives $2C(B) + 3H_2(g) = C_2H_6(g)$. Heat of reaction, $\Delta H_R = -2(94.05) - 204.95 - (-372.82) = -20.23$ kcal/gmol ΔH_f^o of ethane $= \Delta H_R - (\Delta H_f^o)_{C(B)} - (\Delta H_f^o)_{H_2} = -20.23$ kcal/gmol

Example 3.8

(Calculation of Heat of Reaction from Heats of Formation)

Calculate the standard heat of reaction for the following reaction (a) from the heats of formation and (b) from the heats of combustion:

$$C_6H_6(g) \to 3C_2H_2(g)$$

Solution

(a) The heats of formation of benzene (g) and acetylene (g) are

Benzene: $\Delta H_f^o = 48.66$ kJ/g mol at standard condition

Acetylene: $\Delta H_f^o = 226.75$ kJ/g mol at standard condition

$$\Delta H_R = \sum \Delta H_{f(products)} - \sum \Delta H_{f(reactants)}$$

$$\Delta H_R = 3 \times 226.75 - 48.66 = 631.59 \text{ kJ/g mol.}$$

(b) The heats of combustion of benzene and acetylene are as follows:

Benzene: $\Delta H_c = -3267.6$ kJ/g mol

Acetylene: $\Delta H_c = -1299.61$ kJ/g mol

$$\Delta H_R = \sum \Delta H_{c(reactants)} - \sum \Delta H_{c(products)} \quad \text{at } 25°C$$
$$= -3267.6 - 3 \times -1299.6$$
$$= 631.2 \text{ kJ/g mol.}$$

Example 3.9

The standard heat of reaction for NH_3 synthesis from nitrogen and hydrogen is -92.22 kJ/g per mol N_2. What is the heat of reaction per mole of $H_2(g)$ at a temperature of 430 K if the average molal heat capacities of the three components are as given below?

$$C_{PH_2} = 29.01 \text{ J/g mol·K} \quad C_{PN_2} = 29.23 \text{ J/g mol·K} \quad C_{PNH_3} = 37.32 \text{ J/g mol·K}$$

Solution

The reaction equation is $N_2(g) + 3H_2(g) \to 2NH_3(g)$.

By Equation 3.20, $\Delta H_T = \Delta H_{T_0} + [\sum(n_i C_{Pi})_{products} - \sum(n_i C_{Pi})_{reactants}](T - T_o)$

$$\Delta C_P = 2 \times 37.32 - 3 \times 29.01 - 29.23 = -41.62 \text{ J/g mol · K}$$

$$(\Delta H_R)_{430K} = -92220 + [-41.62(430 - 298)\}$$
$$= -97714 \text{ J/g mol} = -97.7 \text{ kJ/g mol}$$
$$= -97.7/3 = -32.57 \text{ kJ/g mol of } H_2(g)$$

Example 3.10

The reaction between CO_2 and H_2 can be represented as follows:

$$CO_2 + H_2(g) = CO(g) + H_2(g) \qquad \Delta H_R^o = 41166 \text{ J/gmol}$$

ΔC_P for products and reactants is calculated as given by the following expression:

$$\Delta C_P = -2.54 - 3.141 \times 10^{-2} T + 3.6374 \times 10^{-5} T^2$$

Calculate the heat of reaction at 398K.

Solution

$$\Delta H_T = \Delta H_{T_o} + \int_{T_o}^{T} \Delta Cp \, dT$$

$$\Delta H_{398} = 41166 + \int_{298}^{398} (-2.54 - 3.141 \times 10^{-2} T + 3.6374 \times 10^{-5} T^2) \, dT$$

$$= 41166 + \left[-2.54T - 3.141 \times 10^{-2} \frac{T^2}{2} + 3.6374 \times 10^{-5} \frac{T^3}{3} \right]_{298}^{398}$$

$$= 41166 + (-254 - 1093 + 443.6) = 40763 \text{ J/gmol}$$
$$= 40.76 \text{ kJ/gmol}$$

Energy Balance with Accounting for Heat of Reaction

In applying energy balance for a process with chemical reaction, the first thing to do is a material balance. Standard heat of reaction and heat of formation are known at standard condition of 25°C and 1 atm. In most cases this will be chosen as the reference state. Enthalpies of all components entering and leaving the system should be calculated relative to the reference state. Account for the heat of formation, sensible heats, and latent heats due to phase transition.

For a flow reaction with negligible changes in potential and kinetic energies and with no work, the energy balance reduces to

$$0 = \sum M_I \hat{H}_I - \sum M_o H_o \quad \text{or} \quad Q = \Delta H \qquad (3.27)$$

For a non-flow reaction taking place at constant pressure,

$$Q = \Delta H \qquad (3.27a)$$

For a non-flow reaction at constant volume, the energy balance reduces to

$$Q = \Delta U \qquad (3.27b)$$

Example 3.11

Sulfur dioxide gas is converted to sulfur trioxide in 100 percent excess air with 80 percent conversion to SO_3. The gases enter the converter at 400°C and leave it at 450°C. How much heat has to be removed from the converter per gmol of SO_2 converted by the heat exchangers?

Data: specific heats of components

SO$_2$: $C_P = 38.91 + 3.904 \times 10^{-2} T - 3.104 \times 10^{-5} T^2$
$+ 8.606 \times 10^{-9} T^3$ J/g mol, T in °C

SO$_3$: $C_P = 49.5 + 9.188 \times 10^{-2} T - 8.540 \times 10^{-5} T^2$
$+ 32.4 \times 10^{-9} T^3$ J/g mol, T in °C

O$_2$: $C_P = 29.1 + 1.158 \times 10^{-2} T - 0.6076 \times 10^{-5} T^2$
$+ 1.311 \times 10^{-9} T^3$ J/g mol, T in °C

N$_2$: $C_P = 29.0 + 0.2199 \times 10^{-2} T + 0.5723 \times 10^{-5} T^2$
$- 2.871 \times 10^{-9} T^3$ J/g mol, T in °C

Heats of formation:

SO$_2$: -296.9 kJ/gmol at 25°C and 1 atm
SO$_3$: -395.18 kJ/gmol at 25°C and 1 atm

Solution 1

This is a flow process. No accumulation. We first prepare a sketch and complete material balance. Since heat removal per gmol is desired, we select 1 gmol as the basis for the calculation, as C_P data are on gmol basis.

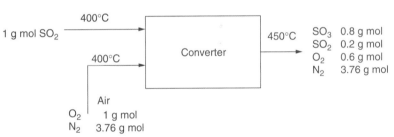

Exhibit 4 Process sketch for Example 3.11

Reaction is   SO$_2$(g) + ½O$_2$(g) → SO$_3$(g)

For the conversion of 1 mol of SO$_2$, 0.5 mol of O$_2$ is required. Excess oxygen is 100 percent. Oxygen to the converter for 100% excess = 1 g mol.
Then nitrogen to the converter = (0.79/0.21) = 3.76 mol.
The material balance is given in Exhibit 4. Since $\Delta U = 0$ (flow process, $W = 0$ as no work is done, and ΔPE and ΔKE negligible), the energy balance with converter as the system reduces to

$$0 = M_I \hat{H}_I - M_o \hat{H}_o + Q$$

Now we calculate the enthalpies of the components at 400°C and 450°C. This can be done by using (1) C_P data for the components in the form of polynomials, (2) tabulated enthalpy data, or (3) average or mean specific data. In the following calculation we use C_P data in the form of polynomials and then solve the same problem with the use of enthalpy data for the components. Solving the problem using the mean specific heats will be left to the reader.

Use of polynomial expressions of C_P data.
Choose reference temperature of 25°C and 1 atm.
Reactants:

Component	gmol	a	$b \times 10^2$	$c \times 10^5$	$d \times 10^9$	ΔH_f^o J/gmol
SO_2	1.0	38.91	3.904	−3.104	8.606	−296900
SO_3	0					
O_2	1.0	29.1	1.158	−0.6076	1.311	0
N_2	3.76	29.0	0.22	0.5723	−2.871	0
		177.05	5.900	−1.56	11.67	−296900

Integrations can be carried out with individual expressions for each component but we can add the respective coefficients for all components and integrate only one expression.

If a computer program is available, this method is more convenient.

Enthalpy at 400°C (Inlet components) using Equation 3.19,

$(\Delta H)_{400}$

$$= -296900 + \int_{25}^{400} (177.05 + 5.9 \times 10^{-2} T - 1.56 \times 10^{-5} T^2 - 11.67 \times 10^{-9} T^3) dT$$

$$= -296900 + \left(177.05 T + 5.9 \times 10^{-2} \frac{T^2}{2} - 1.56 \times 10^{-5} \frac{T^3}{3} - 11.67 \times 10^{-9} \frac{T^4}{4} \right)_{25}^{400}$$

$$= -296900 + (66394 + 4702 - 333 - 75)$$
$$= -226212 \text{ J}$$

Products:

Component	gmol	a	$b \times 10^2$	$c \times 10^5$	$d \times 10^9$	ΔH_f^o J/gmol
SO_2	0.2	38.91	3.904	−3.104	8.606	−59380
SO_3	0.8	48.5	9.188	−8.54	32.4	−316144
O_2	0.6	29.1	1.158	−0.6076	1.311	0
N_2	3.76	29.0	0.22	0.5723	−2.871	0
		173.1	9.653	−5.67	17.63	−375524

Enthalpy at 450°C (outlet components)

$(\Delta H)_{450}$

$$= -375524 + \int_{25}^{450} (173.51 + 9.655 \times 10^{-2} T - 5.67 \times 10^{-5} T^2 + 17.63 \times 10^{-9} T^3) dT$$

$$= -375524 + \left(173.1 T + 9.653 \times 10^{-2} \frac{T^2}{2} - 5.67 \times 10^{-5} \frac{T^3}{3} + 17.63 \times 10^{-9} \frac{T^4}{4} \right)_{25}^{450}$$

$$= -375524 + (73568 + 9744 - 1722 + 181)$$
$$= -291753 \text{ J}$$

$Q = -291753 - (-226212) = -65541$ J/gmol per 0.8 gmol SO_2 converted.
Therefore $Q = -65541/0.8 = -81926$ J/gmol of SO_2 converted.

The problem can be solved more easily with a spreadsheet or a computer program.

Solution 2

If we use enthalpy data from tables (see Himmelblau), arithmetic is rather simple.

In the following table the values of ΔH were obtained by interpolation from Tables D.4, D.5 or D.6 (Appendix D, Himmelblau).

Enthalpy at inlet of converter

Component	gmol	ΔH_f^o J/gmol	ΔH @400°C	$n_i \hat{H}_i$	$n_i \Delta H_f^o$ J
SO_2	1.0	−296900	18155	18155	−296900
SO_3	—				
O_2	1.0	—	12346	12346	
N_2	3.76	—	11835	44500	
				75001	

Net enthalpy in = −296900 + 75001 = −221900 J

Enthalpy at outlet of converter

Component	gmol	ΔH_f^o J/gmol	ΔH @450°C	$n_i \hat{H}_i$	$n_i \Delta H_f^o$ J
SO_2	0.2	−296900	20685	4137	−59380
SO_3	0.8	−316144	28901	23121	−316144
O_2	0.6	—	17347	10408	
N_2	3.76	—	13366	50256	
			—	87922	−375524

Net enthalpy out = −375524 + 87922 = −287622 J
Q = −287622 − (−221900) = −65702 J
Q = −65702 J/gmol SO_2 converted
PER 0.8 gmol

Adiabatic Flame Temperature

When a reaction is carried out without exchanging heat with the surroundings, it is called an adiabatically conducted reaction. In such a reaction, if all the products remain together they will attain a definite temperature, which is called adiabatic reaction temperature or flame temperature. In calculating flame temperature in combustion reactions, complete combustion is assumed. All materials such as inerts or excess reactants present in the products are to be taken into account when calculating adiabatic flame temperature. For steady state energy balance with $Q = 0$, $Ws = 0$ the energy balance equation reduces to

$$\Delta H = 0$$

Temperature T is determined by a trial-and-error procedure so that $\Delta H = 0$.

Example 3.12

(Calculation of Flame Temperature)

Methane is burned adiabatically in 30 percent excess air. Both gas and air are at 25°C. What is the adiabatic flame temperature?

Solution

The equation for the reaction is $CH_4(g) + 2O_2(g) = CO_2(g) + 2H_2O(g)$.

Complete combustion requires 2 mol of oxygen per mol of CH_4. Hence 30 percent excess air means 0.6 mol of oxygen are unreacted and total mol of oxygen used = 2.6 mol.

$$\text{Nitrogen in air} = (0.79/0.21)(2.6) = 9.78 \text{ mol}$$

Inlet gas: $CH_4 = 1$ mol, $O_2 = 2.6$ mol, $N_2 = 9.78$ mol
Product gas: $CO_2 = 1$ mol, $H_2O = 2$ mol, $O_2 = 0.6$ mol and $N_2 = 9.78$ mol

CH_4: $C_p/R = 1.702 + 9.081 \times 10^{-3}T - 2.164 \times 10^{-6}T^2$

O_2: $C_p/R = 3.639 + 0.506 \times 10^{-3}T - 0.227 \times 10^{+5}T^{-2}$

N_2: $C_p/R = 3.28 + 0.593 \times 10^{-3}T + 0.04 \times 10^{+5}T^{-2}$

H_2O: $C_p/R = 3.47 + 1.45 \times 10^{-3}T + 0.121 \times 10^{+5}T^{-2}$

CO_2: $C_p/R = 5.457 + 1.045 \times 10^{-3}T - 1.157 \times 10^{+5}T^{-2}$

(Note: In preceding correlations, R = ideal gas law constant = 8.314 J/mol·K

We simplify the calculation by summing up the coefficients of various corresponding terms times the number of moles of each species as done in Example 3.11.

$$\text{Basis: 1 gmol of } CH_4$$

Material balance is shown in Exhibit 5.

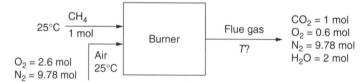

Exhibit 5 Sketch for Example 3.12

$$\Delta H_f^o \text{ of } CH_4 = -74520 \text{ J/g mol}$$

Enthalpy of reactants

$$\Delta H_{CH_4} = (1.00)C_p(298 - 298) - 74520 = -74520$$

$$\Delta H_{O_2} = (2.6)C_p(298 - 298) + 0 = 0$$

$$\Delta H_{N_2} = (9.78)C_p(298 - 298) + 0 = 0 \text{ (from Equation 3.19)}$$

Enthalpy of products

Component	g mol	a	$b \times 10^3$	$c \times 10^6$	$d \times 10^{-5}$	ΔH_f^o J/g mol
CO_2	1.00	5.457	1.045	—	-1.157	-393509
H_2O	2.00	6.94	2.9	—	-0.242	-2×241818
O_2	0.6	2.183	0.3036	—	-0.1362	0
N_2	9.78	32.08	5.8	—	+0.3912	0
		$\Sigma a = 46.66$	$\Sigma b = 10.05$	—	$\Sigma d = -0.66$	-877145

Energy balance gives $\Delta H = 0$.

$$0 = [-877145 + (8.314)\int_{298}^{T}(46.66 + 10.05\times 10^{-3}T - 0.66\times 10^{5}T^{-2})dT - (-74520)$$

$$= -802625 + 8.314\left(46.66(T-298) + 10.05\times 10^{-3}\frac{T^2 - 298^2}{2}\right.$$

$$\left.+ 0.66\times 10^{5}\left[\frac{1}{T} - \frac{1}{298}\right]\right)$$

T is to be calculated by trial and error so that $\Delta H = 0$ is satisfied. When this is done conveniently with the help of a computer spread sheet program, the following results are obtained:

Temperature Assumed, K	Calculated ΔH	Deviation from Zero
2000	822093	19468
1950	794453	−8172
1965	802723	97.9
1964.84	794453	−0.9 (very close to zero)

So the adiabatic flame temperature is 1964.8 K or 1691.7°C. You can also plot a graph of deviation versus temperature and quickly locate the adiabatic flame temperature on the graph rather than making many trials.

Heats of Mixing and Enthalpy of a Solution

Enthalpy change that takes place on dissolving a component in another substance is called the *heat of mixing*. The enthalpy change of mixing depends upon the nature of the solute and solvent, temperature, and initial and final concentrations of the solution.

Standard integral heat of solution is the change in the enthalpy of the system when 1 mol of a solute is mixed with n_1 moles of a solvent at constant temperature of 25°C and constant pressure of 1 atm. The enthalpy of the solution H_s is given by

$$H_s = n_1\hat{H}_1 + n_2\hat{H}_2 + n_2\Delta\hat{H}_{s2} \tag{3.28}$$

where

H_s = enthalpy of $n_1 + n_2$ moles of solution of components 1 and 2 at temperature T relative to temperature T_0 (reference temperature)

$\hat{H}_1, \hat{H}_2$ = molal enthalpies of components 1 and 2 at temperature T relative to the temperature T_0

$\Delta\hat{H}_{s2}$ = integral heat of solution of component 2 at temperature T

Integral heats of solution are available elsewhere. Heats of solution involved in the mixing of liquids are called heats of mixing.

Example 3.13

How much heating or cooling must be done if 50 gpm of 20 wt% $CaCl_2$ brine is to be produced at 90°F by feeding to the dissolver solid anhydrous $CaCl_2$ at 60°F and water at 86°F? The dissolver is equipped with a 25-hp agitator, which has an efficiency of 80 percent. See Exhibit 6.

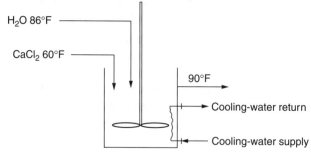

Exhibit 6 Sketch for Example 3.13

Solution

Energy balance on the contents of the tank results in

$$0 = n_1 \hat{H}_1 + n_2 \hat{H}_2 - (n_1 + n_2)\Delta H_s + Q_s - Q_t$$

where Q_s is the heat generated because of the agitator and Q_t is the heat caused by heating or cooling. It is required first to determine $\hat{H}_1$, $\hat{H}_2$, and $\hat{H}_s$. For calculation of enthalpies, assume 25°C as the reference temperature. Obtain specific heats of solution of $CaCl_2$ and enthalpy of water. For each 100 lb of 20 percent $CaCl_2$ solution,

$$\text{Moles of } CaCl_2 = \frac{20}{111} = 0.18 \qquad \text{Moles of water} = \frac{80}{18} = 4.44$$

$$\text{Molecular weight of solution} = \frac{100}{4.44 + 0.18} = 21.645$$

$$\frac{\text{Moles of water}}{\text{Moles of } CaCl_2} = \frac{4.44}{0.18} = 24.7 \text{ mol}$$

From *Chemical Process Principles*, Part I, (Hougen and Watson),

$$\hat{H}_s = -17{,}500 \text{ cal/g mol at } 25°C = -31{,}500 \text{ Btu/lb mol}$$

From steam tables, enthalpy of water at 77°F(25°C) is

$$45.03(18) = 810.5 \text{ Btu/lb mol}$$

(Datum temperature for enthalpies of water is 32°F.) The specific heat of $CaCl_2$ solid is

$$0.164 \text{ Btu/lb} \cdot °F = 18.20 \text{ Btu/lb mol} \cdot °F$$

The enthalpy of $CaCl_2$ at 25°C(77°F) is

$$C_p \Delta t = 18.2(77 - 32) = 819 \text{ Btu/lb mol}$$

The enthalpy of $CaCl_2$ solution at 60°F is

$$819 - 17(18.2) = 509.6 \text{ Btu/lb mol}$$

(Datum temperature for $CaCl_2$ is also taken as 32°F.) Enthalpy of $CaCl_2$ solution at 25°C or 77°F equals

$$(n_1 + n_2)\hat{H}_s = n_1\hat{H}_1 + n_2\hat{H}_2 + n_2\Delta\hat{H}_s$$
$$= 4.44(810.5) + 0.18(819) + 0.18(-31,500) = -1924 \text{ Btu}$$

or

$$\hat{H}_s = \frac{-1924}{0.18 + 4.44} = -416.4 \text{ Btu/lb mol solution.}$$

Specific heat of $CaCl_2$ solution equals

$$0.77 \text{ Btu/lb} \cdot °F = 0.77(21.645)$$
$$= 16.7 \text{ Btu/lb mol of solution} \cdot °F$$

Therefore $(\hat{H}_s)_{90} = -416.4 + \int_{77}^{90} C_p dT$

$$= -416.4 + 16.7(90 - 77) = -199.3 \text{ Btu/lb mol}$$

50 gal of 20 percent solution of specific gravity 1.1721 at 90°F is to be prepared per minute.

$$\text{Solution to be prepared} = 50 \text{ gpm} \times (8.33)(1.1721 \text{ lb/min})$$
$$= 488.18 \text{ lb/min}$$

$$= \frac{488.18}{21.645} = 22.6 \text{ lb mol/min}$$

$$\text{Moles of water} = \frac{4.44}{4.62}(22.6) = 21.71 \text{ lb mol}$$

Moles of $CaCl_2$ = 0.88 lb mol

The enthalpy of water at 86°F is

$$54.03(18) = 972.5 \text{ Btu/lb mol}$$

By heat balance

$$0 = 21.71(972.5) + 0.88(509.6) - 22.6(-199.3) + 42.4(25)(0.8) + Q_t$$
$$Q_t = -26,914 \text{ Btu/min}$$

The minus sign indicates that the heat has to be removed and cooling is required.

REVERSIBLE AND IRREVERSIBLE PROCESSES

A process is said to be *reversible* if it proceeds under conditions of balanced forces. Its direction can be reversed at any point by an infinitesimal change in external conditions. In a reversible process there is no friction. It is not far removed more than differentially from equilibrium. When reversed it returns to its forward path and restores the initial state of the system and its surroundings, so that there is no degradation of energy and the availability of the energy of the combined system and its surroundings is constant. A reversible process is a concept only. Actual processes are always *irreversible* to some extent and they involve degradation in available energy, for example, through frictional losses.

A reversible process, although only a concept, is very useful to help understand the maximum potential of a process and how the performance of the actual process compares with it. This concept is the basis of the second law of thermodynamics.

Entropy and Second Law of Thermodynamics

Mass and energy balances alone are not always sufficient to solve all the problems involving energy flows. In energy balance equations, both work and heat are considered to be forms of energy. However, they are energies of different quality. Whereas work can be transformed into another form of work or heat quantitatively, it is known from practical experience that heat cannot be converted to work quantitatively and that heat cannot be transferred from a lower temperature to a higher temperature without the aid of an external agency. Heat is therefore energy of a lower or degraded quality when compared to work, and it depends on temperature. The higher the temperature at which it is available in reference to a given lower temperature (usually that of surroundings), the higher is its quality. An additional state variable to account for the unidirectional nature of the heat flow and the less efficient conversion of heat into work is provided by the postulation of the second law of thermodynamics.

Second Law of Thermodynamics

No practical engine (not even a reversible one) can convert heat into work quantitatively, or it is impossible for a self-acting machine unaided by an external agency to transfer heat from a lower temperature to a higher one.

The second law of thermodynamics limits the efficiency of a process involving the conversion of heat into work. The Carnot principle states that the efficiency of a reversible engine depends only on the temperature levels at which the engine absorbs and rejects heat and is independent of the medium used. Thus, if a reversible engine abstracts heat Q_H at a higher temperature T_H and rejects heat Q_C at the lower temperature T_C while doing a net amount of work W_{rev} in the process, the efficiency of the engine is given by

$$\text{Efficiency } \eta_{rev} = \frac{W_{rev}}{Q_H} = \frac{Q_H - Q_C}{Q_H} = \frac{T_H - T_C}{T_H} \qquad (3.29)$$

from which also follows the relation

$$\frac{Q_C}{Q_H} = \frac{T_C}{T_H} \qquad (3.29a)$$

If $Q_C = 0$, $T_C = 0$ since $Q_H \neq 0$ and $T_H \neq 0$, which means the zero on the absolute scale of the temperature is the temperature of a heat reservoir to which a reversible heat engine rejects no heat because all the absorbed heat is converted to work. (This is true only of the Carnot reversible idealized heat engine.) The Carnot principle thus enables one to define zero on the temperature scale. It can also be deduced from the preceding that all reversible engines operating between the same temperature limits have the same efficiencies.

For an engine undergoing reversible Carnot cycle, it can be shown that

$$\frac{Q_H}{T_H} - \frac{Q_C}{T_C} = 0 \qquad (3.30)$$

and for a differential reversible cycle, regardless of path,

$$\oint \frac{dQ_R}{T} = 0 \qquad (3.31)$$

This integral $\int dQ_R/T$ is the same for any reversible path and is only dependent on the initial and final states. This integral is termed the change in entropy, which is a state property and was first introduced by Clausius. Thus

$$dS = \frac{dQ_R}{T} \quad (3.32)$$

where S is entropy $= \int dQ_R/T$.

In an engine that uses a finite temperature difference for the transfer of heat, the work done by the engine is given by

$$Q_1 - Q_2 + W = 0 \quad (3.33)$$

and

$$\frac{Q_H}{T_H} - \frac{Q_C}{T_C} + S_p = 0 \quad (3.34)$$

where S_p is the entropy production (increase in entropy). The work done by the engine is then given by

$$W = Q_H \left(\frac{T_H - T_C}{T_H} \right) - T_C S_p \quad (3.35)$$

and, therefore, to obtain maximum work from an engine operating between two temperature limits T_H and T_C, all the processes must be carried out reversibly so that $S_p = 0$. In other words, the reversible process produces the maximum work and no actual process can be as efficient as the reversible one. By solving Equation 3.34 for Q_C, one gets

$$Q_C = Q_H \frac{T_C}{T_H} + T_C S_p \quad (3.36)$$

In the preceding equation, Q_C can be zero only if $T_C = 0$, and this leads to the principle of the second law of thermodynamics that it is impossible to convert heat quantitatively into work.

$$S_p \begin{cases} = 0 & \text{for a reversible process} \\ > 0 & \text{for an irreversible process} \end{cases} \quad (3.37)$$

These two relations are combined as one other statement of the second law of thermodynamics, viz.,

$$S_p = \Delta S \geq 0 \quad (3.38)$$

Lost work is given by

$$W_{\text{lost}} = T_s \Delta S \quad (3.39)$$

Since the entropy is an extensive state property and is characteristic of the state of the system, the principle of the balance or accounting can be applied to it over a selected accounting period. The entropy balance is given by

$$S_E - S_B = \sum S_t + \sum \frac{Q}{T_B} + S_p - S_c \quad (3.40)$$

$$S_p > 0$$

where

S_E = Entropy of system at end of accounting period
S_B = Entropy of system at beginning of accounting period
$\sum S_t$ = Entropy associated with mass transfer across boundaries of system during accounting period
$\sum Q/T_B$ = Entropy transfer not associated with mass transfer
S_p = Entropy increase in a nonreversible process
S_p = 0 for a reversible process
S_c = Entropy production associated with atomic transmutations.

By application of the energy and entropy balances to a constant mass and composition system undergoing a reversible process with the pressure as the only external force, the following relation results:

$$U = T\,dS - P\,dV \tag{3.41}$$

Using this relation and the definition of enthalpy H, the following relation is obtained:

$$dH = d(U + PV) = T\,dS + V\,dP \tag{3.42}$$

Two more thermodynamic properties are defined for convenience. The work function or Helmholtz function is

$$A = U - TS \tag{3.43}$$

Gibb's free energy is

$$G = H - TS \tag{3.44}$$

Differentiation of Equations 3.43 and 3.44, and the use of Equations 3.41 and 3.42 for substitution yields the relations

$$dA = -P\,dV - S\,dT \tag{3.45}$$
$$dG = V\,dP - S\,dT \tag{3.46}$$

Example 3.14

A reversible heat engine absorbs 1200 kJ at 540 K, does work, and then rejects heat to a heat sink maintained at 310 K. What is the entropy change for the heat source, the heat sink, and the total entropy change resulting from the process? What is the amount of work done by the engine?

Solution

Work done by the engine: The energy balance reduces to $Q + W = 0$ or $W = -Q$.

$$W = -Q_1 \frac{T_H - T_C}{T_H} = 1200 \frac{540 - 310}{540} = 511 \text{ kJ}$$

With respect to the heat reservoir, Q is negative. $\therefore \Delta S = -\dfrac{1200}{540} = -2.222 \dfrac{\text{kJ}}{\text{K}}$

Now $\dfrac{Q_H}{T_H} = \dfrac{1200}{540} = -\dfrac{Q_c}{T_c} = \dfrac{-Q_C}{310}$ Then $Q_2 = -688.9$ kJ

Since heat sink is receiving the heat, it is positive with heat sink as system.

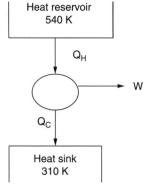

Exhibit 7 Sketch for Example 3.14

Hence

$$\Delta S(\text{heat sink}) = (+Q_C/T_C) = (688.9/310) = +2.222 \text{ kJ/K}$$

ΔS for the whole process is $-2.222 + 2.222 = 0$ since the process is reversible.

Example 3.15

Calculate the entropy change when 1 lb of SO_2 gas is heated from 70 to 2000°F at constant pressure. The molar heat capacity of SO_2 is given by

$$C_p = 6.157 + (1.38 \times 10^{-2})T + 0.9103 \times 10^{-5}T^2 + 2.057 \times 10^{-9}T^3$$

where T is in K.

Solution

Calculate first the entropy change for 1 gmol using temperatures in Kelvins.

$$T_1 = \frac{460 + 70}{1.8} = 294.44 \text{ K}$$

$$T_2 = \frac{460 + 2000}{1.8} = 1366.7 \text{ K}$$

$$\Delta \bar{S} = \int_{T_1}^{T_2} \frac{C_p \, dT}{T}$$

$$= \int_{294.44}^{1366.7} \left(\frac{6.157}{T} + 1.384 \times 10^{-2} - 0.9103 \times 10^{-5} T + 2.057 \times 10^{-9} T^2 \right) dT$$

$$= 23.32 \text{ g cal/gmol} \cdot \text{K}$$

$$= 23.32 \text{ Btu/lb mol} \cdot °R \quad \text{or} \quad \frac{23.32}{64} = 0.3644 \text{ Btu/lb} \cdot °R$$

Example 3.16

A steam turbine is operating under the following conditions: steam to turbine at 900°F and 120 psia, velocity = 250 ft/s; steam exiting turbine at 700°F and 1 atm, velocity = 100 ft/s. Calculate the rate at which work can be obtained from this turbine if the steam flow is 25,000 lb/h and the turbine operation is steady state adiabatic.

Solution

The system is space bounded by turbine (i.e., turbine and its content). See Exhibit 8. Mass balance is

$$\overset{3}{M_E} - \overset{3}{M_B} = M_I - M_O + \overset{2}{\Sigma M_P} - \overset{1}{\Sigma M_C}$$

1. No atomic transmutations
2. No chemical transformation
3. Steady state

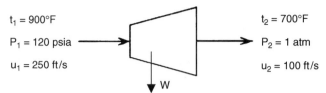

Exhibit 8 Sketch for Example 3.16

Therefore, mass balance reduces to

$$M_I - M_O = 0 \quad \text{or} \quad M_I = M_O$$

Energy balance also reduces to

$$0 = (H + KE)_I - (H + KE)_O + W$$
$$W = (H + KE)_O - (H + KE)_I$$

or

$$W = (H + KE)_O - (H + KE)_I$$

From the steam tables, $\hat{H}_I = 1478.8$ Btu/lb, $\hat{H}_O = 1383.2$ Btu/lb. Since the steam is still superheated (i.e., it is above its saturation temperature), there is no condensation. Now since $M_I = M_O = 25{,}000$ lb, the work W can be obtained as

$$W = (H_O - H_I) + (KE_O - KE_I)$$

$$= M(\hat{H}_I - \hat{H}_O) + \frac{Mu_2^2}{2g_c} - \frac{Mu_1^2}{2g_c}$$

$$= 25{,}000 \left[(1383.2 - 1478.8)\frac{\text{Btu}}{\text{lb}} + \frac{100^2 - 250^2}{2(32.2)} \frac{\text{ft}^2/\text{s}^2 (1 \text{ Btu})}{(\text{ft} \cdot \text{lb}/\text{lb}_f \cdot \text{s}^2) 778 \text{ ft} \cdot \text{lb}_f} \right]$$

$$= 25{,}000(-95.6 - 1.05) \text{ Btu/h} = -25{,}000(96.65) = -2412500 \text{ Btu/h}$$

The minus sign indicates that work is transferred out from the system.

$$|W| = \frac{25{,}000(96.65) \text{ Btu/h}}{2545 \text{ (Btu/h)/hp}} = 949.4 \text{ hp}$$

Example 3.17

Superheated steam at 200 psia and 50°F superheat expands adiabatically and reversibly to 14.7 psia. Calculate the final enthalpy, specific volume, and entropy.

Solution

The initial conditions are 200 psia and 50°F superheat, i.e., 432°F. From the steam tables at 200 psia and 432°F

$$\hat{S}_1 = 1.58 \text{ Btu/lb} \cdot {}^\circ\text{R} \qquad \hat{V}_1 = 2.45 \text{ ft}^3/\text{lb} \qquad \hat{H}_1 = 1226 \text{ Btu/lb}$$

Final condition

$$\text{Pressure} = 14.7 \text{ psia}$$

$$\text{Initial entropy } \hat{S}_1 = \hat{S}_2 \quad \text{(final entropy)} = 1.58 \text{ Btu/lb} \cdot {}^\circ\text{R}$$

At 14.7 psia, this value lies between the entropies of the liquid and the vapor, and therefore the final condition of the steam is a mixture of the liquid and vapor, since $\hat{S}_l = 0.3121$ Btu/lb · °R and $\hat{S}_g = 1.7568$ Btu/lb · °R.

Let x be quality of the final steam. Then the entropy of the steam in the final condition

$$\hat{S}_f = 1.7568x + 0.3121(1-x) = 1.58 \text{ Btu/lb} \cdot °R$$

from which $\quad x = 0.878$

$$\hat{V}_2 = x(\hat{V}_g) + (1-x)\hat{V}_l$$

$V_2 = 0.878(26.78) + (1-0.878)(0.01672) = 23.515 \text{ ft}^3/\text{lb mixture}$

$\hat{H}_2 = 0.878(1150.4) + (1-0.878)(180.1) = 1032 \text{ Btu/lb mixture}$.

Note: In throttling expansion, the initial and final enthalpies are the same. This fact can be used to calculate the quality of a mixture after an isenthalpic expansion similar to one through a throttling valve.

Example 3.18

Ethane is compressed adiabatically from 200 psia and 80°F to 1050 psia. Calculate the reversible work for this process given that flow is 150 lb/min.

Solution

Choose as a system the space within the boundaries of the compressor and its inlet and outlet at the specified conditions. For the steady-state compression, the energy balance reduces to

$$W = H_O - H_I = M(\hat{H}_O - \hat{H}_I) = 150(\hat{H}_O - \hat{H}_I)$$

Similarly, the entropy balance reduces to

$$S_I = S_O \quad \text{or} \quad M\hat{S}_I = M\hat{S}_O$$

i.e., $S_I = S_O$. $\hat{S}_I = 1.6395$ Btu/lb·°R. Use tables of ethane properties to obtain the required values: at inlet, $t = 80°F$, $P = 200$ psia, $\hat{H}_I = 446.5$ Btu/lb; at outlet $P = 1050$ psia, $\hat{S}_O = 1.6395$ Btu/lb·°R. Since there are no data for $P = 1050$ psia, a cross-interpolation is required to determine the outlet enthalpy and temperature. Reading entropy columns at 1000 psia, it is seen that temperature lies between 240°F and 260°F. Obtain by interpolation $\hat{H}$ and $\hat{S}$ at 1050 psia.

$$\hat{S}_{1050} = (-1.6281 + 1.5735)\frac{50}{500} + 1.6281 = 1.62264 \text{ at } 240°F$$

$$\hat{S}_{1050} = (-1.6472 + 1.5964)\frac{50}{500} + 1.6472 = 1.64212 \text{ at } 260°F$$

$$\hat{H}_{1050} = (-493.5 + 470.3)\frac{50}{500} + 493.5 = 491.2 \text{ at } 240°F$$

$$\hat{H}_{1050} = (-507.2 + 486.4)\frac{50}{500} + 507.2 = 505.12 \text{ at } 260°F$$

Now obtain values of $\hat{H}$ and t at $\hat{S} = 1.6395 = \underline{S}_O$.

t, °F	$\hat{H}$, Btu/lb	$\hat{S}$, Btu/lb·°R
240	491.2	1.62284
260	505.12	1.64212

$$\hat{H}_O = \frac{1.6395 - 1.6228}{1.6421 - 1.6228}(505.12 - 491.2) + 491.2 = 503.23 \text{ Btu/lb}$$

$$t_o = \frac{1.6395 - 1.6228}{1.6421 - 1.6228}(260 - 240) + 240 = 257.3°F$$

$$-W = M(\hat{H}_O - \hat{H}_I) = 150(503.23 - 446.5) = 8509.5 \text{ Btu/min}$$

$$W = -8509.5 \text{ Btu/min reversible work}$$

Note: The sign is negative because work is transferred out from the system.

THERMODYNAMIC PROPERTIES OF MATTER

The mass, energy, and entropy balances relate the mass, heat, and work to the changes in the thermodynamic state of a system. To make effective thermodynamic analysis of the practical systems and processes, values of the thermodynamic state properties such as internal energy, enthalpy, and entropy are required. However, these are not easily measurable properties. For some fluids, thermodynamic property data in the form of charts or tables are available. In many situations, only the *PVT* and heat capacity data are available. The thermodynamic analysis thus involves computation of the required state properties using the available *PVT* and heat capacity data, equations of state, and graphical as well as tabular data.

Useful property relations for enthalpy and entropy of a homogeneous phase are obtained when the properties are expressed as functions of temperature and pressure. Equations 3.41, 3.42, 3.45, and 3.46 are the basis from which relations are developed expressing changes in enthalpy and entropy as functions of temperature and pressure. For example, the following two relations are obtained:

$$dH = C_P dT + \left[V - T\left(\frac{\delta V}{\delta T}\right)_P\right] dP \qquad (3.46\text{a})$$

$$dS = C_P \frac{dT}{T} - \left(\frac{\delta V}{\delta T}\right)_P dP \qquad (3.46\text{b})$$

Gibbs free energy ($G = G(P, T)$) function is especially useful for generating other properties. The following relations are obtained using this function (for derivation, see Smith et al.):

$$\frac{S}{R} = \frac{H}{RT} - \frac{G}{RT} \quad \text{and} \quad \frac{U}{RT} = \frac{H}{RT} - \frac{PV}{RT} \qquad (3.47)$$

Thus if G is available as a function of T and P, S, U, H can be easily calculated. However, G is of little practical use because there is no method available to experimentally measure G directly.

Residual Properties

To get some practical functions for evaluation of properties, the concept of *residual property* is introduced. It is convenient to estimate the difference between the value of a property for an actual fluid and that for an ideal gas at the same temperature and pressure.

Residual Gibbs free energy is defined as $G^R = G - G^{ig}$ where G^R is residual free energy and G and G^{ig} are actual and ideal-gas values of G at the same temperature and pressure. Other residual properties such as residual volume, residual enthalpy, or entropy are defined in a similar manner. For example, residual volume is given by

$$V^R = V - V^{ig} = V - \frac{RT}{P} \quad (3.48)$$

In general, if M is the actual property value and M^{ig} is the corresponding ideal gas value at the same temperature and pressure, the residual molar value of any thermodynamic property such as $V, H, U, S,$ or G is given by

$$M^R = M - M^{ig} \quad (3.49)$$

With the definition of residual property, a number of useful functions are derived (see Smith et al.) as follows:

$$\frac{G^R}{RT} = \int_0^P (z-1)\frac{dP}{P} \quad \text{(constant } T\text{)} \quad (3.50)$$

$$\frac{H^R}{RT} = -T\int_0^P \left(\frac{\delta z}{\delta T}\right)_P \frac{dP}{P} \quad \text{(constant } T\text{)} \quad (3.51)$$

$$\frac{S^R}{R} = -T\int_0^P \left(\frac{\delta z}{\delta T}\right)_P \frac{dP}{P} - \int_0^P (z-1)\frac{dP}{P} \quad \text{(constant } T\text{)} \quad (3.52)$$

Values of z and the derivative $\left(\frac{\delta z}{\delta T}\right)_P$ are directly obtained from experimental data and therefore the integrals in Equations 3.50 to 3.52 can be evaluated by graphical or numerical method. The integrals can be evaluated analytically if z is expressed as a function of T and P by an equation explicit in volume. Thus when PVT data or an equation of state is available, H^R, S^R, and hence other properties, can be evaluated. For details and examples please refer to texts on thermodynamics (Smith et al.). Hougen and Watson have given generalized property departure charts. The Lee-Kessler generalized correlation tables for compressibility factor, residual properties, enthalpy, and entropy, and a correlation for fugacity coefficients are also available.

PVT Relationships—Equations of State for Ideal Gases

In Chapter 2, we reviewed the ideal gases and their *PVT* behavior as well as modified equations of state such as that of van der Waals, and the use of compressibility factor and generalized correlations to obtain *PVT* properties of actual or non-ideal gases. In the next section, we review thermodynamic relations for ideal gases.

Thermodynamic Property Relations for an Ideal Gas

To recall, for ideal gases, $PV = nRT$, $U = U(T)$, and $H = H(T)$.

Application of the definition of the heat capacities at constant volume and that at constant pressure to an ideal gas yields

$$C_P = \left(\frac{dH}{dT}\right)_P = \left(\frac{d(U+PV)}{dT}\right) = \frac{d(U+RT)}{dT} = \frac{dU}{dT} + R = C_V + R \quad (3.53)$$

By the application of the first law of thermodynamics to a reversible closed system of 1 mol of an ideal gas, the following relations result:

$$dU = dQ + dW = C_V dT \quad (3.54)$$

The work for a reversible closed system process is given by Equation 3.3a and is

$$dW = -PdV \quad (3.54a)$$

Then
$$dQ = C_V dT + PdV \quad (3.54b)$$

By using the relation $PV = RT$, and expressing one of the variables in terms of the other two, a number of relations can be established as follows:

With $P = RT/V$

$$dQ = C_V dT + RT \frac{dV}{V} \quad (3.55)$$

and
$$dW = -RT \frac{dV}{V} \quad (3.56)$$

With $V = RT/P$

$$dQ = C_V dT + P\left(\frac{R}{P} dT - \frac{RT}{P^2} dP\right)$$

Using the relationship of Equation 3.47, this reduces to

$$dQ = C_P dT - RT \frac{dP}{P} \quad (3.57)$$

similarly,
$$dW = -RdT + RT \frac{dP}{P} \quad (3.58)$$

With $T = PV/R$

$$dQ = C_V \left(\frac{V}{R} dP + \frac{P}{R} dV\right) + PdV$$

and with the relationship of Equation 3.47, this becomes

$$dQ = \frac{C_V}{R} VdP + \frac{C_P}{R} PdV \quad (3.59)$$

All these equations are applicable to closed system reversible process involving ideal gases only. They can be applied to various processes as follows:

Constant volume (isochoric process):

$$dU = dQ = C_V dT \quad \text{or} \quad \Delta U = Q = \int C_V dT \tag{3.60}$$

Constant pressure (isobaric process):

$$d\hat{H} = dQ = C_p dT \quad \Delta \hat{H} = Q = \int C_p dT \tag{3.61}$$

Constant temperature (isothermal process):

$$d\hat{U} = dQ + dW = 0 \quad \text{since} \quad d\hat{U} = C_V dT = 0$$

Thus

$$W = -\int P \, d\hat{V} \tag{3.62}$$

and integration results in

$$Q = -W = -RT \ln \frac{\hat{V}_2}{\hat{V}_1} = RT \ln \frac{P_2}{P_1} \quad \text{(Constant } T\text{)} \tag{3.63}$$

Adiabatic Process

For an adiabatic process,

$$dQ = 0 \quad \text{and} \quad d\hat{U} = dW = -P \, dV$$

In this case the following relations can be developed from Equations 3.55, 3.57, and 3.59 by putting $dQ = 0$:

$$\frac{T_2}{T_1} = \left(\frac{V_1}{V_2}\right)^{R/C_V} \quad \frac{T_2}{T_1} = \left(\frac{P_2}{P_1}\right)^{R/C_p} \quad \frac{P_2}{P_1} = \left(\frac{V_1}{V_2}\right)^{C_p/C_V} \tag{3.64}$$

These relations can be expressed in terms of k ratio of specific heats as follows:

$$\frac{T_2}{T_1} = \left(\frac{V_1}{V_2}\right)^{k-1} \quad \frac{T_2}{T_1} = \left(\frac{P_2}{P_1}\right)^{\frac{k-1}{k}} \quad \frac{P_2}{P_1} = \left(\frac{V_1}{V_2}\right)^{k} \tag{3.64a}$$

They may also be expressed as

$$TV^{k-1} = \text{Constant} \quad TP^{\frac{k-1}{k}} = \text{Constant} \quad PV^k = \text{Constant} \tag{3.64b}$$

where k is the ratio of the specific heats, C_p/C_V. Approximate values of k are $k = 1.67$ for monatomic gases, 1.4 for diatomic gases and 1.3 for simple polyatomic gases.

The following relation is useful to calculate the work of an adiabatic process:

$$W = \frac{RT_1}{k-1}\left[\left(\frac{P_2}{P_1}\right)^{(k-1)/k} - 1\right] = \frac{P_1 V_1}{k-1}\left[\left(\frac{P_2}{P_1}\right)^{(k-1)/k} - 1\right] \tag{3.65}$$

Polytropic Compression

The polytropic compression is neither adiabatic nor isothermal. For this type of compression, the equation of state PV^n = constant applies, where n is given by

$$\frac{n-1}{n} = \frac{k-1}{k\eta_p} \tag{3.65a}$$

polytropic work of compression of an ideal gas is given by

$$W_p = \frac{RT_1}{n-1}\left[1 - \left(\frac{P_2}{P_1}\right)^{(n-1)/n}\right] \tag{3.66}$$

Also

$$\frac{T_2}{T_1} = \left(\frac{P_2}{P_1}\right)^{(n-1)/n} = \left(\frac{V_1}{V_2}\right)^{n-1} \tag{3.67}$$

Calculation of Entropy Changes for Ideal Gas

It can be easily shown that when an ideal gas at P_1, T_1, V_1, is compressed to P_2, V_2, T_2, the total entropy change is given by

$$\Delta S_{\text{total}} = \int_{T_1}^{T_2} \frac{C_p dT}{T} - R \ln \frac{P_2}{P_1} \tag{3.68}$$

If the equation for heat capacity is in the form $C_p^o/R = a + bT + cT^2 + \cdots$, Equation 3.62 takes the form

$$\Delta S_{\text{total}} = \int_{T_1}^{T_2} \frac{C_p^o dT}{RT} - R\ln \frac{P_2}{P_1} \tag{3.68a}$$

Graphical Presentation of Property Data

In presenting the thermodynamic property data in graphical form, two variables are selected arbitrarily as the independent variables and the third variable is treated as a parameter. Most important charts are: (1) *PV* diagram with *T* as a parameter, (2) temperature-entropy (*TS*) diagram, (3) pressure-enthalpy (*PH*) diagram, and (4) enthalpy-entropy (*HS*) diagram. The enthalpy-entropy diagram is also called the *Mollier diagram* and is very useful in solving turbine and compressor problems. Pressure-enthalpy or *PH* diagrams are useful in solving refrigeration problems; temperature-entropy (*TS*) diagrams are useful in the analysis of engines.

Third Law of Thermodynamics

Nernst and Planck postulated that at absolute zero temperature, the absolute entropy of a pure, perfectly crystalline substance, free of any random arrangement is equal to zero. This was subsequently proved to be true by experimentation and is considered the third law of thermodynamics. The absolute entropy of a substance at constant pressure can be calculated by the equation

$$\hat{S} = \int_0^{T_f} \frac{C_{ps}}{T} dT + \frac{\Delta \hat{H}_f}{T_f} + \int_{T_f}^{T_b} \frac{C_{pl}}{T} dT + \frac{\Delta \hat{H}_v}{T_b} + \int_{T_b}^{T} \frac{C_{pg}}{T} dT \tag{3.69}$$

where
$\hat{S}$ = absolute entropy
T_f = temperature of fusion
$\Delta \hat{H}_f$ = heat of fusion
C_{pl} = specific heat of liquid
T_b = boiling point of liquid
$\Delta \hat{H}_v$ = heat of vaporization
C_{pg} = specific heat of vapor

In the preceding equation consistent units should be used. Temperatures are absolute temperatures.

Example 3.19

Methyl alcohol melts at 175.3 K with heat of fusion of 3.17 kJ/gmol. Its normal boiling point is 337.9 K and its heat of vaporization at normal boiling point is 35.3 kJ/gmol. Its average specific heat between its melting and boiling points can be taken as 0.61 cal/g·K.

Calculate the entropy changes that take place during melting, sensible heating up to boiling point and during vaporization to saturated vapor.

Solution

$$\Delta S_{\text{fusion}} = \frac{\lambda_f}{T_f} = \frac{3.15 \text{ kJ/g mol}}{175.3 \text{ K}} = 0.018 \frac{\text{kJ}}{\text{g mol} \cdot \text{K}}$$

$$\Delta S_{\text{heating}} = \int_{175.3}^{337.9} \frac{C_P dT}{T} = (C_P) \int_{175.3}^{337.9} \frac{dT}{T} = C_P \ln \frac{T}{T_o}$$

$$= 0.61 \times 32 \times 4.184 [\ln(337.9/175.3)]$$
$$= (53.55 \text{ J/gmol} \cdot \text{K}) \ln(1.9276)$$
$$= 53.6 \text{ J/gmol} \cdot \text{K} = 0.0536 \text{ kJ/gmol} \cdot \text{K}$$

$$\Delta S_{\text{vaporization}} = \frac{\lambda_v}{T_b} = \frac{35.3}{337.9} = 0.1045 \frac{\text{kJ}}{\text{g mol} \cdot \text{K}}$$

Example 3.20

Calculate Q, W, ΔU, ΔH, and ΔS for (a) the system, (b) the surroundings, and (c) the universe for freezing of 0.025 kg of super-cooled liquid SO_2 at 185.7 K and 0.1 MPa. Data for SO_2 at 0.1 MPa are as follows: melting point = 197.7 K, latent heat of fusion = 7.4×10^3 kJ/kg mol at melting point; c_p of liquid SO_2 = 1.3 kJ/kg·K; c_p of solid SO_2 = 0.96 kJ/kg·K.

Solution

Assume 1 kg mol of SO_2 as a basis. Visualize the freezing process as follows:

Liquid SO_2 →   liquid SO_2 →   solid SO_2 → solid SO_2
At 185.7 K(T_1)   at 197.7 K(T_2)   at 197.7 K(T_2)   at 185.7 K(T_3)

For a constant-pressure process, the energy balance reduces to

$$\Delta U = Q + W = Q - P\,\Delta V$$

or

$$Q = \Delta(U + PV) = \Delta H$$

Since ΔV is negligible for this cooling process

$$Q = \Delta U = \Delta H \quad \text{and} \quad W = P\,\Delta V$$

Now

$$Q = \Delta \hat{H} = \int_{T_1}^{T_2} C_{p\ell}\,dT + \Delta \hat{H}_f + \int_{T_2}^{T_3} C_{ps}\,dT = \Delta \hat{U}$$

$$= C_{p\ell}(T_2 - T_1) + \Delta \hat{H}_f^o + C_{ps}(T_3 - T_2)$$

$$= 64(1.3)(197.7 - 185.7) - 7.4 \times 10^3 + 64(0.96)(185.7 - 197.7)$$

$$= -7138.88 \text{ kJ/kg mol} = -7.138 \times 10^6 \text{ J/kg mol}$$

$$\Delta \hat{S} = \int \frac{dQ}{T} = \int_{T_1}^{T_2} C_{p\ell}\frac{dT}{T} + \frac{\Delta H_f}{T_f} + \int_{T_2}^{T_3} \frac{C_{ps}\,dT}{T}$$

$$= C_{p\ell}\ln\frac{T_2}{T_1} + \frac{\Delta H_f}{T_f} + C_{ps}\ln\frac{T_3}{T_2}$$

$$= 64(1.3)\ln\frac{197.7}{185.7} - \frac{7.4 \times 10^3}{197.7} + 64(0.96)\ln\frac{185.7}{197.7}$$

$$= -36.068 \text{ kJ/kg mol} \cdot \text{K}$$

For 0.025 kg of SO_2

$$S = -36.068\left(\frac{0.025}{64}\right) = -0.014 \text{ kJ/K}$$

$$Q = \frac{-7.138 \times 10^6 (0.025)}{64} = -2788.2 \text{ J}$$

or the surroundings

$$\Delta S = +\frac{2788.2}{185.7} = 0.015 \text{ kJ/K}$$

For the universe

$$S = S(\text{system}) + S(\text{surroundings})$$
$$= -0.014 + 0.015 = 0.001 \text{ kJ/K}$$

Solution Thermodynamics

So far our review has covered thermodynamic properties of constant composition fluids. However, in practical systems, in addition to temperature and pressure, composition is a very important variable. Therefore new properties are defined for homogeneous solutions. For example, chemical potential for phase and chemical equilibria and partial molal properties are defined for solutions. Properties of ideal gas mixtures are important in the treatment of real gas mixtures. A new function called the fugacity is related to chemical potential and is very useful in developing relations in phase and reaction equilibria. The concepts of ideal solution and excess property are also important. We will review here only the fugacity, an important property useful in the formulation of phase and chemical reaction equilibria.

Fugacity and Fugacity Coefficients

At constant T, $\quad dG_T = \hat{V}dP = RT\dfrac{dP}{P} = RTd\ln P \quad$ for ideal gas. $\hfill$ (3.70)

For nonideal substances, a fugacity for a pure component i is defined by

$$d\hat{G}_i = RT\, d\ln f_i \quad \text{at const. } T \tag{3.71}$$

in which P is replaced by f_i which has the units of pressure. Equation 3.71 partially defines f_i, the fugacity of a pure species i. For ideal gases this equation becomes

$$(dG)_T = \hat{V}dP = RT\dfrac{dP}{P} = RTd\ln P \tag{3.71a}$$

A further restriction given below completes the definition of fugacity.

$$\lim_{P \to 0} \dfrac{f_i^V}{P} = 1 \tag{3.71b}$$

Thus as $P \to 0$ the properties of nonideal gas approach those of an ideal gas. The ratio f_i^V/P is called the fugacity coefficient for vapor phase, for which a generalized chart is available. It is denoted by ϕ_i^V.

Likewise the fugacity $\dfrac{\hat{f}_i^L}{P}$ of a component in solution is defined by analogous equations as follows:

$$dG_i = RTd\ln \hat{f}_i^L \quad \text{at constant temperature} \tag{3.72}$$

with the restriction $\quad \lim_{P \to 0} \dfrac{\hat{f}_i^L}{P} = 1 \hfill$ (3.72a)

For a component in solution, the fugacity coefficient is defined in the same manner as that for vapor phase and is given by

$$\hat{\phi}_i^L = \dfrac{\hat{f}_i^L}{x_i P} \tag{3.72b}$$

The fugacity f of a solution as a whole is also defined in the same manner as the fugacity of a pure material.

$$dG = RTd\ln f \quad \text{at constant temperature} \tag{3.73}$$

and $\quad \lim_{P \to 0} \dfrac{f}{P} = 1 \hfill$ (3.73a)

In this case G is the molar Gibbs free energy of the solution.

By combining the equation for fugacity with that for ideal gas and noting that at constant temperature $\left(\frac{d \ln f_i^V}{d \ln P}\right)_T = z$ (compressibility), one can derive the following relation:

$$d \ln \frac{f_i^V}{P} = [(z-1) d \ln P] \quad \text{at constant } T \tag{3.74}$$

At $P = 0$, fugacity and pressure are equal by definition. Using this fact and writing the previous equation in terms of reduced conditions and by integration from $P = 0$ to P_r, we get

$$\ln \frac{f_i^V}{P} = \left[\int_0^{P_r} \frac{(z-1) dP_r}{P_r}\right] \tag{3.75}$$

Fugacity can be determined using *PVT* data or an equation of state. If the law of corresponding states is used

$$dG = RT d \ln f_i^V = \frac{zRT}{P} dP \quad \text{at constant } T \tag{3.76}$$

or

$$\ln f_i^V = \int_0^P z \frac{dP}{P} \tag{3.77}$$

The integral on the righthand side of the equation can be evaluated by numerical integration using generalized z charts. Hougen et al. have given a table of fugacity coefficients, ϕ_i versus P_r with T_r as parameter for substances having critical compressibility $z_c = 0.27$. A table of data and a correction equation allow the computation of the fugacity coefficients for other materials having different critical compressibility factors.

Fugacities in Ideal Solutions and Standard States

The fugacity coefficients of a component in a solution and its pure state are related by

$$\ln \frac{\hat{\phi}_i}{\phi_i} = \frac{1}{RT} \int_0^P (\hat{V}_i - V_i) dP \tag{3.78}$$

If $\hat{\phi}_i$ and ϕ_i are replaced by their definitions, one obtains

$$\ln \frac{\hat{f}_i}{f_i x_i} = \frac{1}{RT} \int_0^P (\hat{V}_i - V_i) dP \tag{3.79}$$

For an ideal solution, $\hat{V}_i = V_i$, therefore $\ln \frac{\hat{f}_i^{id}}{f_i x_i} = 0$ and then $\hat{f}_i^{id} = f_i x_i$.

This expression shows that for an ideal solution, the fugacity of a component in solution is equal to its mol fraction times its fugacity in pure state. This is known as the *Lewis-Randall rule*.

A more general definition of an ideal solution is based on the concept of *standard states*. Based on a standard state, the fugacity of component i in an ideal solution is defined by

$$\hat{f}_i^{id} = x_i f_i^o \tag{3.80}$$

where f_i^o is the fugacity of component i in a standard state at the same temperature and pressure as that of the mixture. The standard state can be either the actual state of pure i or it could be a hypothetical state. If the actual state of pure i is chosen, $f_i^o = f_i$. Two convenient states, one based on the Lewis-Randall rule and the other based on Henry's law, are defined as follows:

Based on the Lewis-Randall rule

$$\lim_{x_i \to 1} \frac{\hat{f}_i}{x_i} = f_i \quad \text{where standard state fugacity is } f_i \tag{3.81}$$

Based on Henry's law

$$\lim_{x_i \to 1} \frac{\hat{f}_i}{x_i} = k_i \quad \text{where standard state fugacity is } k_i \tag{3.82}$$

With standard states defined in the previous manner, the fugacity of ideal solution will be given by

$$\hat{f}_i^{id} = x_i f_i \quad \text{(based on Lewis-Randall rule)} \tag{3.83}$$

where, f_i is the fugacity of component i in the standard state. The other standard state results in the following expression for the fugacity of a component in solution:

$$\hat{f}_i^{id} = x_i k_i \quad \text{(based on Henry's law)} \tag{3.84}$$

This relation is valid for values of x_i near zero. In this case, the standard state fugacity is k_i and $k_i = f_i$.

Isothermal Effect of Pressure on $\hat{G}$, $\hat{S}$, $\hat{U}$

The isothermal effect of pressure on $\hat{G}$, $\hat{S}$, and $\hat{U}$ is given by the following equations:

$$(\hat{G}_2 - \hat{G}_1)_T = RT \ln \frac{f_2}{f_1} = RT \ln \frac{v_2 P_2}{v_1 P_1} \tag{3.85}$$

$$(\hat{S}_2 - \hat{S}_1)_T = \frac{(\hat{H}_2 - \hat{H}_1)_T - (\hat{G}_2 - \hat{G}_1)_T}{T} \tag{3.86}$$

$$(\hat{U}_2 - \hat{U}_1)_T = (\hat{H}_2 - \hat{H}_1)_T - (P_2 \hat{V}_2 - P_1 \hat{V}_1)_T \tag{3.87}$$

$$= (\hat{H}_2 - \hat{H}_1)_T - RT(Z_2 - Z_1)_T \tag{3.88}$$

Example 3.21

Estimate the fugacity of SO_2 at 1000 psia and 460° F. Use the fugacity coefficient chart for SO_2. $P_c = 1143.7$ psia, $T_c = 775.3°$ R.

Solution

$$P_r = \frac{1000}{1143.7} = 0.8744 \qquad T_r = \frac{460+460}{775.3} = 1.187$$

From the fugacity coefficient chart,

$$\phi = 0.86 = \frac{f}{P}$$

Hence

$$f = 0.86(1000) = 860 \text{ psi}$$

Example 3.22

Determine the fugacity and fugacity coefficient for SO_2 at 400° F and 800 psia from the enthalpy and entropy values. The enthalpy and entropy data for SO_2 are given below.

Properties of SO_2 at 400° F

P (pressure), psi	$\hat{V}$ (volume), ft³/lb	$\hat{H}$ (enthalpy), Btu/lb	$\hat{S}$ (entropy), Btu/lb · °R
10	14.35	266.7	0.5506
800	0.1446	249.5	0.4004

Solution

$$d\hat{G}_T = RT\, d\ln f_T \quad \text{or} \quad d\ln f_T = \frac{d\hat{G}_T}{RT}$$

Integration between high P and a low pressure (reference state) gives

$$RT \ln \frac{f}{f^*_T} = (\hat{G} - \hat{G}^*)$$

and since $\hat{G} = \hat{H} - T\hat{S}$

$$\ln \frac{f}{f^*} = \frac{1}{R}\left[\frac{\hat{H} - \hat{H}^*}{T} - (\hat{S} - \hat{S}^*)\right]$$

If the reference state is a low pressure, then $f^* = P^*$ and

$$\ln \frac{f}{P^*} = \frac{1}{R}\left[\frac{\hat{H} - \hat{H}^*}{T} - (\hat{S} - \hat{S}^*)\right]$$

Assuming SO_2 behaves as an ideal gas at 10 psia

$$P^* = 10 \text{ psia} \qquad \hat{H}^* = 266.7 \text{ Btu/lb} \qquad \hat{S}^* = 0.5506 \text{ Btu/lb} \cdot °R$$

For superheated SO_2 at 800 psi and 400°F

$$\hat{H} = 249.5 \text{ Btu/lb} \quad \hat{S} = 0.4004 \text{ Btu/lb·°R}$$

Hence

$$\ln \frac{f}{P^*} = \frac{64}{1.987}\left(\frac{249.5 - 266.7}{400 + 460} - 0.4004 + 0.5506\right) = 4.194$$

Thus

$$\frac{f}{P^*} = e^{4.194} = 66.3.$$

Therefore

$$f = 66.3(10) = 663 \text{ psia}$$

and

$$\text{fugacity coefficient} = \frac{f}{P} = \frac{663}{800} = 0.829$$

POWER CYCLES AND REFRIGERATION

A cyclic process or a cycle consists of a series of operations repeated in the same order. Cycles of heat engines used to convert heat into work consist of the following elements: (1) working fluid that receives heat at a higher temperature and rejects it at a lower temperature, (2) a high-temperature reservoir from which heat is added to the working fluid, (3) a low-temperature receiver to which heat is rejected, and (4) a heat engine or heat pump to convert heat into work or work into heat. During each cycle, the working substance goes through a series of operations and returns to its initial condition. In many cases, the same mass of the fluid may not be involved although the states of the cyclic operations are repeated. A reversible cyclic process is composed of individual operations, which themselves are reversible.

The Carnot reversible cycle is a hypothetical device that uses a perfect gas as the working fluid. It is taken as a standard in evaluating the efficiency of the actual power cycles used to convert heat into work. The Carnot cycle consists of the following four reversible steps (see Fig. 3.1):

1. Reversible isothermal expansion from V_1 to V_2 of the working fluid by absorbing heat Q_H at constant temperature T_H, from the heat source. Work obtained in this step is

$$W = -\int_{V_1}^{V_2} P\, dV$$

Negative sign is used because here work is done by the gas on the piston.

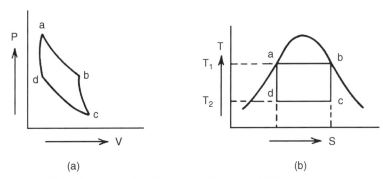

Figure 3.1 Carnot cycle (a) on a *PV* diagram, (b) *TS* diagram

2. This step comprises a reversible adiabatic expansion of the fluid from V_2, P_2, T_1 to the state P_3, V_3, T_2. Work obtained in this expansion is

$$-\int_{V_2}^{V_3} P\,dV$$

This is obtained by direct conversion of the internal energy into work.

3. This step consists of a reversible isothermal compression from V_3 to V_4 at T_2. Work of compression is

$$\int_{V_3}^{V_4} P\,dV$$

and amount $-Q_2$ flows into the low-temperature (T_2) receiver.

4. Isentropic (adiabatic) compression to P_1, T_1, and V_1. The reversible work done on the gas in this step is

$$\int_{V_4}^{V_1} P\,dV$$

As the fluid returns to its original state after completing the reversible cycle, $\Delta U = 0$. Because no heat was gained or lost during the adiabatic steps, the net work $(W_{rev})_{net}$ is given by

$$W_{rev} = \sum \oint P\,dV = Q_1 - Q_2$$

Since total entropy change during the reversible cycle is zero, one obtains

$$\Delta S_T = S_1 + S_2 = 0$$

from which

$$\frac{Q_H - Q_C}{Q_H} = \frac{T_H - T_C}{T_H} = \frac{W_{net}}{Q_H} \qquad (3.89)$$

W_{net}/Q_H is called the *thermodynamic ideal efficiency* of the cycle. The 100 percent efficiency for the Carnot cycle is possible only if the temperature of the receiver is absolute zero or if the engine absorbs heat at an infinite temperature.

Example 3.23

A working fluid goes through a Carnot cycle between the temperature limits of 533.2 and 333.2 K. At higher temperature 527 kJ are supplied to the cycle. Find the thermal efficiency, work done, amount of heat rejected, change in entropy during the isothermal process, and total entropy change.

Solution

$$T_H = 533.2 \text{ K}, \qquad T_C = 333.2 \text{ K}$$

1. Since the cycle is Carnot reversible, thermal efficiency e is

$$e = \frac{T_H - T_C}{T_H} = \frac{533.2 - 333.2}{533.2}(100) = 37.5 \text{ percent}$$

2. Work done and amount of heat rejected

$$\frac{Q_H - Q_C}{Q_H} = \frac{T_H - T_C}{T_H}$$

$$W_{net} = Q_H - Q_C = \frac{T_H - T_C}{T_H} Q_H$$

$$= 0.375(527) = 197.6 \text{ kJ}$$

Q_C = heat rejected = $Q_H - W_{net}$ = 527 − 197.6 = 329.4 kJ

3. Entropy change in isothermal step

$$\Delta S = \frac{Q_H}{T_H} = \frac{527}{533.2} = 0.988 \text{ kJ/K}$$

for total mass at higher temperature

$$\Delta S = \frac{Q_C}{T_C} = -\frac{329.3}{333.2} = -0.988 \text{ kJ/K} \quad \text{at lower temperature.}$$

4. Because cycle is reversible, $\Delta S_T = 0$ or

$$\frac{Q_H}{T_1} - \frac{Q_C}{T_C} = 0.988 - 0.988 = 0$$

Refrigeration

The purpose of refrigeration is to maintain a given system at a temperature below that of its surroundings. Heat is abstracted from the low-temperature region and rejected at a higher temperature. The refrigeration methods of more importance are (1) vapor-compression refrigeration and (2) absorption refrigeration.

Definitions

A ton of refrigeration is the removal of heat at a rate of 200 Btu/min, 12,000 Btu/h, or 288,000 Btu/d. Coefficient of performance is the ratio of heat absorbed at low temperature divided by the work done in compression.

$$\beta = \frac{\text{heat removed from low} - \text{temperature region}}{\text{work of compression}} \tag{3.90}$$

Carnot Refrigerator (Reversed Carnot Cycle). An ideal refrigeration cycle is the reversed ideal Carnot cycle, and its performance depends only on the temperature at which heat is absorbed and rejected.

Carnot refrigeration cycle steps are shown on *PV* and *TS* diagrams in Figure 3.2. For a Carnot cycle, the coefficient of performance β is given by

$$\beta = \frac{\text{heat absorbed in evaporator}}{\text{work done on the fluid}} = \frac{Q_C}{W} = \frac{Q_C}{Q_C - Q_H} \tag{3.91}$$

$$= \frac{T_C \Delta S}{(T_H - T_C) \Delta S} = \frac{T_C}{T_H - T_C} \tag{3.91a}$$

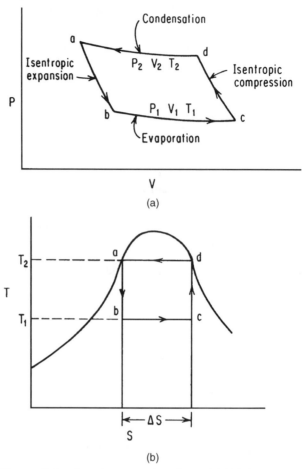

Figure 3.2 Carnot reversed cycle on (a) PV diagram, (b) TS diagram

The actual refrigeration cycles are not as efficient as Carnot's reversed cycle because of the irreversibilities in the actual processes. They are carried out in two ways. One uses *free expansion* through a throttle valve and the other *adiabatic expansion* through a turbine. Free expansion is isenthalpic while the turbine expansion is more or less *isentropic*.

Vapor-Compression Cycle with Turbine Expansion

The vapor-compression cycle using turbine expansion is shown in Figure 3.3. Both isentropic compression and expansion are less than theoretically efficient in practice, and hence *AB* and *CD* (Figure 3.3) will not be vertical on the *TS* diagram. In the ideal case,

$$\hat{S}_l = \hat{S}_{gl} \quad \text{and} \quad \hat{S}_d = \hat{S}_g$$

where subscripts indicate the following:
- g = vapor at compressor inlet
- d = vapor at compressor discharge
- gl = liquid-vapor mixture entering evaporator
- l = cooled liquid leaving condenser.

Heat absorbed at low temperature

$$Q_C = \Delta H = \hat{H}_g - \hat{H}_{gl}$$

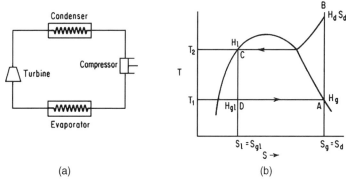

Figure 3.3 Vapor compression refrigeration cycle with turbine expansion: (a) system representation (b) TS diagram

Heat rejected at condenser

$$\text{Work done} = (\hat{H}_d - \hat{H}_l) - (\hat{H}_g - \hat{H}_{gl})$$

and therefore β, the coefficient of performance, is given by

$$\beta = \frac{\hat{H}_g - \hat{H}_{g\ell}}{(\hat{H}_d - \hat{H}_\ell) - (\hat{H}_g - \hat{H}_{g\ell})}$$

After turbine expansion, the refrigerant is a liquid-vapor mixture. Its entropy is given by

$$\hat{S} = \hat{S}_{g\ell} = \hat{S}_g - (\Delta\hat{S}_{\text{vap}})x$$

where x is the quality of refrigerant after turbine expansion.

Similarly, $$\hat{H}_{gl} = \hat{H}_g - x(\Delta\hat{H}_{\text{vap}})$$

where $\Delta\hat{S}_{VAP}$ and $\Delta\hat{H}_{VAP}$ represent the changes of the entropy and enthalpy per unit mass at the low pressure and temperature. If x is eliminated from the preceding equations

$$\hat{H}_{g\ell} = \hat{H}_g - \frac{\Delta\hat{H}}{\Delta\hat{S}_{\text{vap}}}(\hat{S}_g - \hat{S}_{g\ell})$$

$$= \hat{H}_g - \left(\frac{\Delta\hat{H}}{\Delta\hat{S}}\right)_{\text{vap}}(\hat{S}_g - \hat{S}_\ell)$$

from which

$$\hat{H}_{g\ell} = \hat{H}_g - T_1(\hat{S}_g - \hat{S}_{g\ell})$$

This equation allows calculation of $\hat{H}_{gl}$ provided $\hat{S}_g$ is saturation vapor entropy. $\hat{H}_{gl}$ could also be obtained by a trial-and-error procedure from

$$(x\hat{S}'_\ell) + (1-x)\hat{S}'_g = \hat{S}_{g\ell} = \hat{S}_\ell$$

where $\hat{S}'_l$ and $\hat{S}'_g$ are the entropies of the liquid and vapor in the liquid-vapor mixture after turbine expansion.

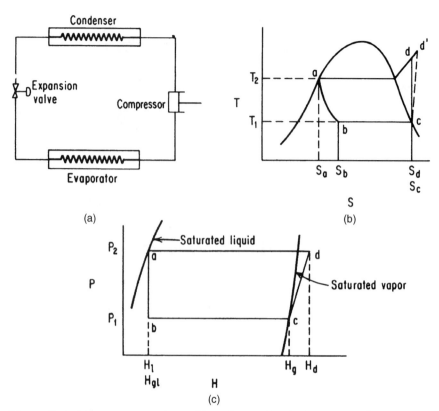

Figure 3.4 Vapor compression refrigeration cycle with free expansion: (a) system sketch; (b) TS diagram; (c) PH diagram

Vapor Compression with Free Expansion

The basic single-stage cycle consists of steps, as shown in Figure 3.4. The calculation steps are as follows:

1. Net refrigeration effect is given by

$$Q_C = \hat{H}_g - \hat{H}_{gl} = \hat{H}_g - \hat{H}_l \quad \text{because } \hat{H}_l = \hat{H}_{gl}$$

where
$\hat{H}_g$ = enthalpy of vapor leaving evaporator
$\hat{H}_l$ = enthalpy of liquid leaving condenser, Btu/lb
$\hat{Q}_C$ = refrigeration effect, Btu/lb.

2. Weight rate of flow in lb/min·ton is

$$W = \frac{200 \text{ Btu/min·ton}}{Q_C \text{ Btu/lb}}$$

$$\text{Volume of vapor to be handled} = W\hat{V}_g$$

where $\hat{V}_g$ is the specific volume of vapor entering compressor.

$$\text{Heat of compression} = \hat{H}_d - \hat{H}_g$$

where $\hat{H}_d$ is the enthalpy of vapor discharged by compressor, Btu/lb, and $\hat{H}_g$ is the enthalpy of vapor at compressor suction, Btu/lb.

Work of compression = W_s

$$(\hat{H}_d - \hat{H}_g)W = \text{Btu/min·ton}$$
$$1 \text{ hp} = 42.4 \text{ Btu/min}$$

$$\frac{\text{hp}}{\text{ton}} = \frac{\text{work of compression (Btu/min·ton)}}{42.4 \text{ Btu/min}} = \frac{4.713}{\beta}$$

Condenser heat load = $Q_H = \hat{H}_d - \hat{H}_\ell$ Btu/lb

This is also equal to the refrigeration effect plus the heat of compression:

$$\hat{H}_g - \hat{H}_\ell + \hat{H}_d - \hat{H}_g = \hat{H}_d - \hat{H}_\ell \text{ Btu/lb}$$

$$\beta = \frac{\text{net refrigeration effect, Btu/lb}}{\text{heat of compression, Btu/lb}}$$

$$= \frac{\hat{H}_g - \hat{H}_\ell}{\hat{H}_d - \hat{H}_g}$$

Example 3.24

A refrigerator is to be maintained at 0°F. Cooling water is available at 86°F. The evaporator and condenser are of sufficient capacity so that a 10°F approach is possible in each unit. Freon 12 is to be used as the refrigerant. Capacity of the unit is to be 5 tons. Assume the compression to be isentropic. Compute (a) the pounds of refrigerant to be circulated, (b) the coefficient of performance β and (c) volumetric displacement of compressor, for (*i*) vapor compression cycle with turbine expansion and (*ii*) vapor compression cycle with free expansion. (*iii*) What is β for Carnot reversed cycle?

Solution

Properties of Freon 12 required for the example are given in Table 3.1.

Table 3.1 Properties of Freon 12

Temp. (°F)	Pressure (psi)	Saturated Liquid $\hat{H}$ Btu/lb	Saturated Liquid $\hat{S}$ Btu/lb·°R	Saturated Vapor $\hat{H}$ Btu/lb	Saturated Vapor $\hat{S}$ Btu/lb·°R	$\hat{V}_g$ ft³/lb
0	23.95	8.52	0.019323	77.27	0.16888	1.6089
96	124.7	30.14	0.061536	86.69	0.16333	
110	—	—	—	89.23	0.16776	
120	—	—	—	91.01	0.17087	

Case i. With a condensed liquid,

$$\hat{H}_\ell = 30.14 \text{ Btu/lb} \quad \hat{S}_\ell = 0.061536 \text{ Btu/lb·°R}$$

Liquid vapor mixture after turbine expansion is isentropic expansion; therefore $\hat{S}_\ell = \hat{S}_{g\ell} = 0.061536$ Btu/lb·°R. Hence

$$\hat{H}_{g\ell} = \hat{H}_g - T_H(\hat{S}_g - \hat{S}_{g\ell})$$
$$= 77.27 - 460(0.16888 - 0.061536) = 27.89 \text{ Btu/lb}$$

Since the evaporator temperature is known, $\hat{H}_{g\ell}$ can also be obtained by calculating x, the quality of refrigerant after expansion and using x to calculate the enthalpy of the liquid-vapor mixture.

Compressor Discharge. By following the *PH* diagram at constant entropy = 0.16888 (saturated vapor at 0°F), the compressor discharge temperature is closer to 110°F but less than 120°F. *H* can be read from the *PH* diagram, but a more accurate value can be obtained by interpolation from the tables. Thus

$$\hat{H}_d = 89.23 + \frac{0.16888 - 0.16776}{0.17087 - 0.16776}(91.01 - 89.23)$$
$$= 89.23 \text{ Btu/lb}$$

$$\beta = \frac{\hat{H}_g - \hat{H}_{g\ell}}{(\hat{H}_d - \hat{H}_\ell) - (\hat{H}_g - \hat{H}_{g\ell})}$$
$$= \frac{77.27 - 27.89}{(89.87 - 30.14) - (77.27 - 27.89)} = 4.8$$

$$\text{Refrigerant circulation} = \frac{12{,}000 \text{ Btu/h} \cdot \text{ton (5 ton)}}{\hat{H}_g - \hat{H}_{g\ell}}$$
$$= \frac{60{,}000}{77.27 - 27.89} = 1215 \text{ lb/h}$$

$$\text{Refrigerant circulation} = \frac{12{,}000 \text{ Btu/h} \cdot \text{ton (5 ton)}}{\hat{H}_g - \hat{H}_{g\ell}}$$
$$= \frac{60{,}000}{77.27 - 27.89} = 1215 \text{ lb/h}$$

$$\text{Volumetric displacement of compressor} = \frac{1215}{60}(1.6089) = 32.5 \text{ ft}^3/\text{min}$$

Case ii. Vapor compression cycle with free expansion: In this case $\hat{H}_g = 77.27$ Btu/lb.

$$\hat{H}_{g\ell} = \hat{H}_\ell = 30.14 \text{ Btu/lb}$$

$$\beta = \frac{\hat{H}_g - \hat{H}_\ell}{\hat{H}_d - \hat{H}_g} = \frac{77.27 - 30.14}{89.86 - 77.27} = 3.74$$

$$\text{Refrigerant circulation} = \frac{60{,}000}{77.27 - 30.14} = 1273 \text{ lb/h}$$

$$\text{Volumetric displacement} = \frac{1273(\hat{V}_g)}{60} = \frac{1273(1.6089)}{60} = 34.14 \text{ ft}^3/\text{min}$$

Case iii. Carnot refrigerator. In this case the coefficient of performance is

$$\beta = \frac{460}{(460+96)-460} = 4.792$$

RECOMMENDED REFERENCES

1. Perry and Green (eds.), *Chemical Engineers' Handbook,* platinum ed., McGraw-Hill, 2000.
2. Hougen, Watson, and Ragatz, *Chemical Process Principles Part II,* John Wiley & Sons, 1947.
3. Hougen, Watson, and Ragatz, *Chemical Process Principles Part 1,* John Wiley & Sons, 1943.
4. Himmelblau, *Basic Principles and Calculations in Chemical Engineering,* 6th ed., Prentice Hall, 2004.
5. Smith, Van Ness, and Abbott, *Introduction to Chemical Engineering Thermodynamics,* 6th ed., McGraw-Hill, 2001.

CHAPTER 4

Fluid Mechanics

OUTLINE

VISCOSITY 132

FLUID STATICS 132
Static Head ■ Liquid Manometers ■ Bernoulli's Theorem

FLUID MEASUREMENTS 138
Static Pressure ■ Stagnation Pressure ■ Venturimeters ■ Orificemeters ■ Mass Flowmeters–Operating Principle ■ Vortex Shedding Flowmeter-Operating Principle

PERMANENT PRESSURE LOSS 143

FLOW OF FLUIDS IN PIPES 147
Consistent Units of Variables to Calculate Dimensionless Reynolds Number ■ Distribution of Velocities ■ Laminar Flow of Newtonian Fluids in Cylindrical Pipes ■ Turbulent Flow ■ Frictional Losses in Circular Pipes ■ Hydrodynamic Entry Length ■ Application of Surface Tension in Fluid Mechanics ■ Equivalent Diameters for Noncircular Conduits ■ Frictional Losses through Fittings and Values ■ Resistance Coefficient K ■ Equivalent Length L_e ■ Flow Coefficient

PUMP CALCULATIONS 168
Pump Affinity Laws

CONTROL VALVES 187
Capacity ■ Characteristics ■ Rangeability ■ Control Valve Sizing

PARALLEL AND BRANCHED SYSTEMS 192

REPLACEMENT OF ONE PIPELINE WITH n PARALLEL PIPELINES OF EQUAL EQUIVALENT LENGTHS AS THE ORIGINAL PIPE 192
Economic Pipe diameter

FLOW OF COMPRESSIBLE FLUIDS 193
Pressure Drop Calculation for Compressible Fluid Obeying Ideal-Gas Law ■ Compression Equipment ■ Ratio of Specific Heats k ■ Multistage Compression ■ Fans ■ Fan Static Pressure P_s

TWO-PHASE (VAPOR/LIQUID) FLOW CONSIDERATION 201
Lockhart-Martinelli Method ■ Duckler's Homogeneous Method

RECOMMENDED REFERENCES 207

The following topics are reviewed in this chapter: fluid statics and pressure measurement, flow measurement, Bernoulli's theorem, and fluid flow and transportation.

VISCOSITY

Viscosity is a measure of the resistance of a fluid to shear or angular deformation. The shear stress between two adjacent layers of a Newtonian fluid varies as the rate of shear at constant temperature and pressure:

$$\tau \propto \frac{du}{dy} \quad \text{hence} \quad \tau = \frac{\mu}{g_c}\frac{du}{dy} \tag{4.1}$$

where
  τ = shear stress, lb_f/ft^2, [N/m^2, or Pa]
  g_c = Newton's law proportionality factor, 32.17 (lb · ft/s^2)/lb$_f$), [1(kg · m/s^2)/N]
  u = linear velocity, ft/s, [m/s]
  y = distance perpendicular to the direction of flow, ft [m]
  μ = viscosity, lb/(ft · s), [kg/(m · s)]

The conversion factors for viscosity from one system of units into the others are given in Table 1.4. The first unit of the variables explained above follows the American Engineering System (AES). The units shown in brackets [ ] refer to SI.

FLUID STATICS

Static Head

The static head of a fluid is the pressure at a given point in a fluid that is exerted by a vertical column of the fluid above the point. Because liquids are more or less incompressible,

$$P_h = \frac{h\rho g}{g_c}$$

$$\Delta P_h = \frac{(\Delta h)\rho g}{g_c} \tag{4.2a}$$

$$\frac{P_h}{\rho} = h\left[\frac{g}{g_c}\right] \tag{4.2b}$$

If the column of fluid is open to atmosphere, then the atmospheric pressure must be added in appropriate unit to get P_h in absolute scale of pressure.

where
  P_h = static pressure, lb_f/ft^2 [N/m^2 or Pa], due to the column of fluid only of height h
  h = Static head of liquid above the point, ft [m]
  ρ = liquid density, lb/ft^3 [kg/m^3]
  g = local acceleration due to gravity, ft/s^2 [m/s^2]
  g_c = 32.17 lb · ft/(lb$_f$ · s^2), [1 (kg · m/s^2)/N]: dimensional constant.

$$\frac{g}{g_c} \simeq \begin{cases} 1\,\dfrac{lb_f}{lb} & \text{on earth in AES unit} \tag{4.2c} \\[2ex] 9.81\,\dfrac{N}{kg} & \text{on earth in SI unit} \end{cases} \tag{4.2d}$$

Typical value of g on earth is 32.17 ft/s^2, or 9.81 m/s^2.

When the value of $g = 32.17$ ft/s^2 (typically on earth), $g/g_c = 1$ (lb$_f$/lb) or 9.81 N/kg on earth. Therefore, substituting appropriate values of g/g_c, and $\rho = s$ (62.4) in AES units in Equation 4.2a gives on earth:

$$P_h = \begin{cases} h(62.4s)\dfrac{\text{lb}_f}{\text{ft}^2} = \dfrac{h(62.4s)}{144} = 0.433hs \text{ psi in AES unit;} \\ \qquad\qquad (h = \text{ft}, s = \text{specific gravity}) \quad (4.3a) \\ h\rho(9.81)\dfrac{\text{N}}{\text{m}^2} \text{ in SI unit;} \left(h = \text{meter}, \rho = \dfrac{\text{kg}}{\text{m}^3}\right) \quad (4.3b) \end{cases}$$

Liquid Manometers

The height of a liquid in an open manometer tube connected to a pressure source (e.g., a liquid in a vessel) is a direct measure of the static pressure at the point of connection. The tubes may be straight or U-type. The pressure-measuring fluid may be different from the fluid whose pressure is to be measured. Both the open and differential tube gauges are used. The height-measuring fluid must be immiscible with the test fluid.

In an open manometer of the type shown in Figure 4.1a, the pressure at point A (connecting point) is

$$P_A = (h_m \rho_m - h_A \rho_l)\dfrac{g}{g_c} \quad (4.4)$$

Since $\dfrac{g}{g_c} = \dfrac{1\,\text{lb}_f}{\text{lb}}$ on earth in AES unit, g/g_c may be dropped from the second half of the equation when solving problems in AES units. However, $\dfrac{g}{g_c} = \dfrac{9.81\,\text{N}}{\text{kg}}$ must be substituted for solving problems in SI unit, P_A is $\dfrac{\text{lb}_f}{\text{ft}^2}$ in AES and $\dfrac{\text{N}}{m^2}$ in SI,

where
- h_m = differential height of manometric fluid, ft [m]
- h_A = distance between point of attachment and the interface between the test fluid and the manometric fluid, ft [m]
- ρ_m = density of manometric fluid, lb/ft^3 [kg/m^3]
- ρ_ℓ = density of test fluid, lb/ft^3 [kg/m^3].

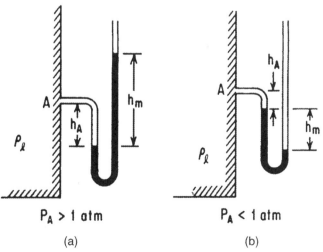

Figure 4.1 Open manometers

Therefore the head at point A is obtained by dividing Equation 4.4 by ρ_L.

$$h_\ell = \frac{P_A}{\rho_L} = \left[h_m \frac{\rho_m}{\rho_l} - h_A \right] \frac{g}{g_c} \quad \frac{\text{ft} \cdot \text{lb}_f}{\text{lb}} \quad \text{or} \quad \frac{\text{N} \cdot \text{m}}{\text{kg}} \tag{4.5}$$

In the case of gases, ρ_ℓ is so small that the terms containing h_A may be neglected in the preceding equation. In the case of the differential U-tube manometer shown in Figure 4.2b,

$$P_A - P_B = [h_m(\rho_m - \rho_A) + h_A \rho_A - h_B \rho_B] \frac{g}{g_c} \tag{4.6}$$

where ρ_A and ρ_B are the densities at points A and B.

As a special case when $\rho_A = \rho_B = \rho$, $h_A = h_B$, and s_m and s are the specific gravities of the manometric fluid and process fluid, respectively Equation 4.6 reduces to

$$\frac{P_A - P_B}{\rho} = \Delta H \left(\frac{g}{g_c} \right) = h_m \left(\frac{s_m}{s} - 1 \right) \left(\frac{g}{g_c} \right) \tag{4.7}$$

where, ΔH is the static head difference in ft [m] of appropriate fluid.

The diameter of a manometer tube should be at least 0.5 in. to avoid capillary error. For U tubes, small diameters are permissible because the capillary rises in the two legs tend to compensate for each other.

For a simple manometer (Fig. 4.3a), the following relation can be established:

$$\Delta P = P_A - P_B = h_m(\rho_A - \rho_B) \frac{g}{g_c} \tag{4.8}$$

To obtain a better accuracy in measurements, multiplying manometers are used. To multiply the reading of height, a fluid of lower density is used in an open tube. In differential manometers, the difference between the density of the manometric fluid and that of the test fluid should be as small as possible. In an inclined manometer (Fig. 4.3b), the fluid head is given by

$$h_m = R \sin\theta \tag{4.9}$$

where R is the distance between the two meniscuses along the tube and θ is the angle of inclination of the manometer in radians.

A draft gauge (Fig. 4.3c) consists of an inclined tube connected to a reservoir. If the gauge liquid is other than water, the reading must be multiplied by ρ/ρ_w, where ρ_w is the density of water and ρ is the density of gauge fluid.

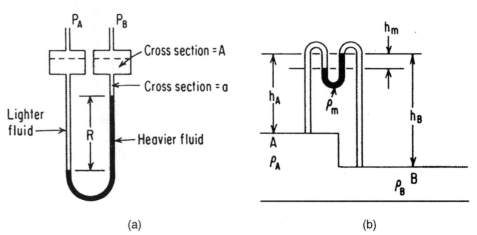

Figure 4.2 U-tube manometers: (a) two-fluid; (b) differential

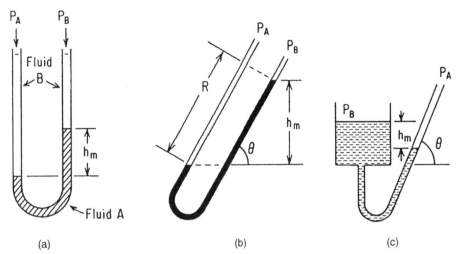

Figure 4.3 (a) Simple manometer, (b) inclined manometer, (c) draft gauge

The two-fluid U-tube manometer (Fig. 4.2a) is very useful in measuring small heads of gases. In this case, the differential pressure is given by

$$\Delta P = P_A - P_B = (R - R_0)\left(\rho_2 - \rho_1 + \frac{a}{A}\rho_1\right)\left(\frac{g}{g_c}\right) \text{lb}_f/\text{ft}^2 \text{ or } \text{N/m}^2 \quad (4.10)$$

where
- R = reading of heavier manometric fluid, ft [m]
- R_0 = reading of heavier manometric fluid, ft [m] when $P_A = P_B$
- ρ_1 = density of lighter fluid, lb/ft³ [kg/m³]
- ρ_2 = density of heavier fluid, lb/ft³ [kg/m³]

If A/a is very large, the term $(a/A)\rho_1$ is negligible, but it may not be omitted without a check.

Bernoulli's Theorem

Applying the principle of conservation of energy (first law of thermodynamics) to a flowing fluid (Fig. 4.4), the following Bernoulli equation is obtained for the steady flow of a fluid:

$$Z_B\left(\frac{g}{g_c}\right) + \frac{u_B^2}{2\alpha_B g_c} + \frac{P_B}{\rho_B} = Z_A\left(\frac{g}{g_c}\right) + \frac{u_A^2}{2\alpha_A g_c} + \frac{P_A}{\rho_A} - F + Q + W \quad (4.11)$$

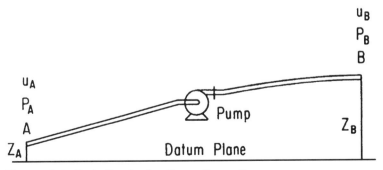

Figure 4.4 Illustration for the Bernoulli equation

where

- u_A, u_B = velocities of fluid, ft/s [m/s] at A and B, respectively
- Z_A, Z_B = liquid heights at A and B with respect to a datum plane, ft[m]
- P_A, P_B = pressures at A and B, lb$_f$/ft^2 [N/m^2]
- ρ_A, ρ_B = densities of fluids at A and B, lb/ft^3 [kg/m^3]
- $u^2/2g_c$ = kinetic head, ft · lb$_f$/lb [N · m/kg]
- F = energy loss due to friction, ft · lb$_f$/lb [N · m/kg]
- W & Q = work done by the pump on the fluid, and heat added respectively, ft · lb$_f$/lb [N · m/kg]
- α = Kinetic energy correction factor

The various terms in the Bernoulli equation must be expressed in the same units as energy per unit mass (ft · lb$_f$/lb or N · m/kg). Normally $g/g_c \doteq 1$ in AES units on earth, and therefore g/g_c can be dropped from the terms containing Z for problem solving on earth using AES units, but not SI units, because the numerical value of g/g_c in SI is 9.81 on earth. Also the kinetic energy correction factors α_A and α_B [both dimensionless] are approximately 0.5 for laminar flow of Newtonian fluid in circular pipe and 0.94 for highly turbulent flow. The term F accounts for the irreversible frictional pressure loss for the entire flow path, ft · lb$_f$/lb in AES units. Unless a problem is solved in SI units, the factor (g/g_c) will be taken as 1 (lb$_f$/lb) in the rest of the book. Use the value of (g/g_c) = 9.81 N/kg for solving problems in SI units. The conversion factor for heat and work in AES: 1 Btu = 778.2 ft · lb$_f$, in SI: 1 J = 1 N · m.

The kinetic energy correction factor, α, may be defined as

$$\alpha = \frac{\text{Plug flow kinetic energy transport rate}}{\text{True kinetic energy transport rate}} \quad (4.11a)$$

Example 4.1

As shown in Exhibit 1, water flows through the pipe. If the contraction loss is half a velocity head based on the velocity at B, calculate the velocity in feet per second and diameter in inches at B. Neglect the pressure drop due to friction.

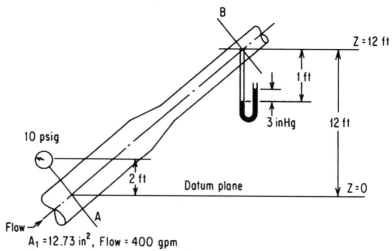

Exhibit 1 Data illustration for Example 4.1

Solution

Apply the Bernoulli equation between the points A and B. There is no pump in the line; therefore $W = 0$. Assume α_A and $\alpha_B = 1$. The frictional drop is given to be negligible. Also assume $F = 0$. Hence

$$Z_A + \frac{u_A^2}{2g_c} + \frac{P_A}{\rho} = Z_B + \frac{u_B^2}{2g_c} + \frac{P_B}{\rho} + \frac{0.5u_B^2}{2g_c}$$

$Z_A = 0$ (taken as datum plane).

The cross-sectional area at point A is computed to be

$$\frac{12.73 \text{ in.}^2}{144 \text{ in.}^2/\text{ft}^2} = 0.0884 \text{ ft}^2$$

Flow = 400 gpm

$$u_A = \frac{400 \text{ gpm}}{(7.48 \text{ gal/ft}^3)(60 \text{ s/min})} \frac{1}{0.0884 \text{ ft}^2} = 10.08 \text{ ft/s}$$

$P_A = 10$ psig + 2 ft
$= 24.7(2.31) + 2$ ft $= 59.06$ ft of water
$Z_B = 12$ ft u_B = required

$$3 \text{ in. Hg} = 13.6\left(\frac{3}{12}\right) = 3.4 \text{ ftH}_2\text{O}$$

From fluid statics, the pressure head at point B is equal to

$$P_B = 3 \text{ in. Hg} + 14.7 \text{ psi} - 1 \text{ ftH}_2\text{O}$$
$$= 3.4 + 33.96 - 1 = 36.36 \text{ ft}$$

Take $g_c = 32.17 \doteq 32.2$ for calculation. Substitution in the Bernoulli equation gives

$$0 + \frac{10.08^2}{64.4} + 59.06 = 12 + \frac{1.5u_B^2}{64.4} + 36.36$$

$$\frac{1.5u_B^2}{64.4} = 12.28 \quad u_B = 22.96 \text{ ft/s}$$

$$\text{volumetric flow rate} = \frac{400}{60(7.48)} = 0.8913 \text{ ft}^3/\text{s}$$

$$\text{Cross section at } B = \frac{0.8913}{22.96} = 0.0388 \text{ ft}^2 = 5.59 \text{ in}^2$$

$$\text{Diameter at } B = \left(\frac{5.59}{0.785}\right)^{0.5} = 2.67 \text{ in}$$

FLUID MEASUREMENTS

Static Pressure

The static pressure is measured by a piezometer opening or a pressure tap. The piezometer opening in the side of the conduit should be normal to and flush with the surface. A *piezometer ring* is a manifold into which are connected several sidewall static taps. Its advantages are (1) it gives average pressure; (2) it reduces the possibility of completely plugging all the static openings. The specifications for pressure tap holes are given by Perry.[1a]

Stagnation Pressure

The stagnation pressure is the pressure of a fluid attained when the fluid is decelerated to zero velocity in a reversible and adiabatic process (isentropic) with no elevation change. Because the stagnation pressure is used as the inlet pressure of a relief device in the sizing of the device, it is important to understand the definition.

$$P_{stagnation} = \begin{cases} P_{static} + \dfrac{\rho u_0^2}{2g_c} \text{ for incompressible flow \& compressible flow with } N_{Ma} < 0.4 \\[2ex] P_{static} + \dfrac{\rho u_0^2}{2g_c}\left[1 + \dfrac{N_{Ma}^2}{4} + \left(\dfrac{2-k}{24}\right)(N_{Ma}^4) + \cdots\right] \\[2ex] \text{for compressible flow with } N_{Ma} \geq 0.4. \text{ Where} \\[2ex] k = \dfrac{C_p}{C_v} = \dfrac{\text{molar specific heat of fluid at constant pressure}}{\text{molar specific heat of fluid at constant volume}} \end{cases} \quad (4.12)$$

In Equation 4.12, $P_{stagnation}$ is in lb_f/ft^2 or N/m^2 when appropriate units are used for the associated variables.

Measurement of Local Velocities, Pitot Tubes

Consider a Pitot tube (named after it's inventor Henry Pitot) of the type shown in Figure 4.5a. Apply the Bernoulli equation to the points A and B with the kinetic enrgy correction factor as unity and no heat effect.

$$\frac{u_A^2}{2g_c} + \frac{P_A}{\rho_A} + Z_A \frac{g}{g_c} + W - F = \frac{u_B^2}{2g_c} + \frac{P_B}{\rho_B} + Z_B \frac{g}{g_c}$$

In the above equation, $Z_A \doteq Z_B$, $u_A = 0$, $W = 0$, $F = 0$.

[*Note:* In Fig. 4.5a, the point A is at the tip of the Pitot tube opposing the direction of the flow. Because the streamline-point hitting the tip of the tube has no way out through the tube, it is decelerated to zero velocity. The additional pressure created at the expense of the velocity pressure adds to the local static pressure, and their sum is called the impact pressure or stagnation pressure as defined in Equation 4.12. By subtracting the local static pressure from the generated stagnation pressure, Equation 4.14 computes the local velocity.]

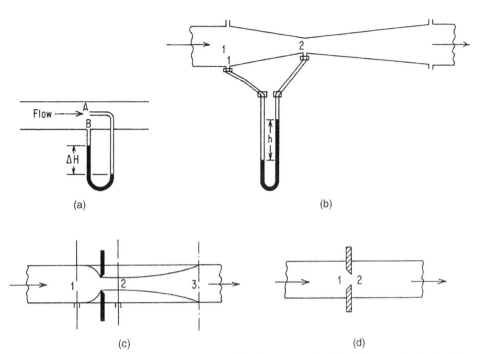

Figure 4.5 (a) Pitot tube, (b) venturimeter, (c) sharp-edged orifice, (d) round-edged orifice

Therefore

$$u_B^2 = 2g_c\left(\frac{P_A}{\rho_A} - \frac{P_B}{\rho_B}\right)$$

or

$$u_B = C\sqrt{2g_c\left(\frac{P_A}{\rho_A} - \frac{P_B}{\rho_B}\right)} \qquad (4.13)$$

The constant C is introduced in the preceding equation to account for the fluctuations in the velocity, which cause errors in the measurement. For the incompressible fluids, $\rho_A = \rho_B = \rho$, and the preceding equation becomes

$$u_B = C\sqrt{2g_c(\Delta H)\frac{g}{g_c}} = C\sqrt{2g(\Delta H)} = C\sqrt{2g_c\frac{P_i - P_{static}}{\rho}} \qquad (4.14)$$

where
    u_B = local velocity at the point where tip is located, ft/s [m/s]
    C = correction coefficient, dimensionless
    ΔH = differential head, ft · lb_f/lb [N · m/kg] = $P_i/\rho - P_{static}/\rho$
    $P_A = P_i$ = impact pressure, lb_f/ft² [N/m²]
    $P_B = P_{static}$ = local static pressure, lb_f/ft² [kg/m³]
    ρ = density of fluid, lb/ft³ [kg/m³]

The coefficient $C = 1 \pm 0.01$ for simple Pitot tubes and $C = 0.98$ to 1 for Pitot static tubes.

For gases at velocities >200 ft/s, the compressibility is important and the following equation is to be used:

$$u_0 = C\left\{\frac{2g_c k}{k-1}\frac{P_{static}}{\rho_{static}}\left[\left(\frac{P_i}{P_{static}}\right)^{(k-1)/k} - 1\right]\right\}^{1/2} \quad (4.15)$$

where u_0 is the velocity at the tip of the Pitot tube, ft/s [m/s], k is (molar specific heat of the gas at constant pressure/molar specific heat of the gas at constant volume), P_{static} is local static pressure, lb_f/ft^2 [N/m^2], ρ_{static} is the fluid density in lb/ft^3 [kg/m^3] at P_{static} and the local temperature.

Venturimeters

A venturimeter (Fig. 4.5b) consists of a tube with a constricted throat, which increases the velocity of the fluid at the expense of pressure. The constriction is followed by a gradually diverging portion where the velocity is decreased, with an increase in the pressure accompanied by slight friction losses.

Writing the Bernoulli equation for an ideal case (no friction losses and heat addition or compression work) gives

$$\frac{P_1}{\rho_1} + Z_1\frac{g}{g_c} + \frac{u_1^2}{2g_c} = \frac{P_2}{\rho_2} + \frac{u_2^2}{2g_c} + Z_2\frac{g}{g_c} \quad \text{and} \quad Z_1 = Z_2$$

By the continuity equation, $u_1 = (A_2/A_1)(\rho_2/\rho_1)u_2$. Then ideal u_2 becomes

$$u_2 = \left[\frac{2g_c(P_1/\rho_1 - P_2/\rho_2)}{1 - (A_2/A_1)^2(\rho_2/\rho_1)^2}\right]^{1/2} \quad (4.16)$$

where A_1 and A_2 are the cross sections at points 1 and 2, respectively. Because of the frictional losses, the actual velocity will be smaller, and therefore a discharge coefficient is introduced in the preceding equation. In addition, if ρ is constant, the preceding equation reduces to

$$u_2 = C\left[\frac{2g_c(P_1 - P_2)/\rho}{1 - (A_2/A_1)^2}\right]^{1/2}$$

$$u_2 = C\sqrt{\frac{2g_c \Delta H\left(\frac{g}{g_c}\right)}{1 - \beta^4}} = C\sqrt{\frac{2g\Delta H}{1 - \beta^4}} \quad (4.17a)$$

where
$\beta^2 = A_2/A_1 = D_2^2/D_1^2$
$D_1, D_2 = $ diameters at points 1 and 2, respectively
$\Delta H = $ differential head, ft[m]

Please note how g_c changes to g when pressure head is changed to static head of fluid. While it numerically makes no difference whether one uses g_c or g in solving problems on earth's surface in AES units, it does make a difference in SI units because the two terms g_c and g are not numerically the same in SI units. The velocity u_2 is in ft/s [m/s].

The volumetric flow rate through a venturi throat is given by

$$q = CYA_T \sqrt{\frac{2g\Delta H}{1-\beta^4}} \text{ ft}^3/\text{s, [m}^3/\text{s]} \tag{4.17b}$$

where Y is an expansion factor and A_T is the throat area, ft² [m²]. For the flow of liquids, $Y = 1.0$.

The change in the potential energy for the inclined pipes needs to be allowed for. The flow equation is then to be modified to

$$q = CA_T \sqrt{\frac{2g\Delta H + 2g(Z_1 - Z_2)}{1-\beta^4}} \text{ ft}^3/\text{s, [m}^3/\text{s]} \tag{4.18}$$

where
Z_1 = Vertical height of upstream pressure tap point from datum line, ft, [m] of fluid
Z_2 = Vertical height of downstream pressure tap point from datum line, ft, [m] of fluid

Mass flow rate can be obtained by using the equation

$$W = q\rho \text{ lb/s, [kg/s]} \tag{4.19}$$

where ρ is the density of fluid at the flowing condition, lb/ft³ [kg/m³].

For gases, the expansion factor Y is not unity and the values of Y for subsonic flow are best obtained from

$$Y = \left[\left(r^{\left(\frac{2}{k}\right)}\right)\left(\frac{k}{k-1}\right)\left(\frac{1-r^{\left(\frac{k-1}{k}\right)}}{1-r}\right)\left(\frac{1-\beta^4}{1-\beta^4\left(r^{\frac{2}{k}}\right)}\right)\right]^{0.5} \tag{4.20}$$

where

$$r = \frac{\text{downstream pressure in absolute}}{\text{upstream pressure in absolute}}$$

$$k = \frac{\text{molar specific heat at constant presure}}{\text{molar specific heat at constant volume}}$$

For compressible flow, if the value of r is lower than r_c, then r_c should be used from Equation 4.104.

For the venturi tubes, $C = 0.98$ in most cases if Reynolds number, $N_{Re} >$ 10,000. For definition of Reynolds number, refer to Equation 4.27.

Orificemeters

An orifice (Fig. 4.5c and 4.5d) is a simple, flat plate with a central opening. The contraction of a stream flowing through an orifice is quite large. The point of the minimum cross section, *vena contracta,* is one or two diameters downstream from the orifice plate. For an orifice plate, the velocity through the orifice is given by

$$u_0 = C_0 Y \sqrt{\frac{2g(\Delta H)}{1-\beta^4}} \text{ ft/s, [m/s]} \tag{4.21}$$

and the mass flow rate lb/s [kg/s] can be calculated by

$$W = u_0 \rho A_0 = \rho A_0 C_0 Y \sqrt{\frac{2g(\Delta H)}{1-\beta^4}} \tag{4.22}$$

where ρ is the density of fluid at the flowing condition, lb/ft^3 [kg/m^3] and β is the usual diameter ratio and Y is the expansion factor. The factor Y is unity for liquids. For gases, Y can be obtained from a plot.[1b] Alternatively, Y may be approximately calculated for subsonic flow ($r_c < r < 1.0$) from[1c]

$$Y = 1 - \frac{0.41 + 0.35\beta^4}{k}(1-r) \tag{4.23}$$

The coefficient C_0 is a function of N_{Re} and β. These values can be obtained from a plot[1c] of C_0 versus N_{Re} with β as a parameter.

Mass Flowmeters–Operating Principle

The principle of the mass flowmeter is based on Newton's second law of angular motion,

$$\frac{dM_a}{dt} = \frac{dm}{dt}\left(r_g^2 \omega\right) \tag{4.24}$$

where

$\dfrac{dM_a}{dt}$ = rate of change of angular momentum

$\dfrac{dm}{dt}$ = mass flow rate

r_g = radius of gyration

ω = angular velocity

Thus if an angular momentum is introduced into a fluid, and the resultant torque and angular velocity are measured, the mass flow rate can be determined. One such mass flow meter belongs to the Coriolis/Gyroscopic flowmeter. This meter uses a C-shaped pipe and a T-shaped leaf spring as the opposite ends of a tuning fork. In the meter, the angular momentum is transmitted by a harmonic vibration of the tube. This results in forces that are proportional to the product of fluid density and mass velocity. This force causes a measurable effect on the tube wall. Each straight section of the tube has a detector. Fluid travels in two opposite directions in the straight sections of the tube, thereby causing an oscillatory twisting around the horizontal axis. This twist along the horizontal axis is small compared with the main vibration around the vertical axis, but it causes a slight difference in phase between the detectors. This phase difference is proportional to the mass flow rate through the tube. The meter is accurate (inaccuracy is ½ to

1% of actual flow rate), but the accuracy is not as good if air bubbles are present. Careful calibration is required for shear-thinning materials.

Vortex Shedding Flowmeter—Operating Principle

When a bluff body (a blunt body with sharp corners) is placed as an obstacle in a moving fluid, the boundary layer separation takes place right behind the body. The separated layers become detached from the main streams, and form vortices in the low-pressure area behind the body. When a vortex is shed from one side of the body, the fluid velocity on that side increases, but the pressure decreases. On the opposite side, a reverse phenomenon takes place: Fluid velocity decreases, but pressure increases. This causes a net pressure change across the bluff body. The process of vortex shedding is reversed, and the shedding takes place from the opposite side. A good example of vortex shedding is the fluttering of a flag.

Detectors are used to measure the oscillating flow (a heated thermistor) or the oscillating pressure (metal diaphragm) across the sides, and convert the signal into measurable flow rate. The meter may be used for gas and liquid services. Inaccuracy is +/–0.75 percent of the rate.

PERMANENT PRESSURE LOSS

Venturi

Permanent pressure loss through venturis depends on the diameter ratio β and the discharge cone angle α. The pressure loss for smaller angles (5 to 7°) is 10 to 15 percent of the total pressure differential $(P_1 - P_2)$. It is 10 to 30 percent for large angles (>15°).

Subsonic Flow Nozzles

For subsonic flow nozzles, the pressure loss is given by

$$\text{Pressure loss} = \frac{1-\beta^2}{1+\beta^2}(P_1 - P_2) \qquad (4.25)$$

where β is the nozzle throat diameter divided by the pipe diameter.

Concentric Circular Nozzles

In the case of the concentric circular nozzle,

$$\text{Permanent pressure loss} = (1 - \beta^2)(P_1 - P_2) \qquad (4.26)$$

Example 4.2

A venturimeter is to be installed in a schedule 40, 6-in. line to measure the flow of water. The maximum rate is expected to be 800 gpm at 86°F. A 50-in. Hg manometer is to be used. Specify the throat diameter of the venturi and calculate the power required to operate it. The discharge cone angle is 5°.

Solution

From Equation 4.16,

$$u_T = C\sqrt{\frac{2g(\Delta H)}{1-\beta^4}}$$

$C = 0.98$ for venturis if $N_{Re} > 10{,}000$ (i.e., fully turbulent flow). Assume fully turbulent flow. Then

$$\text{Flow} = \frac{800}{60(7.48)} = 1.7825 \text{ ft}^3/\text{s}$$

$$\text{ID} = 6.6025 \text{ in} = 0.5052 \text{ ft}$$

$$\Delta H = h_m \left(\frac{s_m}{s} - 1\right) \text{ from Eq. (4-7)}$$

$$= \frac{50}{12}(13.6 - 1) = 52.5 \text{ ft}$$

$$u_T = 0.98 \sqrt{\frac{64.4(52.5)}{1 - \beta^4}}$$

$$\text{Flow } u_T A_T = 0.98 \, A_T \sqrt{\frac{64.4(52.5)}{1 - \beta^4}} = 1.7825$$

$$\frac{A_T}{\sqrt{1 - \beta^4}} = \frac{1.7825}{0.98\sqrt{64.4(52.5)}} = 0.03128$$

$$A_T = \frac{1}{4}\pi D_T^2 = 0.7854 D_T^2$$

$$\frac{0.7854 \, D_T^2}{\sqrt{1 - (D_T/D_P)^4}} = 0.03128$$

where D_P is the inside diameter of pipe, ft. Squaring both sides, one obtains

$$\frac{0.617 D_T^4}{1 - (D_T/D_P)^4} = 0.0009784$$

Simplifying and solving for D_T gives

$$D_T = 0.198 \text{ ft} = 2.38 \text{ in}$$

Use 2.375 in. = 0.1979 ft (a standard size). Check the Reynolds number through the throat of the venturi.

$$u_T = \frac{1.7825}{0.7854(0.1979)^2} = 57.95 \text{ ft/s}$$

$$N_{Re} = \frac{D_T u_T \rho}{\mu} = \frac{0.1979(57.95)(62.4)}{0.85(0.000672)} = 1.253 \times 10^6$$

This N_{Re} is greater than 10,000 and therefore the flow is in the fully turbulent region. Therefore, the size of the venturi is adequate.

Next calculate the permanent pressure loss and the power for operation:

$$\beta = \frac{D_T}{D_P} = \frac{2.375}{6.0625} = 0.3918 \qquad \beta^4 = 0.02355$$

Substitution in the equation for u_T gives

$$57.95 = 0.98\sqrt{\frac{64.4\,\Delta H}{1-0.02355}}$$

$$\Delta H = \frac{(57.95)^2(1-0.02355)}{(0.98)^2(64.4)} = 53.02 \text{ ft of } H_2O$$

From Equation 4.7,

$$\Delta P = \rho(\Delta H)\frac{g}{g_c} = 62.3\,\frac{\text{lb}}{\text{ft}^3}(53.02\text{ ft})\left(\frac{\text{lb}_f}{\text{lb}}\right)$$

$$= 3305\,\frac{\text{lb}_f}{\text{ft}^2}$$

The discharge cone angle is 5°, and the permanent pressure loss for the venturi may be taken as 10 percent of the ΔP.

$$\text{Pressure loss} = 0.1(3305) \doteq 331 \text{ lb}_f/\text{ft}^2$$

and

$$\text{Flow rate} = 800/7.48 = 106.95 \text{ ft}^3/\text{min}$$

$$\text{Power required to operate venturi} = \frac{106.95\,\frac{\text{ft}^3}{\text{min}}(331)\,\frac{\text{lb}_f}{\text{ft}^2}}{33{,}000\,\frac{\text{ft}\cdot\text{lb}_f/\text{min}}{\text{hp}}}$$

$$= 1.07 \text{ hp} = 0.8 \text{ kW}$$

Example 4.3

Natural gas (viscosity = 0.011 cP) is flowing through a 6-in. schedule 40 pipe equipped with a 2-in. orifice with flanged taps. The gas is at 90°F and 20 psia at the upstream tap. The manometer reading is 50 in. H_2O at 60°F. k for natural gas is 1.3. Calculate the rate of flow of the gas through the line in pounds per hour. Assume that the molecular weight of the gas is 16.

Solution

As defined in Equation 4.21, the velocity of gas through the orifice is

$$u_0 = C_0 Y\sqrt{\frac{2g\Delta H}{1-\beta^4}}$$

Because C_0 and Y are not known, a trial-and-error solution is required. Assume $C_0 = 0.61$ (based on the assumption of fully turbulent flow). Also

$$\beta = \frac{2}{6.065} = 0.3229 \qquad \beta^4 = 0.01184 \qquad \beta^2 = 0.109 \doteq 0.11$$

The average density of the gas is

$$\rho_1 = \frac{16}{359}\left(\frac{20}{14.7}\right)\left(\frac{492}{460+90}\right) = 0.0543\,\text{lb/ft}^3$$

From Equation 4.7,

$$\Delta H = h_m\left(\frac{S_m}{S}-1\right) = \frac{50}{12}\left(\frac{62.4}{0.0543}-1\right) = 4784\,\text{ft}$$

$$\Delta P = \rho_1 \Delta H \frac{1}{144} = 1.8\,\text{psi}$$

$$P_2 = 20 - 1.8 = 18.2\,\text{psia}$$

$$r = \frac{P_2}{P_1} = \frac{18.2}{20} = 0.91$$

$$\text{Critical pressure ratio} = r_c = \left(\frac{2}{k+1}\right)^{\frac{k}{k-1}} = \left(\frac{2}{1.3+1}\right)^{\frac{1.3}{0.3}} = 0.55$$

Because $r > r_c$, flow is subcritical.

Y can be calculated approximately by Equation 4.23 as follows:

$$Y = 1 - \frac{0.41 + 0.35\beta^4}{k}(1-r)$$

$$= 1 - \frac{0.41 + 0.35(0.01184)}{1.3}(1-0.91) = 0.9713$$

$$u_0 = 0.61(0.9713)\sqrt{\frac{64.4(4784)}{1-0.01184}} = 330.9\,\text{ft/s}$$

$$N_{Re} = \frac{Du\rho}{\mu} = \frac{\frac{2}{12}(330.8)(0.0543)}{0.011(0.000672)} = 404{,}997$$

Because the Reynolds number is very high and in the fully turbulent region,[1b]

$$C_0 = 0.61$$

and

$$T_2 = T_1\left(\frac{P_2}{P_1}\right)^{(k-1)/k} = 550(0.91)^{0.3/1.3} = 538°R$$

$$\rho_2 = 0.0543\left(\frac{550}{538}\right)\left(\frac{18.20}{20}\right) = 0.505\,\text{lb/ft}^3$$

$$\text{Average } \rho = \frac{1}{2}(0.0543 + 0.0505) = 0.0524\,\text{lb/ft}^3$$

$$\text{Flow of gas} = 330.8(0.7854)\left(\frac{2}{12}\right)^2(0.0524)(3600) = 1361\,\text{lb/h}$$

Example 4.4

A pitot tube is inserted in a 30-in.-ID duct carrying air so that the tip is located at the center of the duct and is aimed in the direction of the flow. The manometer reading is 1.1 in. H_2O, and the static pressure at the point of measurement is 33 in. H_2O. Calculate the flow of the air in cubic feet per minute if the temperature of the air is 100°F and the viscosity of air is 0.02 cP.

Solution

For a pitot tube, velocity in the pipe is given by Equation 4.14.

$$u = \sqrt{2g(\Delta H)}$$

$$\text{Air pressure} = 14.7 + \frac{33}{407}(14.7) = 15.9 \text{ psia}$$

$$1 \text{ atm} = 14.7 \text{ psia} \doteq 407 \text{ inH}_2\text{O}$$

$$\text{Temperature of air} = 460 + 110 = 570°\text{R}$$

$$\text{Density}^* \text{ of air } \rho = \frac{29}{359}\left(\frac{15.9}{14.7}\right)\left(\frac{492}{570}\right) = 0.0754 \text{ lb/ft}^3$$

From the manometer reading, $\Delta P = (1.1/12)(62.4)(g/g_c) = 5.72 \text{ lb}_f/\text{ft}^2$ where the height of the air column is ignored. Because the tip of the pitot tube is at the center,

$$u_{max} = 0.98\sqrt{2g_c \frac{\Delta P}{\rho}} = 0.98\sqrt{\frac{(64.4)(5.72)}{0.0754}} = 68.5 \text{ ft/s}$$

Neglecting the compressibility correction, the Reynolds number is given by

$$N_{Re} = \frac{\frac{30}{12}(68.5)(0.0754)}{0.02(0.000672)} = 960{,}733$$

At this value of the Reynolds number,[1d]

$$\frac{u}{u_{max}} = 0.82$$

Therefore $\quad u_{av} = 0.82(68.5) = 56.17 \text{ ft/s}$

Then $\quad \text{Airflow} = 0.7854(2.5)^2 (56.17)(60) = 16{,}544 \text{ ft}^3/\text{min}$

FLOW OF FLUIDS IN PIPES

The nature of the flow of a fluid in a pipe depends on the Reynolds number, which is defined as

$$N_{Re} = \frac{Du\rho}{\mu} = \frac{DG}{\mu} \qquad (4.27)$$

where
- D = inside diameter of pipe, ft
- u = velocity of fluid, ft/s
- ρ = density of fluid, lb/ft^3
- G = mass velocity, lb/h · ft^2
- μ = viscosity, lb/ft · s for $Du\rho/\mu$ and lb/ft · h for DG/μ

---

*The density can also be calculated as $\rho = PM_w/RT = 15.9(29)/10.73(570) = 0.0754 \text{ lb/ft}^3$.

Consistent Units of Variables to Calculate Dimensionless Reynolds Number

The following table shows how to select the appropriate units of variables so that the Reynolds number is dimensionless.

Variable	$N_{Re} = \dfrac{Du\rho}{\mu}$		$N_{Re} = \dfrac{DG}{\mu}$	
	AES Unit	SI Unit	AES Unit	SI Unit
D	ft	m	ft	M
u	ft/s	m/s	—	—
ρ	lb/ft³	kg/m³	—	—
G	—	—	lb/(ft² · h)	kg/(m² · s)
μ	lb/(ft · s)	Pa · s or kg/(m · s)	lb/(ft · h)	Pa · s or kg/(m · s)

Some useful conversion factors: 1 cP = 0.000672 lb/(ft · s) = 2.42 lb/(ft · h) = 1 mPa · s
1 Pa · s = 1 kg/(m · s)

Flow regimes are defined as follows:

$$N_{Re} \begin{cases} <2100 & \text{viscous flow} \\ >4000 & \text{turbulent flow} \\ =2100-4000 & \text{transition region} \end{cases}$$

Distribution of Velocities

For fluids flowing through a pipe, the velocity distribution will depend on the type of flow. For the laminar or viscous flow, the velocity distribution is truly parabolic.

Laminar Flow of Newtonian Fluids in Cylindrical Pipes

Starting from the definition of viscosity and considering the shear stress of an incompressible fluid through the tube, it can be shown that

$$\frac{u}{u_{max}} = 1 - \left(\frac{r}{r_w}\right)^2 \qquad (4.28)$$

where u is the local velocity at r and u_{max} is the maximum velocity at $r = 0$ (center of tube). For the laminar flow of a fluid in a tube, the average velocity is given by[1d]

$$\frac{u_{av}}{u_{max}} = 0.5 \qquad (4.29)$$

Turbulent Flow

For turbulent flow,[1d]

$$\frac{u_{av}}{u_{max}} \doteq 0.82 \qquad (4.29\text{a})$$

The relationship between u_{av}/u_{max} versus Reynolds number $Du\rho/\mu$ is available in a graphical form.[1d]

Frictional Losses in Circular Pipes

The frictional losses for flowing fluids are a function of Reynolds number. ΔP and the head loss from friction are given by

$$\Delta P_f = \frac{4u^2 L \rho}{2g_c D} \phi\left(\frac{Du\rho}{\mu}\right)$$

$$\Delta H_f = \frac{4u^2 L}{2g_c D} \phi\left(\frac{Du\rho}{\mu}\right)$$

(4)

A more common method is to use the Fanning equation, which expresses pressure drop in terms of a friction factor.

$$\Delta P_f = \frac{4fu^2 L \rho}{2g_c D} = \frac{2fu^2 L \rho}{g_c D} \quad \text{lb}_f/\text{ft}^2, [\text{N/m}^2]$$

$$\Delta H_f = \frac{2fu^2 L}{g_c D} = h_L \quad \frac{\text{ft} \cdot \text{lb}_f}{\text{lb}}, \left[\frac{\text{N} \cdot \text{m}}{\text{kg}}\right]$$

(4.31)

where f, the Fanning friction factor, may be obtained from Figure 4.6 and L is the equivalent length of the pipe in feet [m].

For the laminar flow, if the Fanning equation is combined with the Hagen-Poiseuille equation, the following relation for the friction factor results:

$$f = \frac{16}{N_{Re}}$$

(4.32)

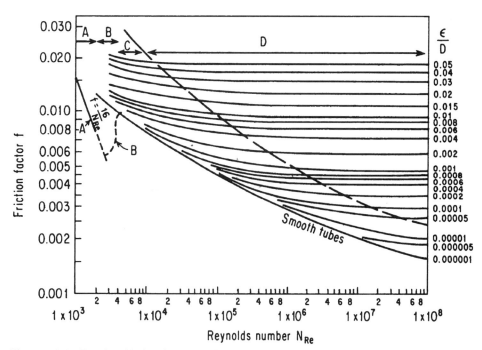

Figure 4.6 Fanning friction factor vs. Reynolds number. (Reprinted from *Chemical Engineer's Handbook*, 5th ed., McGraw-Hill Book Company, New York, 1973, with permission.) (A) Laminar region; (B) critical region; (C) transition region; (D) region of complete turbulence

For both the turbulent flow and the laminar flow, the friction factor f may be obtained from the Fanning friction factor chart given in Figure 4.6.

If the upper boundary of the streamline flow is given by $Du\rho/\mu = 2100$, the critical velocity is given by

$$u_{cr} = 2100 \frac{\mu}{\rho D} \quad (4.33)$$

The Fanning friction factor f also depends on the roughness of pipe ε, in ft or m. The roughness of the pipe wall is the texture of the surface that causes more or less friction. The roughness factor has the dimension of length (ft or m). The effect of the roughness of the pipe on the friction factor is shown in the Fanning friction chart by giving the values of f with ε/D [dimensionless] as parameter.

The following explicit form of the Fanning friction factor for complete turbulent flow is independent of Reynolds number, and may be obtained from

$$f_{\text{fullturbulent}} = \frac{1}{16\left[\log\left(\frac{3.7D}{\varepsilon}\right)\right]^2} \quad (4.34)$$

As a rule of thumb, when the Reynolds number exceeds [(nominal diameter of pipe in inch) 10^6], full turbulence is expected.

The Fanning friction factor for turbulent flow is a function of Reynolds number, and, in general, may be obtained from

$$f_{\text{turbelent}} = \frac{1}{2.444\left[\ln\left(\frac{1.35\varepsilon}{D}\right) + \frac{6.5}{N_{Re}}\right]^2} \quad (4.35)$$

The Altshul[7] Fanning friction factor equation is the simplest to use for manual computation. It may be used in laminar flow for Reynolds number above 1000, and in turbulent flow. The value of friction factor obtained from the Altshul equation below the Reynolds number of 1000 is not conservative.

$$f = 0.0275\left[\frac{\varepsilon}{D} + \frac{68}{N_{Re}}\right]^{0.25} \quad (4.36)$$

Finally, the Churchill equation of the Fanning friction factor is valid for both laminar and turbulent conditions. The equation is tedious, but is an excellent form for computer-aided calculation. First, parameters A and B are determined.

$$A = \left[2.457\ln\left(\frac{1}{\left(\frac{7}{N_{Re}}\right)^{0.9} + \frac{0.27\varepsilon}{D}}\right)\right]^{16}$$

$$B = \left(\frac{37530}{N_{Re}}\right)^{16} \quad (4.37)$$

$$f = 2\left[\left(\frac{8}{N_{Re}}\right)^{12} + \frac{1}{(A+B)^{1.5}}\right]^{\frac{1}{12}}$$

The Churchill equation may be used to benchmark any new equation for friction factor.

Hydrodynamic Entry Length

The hydrodynamic entry length is the minimum straight length, L, of pipe required to develop the parabolic velocity profile, and it is a function of Reynolds number.

$$\frac{L}{D} = \begin{cases} 0.05 N_{Re} & \text{for laminar flow} \\ 0.623 (N_{Re})^{0.25} & \text{for turbulent flow} \end{cases} \quad (4.38)$$

Note that (L/D) in the preceding equation is dimensionless.

Application of Surface Tension in Fluid Mechanics

The value of surface tension is needed to figure out the droplet size of liquid, bubble rise velocity through a pool of liquid, and wetting rate of a liquid in a vertical surface.

1. To estimate the maximum diameter, medium volume diameter, and the Sauter mean diameter of droplets, the following equations may be used.[11]

$$\frac{x_m}{D} = 57 \left(\frac{Dv\rho}{\mu}\right)^{-0.48} \left(\frac{\mu v}{\sigma}\right)^{-0.18} \quad (4.39)$$

$$x_{mvd} = \frac{x_m}{1+a}$$

Sauter mean diameter = $0.85(x_{mvd})$
where
  x_m = maximum drop diameter, m
  x_{mvd} = diameter which corresponds to 50% of droplet volume, m
  D = nozzle nominal bore diameter, m
  v = velocity through nozzle orifice, m/s
  ρ = liquid density, kg/m^3
  μ = vicosity of liquid, kg/m · s
  σ = surfsce tension, N/m
  a = constant = 0.8803

The preceding equations return diameter in meters; multiply it with 10^6 to get the droplet diameter in microns. The nozzle orifice velocity may be obtained from the manufacturers of the nozzle. For example, the water velocity at the exit of the spray orifice of the Angus Thermospray system ranges from 15 m/s to 25 m/s when used within an operating pressure of 1.4 bar to 3.5 bar. The nozzle orifice may also be sized to give the desired velocity.

The Sauter mean diameter is useful in sizing spray condensers and spray driers.

The droplet diameter may also be used to size the vapor liquid knock-out pot. In this case, the maximum droplet diameter may be determined by

$$d_m = 1600 \left(\frac{\sigma}{\rho_L - \rho_G}\right)^{0.5} \quad (4.40)$$

where

d_m = maximum droplet diameter, micron
σ = surface tension, dyne/cm
ρ_L = liquid density, lb/ft^3
ρ_G = vapor density, lb/ft^3

The average droplet diameter may be estimated by $(0.45)d_m$. Commercial vapor liquid knock-out pot may be designed for the removal of droplets of 400 microns in diameter by proper control of superficial velocity without any mist-eliminating device.

When droplets of lower particle size are to be removed, mist-eliminating devices should be used.

2. The following equation may be used to determine the bubble rise velocity in a pool of liquid.

$$v_\infty = \frac{K[32.17(2.2046)(10^{-3})\sigma(\rho_L - \rho_G)]^{0.25}}{\rho_L^{0.5}}$$

σ = surface tension of liquid, dyne/cm
ρ_L = liquid density, lb/ft^3 (4.41)
ρ_G = vapor density, lb/ft^3
v_∞ = bubble rise velocity, ft/s

The factor K may be determined from the following table for a non-foamy system.

Fluid Model	Characteristics	K
Churn-turbulent	Single component or close-boiling homologues (boiling point difference <80°C), and clean liquid, and viscosity <100 cP	1.53
Bubbly	Dissolved gas or multicomponent with b.p. difference >80°C and clean liquid and viscosity >100 cP	1.18

When a liquid is boiled under fire, the preceding equation is useful to determine the possibility of two-phase flow.

3. The following equation may be used to determine the minimum wetting rate to keep the inside surface of a vertical tube wet.

$$W_{min} = 19.5[\mu s \sigma^3]^{0.2} \quad (4.42)$$

where

W_{min} = minimum wetting rate, lb/h.ft of perimeter
μ = viscosity of liquid, cP
s = specific gravity of liquid
σ = surface tension of liquid, dyne/cm

The preceding equation is used in the design of a falling film evaporator or to determine the minimum spray water density for fire-protection of a piece of equipment.

Equivalent Diameters for Noncircular Conduits

For noncircular conduits, an equivalent diameter for fluid flow is defined as

$$D_e = \frac{4 \text{ (cross-sectional area of flow)}}{\text{wetted perimeter of channel}} = 4r_h \quad (4.43)$$

where the hydraulic radius is

$$r_h = \frac{\text{cross-sectional area of flow}}{\text{wetted perimeter of channel}} \quad (4.44)$$

For a square channel,

$$r_h = \frac{b}{4} \quad \text{and} \quad D_e = b \quad (4.45)$$

where b is the side of the square. For a rectangular channel of sides a and b,

$$r_h = \frac{ab}{2(a+b)} \quad \text{and} \quad D_e = \frac{2ab}{a+b} \quad (4.46)$$

For annular spaces, area $A_c = \frac{1}{4}\pi(D_2^2 - D_1^2)$, wetted perimeter $= D_2 + D_1$, and hence

$$D_e = \frac{4\pi(D_2^2 - D_1^2)}{4\pi(D_2 + D_1)} = D_2 - D_1 \quad (4.47)$$

where D_1 is the outside diameter of the inner pipe and D_2 is the inside diameter of the outside pipe.

The friction-factor relationships for the circular conduits apply also to the noncircular conduits, but the equivalent diameter must be used in calculating the Reynolds number.

Example 4.5

Find the hydraulic radius and equivalent diameter for the section shown in Exhibit 2 if the water is flowing 4 ft 6 in. deep in the channel.

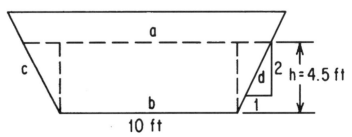

Exhibit 2 Data sketch for Example 4.5

Solution

The hydraulic radius is

$$r_h = \frac{\text{area of cross section of flow}}{\text{wetted perimeter}}$$

$$\text{Area of cross section} = \frac{1}{2}(a+b)h$$

$$\text{Width of top at liquid level} = 10 + 2(4.5)\left(\frac{1}{2}\right) = 14.5 \text{ ft}$$

$$\text{Length of each side } (c \text{ and } d) = \sqrt{4.5^2 + \left(\frac{4.5}{2}\right)^2} = 5.031 \text{ ft}$$

$$\text{Wetted perimeter} = 2(5.031) + 10 = 20.062 \text{ ft}$$

$$\text{Area of cross section} = \frac{1}{2}(14.5+10)(4.5) = 55.125 \text{ ft}^2$$

Then

$$r_h = \frac{55.125}{20.062} = 2.75 \text{ ft}$$

and the equivalent diameter is

$$D_e = 4r_h = 4(2.75) = 11 \text{ ft}$$

Note: The friction factor used in the previous equations is the Fanning friction factor, which is an arbitrary constant and should be used only in conjunction with the Fanning friction-factor chart of Figure 4.6. Other charts are also in use. The most common of these is the Moody-Darcy chart. The Moody friction factor f' is related to the Fanning friction factor f by the following relation:

$$f' = 4f \tag{4.48}$$

A quick way to distinguish between the two friction factor values is that the Fanning friction factor in turbulent flow is approximately 0.005, and the Moody friction factor for turbulent flow is approximately 0.02. All friction factor equations used in this book give the Fanning friction factor.

Frictional Losses through Fittings and Values

Frictional losses occur because of local disturbances in the flow-through conduits. The disturbances are caused by fittings such as bends and elbows and by sudden changes in the direction of the flow caused by obstructions. These losses are generally expressed in terms of a resistance coefficient K, an equivalent length L_e, and a flow coefficient C_v.

Resistance Coefficient *K*

The resistance coefficient K is defined as the number of the velocity heads that are lost because of a fitting or an obstruction. It is assumed to be independent of the friction factor irrespective of either the laminar or turbulent flow. Data on K values for the various fittings, obstructions, and valves are available,[2,1e] and some are given in Table 4.1.

Equivalent Length *Lₑ*

The equivalent length of a fitting or an obstruction is the length of a pipe that causes the same pressure drop due to friction as the fitting or obstruction under consideration. It is usually expressed in terms of the equivalent pipe diameter

Table 4.1 Equivalent lengths and K values

Type	Comment Valves	L_e/D*	L_e/d*
Gate (disk or plug)	Fully open	13	1.1
Globe	Seat flat, bevel or plug	340	28.3
(conventional)	Wing or pin guided disc	450	37.5
Angle	With no obstruction		
(conventional)	(seat flat, bevel, or plug)	145	12.1
	Wing or pin guided disk	200	16.7
Y-pattern globe	Stem at 60° from run of pipe	175	14.6
	Stem at 45° from run of pipe	145	12.1
Ball valve	Fully open (full port)	3	0.25
Butterfly	Sizes 2 in. to 8 in.	45	3.8
	Sizes 10 in. to 14 in.	35	2.9
	Sizes 16 in. to 24 in.	25	2.1
Plug	Straight-way	18	1.5
	3-way (straight run)	30	2.5
	3-way through branch	90	7.5
Foot valve with	Poppet disk	420	35
Strainer	Hinged disk	75	6.25
Swing check	Clearway	50	4.2
	Tilting seat	100	8.34
Tilting disk check	Disk angle, 5° (15°) 2 in. to 8 in.	40 (120)	3.3 (10)
	10 in. to 14 in.	20 (90)	2.5 (7.5)
	16 in. to 48 in.	20 (60)	1.7 (5)
Lift or stop check	Globe lift or stop	450	37.5
	Angle lift or stop	200	16.7

* L_e and D are in feet and d is in inches.

Standard elbow	90°	30	2.5
	45°	16	1.33
	90° long radius	20	1.67
Return bend	180° close pattern	50	4.17
Standard tee	Flow-through run	20	1.67
	Flow-through branch	60	5.0
Elbow 90°	90° $R/D = 1.0$	18	1.5
	$R/D = 1.5$	12	1.0
	$R/D = 2.0$	10	0.83

Welded Fittings			
Elbow 45°	Multiply L_e/D for 90° elbow by 0.64		
Return bend, 180°	Multiply L_e/D for 90° elbow by 1.34		
Tees	100% flow-through run	18	0.7
	100% flow-in through branch	58	4.83
	100% flow-in through branch	43	3.6
Reducer		30	2.5

Miscellaneous Obstructions		
Type	Comment	K
Pipe entrance	With inward projection	0.78
Pipe entrance	Sharp-edged	0.50

(*Continued*)

Table 4.1 Equivalent lengths and K values (*Continued*)

Type	Comment	K
Pipe entrance	Slightly rounded	0.23
Pipe entrance	Well rounded	0.04
Exit from pipe	Projecting type, sharp edged, or rounded	1.0
Sudden contraction	β = small diam./large diam.	$0.5(1-\beta^2)$
Sudden expansion	β = small diam./large diam.	$(1-\beta^2)^2$

Note: The K factor for exit ($K = 1.0$) should be included only when the fluid exits to a confined space because the kinetic energy is dissipated in the confined space. When the fluid exits to unconfined space, such as atmosphere, $K = 0$ because the velocity of the fluid exiting the pipe (free jet velocity) is the same as the velocity of the fluid inside the pipe.

Table 4.1A Roughness factor, ε, ft

Material	Roughness factor, ε, ft
Commercial steel pipe, new	0.00015
Sheet metal	0.0001
Glass/plastic lined pipe	0.00006

Note: average thickness of glass lining = 0.066 in.

L_e/D. Equivalent lengths are available from various sources.[3a] Some values are reproduced in Table 4.1.

Calculation of Equivalent Length of a Series of Piping of Different Diameters and Lengths

If pipes of inside diameter (d) and length (L), such as (d_1, L_1), (d_2, L_2), (d_3, L_3) are put in series, then the equivalent length of the combination may be approximated in terms of single diameter d_1.

Equivalent length in terms of diameter d_1

$$= L_1 + L_2 \left(\frac{d_1}{d_2}\right)^5 \left(\frac{f_2}{f_1}\right) + L_3 \left(\frac{d_1}{d_3}\right)^5 \left(\frac{f_3}{f_1}\right) + \cdots \quad (4.48a)$$

For complete turbulent conditions, the correction for friction factor, f, may be ignored.

Table 4.1B Inside diameter of lined schedule 40 metallic pipe

	Inside Diameter, in	
Nominal Size, in	SL, PPL, KL	TFE
1	0.735	0.785
1.5	1.272	1.342
2	1.723	1.817
3	2.744	2.844
4	3.634	3.798
6	5.629	5.785
8	7.619	7.755

Note: SL = Saran-lined, PPL = Polypropylene-lined, KL = Kynar lined, TFE = Teflon-lined.

Flow Coefficient

The flow coefficient C_v of a control valve is defined as the flow of water at $60°\,F$ in gallons per minute at a pressure drop of 1 $lb_f/in.^2$ across the valve when the valve is completely open. (See Equation 4.92.)

Relation between Flow Coefficient of a Valve and Resistance Coefficient

$$K = \frac{891 d^4}{(C_v)^2} \tag{4.48b}$$

where, d = nominal diameter, in.

Relation between K and L_e

The resistance coefficient K and the equivalent length are related by

$$K = 4f \frac{L_e}{D} \quad \text{or} \quad K = f' \frac{L_e}{D} \tag{4.49}$$

where f is the Fanning friction factor and f' the Moody-Darcy friction factor.

The flow resistance coefficient K is considered independent of friction factor or Reynolds number, and may be treated as constant for a specific valve or fitting of a given size under all flow conditions, laminar or turbulent. The resistance coefficient K, however, varies with size. The variation of K with size is given by

$$K_a = K_b \left(\frac{d_a}{d_b} \right)^4 \tag{4.49a}$$

Because K is constant, (L/D) varies with friction factor as shown in the preceding equation.

Variation of (L/D) factor with size for turbulent flow is as follows:

$$\left(\frac{L}{D} \right)_a = \left(\frac{L}{D} \right)_b \left(\frac{d_a}{d_b} \right)^4 \tag{4.49b}$$

When a rupture disk is a piping component, its flow resistance coefficient, K_R, may be included in the total resistance of the system.

The single-K flow resistance coefficient assumes that K values are constant in all flow conditions: laminar or turbulent.

The single-K method of computing pressure drop, therefore, becomes less accurate at low Reynolds number. To take care of this situation, 2-K (Hooper) and 3-K (Darby) methods have been published.

$$K = \begin{cases} \dfrac{K_1}{N_{Re}} + K_\infty \left(1 + \dfrac{1}{d}\right) & \text{2-}K\text{ (Hooper)} \qquad (4.50) \\[2ex] \dfrac{K_1}{N_{Re}} + K_i \left(1 + \dfrac{K_d}{d_n^{0.3}}\right) & \text{3-}K\text{ (Darby)} \qquad (4.51) \end{cases}$$

where
> d = inside diameter of fitting, in
> d_n = nominal diameter of fitting, in

The 2-K values (K_1 & K_∞) of fittings[10] and 3-K values (K_1, K_i & K_d) of fittings[5] are shown in Tables 4.1C & 4.1D.

The Darby method is the most recent one, and this method has improved the effect of the scaling term ($1/d$) of the 2-K method over a wider range of the sizes of valves and fittings. The 3-K values of fittings, if available, should be used; the 2-K values should be used only when the 3-K values are not available.

Table 4.1C Darby 3-K constants for loss coefficients for valves and fittings[5]

Fittings	Specification	K_1	K_i	K_d
Elbow, 90°	Threaded, standard, ($r/D = 1$)	800	0.14	4.0
	Threaded, long radius, ($r/D = 1.5$)	800	0.071	4.2
	Flanged, welded, ($r/D = 1$)	800	0.091	4.0
	Flanged, welded, ($r/D = 2$)	800	0.056	3.9
	Flanged, welded, ($r/D = 4$)	800	0.066	3.9
	Flanged, welded, ($r/D = 6$)	800	0.075	4.2
Elbow, Mitered	90° (1 weld)	1000	0.27	4.0
	45° (2 welds)	800	0.068	4.1
	30° (3 welds)	800	0.035	4.2
Elbow, 45°	Threaded standard, ($r/D = 1$)	500	0.071	4.2
	Long radius, ($r/D = 1.5$)	500	0.052	4.0
	Mitered, 1 weld (45°)	500	0.086	4.0
	Mitered, 2 welds (22 1/2°)	500	0.052	4.0
Elbow, 180°	Threaded, close return bend, ($r/D = 1$)	1000	0.23	4.0
	Flanged ($r/D = 1$)	1000	0.12	4.0
	All types ($r/D = 1.5$)	1000	0.10	4.0
Tee, through branch	Threaded, ($r/D = 1$)	500	0.274	4.0
	Threaded, ($r/D = 1.5$)	800	0.14	4.0
	Flanged, ($r/D = 1$)	800	0.28	4.0
	Stub-in branch	1000	0.34	4.0
Tee, through run	Threaded, ($r/D = 1$)	200	0.091	4.0
	Flanged, ($r/D = 1$)	150	0.5	4.0
	Stub-in branch	100	0	0
Valve, Angle	45° Full line size, $\beta = 1$	950	0.25	4.0
	90° Full line size, $\beta = 1$	1000	0.69	4.0
Valve, Globe	Standard, $\beta = 1$	1500	1.7	3.6
Valve, Plug	Branch flow	500	0.41	4.0
	Straight through	300	0.084	3.9
	Three-way (flow through)	300	0.14	4.0
Valve, Gate	Standard, $\beta = 1$	300	0.037	3.9
Valve, Ball	Standard, $\beta = 1$	300	0.017	4.0
Valve, Diaphragm	Dam type	1000	0.69	4.9
Valve, Check	Swing check	1500	0.46	4.0
	Lift check	2000	2.85	3.8

r = radius of the arc formed by an elbow or turn. D = internal diameter of fitting, r/D is dimensionless.
β = ratio of port diameter to pipe inside diameter

Table 4.1D Hooper K-factors for contractions, expansions, entrance, and exit
[*Chemical Engineering*, November 7, 1988] [See Fig. 4.7 for illustration of fittings]

Fittings	Explanation	Inlet N_{Re1}	K to be Used with Inlet Velocity Head, $v_{inlet}^2/2g_c$
A. Square Reduction	D_1 = bigger diameter (inlet) D_2 = smaller diameter (outlet) f_1 = Fanning friction factor based on D_1	≤ 2500	$K = \left[1.2 + \dfrac{160}{N_{Re1}}\right]\left[\left(\dfrac{D_1}{D_2}\right)^4 - 1\right]$
		> 2500	$K = [0.6 + 1.92 f_1]\left(\dfrac{D_1}{D_2}\right)^2\left[\left(\dfrac{D_1}{D_2}\right)^2 - 1\right]$
B. Tapered Reduction	θ = tapered cone angle In a tapered reduction, the reduction is gradual, and not sharp as in square reduction.	All	Multiply K from case A by: $\begin{cases} \sqrt{\sin\left(\dfrac{\theta}{2}\right)} & \text{for } 45° < \theta \leq 180° \text{ or} \\ \left[1.6\sin\left(\dfrac{\theta}{2}\right)\right] & \text{for } 0° < \theta \leq 45° \end{cases}$
C. Pipe reducer	A pipe reducer has rounded reduction.	All	$K = \left[0.1 + \dfrac{50}{N_{Re1}}\right]\left[\left(\dfrac{D_1}{D_2}\right)^4 - 1\right]$
D. Thin, sharp orifice	D_1 = Pipe diameter D_2 = Orifice diameter	≤ 2500	$K = \left[2.72 + \left(\dfrac{D_2}{D_1}\right)^2\left(\dfrac{120}{N_{Re1}} - 1\right)\right]\left[1 - \left(\dfrac{D_2}{D_1}\right)^2\right]\left[\left(\dfrac{D_1}{D_2}\right)^4 - 1\right]$
		> 2500	$K = \left[2.72 - \left(\dfrac{D_2}{D_1}\right)^2\left(\dfrac{4000}{N_{Re1}}\right)\right]\left[1 - \left(\dfrac{D_2}{D_1}\right)^2\right]\left[\left(\dfrac{D_1}{D_2}\right)^4 - 1\right]$
E. Thick orifice	D_1 = Pipe diameter D_2 = Orifice diameter L = Orifice plate thickness	All	If $L/D_2 > 5$, K = sum of K in case A & Case F, otherwise multiply K from case D by: $K = \left\{0.584 + \left[\dfrac{0.0936}{(L/D_2)^{1.5} + 0.225}\right]\right\}$
F. Square expansion and pipe expansion with rounded expansion	D_1 = smaller diameter (inlet) D_2 = bigger diameter (outlet) f_1 = Fanning friction factor based on D_1	≤ 2500	$K = 2\left[1 - \left(\dfrac{D_1}{D_2}\right)^4\right]$
		> 4000	$K = [1 + 3.2 f_1]\left\{\left[1 - \left(\dfrac{D_1}{D_2}\right)^2\right]^2\right\}$
G. Tapered Expansion	θ = tapered cone angle In a tapered expansion, the expansion is gradual, and not sharp as in square expansion.	All	If $\theta > 45°$ then use K from case F, otherwise multiply K from case F by $\left[2.6\sin\left(\dfrac{\theta}{2}\right)\right]$
H. Pipe entrance	K_∞ for flush, rounded entrance: r/D: 0.02 0.04 0.06 0.10 0.15&up K_∞: 0.28 0.24 0.15 0.09 0.04	All	$K = \dfrac{160}{N_{Re}} + K_\infty$, $K_\infty = \begin{cases} 1.0 \text{ for inward projecting (Borda)} \\ 0.5 \text{ for flush, sharp, no rounding} \end{cases}$
I. Pipe exit		All	$K = 1.0$

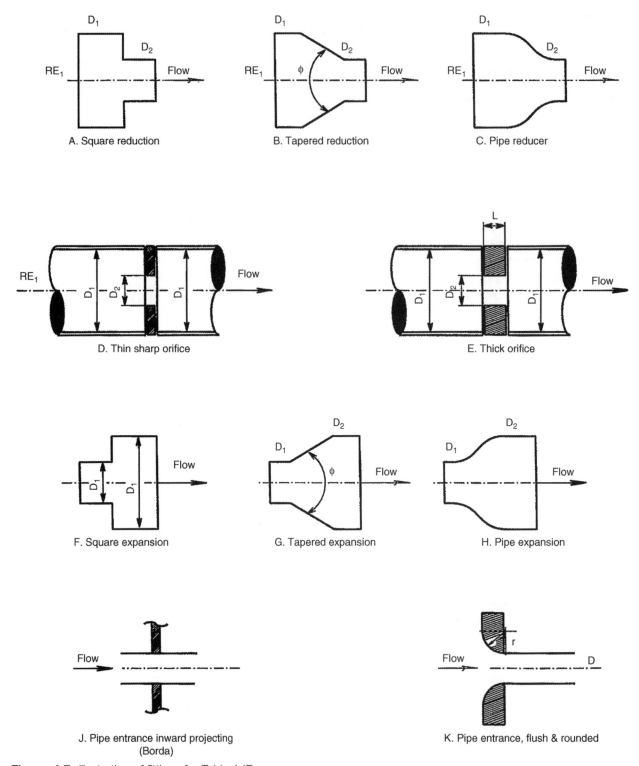

Figure 4.7 Illustration of fittings for Table 4.1D

In general, the irreversible pressure drop due to friction in a pipeline may be computed by

$$\left(\frac{\Delta P_f}{\rho}\right) = \left[\sum K + \frac{4fL}{D}\right]\frac{u^2}{2g_c} \qquad (4.52)$$

where

Variable	Name	AES Unit	SI Unit
ΔP_f	Frictional pressure drop	lb_f/ft^2	N/m^2
K	Sum of flow resistance coefficients	Dimensionless	dimensionless
F	Fanning friction factor	Dimensionless	dimensionless
L	Straight length of pipe	ft	m
D	Inside diameter of pipe	ft	m
u	Velocity of fluid in the pipe	ft/s	m/s
α	Velocity correction factor	See Equation 4.11a, dimensionless	See Equation 4.11a, dimensionless
g_c	Dimensional constant	$32.17\ lb \cdot ft/(s^2 \cdot lb_f)$	$1\ kg \cdot m/(s^2 \cdot N)$
ρ	Density of fluid	lb/ft^3	kg/m^3

Convenient Equations for Reynolds Number and Pressure Drop for Newtonian Fluid

Some convenient forms of equation in terms of commonly used units in the USA [AES units] are shown in the following table.

Reynolds number and pressure drop equations

Reynolds Number	Pressure Drop, psi	
$N_{re} = \dfrac{123 du\rho}{\mu}$	$\Delta P_f = \dfrac{0.005176 fL\rho u^2}{d}$	(4.53a)
$N_{re} = \dfrac{50.65 Q\rho}{\mu d}$	$\Delta P_f = \dfrac{0.000864 fL\rho Q^2}{d^5}$	(4.53b)
$N_{Re} = \dfrac{379\rho q}{\mu d}$	$\Delta P_f = \dfrac{0.0484 fL\rho q^2}{d^5}$	(4.53c)
$N_{Re} = \dfrac{6.32 W}{\mu d}$	$\Delta P_f = \dfrac{0.0000134 fLW^2}{\rho d^5}$	(4.53d)

The calculation of flows through nozzles and orifices may be done by

$$W = 1891\left(d_1^2\right)(C)\sqrt{\Delta P_f(\rho)} \quad \text{for incompressible flow} \quad (4.53e)$$

In the preceding table, d = internal diameter, in., d_1 = orifice diameter, in.; f = Fanning friction factor; L = pipe equivalent length, ft; ΔP_f = frictional pressure loss, psi. It should be the critical pressure drop, psi, that is used for compressible flow when flow is critical or choked. For orifice flow formula, this pressure drop is measured between taps located 1 diameter upstream and 5 diameters downstream. q = flow rate, cuft/min; Q = flow rate, gpm; N_{Re} = Reynolds number, dimensionless; u = velocity, ft/s; W = flow rate, lb/h; μ = viscosity, cP; ρ = fluid density, $lb/ft^3 \cdot \rho_1$ = fluid density at upstream condition, lb/ft^3. Do not use Moody-Darcy friction factor in preceding equations.

For parameter C (orifice coefficient), please see Crane[9] (p. A-20).

Example 4.6

Water is being pumped through the system shown in Exhibit 3. The temperature of the water is 86° F and the flow rate is 110 gpm. Calculate the total frictional pressure drop through the system if the viscosity of water is 0.85 cP.

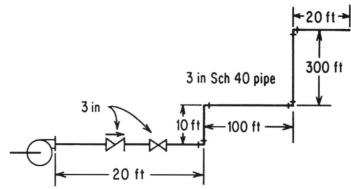

Exhibit 3 Pumping system (Example 4.6). Distances covered by fittings are not included.

Use the K-values for the fittings in the following table.

K-Values	
90° bends	$K = 1.36$
3-in. check valve	$K = 2.00$
3-in. gate valve	$K = 0.23$
Exit enlargement	$K = 1.00$

Solution

$$ID = 3.068 \text{ in.} = 0.2557 \text{ ft}$$
Total length of pipe $= 20 + 10 + 100 + 300 + 20 = 450$ ft

K-Values	
90° bends	$K = 4(0.34) = 1.36$
3-in. check valve	$K = 2.00$
3-in. gate valve	$K = 0.23$
Exit enlargement	$K = 1.00$

$$\frac{\Delta P_f}{\rho} = \frac{\Sigma K u^2}{2g_c} + \frac{2fLu^2}{g_c D}$$

$$= \left(\Sigma K + \frac{4Lf}{D} \right) \frac{u^2}{2g_c}$$

$$\Sigma K = 1.36 + 2 + 0.23 + 1 = 4.59$$

$$u = \frac{110}{7.48(60)(0.7854)(0.2557)^2} = 4.773 \text{ ft/s}$$

$$\frac{Du\rho}{\mu} = \frac{0.2557(4.773)(62.4)}{0.85(0.000672)} = 133{,}327$$

For steel pipe

$$\varepsilon = 0.00015 \text{ ft}, \quad \frac{\varepsilon}{D} = \frac{0.00015}{0.2557} = 0.00059$$

$$f = 0.0048 \quad \text{from Figure 4.6}$$

Alternatively, the friction factor may be computed by any of the explicit equations presented before. Thus,

$$f = \begin{cases} 0.0043 \text{ from Equation 4.34} \\ 0.0046 \text{ from Equation 4.35} \\ 0.0050 \text{ from Equation 4.36} \\ 0.0050 \text{ from Equation 4.37} \end{cases}$$

$$\frac{\Delta P_f}{\rho} = \left[4.59 + \frac{4(450)(0.0048)}{0.2557} \right] \frac{4.773^2}{64.4} = 13.58 \ \frac{\text{ft} \cdot \text{lb}_f}{\text{lb}}$$

$$\Delta P_f = 13.58 \frac{\text{ft} \cdot \text{lb}_f}{\text{lb}} \left(62.4 \frac{\text{lb}}{\text{ft}^3} \right) \frac{1 \text{ ft}^2}{144 \text{ in}^2} = 5.88 \text{ psi}$$

Example 4.7

Solve Example 4.6 with the use of the equivalent lengths. Use the equivalent lengths as shown in the following table.

Item	Equivalent Length, L_e, ft
3-in. elbow	7.5
3-in. check valve	35
3-in. gate valve	3.5
Sudden enlargement	15
Straight pipe length	450

Solution

Calculate the total equivalent length.

Item	Equivalent Length, L_e, ft
Four 3-in. elbows	(4)(7.5) = 30
One 3-in. check valve	35
One 3-in. gate valve	3.5
Sudden enlargement	15
Straight pipe length	450
Total equivalent length	30 + 35 + 3.5 + 15 + 450 = 533.5

$$f = 0.0048 \quad \text{as in Example 4.6}$$

$$\Delta P = \frac{2 f L_e u^2}{g_c D} \left(\frac{\rho}{144} \right) = \frac{2(0.0048)(533.5)(4.773)^2(62.4)}{(32.17)(0.2557)(144)} = 6.10 \text{ psi}$$

Example 4.8

Soda ash liquor having 1250 kg/m³ density and 1.2 cP viscosity flows through a 215 m length of a 150-mm ID steel pipe. The equivalent length of the fittings may be taken as 70 pipe diameters. A venturimeter with a throat diameter of 75 mm installed in the line shows a differential column height of 26 mm on a mercury manometer. What is the flow rate in kilograms per second? What is the total pressure drop through the line if the roughness of the pipe inside surface is 0.0000457 m? Assume the flow coefficient of the venturi is 0.985.

Solution

$$\rho = 1250 \text{ kg/m}^3 \quad \text{Specific gravity} = 1.25$$
$$\mu = 1.2 \text{ cP} = 0.0012 \text{ N} \cdot \text{s/m}^2 \text{ or kg/m} \cdot \text{s}$$

From Equation 4.7,

$$\Delta H = h_m \left(\frac{S_m}{S} - 1 \right)$$

Therefore,

$$\Delta H = 26 \left(\frac{13.6}{1.25} - 1 \right) = 257 \text{ mm of liquid} = 0.257 \text{ m}$$

$$\beta = \frac{D_T}{D_P} = \frac{75}{150} = 0.5 \quad \beta^4 = 0.0625 \quad 1 - \beta^4 = 0.9375$$

u_T = velocity through the throat of the venturi

$$= C \sqrt{\frac{2g\Delta H}{1 - \beta^4}}$$

$$= 0.985 \sqrt{\frac{2(9.81)(0.257)}{0.9375}} = 2.284 \text{ m/s}$$

$$A_T = \text{cross section of venturi} = \frac{1}{4}\pi(0.075)^2 = 0.004418 \text{ m}^2$$

Flow rate = $u_T A_T$
= (2.284 m/s)(0.004418 m²)(1250 kg/m³)
= 12.62 kg/s

Volumetric flow rate = 2.284 (0.004418) = 0.0101 m³/s

Flow cross section of pipe = $\frac{1}{4}\pi(0.15)^2 = 0.01767 \text{ m}^2$

Velocity through pipe = $\frac{0.0101 \text{ m}^3/\text{s}}{0.01767 \text{ m}^2} = 0.572 \text{ m/s}$

$$N_{Re} = \frac{Du\rho}{\mu} = \frac{0.15(0.572)(1250)}{0.0012} = 89,375$$

Roughness factor = $\frac{\varepsilon}{D} = \frac{0.0000457}{0.15} = 0.0003$

From Figure 4.6,

Fanning friction factor = 0.0048
Equivalent length of fittings = 70(150) = 10,500 mm = 10.5 m
Total equivalent length = 215 + 10.5 = 225.5 m

$$\Delta P \text{ through pipe and fittings} = \frac{2fL_e u^2 \rho}{g_c D}$$

$$= \frac{2(0.0048)(225.5)(0.572)^2(1250)}{1(0.15)}$$

$$= 5902 \text{ N/m}^2 = 5.9 \text{ kN/m}^2$$

Assume that the permanent ΔP through the venturi is 15 percent of the pressure differential across the venturi. On this basis, the differential head across the venturi is

$$\Delta P = (\Delta H)\left(\frac{g}{g_c}\right)\rho$$

$$= (0.257 \text{ m})(9.81 \text{ N/kg})(1250 \text{ kg/m}^3)$$

$$= 3151.5 \text{ N/m}^2 = 3.15 \text{ kN/m}^2$$

ΔP through venturi = 0.15(3.15) = 0.47 kN/m²
Total ΔP in the line = 5.9 + 0.47 = 6.37 kN/m²

$$= \frac{6.37}{6.896} = 0.924 \text{ psi}$$

Example 4.9

Water at 86°F is to flow through a horizontal pipe at the rate of 175 gpm. μ = 0.85 cP. A 25-ft head is available. What should be the pipe diameter? Assume ε, the roughness of the pipe, is 0.00015 ft and the length of pipe is 1000 ft.

Solution

From Equation 4.32,

$$h_L = \frac{2fLu^2}{g_c D} \frac{\text{ft} \cdot \text{lb}_f}{\text{lb}} \quad \text{or} \quad D = \frac{2fLu^2}{g_c h_L}$$

u can be calculated in terms of D as

$$u = \frac{175 \text{ gpm}}{(7.48 \text{ gal/ft}^3)(\frac{1}{4}\pi D^2)(60 \text{ s/min})} = \frac{0.497}{D^2} \text{ ft/s}$$

Substitution in the equation for D gives

$$D = \frac{2fL}{g_c h_L}\left(\frac{0.497}{D^2}\right)^2 \quad \text{or} \quad D^5 = \frac{2(0.497)^2(1000)f}{32.2(25)} = 0.6137f$$

Because f is not known, the solution for D is to be obtained by trial and error. A first estimate could be obtained by using the guidelines in Table 4.2. Thus, the

Table 4.2 Line sizing guideline

	Typical ΔP_{100}, psi/100 ft	Typical Velocity, ft/s	Remarks
1. Liquid service			
a. Pump suction	0.05 to 0.5	1 to 6	Use lower ΔP_{100} and velocity for hydrocarbons and boiling liquids. For water and similar service $\Delta P_{100} = 0.5$ to 1 may be used. (Line size is often dictated by the NPSH required by the pump.) Rule of thumb: line size (in.) = $\sqrt{\text{(gallons per min)}/10}$
b. Pump discharge	2 to 6	3 to 14	Final line size should be selected from an economic analysis, which may sometimes lead to $\Delta P_{100} \doteq 8$. Rule of thumb: line size (in.) = $0.25\sqrt{\text{gpm}}$
c. Water header	0.5 to 1	2 to 10	
d. Water lateral	0.5 to 2	2 to 12	
e. Reboiler inlet			
i. Once-through	0.1 to 0.2		Usually dictated by the static head available and flow rate.
ii. Thermosyphon recirculation	0.8 to 1		Usually dictated by the static head available and the required circulation ration. Rule of thumb: cross section of inlet pipe = (0.5) (cross section of all tubes).
f. Gravity flow			For self-venting, line size, in = $0.92 Q^{0.4}$ where Q = flow in gpm.
2. Vapors and gases			
a. Steam			
15–28 in. Hg vacuum	0.05 to 0.2		Use lower ΔP_{100} at lower operating pressure.
15 in. Hg vacuum to 0 psig	0.2 to 0.5		
50–150 psig	1 to 1.5		
150–300 psig	1.5 to 2		
Over 300 psig	3		
b. Vapor and gas			
15–28 in. Hg vacuum	0.025 to 0.05		Use lower ΔP_{100} at lower operating pressure. For operating pressure below 15 in. Hg vacuum, total system pressure drop should not exceed 10% operating pressure (absolute). Line size is often dictated by the total allowable system pressure drop and economics.
15 in. Hg vacuum to 0 psig	0.05 to 0.1		
0–50 psig	0.1 to 0.25		
50–150 psig	0.25 to 0.75		
150–300 psig	0.75 to 1.5		
c. Kettle type reboiler outlet	0.1 to 0.2		
d. Compressor suction		25 to 100	Final line size should be selected from a pressure profile study and economics.
e. Compressor discharge	1 to 2	100 to 200	
3. Two-phase flow			
a. Slurry	—	4 to 10	Never size line on the basis of ΔP_{100}. Velocity should be greater than deposit velocity. For slurries containing solids that tend to stick to the surface when settled, velocities like 16 ft/s are not unusual.

Table 4.2 Line sizing guideline

	Typical ΔP_{100}, psi/100 ft	Typical Velocity, ft/s	Remarks
b. Gas-liquid	—	35 to 75 maximum $u = 100/\sqrt{\rho_{mix}}$	ρ_{mix} = density of mixture in lb/ft³. Prefer dispersed flow. Avoid slug flow.
			Rule of thumb: condensate line size is two sizes smaller than steam line size.
c. Air-solid air/solid (ft³/lb)	—		
10 to 40		90 to 70	Use higher velocity for lower air/solid ratio.
40 to 100	—	70 to 60	
d. Thermosyphon reboiler outlet	0.1 to 1.0	—	Rule of thumb: cross section of outlet pipe = cross section of all tubes.

estimated line size = $0.25\sqrt{175} = 3.3$ in. Because 3.3 in. is not a standard pipe size, assume 4-in. schedule 40 pipe. Alternatively, assume a trial value of f in turbulent flow = 0.005, and calculate D from the preceding equation as 3.77-in. or rounded to 4-in. nominal pipe

$$\text{ID} = 4.026 \text{ in.} = 0.3355 \text{ ft}$$

$$u = \frac{0.497}{(0.3355)^2} = 4.415 \text{ ft/s}$$

$$\frac{Du\rho}{\mu} = \frac{0.3355(4.415)(62.4)}{0.85(0.000672)} = 1.62 \times 10^5$$

$$\frac{\varepsilon}{D} = \frac{0.00015}{0.3355} \doteq 0.00045$$

$$f = 0.0047 \text{ from Fig. 4.7}$$

$$D^5 = 0.6137 (0.0047) = 0.002884$$

$$D = 0.3105 \text{ ft} = 3.73 \text{ in}$$

which is close to assumed 4.026-in. ID. A 4-in.-diam. schedule 40 pipe should be specified.

Example 4.10

Water is flowing through an annular channel at a rate of 25 gpm. The channel is made of $\frac{1}{2}$- and $1\frac{1}{4}$-in. schedule 40 pipes. Calculate ΔP through an annular channel of length 20 ft, assuming $\varepsilon/D = 0.0014$:

$$\mu = 0.9 \text{ cP} \quad \rho = 62.4 \text{ lb/ft}^3$$

Solution

First calculate equivalent diameter. The outside diameter of the inner pipe is $D_1 = 0.84$ in. $= 0.07$ ft, and the inside diameter of the outer pipe is $D_2 = 1.38$ in. $= 0.115$ ft.

$$\text{Equivalent diameter} = \frac{4(\text{cross-sectional area of flow})}{\text{wetted perimeter}} = \frac{4\left(\frac{\pi}{4}\right)\left(D_2^2 - D_1^2\right)}{\pi(D_2 + D_1)}$$

$$= D_2 - D_1 = 0.115 - 0.07 = 0.045 \text{ ft}$$

$$\text{Cross section} = \frac{1}{4}\pi\left(D_2^2 - D_1^2\right) = 0.785(0.115^2 - 0.07^2) = 0.006535 \text{ ft}^2$$

$$\text{Velocity } u = \frac{25}{7.48(60)(0.006535)} = 8.52 \text{ ft/s}$$

$$N_{Re} = \frac{D_e u \rho}{\mu} = \frac{0.045(8.52)(62.4)}{0.9(0.000672)} = 39{,}557$$

$$\frac{\varepsilon}{D} = 0.0014 \quad \text{(given)}$$

$$f = 0.0068$$

From the Fanning friction chart (Fig. 4.6),

$$\Delta P = \frac{2fLu^2\rho}{g_c D_e (144)} = \frac{2(0.0068)(20)(8.52)^2(62.4)}{32.17(0.045)(144)} = 5.91 \text{ psi}$$

PUMP CALCULATIONS

Various terms in connection with the pump calculations (Fig. 4.8) are defined in the following paragraphs.

Capacity is the quantity of fluid discharged per unit time. In the fps system, this is expressed in gallons per minute (gpm) for liquids.

Static Head for a liquid being pumped is the difference in elevation, in feet, between the datum line and the liquid surface or the point of free delivery. For

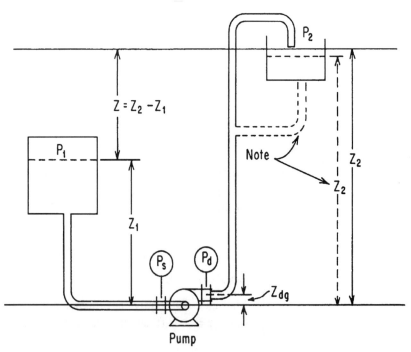

Figure 4.8 Pumping system terminology. (*Note:* If the discharge line of the pump is connected to the bottom of the receiver tank as shown by the dashed line, then Z_2 is the distance from the centerline of the pump to the liquid surface.)

the horizontal centrifugal pumps, the datum line is the pump centerline. For the vertical pumps, the datum line is taken at the eye of the first-stage impeller.

In Figure 4.8, Z_1 and Z_2 are the static heads. Note that Z_2 is measured up to the end of the discharge pipe because the pipe end is the point of free delivery. If the pipe were connected at the bottom, Z_2 would be measured up to the liquid surface.

Pressure Head is given in feet and defined as follows:

$$\text{Pressure head, } \left(\frac{\text{ft} \cdot \text{lb}_f}{\text{lb}}\right) = \frac{144 P_1}{\rho} = \frac{2.31 P_1}{s} \tag{4.54}$$

where
P_1 = absolute pressure, psia
ρ = density of the fluid, lb/ft^3
s = specify gravity of the fluid with respect to water.

Velocity Head is given in feet and defined as follows:

$$\text{Velocity head, } (\text{ft} \cdot \text{lb}_f/\text{lb}) = \frac{u^2}{2\alpha g_c}$$

where u is the velocity, ft/s, and g_c is Newton's law proportionality factor = 32.17 ft · lb/(lb$_f$ · s^2), and α is the kinetic energy correction factor = 0.5 for laminar flow, approximately 0.94 for turbulent flow. Generally, α is taken as 1 for turbulent flow.

Static Suction Head is the difference in elevation, in feet, between the centerline (or impeller eye) of the pump and the liquid surface in the suction vessel. The liquid surface is above the pump centerline. In Figure 4.8, $Z_1(g/g_c)$ is the static suction head.

Static Suction Lift is the difference in elevation, in feet, between the liquid surface of the suction vessel and the centerline of the pump when the liquid level in the suction vessel is below the centerline (or impeller eye) of the pump. In Figure 4.9c, $Z_1(g/g_c)$ is the static suction lift.

Total Suction Head or *Lift* is defined as total suction head (Fig. 4.9b and 4.9c) and is the absolute pressure head in the supply vessel plus the static suction head minus the friction head, or

$$\text{Total suction head} = (Z_{pt} \pm Z_1 - Z_f)(g/g_c) \tag{4.55}$$

Use the minus sign for Z_1 in the case of the suction lift.

If the total suction head (or lift) is measured from the reading of a pressure gauge at the suction flange of the pump, then

$$\text{Total suction head} = \frac{P_s(144)}{\rho} + \frac{u_s^2}{2\alpha g_c} \tag{4.56}$$

where
$P_s = \pm^*$ gauge reading + barometric pressure, psia
ρ = density of liquid, lb/ft^3
u_s = suction velocity, ft/s
g_c = Newton's law proportionality factor
= 32.17 ft · lb/(lb$_f$ · s^2)

---

*Use the minus sign when using the vacuum-gauge reading.

Static Discharge Head is the difference in elevation, in feet, between the point of the free delivery or the liquid surface in the discharge vessel and the centerline of the pump. In Figure 4.8, Z_2 is the static discharge head in ft of fluid.

Total Discharge Head is defined as the absolute pressure head in the discharge vessel plus the static discharge head plus the friction head. If the total discharge head is determined from the reading of the pressure gauge at the discharge flange of the pump as shown in Figure 4.8, then

$$\text{Total discharge head} = \frac{144 P_d}{\rho} + Z_{dg}\left(\frac{g}{g_c}\right) + \frac{u_d^2}{2\alpha g_c} \quad (4.57)$$

where
  P_d = barometric pressure +gauge reading, psia
  Z_{dg} = elevation of the discharge flange from the datum line, ft
  u_d = discharge velocity, ft/s

Total Dynamic Head or *Total Head (TDH)* is the energy, in ft · lb$_f$/lb, of liquid that the pump has to impart to the liquid in order to transport it to the desired location. It can be calculated from

$$\text{TDH} = \text{total discharge head} - \text{total suction head}$$

or from

$$\begin{aligned}\text{TDH} &= \frac{144 P_2}{\rho} - \frac{144 P_1}{\rho} + \frac{\Delta P_{f1}(144)}{\rho} + \frac{(144)\Delta P_{f2}}{\rho} \\ &\quad + Z_2\left(\frac{g}{g_c}\right) - Z_1\left(\frac{g}{g_c}\right) + \frac{u_2^2}{2\alpha g_c} - \frac{u_1^2}{2\alpha g_c} \\ &= (P_2 - P_1 + \Delta P_{f1} + \Delta P_{f2})\frac{2.31}{s} + Z_2 - Z_1 \end{aligned} \quad (4.58)$$

where
  P_2 = pressure in discharge vessel, psia
  P_1 = pressure in suction vessel, psia
  ΔP_{f1} = pressure drop in suction line, psi
  ΔP_{f2} = pressure drop in discharge line, psi
  S = gravity of liquid at pumping temperature with respect to water
  Z_2 = elevation of liquid discharge point above pump centerline, ft
  Z_1 = Elevation of liquid level in suction vessel above pump centerline, ft
    (Z_1 is negative for suction lift)

Note: The second part of Equation 4.58 is a simplified form, which ignores the velocity heads, and is applicable in AES unit on earth because the (g/g_c)-factor is 1 lb$_f$/lb.

Shut-off Head is the head developed by a pump with the discharge valve closed. It is added to the suction head to determine the maximum discharge pressure of the pump.

Net Positive Suction Head (NPSH$_A$) available is the total suction head, in feet, of liquid (absolute) that is available in excess of the liquid vapor pressure, also expressed in feet, of liquid at the pump suction flange. It is required to move the liquid into the eye of the impeller, for which the pump itself is not responsible. The liquid should be brought into the pump in the liquid state without vaporization.

Each pump requires a particular NPSH depending on its design. Typical suction systems with formulas for NPSH$_A$ in each case are given below.

System 1A

In this case, the suction supply is open to atmosphere, the liquid level of the supply is above the pump centerline (Fig. 4.9a), and the NPSH$_A$ is

$$\text{NPSH}_A = Z_1 + Z_a - Z_v - Z_f \tag{4.59}$$

where

Z_1 = static suction head, ft of fluid
Z_a = absolute atmospheric pressure over the liquid level, ft of fluid
Z_v = absolute vapor pressure of liquid at the pumping temperature, ft of fluid
Z_f = frictional losses, ft of fluid

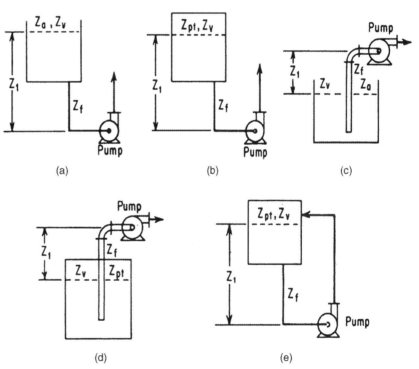

Figure 4.9 NPSH$_A$ for various suction conditions: (a) suction vessel open to atmosphere and liquid level above pump centerline; (b) suction vessel closed to atmosphere and above the centerline of pump; (c) suction vessel open to atmosphere but below the centerline of pump; (d) suction vessel closed to atmosphere and below the centerline of pump; (e) boiling liquid in a closed tank located above the pump

System 1B

In this case, the suction vessel is closed to the atmosphere and is located above the pump centerline (Fig. 4.9b). For this system

$$\text{NPSH}_A = Z_1 + Z_{pt} - (Z_v + Z_f) \tag{4.60}$$

where Z_{pt} is the total absolute pressure on the surface of the liquid, ft. If $Z_{pt} = Z_v$, the above reduces to

$$\text{NPSH}_A = Z_1 - Z_f \tag{4.61}$$

System 2A

In this case, the suction vessel is open to atmosphere but is located below the centerline of the pump (Fig. 4.9c). For this system

$$\text{NPSH}_A = Z_a - (Z_1 + Z_v + Z_f) \tag{4.62}$$

in which Z_1 is the suction lift, ft.

System 2B

In this case, the suction vessel is closed to atmosphere and below the centerline of the pump (Fig. 4.9d). For this case:

$$\text{NPSH}_A = Z_{pt} - (Z_1 + Z_v + Z_f) \tag{4.63}$$

System 3

For a boiling liquid in a closed tank located above the pump suction line (e.g., liquid refrigerant),

$$\begin{aligned}\text{NPSH}_A &= Z_1 + Z_{pt} - (Z_v + Z_f) \\ &= Z_1 + Z_{pt} - Z_v + Z_f \\ &= Z_1 - Z_f\end{aligned} \tag{4.64}$$

since $Z_{pt} = Z_v$ in this case. Requirements of NPSH are usually determined on the basis of handling water.

In an existing system, the NPSH_A is determined by gauge reading at the pump suction and with the use of the following formula:

$$\text{NPSH}_A = Z_{pt} - Z_v \pm R_G + Z_k \pm Z_{1p} \tag{4.65}$$

where
- R_G = gauge reading, ft (Use minus sign for vacuum gauge reading.)
- Z_k = velocity head in suction pipe at the gauge connection, ft
- Z_{1p} = static suction head or lift of pressure gauge with respect to centerline of the pump. (Use plus sign for the suction head and minus sign for the suction lift.)

Cavitation

When there is no sufficient NPSH at the pump suction, the pressure of the liquid reduces to a value equal to or below its vapor pressure, which causes the liquid to vaporize resulting in the formation of small vapor bubbles. These bubbles collapse, when they reach a high-pressure area as they move along the impeller vanes. This is called cavitation of the pump. To prevent the adverse effects of cavitation (such as pump noise, loss of head, and impeller damage), it must be ensured that the available NPSH in the system is greater than the NPSH required by the pump.

Specific Speed

The specific speed of an impeller is defined as the revolutions per minute needed to produce 1 gpm at 1 ft head. The specific speed is related to the capacity, head, and impeller speed by

$$N_s = \frac{nQ^{1/2}}{H^{3/4}} \qquad (4.66)$$

$$n_s = \frac{nQ_v^{1/2}}{H^{3/4}} \qquad (4.67)$$

where
   N_s = specific speed of the pump impeller, rpm
   n_s = specific speed of the blower or fan impeller, rpm
   n = impeller speed, rpm
   Q = flow, gpm
   H = total dynamic head, ft of fluid flowing
   Q_v = flow, cfm, in case of fans.

Cavitation Parameter σ

The cavitation in the pump must be avoided for the sake of efficiency and for the prevention of impeller damage. For pumps, a cavitation parameter σ is given by

$$\sigma = \frac{P_1/\rho - P_v/\rho + Z_1 - Z_f}{H} \qquad (4.68)$$

where
   H = total dynamic head of the pump, ft
   P_v = vapor pressure, ft
   P_1 = pressure upon the liquid surface in the suction vessel, ft.

The critical value of the cavitation parameter σ_c is the value at which there is an observed change in the efficiency. σ_c and N_s are related as follows:

$$\sigma_c = \begin{cases} 6.3 \times 10^{-6} N_s^{1.33} & \text{for single suction pumps} \quad (4.69) \\ 4 \times 10^{-6} N_s^{1.33} & \text{for double suction pumps} \quad (4.70) \end{cases}$$

Suction Specific Speed

The suction specific speed S is given by

$$S = \frac{nQ^{1/2}}{\text{NPSH}^{3/4}} \qquad (4.71)$$

σ_c depends on both N_s and S. A relation among σ_c, N_s, and S is

$$\sigma_c = \left(\frac{N_s}{S_c}\right)^{4/3} \qquad (4.72)$$

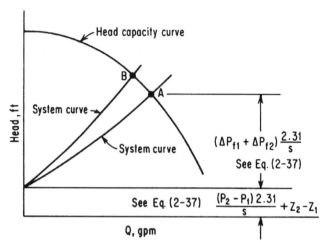

Figure 4.10 Operating characteristics and operating points for a centrifugal pump

where S_c is the critical suction speed of the pump, rpm. The critical net positive suction head, $NPSH_c$ is given by

$$NPSH_c = \sigma_c H \tag{4.73}$$

Performance Curve is a plot of the head developed vs. the pump capacity. It is also called the *head-capacity* or the *HQ* curve (Fig. 4.10). The head developed is the net head obtained after subtraction of the vane and shock losses. For a given pump, the head-capacity curve is unique.

System Head, or the total head a pump has to produce, is the sum of all the work from the liquid source to the discharge. This includes the frictional resistance, the static head difference, and the pressure head difference that have to be overcome. For a given pump and piping arrangement, a system curve can be prepared and superimposed on the head-capacity curve (Fig. 4.10). The system head curve is a function of the system static head, pressure head, and frictional head. Only the frictional head varies with the flow.

The point of intersection of the system curve with the *HQ* curve is the operating point (*A* in Fig. 4.10). This is the only flow rate the pump will deliver. If a change in the flow rate delivered is desired, characteristics for the system must be changed (point *B* in Fig. 4.10). This is usually accomplished by throttling the discharge valve. Head efficiency and horsepower curves vary with specific speed.

Brake Horsepower (bhp) is the actual horsepower consumed by the pump in generating the required head and volumetric flow rate.

Centrifugal Pump Calculations

The expressions for calculation of various quantities in the case of the centrifugal pumps in AES units are summarized next.

$$Z = \Delta P \frac{2.31}{s} \quad \text{and} \quad \Delta P = 0.433 s Z \tag{4.74}$$

$$Q = \frac{W}{500 s} \tag{4.75}$$

$$\text{TDH} = (P_2 - P_1 + \Delta P_{f1} + \Delta P_{f2})\frac{2.31}{s} + Z_2 - Z_1 \qquad (4.76)$$

Net Positive Suction Head

$$\text{NPSH} = (P_1 - P_v - \Delta P_{f1})\frac{2.31}{s} \pm Z_1 \, \text{ft} \qquad (4.77)$$

$$P_s = P_1 \pm 0.433 Z_1 s - \Delta P_{f1} \qquad (4.78)$$

$$P_d = P_s + 0.433(\text{TDH})s \qquad (4.79)$$

$$\text{hhp} = \frac{W(\text{TDH})}{1.98 \times 10^6} = \frac{Q(\text{TDH})s}{3960} = \frac{Q(P_d - P_s)}{1714} \qquad (4.80)$$

$$\text{bhp} = \frac{W(\text{TDH})}{1.98 \times 10^6 \varepsilon} = \frac{Q(\text{TDH})s}{3960\varepsilon} = \frac{Q(P_d - P_s)}{1714\varepsilon} \qquad (4.81)$$

$$\text{hp} = \frac{\text{bhp}}{\varepsilon_m} \qquad (4.82)$$

$$kW = \frac{0.745(\text{bhp})}{\varepsilon_m} \qquad (4.83)$$

where
W = weight rate of flow, lb/h
TDH = required total dynamic head of the pump, ft
P_1 = pressure in suction vessel, psia
P_2 = pressure in discharge vessel, psia
ΔP_{f1} = total pressure drop in suction line, psi
ΔP_{f2} = total pressure drop in discharge line, psi
P_v = vapor pressure at pumping temperature, psia
Q = pump capacity, gpm
Z_1 = static suction head, ft
Z_2 = static discharge head, ft
hp = motor horsepower
hhp = hydraulic horsepower
bhp = brake horsepower
ε = efficiency of pump
ε_m = motor efficiency
P_s = suction pressure, psia
P_d = discharge pressure, psia
s = specific gravity

Actual motor size selection depends on available standard motor sizes. This results in having motors of ratings 110% to 125% of the rated brake horsepower of the pump.

Pump Affinity Laws

These are the relationships among the capacity Q, head H, power bhp, impeller diameter D, and speed of revolution (in revolutions per minute) of centrifugal pumps and fans.

1. Effect of Speed Change When D is Constant

Capacity
$$\frac{Q_2}{Q_1} = \frac{N_2}{N_1} \tag{4.84}$$

Head
$$\frac{H_2}{H_1} = \left(\frac{N_2}{N_1}\right)^2 \tag{4.85}$$

Power
$$\frac{bhp_2}{bhp_1} = \left(\frac{N_2}{N_1}\right)^3 \tag{4.86}$$

NPSH required
$$\frac{NPSH_2}{NPSH_1} = \left(\frac{N_2}{N_1}\right)^2 \tag{4.87}$$

These laws can be used to determine the performance curve at another rpm level if the performance curve is available at a known rpm.

2. Effect of Impeller Diameter Change
Within the same pump at a constant speed, the following relations apply when the impeller diameter is changed.

Capacity
$$\frac{Q_2}{Q_1} = \frac{D_2}{D_1} \tag{4.88}$$

Head
$$\frac{H_2}{H_1} = \left(\frac{D_2}{D_1}\right)^2 \tag{4.89}$$

bhp
$$\frac{bhp_2}{bhp_1} = \left(\frac{D_2}{D_1}\right)^3 \tag{4.90}$$

The laws relating to the impeller diameter are accurate within a certain range of the change of impeller diameter. In general, these laws are not as accurate as the laws relating to the rpm.

3. For Geometrically Similar Pumps with Different Impeller Diameter But Same Speed
Pumps are geometrically similar if they are of different sizes but of the same style and the relationship between the casing and impeller dimensions is the same. For such pumps,

$$Q_2 = Q_1 \left(\frac{D_2}{D_1}\right)^3, \quad N = \text{constant}$$

$$H_2 = H_1 \left(\frac{D_2}{D_1}\right)^2, \quad N = \text{constant} \tag{4.91}$$

$$bhp_2 = bhp_1 \left(\frac{D_2}{D_1}\right)^5, \quad N = \text{constant}$$

These laws can be applied to develop a performance curve at a different diameter from a given performance curve at a known diameter.

When the diameter and rpm are both changed, the theorem of joint variation can be applied to estimate the effect of change. The above affinity laws apply to centrifugal fans and blowers, too. In the case of this equipment, NPSH does not apply.

Example 4.11

Determine the available NPSH from Exhibit 4a.

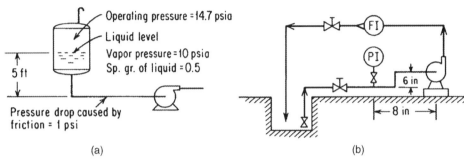

Exhibit 4 (a) Data sketch for Example 4.11; (b) data sketch for Example 4.12

Solution

$$\text{NPSH} = Z_l + Z_a - Z_v - Z_f$$

$$= 5 + \frac{(14.7 - 10 - 1)(2.31)}{0.5} = 22.1 \text{ ft}$$

Example 4.12

The adjoining Exhibit 4b shows a typical setup for NPSH requirement test. One set of data in one of the tests was

Pressure gauge reading = 9.12 psi (vacuum)
Vapor pressure of liquid = 0.507 psia
Specific gravity of liquid = 1
Velocity in suction line = 8.03 ft/s

Calculate the NPSH required for the specified flow.

Solution

From Equation 4.65,

$$\text{NPSH} = Z_{pt} \pm R_G - Z_v + Z_k \pm Z_{1p}$$

$$= \frac{14.7 - 9.12 - 0.507}{0.433(1)} + \frac{8.03^2}{2(32.2)} - 0.5 = 12.22 \text{ ft}$$

Example 4.13

A solution (specific gravity = 1.25, viscosity = 1.2 cP) is pumped through a 100-mm-ID stainless steel pipe of total length 200 m in the horizontal and vertical directions. The net elevation is 15 m. In the line, there are fifteen 90° standard elbows, five ball valves, a control valve, and a filter. The roughness factor of the inside surface is 0.0004 m. Flow rate is 0.0285 m³/s. Maximum pressure drop through the filter is 210 cmHg with a Hg density of 13,558 kg/m³, and the equivalent length of the control valve can be taken as 250 pipe diameters. The pipe discharges to an open tank.

Calculate (a) pressure loss through the pipes and fittings, (b) total head loss in meters, (c) total head to be developed by the pump, and (d) power requirements of the pump if it is 65% efficient. Use the following table to find equivalent length.

Item	Equivalent Length
90° std. elbow	3 m
Ball valve	0.7 m
Control valve	25 m
Pipe exit	4.5 m

Solution

(a) Flow through pipe = 0.0285 m³/s

$$\text{Area of pipe-flow cross section} = \frac{1}{4}\pi(0.1)^2 = 0.007854 \text{ m}^2$$

$$\text{Velocity through pipe} = \frac{0.0285}{0.007854} = 3.63 \text{ m/s}$$

Viscosity of liquid $\mu = 1.2$ cP $= 0.0012$ N · s/m² or kg/m · s

$$N_{Re} = \frac{Du\rho}{\mu} = \frac{0.1(3.63)(1250)}{0.0012} = 3.78 \times 10^5$$

$$\text{Roughness factor} = \frac{\varepsilon}{D} = \frac{0.00004}{0.1} = 0.0004$$

From Figure 4.6, Fanning friction factor $f \doteq 0.0043$.
Calculate total equivalent length.

Item	Equivalent Length, m
Straight pipe	200
Fifteen 90° std. elbows	45 (3 m each)
Five ball valves	3.5 (0.7 m each)
Control valve	25
Pipe exit	4.5
Total	**278.0**

ΔP through pipe, fittings, and control valve is

$$\Delta P = \frac{2fL_e u^2 \rho}{g_c D}$$

$$= \frac{2(0.0043)(278)(\text{m})(3.63)^2 \left(\frac{\text{m}}{\text{s}}\right)^2 (1250)\left(\frac{\text{kg}}{\text{m}^3}\right)}{\left(\frac{1 \text{ kg} \cdot \frac{\text{m}}{\text{s}^2}}{\text{N}}\right) 0.1(\text{m})}$$

$$= 393{,}792 \text{ N/m}^2 = 393.8 \text{ kN/m}^2$$

(b) ΔP through line including control valve = 393.8 kN/m^2
From Equation 4.2a,

$$\Delta P \text{ through filter} = \frac{\Delta h \rho g}{g_c} = 2.1 \text{ m}\left(13558 \frac{\text{kg}}{\text{m}^3}\right)\left(9.81 \frac{\text{N}}{\text{kg}}\right) = 279.3 \frac{\text{kN}}{\text{m}^2}$$

Then

$$\text{Total } \Delta P = 393.8 + 279.3 = 673.1 \text{ kN/m}^2 = 673,100 \text{ N/m}^2$$

From Equation 4.2a,

$$\text{Head loss, m} = \frac{\Delta P}{\rho \left[\frac{g}{g_c}\right]}$$

$$\text{Head loss} = \frac{673,100 \frac{\text{N}}{\text{m}^2}}{1250 \left(\frac{\text{kg}}{\text{m}^3}\right)(9.81) \frac{\text{N}}{\text{kg}}} = 54.9 \text{ m}$$

(c) Total head to be developed is the head loss plus the net elevation:

$$54.9 + 15 = 69.9 \text{ m}$$

(d) Mass flow = $(0.0285 \text{ m}^3/\text{s})(1250 \text{ kg/m}^3) = 35.625$ kg/s

$$\begin{aligned}
\text{Power required} &= (35.625 \text{ kg/s})(69.9 \text{ m})(9.81 \text{ m/s}^2) \\
&= 24,429 \text{ (kg} \cdot \text{m}^2/\text{s}^2)/\text{s} = 24,429 \text{ J/s} = 24,429 \text{ W} \\
&= 24.43 \text{ kW} \\
&= \frac{24.43}{0.746} = 32.8 \text{ hp}
\end{aligned}$$

Example 4.14

A pump takes brine from a tank and transports it to another tank through a 6-in. schedule 40 line. A sketch of the system is shown in Exhibit 5. The flow rate is 825 gpm. In the suction line there are one gate valve, one strainer, two standard tees, and two 90° elbows. In the discharge line, there are six standard tees, one gate valve, one check valve, and five 90° short elbows (not all shown in the figure). Two pressure gauges are installed at the suction and discharge of the pump as

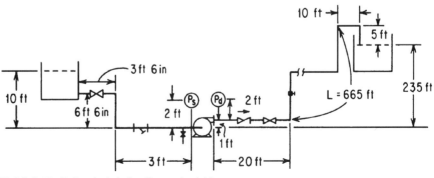

Exhibit 5 Data sketch for Example 4.14

shown. Specific gravity of brine is 1.2, and its viscosity is 1.2 cP. Calculate the frictional pressure drops in the suction and discharge lines. What pressures are registered by the gauges?

Use *K*-values from the following table:

	K
Sudden contraction	= 0.50
Gate valve	= 0.09
Strainer	= 0.88
Standard tees	= 0.6
90° short elbows	= 0.3

Solution

Calculate the pressure drop in the suction line. The line contains the items shown in Table 4.3.

$$ID = 6.065 \text{ in.} = 0.5054 \text{ ft}$$

$$\text{Velocity of brine} = \frac{825 \text{ gpm}}{7.48(60)(0.785)(0.5054)^2} = 9.163 \text{ ft/s}$$

$$N_{Re} = \frac{Du\rho}{\mu} = \frac{0.5054(9.163)(1.2)(62.4)}{1.2(0.000672)} = 4.3 \times 10^5$$

$$\frac{\varepsilon}{D} = \frac{0.00015}{0.5054} = 0.0003$$

f from Fanning friction factor chart (Fig. 4.6) is 0.0041. The pressure drop on the suction side is

$$\frac{\Delta P_f}{\rho} = \left(\Sigma K + \frac{4fL}{D}\right)\frac{u^2}{2g_c}$$

$$= \left[3.27 + \frac{4(0.0041)(13)}{0.5054}\right]\frac{9.163^2}{64.4}$$

$$= 4.8 \text{ ft-lb}_f/\text{lb}.$$

$$\Delta P_f = 4.8 \frac{\text{ft} \cdot \text{lb}_f}{\text{lb}}\left(1.2 \times 62.4 \frac{\text{lb}}{\text{ft}^3}\right)\frac{1 \text{ ft}^2}{144 \text{ in}^2} = 2.5 \text{ psi}$$

Table 4.3 Calculate total *K* value

	K
One sudden contraction	1(0.5) = 0.50
One gate valve	1(0.09) = 0.09
One strainer	1(0.88) = 0.88
Two standard tees	2(0.6) = 1.2
Two 90° short elbows	2(0.3) = 0.6
	$\Sigma K = 3.27$

Straight pipe length = 13 ft.

Table 4.4

	K
Six standard tees	6(0.6) = 3.6
One gate valve	0.09
Five check valve	2.0
Five 90° short elbows	5(0.3) = 1.5
One sudden expansion	1.0
	$\Sigma K = 8.19$

Total pipe length = 700 ft.

The suction static head is

$$10(0.433)(1.2) = 5.2 \text{ psi}$$

The suction velocity head is

$$\frac{9.163^2}{64.4}(0.433)(1.2) = 0.678 \text{ psi}$$

Therefore the pressure at inlet of nozzle is

$$14.7 + 5.2 - 2.5 - 0.678 = 16.7 \text{ psia}$$

The pressure indicated by gauge is 16.7 minus the elevation of the gauge from the centerline at the suction nozzle and equals

$$16.7 - 2(0.433)(1.2) \text{ B } 15.7 \text{ psia} = 1 \text{ psig}$$

Calculate pressure drop in the discharge line. The line contains the data listed in Table 4.4.
  The friction loss in the discharge line is therefore

$$\Delta P_{fd} = \left[8.19 + \frac{4(0.0041)(700)}{0.5054} \right] \frac{9.163^2}{64.4} = 40.3 \text{ ft of fluid} = 20.8 \text{ psi}$$

The discharge static head is

$$(235 \text{ ft})(0.433)(1.2) = 122.1 \text{ psi}$$

The pressure indicated by the gauge (neglecting the velocity head) is calculated to be

$$14.7 + 122.1 + 20.8 - \text{elevation}(0.433)s = 157.6 - 2(0.433)(1.2)$$
$$= 156.56 \text{ psia} = 141.9 \text{ psig}$$

where the elevation of the gauge from the centerline is in ft.

Example 4.15

In Example 4.14, calculate (a) NPSH available, (b) total head, (c) hydraulic horsepower, and (d) brake horsepower if efficiency is 74.7 percent.
  The vapor pressure of water over the brine solution at 86°F is 0.6 psia.

Solution

(a) From Equation 4.59 and considering the velocity head negligible,

$$\text{NPSH}_A = Z_1 + Z_a - Z_v - Z_f$$

From the previous problem solution and the data of this example,

$$Z_1 = 10 \text{ ft}$$

$$Z_a = 14.7 \text{ psia} = 14.7 \frac{2.31}{1.2} = 28.3 \text{ ft}$$

$$Z_f = 4.8 \text{ ft} \quad \text{(calculated in Example 4.14)}$$

$$Z_v = (0.6 \text{ psi}) \frac{2.31}{1.2} \doteq 1.2 \text{ ft}$$

$$\text{NPSH}_A = 10 + 28.3 - 1.2 - 4.8 = 32.3 \text{ ft}$$

(b) *Total Head*

$$\text{TDH} = (P_2 - P_1 + \Delta P_{f1} + \Delta P_{f2}) \frac{2.31}{s} + Z_2 - Z_1$$

$$= (14.7 - 14.7 + 2.5 + 20.8) \frac{2.31}{1.2} + 235 - 10$$

$$= 269.9 \text{ ft} = 140.2 \text{ psi}$$

(c) *Hydraulic Horsepower*

$$\text{hhp} = \frac{(\text{gpm})(\text{psi})}{1714} = \frac{825(140.2)}{1714} = 67.5$$

(d) *Brake Horsepower*

$$\text{bhp} = \frac{\text{hhp}}{\text{efficiency}} = \frac{67.5}{0.747} = 90.4$$

Example 4.16

Water is pumped at 86°F through a piping system from one tank to another as shown in Exhibit 6. The suction line consists of 30 ft of 3-in. schedule 40 straight pipe and 260 ft of 2-in. schedule 40 pipe in the discharge line. If the globe valves are completely open, determine the flow through piping in gallons per minute.

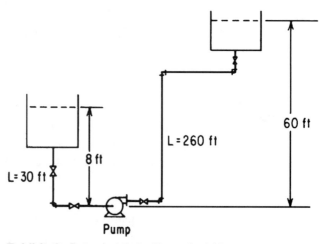

Exhibit 6 Data sketch for Example 4.16

Assume the viscosity of water is 0.85 cP. Data for the centrifugal pump to be used are given below. Assume a roughness factor of 0.00015 ft.

Capacity, gpm	0	10	20	30	40	50	60	70	80
Developed head, ft	120	119.5	117	113	107.5	100	90	82	75
Efficiency, %	0	13	23.5	31.6	37.5	42.2	42.5	41.7	39.5

Use the equivalent lengths from the following table:

Suction Line	Length, ft
Contraction	7.5
Open glove valve	90
One 90° elbow	7.5
Discharge Line	
Elbow	5.33
Globe valve	60
Sudden expansion	1.3

Calculate the brake horsepower.

Solution

Plot the pump-head curve as in Exhibit 7. The flow rate through the piping will be given by the intersection of the system curve with the pump-head curve. To construct

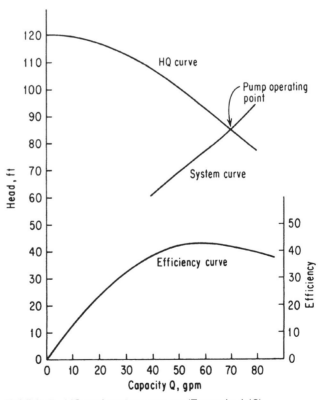

Exhibit 7 *HQ* and system curves (Example 4.16)

the system curve, the total head for a given flow through the piping system is given by

$$H = P_2 - P_1 + Z_2 - Z_1 + \Delta P_{f1} + \Delta P_{f2}$$
$$P_2 - P_1 = 0$$
$$Z_2 - Z_1 = 60 - 8 = 52 \text{ ft}$$

$$\Delta P_{f1} = h_1 = \frac{2 f_1 L_1 u_1^2}{g_c D_1}$$

$$\Delta P_{f2} = h_2 = \frac{2 f_2 L_2 u_2^2}{g_c D_2} \quad \text{(both in ft·lb}_f\text{/lb)}$$

where L_1 and L_2 are total equivalent lengths. See Table 4.5.

$$D_1 = \frac{3.068}{12} = 0.2557 \text{ ft} \quad D_2 = \frac{2.067}{12} = 0.1723 \text{ ft}$$

Prepare a table of total head vs. capacity for the pump.

$$\Delta P_{f1} = \frac{2(225) f_1 u_1^2}{32.17(0.2557)} = 54.7 f_1 u_1^2$$

$$\Delta P_{f2} = \frac{2(397.3) f_2 u_2^2}{32.17(0.1723)} = 143.4 f_2 u_2^2$$

$$u_1 = \frac{\text{gpm}}{7.48(60)(0.785)(0.2557)^2} = 0.044 \text{ (gpm)}$$

$$u_2 = \frac{\text{gpm}}{7.48(60)(0.785)(0.1723)^2} = 0.096 \text{ (gpm)}$$

$$\frac{\varepsilon}{D_1} = \frac{0.00015}{0.2557} = 0.00059 \quad \frac{\varepsilon}{D_2} = \frac{0.00015}{0.1723} = 0.0009$$

Table 4.5

Suction Line	Length, ft
3-in. pipe length	30
Contraction	7.5
Two open glove valves	180.0
One 90° elbow	7.5
Total	225.0
Discharge Line	
2-in. pipe length	260
Three elbows	16
Two globe valves	120
Sudden expansion	1.3
Total	397.3

$$N_{Re1} = \frac{50.65\rho Q}{\mu d_1} = \frac{50.65(62.16)(\text{gpm})}{0.85(3.068)}$$

$$N_{Re2} = \frac{50.65\rho Q}{\mu d_2} = \frac{50.65(62.16)(\text{gpm})}{0.85(2.067)}$$

$$f_1 = \frac{1}{2.444\left[\ln\left(1.35\frac{\varepsilon}{D_1}\right) + \frac{6.5}{N_{Re1}}\right]^2}$$

$$f_2 = \frac{1}{2.444\left[\ln\left(1.35\frac{\varepsilon}{D_2}\right) + \frac{6.5}{N_{Re2}}\right]^2}$$

$$H = 52 + 54.7 f_1 u_1^2 + 143.4 f_2 u_2^2$$

With assumed gpm, calculate velocity, Reynolds number, friction factors, and head.

The head H is plotted vs. capacity in gallons per minute in Exhibit 7 to get the system curve. The system curve cuts the pump head curve at 70 gpm and 85 ft head. At 70 gpm, efficiency is 41.5 percent. Thus,

$$\Delta P = 85(0.433) = 36.84 \text{ psi}$$

$$\text{bhp} = \frac{70(36.84)}{1714(0.415)} = 3.62$$

Assuming 125 percent of the rated brake horsepower, the motor hp is

$$\text{hp} = 3.62(1.25) = 4.53$$

A 5-hp standard size motor will have to be specified.

Example 4.17

A liquid pumping system has the following data:

Capacity	42 gpm
TDH	50 ft
NPSH available	10 ft
Specific gravity	0.5

A pump with a $5\frac{1}{2}$-in. impeller is available in storage. However, the performance curve is lost. The only information from the vendor is the performance curves of a 6-in. impeller pump of the identical model number at various rpm as shown in Exhibit 8. Also available in stock are electric motors with 80 percent efficiency as given in Table 4.6. Justify whether the pump in storage can be used for the specified duty with or without any modification of the pump dimensions. Also determine which one of the above motors will meet the pumping requirement.

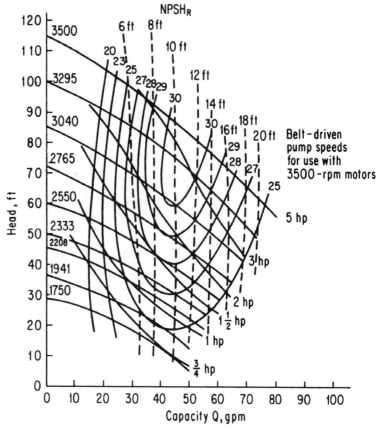

Exhibit 8 Pump characteristics (Example 4.17). Pump size = 1 × 1 – 6; eye area = 1.23 in.^{2}

Solution

The brake horsepower is given by

$$\text{bhp} = \frac{(\text{gpm})(\text{psi})}{1714 \,(\text{pump efficiency})} = \frac{(\text{gpm})(\text{TDH})(0.433)(s)}{1714 \,(\text{pump efficiency})}$$

$$= \frac{42(50)(0.433)(0.5)}{1714(\text{pump efficiency})} = \frac{0.265}{\text{pump efficiency}}$$

We do not know the pump efficiency. We ask the question: If a $5\frac{1}{2}$-in. impeller pump has a capacity of 42 gpm and TDH of 50 ft, what are the corresponding capacity and TDH of a 6-in. impeller pump at the same efficiency and same rpm?

Table 4.6 Revolution per minute of stock electric motors

	Hp	
1	$1\frac{1}{2}$	2
3500	3500	3500
3295	3295	3295
3040	3040	3040
2500	2500	2500
1750	1750	1750

From the affinity laws, *at an equal efficiency,* and *equal rpm*,

$$\frac{Q_6}{Q_{5(1/2)}} = \frac{D_6}{D_{5(1/2)}} = \frac{6}{5.5} = 1.091$$

$$Q_6 = 1.091(42) = 46 \text{ gpm}$$

$$\frac{H_6}{H_{5(1/2)}} = \left(\frac{D_6}{D_{5(1/2)}}\right)^2 = 1.19$$

$$H_6 = 1.19(50) = 59.5 \text{ ft}$$

Locating $Q_6 = 46$ gpm and TDH = 59.5 ft on the performance curve gives the impeller 3040 rpm and a pump efficiency of 30 percent. This is also the efficiency of a $5\frac{1}{2}$-in. impeller at 42 gpm and 50 ft TDH and operating at 3040 rpm. Therefore the brake horsepower of the pump in storage is

$$\frac{0.265}{0.30} = 0.883 \quad \text{Motor hp} = \frac{0.883}{0.8} = 1.10$$

Hence, choose a $1\frac{1}{2}$-hp, 3040-rpm motor, and the existing pump will work. Next, check the NPSH requirement. From the performance curve, the NPSH required is 10 ft, when the impeller diameter is 6 in. at 3040 rpm. At the same rpm but with an impeller diameter of $5\frac{1}{2}$-in, estimate the NPSH required.

$$\frac{H_2}{H_1} = \left(\frac{D_2}{D_1}\right)^2 \quad \text{when } N = \text{const}$$

$$H_2 = H_1 \left(\frac{D_2}{D_1}\right)^2 = 10\left(\frac{5.5}{6}\right)^2 = 8.4 \text{ ft}$$

The motor hp can also be estimated alternatively. From Exhibit 8, the bhp for a 6-in. impeller at the point of equivalence (i.e., at 46 gpm and 59.5 ft TDH) is 2.4 hp approximately. Hence bhp for a $5\frac{1}{2}$-in. impeller at the same rpm (3040) by the affinity law is

$$\text{bhp}\left(\frac{D_{5(1/2)}}{D_6}\right)^3 (s) = 2.4\left(\frac{5.5}{6}\right)^3 (0.5) = 0.92$$

$$\text{Motor hp} = \frac{0.92}{0.8} = 1.15$$

So, choose a $1\frac{1}{2}$-hp motor of 3040 rpm.

Note: The bhp's shown on the pump performance curves are generally based on water. Hence, when estimating the bhp for any other liquid, correction of specific gravity needs to be made, as shown previously.

CONTROL VALVES

A control valve is a variable opening used to regulate the flow of a process fluid as required by the process. The three important aspects of a control valve are capacity, characteristics, and rangeability.

Capacity

This is expressed in terms of C_v, the flow coefficient, which is the flow of water at 60°F in gallons per minute at a pressure drop of 1 psi across the valve when it is completely open.

Characteristics

This is the relationship between the change in the valve opening and the change in the flow through the valve, viz., equal percentage, linear, and nonlinear.

Rangeability

This is the ratio of the maximum controllable flow to the minimum controllable flow. In terms of C_v, the rangeability R is given by

$$R = \frac{\text{rated } C_v}{\text{minimum controllable } C_v}$$

The formulas used for control valve sizing are as follows:

Volumetric flow:
$$C_v = \begin{cases} Q\sqrt{\dfrac{s}{\Delta P}} & \text{liquid service} \\ \dfrac{Q}{963}\sqrt{\dfrac{s_g T}{\Delta P(P_1+P_2)}} & \text{gas service} \end{cases} \quad (4.92)$$

Mass flow:
$$C_v = \begin{cases} \dfrac{W}{500\sqrt{s\,\Delta P}} & \text{liquid service} \\ \dfrac{W}{3.22\sqrt{\Delta P(P_1+P_2)s_{gf}}} & \text{gas service} \end{cases} \quad (4.93)$$

where
- C_v = valve coefficient
- Q = flow rate, gpm for liquid, ft³/min for gases and vapors
- s = specific gravity at flowing temperature (water s at 60°F = 1)
- ΔP = pressure drop across valve, psi
- W = flow rate, lb/h
- s_g = specific gravity of gas relative to air
   = $\frac{1}{29}$ the molecular weight of gas
- $s_{gf} = s_g\,(520/T)$
- T = flow temperature, °R
- P_1 = upstream pressure, psia
- P_2 = downstream pressure, psia

In terms of ΔP and Q (gpm) the rangeability is given by

$$R = \frac{Q_{\max}}{Q_{\min}}\sqrt{\frac{\Delta P_{\min}}{\Delta P_{\max}}} \quad (4.94)$$

where

Q_{max} = maximum flow
Q_{min} = minimum flow
ΔP_{min} = pressure drop at the minimum flow (not the minimum pressure drop)
ΔP_{max} = pressure drop at the maximum flow (not the maximum pressure drop)

Equations 4.92 and 4.93 apply to subcritical flow. For liquids, when the ΔP across the control valve is less than the ΔP that would cause flashing, the flow is subcritical. For gases, the flow is subcritical when the ΔP across the control valve is roughly less than 0.5 times upstream pressure in absolute units. The sizing calculation of the control valve for critical and supercritical flows requires additional information beyond the scope of this book.

Control Valve Sizing

For sizing a control valve for liquid service, a flow coefficient C_{vc} is calculated with the design flow rate in gallons per minute from

$$C_{vc} = Q\sqrt{\frac{s}{\Delta P}} \qquad (4.95)$$

For a good range of control, a value of $C_v > C_{vc}$ is required. The normal practice is to select a valve such that[3b]

$$\frac{C_v}{C_{vc}} = 1.25 \text{ to } 2 \qquad (4.96)$$

where C_v is the coefficient of the selected valve from the manufacturer's catalog. The pressure drop across the control valve is generally taken as 50 percent of the frictional drop in the system excluding the control valve or 5 psi to 10 psi minimum. Thus if the frictional drop in the line, heat exchanger, fittings, etc., at the design flow rate is 30 psi, the drop across the control valve should be 15 psi. For a very high-pressure drop line, the percentage drop across the valve may be 15 to 20 percent of the frictional drop through the system.

The minimum pressure drop at the fully open condition is given by

$$\Delta P_{min} = \left(\frac{Q}{C_v}\right)^2 s \text{ psi} \qquad (4.97)$$

For intermediate positions, the pressure drop can be calculated by

$$\Delta P = \left[\frac{Q}{(C_{vc}/C_v)C_v}\right]^2 s \text{ psi} \qquad (4.98)$$

where C_v is the valve coefficient from the manufacturer's catalog and C_{vc}/C_v is the intermediate flow condition.

When sizing control valves for a gas, steam, or vapor service, the valve coefficients may be calculated with the use of either Equation 4.92 or 4.93. For good control, the ratio of the selected valve coefficient to the calculated coefficient should be according to Equation 4.96 in the case of gases also.

Example 4.18

Select the control valve size from the following data assuming subcritical flow of 100 gpm, an available pressure drop of 10 psi, and a specific gravity of 2.5.

Valve size, in.	1	$1\frac{1}{2}$	2	3	4
C_v	7.6	19	35	76	130

After the final selection, estimate the maximum flow through the valve for the specified pressure drop.

Solution

$$C_{vc} = Q\sqrt{\frac{s}{\Delta P}} = 100\sqrt{\frac{2.5}{10}} = 50$$

With a 3-in. valve,

$$C_v = 76$$

$$\frac{C_v}{C_{vc}} = \frac{76}{50} = 1.52$$

This ratio is between 1.25 and 2 and acceptable. Therefore,

$$76 = Q\sqrt{\frac{2.5}{10}} = Q(0.5)$$

$$Q = 152 \text{ gpm (maximum)}$$

Example 4.19

An existing pumping system has a flow control valve working at 50 percent of the rated flow coefficient. The system study indicates that the valve is working with a pressure drop of 20 psi and specific gravity of 0.9. Because of the expansion of the plant capacity, the flow from the pump is to be increased from 100 to 130 gpm and, consequently, the pressure drop available for the control valve drops to 15 psi. If the pump is retained, will the control valve need replacement?

Solution

For existing operation,

$$C_{vc} = Q\sqrt{\frac{s}{\Delta P}} = 100\sqrt{\frac{0.9}{20}} = 21.2$$

$$\text{Rated } C_v = \frac{21.2}{0.5} = 42.4$$

For the new case,

$$C_{vc} = 130\sqrt{\frac{0.9}{15}} = 31.8 \qquad \frac{C_v}{C_{vc}} = \frac{42.4}{31.8} = 1.33$$

This ratio is within the range of 1.25 to 2, and hence the control valve would work and would need no replacement.

Example 4.20

In a system, the frictional drop excluding the pressure drop across the control valve for the maximum controllable flow of 100 gpm is 30 psi. The pressure drop available for the control valve for this maximum flow is 5 psi. If the rangeability of the control valve is 50 and the terminal pressures do not change, estimate the minimum controllable flow of the valve.

Solution

$$R = \frac{Q_{max}}{Q_{min}} \sqrt{\frac{\Delta P_{min}}{\Delta P_{max}}}$$

ΔP_f = frictional drop = $\Delta P_1 + \Delta P_2$
$P_1 - P_2$ = constant = $\Delta P_f + \Delta P$
ΔP = control valve pressure drop
ΔP_1 = frictional drop upstream of the control valve
ΔP_2 = frictional drop downstream of the control valve
P_1 = upstream terminal pressure
P_2 = downstream terminal pressure

Any lowering in the frictional drop would increase the available ΔP for the control valve:

$$(\Delta P_f)_{max} = 30 \text{ psi}$$

when the flow is

$$Q_{max} = 100 \text{ gpm} \quad \text{and} \quad \Delta P_{max} = 5 \text{ psi}$$

$$(\Delta P_f)_{min} = \left(\frac{Q_{min}}{Q_{max}}\right)^2 (30)$$

The decrease in the frictional drop at the minimum flow is

$$30 - \left(\frac{Q_{min}}{Q_{max}}\right)^2 (30) = 30\left[1 - \left(\frac{Q_{min}}{Q_{max}}\right)^2\right]$$

$$\Delta P_{min} = 5 + 30\left[1 - \left(\frac{Q_{min}}{Q_{max}}\right)^2\right]$$

$$R = \frac{Q_{max}}{Q_{min}} \sqrt{\frac{\Delta P_{min}}{\Delta P_{max}}}$$

$$50 = \frac{100}{Q_{min}} \sqrt{\frac{5 + 30[1 - (Q_{min}/100)^2]}{5}}$$

$$1.25 \, Q_{min}^2 = 5 + 30 - 0.003 \, Q_{min}^2$$

Solving for Q_{min} gives Q_{min} = 5.3 gpm.

PARALLEL AND BRANCHED SYSTEMS

Fluid distribution systems consist of piping networks with series and parallel connections. The pressure drop and flow calculations involve node analysis. A node is a point where two or more lines meet. Each node is assigned a number. The flow rate from node i to node j is denoted by Q_{ij}. Five items are established for each segment:

1. The pressure at the node, P_i: P_1, P_2 ...

2. The material balance at each node: total flow in = total flow out

3. The overall material balance

4. The Reynolds number and friction factor equations for flow between nodes i and j

$$N_{Re\,ij} = \frac{6.32 W_{ij}}{\mu_{ij} d_{ij}}$$

$$f_{ij} = \frac{1}{2.44 \left[\ln\left(\frac{0.135(\varepsilon)(12)}{d_{ij}} \right) + \frac{6.5}{N_{Re\,ij}} \right]^2}$$

5. The frictional pressure drop equation between nodes i and j

$$\Delta P_{ij} = \frac{0.0000134 f_{ij} L_{ij} W_{ij}^2}{\rho_{ij} d_{ij}^5}$$

For the same fluid at the same temperature, as in a water distribution system, the physical properties may be assumed the same and initial estimate can be done with Fanning friction factor as 0.005.

In general, the terminal pressures, line lengths, and fittings and diameters are known, and the flow rates through various branches are to be found out. The number of unknowns must match the number of independent equations.

REPLACEMENT OF ONE PIPELINE WITH n PARALLEL PIPELINES OF EQUAL EQUIVALENT LENGTHS AS THE ORIGINAL PIPE

This is a simple application of piping network.

The original pipe of internal diameter d_o in., length L ft, and flow, Q gpm.

The pressure drop from node 1 to node 2 may be given as

$$\Delta P = \frac{0.000864 f_o L \rho Q_o^2}{d_o^5} = \frac{k f_o Q_o^2}{d_o^5}$$

where k is a constant for the same fluid and length of pipe.

Because the flow will be divided by n equal parts,

$$\Delta P = \frac{0.000864 f_n L \rho \left(\frac{Q_o}{n} \right)^2}{d_n^5} = \frac{k f_n \left(\frac{Q_o}{n} \right)^2}{d_n^5}$$

where k is a constant for the same fluid and length of pipe.

If the flow is fully turbulent in both the cases, $f_o = f_n$. Equating the two preceding expressions, the following two equations may be derived:

1. To replace one big pipe of internal diameter, d_o, with n small pipes of internal diameter, d_n:

$$d_n = \frac{d_o}{n^{0.4}} \tag{4.99}$$

2. To replace n small pipes of internal diameter, d_n, with one big pipe of internal diameter, d_o:

$$d_o = (n^{0.4})d_n \tag{4.100}$$

Economic Pipe Diameter

A simple approximate estimate[8] may be made from the following equations:

1. For turbulent flow in steel pipes:

$$d_{optimum} = \begin{cases} 3.9(q^{0.49})(\rho^{0.14}) & \text{for} \quad d_{internal} \geq 1\,\text{in} \\ 4.7(q^{0.49})(\rho^{0.14}) & \text{for} \quad d_{internal} < 1\,\text{in} \end{cases} \tag{4.101}$$

2. For viscous flow in steel pipes:

$$d_{optimum} = \begin{cases} 3.0(q^{0.36})(\mu^{0.18}) & \text{for} \quad d_{internal} \geq 1\,\text{in} \\ 3.6(q^{0.40})(\mu^{0.20}) & \text{for} \quad d_{internal} < 1\,\text{in} \end{cases} \tag{4.102}$$

where

$d_{internal}$ = internal diameter, in
ρ = fluid density, lb/ft^3
μ = fluid viscosity, cP
q = fluid flow rate, ft^3/s

FLOW OF COMPRESSIBLE FLUIDS

A fluid is compressible when its density changes with pressure. The following terms are often used in connection with the flow of compressible fluids.

Mach Number N_{Ma}

This is the ratio of the fluid velocity to the velocity of sound in the same material under identical temperature and pressure.

$$N_{Ma} = \frac{u}{u_s} \tag{4.103a}$$

where
u = fluid velocity, ft/s

$$u_s = \begin{cases} \sqrt{\frac{Bg_c}{\rho}} & \text{for solids, liquids, and gases, ft/s} \\ \sqrt{kg_c RT} & \text{for gases only, ft/s} \end{cases} \tag{4.103b}$$

u_s = velocity of sound in the same material
B = bulk modulus of elasticity of the material, lb_f/ft^2
ρ = density of material, lb/ft^3

$$N_{Ma} \begin{cases} < 1, & \text{subsonic flow} \\ = 1, & \text{sonic flow} \\ > 1, & \text{supersonic flow} \end{cases}$$

For explanation of k, g_c, R, T, see Equation 4.106.

Critical Pressure Ratio

For a constant pressure, the flow rate of a compressible fluid increases as the downstream pressure is lowered until a downstream pressure is reached at which the flow rate is the maximum and the fluid velocity equals the sonic velocity. At this condition, the ratio of downstream pressure to the upstream pressure in absolute scale is called the *critical pressure ratio* (r_c). For gases obeying the ideal-gas law, the critical pressure ratio is given by

$$r_c = \frac{P_{downstream}}{P_{upstream}} = \left(\frac{2}{k+1}\right)^{k/(k-1)} \quad (4.104)$$

$$k = \frac{C_p}{C_v} = \frac{C_p}{C_p - 1.987} \quad (4.105)$$

where
 k = specific heat ratio
 C_p = molar specific heat at constant pressure, Btu/lb mol · °F
 C_v = molar specific heat at constant volume, Btu/lb mol · °F

If the downstream pressure is less than what is calculated from the preceding equation, the flow rate becomes independent of the downstream pressure.

Critical Velocity

The critical velocity of a gas through a pipe or an orifice is the sonic velocity. It is given by

$$u_s = \sqrt{kg_c RT} \text{ ft/s, [m/s]} \quad (4.106)$$

where
 k = ratio of specific heats at constant pressure and volume, respectively
 g_c = 32.2 lb · ft/(lb$_f$ · s^2), [(1 kg·m/s^2)/N]
 $R = \frac{1545}{M}$ ft · lb$_f$/lb · °R, $\left[\frac{8314 \text{ N} \cdot \text{m}}{M \text{ kg} \cdot \text{K}}\right]$
 T = temperature, °R, K
 M = molecular weight

For the estimation of the flow of compressible fluids, the following rules may be used:

1. When the pressure drop is less than 10 percent of inlet pressure, use the density based on either the inlet or outlet conditions.

2. When the pressure drop is more than 10 percent but less than 40 percent of the inlet pressure, use the average density based on the inlet and outlet conditions.

3. For a pressure drop more than 40 percent of the inlet pressure, both the inlet and outlet densities should be considered.

4. At a very high pressure, the flow of gases may be treated the same way as the flow of incompressible fluid since the change in the density is small.

Pressure Drop Calculation for Compressible Fluid Obeying Ideal-Gas Law

In the practical application of compressible fluid flow, when the downstream pressure is lowered with a fixed upstream pressure and a fixed uniform geometry, the velocity of the compressible fluid increases and so does the flow until the velocity reaches the velocity of sound in the fluid ($N_{Ma} = 1$). If the downstream pressure is lowered further, flow does not increase. This is also known as chocked flow. Thus the knowledge of Mach number allows one to determine the maximum possible flow for a given inlet pressure and geometry of the hardware for a compressible fluid. All compressible flows, whether subsonic or supersonic initially in a pipe of uniform diameter, will reach sonic velocity as the length is increased. However, a subsonic flow can be transformed into a supersonic flow through a convergent-divergent nozzle if the sonic velocity is reached at the throat and the pressure at the outlet is lower than the throat pressure.

If downstream pressure is less than what is calculated from Equation 4.104, the flow rate becomes independent of the downstream pressure. As a rule of thumb, when the absolute pressure ratio $P_{downstream}/P_{upstream}$ is less than 0.5, critical flow (choked flow) is anticipated.

A simple manual method of computing compressible fluid flow rate can be done using the Crane[9] formulae as follows:

$$W = \begin{cases} 1891(Y)(d^2)\sqrt{\frac{(\Delta P)\rho_1}{\Sigma K}} & \text{for flow through lines, valves, and fittings} \quad (4.106a) \\ 1891(Y)(d_1^2)(C)\sqrt{(\Delta P)\rho_1} & \text{for flow through nozzles amd orifices} \quad (4.106b) \end{cases}$$

where

W = flow, lb/h
Y = expansion factor, dimensionless, for lines see Crane A-22; for orifice, A-21
d = pipe inside diameter, in
d_1 = orifice or nozzle diameter, in
$\Delta P = \begin{cases} P_{upstream} - P_{downstream} & \text{for sub-critical flow, psi} \\ \text{Limiting } \Delta P \text{ for critical flow, psi} \end{cases}$
$K = \Sigma K_{\text{fittings \& valves}} + \frac{4fL}{D}$
ρ_1 = fluid density at upstream conditions, lb/ft^3
C = flow coefficient for orifice, see Crane A-20
f = Fanning friction factor [Do not use Crane friction factor!]
L = length of pipe, ft
D = inside diameter of pipe, ft

For compressible flow calculation, the following steps may be used:

1. For the system, calculate the total flow resistance factor K

2. From the given upstream pressure and downstream pressure (pressure at the exit of pipe), calculate

$$\left[\frac{\Delta P}{P'_1}\right]_{actual} = \frac{P_{upstream} - P_{downstream}}{P_{upstream}}$$

3. For a given specific heat ratio and calculated K factor, examine if

   (a) $$\left[\frac{\Delta P}{P'_1}\right]_{actual} \geq \left[\frac{\Delta P}{P'_1}\right]_{sonic}$$

   (b) $$\left[\frac{\Delta P}{P'_1}\right]_{actual} < \left[\frac{\Delta P}{P'_1}\right]_{sonic}$$

4. If 3 (a) is true, then calculate

$$\Delta P_{limiting} = \left[\frac{\Delta P}{P'_1}\right]_{sonic} \text{ value form table} \times (P'_1)$$

Use this and corresponding value of Y to compute the flow rate.

5. If 3(b) is true, then use actual pressure drop and corresponding Y to compute flow rate.

An alternative iterative method uses Mach number as follows:

Calculate the friction factor using Equation 4.34, and then estimate the upstream Mach number (Ma1) by trial and error using Equation 4.107, followed by the calculation of remaining parameters. The following equations are valid for adiabatic compressible flow:

$$\frac{(k+1)}{2}\ln\frac{2+(k-1)N^2_{Ma1}}{(k+1)N^2_{Ma1}} - \left(\frac{1}{N^2_{Ma1}}-1\right) + k(4fL/D) = 0 \qquad (4.107)$$

$$Y_1 = 1 + \frac{(k-1)N^2_{Ma1}}{2} \quad \text{dimensionless} \qquad (4.108)$$

$$T_{choked} = 2Y_1 T_1/(k+1), \, °R[K] \qquad (4.109)$$

$$P_{choked} = P_1 N_{Ma1}[2Y_1/(k+1)]^{0.5}, \, lb_f/ft^2 [N/m^2] \qquad (4.110)$$

$$G_{choked} = P_{choked}[kg_c M/(RT_{choked})]^{0.5}, \, lb/ft^2 \cdot s \, [kg/m^2 \cdot s] \qquad (4.111)$$

$$\text{Mass flow} = G_{choked} \pi D^2/4, \, lb/s[kg/s] \qquad (4.112)$$

k = ratio of specific heat at constant pressure to that at constant volume; f = Fanning friction factor; L = equivalent length of pipe, ft; D = internal diameter of pipe, ft

Compression Equipment

The objective of compression is to deliver a gas in a required quantity at a pressure higher than the initial. The compression equipment is basically of two types:

1. Positive displacement, e.g., reciprocating
2. Velocity or dynamic compressor, e.g., centrifugal

Compression is the following three types: (1) adiabatic (2) isothermal, and (3) polytropic. In the adiabatic compression there is no heat exchange with the surroundings and the equation $PV^k = C_1$ applies, where k is the ratio of specific heats C_p/C_v for the gas.

During an isothermal compression, the heat of compression is removed from the gas by cooling to maintain the gas temperature constant. The equation $PV = C_2$ applies in this case.

The polytropic compression is neither adiabatic nor isothermal. In this case the equation $PV^n = C_3$ applies. If ε_p is the polytropic compression efficiency, k and n are related by the equation

$$\frac{n}{n-1} = \varepsilon_p \frac{k}{k-1} \tag{4.113}$$

where n is the polytropic coefficient. An average value of 0.725 for ε_p is usually assumed for estimation purposes. The actual efficiency will be different depending on the speed, wheel design, compression ratio, and other factors.

A compressor operates on a predetermined performance curve that shows the relationship between the total head and horsepower requirement as a function of the volumetric flow rate. As in the case of pumps, the exact operating point of a compressor is determined by the intersection of the system and performance curves. Compressor calculations are done either with the use of simplified equations (Table 4.7) for the adiabatic or polytropic paths or by the use of the Mollier diagrams where such diagrams are available or by using the thermodynamic property data. The Mollier diagrams are easier to use but they are available mostly for the pure components, especially for the refrigerants.

Ratio of Specific Heats *k*

Values of the specific heat ratios k are available for some gases at 1 atm. When the k value is not available, it can be calculated from molar specific heats at constant pressure by the formula

$$k = \frac{C_p}{C_p - 1.987} \tag{4.114}$$

For gas mixtures, an average k_{av} can be calculated on the basis of their mole fractions.

Multistage Compression

The multistage operation allows interstage cooling of the gas, which reduces work of compression. The minimum work is obtained when the compression ratio in

Table 4.7 Calculation of power requirements in compression

Type of Compression	Adiabatic
Equation of state	$PV^k = C_1$
Exponent	$k = \dfrac{C_p}{C_v}$
Theoretical discharge temperature	$T_2 = T_1 \left(\dfrac{P_2}{P_1}\right)^{(k-1)/k}$
Actual discharge temperature	$T_2 = T_1 + \dfrac{T_1[(P_2/P_1)^{(k-1)/k} - 1]}{\varepsilon_{ad}}$
Head, ft · lb$_f$/lb or ft	$H_{ad} = \dfrac{k}{k-1}\dfrac{RT_1}{M}\left[\left(\dfrac{P_2}{P_1}\right)^{(k-1)/k} - 1\right]Z_{av}$
Gas horsepower, ghp (using cfm)	$\text{ghp} = \dfrac{0.00437 Q_1 P_1 [k/(k-1)][(P_2/P_1)^{(k-1)/k} - 1]Z_{av}}{\varepsilon_{ad}}$
Gas horsepower, ghp (using weight)	$\text{ghp} = \dfrac{W H_{ad}}{33{,}000\,\varepsilon_{ad}}$
Brake horsepower, bhp	$\text{bhp} = \dfrac{\text{ghp}}{\varepsilon_m}$

C_p, C_v = molar specific heats at constant pressure and constant volume, respectively, Btu/lb mol · °R
T_1, T_2 = inlet and outlet temperature, °R
P_1, P_2 = inlet and discharge pressure, psia
ε = efficiency (ad = adiabatic, P = polytropic, iso = isothermal, m = mechanical)
W = lb/min, Q_1 = cfm at suction conditions
H_{ad} = adiabatic head, ft; H_p = polytropic head, ft; H_{iso} = isothermal head, ft
R = gas constant = 1545 ft · lb$_f$/lb mol · °R
M = molecular weight
$Z_{av} = \tfrac{1}{2}(Z_1 + Z_2)$, Z_1, Z_2 compressibilities at suction and discharge

Polytropic	Isothermal
$PV^n = C_2$	$PV = C_3$
$\dfrac{n-1}{n} = \dfrac{k-1}{k}\dfrac{1}{\varepsilon_P}$	1
$T_2 = T_1\left(\dfrac{P_2}{P_1}\right)^{(n-1)/n}$	$T_2 = T_1$
$T_2 = T_1\left(\dfrac{P_2}{P_1}\right)^{(n-1)/n}$	$T_2 = T_1$
$H_p = \dfrac{n}{n-1}\dfrac{RT_1}{M}\left[\left(\dfrac{P_2}{P_1}\right)^{(n-1)/n} - 1\right]Z_{av}$	$H_{iso} = Z_{av}\dfrac{RT_1}{M}\ln\dfrac{P_2}{P_1}$
$\text{ghp} = \dfrac{0.00437 Q_1 P_1 [(P_2/P_1)^{(n-1)/n} - 1]Z_{av}}{\varepsilon_P}$	$\text{ghp} = \dfrac{0.00437 Q_1 P_1 \ln(P_2/P_1)}{\varepsilon_{iso}}$

Table 4.7 Calculation of power requirements in compression (*Continued*)

Polytropic	Isothermal
$\text{ghp} = \dfrac{WH_P}{33,000}$	$\text{ghp} = \dfrac{WH_{iso}}{33,000\, \varepsilon_{iso}}$
$\text{bhp} = \dfrac{\text{ghp}}{\varepsilon_m}$	$\text{bhp} = \dfrac{\text{ghp}}{\varepsilon_m}$

each stage is equal to

$$\frac{P_n}{P_{n-1}} = \left(\frac{P_n}{P_1}\right)^{1/N_s} \qquad (4.115)$$

where
 N_s = Number of stages
 P_n = pressure after n stages
 P_1 = pressure to first stage

Fans

These are used for low pressure and usually for pressure heads less than 1.5 psi. They are either centrifugal or axial flow type. Air horsepower of a fan is given by

$$\text{Air hp} = \frac{144(P_2 - P_1)Q_V}{33,000} = 0.00436(P_2 - P_1)Q_V \qquad (4.116)$$

where P_2 and P_1 are in psia. If $P_2 - P_1$ is expressed in inches of water, the air horsepower is given by

$$\text{Air hp} = 0.0001575 Q_V\, \Delta P \qquad (4.117)$$

where ΔP is the developed head across the fan in inches of water and Q_V is flow in cubic feet per minute at the inlet conditions.

Fan Static Pressure P_s

This is the total pressure rise ΔP minus the velocity pressure in the fan outlet. Thus $P_s = \Delta P - P_v$. P_v = velocity head expressed in inches of water. It is given by

$$P_v = \frac{u_2^2}{2g_c}\left[\frac{\rho_{air}(12)}{62.4}\right] \text{ in } H_2O$$

where u_2 is the fan outlet velocity, ft/s.

$$u_2 = 18.3\sqrt{\frac{P_v}{\rho_{air}}}, \quad \rho_{air} = \text{air density lb/ft}^3 \qquad (4.118)$$

Example 4.21

A centrifugal fan operating at 1740 rpm has characteristics as shown in Exhibit 9. It is connected to a duct system that offers a static resistance of 1.5 in. H$_2$O when handling 2500 cfm of air. (a) At what flow, static pressure, and bhp will the fan and the duct system operate when connected together? (b) The flow through the duct is to be 4500 cfm of air by changing pulley ratios. What speed, static pressure, and bhp would be required to do this?

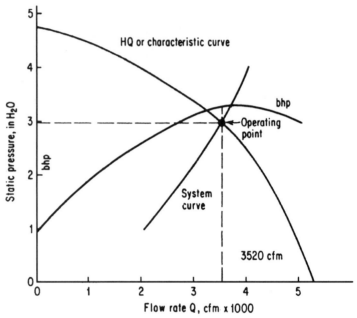

Exhibit 9 Data and solution

Solution

(a) A fan or a pump will always operate at the intersection of its headcapacity curve and the system curve. Therefore, a system curve has to be prepared first.

ΔP at 2500 cfm is given to be 1.5 in. H$_2$O. Assume turbulent flow and f, the friction factor, constant. Then $\Delta P \propto u^2$, i.e., as Q^2, if the duct cross-section is constant. Calculate values as follows:

Cfm	2500	3000	3250	3500	4000
ΔP, in. H$_2$O	1.5	2.16	2.54	2.94	3.84

These points are plotted on the same graph to give the system curve that cuts the HQ curve. This is the operating point for the system (Exhibit 9). At this point,

$$\text{Flow} = 3520 \text{ cfm}$$
$$\text{bhp} = 3.28$$
$$\text{Static pressure} = 2.98 \text{ in. H}_2\text{O}$$

(b) To obtain the solution to part b, make use of the affinity laws.

$$\Delta P_s \propto Q^2 \quad \text{New } \Delta P_s = 2.98 \left(\frac{4500}{3520}\right)^2 = 4.87 \text{ in } H_2O$$

$$\text{bhp} \propto Q^3 \quad \text{New bhp} = 3.28 \left(\frac{4500}{3520}\right)^3 = 6.85 \text{ bhp}$$

$$\text{rpm} \propto Q \quad \text{New rpm} = 1740 \frac{4500}{3520} = 2225 \text{ rpm}$$

TWO-PHASE (VAPOR/LIQUID) FLOW CONSIDERATION

The two-phase pressure drop consists of the sum of three items:

1. Irreversible frictional loss
2. Reversible loss of pressure due to change in elevation of the fluid
3. Reversible loss of pressure due to acceleration.

The pressure drop calculation is complex due to the following:

1. The density, velocity, and friction factors are often different in each phase
2. The gas phase generally slips from the liquid phase.
3. Liquid phase composition may change as a result of pressure drop.

Various types of flow regimes have been recognized: bubbly, plug, slug, stratified, wavy, annular, and dispersed. Of these, the slug flow should be avoided because it induces instability.

One empirical relation to avoid slug flow is to have a minimum two-phase velocity, m/s, exceeding [3.05 + 0.024(inside pipe diameter in mm)].

Use the following procedures to determine the flow regimes.

Property	Air at 68°F & 1 atm		Water at 68°F & 1 atm	
Viscosity (μ_{water})			1.033 mPa.s	1.033 cP
Density (ρ_{air}, ρ_{water})	1.2 kg/m³	0.075 lb/ft³	1009 kg/m³	62.32 lb/ft³
Surface tension (σ_{water})			0.073 N/m	72.966 Dyn/cm

Step 1: Calculate parameters λ and Ψ as follows using the physical properties of water and air at standard conditions from the preceding table, and the corresponding physical properties of the liquid and gas of the two-phase mixture under consideration in consistent units.

$$\lambda = \left[\frac{\rho_G \rho_L}{\rho_{air} \rho_{water}}\right]^{0.5} \qquad (4.119)$$

$$\Psi = \left(\frac{\sigma_{water}}{\sigma_L}\right) \left[\frac{\mu_L}{\mu_{water}} \left(\frac{\rho_{water}}{\rho_L}\right)^2\right]^{0.33} \qquad (4.120)$$

Step 2: Calculate the X- and Y-parameters of Baker's flow regime map as follows.

$$Y - \text{parameter} = \frac{G'}{\lambda} \quad (4.121)$$

$$X - \text{parameter} = \frac{(L')(\lambda)(\Psi)}{G'} \quad (4.122)$$

where

G' = gas mass velocity, $\dfrac{\text{lb}}{\text{ft}^2 \cdot \text{h}}$, or $\dfrac{\text{kg}}{\text{m}^2 \cdot \text{s}}$

L' = liquid mass velocity, $\dfrac{\text{lb}}{\text{ft}^2 \cdot \text{h}}$, or $\dfrac{\text{kg}}{\text{m}^2 \cdot \text{s}}$

The mass velocities are obtained from corresponding mass flow rates and the internal flow cross-section of the pipe. From these parameters, read the flow regime from Baker's Flow Regime Map [Fig. 4.11].

Two simplified methods are presented here.
Symbols used in the methods are as follows:

W_L = liquid flow rate, lb/h
W_G = gas flow rate, lb/h
P_{in} = inlet pressure, psig
P_{out} = outlet pressure, psig
d = inside diameter, in.
μ_L = viscosity of liquid phase, cP
μ_G = viscosity of gas phase, cP
ρ_L = density of liquid phase, lb/ft^3
ρ_G = density of gas phase, lb/ft^3

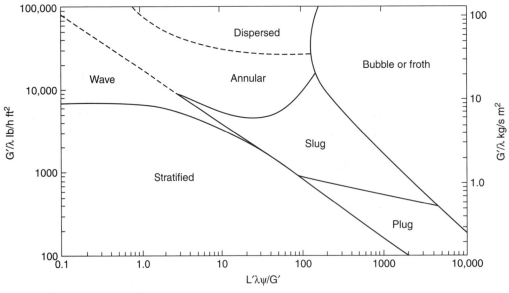

Figure 4.11 Baker's Flow Regime Map

$(\rho_G)_{avg}$ = average of inlet and outlet gas phase density, lb/ft^3
ε = roughness factor, ft
L = pipe length, ft
F = Fanning friction factor

The two methods presented here apply for frozen flows (i.e., flows in which the liquid and vapor flow rates do not change due to vaporization caused by pressure drop and friction), and the two phases have no slip between them. Calculation of flashing flows involving vapor-liquid phase slip is computationally intensive, and may be studied in the cited text.[5]

Lockhart-Martinelli Method

In this method, pressure drops (ΔP_L and ΔP_G) are calculated using each phase (L and G) as if each phase were flowing by itself. Then the Lockhart-Martinelli parameter, X, is determined as follows:

$$X = \left(\frac{\Delta P_L}{\Delta P_G}\right)^{0.5} \quad (4.123)$$

In the computation of the gas-phase pressure drop, an initial estimate of pressure drop is made to calculate the average gas density. If this estimate is too far off, then an iterative process is used to converge. Alternatively, the calculation could be done on small incremental lengths in which the inlet pressure of a segment becomes the outlet pressure of the previous segment. This method is tedious and requires a spreadsheet or a computer. This method does not include the acceleration loss.

Next the two-phase flow multiplier is calculated using the value of C from the following table.

The turbulent or viscous condition is determined for each phase from the Reynolds number.

$$\Phi_{Lm}^2 = 1 + \frac{C}{X} + \frac{1}{X^2} \quad (4.124)$$

$$\Phi_{Gm}^2 = 1 + CX + X^2 \quad (4.125)$$

$$\Delta P_{\text{two-phase}} = \Phi_{Lm}^2(\Delta P_L) = \Phi_{Gm}^2(\Delta P_G) \quad (4.126)$$

Table 4.8 Lockhart-Martinelli two-phase flow multiplier

Flow State Combination	Liquid	Gas	C
Tt	Turbulent	turbulent	20
Vt	Viscous (laminar)	turbulent	12
Tv	Turbulent	viscous (laminar)	10
Vv	Viscous (laminar)	viscous (laminar)	5

Duckler's Homogeneous Method

Duckler's method is based on the assumption that two phases move without a slip and the properties of the combined phase can be suitably averaged. The following method of averaging is proposed:

$$\rho_{NS} = \rho_L \lambda_L + \rho_G (1 - \lambda_L) \qquad (4.127)$$

$$\mu_{NS} = \mu_L \lambda_L + \mu_G (1 - \lambda_L) \qquad (4.128)$$

$$\lambda_L = \frac{Q_L}{Q_L + Q_G} \qquad (4.129)$$

where

ρ = density, lb/ft^3
μ = viscosity, cP
Q = volumetric flow rate Subscripts: NS = no slip
L = liquid
G = gas

The frictional pressure drop is calculated on the basis of total flow as follows:

$$\Delta P_f = \frac{0.0000134(f)(L)(W_L + W_G)^2}{(\rho_{NS}) d^5} \qquad (4.130)$$

The acceleration loss and two-phase pressure drop are calculated as follows:

$$A_{cc} = \left(\frac{16}{\pi^2}\right) \left[\frac{(W_G + W_L) W_G}{(3600)^2 (g_c)}\right] \left[\frac{12}{d}\right]^4 \left[\frac{\left(\frac{P_{in} + P_{out}}{2}\right) + 14.7}{(P_{in} + 14.7)(P_{out} + 14.7)(\rho_G)_{avg}}\right] \qquad (4.131)$$

$$\Delta P_{\text{two-phase}} = \frac{\Delta P_f}{A_{cc}} \qquad (4.132)$$

Of these two cited methods, the Lockhart-Martinelli is more conservative.

Example 4.22

Calculate the pressure drop in the horizontal section of a pipe carrying a mixture of water and air using (a) Lockhart-Martinelli method, (b) Duckler's homogeneous method. The following information is available:

Inlet pressure: 110 psig, inlet temperature: 70°F

Pipe inside diameter: 4.026 in.

Pipe inside roughness factor: 0.00015 ft

Pipe equivalent length: 150 ft.

Fluid	Water	Air
Flow rate, lb/h	55000	6000
Density	62.3	calculate
Viscosity, cP	1	0.018

Solution

The flows of the two streams are frozen in rates: their rates do not change due to pressure drop.

(a) Lockhart-Martinelli method.
   Step 1: Assume a pressure drop of 10 psi and calculate average gas density.

$$\rho_{in} = \frac{PM}{RT} = \frac{(124.7)(29)}{10.731(460+70)} = 0.636 \frac{\text{lb}}{\text{ft}^3}$$

$$\rho_{out} = \frac{PM}{RT} = \frac{(114.7)(29)}{10.731(460+70)} = 0.585 \frac{\text{lb}}{\text{ft}^3}$$

$$\rho_{avg} = \frac{\rho_{in} + \rho_{out}}{2} = 0.61 \frac{\text{lb}}{\text{ft}^3}$$

Step 2: Calculate the Reynolds numbers using liquid alone and gas alone flowing through pipe and determine the viscous/turbulent flow condition due to each phase.

$$N_{Re\,L} = \frac{6.32 W_L}{(\mu_L)d} = \frac{6.32(55000)}{(1)(4.026)} = 8.634(10^4)$$

$$N_{Re\,G} = \frac{6.32 W_G}{(\mu_G)d} = \frac{6.32(6000)}{(0.018)(4.026)} = 5.233(10^5)$$

Flow is turbulent-turbulent with respect to liquid and gas.
Step 3: Calculate friction factors due to liquid and gas phase.

$$f_L = \frac{1}{2.44\left[\ln\left(\frac{0.135(\varepsilon)(12)}{d}\right) + \frac{6.5}{N_{Re\,L}}\right]^2}$$

$$= \frac{1}{2.44\left[\ln\left(\frac{0.135(0.00015)(12)}{4.026}\right) + \frac{6.5}{8.634(10^4)}\right]^2} = 0.004335$$

$$f_G = \frac{1}{2.44\left[\ln\left(\frac{0.135(\varepsilon)(12)}{d}\right) + \frac{6.5}{N_{Re\,G}}\right]^2}$$

$$= \frac{1}{2.44\left[\ln\left(\frac{0.135(0.00015)(12)}{4.026}\right) + \frac{6.5}{5.233(10^5)}\right]^2} = 0.004335$$

Step 4: Calculate frictional pressure drop based on liquid alone and gas alone.

$$\Delta P_L = 0.0000134(f_L)(L)\left[\frac{W_L}{\rho_L(d^5)}\right]$$

$$= 0.0000134(0.004335)(150)\left[\frac{55000}{62.3(4.026)^5}\right]$$

$$= 0.4 \text{ psi}$$

$$\Delta P_G = 0.0000134(f_L)(L)\left[\frac{W_G}{\rho_{avg}(d^5)}\right]$$

$$= 0.0000134(0.004335)(150)\left[\frac{6000}{0.61(4.026)^5}\right]$$

$$= 0.486 \text{ psi}$$

Step 5: Calculate Lockhart-Martinelli Parameter.

$$X = \left[\frac{\Delta P_L}{\Delta P_G}\right]^{0.5} = \left[\frac{0.4}{0.486}\right]^{0.5} = 0.907$$

Step 6: Pick appropriate C-value from Table 4.8 and compute two-phase pressure drop.

Corresponding to turbulent-turbulent flow, $C = 20$.

$$\Phi_{Lm}^2 = 1 + \frac{C}{X} + \frac{1}{X^2} = 1 + \frac{20}{0.907} + \frac{1}{(0.907)^2} = 24.258$$

$$\Phi_{Gm}^2 = 1 + CX + X^2 = 1 + 20(0.907) + (0.907)^2 = 19.97$$

$$\Delta P_{\text{two-phase}} = \Phi_{Lm}^2[\Delta P_L] = 24.258[0.4] = 9.703 \text{ psi}$$

$$= \Phi_{Gm}^2[\Delta P_G] = 19.97[0.486] = 9.703 \text{ psi}$$

(b) Duckler's homogeneous method

Step 1: Calculate volume fraction of liquid phase, no-slip density, and no-slip viscosity.

$$Q_L = \frac{W_L}{\rho_L} = \frac{55000}{62.3} = 882.825 \ \frac{\text{ft}^3}{\text{h}}$$

$$Q_G = \frac{W_G}{\rho_{avg}} = \frac{6000}{0.61} = 9836 \ \frac{\text{ft}^3}{\text{h}}$$

$$\lambda_L = \frac{Q_L}{Q_L + Q_G} = 0.082$$

$$\rho_{NS} = \rho_L \lambda_L + \rho_{avg}(1 - \lambda_L) = 62.3(0.082) + 0.61(1 - 0.082) = 5.694 \ \frac{\text{lb}}{\text{ft}^3}$$

$$\mu_{NS} = \mu_L \lambda_L + \mu_G(1 - \lambda_L) = 1(0.082) + (0.018)(1 - 0.082) = 0.099 \text{ cP}$$

Step 2: Calculate the no-slip Reynolds number based on total flow, no-slip friction factor, and frictional pressure drop without acceleration loss.

$$N_{\text{Re }NS} = \frac{6.32(W_L + W_G)}{\mu_{NS} d} = \frac{6.32(61000)}{0.099(4.026)} = 9.68(10^5)$$

$$f_{NS} = \frac{1}{2.44\left[\ln\left(\frac{0.135(\varepsilon)(12)}{d}\right) + \frac{6.5}{N_{\text{Re }NS}}\right]^2}$$

$$= \frac{1}{2.44\left[\ln\left(\frac{0.135(0.00015)(12)}{4.026}\right) + \frac{6.5}{9.68(10^5)}\right]^2} = 0.004335$$

$$\Delta P_{NS} = 0.0000134(f_L)(L)\left[\frac{W_L + W_G}{\rho_{NS}(d^5)}\right]$$

$$= 0.0000134(0.004335)(150)\left[\frac{61000}{5.694(4.026)^5}\right]$$

$$= 5.384 \text{ psi}$$

Step 3: Calculate the acceleration factor, and compute two-phase pressure drop including acceleration loss.

$P_{in} = 110$ psig, $P_{out} = 100$ psig

$$A_{cc} = \frac{\left[16(W_L + W_G)(W_G)\left[\left(\frac{P_{in} + P_{out}}{2}\right) + 14.7\right]\right]}{\left[\pi^2(3600)^2(32.174)\left(\frac{d}{12}\right)^4 (P_{in} + 14.7)(P_{out} + 14.7)(144)(\rho_{avg})\right]}$$

$$= \frac{\left[16(61000)(6000)\left[\left(\frac{110+100}{2}\right) + 14.7\right]\right]}{\left[\pi^2(3600)^2(32.174)\left(\frac{4.026}{12}\right)^4 (110 + 14.7)(100 + 14.7)(144)(0.61)\right]} = 0.011$$

$$\Delta P_{\text{two-phase}} = \frac{\Delta P_{NS}}{1 - A_{cc}} = \frac{5.384}{1 - 0.011} = 5.442 \text{ psi}$$

RECOMMENDED REFERENCES

1. *Chemical Engineers' Handbook,* 5th ed., Perry (ed.), McGraw-Hill, New York, 1973; (*a*) p. 5–7; (*b*) p. 5–11; (*c*) p. 5–13; (*d*) p. 5–10; (*e*) p. 5–36.
2. Simpson and Weirich, *Chem. Eng.,* vol. 85, desk book issue/April 3, 1978, pp. 35–42.
3. Kern, *Chem. Eng.,* vol. 82, 1975: (*a*) Jan. 6, pp. 115–120; (*b*) April 14, p. 88.
4. Simpson et al, *Chemical Engineering,* June 17, 1968, pp. 192–214.

5. Darby, *Chemical Engineering Fluid Mechanics*, 2nd ed., Marcel Dekker, Inc., NY.
6. McCabe and Smith, *Unit Operations of Chemical Engineering*, 3rd ed., McGraw Hill Book Company, NY.
7. Altshul and Kiselev, *Gidravlika i Aerodinamika*, Moscow.
8. Peters and Timmerhaus, *Plant Design and Economics for Chemical Engineers*, McGraw Hill Book Company, NY.
9. Crane Technical Paper 10, CRANE, Chicago, IL.
10. Hooper, "The two-K method predicts head losses in pipe fittings," *Chemical Engineering,* August 1981.
11. Murty, "Quick estimate of spray-nozzle mean drop size," *Chemical Engineering,* July 1981.

CHAPTER 5

Heat Transfer

OUTLINE

THERMAL CONDUCTION 210
Composite Flat Walls

THERMAL CONVECTION 213
Simultaneous Convection and Radiation

THERMAL RADIATION 214
Composite Walls ■ Kirchhoff's Law ■ Radiation to a Completely Absorbing Receiver ■ Heat Exchange between Source and Receiver ■ Nongray Enclosures ■ Heat Transfer by Convection to Fluids Flowing Inside and Outside of Pipes ■ Film Coefficients for Fluids in Pipes and Tubes (No Change of Phase) ■ Free Convection of Fluids outside Horizontal Pipes ■ Boiling ■ Condensation Outside Tubes ■ Condensation Inside Tubes ■ Shell and Tube Exchangers ■ Logarithmic Temperature Differences

PRESSURE DROP IN EXCHANGER 251
Double Pipe Exchanger: Annulus ■ Double Pipe Exchanger: Inner Pipe ■ Shell-and-Tube Exchanger: Tube Side ■ Shell-and-Tube Exchanger: Shell Side ■ Pressure Drop in Condenser

THE EFFECTIVENESS-NTU METHOD 260

REBOILERS 262
Design Guidelines for Reboilers

SUBLIMATION 263

CRYSTALLIZATION 263

EXTENDED SURFACE HEAT EXCHANGER 264

HEATING AND COOLING OF LIQUID BATCHES 268

NONMETALLIC HEAT EXCHANGERS 269
Teflon ■ Graphite ■ Glass Exchanger

HEAT PUMP 271

RECOMMENDED REFERENCES 276

There are three modes of heat transfer: conduction, convection, and radiation. The basic laws governing these are covered in this chapter.

THERMAL CONDUCTION

Thermal conduction is a mode of heat transfer in which the vibrational energy from one molecule to another migrates across a medium. This mode of heat transfer is caused by a temperature difference with no intermixing or flow of material. Fourier's law formulates the steady-state conduction of heat.

Fourier's law states that

$$Q_k = -kA \frac{dT}{dx} \qquad (5.1)$$

where

Q_k = heat transferred by conduction, Btu/h [W or J/s]
k = thermal conductivity of material, Btu/(h · ft² · °F/ft) [W/(m.K)]
A = area of cross section perpendicular to direction of heat flow, ft² [m²]
dT/dx = temperature gradient in the direction of heat flow, °F/ft [K/m].

If k is independent of temperature,

$$Q_k = \frac{A(T_{hot} - T_{cold})}{\frac{X}{k}} = \frac{A(T_{hot} - T_{cold})}{R_F} = \frac{(T_{hot} - T_{cold})}{\frac{R_F}{A}} = \frac{(\Delta T)}{\frac{X}{kA}} = \frac{(\Delta T)}{R_k} \qquad (5.2)$$

Analogy between heat transfer rate and electrical current is

$$\text{Electrical current} = \frac{\text{Electrical potental difference}}{\text{Electrical resistance}}$$

$$\text{Heat transfer rate} = \frac{\text{Temperature difference}}{\text{Thermal resistance}}$$

$$R_k = \frac{X}{kA} = \text{resistance to thermal conduction} = \frac{R_F}{A} \qquad (5.3)$$

$$K_k = \frac{kA}{X} = \text{thermal conductance} \qquad (5.4)$$

$R_F = \frac{X}{k}$ = R-factor used to characterize insulation properties (thermal resistance of a slab of insulation per unit area) of building materials in the USA, where X is the distance between the hot and cold surfaces in the direction of heat flow, in feet [m]. Please note that the unit of thermal resistance is (h·°F/Btu) [(K)/W] and the unit of thermal conductance is Btu/(h·°F) [W/K]. The unit of R-factor is h·ft²·°F/Btu. Thermal conductivity is the property of a *material;* however, the resistance or the R-factor is the property of a *particular piece of material*. In the USA, the R-factor of a building insulation is shown by a number to indicate its thermal resistance, generally with no specification of unit. The higher the R-factor, the better the insulation. Thus R-19 means that the piece of insulation material carrying the number has the *resistance* of 19 h·ft²·°F/Btu per square foot area. In other words, it will lose heat at the rate of (1/19) Btu/h through 1 square foot of area when the temperature difference between the hot and cold surface is 1°F. The relation between resistance and R-factor is: Thermal resistance = R-factor per square foot. It may be observed later in the chapter that the R-factor may be added to the reciprocal of the heat transfer coefficient, and fouling factor when they are in consistent units.

Composite Flat Walls

The rate of heat transfer through a series of flat plates forming a composite wall is given by

$$Q_k = \frac{(T_0 - T_1)}{R_{k1}} = \frac{(T_1 - T_2)}{R_{k2}} = \cdots = \frac{(T_{n-1} - T_n)}{R_{kn}} = \frac{(T_0 - T_n)}{R_{kt}} = \frac{(T_0 - T_n)}{\sum_1^n \frac{X_i}{k_i A_i}} \quad (5.5)$$

where $T_0 > T_1 > \cdots > T_n$, R_{kn} = resistance of nth plate = $X_n/(k_n A_n)$, and R_{kt} = total resistance = $(R_{k1} + R_{k2} + \cdots + R_{kn})$.

Thus, the equivalent thermal resistance of a series of resistances may be computed by electrical resistance analogy.

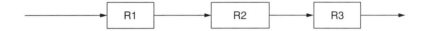

The equivalent resistance of the preceding combination would be

$$R_{kt} = R1 + R2 + R3 \quad (5.6)$$

In the case of parallel resistances, such as shown below, the computation method differs.

The equivalent resistance of the preceding parallel resistances may be given by:

$$\frac{1}{R_{kt}} = \frac{1}{R1} + \frac{1}{R2}$$

$$R_{kt} = \frac{(R1)(R2)}{R1 + R2} \quad (5.7)$$

When a combination of parallel and series resistances is required, first the equivalent resistances between parallel groups are determined, and then the equivalent resistance of the series groups is calculated.

Calculation of equivalent resistance of parallel thermal resistances requires consideration of (individual heat transfer area/total heat transfer area). This can be done through the R-factor concept. Consider a parallel resistance as follows: Temperature $T1 > T2$. Two slabs of thickness X are on top of each other. The height of top slab is h_1 and bottom slab, h_2. The thermal conductivity of top slab is k_1 and bottom slab, k_2.

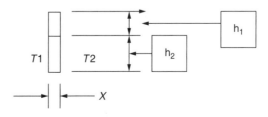

Considering the width of unit length,

Heat flow area of top slab = $a_1 = h_1 \times 1 = h_1$

Heat flow area of bottom slab = $a_2 = h_2 \times 1 = h_2$

Total area of heat transfer = $a_1 + a_2 = h_1 + h_2$

$$Q_1 = a_1 \frac{\Delta T}{\frac{X}{k_1}} = a_1 \frac{\Delta T}{R_{F1}} = \frac{\Delta T}{\frac{R_{F1}}{a_1}}$$

$$Q_2 = a_2 \frac{\Delta T}{\frac{X}{k_2}} = a_2 \frac{\Delta T}{R_{F2}} = \frac{\Delta T}{\frac{R_{F2}}{a_2}}$$

Total heat flow = $Q_1 + Q_2 = a_1 \dfrac{\Delta T}{R_{F1}} + a_2 \dfrac{\Delta T}{R_{F2}} = \left(\dfrac{a_1}{R_{F1}} + \dfrac{a_2}{R_{F2}}\right)(\Delta T)$

$$= \frac{\Delta T}{\dfrac{1}{\dfrac{a_1}{R_{F1}} + \dfrac{a_2}{R_{F2}}}} = \frac{\Delta T}{\dfrac{R_{F-\text{equivalent}}}{a_1 + a_2}} = \frac{\Delta T}{\dfrac{1}{\dfrac{a_1 + a_2}{R_{F-\text{equivalent}}}}}$$

It follows from the preceding equation

$$\frac{1}{R_{F-\text{equivalent}}} = \frac{a_1}{a_1 + a_2}\left(\frac{1}{R_{F1}}\right) + \frac{a_2}{a_1 + a_2}\left(\frac{1}{R_{F2}}\right) = \frac{h_1}{h_1 + h_2}\left(\frac{1}{R_{F1}}\right) + \frac{h_2}{h_1 + h_2}\left(\frac{1}{R_{F2}}\right)$$

(5.8)

Insulation should be securely placed in chemical process equipment using metal band and caulking to avoid condensation indoors. For outdoor applications, jacketing is necessary to protect insulation from rain. For fire-proof insulation, the insulation must be held by stainless steel bands, and secured by stainless steel jackets.

Table 5.1 Insulation type and placement

Insulation Name	ASTM Specification	Maximum Temperature, °F	Equation $k = \text{Btu-in/h} \cdot \text{ft}^2 \cdot {}°\text{F}$, t_m = mean temperature, °F
Mineral Fiber Pipe	C547-93 Class 3	1200	$k = 0.3016 + 1.037(10^{-3})t_m - 4.976(10^{-7})t_m^2$
Mineral Fiber Block	C612-83 Class 3	850	$k = 0.24 + 6.0(10^{-4})t_m + 2.316(10^{-11})t_m^2$
Mineral Fiber Block	C612-83 Class 4	1000	$k = 0.3525 - 8.25(10^{-4})t_m + 3.75(10^{-6})t_m^2$
Elastomer	C534-88	200	$k = 0.28 + 1.556(10^{-4})t_m + 1.481(10^{-6})t_m^2$
Polystyrene	C578-87 Type 1	165	$k = 0.2744 + 6.466(10^{-4})t_m - 4.635(10^{-7})t_m^2$
Calcium Silicate	C533-85 Type 1	1200	$k = 0.31 + 4.676(10^{-4})t_m + 6.386(10^{-8})t_m^2$
Cellular Glass	C552-91	800	$k = 0.3019 + 5.240(10^{-4})t_m + 8.692(10^{-7})t_m^2$
Perlite	C610-85 Type 2	1200	$k = 0.4519 + 4.246(10^{-4})t_m + 2.679(10^{-7})t_m^2$

Application of Thermal Conductance in Thermometry

The principle of resistance thermometer is based on the fact that the electrical resistance of metals varies according to the following equation:

$$R_t = R_o(1 + at + bt^2) \tag{5.9}$$

where
R_t = electrical resistance at temperature, t
R_o = electrical resistance at reference temperature, t_o
a, b = characteristic coefficients for the metal.

A resistance thermometer is an instrument to measure the electrical resistance at a temperature, but it is calibrated in units of temperature. The Wheatstone bridge is the most common bridge used in the design of resistance thermometers. Platinum, nickel, and copper are the most common metals used to manufacture the resistance thermometers. The resistance thermometers must come in contact with the body whose temperature is to be measured. This sets an upper limit of the temperature measurement at the melting point of material of construction of the resistance thermometer element.

THERMAL CONVECTION

Thermal convection is a mode of heat transfer between a surface and a moving fluid when they are at different temperatures. When the fluid moves naturally, induced by buoyancy, the mode of heat transfer is called natural convection; otherwise it is known as forced convection.

The convection heat transfer rate can be expressed as

$$Q_c = h_c A \Delta T \tag{5.10}$$

where
Q_c = rate of heat transfer by convection, Btu/h [W]
A = area of heat transfer, ft² [m²]
ΔT = temperature difference between the surface and the fluid, $t_s - t_f$, °F [K]
h_c = convection coefficient of heat transfer, Btu/h·ft²·°F [W/(m²·K)]

Simultaneous Convection and Radiation

The radiation and convection from a point source go in parallel.

$$Q_r = h_r A(\Delta T)$$
$$Q_c = h_c A(\Delta T)$$

$$Q_{\text{total}} = Q_r + Q_c = (h_r + h_c) A(\Delta T) = \frac{\Delta T}{\frac{1}{(h_r + h_c)A}} \tag{5.11}$$

So, the radiation coefficient and convection coefficient in the same unit may be added together to find the overall coefficient or resistance.

THERMAL RADIATION

Radiation involves some mysticism in the sense that physicists have ascribed duality, meaning both wave and particulate (photons) properties, to it in order to explain all of its phenomena. Contrary to convection and conduction, which require a medium and a temperature gradient for heat transfer, radiation needs neither a medium nor a temperature gradient for the transfer of heat. As long as the body is above absolute zero of temperature, it radiates heat irrespective of the presence or absence of matter surrounding it and regardless of whether the body is colder or hotter than the surroundings. It is, however, the net exchange of radiant heat that affects the net heat transfer. Thermal radiation involves an intermediate portion of the complete spectrum of radiation, and extends approximately from 0.1 to 100 micrometer in the range of wave length.

By the Stefan-Boltzmann law, the heat transfer by radiation for a blackbody is

$$Q_r = \sigma A_1 T_1^4 \tag{5.12}$$

where
- Q_r = heat transfer by radiation, Btu/h [W]
- σ = Stefan-Boltzmann constant
  - $= 0.1713 \times 10^{-8}$ Btu/h · ft² · °R⁴
  - $= 5.67 \times 10^{-8}$ W/m² · K⁴
- A_1 = area of radiating surface, ft² [m²]
- T_1 = radiating surface temperature, °R[K]

Real bodies emit radiation at lower rates than blackbodies. Gray bodies emit a constant fraction of the blackbody emission at each wavelength at a given temperature. The net rate of the radiation heat transfer from a gray body at T_1 to a surrounding blackbody at T_2 is given by

$$Q_r = \sigma A_1 \varepsilon_1 \left(T_1^4 - T_2^4\right) \tag{5.13}$$

where ε_1 is the emissivity of the gray surface.

Heat transfer between two gray bodies is given by

$$Q_r = \sigma A_1 F_{1 \to 2} \left(T_1^4 - T_2^4\right) \tag{5.14}$$

where $F_{1 \to 2}$ is a view factor that accounts for the emissivity and relative geometry of the gray bodies. Radiation coefficient is given by

$$h_r = \frac{\sigma F_{1 \to 2} \left(T_1^4 - T_2^4\right)}{T_1 - T_2'} \tag{5.15}$$

where T_2' is a convenient reference temperature. Usually the surface temperature is taken as the reference temperature.

Application of Radiation Principle in Thermometry

The resistance thermometer has the limitation to measure temperature, which may be high enough to melt the element. It is also difficult to measure the temperature of a moving body with a resistance thermometer. At these unusual conditions, the radiation principle is used to make a special kind of thermometer, called the radiation thermometer.

Example 5.1

A 6-in. thick wall is 12 ft high and 16 ft long. One face is at 1500°F and the other at 300°F. Find the heat loss in Btu/h.

Solution

$$A = 12(16) = 192 \text{ ft}^2$$
$$\Delta T = 1500 - 300 = 1200°F$$
$$\Delta X = \frac{6}{12} = 0.5 \text{ ft}$$
$$Q = kA\frac{\delta T}{\delta X} = -kA\frac{\Delta T}{\Delta X} = -0.15(192)\frac{1200}{0.5} = -69,120 \text{ Btu/h}$$

Composite Walls

For resistances in series (Figure 5.1a), the rate of heat transfer is given by

$$Q = \frac{A(\Delta T)}{R_{FT}} = \frac{A(t_0 - t_n)}{(X_A/k_a) + (X_B/k_b) + (X_C/k_c) + \cdots + (X_n/k_n)} \quad (5.16)$$

where ΔT is the overall temperature difference, °F.

Example 5.2a

The inside temperature of a composite wall is maintained at 2000°F, and the outside temperature is 70°F. The thicknesses from the hotter to colder surfaces are 12, 12, and 10 in., respectively, and the corresponding thermal conductivities are 0.4, 0.2, 0.1 Btu/(h·ft²·°F/ft), respectively. The outside surface coefficient of heat transfer is 2 Btu/h·ft²·°F. Calculate the heat loss and outside surface temperature. (Refer to Figure 5.1b.)

Solution

Heat loss Q_L is given by

$$Q_L = \frac{A(t_0 - t_1)}{R_{FA}} = \frac{A(t_1 - t_2)}{R_{FB}} = \frac{A(t_2 - t_3)}{R_{FC}} = \frac{A(t_3 - t_a)}{R_a} = \frac{A(t_0 - t_a)}{R_{FT}}$$

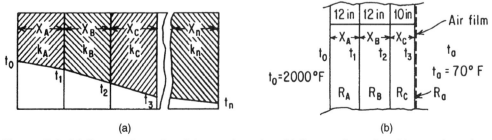

Figure 5.1 (a) Composite wall resistances in series; (b) Composite wall resistances in series for Example 5.2a

$$R_{FA} = \frac{X_A}{k_a} = \frac{12/12}{0.4} = 2.5 \frac{\text{h} \cdot \text{ft}^2 \cdot °\text{F}}{\text{Btu}}$$

$$R_{FB} = \frac{X_B}{k_b} = \frac{12/12}{0.2} = 5.0 \frac{\text{h} \cdot \text{ft}^2 \cdot °\text{F}}{\text{Btu}}$$

$$R_{FC} = \frac{X_C}{k_c} = \frac{10/12}{0.1} = 8.33 \frac{\text{h} \cdot \text{ft}^2 \cdot °\text{F}}{\text{Btu}}$$

$$R_A = \frac{1}{h_a} = \frac{1}{2} = 0.5 \frac{\text{h} \cdot \text{ft}^2 \cdot °\text{F}}{\text{Btu}}$$

$$R_{FT} = \Sigma R = 16.33 \ (\text{h} \cdot \text{ft}^2 \cdot °\text{F})/\text{Btu}, \quad \Delta T = 2000 - 70 = 1930°\text{F}, \ A = 1 \ \text{ft}^2$$

$$\text{Heat loss} = Q_L = \frac{(1 \ \text{ft}^2) 1930(°\text{F})}{16.33 \left(\frac{\text{h} \cdot \text{ft}^2 \cdot °\text{F}}{\text{Btu}} \right)} = 118 \ \text{Btu/h}$$

$$t_3 - t_a = \left(118 \frac{\text{Btu}}{\text{h}} \right) \left(0.5 \frac{\text{h} \cdot \text{ft}^2 \cdot °\text{F}}{\text{Btu}} \right) \left(\frac{1}{1 \ \text{ft}^2} \right) = 59°\text{F}$$

Therefore, $t_3 = t_a + 59 = 70 + 59 = 129°\text{F}$.

Example 5.2b

A furnace is constructed with 0.3 m of fire brick, 0.15 m of insulating brick, and 0.25 m of ordinary building brick. The inside surface temperature is 1530 K, and the outside surface temperature is 325 K. Calculate the heat loss per unit area and the temperatures at the junctions of the bricks. The thermal conductivities of the fire, insulating, and building bricks are 1.4, 0.21, and 0.7 W/m · K, respectively.

Solution

Calculate resistance based on 1 m² area.

$$\text{Total resistance} = \frac{0.3}{1.4(1)} + \frac{0.15}{0.21(1)} + \frac{0.25}{0.7(1)}$$
$$= 0.2143 + 0.7143 + 0.3571$$
$$= 1.2857 \ \text{K/W}$$

Therefore
$$\text{Heat loss} = \frac{1530 - 325}{1.2857} = 937.2 \ \text{W}$$

$$\text{Total temperature drop} = 1530 - 325 = 1205 \ \text{K}$$

$$\text{Temperature drop over fire brick} = \frac{0.2143}{1.2857}(1205) = 201 \ \text{K}$$

$$\text{Temperature drop over insulating brick} = \frac{0.7143}{1.2857}(1205) = 669 \ \text{K}$$

$$\text{Temperature drop over building brick} = \frac{0.3571}{1.2857}(1205) = 335 \ \text{K}$$

Temperature at junction of fire brick and insulating brick

$$= 1530 - 201 = 1329 \text{ K}$$

Temperature at junction of insulating and building bricks

$$= 1329 - 669 = 660 \text{ K}$$

Heat Flow Through a Pipe Wall
The differential equation for heat flow through a pipe wall (Fig. 5.2a) is given by

$$q = 2\pi r k \left(-\frac{dt}{dr} \right) \quad \text{Btu/h·ft} \tag{5.17}$$

The boundary conditions are when $r = r_i$, $t = t_i$, and when $r = r_o$, $t = t_o$, where i and o refer to the inside and outside surfaces. The solution of the differential equation is

$$q = \frac{2\pi k(t_i - t_o)}{2.3 \log(r_o/r_i)} = \frac{2\pi k(t_i - t_o)}{\ln(r_o/r_i)} = \frac{2\pi k(t_i - t_o)}{\ln(D_o/D_i)} \quad \text{Btu/h·ft} \tag{5.18}$$

If $D_i > 0.75 D_o$, the arithmetic mean of the two areas can be used without much error.

Then

$$q = \frac{A_m(\Delta T)}{\frac{l_w}{k_a}} = \frac{\Delta T}{\frac{l_w}{k_a A_m}} = \frac{t_i - t_o}{\frac{\frac{1}{2}(D_o - D_i)}{\frac{1}{2}\pi k_a (D_o + D_i)}} = \frac{t_i - t_o}{\frac{1}{\pi k_a} \frac{D_o - D_i}{D_o + D_i}} \quad \text{Btu/h·ft} \tag{5.19}$$

where $l_w = \tfrac{1}{2}(D_o - D_i)$ is the wall thickness of the pipe.

Composite Cylindrical Resistance
Referring to Figure 5.2b, the temperature drop between the inside surface of pipe and the outside surface of insulation is given by

$$t_1 - t_3 = \frac{q}{2\pi k_p} \ln \frac{D_2}{D_1} + \frac{q}{2\pi k_b} \ln \frac{D_3}{D_2} \tag{5.20}$$

Heat Loss from an Insulated Pipe to Air
For an insulated pipe in which steam is flowing, there are four resistances to be considered. Since the resistances are series, the following equation may be written.

$$q = \frac{t_s - t_1}{\frac{1}{h_s \pi D_1}} = \frac{t_1 - t_2}{\frac{\ln(D_2/D_1)}{2\pi k_p}} = \frac{t_2 - t_3}{\frac{\ln(D_3/D_2)}{2\pi k_b}} = \frac{t_3 - t_a}{\frac{1}{h_a \pi D_3}} \tag{5.21}$$

$$q = \frac{t_s - t_a}{\frac{1}{h_s \pi D_1} + \frac{\ln(D_2/D_1)}{2\pi k_p} + \frac{\ln(D_3/D_2)}{2\pi k_b} + \frac{1}{h_a \pi D_3}} \tag{5.22}$$

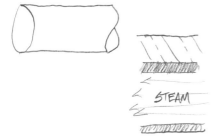

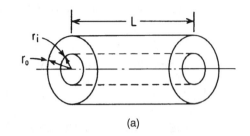

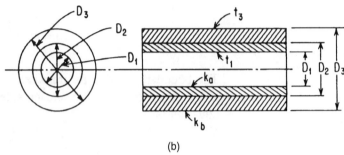

Figure 5.2 (a) Heat flow through pipe wall; (b) Composite cylindrical resistances

Note that the last term of the equation is obtained from the algebraic law of proportionality

$$\frac{a}{b} = \frac{c}{d} = \frac{e}{f} = \frac{a+c+e}{b+d+f} = k = \text{constant} \quad (5.23)$$

By transposing

$$t_s - t_a = q\left[\frac{1}{h_s \pi D_1} + \frac{\ln(D_2/D_1)}{2\pi k_p} + \frac{\ln(D_3/D_2)}{2\pi k_b} + \frac{1}{h_a \pi D_3}\right] \quad (5.24)$$

where
- q = heat loss, Btu/h·ft
- t_s = temperature of steam, °F, t_1 = temperature of inside of pipe wall, °F, t_2 = temperature of outside of pipe wall, °F
- t_a = temperature of air, °F
- k_p = thermal conductivity of pipe wall, Btu/(h·ft²·°F/ft)
- k_b = thermal conductivity of insulation, Btu/(h·ft²·°F/ft)
- h_a = surface coefficient of heat transfer, Btu/h·ft²·°F, on the outside surface of insulation.
- h_s = steam film coefficient of heat transfer, Btu/h·ft²·°F

Generally the steam-film and pipe-wall resistances are negligible and can be ignored in comparison with the other resistances, when the latter are comparatively high.

Critical Radius and Maximum Heat Loss

The addition of the insulation to the outside of small-diameter tubes or pipes does not always reduce the heat loss. The expression for the value of the critical radius at maximum heat loss can be derived from Equation 5.23 and is given by

$$(r_s)_c = \frac{k_b}{h_a} \quad (5.25)$$

where $(r_s)_c$ is the critical radius. It should be noted that when the insulation thickness is such that the outside radius of insulation is less than $(r_s)_c$, adding insulation of that particular thickness actually increases heat loss.

Long, Hollow Cylinder

Heat-transfer rate through a long, hollow cylinder is given as follows:

For conductive resistance alone

$$Q = \frac{\Delta t}{\frac{X_{wall}}{k_w A_m}} = \frac{\Delta t}{\frac{(r_2 - r_1)}{k_w A_m}} = \frac{k_w A_m (\Delta t)}{(r_2 - r_1)} \quad (5.26)$$

and the log mean area is defined as

$$A_m = \frac{A_2 - A_1}{\ln(A_2/A_1)} = \frac{2\pi r_2 L - 2\pi r_1 L}{\ln(2\pi r_2 L / 2\pi r_1 L)} = \frac{2\pi (r_2 - r_1) L}{\ln(r_2/r_1)} \quad (5.27)$$

Substituting Equation 5.27 in Equation 5.26

$$Q = \frac{k_w 2\pi L (\Delta t)}{\ln(r_2/r_1)} \quad (5.28)$$

where $\Delta t = t_i - t_o$ and t_i and t_o are the inside and outside surface temperatures, respectively.

r_2 and r_1 are outside and inside radius respectively. L = length of pipe.

The arithmetic mean area is given by

$$A_a = \frac{A_1 + A_2}{2} \quad (5.29)$$

When A_2/A_1 is not > 2, A_a is within 4 percent of A_m. If the log mean areas are used, the expression for Q in Equation 5.23 can be written as

$$Q = \frac{t_s - t_a}{1/(h_i A_i) + l_w/(k_w A_{mp}) + x_i/(k_b A_{mb}) + 1/(h_o + h_r) A_o} \quad (5.30)$$

where
A_{mp} = log mean area for pipe
A_{mb} = log mean area for insulation
x_i = thickness of insulation
Q = heat transfer, Btu/h, [W].

Hollow Sphere

For heat transfer through a hollow sphere, the mean area is

$$A_m = 4\pi r_1 r_2 = \sqrt{A_1 A_2} \quad (5.31)$$

Radiation

The relation among the absorbed, reflected, and transmitted energies is

$$\alpha + R + \tau = 1 \quad (5.32)$$

where α, R, and τ are the absorptivity, reflectivity, and transmittivity, respectively.

For opaque bodies $\tau = 0$, and therefore

$$\alpha + R = 1. \quad (5.33)$$

Kirchhoff's Law

At thermal equilibrium, the ratio of the total emissive power to the absorptivity for all bodies is the same, or

$$\frac{\varepsilon_1}{\alpha_1} = \frac{\varepsilon_2}{\alpha_2} = \varepsilon_b \tag{5.34}$$

where subscript b refers to blackbody. Emissivity of an actual body is given by

$$\varepsilon = \frac{\text{emissive power of body } \varepsilon_a}{\text{blackbody emissive power } \varepsilon_b} \tag{5.35}$$

Radiation to a Completely Absorbing Receiver

When a radiating source is small by comparison with the receiving enclosure, it is assumed for simplification that the heat radiated by the source is not reflected back to it. In this case, the heat loss by the source is given by

$$\frac{Q_r}{A_1} = \varepsilon_1 \sigma \left(T_1^4 - T_2^4 \right) \tag{5.36}$$

where
Q_r = heat loss due to radiation, Btu/h
ε_1 = emissivity of radiating sources
σ = Stefan-Boltzmann constant
T_1 = temperature of the radiating source, °R
T_2 = temperature of the receiving surface, °R.

Example 5.3

A 2-in. IPS (OD = 2.38-in.) steel pipe carries steam at 325° F through a room at 70° F. What decrease in radiation occurs if the bare pipe is coated with 26 percent aluminum paint?
Emissivity of oxidized steel = 0.79, emissivity of painted steel (26% Al) = 0.3.

Solution

Assume negligible resistance of the steam film and metal wall.

Surface temperature t_s	= 325° F = 785° R
Air temperature t_a	= 70° F = 530° R
OD of pipe	= 2.38-in.
Area A_0	= 0.622 ft²/ft of pipe
Emissivity of oxidized steel	= 0.79
Emissivity of painted steel (26% Al)	= 0.3
Heat loss from bare pipe	= $0.1713 \times 10^{-8}(0.79)[(785)^4 - (530)^4]$
	= $0.1713(0.79)[(\frac{785}{100})^4 - (\frac{530}{100})^4]$
	= 407.1 Btu/h · ft²
Heat loss from painted pipe	= $0.1713(0.3)[(\frac{785}{100})^4 - (\frac{530}{100})^4]$
	= 154.6 Btu/h · ft²
Decrease in radiation	= 407.1 − 154.6 = 252.5 Btu/h · ft²

Heat Exchange between Source and Receiver

In general, the radiation received by a receiver from a source is given by

$$Q = \begin{cases} F_A A_1 \sigma \left(T_1^4 - T_2^4\right) \text{ for blackbodies} & (5.37) \\ F_A F_\varepsilon A_1 \sigma \left(T_1^4 - T_2^4\right) \text{ for gray surfaces} & (5.38) \end{cases}$$

Where
F_ε = emissivity correction[5j]
F_A = correction for relative geometry of the bodies or view factor[5j]
σ = Stefan-Boltzmann constant
T_1 = temperature of source, °R
T_2 = temperature of receiver, °R

Nongray Enclosures

For a small nongray body of area A_1 and at T_1 in black surroundings at T_2, the radiation interchange is given by

$$Q_{1 \to 2} = A_1 \sigma \left(\varepsilon_1 T_1^4 - \alpha_{1 \to 2} T_2^4\right) \quad (5.39)$$

where ε_1 is the emissivity of small nongray body of area A_1 and $\alpha_{1 \to 2}$ is the absorptivity of surface A_1 at T_1 for blackbody radiation at T_2.

Example 5.4

Calculate the heat transfer by radiation between an oxidized nickel tube of 3-in. OD at a temperature of 1000°F and an enclosing chamber maintained at 2000°F if (a) the chamber is very large relative to the tube diameter and (b) 12-in.-square inside. The chamber is lined with glazed silica brick having an emissivity of 0.85.

Solution

(a) If the chamber is very large, it is not necessary to correct for emissivity of silica brick, because the surroundings from the enclosed small body appear black.

ε of nickel at 1000°F = 0.463 (interpolated value)
ε of nickel at 2000°F = 0.62 (extrapolated value)
= absorptivity at 2000°F

$$\frac{\text{Surface area}}{\text{Tube length (ft)}} = \pi D(1) = \tfrac{3}{12}\pi(1) = 0.7854 \text{ ft}^2/\text{ft}$$

$$q_r = 0.1713(0.7854)\left[0.463\left(\frac{1000+460}{100}\right)^4 - 0.62\left(\frac{2000+460}{100}\right)^4\right]$$

$$= -27{,}718 \text{ Btu/h} \cdot \text{ft}.$$

(b) In this case, it is necessary to allow for ε of the silica brick, since the enclosure is small. For this, use the following equation given for two nonblack source-sink surfaces of areas A_1 and A_2:

$$\frac{1}{A_1 F_{1 \to 2}} = \frac{1}{A_1}\left(\frac{1}{\varepsilon_1} - 1\right) + \frac{1}{A_2}\left(\frac{1}{\varepsilon_2} - 1\right) + \frac{1}{A_1 \overline{F}_{1 \to 2}} \quad (5.40)$$

where $F_{1\to 2}$ is the overall interchange factor for gray surfaces and $\overline{F}_{1\to 2}$ is the view factor with allowance for the refractory surface. Since the temperature of the brick is higher than that of the tube, $F_{1\to 2}$ is more important than $\overline{F}_{1\to 2}$. Therefore, use $\alpha_{1\to 2}$ instead of ε_1 of the tube. Thus $\varepsilon_1 = \alpha_{1\to 2} = 0.62$. $= 1$, since all radiation emitted by A_1 is received by A_2. $\alpha_{1\to 2} = 0.62$ and $\varepsilon_2 = 0.85$.

$$A_1 = 0.7854 \text{ ft}^2/\text{ft of tube}$$
$$A_2 = (4)(1)(1) = 4 \text{ ft}^2/\text{ft of enclosing chamber}$$

From Equation 5.40

$$\frac{1}{F_{1\to 2}} = \left(\frac{1}{\varepsilon_1} - 1\right) + \frac{A_1}{A_2}\left(\frac{1}{\varepsilon_2} - 1\right) + 1$$

$$= \left(\frac{1}{0.62} - 1\right) + \frac{0.7854}{4}\left(\frac{1}{0.85} - 1\right) + 1 = 1.6476$$

from which

$$F_{1\to 2} = 0.607$$

Therefore

$$q_r = 0.1713(0.7854)(0.607)\left[\left(\frac{1460}{100}\right)^4 - \left(\frac{2460}{100}\right)^4\right]$$

$$= -26{,}197 \text{ Btu/h} \cdot \text{ft}$$

Example 5.5

A 4-in. NPS steel pipe carrying steam at 450°F is insulated with 1-in. of kapok surrounded by 1-in. of magnesite. The surrounding air is at 70°F. What is the heat loss per linear foot given the following data?

k of kapok $= 0.02$ Btu/(h · ft² · °F/ft)
k of magnesite $= 0.35$ Btu/(h · ft² · °F/ft)
ε of plaster $= 0.9$

Solution

Neglect the steam-film and pipe-wall resistances. Here t_s, the surface temperature, is not known. Therefore, a trial-and-error solution is required. The heat loss in Btu/h·ft, q, is given by

$$q = \frac{t_i - t_a}{(1/2\pi k_k)\ln(D_3/D_2) + (1/2k_m\pi)\ln(D_s/D_3) + (1/h_a\pi D_s)}$$

Since the outside diameter of 4-in. NPS pipe is 4.5-in.,

$$D_2 = \frac{4.5}{12} = 0.375 \text{ ft}$$

$$D_3 = \frac{6.5}{12} = 0.5417 \text{ ft}$$

$$D_s = \frac{8.5}{12} = 0.7084 \text{ ft}$$

$$t_a = 70°F = 530°R$$

Assuming $t_s = 130°F = 590°R$,

$$h_r = 0.1713(0.9)\frac{\left(\frac{590}{100}\right)^4 - \left(\frac{530}{100}\right)^4}{130-70} = 1.09 \, \text{Btu/h} \cdot \text{ft}^2 \cdot {}^\circ\text{F}$$

$$h_c = 0.5\left(\frac{\Delta t}{d_0}\right)^{0.25} = 0.5\left(\frac{60}{8.5}\right)^{0.25} = 0.82 \, \text{Btu/h} \cdot \text{ft}^2 \cdot {}^\circ\text{F}$$

Note: d_0 is in inches in the formula for h_c.

$$h_a = h_c + h_r = 1.91 \, \text{Btu/h} \cdot \text{ft}^2 \cdot {}^\circ\text{F}$$

$$q = \frac{450-70}{\frac{1}{2\pi(0.02)}\ln\left(\frac{0.5417}{0.375}\right) + \frac{1}{2\pi(0.35)}\ln\left(\frac{0.7084}{0.5417}\right) + \frac{1}{1.91\pi(0.7084)}}$$

$$= \frac{380}{3.049 + 0.235} = 115.8 \, \text{Btu/h} \cdot \text{ft}$$

Checking t_s for correctness,

$$q = \frac{t_s - 70}{1/1.92\pi(0.7084)} = 115.8 \, \text{Btu/h} \cdot \text{ft}$$

$$t_s = 97.1{}^\circ\text{F no check}$$

Another trial with $t_s = 100{}^\circ$F, $h_r = 1.01$, $h_c = 0.69$, and $h_a = 1.7$ gives $t_s = 100.3{}^\circ$F which is close to the assumed value of $100{}^\circ$F. Hence $t_s = 100.3{}^\circ$F, and the heat loss = 114.7 Btu/h · ft.

Heat Transfer by Convection to Fluids Flowing Inside and Outside of Pipes

Overall Coefficient of Heat Transfer

The overall coefficient of heat transfer is expressed in terms of the individual coefficients of heat transfer. U_o, the general overall coefficient based on the outside area, is given by

$$U_0 = \frac{1}{\left[\left(\frac{1}{h_o} + f_{do}\right)\frac{1}{E_f} + r_w + f_{di}\left(\frac{A_o}{A_i}\right) + \frac{1}{h_i}\left(\frac{A_o}{A_i}\right)\right]} \quad (5.41)$$

Where

$$r_w = \begin{cases} \dfrac{d_o}{24k_w}\left[\ln\left(\dfrac{d_o}{d_o-2t}\right)\right] & \text{for bare tube} \\[1em] \dfrac{t}{12k_w}\left[\dfrac{d_o + 2Nw(d_o+w)}{d_o-t}\right] & \text{for finned tube} \end{cases}$$

The simplified form of the overall heat transfer coefficient for shell-and-bare-tube exchanger is given by

$$\frac{1}{U_o} = \frac{1}{h_o} + f_{do} + \frac{l_w}{k_w}\frac{D_o}{D_{av}} + f_{di}\frac{D_o}{D_i} + \frac{1}{h_1}\frac{D_o}{D_i} \quad (5.42a)$$

where
- U_o = overall heat transfer coefficient, Btu/h· ft²· °F, based on the outside area
- h_i = inside film coefficient of heat transfer, Btu/h· ft²· °F
- f_{di} = fouling resistance on the inside, h · ft² · °F/Btu
- l_w = tube wall thickness of pipe, ft
- k_w = thermal conductivity of pipe wall, Btu/(h · ft² · °F/ft)
- f_{do} = fouling resistance on outside of pipe, h · ft² · °F/Btu
- h_o = outside film coefficient of heat transfer, Btu/h · ft² · °F
- D_i, D_o, D_{av} = inside, outside and average diameters respectively, ft
- r_w = Tube wall resistance, h ·ft² · °F/Btu
- d_o = Outside diameter of tube, or root diameter of integrally finned tube, inch
- t = Thickness of tube wall, inch

Additional for finned tubes:
- E_f = Fin efficiency, dimensionless
- w = Fin height, inch
- N = Number of fins per inch of tube length

Based on inside area, for bare tube exchangers, U_i is given by

$$\frac{1}{U_i} = \frac{1}{h_i} + f_{di} + \frac{l_w}{k_w}\frac{D_i}{D_{av}} + f_{do}\frac{D_i}{D_o} + \frac{1}{h_o}\frac{D_i}{D_o} \tag{5.42b}$$

Fouling in Heat Exchanger and Fouling factor, f_d

Fouling is created on heat exchanger process surfaces, both inside and outside, due to the scales formed by solids that settle due to low velocity, reverse solubility, and corrosion products. Vaporization-induced fouling is severe in kettle reboilers because they operate near total vaporization. Fouling has two major effects: reduction of heat transfer rate and increase of pressure drop. At the extreme, one of these factors, or both, may require shutdown and cleaning of the equipment. In the design of exchangers in petrochemical industries, therefore, additional resistance, called the fouling factor, is included. The factor has the unit of the reciprocal of heat transfer coefficient, and is not a dimensionless number. Therefore, attention is required when a fouling factor from information in SI units is used in an equation in AES unit, and vice versa.

Fouling is an unsteady state phenomenon; therefore, the fouling factor is the limit of an unsteady state resistance, which is added to steady state resistances to calculate the heat transfer area. The fouling factors, therefore, should be associated with a continuous period after which the equipment requires cleaning. An uneducated guess in fouling factors may invite the problem the designer is trying to avoid in the first place. For example, when no phase change is involved, such as in cooling water service, the exchanger may be designed with no fouling factor first, then the tube length is extended, rather than increasing the number of tubes, to account for the fouling factor. If we were to increase the number of tubes in this case, the velocity would be lowered, thereby inviting fouling, which the designer wants to avoid. In the same context of cooling water service, the outlet temperature should be limited to $\leq 120°F$ to avoid temperature-induced scaling. Forced circulation reboilers are less prone to fouling because of higher velocity.

By observing the performance of an exchanger from the time it is installed, the fouling correlation may be developed[16] as follows:

$$R_{f,\theta} = R_f^*(1 - e^{-B\theta}) \tag{5.43}$$

where

$R_{f,\theta}$ = instantaneous fouling factor (resistance) at time, θ
R_f^* = an asymptotic value of R_f which the fouling curve tends to appproach in time
B = a constant to be determined be experiment

The asymptotic fouling resistance of a liquid service in a tube is related to the tube diameter and liquid velocity as follows:

$$R_f^* = \frac{k_1}{v^{k_2} d^{k_3}} \tag{5.44}$$

v = tube side velocity
d = tube inside diameter
k_1, k_2, k_3 are constants.

A good source of fouling factors is the Tubular Exchanger Manufacturers Association[11a] (TEMA) standard. The fouling factors in TEMA are applicable in shell-and-tube heat exchangers only, and should not be used in plate exchangers or on the fin-side of air coolers. Plate exchangers generally use 10 to 15 percent oversurface to account for the fouling.

Logarithmic Mean Temperature Difference (LMTD)

The logarithmic mean temperature difference is given by the relation

$$\Delta T_{LM} = \frac{\Delta T_2 - \Delta T_1}{\ln(\Delta T_2/\Delta T_1)} \tag{5.45}$$

ΔT_1 and ΔT_2 are the terminal temperature differences.

The expression for log mean difference is the same for both the parallel and countercurrent flows. When one of the fluids is isothermal, the numerical value of ΔT_{LM} is the same for both the parallel and countercurrent flows. (See Eq. 5.112 for the correction factor.) It must be remembered that the concept of LMTD becomes meaningless under some circumstances:

1. When there is a large variation of overall heat transfer coefficient at the terminals. Under this condition use the following method:

$$Q = A\left[\frac{U_2(\Delta T_1) - U_1(\Delta T_2)}{\ln \frac{U_2(\Delta T_1)}{U_1(\Delta T_2)}}\right] \tag{5.46}$$

2. When dealing with a fluid that undergoes subcooling and phase change such as subcooling superheated steam, followed by condensation, and subcooling of the condensate, or condensation of mixed vapor when the condensation curve is non-linear. In these cases, a weighted temperature difference should

be used. For example, in the case of desuperheating, condensing, and subcooling, the LMTD concept may be applied at each section of desuperheating, condensing, and subcooling followed by weighting. Thus,

$$\Delta T_{\text{weighted}} = \frac{Q_{\text{total}}}{\dfrac{Q_{\text{desuperheating}}}{\text{LMTD}_{\text{desuperheating}}} + \dfrac{Q_{\text{condensing}}}{\text{LMTD}_{\text{condensing}}} + \dfrac{Q_{\text{subcooling}}}{\text{LMTD}_{\text{subcooling}}}} \quad (5.47)$$

When the condensation curve is nonlinear, several segments may be used to calculate weighted temperature difference. The corresponding U_D must also be weighted.

$$U_{D,\text{weighted}} = \frac{Q_{\text{total}}}{\dfrac{Q_{\text{desuperheating}}}{U_{D,\text{desuperheating}}} + \dfrac{Q_{\text{condensing}}}{U_{D,\text{condensing}}} + \dfrac{Q_{\text{subcooling}}}{U_{D,\text{subcooling}}}} \quad (5.48)$$

Caloric Fluid Temperature[5h]

When U is not constant along the length of the exchanger because of a change in the viscosity of the fluid, the fluid properties should be evaluated at the caloric temperature for the calculation of the film coefficients.

Pipe-Wall Temperature (When U is not Constant)

If the caloric temperature and the film coefficient h_i and h_o are known, and if the pipe-wall resistance is neglected, the wall temperature t_w can be calculated by the following relations. When hot fluid is on the outside of the pipe

$$t_w = \begin{cases} t_c + \dfrac{h_o}{h_{io} + h_o}(T_c - t_c) & (5.49) \\[2ex] T_c - \dfrac{h_{io}}{h_{io} + h_o}(T_c - t_c) & (5.50) \end{cases}$$

where T_c and t_c are the caloric temperatures of the hot and cold fluids respectively, and $h_{io} = h_i(D_i/D_o)$. When the hot fluid is inside the pipe, t_w is given by

$$t_w = \begin{cases} t_c + \dfrac{h_{io}}{h_{io} + h_o}(T_c - t_c) & (5.51) \\[2ex] T_c - \dfrac{h_o}{h_{io} + h_o}(T_c - t_c) & (5.52) \end{cases}$$

If h_i and h_o are not known, a trial-and-error calculation to fix the caloric temperature is required.

Film Coefficients for Fluids in Pipes and Tubes (No Change of Phase)

Streamline Flow

Sieder and Tate[1] correlation for laminar flow ($N_{\text{Re}} < 2100$) is

$$N_{\text{Nu}} = \frac{h_i D_i}{k} = 1.86 \left[\left(\frac{D_i G}{\mu}\right)\left(\frac{c_p \mu}{k}\right)\left(\frac{D_i}{L}\right)\right]^{1/3} \left(\frac{\mu_b}{\mu_w}\right)^{0.14} \quad (5.53)$$

Equation 5.53 simplifies to

$$\frac{h_i D_i}{k} = 1.86 \left(\frac{4W(c_p)}{\pi k L}\right)^{1/3} \left(\frac{\mu}{\mu_w}\right)^{0.14} \quad (5.54)$$

For turbulent flow ($N_{Re} > 10{,}000$), the Sieder and Tate[1] equation is

$$\frac{h_i D_i}{k} = 0.027 \left(\frac{D_i G}{\mu}\right)^{0.8} \left(\frac{c_p \mu}{k}\right)^{1/3} \left(\frac{\mu}{\mu_w}\right)^{0.14} \quad (5.55)$$

The bulk temperature for streamline flow and the caloric temperature for turbulent flow are to be used to evaluate the physical properties of the fluid, except μ_w, which is taken at the wall temperature. If the caloric temperature factor chart for the fluid is not available, the average bulk fluid temperature is used to evaluate the physical properties when: (1) viscosity at the cold temperature is low (~5 cP), (2) temperature range is moderate (~100°F), and (3) temperature difference is low (~75°F). Also $(\mu/\mu_w)^{0.14}$ is assumed equal to ~1.

At moderate Δt values, the Dittus-Boelter[2] equation for turbulent flow is

$$\frac{h_i D_i}{k_b} = \begin{cases} 0.023 \left(\dfrac{D_i G}{\mu_b}\right)^{0.8} \left(\dfrac{c_p \mu}{k}\right)_b^{0.4} & \text{for heating} \quad (5.56) \\[2ex] 0.023 \left(\dfrac{D_i G}{\mu_b}\right)^{0.8} \left(\dfrac{c_p \mu}{k}\right)_b^{0.3} & \text{for cooling} \quad (5.57) \end{cases}$$

where
h_i = inside heat-transfer coefficient, Btu/h · ft² · °F
D_i = inside diameter, ft
k = thermal conductivity of fluid, Btu · ft/h · ft² · °F
G = mass velocity, lb/h · ft²
m = viscosity, lb/h · ft
c_p = specific heat of fluid = Btu/lb · °F
L = length of pipe, ft
W = flow rate, lb/h

The subscript b indicates that the fluid physical properties are to be evaluated at the bulk temperature of the fluid.

Heat Transfer Coefficient from Colburn-Factor J_H

The Colburn-factor, J_H, is the product of Stanton number, and Prandtl number raised to the power 2/3. From the known Reynolds number, Prandtl number, specific heat, and mass velocity, and the J_H-factor, the heat transfer coefficient can be calculated.

$$J_H = N_{St} \left[N_{Pr}^{2/3} \right] = \frac{h}{c_p G} \left[N_{Pr}^{2/3} \right] \quad (5.58)$$

$$N_{St} = \frac{h}{c_p \rho u} = \frac{h}{c_p G} = \frac{N_{Nu}}{(N_{Re})(N_{Pr})} \quad (5.59)$$

$$N_{Pr} = \frac{c_p \mu}{k}, \quad N_{Nu} = \frac{hD}{k}, \quad N_{Re} = \frac{DG}{\mu} \quad (5.60)$$

Equation 5.61 is applicable in all flow regimes, laminar or turbulent. The J factor is computed from Reynolds number

$$J_1 = \left[\left(\frac{1}{N_{Re}^{9.36}} \right) + \frac{1}{\left\{ \frac{N_{Re}^{1.6}}{7.831(10^{-14})} + \left(\frac{1.969(10^6)}{N_{Re}} \right)^8 \right\}^{3/2}} \right] \quad (5.61)$$

The final Colburn factor is obtained by viscosity correction.

$$J_H = J_1 \left(\frac{\mu}{\mu_w} \right)^{0.14} \quad (5.62)$$

Finally, the heat transfer coefficient is obtained from

$$h = \left\{ J_H \left[\frac{c_p G}{N_{Pr}^{2/3}} \right] \right. \text{ when } J_H \text{ is calculated from Equation 5.62} \quad (5.63)$$

Evaluation of Wall Temperature t_w (When U is Fairly Constant)

When the overall heat-transfer coefficient U is fairly constant over the length of the exchanger, a trial-and-error calculation is required to evaluate t_w.

$$\Delta T_i = \frac{(1/h_i A_i) \Delta T}{(1/h_i A_i) + (1/h_o A_o) + (l_w / k_w A_{av})} \quad (5.64)$$

where ΔT is the difference between the average temperature of the hot fluid and the average temperature of the cold fluid.

Preliminary estimates of h_i and h_o are made, and then the wall temperature is obtained by the equations

$$t_w = \begin{cases} T + \Delta T_i \text{ for heating} & (5.65) \\ T - \Delta T_i \text{ for cooling} & (5.66) \end{cases}$$

where T is the average temperature of the fluid.

Equivalent Diameter for Heat Transfer

When a fluid flows through a circular annulus, the wetted area for the heat transfer is not the same as that for the fluid flow. Therefore, values of the equivalent diameter are different and are given by

$$D_e = \begin{cases} \dfrac{D_2^2 - D_1^2}{D_1} \text{ for heat transfer} & (5.67) \\ D_2 - D_1 \text{ for pressure drop} & (5.68) \end{cases}$$

Flow across Banks of Tubes

The Reynolds number for flow over tube banks is based on the minimum free flow area. For in-line tube arrangement (Figure 5.3a), the minimum free flow area between two adjacent tubes is given by

$$A_{min} = S_T - D_o \text{ ft}^2/\text{unit length of tube} \tag{5.69}$$

where S_T is center-to-center distance of the tubes in adjacent longitudinal rows or the transverse pitch, ft.

For a shell containing N transverse rows of in-line tubes perpendicular to the direction of flow, the minimum flow area (total) is given by

$$A_{min} = (N - 1)(S_T - D_o) + 2C \text{ ft}^2/\text{unit length of bundle} \tag{5.70}$$

where C is the clearance between the shell and the outermost row of tubes, ft.

For the staggered tube arrangement (Fig. 5.3b), the minimum flow area may occur between the adjacent tubes in a row or between the diagonally opposed tubes. In the former case, the minimum flow area is given by

$$A_{min} = S_T - D_o \tag{5.71}$$

S_T is as shown in Figure 5.3b. In the latter case, if S_L/S_T is very small,

$$\sqrt{S_T^2 + S_L^2} < \left(S_T + \frac{1}{2}D_o\right) \tag{5.72}$$

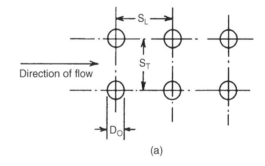

(a)

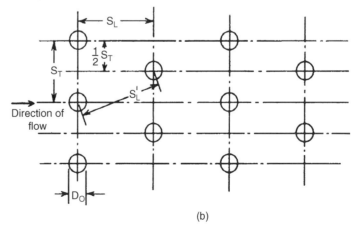

(b)

Figure 5.3 Tube arrangements in cross-flow exchangers: (a) in-line, (b) staggered

For the diagonally opposed tubes, the maximum velocity is then given by

$$V_{max} = V_f \frac{S_T}{\sqrt{S_T^2 + S_L^2} - D_o} \tag{5.73}$$

where
- V_{max} = maximum velocity, ft/s
- V_f = free flow velocity based on shell area without tubes, ft/s
- S_L = center-to-center spacing between transverse tube rows or longitudinal pitch, ft.

For gases flowing normal to the staggered tubes, Colburn[3] recommended the equation

$$\frac{h_m D_o}{k_f} = 0.33 \left(\frac{c_p \mu}{k}\right)_f^{1/3} \left(\frac{D_o G_{max}}{\mu_f}\right)^{0.6} \tag{5.74}$$

The subscript f indicates that the fluid properties are to be evaluated at the film temperature t_f. The units of variables must be consistent to result in dimensionless numbers.

For air, the preceding reduces to

$$\frac{h_m D_o}{k_f} = 0.3 \left(\frac{D_o G_{max}}{\mu_f}\right)^{0.6} \tag{5.75}$$

For the in-line tube arrangement, the constant in the preceding equation should be taken as 0.26 instead of 0.33. The fluid properties are to be evaluated at

$$t_f = \frac{1}{2}(t_s + t) \tag{5.76}$$

where t_s is the surface temperature and t is the bulk temperature.

The ratios of h_m for N rows deep to h_m for 10 rows deep are given in Table 5.2. A simplified dimensional equation for gases flowing normal to a 10-row bank of staggered tubes is

$$h_m = \frac{0.133 c_p G_{max}^{0.6}}{D_o^{0.4}}, \quad \frac{\text{Btu}}{\text{h} \cdot \text{ft}^2 \cdot °\text{F}} \tag{5.77}$$

c_p = fluid specific heat, Btu/lb·°F
G_{max} = Mass velocity through minimum cross section, lb/h·ft^2
D_o = Outside diameter, ft

Table 5.2 Ratio of (h_m for N Rows deep)/(h_m for 10 Rows)[4e]

N	1	2	3	4	5	6	7	8	9	10
Ratio for staggered tubes	0.68	0.75	0.83	0.89	0.92	0.95	0.97	0.98	0.99	1
Ratio for in-line tubes	0.64	0.80	0.87	0.90	0.92	0.94	0.96	0.98	0.99	1

Calculation of Natural Convection Heat Transfer Coefficient

For a vertical plate involving natural convection, Incropera and Dewitt[8a] recommend as follows:

$$Nu = \left\{ 0.825 + \frac{0.387 Ra^{1/6}}{\left[1 + \left(\frac{0.492}{N_{Pr}}\right)^{9/16}\right]^{8/27}} \right\}^2 \qquad (5.78)$$

$$Nu = \text{Nusselt Number} = \frac{h_c L}{k} \qquad (5.79)$$

$$Ra = \text{Rayleigh Number} = \frac{g\beta(T_{hot} - T_{cold})L^3}{\nu\alpha} = \frac{g\beta\rho^2 c_p (T_{hot} - T_{cold})L^3}{\mu k} \qquad (5.80)$$

$$N_{Pr} = \text{Prandtl number} = \frac{c_p \mu}{k} \qquad (5.81)$$

Equation 5.78 is good for the entire range of Rayleigh numbers and discrimination between laminar and turbulent flows is not required.

- g = acceleration due to gravity, $4.17(10^8)$ ft/h² [9.81 m/s²]
- h_c = convection heat transfer coefficient, Btu/hr·ft²·°F [W/m²K]
- k = thermal conductivity of fluid, Btu-ft/hr·ft²·°F [W/m·K]
- L = characteristic length of equipment across which convection is taking place; generally it is the height of a vertical plate, ft [m]
- T_{hot} = hotter temperature, °R [K]
- T_{cold} = colder temperature, °R [K]
- $\alpha = k/(c_p \rho)$ = thermal diffusivity, ft²/h [m²/s]
- c_p = specific heat at constant pressure, Btu/lb·°F [J/kg·K]
- ρ = Fluid density, lb/ft³ [kg/m³]
- β = Coefficient of volumetric thermal expansion of fluid, °R⁻¹ [K⁻¹] = $1/T$ for ideal gas

$$\beta = \frac{1}{V_{avg}}\left[\frac{dV}{dT}\right] = \frac{\left(\frac{1}{\rho_2} - \frac{1}{\rho_1}\right)}{\left(\frac{1}{\rho_{avg}}\right)(T_2 - T_1)} = \frac{s_1^2 - s_2^2}{2(T_2 - T_1)s_1 s_2} \qquad (5.82)$$

For liquids where, ρ_1 and ρ_2 are densities, lb/ft³ [kg/m³], s_1 and s_2 are dimensionless specific gravities at temperatures T_1 and T_2 respectively.

V is the specific volume, volume/mass. ft³/lb [m³/kg]

ν = kinematic viscosity
  = Absolute viscosity (μ)/density (ρ), ft²/h [m²/s]

μ = absolute viscosity, lb/ft·hr [kg/m·s]

For horizontal surface[8b]

$$Nu = 0.54\, Ra^{0.25} \quad \text{for} \quad (10^5 \le Ra \le 10^7) \tag{5.83}$$

$$Nu = 0.15\, Ra^{0.25} \quad \text{for} \quad (10^7 \le Ra \le 10^{10}) \tag{5.84}$$

For horizontal surfaces such as the bottom head of a vertical tank under external fire

$$Nu = 0.27 Ra^{0.25} \quad \text{for} \quad (10^5 \le Ra \le 10^{10}) \tag{5.85}$$

In Equations 5.83 and 5.85, the characteristic length is defined by surface area/perimeter.

Simplified equations have been proposed for natural convection for vertical plates and cylinders based on Rayleigh number (characteristic length >1 m).[19]

$$Nu = \begin{cases} 1.36 Ra^{1/5} & \text{for} \quad Ra < 10^4 \\ 0.55 Ra^{1/4} & \text{for} \quad Ra = 10^4 - 10^9 \\ 0.13 Ra^{1/3} & \text{for} \quad Ra > 10^9 \end{cases} \tag{5.86}$$

More simplified equations have been proposed for natural convection[5] to air.

$$h_c = \begin{cases} 0.28\left(\dfrac{\Delta T}{z}\right)^{0.25} & \text{for vertical plates <2 ft high} \\[2pt] 0.3(\Delta T)^{0.25} & \text{for vertical plates >2ft high} \\[2pt] 0.38(\Delta T)^{0.25} & \text{for horizontal plates facing upward} \\[2pt] 0.2(\Delta T)^{0.25} & \text{for horizontal plates facing downward} \\[2pt] 0.5\left(\dfrac{\Delta T}{d_o}\right)^{0.25} & \text{for horizontal pipe} \\[2pt] 0.4\left(\dfrac{\Delta T}{d_o}\right)^{0.25} & \text{for long vertical pipe} \end{cases} \tag{5.87}$$

where

ΔT = temperature difference between hot fluid and cold air, °F, and z is the height in feet, d_o is the outside diameter in inches, h_c is in Btu/hr · ft² · °F.

For forced convection outside equipment caused by wind velocity, the following equation[20] has been proposed:

$$h_c = 0.296(\Delta T)^{0.25}(1.28V + 1)^{0.5} \tag{5.88}$$

where

ΔT = temperature difference between hot fluid and cold surface, °F, V is the wind velocity, miles per hour, and h_c is in Btu/hr · ft² · °F. It appears that the factor $(1.28V + 1)^{0.5}$ in the equation may be used as a multiplier for the effect of wind velocity.

Free Convection of Fluids outside Horizontal Pipes

Single Horizontal Pipes

The heat-transfer coefficient for both liquids and gases outside a single horizontal pipe is given by

$$\frac{h_c D_o}{k_f} = \alpha \left[\left(\frac{D_o^3 \rho_f^2 g \beta \Delta t}{\mu_f^2} \right) \left(\frac{c_p \mu_f}{k_f} \right) \right]^{0.25} \quad (5.89)$$

where the film temperature t_f is

$$t_f = \frac{t_w + t_b}{2} \quad t_b = \text{bulk temperature} \quad (5.90)$$

The constant α is 0.47 for small pipe and 0.53 for large pipes. Other terms are

h_c = Free convection heat transfer coefficient, Btu/h · ft² · F, [W/m² · K]
D_o = Outside diameter of pipe, ft, [m]
k_f = Thermal conductivity of fluid, Btu · ft/h · ft² · °F, [W/m · k]
ρ_f = Density of fluid, lb/ft³, [kg/m³]
g = Local acceleration due to gravity, ≈ 4.17(10⁸) (ft/h², [9.81 m/s²]
β = Coefficient of volumetric expansion of fluid, 1/°F, [1/K]
Δt = Temperature difference between hot and cold surface, °F, [K]
μ_f = Viscosity of fluid, lb/ft · h, [kg/m · s]
c_p = Specific heat of fluid, Btu/lb · F, [J/kg · K]

In Equation 5.89, the three groups are dimensionless. Therefore consistent units of the variables are to be used. For example, g, acceleration due to gravity, equals 4.17×10^8 ft/h² and β is the coefficient of thermal expansion, 1/°F in the AES system.

Banks of Tubes

For free convection outside a single tube or banks of tubes (liquids and gases),[5a] the convection heat-transfer coefficient is given by the following dimensional equation for use in AES units:

$$h_c = 116 \left[\left(\frac{k_f^3 \rho_f^2 c_f \beta}{\mu_f'} \right) \left(\frac{\Delta t}{d_o} \right) \right]^{0.25} \quad (5.91)$$

For ideal gases, $\beta = 1/T$ where T = average temperature in absolute scale. For liquids,

$$\beta = \frac{1/s_2 - 1/s_1}{(1/s_{av})(t_2 - t_1)} = S_{av} \left[\frac{\frac{1}{s_2} - \frac{1}{s_1}}{t_2 - t_1} \right] \quad (5.92)$$

where s_2 and s_1 are specific gravities at t_2 and t_1 and s_{av} equals $\frac{1}{2}(s_1 + s_2)$. In Equation 5.91, μ_f' is in centipoise and d_o is outside diameter in inches. Properties are evaluated at film temperature. Other terms are

k_f = fluid thermal conductivity, Btu · ft/h · ft^2 · °F
ρ_f = fluid density, lb/ft^3
c_f = fluid specific heat at constant pressure, Btu/lb · °F
Δt = temperature difference between hot surface and cold fluid, °F
h_c = heat transfer coefficient, Btu/h · ft^2 · °F

Boiling

The term boiling refers to evaporation at a solid-liquid interface through the formation of bubbles. As the temperature of the heating surface is increased, heat transfer proceeds through free convection, nucleate boiling, a boiling transition, and film boiling. High heat flux is obtained in the nucleate boiling range, and this happens when the temperature difference between the heating surface and the bubble point is maintained at less than ~100°F. Boiling heat transfer depends on the texture of heat-transfer surface, operating pressure, temperature difference, and physical properties of vapor and liquid.

The nucleate boiling heat transfer flux may be obtained from

$$q_{\text{nucleate}} = (\mu_l h_{fg}) \left[\frac{g(\rho_l - \rho_v)}{g_c \sigma_l} \right]^{0.5} \left[\frac{c_{p,l}(T_s - T_{\text{sat}})}{c_{s,f} h_{fg} N_{\text{Pr},l}^s} \right]^3 \tag{5.93}$$

$$q_{\text{nucleate},Cr} = 0.18 h_{fg} \rho_v \left[\frac{\sigma_l g g_c (\rho_l - \rho_v)}{\rho_v^2} \right]^{1/4} \tag{5.94}$$

where

q_{nucleate} = Nucleate boiling heat flux, Btu/s · ft^2, [W/m^2]
$q_{\text{nucleate},Cr}$ = Critical (highest) nucleate boiling heat flux, Btu/s · ft^2, [W/m^2]
μ_l = viscosity of liquid, lb/ft · s, [N · s/m^2 or kg/(m · s), or Pa · s]
h_{fg} = latent heat of vaporization of liquid, Btu/lb, [J/kg]
g = local acceleration due to gravity, 32.17 ft/s^2, [9.81 m/s^2], on earth
ρ_l = liquid density, lb/ft^3, [kg/m^3]
ρ_v = vapor density, lb/ft^3, [kg/m^3]
g_c = Newton's law proportionality factor, 32.17 (lb · ft/s^2)/lb$_f$, [(1 kg · m/s^2)/N]
σ_l = Surface tension of liquid, lb$_f$/ft, [N/m]
$c_{p,l}$ = Specific heat of liquid, Btu/lb · °F, [J/kg · K]
T_s = Heating surface temperature, °F, [K]
T_{sat} = Saturation temperature of liquid, °F, [K]
$c_{s,f}$ = Liquid-to-heating surface combination factor

$N_{\text{Pr},l} = \frac{c_{p,l} \mu_l}{k_l}$ = Prandtl number for liquid, dimensionless.

k_l = liquid thermal conductivity, Btu.ft/s · ft^2 · °F, [W/m · K or J/(s · m · K)]

The equation of critical nucleate boiling heat flux is known as the Kutateladze correlation. This equation may be used to calculate emergency relief capacity due to control valve failure of a utility stream.

The liquid-to-heating surface combination factor and the exponent of Prandtl number may be obtained from the following table.[18]

Table 5.3 Surface combination factor and Prandtl no. exponent

Surface Combination	$c_{s,f}$	Exponent of Prandtl Number, s
Water-nickel	0.006	1.0
Water-platinum	0.013	1.0
Water-copper	0.013	1.0
Water-brass	0.006	1.0
CCl_4-copper	0.013	1.7
Benzene-chromium	0.010	1.7
n-Pentane-chromium	0.015	1.7
Ethyl alcohol-chromium	0.0027	1.7
Isopropyl alcohol-copper	0.0025	1.7
35% K_2CO_3 –copper	0.0054	1.7
50% K_2CO_3 –copper	0.0027	1.7
n-butyl alcohol-copper	0.0030	1.7

Condensation Outside Tubes

Vertical Tubes

Some correlations for the condensation coefficients are given below. For vertical tubes[5b] (length L of tube is known), the condensation coefficient is given by

$$\bar{h} = 0.943 \left(\frac{k_f^3 \rho_f^2 \lambda g}{\mu_f L \Delta t_f} \right)^{0.25} \quad (5.95)$$

where

$t_f = \frac{1}{2}(t_{sv} + t_w)$
$\Delta t_f = t_f - t_w$
t_{sv} = temperature of saturated vapor, of [K]
t_w = wall temperature of [K]
$\bar{h}$ = condensation heat transfer coefficient, Btu/h · ft² °F, [W/m² · K]
k_f = thermal conductivfity of condensate, Btu · ft/h · ft² · °F, [W/m · K]
ρ_f = density of condensate, lb/ft³, [kg/m³]
λ = latent heat of condensation, Btu/lb, [J/kg]
g = local acceleration due to gravity, 4.17(10^8) ft/h², [9.81 m/s]
μ_f = condensate viscosity, lb/ft · h, [kg/m · s]
L = tube length, ft, [m]
Δt_f = temperature difference between film and cold surface (tube wall), °F, [K]

Inclined Tubes[5c]

For the inclined tubes $\bar{h}$ is given by

$$\bar{h} = 0.943 \left(\frac{k_f^3 \rho_f^2 \lambda g \sin \alpha}{\mu_f L \Delta t_f} \right)^{0.25} \quad (5.96)$$

where α is the angle that the gravity component of the condensate weight makes with the line perpendicular to the tube, and λ is the latent heat. The consistent units are defined under the previous equation.

When the length of the vertical tube is not known, the following equation can be used for the vertical single or multiple tubes.[5d]

$$\bar{h} = 1.47 \left(\frac{4G'}{\mu_f}\right)^{-1/3} \left(\frac{k_f^3 \rho_f^2 g}{\mu_f^2}\right)^{1/3} \quad (5.97)$$

$$G' = \frac{W'}{\pi D_o} \text{ lb/h} \cdot \text{ft, [kg/m} \cdot \text{s]} \quad (5.98)$$

where W' is the condensate loading/tube = W/N_t and N_t is the number of tubes.
$\bar{h}$ = condensation heat transfer coefficient, Btu/h $\cdot$ ft^2 $\cdot$ °F, [W/m^2 $\cdot$ K]
k_f = thermal conductivity of condensate, Btu $\cdot$ ft/h $\cdot$ ft^2 $\cdot$ °F, [W/m $\cdot$ K]
ρ_f = density of condensate, lb/ft^3, [kg/m^3]
g = local acceleration due to gravity, 4.17(10^8) ft/h^2, [9.81 m/s]
G' = condensate loading lb/h $\cdot$ ft, [kg/m $\cdot$ s]
W' = condensate loading per tube
W = condensate loading, lb/h, [kg/s]
D_o = tube outside diameter, ft, [m]
μ_f = condensate viscosity, lb/ft $\cdot$ h, [kg/m $\cdot$ s]

Horizontal Tubes[5d]
For horizontal tubes $\bar{h}$ is given by

$$\bar{h} = \begin{cases} 0.725 \left(\dfrac{k_f^3 \rho_f^2 g \lambda}{\mu_f D_o \Delta t_f}\right)^{0.25} & \text{for single tube} \quad (5.99) \\[2ex] 1.51 \left(\dfrac{4G''}{\mu_f}\right)^{-1/3} \left(\dfrac{k_f^3 \rho_f^2 g}{\mu_f^2}\right)^{1/3} & \text{for multiple tubes} \quad (5.100) \end{cases}$$

where

$$G'' = \frac{W}{LN_t^{2/3}} \quad (5.101)$$

The notations are as in previous equations. L is the length of tube, ft [m].

The equations for estimating the condensation coefficients outside the vertical and horizontal tubes hold good for the streamline flow of the condensate, i.e., for $N_{Re} < 2100$ where $N_{Re} = DG'/\mu$ for vertical tubes and $N_{Re} = DG''/\mu$ for horizontal tubes. The effects of the high velocity, turbulence, etc., can be accounted for.[4b]

Condensation Inside Tubes

Horizontal Tubes
Equation 5.100 is to be used with

$$G'' = \frac{W}{0.5LN_t} \quad (5.102)$$

Vertical Tubes

$\bar{h}$ is given in terms of Colburn *J*-factor by

$$J = \bar{h}\left(\frac{\mu_f^2}{k_f^3 \rho_f^2 g}\right)^{1/3} = \phi\left(\frac{4G'}{\mu_f}\right) \quad (5.103)$$

where *J* is to be obtained from a plot[5e] of *J* versus N_{Re}. The units should be consistent as shown in the previous equations for dimensionless groups shown in parentheses. The *J*-factor can also be determined from Equation 5.62.

Film Temperature for Condensation

Kern[5f] recommends the relation

$$t_f = \frac{1}{2}(t_{sv} + t_w)\,°\text{F, [K]} \quad (5.104)$$

where
 t_{sv} = temperature of saturated vapor, °F [K]
 t_w = wall temperature, °F [K]
 $\Delta t_f = t_f - t_w$.

McAdams[4c] recommends the relation

$$t_f = t_{sv} - 0.75\Delta t \quad (5.105)$$

where

$$\Delta t = t_{sv} - t_w.$$

Tube Wall Temperature

This may be estimated by the equation

$$t_w = \frac{h_i t_i + h_o t_o}{h_i + h_o} \quad (5.106)$$

where t_i is the bulk temperature of the fluid inside the tube and t_o is the bulk temperature of the fluid outside the tube.

Note that if t_f is calculated as suggested by McAdams,[4c] Δt should be used in the place of Δt_f in Equations 5.95, 5.96, and 5.99.

Example 5.6

A bare horizontal 12-in. IPS steam pipe carries saturated steam at 240° F. The temperature of ambient air is 70° F. Calculate the rate of heat loss per foot of pipe length. Steel emissivity = 0.8.

Solution

Assume the steam-film resistance and pipe-wall resistance negligible. Therefore, the surface temperature is 240° F. For horizontal pipe

$$h_c = 0.5\left(\frac{\Delta t}{d_o}\right)^{0.25}$$

$$= 0.5\left(\frac{170}{12.75}\right)^{0.25} = 0.96 \text{ Btu/h·ft}^2\cdot°\text{F}$$

Steel emissivity = 0.8,

$$h_r = \frac{0.1713(0.8)\left[\left(\frac{460+240}{100}\right)^4 - \left(\frac{460+70}{100}\right)^4\right]}{240-70}$$

$$= 1.30 \text{ Btu/h} \cdot \text{ft}^2 \cdot {}^\circ\text{F}$$

$$h_a + h_c + h_r = 0.96 + 1.30 = 2.26 \text{ Btu/h} \cdot \text{ft}^2 \cdot {}^\circ\text{F}$$

$$\frac{\text{Outer area of pipe}}{\text{Foot length}} = \pi D_o L = \pi \frac{12.75}{12}(1) = 3.338 \text{ ft}^2/\text{ft length}$$

Therefore, heat loss is

$$Q_L = h_a(A_o)(\Delta t)$$
$$= 2.26(3.338)(240-70)$$
$$= 1282.5 \text{ Btu/h} \cdot \text{ft}.$$

Example 5.7

An oil is flowing at velocity of 5 ft/s through a 10 ft long, 1-in. diam. 18 BWG tube. On the outer surface of the tube, steam is condensing at 220° F. The tube is clean. Oil enters the tube at 86° F and leaves at 104° F. The physical-property data (all assumed constant) for oil are

Oil density $\rho = 55$ lb/ft^3
Specific heat $c_p = 0.48$ Btu/lb $\cdot$ °F
Thermal conductivity $k = 0.08$ Btu/(h $\cdot$ ft^2 °F/ft)

The viscosity μ varies with temperature as follows:

t, °F	80	90	100	110	130	140	220
μ, cP	20	18	16.2	15	12	11	3.6

Calculate h_i, the inside oil film coefficient of heat transfer.

Solution

The inside diameter of the tube is

$$0.902 \text{ in} = \frac{0.902}{12} = 0.0752 \text{ ft}$$

Neglect the steam-film resistance. The oil leaves the tube at 104° F.

$$t_w = 220°\text{F}$$

$$t_b = \text{bulk temperature of oil} = \frac{86+104}{2} = 95°\text{F}$$

At $t_b = 95°$F, by interpolation $\mu = 17.1$ cP $= 41.38$ lb/ft $\cdot$ h

$$N_{Re} = \frac{Du\rho}{\mu} = \frac{0.0752(5)(55)(3600)}{41.38} = 1799 < 2100$$

Therefore, Equation 5.53 for the streamline flow applies.

$$\frac{h_i D_i}{k} = 1.86 \left[\left(\frac{DG}{\mu} \right) \left(\frac{c_p \mu}{k} \right) \left(\frac{D}{L} \right) \right]^{1/3} \left(\frac{\mu}{\mu_w} \right)^{0.14}$$

$$\frac{DG}{\mu} = 1799 \quad \text{(as calculated before)}$$

$$\frac{c_p \mu}{k} = \frac{0.48(41.38)}{0.08} = 248.3$$

$$\frac{D}{L} = \frac{0.0752}{10} = 0.00752$$

$$\frac{h_i D_i}{k} = 1.86[(1799)(248.3)(0.00752)]^{1/3} \left[\frac{41.38}{(3.6)(2.42)} \right]^{0.14}$$

$$= 34.65 \text{ Btu/h} \cdot \text{ft}^2 \cdot °F$$
$$h_i = 34.65 \times 0.8/0.752$$
$$= 36.86 \text{ Btu/h} \cdot \text{ft}^2 \cdot °F$$

Example 5.8

Aniline is flowing at a velocity of 8 ft/s in a $\frac{3}{4}$-in. 18 BWG tube. On the outer surface, steam is condensing at 220°F. The tube is clean. Using the Sieder-Tate equation, find h_i, the inside film coefficient of heat transfer if aniline enters the tube at 80°F and leaves at 120°F. Also determine the film coefficient using the Dittus-Boelter equation. The physical-property data for aniline are given in Table 5.4.

Solution

The bulk temperature is

$$t_b = \frac{80 + 120}{2} = 100°F$$

Assuming that the caloric temperature = t_b = 100°F,

$\rho_c = 1.001(62.5) = 62.5 \text{ lb/ft}^3$
$(c_p)_c = 0.49 \text{ Btu/lb} \cdot °F$
$k_c = 0.10 \text{ Btu/(h} \cdot \text{ft}^2 \cdot °F/\text{ft})$
$\mu_c = 2.53(2.42) = 6.123 \text{ lb/h} \cdot \text{ft}$
$\mu_{220} = 0.91 - \frac{20}{100}(0.91 - 0.48) = 0.824 \text{ cP}$

Table 5.4 Physical-property data for aniline

Temp.,°F	μ, cP	k, Btu/(h · ft² · °F/ft)	s	c_p, Btu/lb · °F
60	4.84	0.10	1.026	0.43
100	2.53	0.10	1.001	0.49
150	1.44	0.098	0.986	0.505
200	0.91	0.096	0.962	0.515
300	0.48	0.093	0.922	0.540

Assume that the metal wall resistance is negligible. Check for Reynolds number.

$$D_i = \frac{0.652}{12} = 0.05434 \text{ ft}$$

$$\frac{D_i u \rho_c}{\mu_c} = \frac{0.05434(8)(62.5)}{2.53(0.000672)} = 15{,}981$$

The flow is in the turbulent region, and therefore the Sieder-Tate equation for the turbulent flow applies.

$$h_i = 0.027 \frac{k}{D_i} \left(\frac{D_i G}{\mu_c}\right)^{0.8} \left(\frac{c_p \mu_c}{k_c}\right)^{1/3} \left(\frac{\mu_c}{\mu_w}\right)^{0.14}$$

$$\frac{DG}{\mu_c} = 15{,}981$$

$$\frac{c_p \mu_c}{k_c} = \frac{0.49(6.123)}{0.10} = 30$$

$$\frac{\mu_c}{\mu_w} = \frac{2.53}{0.824} = 3.07$$

$$h_i = 0.027 \left(\frac{0.1}{0.05434}\right)(15{,}981)^{0.8}(30)^{1/3}(3.07)^{0.14}$$

$$= 416.6 \text{ Btu/h} \cdot \text{ft}^2 \cdot {}^\circ\text{F}$$

The Dittus-Boelter equation gives

$$h_i = 0.023 \left(\frac{DG}{\mu_b}\right)^{0.8} \left(\frac{c_p \mu_b}{k_b}\right)^{0.4} \frac{k_b}{D_i}$$

$$= 0.023(15{,}981)^{0.8}(30)^{0.4} \frac{0.1}{0.05434} = 380.5 \text{ Btu/h} \cdot \text{ft}^2 \cdot {}^\circ\text{F}$$

Example 5.9

Water flowing inside a 1-in.-diam. 16 BWG horizontal copper tube is being heated from 90°F to 200°F by saturated steam condensing on the outside of the tube at 250°F. If the average tube wall temperature is 210°F, calculate the condensation coefficient based on the inside area.

Solution

$$t_{sv} = 250°\text{F} \quad t_w = 210°\text{F}$$

The condensation film coefficient on the outside of the tube is calculated by Equation 5.99 with Δt calculated as suggested by McAdams.

$$h = 0.725 \left(\frac{k_f^3 \rho_f^2 g \lambda_f}{\mu_f D_o \Delta t}\right)^{0.25}$$

From Equation 5.105

$\Delta t = t_{sv} - t_w = 250 - 210 = 40°F$

$t_f = t_{sv} - 0.75\Delta t$
$= 250 - 0.75(250 - 210)$
$= 220°F$

$\mu_f = 0.654$ lb/h · ft
$\rho_f = 59.63$ lb/ft^3
$k_f = 0.395$ Btu/(h · ft^2 · °F/ft)
$\lambda_f = 965.2$ Btu/lb
$D_o = \frac{1}{12} = 0.0834$ ft

$$h_o = 0.725 \left[\frac{0.395^3 (59.63)^2 (965.2)(4.18 \times 10^8)}{(0.654)(0.0834)(40)} \right]^{0.25}$$

$= 1829$ Btu/h · ft^2 · °F

h_{oi} is the condensation coefficient based on inside area.

$$\frac{h_o D_o}{D_i} = 1829 \left(\frac{1}{0.87} \right) = 2102 \text{ Btu/h·ft}^2 \cdot °F$$

For the calculation of h_o using the Kern's relationship for the film temperature

$$h_o = 0.725 \left(\frac{k_f^3 \rho_f^2 \lambda g}{\mu_f D \Delta t_f} \right)^{0.25}$$

From Equation 5.104,

$t_f = \frac{1}{2}(t_{sv} + t_w) = \frac{1}{2}(250 + 210) = 230°F$
$\Delta t_f = t_f - t_w = 230 - 210 = 20°F$
$\rho_f = 59.38$ lb/ft^3
$\mu_f = 0.6195$ lb/h · ft
$\lambda = 958.8$ Btu/lb
$k_f = 0.3955$ Btu/(h · ft^2 · °F/ft)

$$h_o = 0.725 \left[\frac{(0.3955)^3 (59.38)^2 (958.8)(4.18 \times 10^8)}{0.6195(0.0834)(20)} \right]^{0.25}$$

$= 2199$ Btu/h · ft^2°F

$$h_{oi} = 2199 \left(\frac{1}{0.87} \right) = 2528 \text{ Btu/h·ft}^2 \cdot °F$$

Example 5.10

Water at an average temperature of 160°F is flowing inside a horizontal 1-in.-diam. 14 BWG clean copper tube at a velocity of 6 ft/s. Saturated steam at 15 psig is condensing on the outside of the tube. Calculate the overall coefficient of heat transfer based on the outside area.

Solution

Use the Dittus-Boelter equation (Eq. 5.56) for the calculation of the water film cofficient.

$\rho_b = 61.0$ lb/ft^3 $\mu_b = 0.97$ lb/h · ft $c_p = 1$ Btu/lb · °F

$k_b = 0.385$ Btu/h · ft^2 · °F/ft) $D_i = \dfrac{0.834}{12} = 0.0695$ ft

$$h_i = 0.023 \left(\frac{D_i u \rho}{\mu}\right)_b^{0.8} \left(\frac{c_p \mu}{k}\right)_b^{0.4} \left(\frac{k_b}{D_i}\right)$$

$$= 0.023 \left[\frac{0.0695(6)(61)(3600)}{0.97}\right]^{0.8} \left[\frac{1(0.97)}{0.385}\right]^{0.4} \left(\frac{0.385}{0.0695}\right)$$

$$= 1761 \text{ Btu/h} \cdot \text{ft}^2 \cdot °F$$

Steam Film Coefficient. As a first approximation, assume

$$t_w = \frac{t_{sv} + t_b}{2} = \frac{250 + 160}{2} = 205°F$$

Use t_f as suggested by McAdams.

$$t_f = t_{sv} - 0.75(t_{sv} - t_w)$$
$$= 250 - 0.75(250 - 205)$$
$$= 216.3°F \quad \text{say } 220°F$$

$\Delta t = 250 - 205 = 45°F \quad \rho_f = 59.63 \text{ lb/ft}^3 \quad D_o = 0.0834 \text{ ft}$

$\lambda = 965.2 \text{ Btu/lb} \quad k_f = 0.395 \text{ Btu/h} \cdot \text{ft}^2 \cdot °F/\text{ft}) \quad \mu_f = 0.654 \text{ lb/ft} \cdot \text{h}$

From Equation 5.99

$$h_o = 0.725 \left[\frac{0.395^3 (59.63)^2 (965.2)(4.18 \times 10^8)}{0.654(0.0834)(45)}\right]^{0.25} = 1776 \text{ Btu/h} \cdot \text{ft}^2 \cdot °F$$

Check the assumed wall temperature. Take areas per linear foot. The water-side resistance is

$$\frac{1}{h_i A_i} = \frac{1}{1761 \pi (0.0695)} = 0.002601$$

The metal-wall resistance is

$$\frac{l_w}{k_w} \frac{1}{A_{av}}$$

$$A_{av} = \frac{A_o - A_i}{\ln(A_o/A_i)} = \frac{(D_o - D_i)\pi}{\ln(D_o/D_i)}$$

$$= \frac{(0.0834 - 0.0695)\pi}{\ln(0.0834/0.0695)} = 0.2395 \text{ ft}^2$$

$$\text{Metal-wall resistance} = \frac{0.0834 - 0.0695}{226(2)} \frac{1}{0.2395} = 0.000128$$

$$\text{Steam-side resistance} = \frac{1}{1776 \pi (0.0834)} = 0.00215$$

$$\text{Total resistance} = \frac{1}{U_o A_o}$$

$$= 0.002601 + 0.00215 + 0.000128 = 0.00488$$

$$\Delta T_{\text{total}} = 250 - 160 = 90°F$$

$$\Delta t_i = \frac{[1/(h_i A_i)]\Delta T}{1/(h_i A_i) + l_w/(k_w A_{av}) + 1/(h_o A_o)}$$

$$= \frac{0.002601(90)}{0.00488} = 48°F$$

$$t_w = t_b + \Delta t_i = 160 + 48 = 208°F$$

and $\quad t_f = 250 - 0.75(250 - 208) = 219°F$ (assumed 220°F)

which is close. h_o was calculated with the physical properties at 220°F. Therefore, correct h_o for Δt only.

New $\quad\quad \Delta t = t_{sv} - t_w = 250 - 208 = 42°F$

$$h_o = 1766\left(\frac{45}{42}\right)^{0.25} = 1807 \text{ Btu/h} \cdot \text{ft}^2 \cdot °F$$

The overall coefficient based on the outside area is

$$\frac{1}{U_o} = \frac{1}{h_o} + \frac{l_w}{k_w}\frac{A_o}{A_{av}} + \frac{1}{h_i}\frac{A_o}{A_i}$$

$$= \frac{1}{1807} + \frac{0.00695}{228}\frac{0.262}{0.2395} + \frac{1}{1761}\frac{0.262}{0.2183} = 0.001268$$

$$U_o = 788.5 \text{ Btu/h} \cdot \text{ft}^2 \cdot °F$$

The following example illustrates the calculation for a double pipe exchanger in which the equivalent diameter has to be considered.

Example 5.11

Three thousand pounds per hour of acetic acid is to be cooled from 250°F to 150°F by heating 6000 lb/h of butyl alcohol, which is available at 90°F in a double pipe exchanger with counter-current flow. Acetic acid flows through a 1-in.-diam 14 BWG tube that is surrounded by an outer pipe of 2.067-in. ID. Calculate the length of the tube required. Thermal conductivity of the tube material is 10 Btu/h·ft²·°F/ft. The specific heats of acetic acid and butyl alcohol are 0.55 and 0.65 Btu/lb·°F, respectively. A fouling factor of 0.001 is to be allowed for each stream. Verify the inside film coefficient by Colburn *J*-factor.

Solution

Overall Heat Balance. The heat loss by acetic acid is

$$3000(250 - 150)(0.55) = 165{,}000 \text{ Btu/h}$$

The temperature rise of butyl alcohol is

$$\frac{165{,}000}{0.65(6000)} = 42.3°F$$

The outlet temperature of butyl alcohol is

$$90 + 42.3 = 132.3°F$$

Log mean temperature difference is expressed as follows:

$$\text{Acetic acid} \xrightarrow{250150}$$

$$\xleftarrow{132.390} \text{Butyl alcohol}$$

$$\Delta t_1 = 250 - 132.3 = 117.7°F \quad \Delta t_2 = 150 - 90 = 60°F$$

$$\Delta t_{LM} = \frac{117.7 - 60}{\ln(117.7/60)} = 85.63°F$$

Use the Dittus-Boelter equation, which requires the evaluation of the physical properties at the bulk temperatures. Bulk temperature of acetic acid is

$$t_B = \frac{250 + 150}{2} = 200°F$$

Bulk temperature of butyl alcohol is

$$\frac{90 + 132.3}{2} = 112.2°F$$

The properties of the streams at the bulk temperatures are listed in Table 5.5.
The inside film coefficient of heat transfer when cooling is given by Equation 5.57

$$h_i = 0.023(N_{Re})_b^{0.8}(N_{Pr})_b^{0.3}\left(\frac{k}{D_i}\right)$$

where

$$N_{Re} = \frac{D_i G}{\mu}, \quad N_{Pr} = \frac{c_p \mu}{k}$$

$$D_i = \frac{0.834}{12} = 0.0695 \text{ ft}$$

$$A_c = 0.003794 \text{ ft}^2 \quad G = \frac{3000}{0.003794} = 790{,}722 \text{ lb/h} \cdot \text{ft}^2$$

$$N_{Re} = \frac{(0.0695)(790722)}{0.5(2.42)} = 4.542(10^4)$$

$$N_{Pr} = \frac{0.55(0.5)(2.42)}{0.098} = 6.791$$

$$h_i = 0.023[N_{Re}]^{0.8}[N_{Pr}]^{0.03}\left(\frac{0.098}{0.0695}\right)$$

$$= 306.4 \text{ Btu/h} \cdot \text{ft}^2 \cdot °F$$

Table 5.5

	Acetic Acid	Butyl Alcohol
μ, cP	0.5	2.7
ρ, lb/ft^3	65.5	50.5
c_p, Btu/lb·°F	0.55	0.65
k, Btu/(h·ft^2·°F/ft)	0.098	0.096

Now verify the inside film coefficient by Colburn *J*-factor.

$$J_1 = \left[\left(\frac{1}{N_{Re}^{9.36}} \right) + \frac{1}{\left\{ \frac{N_{Re}^{1.6}}{7.831(10^{-14})} + \left(\frac{1.969(10^6)}{N_{Re}} \right)^8 \right\}^{3/2}} \right]^{1/12}$$

Substituting the value of Reynolds number,

$$J_1 = \left[\left(\frac{1}{(4.542(10^4))^{9.36}} \right) + \frac{1}{\left\{ \frac{(4.542(10^4))^{1.6}}{7.831(10^{-14})} + \left(\frac{1.969(10^6)}{(4.542(10^4))} \right)^8 \right\}^{3/2}} \right]^{1/12}$$

$$= 2.693(10^{-3})$$

Ignoring the viscosity correction:

$$h_i = J_1 \left(\frac{c_p G}{N_{Pr}^{0.6667}} \right) = \frac{(2.693)(10^{-3})(0.55)(790722)}{(6.791)^{0.6667}} = 326.7 \text{ Btu/h·ft}^2 \cdot °F$$

This value is within ~7 percent of the first value, so agreement is good.

Outside Coefficient of Heat Transfer. For heat transfer, equivalent diameter D_e is

$$D_e = \frac{D_2^2 - D_o^2}{D_o}$$

$$= \frac{(2.067^2 - 1^2)}{12} = 0.2727 \text{ ft}$$

$$\text{Flow area} = \frac{\pi(2.067^2 - 1^2)}{4(144)} = 0.01785 \text{ ft}^2$$

$$G = \frac{6000}{0.01785} = 336,135 \text{ lb/h·ft}^2$$

$$\frac{D_e G}{\mu} = \frac{0.2727(336,135)}{(2.7)(2.42)} = 14,029$$

$$\frac{k}{D_e} = \frac{0.096}{0.2727} = 0.352$$

$$\frac{c_p \mu}{k} = \frac{0.65(2.7)(2.42)}{0.096} = 44.23$$

From Equation 5.56 the outside coefficient of heat transfer is

$$h_o = 0.023 \left(\frac{D_e G}{\mu_b} \right)^{0.8} \left(\frac{c_p \mu}{k} \right)^{0.4} \left(\frac{k_b}{D_e} \right)$$

$$= 0.023(14,029)^{0.8}(44.23)^{0.4}(0.352) = 76.6 \text{ Btu/h·ft}^2 \cdot °F$$

$$\frac{1}{U_o} = \frac{1}{h_o} + f_{do} + \frac{l_w}{k_w}\frac{D_o}{D_{av}} + f_{di}\frac{D_o}{D_i} + \frac{1}{h_i}\frac{D_o}{D_i}$$

$$= \frac{1}{76.6} + 0.001 + \frac{0.006917}{10}\frac{1}{0.917} + 0.001\frac{1}{0.834} + \frac{1}{306}\frac{1}{0.834}$$

$$= 0.01993$$

$$U_o = 50.2 \text{ Btu/h} \cdot \text{ft}^2 \cdot °\text{F}$$

$$A_o = \frac{Q}{U\,\Delta T} = \frac{165{,}000}{50.2(85.63)} = 38.38 \text{ ft}^2$$

$$\text{Length of tube} = \frac{A}{\pi D_o} = \frac{38.38}{\pi(0.0834)} \doteq 147 \text{ ft}$$

Example 5.12

The data in Table 5.6 were obtained for a condenser with water flowing through the tubes and steam condensing on the outside. The steam film coefficient on the outside of the tube is constant and equal to 2000 Btu/h·ft²·°F. What is the scale resistance? Neglect the pipe-wall resistance.

Solution

$$\frac{1}{U_o} = \frac{1}{h_o} + R_d + \frac{1}{au^{0.8}}$$

Since $h_i \propto u^{0.8}$, a plot of $1/U_o$ versus $1/u^{0.8}$ will be a straight line with intercept of $(1/h_o) + R_d$ which is constant (Wilson's plot).[4b] Values of $1/U_o$ and $1/u^{0.8}$ are given in the last two columns of Table 5.6. $1/U_o$ is plotted against $1/u^{0.8}$ in Exhibit 1.

From the graph, the intercept is

$$0.00083 = \frac{1}{h_o} + R_d$$

$$R_d = 0.00063 - \frac{1}{h_o} = 0.00083 - \frac{1}{2000} = 0.00033$$

The scale resistance, or fouling factor, is 0.00033 h · ft² · °F/Btu.

Table 5.6 Overall coefficients for a condenser

u, ft/s	U_o, Btu/h·ft²·°F	$\frac{1}{U_o} \times 10^3$	$\frac{1}{u^{0.8}}$
6.91	485.4	2.060	0.213
6.35	473.3	2.113	0.228
5.68	452.1	2.212	0.249
4.90	421.2	2.374	0.280
2.93	333.2	3.001	0.423
7.01	480.5	2.081	0.211
2.95	325.1	3.076	0.421
4.12	364.6	2.743	0.322
6.76	400.3	2.498	0.217
2.86	298.0	3.356	0.431
6.27	452.7	2.209	0.230

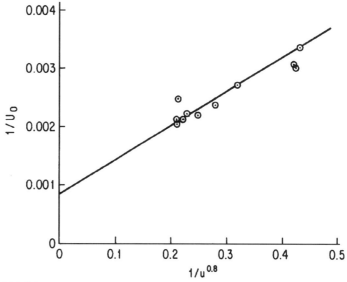

Exhibit 1 Wilson plot (Example 5.12)

Example 5.13

A tube bank consists of 1-in. OD tubes on 2-in. centers in a square in-line arrangement of 16 rows in each direction. The tube wall is at 220°F and air at 80°F flows across the tube bank at a mean velocity of 8000 lb/h·ft² based on the minimum area of flow. Determine the heat-transfer coefficient for air. At what temperature does the air leave the tube bank? Assume $1/2$-in. clearance between the wall and the tubes in the extreme longitudinal rows.

Solution

The surface temperature is 220°F, and the air outlet temperature is not known. Assume the air outlet temperature is 120°F. From Equation 5.76

$$t_f = \frac{1}{2}\left[220 + \frac{1}{2}(120+80)\right] = 160°F$$

The properties of air at 160°F are

$$\mu_f = 0.02 \text{ cP}(2.42) = 0.0484 \text{ lb/h} \cdot \text{ft}$$
$$D_o = 0.0834 \text{ ft}$$
$$k_f = 0.0174 \text{ Btu/(h} \cdot \text{ft}^2 \cdot °F/\text{ft})$$
$$\rho_f = 0.0643 \text{ lb/ft}^3$$

$$h_o = 0.33 \frac{k_f}{D_o}\left(\frac{D_o G_{max}}{\mu_f}\right)^{0.6}\left(\frac{c_p \mu_f}{k_f}\right)^{1/3}$$

$$= 0.33 \frac{0.0174}{0.0834}\left[\frac{0.0834(8000)}{0.0484}\right]^{0.6}\left[\frac{0.24(0.0484)}{0.0174}\right]^{1/3}$$

$$= 18.3 \text{ Btu/h} \cdot \text{ft}^2 \cdot °F$$

$$Q_o = 18.3(220-100) = 2196 \text{ Btu/h} \cdot \text{ft}^2$$

The flow per square foot of minimum area is 8000 lb/h · ft².

$$\Delta T_{LM} = \frac{(220-80)-(220-120)}{\ln(140/100)} = 119°F$$

Consider a 1 ft length of bundle with 16 rows. The flow area is

$$(16-1)\left(\frac{2}{12}-\frac{1}{12}\right)+(2)\left(\frac{1}{2}\right)\left(\frac{1}{12}\right) = \frac{16}{12} \text{ ft}^2/\text{ft length}$$

The flow per foot of bundle is

$$(8000 \text{ lb/h} \cdot \text{ft}^2)\left[\left(\frac{16}{12}\right) \text{ ft}^2/\text{ft}\right] = 10,667 \text{ lb/h} \cdot \text{ft}$$

The surface area per foot of bundle is

$$\frac{\pi(1)}{12}(16)(16) = 67.02 \text{ ft}^2/\text{ft length}$$

Heat balance gives

$$10,667(0.24)(t_o - 80) = 18.3(67.02)(119)$$

from which

$$t_o = 137°F \quad \text{(assumed 120°F)}$$

Reassume

$$t_o = 137°F \quad \text{then} \quad t_f = \frac{1}{2}\left(220 + \frac{80+137}{2}\right)$$

$$= 164°F \quad \text{(previously calculated 160°F)}$$

Because of very small change in temperature, there will not be much change in the physical properties of air.

$$h_o = 18.3 \text{ Btu/h} \cdot \text{ft}^2 \cdot °F \quad \text{as before}$$

$$\Delta T_{LM} = \frac{57}{\ln\frac{140}{83}} \doteq 109°F$$

Then
$$10,667(0.24)(t_o - 80) = 67.02(18.3)\Delta T_{LM} = 67.02(18.3)(109.0)$$
$$t_o = 132.2, \ (137°F \text{ assumed})$$
Another trial with $t_o = 133°F$ yields
$$\Delta T_{LM} = 114.4°F$$
and
$$t_o = 133.4°F, \quad \text{(assumed 133°F)}$$
$$h_i = 18.3 \text{ Btu/h} \cdot \text{ft}^2 \cdot °F \quad \text{and} \quad t_o = 133.4°F$$

Shell and Tube Exchangers

For $2000 < N_{Re} < 10^6$, the heat-transfer coefficient[5g] on the shell side is given by

$$\frac{h_o D_e}{k} = 0.36 \left(\frac{D_e G_s}{\mu}\right)^{0.55} \left(\frac{c_p \mu}{k}\right)^{1/3} \left(\frac{\mu}{\mu_w}\right)^{0.14} \quad (5.107)$$

$$a_s = \frac{D_{si} C' B}{144 P_T} \quad \text{and} \quad G_s = \frac{W}{a_s} \quad (5.108)$$

where*
  d_o = tube outside diameter, in
  D_e = shell side equivalent diameter, ft
  a_s = cross-flow area, ft^2
  D_{si} = inside diameter of shell, in
  P_T = tube pitch, in
  C' = clearance between tubes measured along tube pitch, in
       = $P_T - d_o$
  B = baffle spacing, in
  W = weight flow of fluid, lb/h

The baffle spacing should range from $D_{si}/5$ to D_{si}. For square pitch tube arrangement

$$D_e = \frac{P_T^2 - \frac{1}{4}\pi d_o^2}{3\pi d_o} \text{ ft} \quad (5.109)$$

For 60° triangular equilateral pitch arrangement

$$D_e = \frac{2\left(0.43 p_T^2 - \frac{1}{8}\pi d_o^2\right)}{3\pi d_o} \text{ ft} \quad (5.110)$$

Logarithmic Temperature Differences

In calculating the log mean temperature difference for the shell-tube exchangers, a correction factor F_T must be applied because the flows are not truly countercurrent or parallel, but a combination of the countercurrent and parallel. Thus

$$\Delta T_{\text{true mean}} = \text{LMTD}(F_T) \quad (5.111)$$

Values of F_T are available in the form of graphs in Figures 5.4 and 5.5. Obviously, when terminal temperature differences are identical, LMTD does not apply, and the terminal temperature difference is used as the driving force.

Unless one or both of the fluids exchange heat under isothermal condition, LMTD for countercurrent flow for the same process terminal temperatures will be greater than the LMTD for cocurrent flow, and therefore, LMTD for countercurrent flow has greater potential for heat recovery. When at least one fluid exchanges heat isothermally, LMTD for countercurrent flow is the same as that in the cocurrent flow.

---

*For explanation of the units of other variables, please see nomenclature under Equation 5.56.

In the multipass shell and tube heat exchangers, the flows are not truly countercurrent. Therefore, a correction factor, F_T is introduced. See reference[4a] for evaluation of this parameter. The value of F_T is 1 for true countercurrent exchanger, and is less than 1 for any other design, implying the higher heat recovery by a countercurrent arrangement than any other arrangement for non-isothermal exchange of heat.

$$Q = UAF_T(T)_{lm} \tag{5.112}$$

The correction factor F_T is determined from a graphical plot as a function of two other parameters S (or P) and R defined as follows:

$$S = \frac{t_2 - t_1}{T_1 - t_1} = \frac{\text{cold fluid temperature rise}}{\text{maximum temperature difference between hot \& cold fluid}} \tag{5.113}$$

Some references label the above parameter as P. It is also known as temperature efficiency of the exchanger

$$R = \frac{T_1 - T_2}{t_2 - t_1} = \frac{\text{hot fluid temperature drop}}{\text{cold fluid temperature rise}} \tag{5.114}$$

When R is zero such as when there is isothermal heating medium, $F_T = 1$; when R is infinity or S is zero such as when there is isothermal cooling medium, $F_T = 1$. Such graphs will be provided if the solution of such a problem is required.

Estimation of Outlet Temperatures for Cocurrent and Countercurrent Flows with no Phase Change

Outlet temperatures for cocurrent and countercurrent flows with no phase change, given inlet temperatures (T_1, t_1), heat capacities (C, c), overall heat transfer coefficient (U), heat transfer area (A), and flow rates (W, w), can be estimated beginning with the following equations:

for hot fluid: $\quad Q = WC(T_1 - T_2)$ \hfill (5.115)

for cold fluid: $\quad Q = wc(t_2 - t_1)$ \hfill (5.116)

heat balance: $\quad Q = WC(T_1 - T_2) = wc(t_2 - t_1) = UA(T)_{lm}$ \hfill (5.117)

Define $R = wc/(WC)$ and $K_1 = e^{UR(R-1)/wc}$, $K_2 = e^{UA(R+1)/WC}$, then for countercurrent flow

$$T_2 = \frac{(1-R)T_1 + (1-K)Rt_1}{1 - RK_1} \tag{5.118}$$

For cocurrent flow

$$T_2 = \frac{(R+K_2)T_1 + (K_2 - 1)Rt_1}{(R+1)K_2} \tag{5.119}$$

For both cases, calculate t_2 from Equation 5.117.

$$t_2 = (T_1 - T_2)/R + t_1 \tag{5.120}$$

$$= (T_1 - t_1)(K_1 - 1)/(RK_1 - 1) + t_1 \tag{5.121}$$

Symbols of preceding equations are as follows:

	Hot Fluid	Cold Fluid
Flow rate	W	W
Temperature in	T_1	t_1
Temperature out	T_2	t_2
Heat capacity	C	C

PRESSURE DROP IN EXCHANGER

The following paragraphs show how to calculate pressure drop in exchangers when there is no change of phase.

Double Pipe Exchanger: Annulus

$$\Delta P = \left[\sum K + \frac{4fL}{D_e}\right]\frac{u^2\rho}{2g_c} = \left[\sum K + \frac{4fL}{D_e}\left(\frac{\mu_w}{\mu}\right)^{0.14}\right]\frac{G^2}{2\rho g_{ch}}\frac{\text{lb}_f}{\text{ft}^2} \tag{5.122}$$

where G = mass velocity, lb/h · ft². $g_{ch} = 4.17(10^8)$ (lb · ft/hr²)/lb$_f$. L is the length of the annulus; ft. D_e is equivalent diameter for pressure drop (not heat transfer!) in ft. See Equation 4.52 for explanation of other terms. Appropriate entrance and exit loss coefficient (K) should be incorporated. The viscosity correction factor $(\mu_w/\mu)^{0.14}$ is usually small.

Double Pipe Exchanger: Inner Pipe

$$\Delta P = \left[\sum K + \frac{4fL}{D}\right]\frac{u^2\rho}{2g_c} = \left[\sum K + \frac{4fL}{D}\left(\frac{\mu_w}{\mu}\right)^{0.14}\right]\frac{G^2}{2\rho g_{ch}}\frac{\text{lb}_f}{\text{ft}^2} \tag{5.123}$$

The method in principle is the same as what is used for annulus pipe except that the equivalent diameter is replaced by inner pipe diameter in ft.

Shell-and-Tube Exchanger: Tube Side

$$\Delta P = \left[4N_{tp} + \left(\frac{4fN_{tp}L}{D}\right)\left(\frac{\mu_w}{\mu}\right)^{0.14}\right]\left[\frac{G^2}{2\rho g_{ch}}\right]$$

$$+ \left[K_{ent} + K_{exit}\right]\left[\frac{G_n^2}{2\rho g_{ch}}\right], \frac{\text{lb}_f}{\text{ft}^2} \tag{5.124}$$

where

K_{ent} = entrance resistance factor = 0.5 to 0.78
K_{exit} = exit resistance factor = 1.0
N_{tp} = number of tube pass
f = Fanning friction factor, dimensionless
L = tube length, ft
D = tube inside diameter, ft
$\left(\frac{\mu_w}{\mu}\right)$ = wall to bulk viscosity ratio
G = fluid mass velocity through tubes, $\frac{lb}{h \cdot ft^2}$
G_n = fluid mass velocity through nozzle, lb/h · ft². If the exit and inlet nozzles are different in size, then G_n should be based on the appropriate nozzle size.
ρ = average fluid density, lb/ft³
g_{ch} = Newton's law proportionality factor = $4.17(10^8) \frac{lb \cdot ft}{h^2 \cdot lb_f}$

Shell-and-Tube Exchanger: Shell Side

The determination of shell-side pressure drop requires a number of mechanically induced factors such as leakage through baffle-and-shell clearance, leakage through tube-and-baffle clearance, percent of baffle cut, etc., and is computationally intensive. A conservative estimating method based on Kern,[5m] for 25 percent baffle-cut segmental baffles is presented below.

$$\Delta P_{shell} = \frac{f_{shell} G_s^2 D_s (N+1)}{2 g_{ch} \rho D_e \left(\frac{\mu}{\mu_w}\right)^{0.14}} + [K_{ent} + K_{exit}] \frac{G_{ns}^2}{2(144) \rho g_{ch}}$$

$$= \frac{f_{shell} G_s^2 D_s \left(\frac{12L}{B}\right)}{5.2(10^{10}) s D_e \left(\frac{\mu}{\mu_w}\right)^{0.14}} + [K_{ent} + K_{exit}] \frac{G_{ns}^2}{2(144) \rho g_{oh}} \frac{lb_f}{in^2} \quad (5.125)$$

where

f_{shell} = dimensional friction factor, $\frac{ft^2}{in^2}$
G_s = Shell side mass velocity, lb/h · ft²
G_{ns} = Shell side nozzle mass velocity, lb/h · ft²
    It must be based on the appropriate nozzle area
D_s = Shell internal diameter, ft
D_e = Shell side equivalent diameter, ft
N = Number of baffles
L = Tube length, ft
B = Baffle spacing, in
S = Specific gravity of shell side fluid, dimensionless

The dimensional friction factor, f_{shell}, for shell-side frictional pressure may be read from Kern[5m] or be computed as a function of shell-side Reynolds number, N_{Res}, from

$$f_{shell} = e^{\left[-40.2235 + 5.23522(\ln(N_{Res})) - 0.37497(\ln(N_{Res}))^2 + 0.009650651(\ln(N_{Res}))^3 + \frac{104.528}{\ln(N_{Res})} - \frac{99.3874}{(\ln(N_{Res}))^2}\right]}$$

(5.126)

Pressure Drop in Condenser

The pressure drop in a total condenser may be computed as one-half of the frictional pressure drop excluding the entrance and exit loss for the all-vapor flow based on inlet conditions. For partial condensers, the frictional pressure drop may be computed[21] by

$$\Delta P_{\text{partial cond}} = \frac{1}{2}\left[\frac{1+v_{\text{out}}}{v_{\text{in}}}\right](\Delta P_{\text{all vap inlet}}) \quad (5.127)$$

where

$\Delta P_{\text{partial cond}}$ = pressure drop in a partial condenser
v_{out} = vapor velocity at the outlet
v_{in} = vapor velocity at the inlet
$\Delta P_{\text{all vap inlet}}$ = pressure drop based on all vapor at the inlet conditions

Example 5.14

Seventy thousand pounds per hour of an organic liquid is to be cooled by countercurrent flow of the water in a shell-and-tube exchanger. The data available are listed in Table 5.7. Estimate the required area of the exchanger.

Solution

Step 1. Estimate a preliminary area and assume a geometry for the exchanger. Heat balance gives

$$Q = WC_p(t_2 - t_1) = 70{,}000(0.5)(200 - 115) = 2{,}975{,}000 \text{ Btu/h}$$

Alternatively,

$$Q = 198{,}330(1)(100 - 85)$$
$$= 2{,}974{,}950 \doteq 2{,}975{,}000 \text{ Btu/h}$$

Table 5.7

	Shell Side	Tube Side
Fluid circulated	Water	Organic
Total liquid, lb/h	198,330	70,000
Temperature in, °F	85	200
Temperature out, °F	100	115
Specific gravity	1 at 92°F	1.5 at 158°F
Viscosity, cP	1.0 at 92°F	2 at 158°F
Specific heat, Btu/lb·°F	1 at 92°F	0.5 at 158°F
Thermal conductivity, Btu /h·ft²·°F/ft)	0.36 at 92°F	0.1 at 158°F
Fouling factor, h·ft²·°F/Btu	0.002	0.002
Material of construction	Carbon steel	Carbon steel

Calculate LMTD.

$$\text{LMTD} = \frac{100 - 30}{\ln \frac{100}{30}} = 58.14°F$$

$$\xrightarrow{85100}$$

$$\xleftarrow{115200}$$

For the cooling of organics, U varies between 50 and 150 Btu/h·ft²·°F. Assume for the first trial, $U = 75$ Btu/h·ft²·°F. The estimated area is

$$\frac{2{,}975{,}000}{75(58.14)} = 682 \text{ ft}^2$$

Assume $\frac{3}{4}$-in.-OD 14 BWG tubes and a tube-side velocity of 5 ft/s. From Table 5.12 (p. 270), the flow cross section is

$$0.2679 \text{ in.}^2 = 0.00186 \text{ ft}^2/\text{tube}$$

The number of tubes N_t is

$$N_t = \frac{W}{Au\rho} = \frac{70{,}000}{0.00186(5)(3600)(62.4)(1.5)} = 22.34$$

say, 22 tubes per pass. A tube of $\frac{3}{4}$-in. outside diameter has an outside surface of 0.1963 ft²/ft of tube (Table 5.14 on page 272). Assume a tube length of 20 ft. Then N_p, the number of tube passes, is

$$N_p = \frac{682}{22(0.1963)(20)} = 7.9$$

say, 8 passes. Therefore, the actual area in the exchanger = $8(22)(0.1963)(20) = 691$ ft².

Step 2. Calculate the tube-side coefficient. The total tube-side flow area is $22(0.00186) = 0.04092$ ft².

$$G_t = \frac{70{,}000}{0.04092} = 1.71 \times 10^6 \text{ lb/h} \cdot \text{ft}^2$$

The bulk temperature of fluid is

$$\frac{1}{2}(200 + 115) = 158°F$$

Use the Dittus-Boelter, Equation 5.57, for fluid cooling to calculate h_i.

$$\frac{h_i D_i}{k} = 0.023(N_{Re})^{0.8}(N_{Pr})^{0.3}$$

where N_{Pr} is the Prandtl number $c_p \mu / k$.

$$D_i = \frac{0.584}{12} = 0.04867 \text{ ft}$$

$$h_i = 0.023\left(\frac{0.1}{0.04867}\right)\left[\frac{0.04867(1.71 \times 10^6)}{2(2.42)}\right]^{0.8}\left[\frac{0.5(2)(2.42)}{0.1}\right]^{0.3}$$

$$= 301.6 \text{ Btu/h} \cdot \text{ft}^2 \cdot °F$$

Step 3. Calculate shell-side coefficient. Assume $\frac{3}{4}$-in. OD tubes on $\frac{15}{16}$-in. triangular pitch. With a $17\frac{1}{4}$-in. ID shell, an eight-tube pass exchanger can accommodate 178 tubes (Table 5.15 on page 273). Therefore, choose $17\frac{1}{4}$-in. ID shell. The baffle spacing chosen may vary from 17.25/5 = 3.45-in. to 17.25-in. The shell-side flow area is

$$a_s = \frac{D_{si} C' B}{144 P_T} \text{ ft}^2$$

where
$D_{si} = 17.25$ in.

$C' = \frac{15}{16} - \frac{3}{4} = 0.1875$ in

B = baffle spacing (unknown)

$P_T = \frac{15}{16} = 0.9375$ in

$$\text{Mass velocity} = G_s = \frac{W}{a_s} = u\rho(3600)$$

$$a_s = \frac{W}{u\rho(3600)}$$

Therefore,

$$\frac{D_{si} C' B}{144 P_T} = \frac{W}{u\rho(3600)}$$

from which

$$u = \frac{W(144)P_T}{B\rho(3600)(D_{si})(C')} = \frac{198,330(144)(0.9375)}{B(62.4)(3600)(17.25)(0.1875)} = \frac{36.85}{B}$$

If B is chosen as 6 in., $u = 36.85/6 = 6.14$ ft/s; this is reasonable since B lies between 3.45 and 17.25 in. as required. The equivalent diameter is

$$D_e = \frac{0.86 P_T^2 - \frac{1}{4}\pi d_o^2}{3\pi d_o} = \frac{0.86(0.9375)^2 - \pi(0.75)^2/4}{3\pi(0.75)} = 0.044 \text{ ft}$$

$$G_s = \frac{W}{a_s} = u\rho(3600) = 6.14(62.4)(3600) = 1,379,290 \text{ lb/h} \cdot \text{ft}^2$$

$$N_{Re} = \frac{D_e G_s}{\mu} = \frac{0.044(1.379 \times 10^6)}{1(2.42)} = 2.5 \times 10^4$$

$$N_{Pr} = \frac{c_p \mu}{k} = \frac{1(2.42)}{0.36} = 6.72$$

$$\frac{h_o D_e}{k} = 0.36(N_{Re})^{0.55}(N_{Pr})^{1/3}\left(\frac{\mu}{\mu_w}\right)^{0.14}$$

Assuming $(\mu/\mu_w)^{0.14} = 1.0$,

$$h_o = 0.36 \frac{0.36}{0.044}(2.5 \times 10^4)^{0.55}(6.72)^{1/3} = 1458 \text{ Btu/h} \cdot \text{ft}^2 \cdot °F$$

Step 4. Calculate overall heat transfer coefficient U_o.

$$\frac{1}{U_o} = \frac{1}{h_o} + f_{do} + \frac{l_w}{k_w}\left(\frac{D_o}{D_{av}}\right) + f_{di}\frac{D_o}{D_i} + \frac{1}{h_i}\frac{D_o}{D_i}$$

$$= \frac{1}{1458} + 0.002 + \frac{0.083}{12(26)}\frac{0.75}{0.667} + 0.002\frac{0.75}{0.584}$$

$$+ \frac{1}{301.6}\frac{0.75}{0.584} = 0.00982$$

$$U_o = 101.8 \text{ Btu/h}\cdot\text{ft}^2\cdot\text{°F}$$

Step 5. Calculate corrected LMTD and area.

$$R = \frac{T_1 - T_2}{t_2 - t_1} = \frac{85 - 100}{115 - 200} = 0.176$$

$$S \quad \text{or} \quad P = \frac{t_2 - t_1}{T_1 - t_1} = \frac{115 - 200}{85 - 200} = 0.74$$

From F_T chart for the 1–8 exchanger, Figure 5.4, $F_T = 0.93$. The corrected ΔT_{LM} = $F_T(\Delta T_{LM})$ = 0.93(58.14) = 54.07°F

$$A = \frac{Q}{U_o \Delta T} = \frac{2,975,000}{101.8(54.07)} = 540.5 \text{ ft}^2$$

The area available in the exchanger is 691 ft². So the percent excess area is

$$\frac{691 - 540.5}{540.5}(100) = 27.8\%$$

Therefore, the assumed area is too high an estimate. A second trial with a tube length of 18 ft (other dimensions being the same) gives a more practical estimate of the heat-exchanger surface required. The area of an exchanger with a tube length of 18 ft is $\frac{18}{20}(691) = 622$ ft². It gives [(622 − 540.5)/540.5](100) = 15.0 percent excess area.

In actual practice, the allowable pressure drop through the exchanger may dictate the selection of the geometry of the exchanger. In the preceding example, it is assumed that the allowable pressure drops are not exceeded by the selected geometry of the exchanger.

Example 5.15

A heat-exchanger specification sheet contains the information from Table 5.8.
Check consistency of the data and calculate the percent extra surface included in the specified area.

Solution

$$Q = U_o A_o (\Delta T)_{\text{LMTD}}(F_T)$$
$$= 21.2(91.1)(21.9) = 42,296 \text{ Btu/h}$$

This checks with the specified heat load of 42,300 Btu/h. The calculated dirty transfer rate may be obtained by

Table 5.8

Tubes	OD = $\frac{3}{4}$-in. ID = 0.62-in.
Heat-transfer surface	91.1 ft²
Heat exchanged	42,300 Btu/h
Corrected MTD	21.9°F
Transfer rates:	
Service	21.2 Btu/h · ft² · °F
Clean	25.3 Btu/h · ft² · °F
Fouling resistances:	
Shell side	0.002 h · ft² · °F/Btu
Tube side	0.002 h · ft² · °F/Btu

$$\frac{1}{U_D} = \frac{1}{U_C} + f_{do} + f_{di}\frac{D_o}{D_i}$$

$$\frac{1}{U_D} = \frac{1}{25.3} + 0.002 + 0.002\frac{0.75}{0.62} = 0.043945$$

$$U_D = 22.76 \text{ Btu/h} \cdot \text{ft}^2 \cdot °F$$

$$\text{Calculated required area} = \frac{Q}{U_D \Delta T_{\text{MTD}}}$$

$$= \frac{42,300}{22.76(21.9)}$$

$$= 84.9 \text{ ft}^2$$

$$\text{Actual area} = 91.1 \text{ ft}^2$$

$$\text{Percent excess area} = \frac{91.1 - 84.9}{84.9}(100) = 7.34\%$$

Example 5.16

Benzene is to be condensed in a 1-2 vertical condenser at a rate of 6000 kg/h at essentially atmospheric pressure. Benzene condenses at 353 K on the shell side while the water flows through the tubes. A unit having the following data is available.

Number of tubes	118
Tube OD	25 mm, ID = 22 mm
Length of tube	3 m
Thermal conductivity of tube material	0.045 kW/m · K

Cooling water is available at 303 K. The scale factors may be taken as

Shell side	0.176 m² · K/kW
Water side	0.260 m² · K/kW

Check whether this condenser is adequate for the required duty.

Solution

Latent heat of vaporization of benzene = 394 kJ/kg at 353 K and 1 atm. The density of water is 1000 kg/m³. The condenser heat load is

$$(6000 \text{ kg/h})\left(\frac{1}{3600} \text{ h/s}\right)(394 \text{ kJ/kg}) = 656.7 \text{ kJ/s} = 656.7 \text{ kW}$$

Assume a rise in the temperature of water of 5 K. By heat balance,

$$Mc_p(t_2 - t_1) = 656.7$$

Therefore, mass flow rate of water is

$$M = \frac{656.7}{5(4.1868)}$$

$$= 31.37 \text{ kg/s}$$

The water flow through tubes is

$$\frac{31.37}{1000} = 0.03137 \text{ m}^3/\text{s}$$

and the number of tubes per pass is

$$\frac{118}{2} = 59$$

The area of cross section for flow is

$$59(0.785)(0.022)^2 = 0.02243 \text{ m}^2$$

Water velocity on the tube side is

$$\frac{0.03137}{0.02243} = 1.3986 \text{ m/s}$$

This is a reasonable velocity consistent with the fouling factor suggested, and, therefore, a 5-K rise in the water temperature is adequate for the first trial. (Alternatively, a tube-side velocity of 1.5 m/s consistent with the scale factor could be assumed as a first trial and ΔT computed.)

Log Mean Temperature Difference

$$\Delta T_1 = 353 - 303 = 50 \text{ K}$$
$$\Delta T_2 = 353 - 308 = 45 \text{ K}$$
$$\Delta T_{LM} = \frac{50 - 45}{\ln \frac{50}{45}} = 47.46 \text{ K}$$

Water Film Coefficient
Use the Dittus-Boelter Equation 5.56, since water is heated.

$$h_i = 0.023 \frac{k}{D_i}\left(\frac{D_i G}{\mu}\right)^{0.8}\left(\frac{c_p \mu}{k}\right)^{0.4}_b$$

Bulk temperature $\quad t_b = \dfrac{1}{2}(303 + 308) = 305.5$ K

Properties of water at this temperature are

$\mu = 0.8$ cP $= 0.0008$ kg/m·s
$c_p = 4186.8$ J/kg K $= 4.1868$ kJ/kg K
$k = 0.623$ W/m·K

Mass velocity $G = \dfrac{31.37}{0.02243} = 1398.6$ kg/m^2·s

$$h_i = 0.023 \dfrac{0.623}{0.022}\left[\dfrac{0.022(1398.6)}{0.0008}\right]^{0.8}\left[\dfrac{4.1868(0.8)}{0.623}\right]^{0.4}$$

$= 5943$ W/m^2·K $= 5.943$ kW/m^2·K

Benzene Condensation Coefficient
From Equation 5.97,

$$h_o = 1.47\left(\dfrac{4G'}{\mu_f}\right)^{-0.33}\left(\dfrac{k^3 \rho_f^2 g}{\mu_f^2}\right)^{0.33}$$

The approximate wall temperature is

$$t_w = \dfrac{353 + 305.5}{2} = 329 \text{ K}$$

and the film temperature from Equation 5.104 is

$$t_f = \dfrac{t_{sv} + t_w}{2} = \dfrac{353 + 329}{2} = 341 \text{ K}$$

Properties of Benzene at the Film Temperature

$\mu_f = 0.35$ cP $= 0.00035$ N·s/m^2
$k_f = 0.151$ W/m·K
$\rho_f = 880$ kg/m^3

Benzene condensed $= \dfrac{6000}{3600} = 1.67$ kg/s

Benzene condensed per tube $= \dfrac{1.67}{118} = 0.01415$ kg/s

$$G' = \dfrac{0.01415}{\pi(0.025)} = 0.1802 \text{ kg/m·s}$$

$$h_o = 1.47\left[\dfrac{4(0.1802)}{0.00035}\right]^{-0.33}\left[0.151^3(880)^2 \dfrac{9.81}{0.00035^2}\right]^{0.33}$$

$= 853.04$ W/m^2·K $= 0.853$ kW/m^2·K

The overall coefficient based on outside area from Equation 5.42 is

$$\frac{1}{U_o} = \frac{1}{5.943}\left(\frac{25}{22}\right) + 0.26\left(\frac{25}{22}\right) + \frac{0.0015}{0.045}\left(\frac{25}{23.5}\right) + 0.176 + \frac{1}{0.853}$$

$$= 1.8705$$

Therefore, $\qquad U_o = 0.5346 \text{ kW/m}^2 \cdot \text{K}$

Check on Film Temperature

$$t_w = \frac{5.943(305.5) + (0.853)(353)}{5.943 + 0.853} = 312 \text{ K}$$

$$t_f = 0.5(353 + 312) = 333 \text{ K (assumed 341 K)}$$

There will not be much change in the properties of benzene; hence a recalculation for benzene condensation is not necessary. The calculated value of U_o is therefore adequate. The area of heat transfer required is

$$\frac{656.7}{(0.5346)(47.46)} = 25.9 \text{ m}^2$$

The heat-transfer surface per tube is

$$\pi D_o L = \pi(0.025)(3) = 0.2356 \text{ m}^2$$

The total heat-transfer surface is

$$0.2356(118) = 27.8 \text{ m}^2$$

The percent excess area available is

$$\frac{(27.8 - 25.9)(100)}{25.9} = 7.3\%$$

Therefore, the available condenser is adequate for the specified duty.

THE EFFECTIVENESS-NTU METHOD*

In shell-and-tube exchangers with baffles and multi-pass design, and cross-flow exchangers, where flow patterns are neither truly cocurrent nor truly countercurrent, a consideration of multi-pass correction factor is required. This has been indicated in the text. However, it involves trial and error in applications where performance of a given exchanger is to be evaluated.

Another elegant solution method is the effectiveness-NTU method (Incropera and Dewitt).[8a] This method avoids trial and error solution in evaluating the performance of a given exchanger, which may be cocurrent, countercurrent, cross flow, or shell-and-tube type. An outline of the method involves the utilization of the following equations.

$$\text{Hot fluid heat capacity rate: } C_h = WC \qquad \textbf{(5.128a)}$$

$$\text{Cold fluid heat capacity rate: } C_c = wc \qquad \textbf{(5.128b)}$$

$$\text{Maximum possible heat transfer rate: } Q_{\max} = C_{\min}(T_1 - t_1) \qquad \textbf{(5.129)}$$

where $C_{\min}$ is C_h or C_c, whichever is smaller.

---

*Subscript 1 = inlet, subscript 2 = outlet

The Effectiveness-NTU Method

Actual heat transfer rate: $\quad q = \varepsilon(Q_{max}) = \varepsilon C_{min}(T_1 - t_1)$ (5.130)

The number of transfer units: $NTU = (UA)/C_{min}$ (5.131)

$$C_r = C_{min}/C_{max} \quad (5.132)$$

where, C_{max} is C_h or C_c, whichever is *larger*, and the thermal effectiveness, ε, may be determined from one of the following equations depending on what information is available:

$$\varepsilon = \frac{C_h(T_1 - T_2)}{C_{min}(T_1 - t_1)} \quad (5.133)$$

$$\varepsilon = \frac{C_c(t_2 - t_1)}{C_{min}(T_1 - t_1)} \quad (5.134)$$

$$\varepsilon = \frac{1 - e^{-NTU[1 - C_r]}}{1 - C_r e^{-NTU[1 - C_r]}} \quad \text{for counter flow exchanger} \quad (5.135)$$

$$\varepsilon = \frac{1 - e^{-NTU[1 + C_r]}}{1 + C_r} \quad \text{for parallel flow (concurrent) exchanger} \quad (5.136)$$

$$\varepsilon_1 = 2\left[1 + C_r + \left(1 + C_r^2\right)^{0.5} \left(\frac{1 + e^{-NTU\left(1 + C_r^2\right)^{0.5}}}{1 - e^{-NTU\left(1 + C_r^2\right)^{0.5}}}\right)\right]^{-1} \quad (5.137)$$

for a shell-and-tube exchanger with one shell pass and any multiple of two tube passes

$$\varepsilon_2 = \left[\left(\frac{1 - \varepsilon_1 C_r}{1 - \varepsilon_1}\right)^2 - 1\right]\left[\left(\frac{1 - \varepsilon_1 C_r}{1 - \varepsilon_1}\right)^2 - C_r\right]^{-1} \quad (5.138)$$

for 2-shell passes & 4 or multiple of 4 tube passes in shell- and tube-exchanger

$$\varepsilon = 1 - e^{\left[\left(\frac{1}{C_r}\right)(NTU)^{0.22}\left\{e^{\left[-C_r(NTU)^{0.78}\right]} - 1\right\}\right]} \quad (5.139)$$

for single pass cross flow exchangers with both fluids unmixed

Table 5.9 Suggested maximum values of thermal effectiveness[12]

Exchanger Type	Maximum M
Shell-and-tube	0.9
Plate-and-frame	0.95
Plate-fin	0.98
Printed circuit	0.98

$$\varepsilon = \left(\frac{1}{C_r}\right)\left(1 - e^{\left\{-C_r\left[1-e^{(-\text{NTU})}\right]\right\}}\right) \quad (5.140)$$

for cross flow exchanger with one fluid mixed and the other unmixed, and if C_{max} and C_{min} are associated with mixed and unmixed fluids respectively

$$\varepsilon = 1 - e^{\left(-\frac{1}{C_r}\left\{1-e^{-C_r(\text{NTU})}\right\}\right)} \quad (5.141)$$

for cross flow heat exchanger with one fluid mixed and the other unmixed, and C_{max} and C_{min} are associated with unmixed and mixed fluids respectively.

REBOILERS

Reboilers are shell-and-tube exchangers used as a source of heat/vapor in a distillation column. Usually a portion of bottom stream of a distillation column is vaporized with a small temperature rise. The following are major kinds of reboilers:

- Natural circulation reboilers. Here the bottom stream from a distillation column recirculates through the reboiler. The circulation is caused by thermosyphon, which is created due to the difference in static head between the liquid level in the distillation column and the point of heating (lower tube sheet of a vertical thermosyphon reboiler), and due to the density difference between the fluid entering the reboiler and the heated fluid in the reboiler.

- Forced circulation reboilers. Here a pump moves the fluid through the reboiler, used generally for viscous service.

- Once-through reboilers. Here part of the fluid is vaporized with vapor returned to the column, while the unvaporized part leaves the distillation system. The fluid may or may not be moved by a pump. Once-through reboilers may be falling-film type when material is heat sensitive, and lower residence time is required.

Design Guidelines for Reboilers

- Maintain a temperature difference between the heating medium temperature and process boiling temperature <90°F.

- For a thermosyphon reboiler, cross section of outlet piping should equal to the cross section of all tubes. Cross section of the inlet piping should be half the cross section of the tubes. Minimize the length of inlet piping to reduce heat loss. Set the upper tube sheet even with the liquid level in the distillation column sump for atmospheric or higher pressure operation; for vacuum service, the liquid level in the column sump should be ~50 percent of tube height.

Tube diameter: 1 inch minimum, consider $1/2$ inch to 2 inch.

Maximum velocity at the exit of a horizontal shell side thermosyphon reboiler to avoid instability:

$$V_{\text{exit,max}} = \frac{77.15}{\rho_{\text{two-phase,exit}}}, \text{ m/s} \quad (5.142)$$

where

$\rho_{\text{two-phase,exit}}$ = homogeneous two-phase density, kg/m³, at reboiler exit.

- Install a valve at the inlet line to throttle the flow to the thermosyphon reboiler to reduce the area needed for sensible heat transfer zone.

- Maintain the heat flux and heat transfer coefficient for thermosyphon reboilers according to the following limit:

$$\text{Maximum Heat flux} = \begin{cases} \text{Organics} \begin{cases} \text{Forced circulation: } 20{,}000 - 30{,}000 \text{ Btu/h} \cdot \text{ft}^2 \\ \text{Natural circulation: } 15{,}000 - 20{,}000 \text{ Btu/h} \cdot \text{ft}^2 \end{cases} \\ \text{Water/aq. solution: } 30{,}000 - 40{,}000 \text{ Btu/h} \cdot \text{ft}^2 \end{cases}$$

$$\text{Maximum Heat transfer coefficient} = \begin{cases} \text{Organics: } 500 \text{ Btu/h} \cdot \text{ft}^2 \cdot °F \text{ (typical: } 100\text{–}220) \\ \text{Water/aq. solution: } 2000 \text{ Btu/h} \cdot \text{ft}^2 \text{ (typical : } 220\text{–}350) \end{cases}$$

- Below 4 psia or highly fouling service, consider forced circulation reboiler.

SUBLIMATION

Sublimation is the process in which a solid changes phase directly into vapor without going through the liquid phase. Its commercial application involves recovery of valuable materials at a temperature that will not degrade the materials. By preserving cell structures, for example, in food materials that attribute to the characteristic taste, sublimation is used in food technology to preserve natural taste.

To understand sublimation, one has to understand *triple point*. The triple point temperature is the temperature at which the solid phase, liquid phase, and vapor phase co-exist at a particular pressure, called the triple point pressure. Thus, the triple point temperature of water is 0.0075°C at 1 atmosphere, and the triple point temperature of iodine is 113.5°C at a pressure of 90.5 mm Hg absolute. To make a substance sublime, the operating pressure must be at or below the triple point pressure. For most materials, with the exception of carbon dioxide, the triple point pressure is below atmospheric pressure, and therefore, sublimation for these materials requires vacuum operation. Sometimes an entrainer gas is used, as in the sublimation of salicylic acid with air as an entrainer at 150°C. The effect of entrainer gas on the explosivity of the substance, which is considered safe at its normal states, must be considered.

The sublimation unit operation is analogous to drying, and therefore, the equipment for sublimation includes the same kind of equipment as tray driers, rotary driers, drum driers, etc. The design principles involve the same as used in drying equipment. The condensers are large air-cooled chambers with cleaning mechanisms to keep the wall clean.

CRYSTALLIZATION

Crystallization involves precipitation of a solid from a solution. This happens when the concentration of the dissolved solid exceeds the solubility at a particular temperature and pressure. From the heat transfer point of view, the most common method of precipitation involves evaporation of solvent or change of temperature to a temperature where the solubility is lower. In general, solubility increases with temperature. Knowledge of eutectic temperature and eutectic composition (the composition at which the solution freezes unchanged in composition at a specific temperature) are helpful for separation through crystallization. Some seeding and mild agitation are also necessary.

EXTENDED SURFACE HEAT EXCHANGER

An extended surface heat exchanger is an exchanger whose prime or base surface is extended by appendages intimately attached to it to promote additional heat transfer by conduction, convection, and radiation. The appendages may be in the form of a fin, a cylindrical rod, a conical spine, or a parabolic spine. The fin may be transverse (radial or perpendicular to the length of the tube) or longitudinal (parallel to the length of the tube).

The application of an extended surface heat exchanger includes air coolers or air condensers, refrigeration and cryogenic processes, and in-process heat dissipation. As a rule of thumb, when the outside heat transfer coefficient in a shell-and-bare-tube exchanger is one-fifth or less than the inside heat transfer coefficient, a fin tube should be considered.

Based on bare heat transfer area, for example, the air-cooled exchangers cost 2 to 3 times more. The higher initial cost is offset by the lower operating cost when the coolant water is replaced by natural air. Typically the hot fluid can be cooled to an outlet process temperature approach of 40 – 50°F from the summertime air temperature. When lower temperature is required, a combination of air cooler and water cooler may be used.

Some design guidelines for air-cooled exchanger include

1. The fin surface area to bare tube area is usually 20:1.

2. The air flow pressure drop across the tube bundle usually ranges between 0.4 and 0.8 inches water column. The maximum fan speed is limited to 10,000 ft/min, dictated by noise level.

3. Sound pressure level, SPL, at 3 ft below the cooler bundle is measured by

$$SPL = \begin{cases} 63 + 30\log_{10}(TS) + 10\log_{10}(HP) + 20\log D & \text{for induced draft fan} \\ 66 + 30\log_{10}(TS) + 10\log_{10}(HP) + 20\log D & \text{for forced draft fan} \end{cases} \quad (5.143)$$

where
   TS = fan tip speed in thousand ft/min: divide tip speed in ft/min by 1000
   HP = fan horsepower
   D = fan diameter, ft

where SPL is in decibel with reference level = 0.0002 microbar.
   TS = tip speed of fan, m/s
   HP = fan horsepower
   D = fan diameter, m

4. Common bare tube outside diameters are 5/8-in. to 1-in. (common: 1-in.), maximum: 2-in.

5. Common fin heights are $1/2$-in. to 5/8-in.

6. Number of fins per inch of tube vary from 7 to 11.

7. Fin thickness varies from 0.014-in. to 0.016-in.

8. Fin material of construction is typically aluminum (most common), copper, steel (used when operating temperature exceeds 750°F), or duplex (a combination of steel inside, and copper or aluminum outside).

9. Tube lengths vary from 5 ft to 40 ft, common: 14 ft-24 ft.

10. Use summer dry bulb temperature (not to exceed 2% of the time) as air inlet temperature.

11. The air-side fouling factor may be ignored (or a very low value be used), in comparison with tube side fouling factor.

12. For pressure below 1000 psig, tubes may be expanded to tube sheets with double grooves, for higher pressure tubes should be strength-welded per TEMA R.

13. When tube side heat transfer coefficient based on bare outside diameter is less than 200 Btu/(h·ft²·°F), the total surface of the air-cooled exchanger compares favorably with the water–cooled exchanger.

14. Consult American Petroleum Institute Standard API 661 before you design a finned tube exchanger.

15. When using exposed area for heat load calculation due to fire to estimate the relief device capacity, exclude the fin area, since the fins burn away fast. This applicable if the exchanger is located within estimated flame height. If it is above flame height, assume the condenser load equals the relief load because the condensation will stop due to hot gases. For relief load due to fan failure, a credit due to natural convection may be taken with demonstrated calculation.

Preliminary Estimation Steps of Air-cooled Extended Surface Exchangers

1. For a given service, choose an overall U based on bare surface from the given table.

2. Choose local summer dry bulb temperature, t_1, (not to exceed 2% of the time) as the air inlet temperature.

3. Calculate trial air temperature rise $(t_2 - t_1)$ as follows:

$$(t_2 - t_1)_{trial} = 0.005U\left[\left(\frac{T_1 + T_2}{2}\right) - t_1\right] \qquad (5.144)$$

where
t_1 = air inlet temperature, °F
t_2 = air outlet temperature, °F
T_1 = Hot fluid inlet temperature, °F
T_2 = Hot fluid outlet temperature, °F

4. Correct the air temperature rise obtained from step 3 as follows:

Table 5.10 Types of fins and their application

Type of Fin	Description	Application
1. L-footed fin	Circular tension-wrapped around tubes. They tend to become loose over time thereby creating an air gap resistance.	Marine atmosphere. Upper limit of process temperature <250°F
2. Double L-footed fin	Same as above, but better coverage of the base. More expensive.	Corrosive atmosphere. Upper limit of process temperature <340°F
3. Grooved (embedded) or G-type fin	Fins are embedded in the base metal.	Most commonly used in process industries. Requires heavier base tube wall thickness. May be used to handle process temperature to 750°F
4. Bimetallic fins	G-type fins are embedded in an outer tube, which then slides over the inner base tube.	Corrosive application.
5. Extruded fins	Similar to item 4, but fins are extruded from the outer tube.	Similar to item 4.

Corrected air temperature rise = $F(t_2 - t_1)_{\text{trial}}$, where F is computed from

$$F = \begin{cases} 0.00144(t_2 - t_1)_{\text{trial}} + 0.8832, & \text{when} \quad (t_2 - t_1)_{\text{trial}} > 5°F \\ 0.018(t_2 - t_1)_{\text{trial}} + 0.8, & \text{when} \quad (t_2 - t_1)_{\text{trial}} \leq 5°F \end{cases} \quad (5.145)$$

$$(t_2 - t_1)_{\text{corrected}} = F(t_2 - t_1)_{\text{trial}}, °F$$

Calculate outlet temperature of air.

$$t_2 = (t_2 - t_1)_{\text{corrected}} + t_1 \quad (5.146)$$

5. Calculate logarithm mean temperature difference for true countercurrent flow and determine LMTD correction factor, and then corrected LMTD per procedure suggested before.

6. Calculate exchanger area (preliminary).

$$A = \frac{Q}{U(\text{LMTD})_{\text{corrected}}} \quad (5.147)$$

where
 A = bare surface, ft^2
 Q = heat load, Btu/h
 U = Overall heat transfer coefficient, Btu/h·ft^2·°F
 LMTD$_{\text{corrected}}$ = Corrected logarithm mean temperature difference, °F

Table 5.11 Typical heat transfer coefficient for air cooled exchanger based on outside bare tube surface. (Basis: 1-in. outside diameter tube (0.262 ft²/ft length). Fins are 5/8 inch high, spaced 8 fins per inch, with a fin surface ratio of 16.9 ft²/ft² of bare surface. Divide the overall U by 17.9 to get the overall U based on total surface area of the finned tube exchanger.)

Application	Physical State of Fluid	Condition	Overall U Based on Bare Surface Btu/(h·ft²·°F)
Amine reactivator	Vapor	Condensing	90–100
Ammonia	Vapor	Condensing	100–120
Freon-12	Vapor	Condensing	60–80
Heavy naphtha	Vapor	Condensing	60–70
Light gasoline	Vapor	Condensing	80
Light hydrocarbons	Vapor	Condensing	80–95
Light naphtha	Vapor	Condensing	70–80
Reactor effluent-platformer, rexformers, hydroformers	Vapor	Condensing	60–80
Steam	0-20 psig, vapor	Condensing	130–140
Still overhead-light naphtha, steam & non-condensables	Vapor	Condensing	60–70
Air or flue gas, pressure drop = 1 psi	50 psig, gas	Cooling	10
Air or flue gas, pressure drop = 2 psi	100 psig, gas	Cooling	20
Air or flue gas, pressure drop = 5 psi	100 psig, gas	Cooling	30
Ammonia reactor stream	gas	Cooling	80–90
Hydrocarbon	15-50 psig, gas	Cooling	30–40
Hydrocarbon	50-250 psig, gas	Cooling	50–60
Hydrocarbon	250-1500 psig, gas	Cooling	70–90
Engine jacket water	Liquid	Cooling	120–130
Fuel oil	Liquid	Cooling	20–30
Hydroformer and platformer liquid	Liquid	Cooling	70
Light gas oil	Liquid	Cooling	60–70
Light hydrocarbons	Liquid	Cooling	75–95
Light naphtha	Liquid	Cooling	70
Process water	Liquid	Cooling	105–120
Residuum	Liquid	Cooling	10–20
Tar	Liquid	Cooling	5–10
Kerosene	Liquid	Cooling	55–60
Alcohol and most organic solvents	Liquid	Cooling	70–75
Ethylene glycol/water (50/50 wt)	Liquid	Cooling	100–120
Brine, 25% CaCl2 in 75% water	Liquid	Cooling	90–110

7. Total area of the exchanger may be computed by multiplying the area from step 7 with 17.9, since this is the basis of the fin-tubes in table from which U is chosen.

8. Detail design may be pursued as follows:

   (a) Select a geometry, number of fans, and layout of the exchanger from table II of given reference.[13]

   (b) From the established temperature rise of air, calculate air flow rate from heat balance, and verify the assumed U by cited procedure[7] from known physical properties of the process stream.

HEATING AND COOLING OF LIQUID BATCHES

A detailed treatment of the various cases of batch heating and cooling is available elsewhere.[5h] Example 5.17 illustrates the method of solution.

Example 5.17

A batch of 20,000 lb of a dilute water solution of a salt is to be concentrated from 5 percent to 25 percent solids by evaporation in a batch heater consisting of a coil in a tank that is equipped with an agitator. The heater coil remains submerged in the solution until the end of the evaporation process. The heating medium is a 40,000 lb/h stream of hot oil entering at 300°F. The specific heat of the oil is 0.5 btu/lb·°F and can be assumed constant. U_o is 150 Btu/h·ft^2·°F; assume U_o also is constant. The coil outside heat-transfer surface is 225 ft^2. The initial temperature of the solution is 60°F, and its boiling point is 212°F. The boiling-point elevation caused by the dissolved salts is negligible.

Solution

Since the solution is initially below its boiling point, it must be heated first to its boiling point (212°F). During this period both the batch liquid and the heating medium are nonisothermal.

Initial time θ_1 for heating the solution comes under the following category: coil in a tank heater, nonisothermal heating medium. Therefore, Equation 18.9 of Reference 5i applies.

Now

$$M = 20{,}000 \text{ lb}$$
$$C_c = 1 \quad C_H = 0.5$$
$$T_1 = 300°F \quad t_1 = 60°F \quad t_2 = 212°F$$

$$K_1 = \exp\left(\frac{UA}{WC_H}\right) = \exp\frac{150(225)}{40{,}000(0.5)} = 5.406$$

Then

$$\theta_1 = \frac{20{,}000(1)}{40{,}000(0.5)}\left[\frac{5.406}{5.406-1}\ln\left(\frac{300-60}{300-212}\right)\right] = 1.23 \text{ h}$$

During the evaporation period, the temperature of the solution is constant at 212°F. The heat-transfer rate is constant because U and A are constant. Therefore, the evaporation is a steady-state operation. First calculate water to be evaporated and the heat of evaporation.

The solids in solution are

$$0.05(20{,}000) = 1000 \text{ lb}$$

The water in solution after evaporation is

$$1000\frac{0.75}{0.25} = 3000 \text{ lb}$$

Therefore, water evaporation is

$$19{,}000 - 3000 = 16{,}000 \text{ lb}$$
$$Q_T = \text{heat of evaporation} = 16{,}000(970.3) = 15.525 \times 10^6 \text{ Btu}$$

Now
$$Q = UA\,\Delta T_{LM} = WC_H(T_1 - T_2)$$

$$UA \frac{(T_1 - t) - (T_2 - t)}{\ln[(T_1 - t)/(T_2 - t)]} = WC_H(T_1 - T_2)$$

from which
$$\frac{UA}{WC_H} = \ln \frac{T_1 - t}{T_2 - t}$$

$$\frac{T_1 - t}{T_2 - t} = \exp\left(\frac{UA}{WC_H}\right) = \exp\left[\frac{150(225)}{40{,}000(0.5)}\right] = 5.406$$

$$\frac{300 - 212}{T_2 - 212} = 5.406 \quad \text{and} \quad T_2 = 228.3°F$$

The rate of heat transfer, Q, is

$$40{,}000(0.5)(300 - 228.3) = 1.434 \times 10^6 \text{ Btu/h}$$

Evaporation time is

$$\frac{Q_T}{Q} = \frac{15.525 \times 10^6}{1.434 \times 10^6} = 10.83 \text{ h}$$

Total time for heating and evaporation is $1.23 + 10.83 = 12.06$ h.

NONMETALLIC HEAT EXCHANGERS

The nonmetallic heat exchangers include materials of construction such as Teflon™, graphite, glass, and silicon carbide.

Teflon

Teflon exchangers are available in the form of a tube bundle that can be immersed in the medium. They are also used in the shell-and-tube exchanger, where the shell is generally made of a metal. Because of superior corrosion resistance, Teflon exchangers find application in the chemical industry, and compete with such materials as Carpenter 20, Zirconium, Inconel 625, Hastelloy, Titanium, and Tantalum. Their application in the flue gas heat recovery is an example. Generally, tubes are (1) 0.1-in. OD and 0.08-in. ID, and (2) 0.25-in. OD and 0.2-in. ID.

The following table shows the comparative thermal conductivities of Teflon, glass, graphite, and metals at ambient temperature 32°F-212°F.

There are temperature and pressure limitations. External pressure on tubing varies from 20 psig (300°F) to about 90 psig (60°F). Internal pressure varies from 45 psig (300°F) to 150 psig (75°F). Typical overall heat transfer coefficient Btu/(h· ft². °F), in liquid/liquid non-viscous service, is 50 to 60 in reactor cooling and mild heating in storage tank, and 25 to 32 in condensing service. Liquid velocity is limited to 3 ft/s to avoid failure of Teflon coils. If the corrosive service has small molecules, the effect of permeability of the molecules in Teflon and its impact on the service side should be considered. Same comment applies to Teflon-lined carbon steel tubes.

Table 5.12 Thermal conductivities of materials

Material	Thermal conductivity, Btu.inch/(h.ft^2.°F)
Teflon	1.32
Stainless steel (304–316)	105
Monel	180
Carbon steel	360
Cupro-nickel(70–30)	216
Nickel	420
Admiralty	760
Copper	2680
Glass	(1) whole glass pipe: 8(2) glass in glassed steel: 6.9, steel in glass steel: 360. Note (1)
Impervious graphite	975

Note (1): Glass steel pipe size ranges from $1^1/_2$ inch to 10 inch in 10 ft length. The glass lining is 0.05-in. thick for piping and 0.06-in. for vessels, and the steel wall is 0.154-in. thick.

Graphite

Heat exchangers made of impervious graphite (see Chapter 17) are used in corrosive service. Graphite exchangers can be metallic shell and graphite tube, or all-graphite block and all-graphite plate exchangers. Graphite block exchangers have the advantages of providing unequal surface area to the heat-exchanging streams to match the unequal heat transfer coefficients so that $1/(h_i A_i)$ is close to $1/(h_o A_o)$. Maximum operating pressure for tubes is 50 psig, and maximum operating temperature is 340°F. Typical graphite tube is $1^1/_4$ in. OD and 7/8-in. ID at 1.547-in. triangular pitch. Typical lengths are 6, 9, 12, 14, and 16 ft. Design method for shell and tube is the same as any other shell and tube with proper consideration of tube wall resistance.

Table 5.13 Effect of thermal conductivity on overall heat transfer coefficient

Overall Design Coefficient, U, without Wall Resistance	Type of Operation	Actual Overall U with Wall Resistance			
		Case 1	Case 2	Case 3	Case 4
		Stainless Steel	Graphite	Glass	Glass-lined Steel
300	Heater (Water-steam)	254	284	90	91
200	Cooler (Water-water)	178	193	78	79
150	Condenser (organic vapor-water)	137	146	69	70
100	As above	94	98	56	57
50	High vacuum condenser or oil steam heater	48.5	49.5	36	36
30	Condensate cooler or oil-steam heater	29.5	30	24	24
20	Cooler (gas-water)	19.8	20	17	17
10	As above	10	10	9.3	9.3
5	Cooler (gas-gas)	5	5	4.8	4.8

Case 1: Shell-and-tube exchanger with 16 BWG 304 stainless tube.
Case 2: Carbon steel shell and impervious graphite-tube exchanger with 0.1875-in. wall thickness of tubes.
Case 3: Double pipe glass heat exchanger with 1/16-in. wall thickness of glass tube.
Case 4: 2-in. inner pipe, double pipe glass-lined exchanger 0.05-in. glass thickness, and 0.154-in. steel thickness.

Glass and Silicon Carbide Exchangers

The application of glass is in corrosive environment. The upper limit of pressure and temperature are 1 bar (gage), and 350° F respectively. It may shatter when hot and sprayed with water, and, therefore, not recommended for use in areas handling flammable materials and equipped with a water sprinkler. It may be used in areas handling flammable materials if fire-induced isolation valves are used at all inlets and outlets handling hazardous processes.

Silicon carbide is an excellent nonmetallic material construction for tubes, but is expensive.

Table 5.13 shows a comparative study[17] of the effect of thermal conductivity on overall heat transfer coefficient, U, Btu/(h· ft²· °F).

HEAT PUMP

A heat pump transfers heat from a source at a lower temperature to a sink at a higher temperature. In a broad sense, therefore, all pieces of refrigeration equipment are heat pumps. In industry, the term heat pump, however, is used to indicate machines that can heat and cool, whichever function is needed. A heat pump cycle is identical with the refrigeration cycle in principle, but differs from the latter in purpose: the primary purpose of a heat pump is to heat a space. The most common use of a heat pump is domestic heating and cooling. The heat pump cycle can be reversed through four-way valve switches.

$$Q_H = Q_L + W \tag{5.148}$$

where
Q_H = Heat rejected to high temperature environment.
Q_L = Heat extracted from low temperature environment.
W = input work: energy required to transfer heat from low to high temperature.

Note: All the terms must be in consistent units.

The coefficient of performance (COP) is defined as follows:

$$\text{COP} = \frac{Q_H}{W} = \frac{Q_L + W}{W} = 1 + \frac{Q_L}{W} \tag{5.149}$$

The COP of a heat pump is always greater than 1. Since the heat pump is a reverse heat engine, its COP is limited by the Carnot cycle COP:

$$\text{COP}_{\text{Carnot}} = \frac{T_H}{T_H - T_L} = \frac{1}{1 - \frac{T_L}{T_H}} \tag{5.150}$$

where
T_H = High temperature in cycle, absolute scale
T_L = Low temperature in cycle, absolute scale

For minimum work the heat pump COP may be equated with Carnot COP. Thus,

$$W_{\min} = \frac{Q_H}{1 + \frac{T_L}{T_H}} \tag{5.151}$$

Table 5.14 Characteristics of tubing

Tube O.D. Inches	B.W.G. Gage	Thickness Inches	Internal Area Sq. Inch	Sq. Ft. External Surface Per Foot Length	Sq. Ft. Internal Surface Per Foot Length	Weight Per Ft. Length Steel Lbs.*	Tube I.D. Inches	Moment of Inertia Inches^4	Section Modulus Inches^3	Radius of Gyration Inches	Constant C**	O.D./I.D.	Transverse Metal Area Sq. Inch
1/4	22	0.028	0.0296	0.0654	0.0508	0.066	0.194	0.00012	0.00098	0.0791	46	1.289	0.0195
	24	0.022	0.0333	0.0654	0.0539	0.054	0.206	0.00010	0.00083	0.0810	52	1.214	0.0158
	26	0.018	0.0360	0.0654	0.0560	0.045	0.214	0.00009	0.00071	0.0823	56	1.168	0.0131
	27	0.016	0.0373	0.0654	0.0571	0.040	0.218	0.00008	0.00065	0.0829	58	1.147	0.0118
3/8	18	0.049	0.0603	0.0982	0.0725	0.171	0.277	0.00068	0.0036	0.1166	94	1.354	0.0502
	20	0.035	0.0731	0.0982	0.0798	0.127	0.305	0.00055	0.0029	0.1208	114	1.230	0.0374
	22	0.028	0.0799	0.0982	0.0835	0.104	0.319	0.00046	0.0025	0.1231	125	1.176	0.0305
	24	0.022	0.0860	0.0982	0.0867	0.083	0.331	0.00038	0.0020	0.1250	134	1.133	0.0244
1/2	16	0.065	0.1075	0.1309	0.0969	0.302	0.370	0.0021	0.0086	0.1555	168	1.351	0.0888
	18	0.049	0.1269	0.1309	0.1052	0.236	0.402	0.0018	0.0071	0.1604	198	1.244	0.0694
	20	0.035	0.1452	0.1309	0.1126	0.174	0.430	0.0014	0.0056	0.1649	227	1.163	0.0511
	22	0.028	0.1548	0.1309	0.1162	0.141	0.444	0.0012	0.0046	0.1672	241	1.126	0.0415
5/8	12	0.109	0.1301	0.1636	0.1066	0.601	0.407	0.0061	0.0197	0.1865	203	1.536	0.177
	13	0.095	0.1486	0.1636	0.1139	0.538	0.435	0.0057	0.0183	0.1904	232	1.437	0.158
	14	0.083	0.1655	0.1636	0.1202	0.481	0.459	0.0053	0.0170	0.1939	258	1.362	0.141
	15	0.072	0.1817	0.1636	0.1259	0.426	0.481	0.0049	0.0156	0.1972	283	1.299	0.125
	16	0.065	0.1924	0.1636	0.1296	0.389	0.495	0.0045	0.0145	0.1993	300	1.263	0.114
	17	0.058	0.2035	0.1636	0.1333	0.352	0.509	0.0042	0.0134	0.2015	317	1.228	0.103
	18	0.049	0.2181	0.1636	0.1380	0.302	0.527	0.0037	0.0119	0.2044	340	1.186	0.089
	19	0.042	0.2299	0.1636	0.1416	0.262	0.541	0.0033	0.0105	0.2067	359	1.155	0.077
	20	0.035	0.2419	0.1636	0.1453	0.221	0.555	0.0028	0.0091	0.2090	377	1.126	0.065
3/4	10	0.134	0.1825	0.1963	0.1262	0.833	0.482	0.0129	0.0344	0.2229	285	1.556	0.259
	11	0.120	0.2043	0.1963	0.1335	0.808	0.510	0.0122	0.0326	0.2267	319	1.471	0.238
	12	0.109	0.2223	0.1963	0.1393	0.747	0.532	0.0116	0.0309	0.2299	347	1.410	0.219
	13	0.095	0.2463	0.1963	0.1466	0.665	0.560	0.0107	0.0285	0.2340	384	1.339	0.195
	14	0.083	0.2679	0.1963	0.1529	0.592	0.584	0.0098	0.0262	0.2376	418	1.284	0.174
	15	0.072	0.2884	0.1963	0.1587	0.522	0.606	0.0089	0.0238	0.2411	450	1.239	0.153
	16	0.065	0.3019	0.1963	0.1623	0.476	0.620	0.0083	0.0221	0.2433	471	1.210	0.140
	17	0.058	0.3157	0.1963	0.1660	0.429	0.634	0.0076	0.0203	0.2455	492	1.183	0.126
	18	0.049	0.3339	0.1963	0.1707	0.367	0.652	0.0067	0.0178	0.2484	521	1.150	0.108
	20	0.035	0.3632	0.1963	0.1780	0.268	0.680	0.0050	0.0134	0.2531	567	1.103	0.079
7/8	10	0.134	0.2894	0.2291	0.1589	1.062	0.607	0.0221	0.0505	0.2662	451	1.442	0.312
	11	0.120	0.3167	0.2291	0.1662	0.969	0.635	0.0208	0.0475	0.2703	494	1.378	0.285
	12	0.109	0.3390	0.2291	0.1720	0.893	0.657	0.0196	0.0449	0.2736	529	1.332	0.262
	13	0.095	0.3685	0.2291	0.1793	0.792	0.685	0.0180	0.0411	0.2778	575	1.277	0.233
	14	0.083	0.3948	0.2291	0.1856	0.703	0.709	0.0164	0.0374	0.2815	616	1.234	0.207
	15	0.072	0.4197	0.2291	0.1914	0.618	0.731	0.0148	0.0337	0.2850	655	1.197	0.182
	16	0.065	0.4359	0.2291	0.1950	0.563	0.745	0.0137	0.0312	0.2873	680	1.174	0.165
	17	0.058	0.4525	0.2291	0.1987	0.507	0.759	0.0125	0.0285	0.2896	706	1.153	0.149
	18	0.049	0.4742	0.2291	0.2034	0.433	0.777	0.0109	0.0249	0.2925	740	1.126	0.127
	20	0.035	0.5090	0.2291	0.2107	0.314	0.805	0.0082	0.0187	0.2972	794	1.087	0.092
1	8	0.165	0.3526	0.2618	0.1754	1.473	0.670	0.0392	0.0784	0.3009	550	1.493	0.433
	10	0.134	0.4208	0.2618	0.1916	1.241	0.732	0.0350	0.0700	0.3098	656	1.366	0.365
	11	0.120	0.4536	0.2618	0.1990	1.129	0.760	0.0327	0.0654	0.3140	708	1.316	0.332
	12	0.109	0.4803	0.2618	0.2047	1.038	0.782	0.0307	0.0615	0.3174	749	1.279	0.305
	13	0.095	0.5153	0.2618	0.2121	0.919	0.810	0.0280	0.0559	0.3217	804	1.235	0.270
	14	0.083	0.5463	0.2618	0.2183	0.814	0.834	0.0253	0.0507	0.3255	852	1.199	0.239
	15	0.072	0.5755	0.2618	0.2241	0.714	0.856	0.0227	0.0455	0.3291	898	1.168	0.210
	16	0.065	0.5945	0.2618	0.2278	0.650	0.870	0.0210	0.0419	0.3314	927	1.149	0.191
	18	0.049	0.6390	0.2618	0.2361	0.498	0.902	0.0166	0.0332	0.3367	997	1.109	0.146
	20	0.035	0.6793	0.2618	0.2435	0.361	0.930	0.0124	0.0247	0.3414	1060	1.075	0.106
1-1/4	7	0.180	0.6221	0.3272	0.2330	2.059	0.890	0.0890	0.1425	0.3836	970	1.404	0.605
	8	0.165	0.6648	0.3272	0.2409	1.914	0.920	0.0847	0.1355	0.3880	1037	1.359	0.562
	10	0.134	0.7574	0.3272	0.2571	1.599	0.982	0.0742	0.1187	0.3974	1182	1.273	0.470
	11	0.120	0.8012	0.3272	0.2644	1.450	1.010	0.0688	0.1100	0.4018	1250	1.238	0.426
	12	0.109	0.8365	0.3272	0.2702	1.330	1.032	0.0642	0.1027	0.4052	1305	1.211	0.391
	13	0.095	0.8825	0.3272	0.2775	1.173	1.060	0.0579	0.0926	0.4097	1377	1.179	0.345
	14	0.083	0.9229	0.3272	0.2838	1.036	1.084	0.0521	0.0833	0.4136	1440	1.153	0.304
	16	0.065	0.9852	0.3272	0.2932	0.824	1.120	0.0426	0.0682	0.4196	1537	1.116	0.242
	18	0.049	1.0423	0.3272	0.3016	0.629	1.152	0.0334	0.0534	0.4250	1626	1.085	0.185
	20	0.035	1.0936	0.3272	0.3089	0.455	1.180	0.0247	0.0395	0.4297	1706	1.059	0.134
1-1/2	10	0.134	1.1921	0.3927	0.3225	1.957	1.232	0.1354	0.1806	0.4853	1860	1.218	0.575
	12	0.109	1.2908	0.3927	0.3356	1.621	1.282	0.1159	0.1545	0.4933	2014	1.170	0.476
	14	0.083	1.3977	0.3927	0.3492	1.257	1.334	0.0931	0.1241	0.5018	2180	1.124	0.369
	16	0.065	1.4741	0.3927	0.3587	0.997	1.370	0.0756	0.1008	0.5079	2300	1.095	0.293
2	11	0.120	2.4328	0.5236	0.4608	2.412	1.760	0.3144	0.3144	0.6660	3795	1.136	0.709
	12	0.109	2.4941	0.5236	0.4665	2.204	1.782	0.2904	0.2904	0.6697	3891	1.122	0.648
	13	0.095	2.5730	0.5236	0.4739	1.935	1.810	0.2586	0.2586	0.6744	4014	1.105	0.569
	14	0.083	2.6417	0.5236	0.4801	1.701	1.834	0.2300	0.2300	0.6784	4121	1.091	0.500

* Weights are based on low carbon steel with a density of 0.2836 lbs./cu. in. For other metals multiply by the following factors:

Aluminum 0.35	Aluminum Bronze 1.04	Nickel 1.13
Titanium 0.58	Aluminum Brass 1.06	Nickel-Copper 1.12
A.I.S.I. 400 Series S/Steels 0.99	Nickel-Chrome-Iron 1.07	Copper and Cupro-Nickels 1.14
A.I.S.I. 300 Series S/Steels 1.02	Admiralty 1.09	

** Liquid Velocity = $\dfrac{\text{lbs. Per Tube Hour}}{C \times \text{Sp. Gr. of Liquid}}$ in feet per sec. (Sp. Gr. of Water at 60°F. = 1.0)

Table 5.15 Shell inside diameter vs tube count & tube-passes—triangular pitch

¾ in. OD tubes on ¹⁵⁄₁₆-in. triangular pitch						¾ in. OD tubes on 1-in. triangular pitch					
Shell ID, in.	1-P	2-P	4-P	6-P	8-P	Shell ID, in.	1-P	2-P	4-P	6-P	8-P
8	36	32	26	24	18	8	37	30	24	24	
10	62	56	47	42	36	10	61	52	40	36	
12	109	98	86	82	78	12	92	82	76	74	70
13¼	127	114	96	90	86	13¼	109	106	86	82	74
15¼	170	160	140	136	128	15¼	151	138	122	118	110
17¼	239	224	194	188	178	17¼	203	196	178	172	166
19¼	301	282	252	244	234	19¼	262	250	226	216	210
21¼	361	342	314	306	290	21¼	316	302	278	272	260
23¼	442	420	386	378	364	23¼	384	376	352	342	328
25	532	506	468	446	434	25	470	452	422	394	382
27	637	602	550	536	524	27	559	534	488	474	464
29	721	692	640	620	594	29	630	604	556	538	508
31	847	822	766	722	720	31	745	728	678	666	640
33	974	938	878	852	826	33	856	830	774	760	732
35	1102	1068	1004	988	958	35	970	938	882	864	848
37	1240	1200	1144	1104	1072	37	1074	1044	1012	986	870
39	1377	1330	1258	1248	1212	39	1206	1176	1128	1100	1078

1 in. OD tubes on 1¼-in. triangular pitch						1¼ in. OD tubes on 1⁹⁄₁₆-in. triangular pitch					
8	21	16	16	14							
10	32	32	26	24		10	20	18	14		
12	55	52	48	46	44	12	32	30	26	22	20
13¼	68	66	58	54	50	13¼	38	36	32	28	26
15¼	91	86	80	74	72	15¼	54	51	45	42	38
17¼	131	118	106	104	94	17¼	69	66	62	58	54
19¼	163	152	140	136	128	19¼	95	91	86	78	69
21¼	199	188	170	164	160	21¼	117	112	105	101	95
23¼	241	232	212	212	202	23¼	140	136	130	123	117
25	294	282	256	252	242	25	170	164	155	150	140
27	349	334	302	296	286	27	202	196	185	179	170
29	397	376	338	334	316	29	235	228	217	212	202
31	472	454	430	424	400	31	275	270	255	245	235
33	538	522	486	470	454	33	315	305	297	288	275
35	608	592	562	546	532	35	357	348	335	327	315
37	674	664	632	614	598	37	407	390	380	374	357
39	766	736	700	688	672	39	449	436	425	419	407

1½ in. OD tubes on 1⅞-in. triangular pitch					
12	18	14	14	12	12
13¼	27	22	18	16	14
15¼	36	34	32	30	27
17¼	48	44	42	38	36
19¼	61	58	55	51	48
21¼	76	72	70	66	61
23¼	95	91	86	80	76
25	115	110	105	98	95
27	136	131	125	118	115
29	160	154	147	141	136
31	184	177	172	165	160
33	215	206	200	190	184
35	246	238	230	220	215
37	275	268	260	252	246
39	307	299	290	284	275

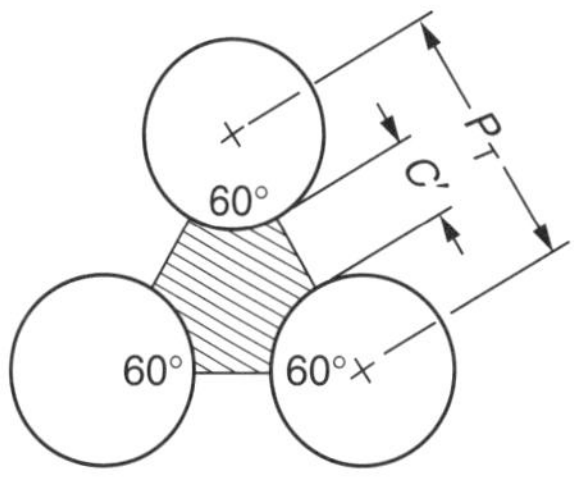

Table 5.16 Shell inside diameter vs tube count & tube-passes—square pitch

¾ in. OD tubes on 1-in. square pitch						1 in. OD tubes on 1¼-in. square pitch					
Shell ID, in.	1-P	2-P	4-P	6-P	8-P	Shell ID, in.	1-P	2-P	4-P	6-P	8-P
8	32	26	20	20		8	21	16	14		
10	52	52	40	36		10	32	32	26	24	
12	81	76	68	68	60	12	48	45	40	38	36
13¼	97	90	82	76	70	13¼	61	56	52	48	44
15¼	137	124	116	108	108	15¼	81	76	68	68	64
17¼	177	166	158	150	142	17¼	112	112	96	90	82
19¼	224	220	204	192	188	19¼	138	132	128	122	116
21¼	277	270	246	240	234	21¼	177	166	158	152	148
23¼	341	324	308	302	292	23¼	213	208	192	184	184
25	413	394	370	356	346	25	260	252	238	226	222
27	481	460	432	420	408	27	300	288	278	268	260
29	553	526	480	468	456	29	341	326	300	294	286
31	657	640	600	580	560	31	406	398	380	368	358
33	749	718	688	676	648	33	465	460	432	420	414
35	845	824	780	766	748	35	522	518	488	484	472
37	934	914	886	866	838	37	596	574	562	544	532
39	1049	1024	982	968	948	39	665	644	624	612	600

1¼ in. OD tubes on 1 9⁄16-in. square pitch						1½ in. OD tubes on 1⅞-in. square pitch					
10	16	12	10								
12	30	24	22	16	16	12	16	16	12	12	
13¼	32	30	30	22	22	13¼	22	22	16	16	
15¼	44	40	37	35	31	15¼	29	29	25	24	22
17¼	56	53	51	48	44	17¼	39	39	34	32	29
19¼	78	73	71	64	56	19¼	50	48	45	43	39
21¼	96	90	86	82	78	21¼	62	60	57	54	50
23¼	127	112	106	102	96	23¼	78	74	70	66	62
25	140	135	127	123	115	25	94	90	86	84	78
27	166	160	151	146	140	27	112	108	102	98	94
29	193	188	178	174	166	29	131	127	120	116	112
31	226	220	209	202	193	31	151	146	141	138	131
33	258	252	244	238	226	33	176	170	164	160	151
35	293	287	275	268	258	35	202	196	188	182	176
37	334	322	311	304	293	37	224	220	217	210	202
39	370	362	348	342	336	39	252	246	237	230	224

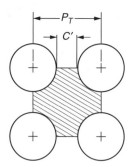

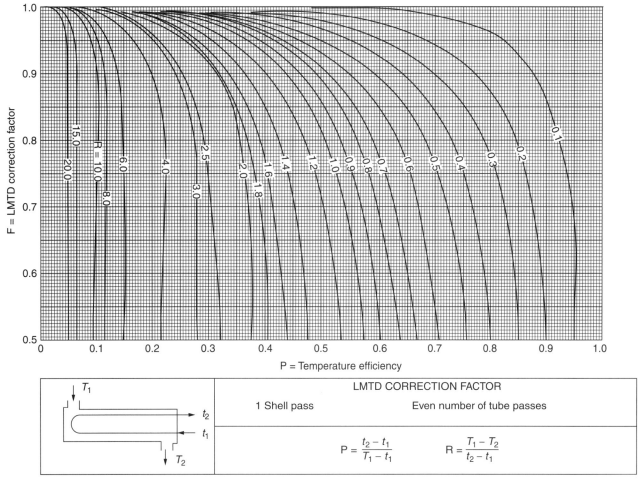

Figure 5.4 LMTD Correction Factor—1 shell pass and even number of tube passes (Source: Tubular Exchanger Manufacturers Association, Inc.)

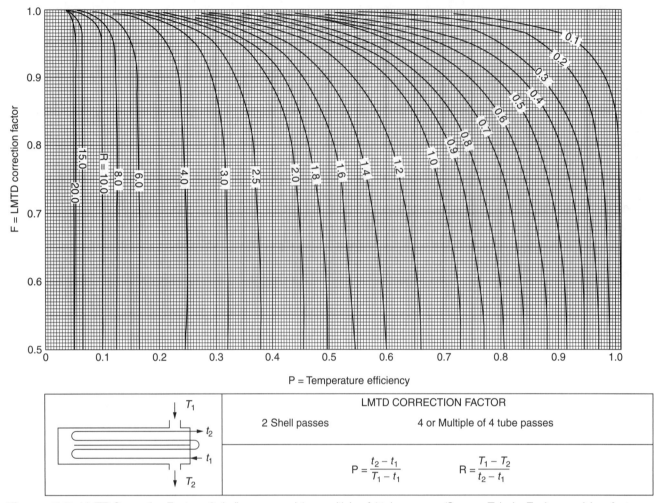

Figure 5.5 LMTD Correction Factor—2 shell passes and 4 or multiple of 4 tube passes (Source: Tubular Exchanger Manufacturers Association, Inc.)

RECOMMENDED REFERENCES

1. Sieder and Tate, *Ind. Eng. Chem.* vol. 28, 1936, pp. 1429–1436.
2. Dittus and Boelter, *Univ. Calif. Pubs. Eng.* vol. 2, 1930, p. 443.
3. Colburn, *Trans. Am. Inst. Chem. Engrs.*, vol. 29, 1933, pp. 174–220.
4. McAdams, *Heat Transmission,* 3rd ed., McGraw-Hill, New York, 1954: (a) pp. 237–246; (b) pp. 335–336; (c) p. 330; (d) p. 345; (e) p. 274.
5. Kern, *Process Heat Transfer,* McGraw-Hill, New York, 1950: (a) p. 217; (b) p. 260; (c) p. 261; (d) p. 265; (e) p. 270; (f) p. 260; (g) pp. 136–139; (h) pp. 624–633; (i) p. 627; (j) p. 82; (k) pp. 841–842; (l) pp. 828–833. (m) Fig-29
6. Perry (ed.), *Chemical Engineers' Handbook,* 5th ed., McGraw-Hill, New York, 1973: pp. 10–24 and 10–25.
7. Kern and Kraus, *Extended Surface Heat Transfer,* McGraw-Hill, New York, 1972: pp. 557–576.
8. Incropera and Dewitt, *Fundamentals of Heat Transfer,* John Wiley & Sons, NY, 1981: pp. 521–536.
9. Ludwig, *Applied Process Design for Chemical and Petrochemical Plants,* vol. 3, Gulf Publishing Company, Houston, 1984.
10. Mukherjee, "Effectively Design Air-cooled Exchanger," *Chemical Engineering Progress,* February, 1987.

11. Tubular Exchangers Manufacturer Association (TEMA) standard, 8th ed.: (*a*) p 285.
12. Kong, "Heat Exchanger Duty: Going for Gold," *Chemical Engineering,* December 2004.
13. Brown, "Design of air-cooled exchangers: A procedure for preliminary estimate," *Chemical Engineering,* March 27, 1978.
14. Ganapathy, "Design of air-cooled exchangers: Process-design criteria," *Chemical Engineering,* March 27, 1978.
15. Yilmaz, "Horizontal Shell Side Thermosyphon Reboilers," *Chemical Engineering Progress,* November 1987.
16. Kern, "Heat Exchanger Design for Fouling Service," *Chemical Engineering Progress,* July 1966.
17. Pfaudler Bulletin 1056.
18. Rohsenow and Harnett, *Handbook of Heat Transfer,* McGraw-Hill Book Company, NY, 1998.
19. Levenspiel, *Engineering Flow and Heat Exchange,* Plenum Press, NY, 1986.
20. Hughes and Deumaga, "Insulation Saves Energy," *Chemical Engineering,* May 1974, pp. 95–100.
21. Chopey, *Handbook of Chemical Engineering Calculations,* McGraw-Hill Book Company, 1994.

CHAPTER 6

Evaporation

OUTLINE

EVAPORATION 279
Types of Evaporators ■ Methods of Operation ■ Evaporator Capacity ■ Heat-Transfer Coefficients ■ Steam Economy in Evaporation

RECOMMENDED REFERENCES 290

EVAPORATION

Evaporation is the removal of a solvent, usually water, by its vaporization from a relatively nonvolatile solute, usually a solid, by application of heat. It is a process of physical separation and involves no phase equilibrium. The evaporators may be single effect or multi effect, the latter for steam economy. The multi-effect evaporators may be classified as forward feed, backward feed, parallel feed, and mixed feed. A further classification of the evaporators is natural circulation and forced circulation.

Typical applications of evaporation include concentration of aqueous solutions of sugar, glycerin, orange juice, tea and coffee extracts, sodium chloride, and hydroxide. In most evaporation applications, concentrated solution is the desired product.

Application of the various types of the evaporators is dependent on the solution characteristics and the capacity required. The pressure and temperature also will be determined by the nature of the product. The factors to consider are:

Liquid Concentration: Initially the solutions are very dilute and less viscous so the heat transfer rates are relatively high. However, as the evaporation proceeds, the solutions become more concentrated and viscous. To obtain higher heat transfer coefficient, agitation or forced circulation may be required.

Solubility: There could be a limit to the degree of concentration that can be achieved by evaporation due to the fact that if the solubility limit is exceeded crystallization may begin.

Temperature-Sensitive Materials: Temperature-sensitive materials need to be evaporated at low temperatures. This may require evaporation under vacuum.

Foam Formation or Frothing: Because of foam formation entrainment losses will occur. This will primarily affect the vapor space to be provided in the evaporator.

Pressure and Temperature and Boiling Point Elevation: The boiling point of a solution depends on both the pressure and the concentration. At higher pressure, the boiling point of solution is higher. Also, as the concentration increases there is a rise in the boiling point. This rise in boiling point is called boiling-point rise or elevation, which reduces the available temperature difference for heat transfer. To maintain a reasonable temperature difference, it may be necessary to operate under vacuum.

The boiling-point elevation is small for dilute solutions but may be very high for the solutions of inorganic salts. Duhring's rule states that the boiling-point elevation of a given solution is a linear function of the boiling point of water. The other factor affecting the boiling point of a solution in an evaporator is the liquid head or height of the liquid level in the tube (hydrostatic head).

Scale Formation and Corrosion: Some solutions may deposit solid materials on the heating surface. The scale affects heat transfer rates and the evaporator needs to be cleaned eventually. Materials of construction to be used for the evaporator must also be selected in such way that preferably there is no corrosion at all or, if it cannot be avoided, it should be at an acceptable low rate.

Types of Evaporators

Many types of evaporator equipment are used for evaporation. These include

- Open kettle or pan
- Vertical natural circulation evaporator
- Horizontal natural circulation evaporator
- Vertical-type long tube evaporator
- Falling film evaporator
- Forced-circulation evaporator
- Agitated film evaporator

Methods of Operation

Single-Effect Evaporators

In a single-effect evaporator, only one evaporator is used. If the capacity is small and/or the cost of steam is relatively low, then a single evaporator system is used. In a single-effect evaporator, 1 kg of steam on condensing will generate approximately 1 kg of water vapor, provided the feed enters the evaporator at its boiling point. The latent heat of steam is all used in vaporization. Maximum steam economy (defined as mass of water vapor generated/unit of steam used) in a single evaporator can be 1 kg/kg (1 lb/lb of steam).

Multiple-Effect Evaporation System

For better steam economy and higher capacities a number (>1) of evaporators are used. In the operation of multiple-effect evaporator system, live steam is fed only to the first evaporator. The vapor from the first evaporator is used as the heating medium in the second evaporator, and so on. This way, live steam usage is reduced. The multiple-effect evaporation system can be operated in various ways.

1. Forward Feed—Fresh feed is added to the first effect (evaporator) and is moved forward to the next effect in the same direction as the steam/vapor flow. This system is used when the final concentrated product is likely to be damaged due to higher temperature. The boiling point decreases from effect to effect since the first effect is operated at 1 atm and the last operated under vacuum.

2. Backward Feed—Backward feed is the reverse of forward feed. Fresh feed enters the coldest effect and the concentrated solution leaves from the first effect. This method is advantageous when the fresh feed is below its boiling point at the pressure in the last effect, as less sensible heat is required to raise feed temperature to the boiling point of solution in the last effect. It is also used when the concentrated solution is highly viscous. Higher temperatures in the first few effects keep the viscosity down whereby reasonable heat transfer coefficients are obtained in these effects.

3. Parallel-Feed Multiple-Effect Evaporators—In this method, fresh feed is added to each effect and concentrated solution withdrawn from the same effect. Vapor from each effect is still used to heat the next effect. The parallel-feed method is generally used when the feed is mostly saturated and solid crystals are the product.

In general, the evaporation problems involve the calculation of the material and heat balances, steam economy, and required heat-transfer surfaces.

Evaporator Capacity

The heat-transfer capacity of an evaporator is given by

$$q = UA \, \Delta T \tag{6.1}$$

where
q = total heat transferred, Btu/h [W]
U = overall heat-transfer coefficient, Btu/h $\cdot$ ft^2 $\cdot$ °F [W/m^2 $\cdot$ K]
ΔT = overall temperature difference, °F[K]
A = heat transfer surface, ft^2[m^2]

For a given heat input, the evaporation capacity is reduced if sensible heat is to be provided to bring the feed to its boiling point. The temperature difference that is available for heat transfer depends on (1) the solution to be evaporated, (2) the liquid head in the tubes, and (3) the difference between pressure in the steam chest and that over the vapor space above the boiling liquid.

Heat-Transfer Coefficients

Heat-transfer coefficients are generally high for evaporators. Steam-side coefficients can be calculated using equations for condensing vapors or special equation for condensing steam. Steam-side condensing coefficient has a high value of about 5700 W/m^2 $\cdot$ K (1000 Btu/h $\cdot$ ft^2).

For tube side also many correlations are available which can be used to calculate the tube-side heat-transfer coefficients provided there is no boiling in the tubes (refer to Chapter 5 for details). The Dittus-Boelter equation with a modified constant is used for forced-circulation evaporators. The equation is

$$\frac{hD}{k} = 0.0278 \left(\frac{Du\rho}{\mu}\right)^{0.8} \left(\frac{C_p \mu}{k}\right)^{0.4} \tag{6.2}$$

where the symbols have their usual meaning as in Chapter 5, Heat Transfer. For long tube vertical natural convection evaporators, it is rather difficult to predict the heat-transfer coefficients because there is a non-boiling zone in the lower portion of the tube and a boiling zone in the upper part of the tube. For the non-boiling zone, Equation 6.2 can be used. For the boiling zone, a number of equations are available (Perry and Green).

Steam Economy in Evaporation

One advantage of the multi-effect evaporation is the steam economy (lb[kg] of water evaporated per lb [kg] of steam) that can be obtained. This is because the live steam is used only in one unit, whereas the vapor from one effect is used in the chest of the next effect. This requires using a vacuum on the system in order to have reasonable temperature differences. An optimum number of the effects for a given application is a function of the economy of steam savings weighted against the investment in the added number of effects.

Calculations

The usual assumptions made in making the calculations of the evaporations or the evaporator heat balances are (1) there is no leakage or entrainment; (2) non-condensable content of the steam or vapor is negligible; and (3) superheating the steam or vapor from each effect, as well as sub-cooling of the condensate, are considered negligible.

Where the enthalpies of the solutions as a function of concentration are available, a heat balance can be made in terms of the enthalpies. When the heat of dilution is negligible, the enthalpies can be calculated with the use of the specific heats of the solution. Usually the temperature of the thick liquor is taken as the reference temperature. The enthalpy of the feed is given by

$$h_F = C_{pg}(t_F - t) \tag{6.3}$$

where
  h_F = enthalpy of feed, Btu/lb[J/kg]
  c_{pf} = specific heat, Btu/lb · °F [J/kg · K]
  t_F, t = temperatures of the feed and thick liquor respectively, °F[K]

Single-Effect Evaporator

Use of the enthalpy composition diagram to solve a problem of a single-effect evaporator is illustrated by the following example.

Example 6.1

(English Units)

A single-effect evaporator is used to concentrate 25,000 lb/h of a NaOH solution from 10 to 50 percent concentration. The evaporator is supplied with 15 psig steam. The feed is at 100°F. It operates under a vacuum of 26 in. Hg referred to a 30-in. barometer. If $U = 450$ Btu/h · ft² · °F, calculate (1) heat-transfer surface, (2) steam consumption, and (3) water requirements of a countercurrent barometric condenser if the water is available at 86°F and if the outlet water temperature is not to exceed 120°F.

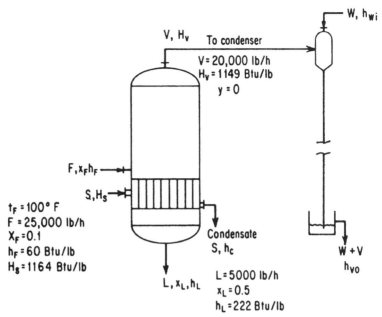

Exhibit 1 Material flows and enthalpies (Example 6.1)

Solution

Refer to Exhibit 1. Given quantities are

$$F = 25,000 \text{ lb/h} \quad x_F = 0.1$$
$$x_L = 0.5 \quad t_F = 100°F$$
$$y = 0$$

where y is the concentration of solute in overhead vapor. From steam tables,

$$\text{Steam pressure} = 14.7 + 15 = 29.7 \text{ psia}$$
$$t_s = 250°F \quad H_s = 1164 \text{ Btu/lb} \quad \lambda_s = 945.5 \text{ Btu/lb}$$

The boiling point of water under a 26-in. Hg vacuum is to be obtained from steam tables.

$$26\text{-in Hg vacuum} = 4\text{-in. Hg pressure} = \frac{4}{30}(14.7) = 1.96 \text{ pisa}$$

From the steam tables, the boiling point of water at 1.96 psia (by interpolation) is

$$\text{bp} = \frac{1.960 - 1.942}{2.222 - 1.942}(130 - 125) + 125 = 125.3°F \not\subset 125°F$$

The enthalpy of saturated steam at 125°F = 1116 Btu/lb (from steam tables). The boiling points of thick liquor (concentrated solution) are to be calculated.

For the enthalpies of NaOH solutions, refer to the enthalpy composition diagram for NaOH. To get the boiling points, refer to the Duhring's lines for NaOH.

$$t_B = 198°F$$
$$h_F = 60 \text{ Btu/lb}$$
$$h_L = 222 \text{ Btu/lb}$$

Material Balance. Amounts of the thick liquor and evaporation are obtained by material balance as

$$Fx_F = Lx_L + Vy$$
$$25{,}000(0.1) = 0.5L + 0$$

Solving,

$$L = 5000 \text{ lb/h}$$
$$V = 20{,}000 \text{ lb/h}$$

Enthalpy Balance. The equation for the enthalpy balance is written as

$$Fh_F + S(H_s - h_c) = VH_v + Lh_L$$

where H_v is the enthalpy of the superheated vapor at 198°F.

$$H_v = h_b + c_p(t_s - t_b)$$
$$= 1116 + 0.46(198 - 125)$$
$$= 1149.0 \text{ Btu/lb}$$

where t_s is the temperature of superheated steam and t_b is the boiling point of the solution, °F.

$$25{,}000(60) + S(1164 - 218.5) = 20{,}000(1149) + 5000(222)$$

from which steam consumption $S = 23{,}892$ lb/h.

Heating Surface

$$Q = UA\,\Delta T \qquad A = \frac{Q}{U\,\Delta T}$$

$$23{,}892(1164 - 218.5) = 450A(250 - 198)$$

$$A = \frac{23{,}892(1164 - 218.5)}{450(250 - 198)} = 965.4 \text{ ft}^2$$

$$\text{Steam economy} = \frac{\text{water evaporated}}{\text{live steam used}} = \frac{20{,}000}{23{,}892}$$
$$= 0.837 \text{ lb/lb of steam}$$

Condenser Water Requirements
An enthalpy balance on the condenser gives (refer to Exhibit 1)

$$VH_v + Wh_{wi} = (W + V)h_{vo}$$
$$20{,}000(1149) + W(54.03) = (W + 20{,}000)(87.97)$$
$$W = 625{,}239 \text{ lb/h} = 1250.5 \text{ gpm}$$

In many cases, the enthalpy composition diagrams are not available and approximate methods are used. The approximations usually made include (1) the latent heat of vaporization of water at the boiling point of the solution rather than the equilibrium temperature is used; (2) heats of dilution are neglected; and (3) the specific heats of the solution, if known, can be used to calculate the enthalpies of the feed and thick liquor. An example is given next.

Example 6.2

(SI Units)

A solution is to be concentrated from 10 to 65 percent solids in a vertical long-tube evaporator. The solution has a negligible elevation of boiling point and its specific heat can be taken to be the same as that of water. Steam is available at 203.6 kPa, and the condenser operates at 13.33 kPa. The feed enters the evaporator at 295 K. The total evaporation is to be 25,000 kg/h of water. Calculate the heat transfer required and the steam consumption in kilograms per hour if the overall coefficient is 2800 W/m² · K.

Solution

From steam tables, the water-vapor temperature at 13.33 kPa is 325 K and the steam temperature at 203.6 kPa is 394 K. The latent heat of vaporization of water at 325 K is 2375 kJ/kg. So

$$\text{Evaporation} = 25{,}000 \text{ kg/h} = \frac{25{,}000}{3600} = 6.94 \text{ kg/s}$$

The feed and thick liquor are denoted by F and B, respectively. By solids balance,

$$F(0.1) = B(0.65)$$

By water balance,

$$F(0.9) = B(0.35) + 6.944$$

By solving the equations simultaneously,

$$F = 8.21 \text{ kg/s} \quad B = 1.26 \text{ kg/s}$$

Taking boiling point in the evaporator as the datum temperature, the evaporator heat load is

$$Q = 6.944(2375) + 8.21(4.1868)(325 - 295)$$
$$= 17{,}523 \text{ kW or kJ/s}.$$

The temperature difference is

$$\Delta T = 394 - 325 = 69 \text{ K}$$

The heat-transfer surface is

$$\frac{17{,}523}{69(2.8)} = 90.7 \text{ m}^2$$

The latent heat of steam (from steam tables) is 2198 kJ/kg, and the steam consumption is

$$\frac{(17{,}523 \text{ kJ/s})(3600 \text{ s/h})}{2198 \text{ kJ/kg}} = 28{,}700 \text{ kg/h}$$

Multi-Effect Evaporator

As stated before, multi-effect evaporation is used to effect steam economy. In practice, the heating areas of evaporators in a multi-effect evaporation system are kept equal to obtain economy in construction.

The heat-transfer rates in each effect are given by

$$q_1 = U_1 A \Delta T_1$$
$$q_2 = U_2 A \Delta T_2 \qquad (6.4)$$
$$q_3 = U_3 A \Delta T_3$$

These are approximately equal, and hence

$$U_1 \Delta T_1 = U_2 \Delta T_2 = U_3 \Delta T_3 = \frac{q}{A} \qquad (6.5)$$

which enables one to get the ΔT distribution in the effects for the first trial.

The multi-effect evaporator calculations are also done as in the case of a single-effect evaporator by solving a number of simultaneous material- and heat-balance equations. However, this procedure is tedious when three or more effects are involved. Therefore, another variation of the method is used. A temperature distribution is assumed and, instead of solving a set of simultaneous equations, a value of the evaporation from one of the effects is also assumed. Then the material- and heat-balance equations are solved one by one for each effect. If the total evaporation calculated does not equal the required evaporation, the value of the assumed evaporation is readjusted, and the trial is repeated. The accuracy of the assumed temperature differences is ascertained by the calculation of the heat-transfer surfaces, which should be equal for all the effects. Example 6.3 illustrates the method.

Example 6.3

(English Units)

A triple effect forced-circulation evaporator is to concentrate NaOH solution from 10 to 50 percent. Feed is at 100°F and 75,000 lb/h. The evaporator operates at 28-in. Hg vacuum in the last effect. The backward feed arrangement is to be used. The overall heat-transfer coefficients are to be: in I, 600; in II, 600; in III, 1000, all in Btu/h · ft² · °F. What are the steam consumption, economy, and heating surfaces needed if live steam at 15 psig is used? (Refer to Exhibit 2.)

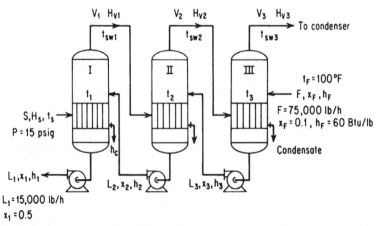

Exhibit 2 Backward-feed triple-effect evaporation system (Example 6.3)

Solution

Given data are

$$U_1 = 600 \quad U_2 = 600 \quad U_3 = 1000$$
$$t_s = 250°F \quad x_F = 0.1 \quad x_1 = 0.5$$
$$t_F = 100°F$$

From the enthalpy composition diagram for NaOH, $h_F = 60$ Btu/lb. From steam tables, $H_s = 1164.1$ Btu/lb and $h_c = 218.9$ Btu/lb. The saturation temperature for water in III is 100°F. Flows of the feed, thick liquor, and evaporation are

$$F = 75{,}000 \text{ lb/h} \quad \text{Solids in feed} = 7500 \text{ lb/h}$$

$$L_1 = \frac{7500}{0.5} = 15{,}000 \text{ lb/h}$$

$$V = 75{,}000 - 15{,}000 = 60{,}000 \text{ lb/h}$$

Total evaporation $V = V_1 + V_2 + V_3$

Trial 1. To know the temperature drop available, the boiling-point elevations are required. To determine these, the concentrations in II and III must be known. For a preliminary estimate, one should divide the total evaporation in three effects based on a judicious guess, or to start with, one can assume equal evaporations in all the effects to calculate the concentrations and boiling-point elevations. In this example, assume equal evaporation in each effect. Therefore,

$$L_3 = 75{,}000 - 20{,}000 = 55{,}000 \quad x_3 = 0.1\left(\frac{75{,}000}{55{,}000}\right) = 0.136$$

$$L_2 = 55{,}000 - 20{,}000 = 35{,}000 \quad x_2 = 0.1\left(\frac{75{,}000}{35{,}000}\right) = 0.214$$

$$L_1 = 35{,}000 - 20{,}000 = 15{,}000 \quad x_1 = 0.1\left(\frac{75{,}000}{15{,}000}\right) = 0.500$$

Note that one can assume unequal evaporations in the effects. The preliminary guess should be based on the considerations of the U values and the probable ΔT that may be available in each effect. To know the boiling points of the solutions and the boiling-point elevations, assume a ΔT of 25°F between the steam and the solution in the first effect and also assume equal ΔP per effect. Pressure in the third effect is known. Therefore, the saturation temperature of water and the boiling point of the solution in each effect can be obtained with the use of the steam tables and Duhring lines for NaOH solutions as given in Table 6.1.

Total boiling-point elevation is 102°F, and therefore the temperature drop available is $(250 - 100) - 102 = 48°F$. Divide this into the ratio of the heat-transfer coefficients using

$\Delta T_1 : \Delta T_2 : \Delta T_3 = 1/U_1 : 1/U_2 : 1/U_3$ as follows:

$$\Delta T_1 = 18 \quad \Delta T_2 = 18 \quad \Delta T_3 = 12$$

288 Chapter 6 Evaporation

Table 6.1 Boiling points of water and NaOH solutions, trial 1, example 6.3

Effect	Conc., wt%	Operating Pressure, psia	bp of Water, °F	bp of Solution, °F	bp Rise, °F
III	13.6	0.98*	100*	110	10
II	21.4	2.35†	131	148	17
I	50.0	3.72‡	150‡	225	75
Total					102°F

*Based on the given 28-inHg vacuum in the third effect and $x_3 = 0.136$.
†Calculated on the assumption of equal ΔP per effect.
‡Based on the assumption of 25°F Δt between the steam and the solution in the first effect and $x_1 = 0.5$.

Using steam tables, now prepare Table 6.2.

Heat Balance Equations
First effect:

$$945.2S + (75{,}000 - V_2 - V_3)90 = 1163.5\ V_1 + 15{,}000(249)$$

Second effect:

$$1003V_1 + (75{,}000 - V_3)65 = 112.2\ V_2 + 90(75{,}000 - V_2 - V_3)$$

Third effect:

$$1024.7\ V_2 + 75{,}000(60) = 1109.2\ V_3 + 65(75{,}000 - V_3)$$

Since the last equation contains only two variables, it is convenient to assume V_3. Let $V_3 = 19{,}000$ lb/h. Then from the third-effect equation

$$V_2 = 19{,}728\ \text{lb/h}$$

Table 6.2 Calculations of enthalpies for first trial

	Evaporator	Temp., °F	H_s or H_v, Btu/lb	h_c, Btu/lb	λ_v, Btu/lb	Flow Rate lb/h
Steam	I	250	1164.1	218.9	945.2	8
ΔT_1	I	18				
bp (t_1)	I	232	1163.5			V_1
bpr*	I	75				
t_{sv1}	to II	157	1129.0	125	1003.0	V_1
ΔT_2	II	18				
bp (t_2)	II	139	1122.2			V_2
bpr	II	17				
t_{sv2}	to III	122	1114.7	92.0	1023.7	V_2
ΔT_3	III	12				
bp (t_3)	III	110	1109.2			V_3
bpr	III	10				
t_{sv3}		100	1105.2	68.0	1037.2	V_3

*Boiling point rise.
†These are obtained by $H_v = H_{tsv} + 0.46$ (bpr) where the subscript sv denotes saturated steam.

From the second-effect equation

$$V_1 = 21{,}677 \text{ lb/h}$$

$$\text{Total } V = 19{,}000 + 19{,}728 + 21{,}677 = 60{,}405 \text{ lb/h} \doteq 60{,}000^*$$

Reassume

$$V_3 = \frac{19{,}000(60{,}000)}{64{,}405} \doteq 18{,}875 \text{ lb/h}$$

Then
$$V_2 = 19{,}600 \quad V_1 \doteq 21{,}550 \text{ lb/h}$$
$$V = V_1 + V_2 + V_3$$
$$= 18{,}875 + 19{,}600 + 21{,}550 = 60{,}025 \text{ lb/h} \doteq 60{,}000 \text{ lb/h}$$

From the first-effect equation,

$$S = 27{,}000 \text{ lb/h}.$$

Values of the enthalpies for the liquid streams are given in Table 6.3. Now check the equality of the heat-transfer surfaces.

$$A_1 = \frac{945.2(27{,}000)}{600(18)} = 2363 \text{ ft}^2$$

$$A_2 = \frac{1004(21{,}550)}{600(18)} = 2003 \text{ ft}^2$$

$$A_3 = \frac{1023.7(19{,}600)}{1000(12)} = 1674 \text{ ft}^2$$

$$\text{Average } A = 2013 \text{ ft}^2$$

Table 6.3 Liquid enthalpies and flow rates, Example 6.3, trial 1

Stream	x_n	Temp., °F	h_t, Btu/lb	Flow Rate, lb/h
F	0.1	100	60	75,000
L_1	0.5	232	249	15,000
L_2	0.214	139	90	$75{,}000 - V_2 - V_3$
L_3	0.136	112	65	$75{,}000 - V_3$

---

*The calculated evaporation in this trial differs only by 0.7 percent from the required evaporation. Therefore, this is a good estimate and no further trial is actually necessary. A second trial, however, is shown here.

The areas are differing too much. Therefore, the assumed ΔT distribution is not correct. Redistribute ΔT's as follows:

$$\Delta T_1 = \frac{2363}{2012.3}(18) \doteq 21$$

$$\Delta T_2 = \frac{2003}{2013.3}(18) \doteq 18$$

$$\Delta T_3 = \frac{1674}{2013.3}(12) \doteq 10$$

Total $\Delta T = \overline{49°F}$

Also readjust concentrations.

$$x_2 = \frac{7500}{75,000 - 18,875 - 19,600} = 0.205$$

$$x_1 = \frac{7500}{75,000 - 18,875} = 0.1336$$

The boiling-point elevation in the second effect will be slightly less than the assumed boiling-point elevation for the first trial.

Trial 2. The calculations completed in a similar manner in trial 1 give $V_1 = 21{,}430$ lb/h, $V_2 = 19{,}650$ lb/h, $V_3 = 18{,}950$ lb/h, and $S = 26{,}840$ lb/h.

Heat-transfer surfaces are now checked again as follows:

	% Deviation from Average A
$A_1 = \dfrac{26,840(945.2)}{600(21)} = 2013.4$ ft^2	+0.26
$A_2 = \dfrac{1005.6(21,430)}{600(18)} = 1995.4$ ft^2	−0.63
$A_3 = \dfrac{1025.8(19,650)}{1000(10)} = 2015.7$ ft^2	+0.38

The average area is 2008 ft^2. The maximum deviation from the average value is 0.63 percent. A more accurate trial will involve fractional ΔT's and this is not necessary since U values are not known more accurately. Hence, the area of each effect is 2008 ft^2, the steam consumption is 26,840 lb/h, and the steam economy is 60,000/26,840 = 2.24 lb/lb.

RECOMMENDED REFERENCES

1. Perry and Green, (Eds.), *Chemical Engineers' Handbook,* platinum ed., McGraw-Hill, 2000.
2. Geankoplis, *Transport Processes and Separation Process Principles,* 4th ed., Prentice-Hall, 2004.
3. Badger and Banchero, *Introduction to Chemical Engineering,* McGraw-Hill, 2004.

CHAPTER 7

Filtration

OUTLINE

FILTRATION EQUIPMENT 291
Types of filters ■ Filter Media and Filter Aids

FILTRATION CALCULATIONS 293
Ruth's Equation ■ Kozeny-Carman Equation

RECOMMENDED REFERENCES 306

Many separation operations involve no mass transfer between two distinct phases and phase equilibrium. The separation is achieved by using mechanical-physical forces acting on the particles, liquids, or their mixtures. This is unlike operations such as distillation, absorption, and adsorption in which diffusion, molecular, or chemical properties, and phase equilibrium play an important role. Examples of operations that involve no phase equilibrium are evaporation, filtration, membrane separations, sedimentation and settling, and centrifugal filtration. In evaporation, heat is used as the medium to physically remove liquid by vaporization. In this chapter, we review the filtration operation by which suspended particles in liquids are physically separated using a *porous medium*. The porous medium retains the particles in the form of a cake and allows the liquid to pass through as a clear filtrate.

FILTRATION EQUIPMENT

Filtration equipment can be classified in many ways. Based on operating cycle, the classification is either batch or continuous. In batch operation, the filter cake is removed after a given amount of slurry is run through the filter. In continuous operation, the cake is removed continuously. Filters can also be classified on the basis of the desired product, namely the filter cake or the clear filtrate.

Types of Filters

Bed Filters
When relatively small amounts of solids are to be removed from large amounts of water, a bed filter is used to clarify the liquid. The bed filter consists of coarse gravel placed on a perforated plate. This forms the bottom layer. On the gravel is

placed a bed of fine sand, which acts as the filter medium. The liquid to be clarified is introduced at the top over a baffle, which acts as a distributor. Filtration is carried out until the bed is clogged. The bed is then washed by water introduced at the bottom, which flows upward and back washes the deposited particles up and out of the filter. Multiple units in parallel can be used to handle large flows.

Plate and Frame Filters

Plate and frame filters are batch-type filters. Plate and frame filters consist of plates and frames assembled alternately. Filter cloth is placed over each side of the plate. The plates have channels through which the filtrate flows. Filtration is stopped when the frames are completely filled with solids. If the cake is to be washed, fresh water is admitted through a separate channel. After washing, the plates and frames are separated and the cake removed manually. The filter is then reassembled and the filtration cycle restarted.

Plate and frame filters are simple to operate and can be designed for high-pressure operation. They are also flexible and versatile in operation; however, they have some disadvantages: their throughput is small and the labor cost is very high.

Leaf Filters

A leaf filter is another type of batch filter. Leaf filters consist of a number of leaves hung together in parallel in a closed tank, and they can handle large volumes of slurry. Each leaf consists of a hollow wire frame covered by filter cloth. Filtrate outlets from the leaves are connected to a common channel. Cake is removed by opening the shell. Sometimes air is used to dislodge the cake. If solids are a waste product, water jets can be used to wash away the cake.

Continuous Rotary Vacuum Filters

A rotary filter consists of a drum mounted on a hollow shaft. The drum is covered with a suitable filtering medium. An automatic valve is used to carry out the filtering, drying, washing, and cake discharge functions in the cycle. A knife scraper is used to remove the cake. Provision can be made to permit the use of compressed air blowback to dislodge the cake. The maximum pressure differential for the vacuum filter can be only 1 atm so this filter is not suitable for viscous fluids or those fluids that need confinement. High pressures could be used if the filter assembly is enclosed in a shell, but the cost of the unit is very high.

Continuous Rotary Disk Filters

A continuous rotary disk filter is a vacuum filter that consists of concentric vertical disks mounted on a rotating shaft. The disks are hollow and covered with filter medium. The disks submerge in the slurry only partly. Washing, drying, and dislodging of the cake is done when the rotating shaft is in the upper half of the disk.

Filter Media and Filter Aids

Filter Media

Filter media must be strong and capable of retaining the solids. Pores should be of such size that the cloth retains the solid particles above its surface and the pores do not get plugged. Several types of cloths can be used, including twill, woven heavy cloth, woolen cloth, and glass cloth.

Filter Aids

Materials such as kieselguhr (diatomaceous earth) are used as filter aids because they are incompressible. These materials are used in two ways: (1) A precoat could be used before the slurry is filtered. The precoat prevents gelatinous-type solids from plugging the filter medium and also give a clear filtrate. (2) The filter aid could be added to the slurry, which gives increased porosity to the cake. Use of this method is limited because it can only be applied when the cake is to be discarded or when the filtered solids can be recovered from the combined cake by chemical leaching.

FILTRATION CALCULATIONS

Given the filtration conditions and the type of filter to be used, common problems to be solved are (1) the amount of the filtrate obtainable in a given time, (2) the amount of the solvent that can be passed through the cake in a given time, and (3) the concentration of the recovered material in the wash solvent.

Filtration can be carried out in two ways: (1) constant-pressure filtration, in which the filtration rate varies, and (2) constant-rate filtration, in which the ΔP varies. Two methods are available to treat filtration problems, Ruth's equation and the Kozeny-Carman relation.

Ruth's Equation

To establish Ruth's equation, the following variables are first defined:

$$V = \text{filtrate, ft}^3 [\text{m}^3]$$
$$\rho = \text{filtrate density, lb/ft}^3 \, [\text{kg/m}^3]$$
$$W = \text{filterable solids, lb [kg]}$$
$$m = \frac{\text{wet cake}}{\text{dry cake}} \quad \text{(washed)}$$
$$x = \text{mass fraction of the solids in a slurry}$$

Using these definitions, the following relations are developed:

$$\text{Liquid in slurry} = \frac{W}{x}(1-x) \text{ lb [kg]} \tag{7.1}$$

$$\text{Liquid retained by cake} = (m-1)W \text{ lb [kg]} \tag{7.1a}$$

from which

$$\text{Filtrate} = V\rho = \frac{W}{x}(1-x) - W(m-1) \text{ lb [kg]} \tag{7.1b}$$

$$W = \frac{V\rho x}{1-mx} \text{ lb [kg]} \tag{7.1c}$$

The Hagen-Poiseuille equation for the laminar flow of fluids in tubes of circular cross section is given by

$$\Delta P = \frac{32Lu\mu}{D^2} \quad \text{(SI)} \qquad \Delta P = \frac{32Lu\mu}{g_c D^2} \quad \text{(English)} \tag{7.2}$$

where
ΔP = pressure drop, lb_f/ft^2 [N/m^2]
u = velocity of fluid flowing through tube, ft/s [m/s] (based on empty cross section)
D = inside diameter of tube, ft [m]
μ = viscosity of fluid, $lb_m/ft \cdot s$ [$Pa \cdot s$ or $kg/m \cdot s$]
g_c = 32.174 $lb_m \cdot ft/lb_f \cdot s^2$ (when English units are used)

Note that in Equation 7.2 SI units, the factor g_c is 1. (Refer to Chapter 1.)

Assuming Equation 7.2 applies to the filtrate flow, it can be shown that the rate of filtration is given by

$$u = \frac{1}{A}\frac{dV}{d\theta} = \frac{\Delta P_C}{32\mu L/D_e^2} \quad \text{or} \quad \frac{dV}{d\theta} = \frac{A}{\mu}\frac{\Delta P_C}{R_C} \quad \text{(SI)} \tag{7.3}$$

$$u = \frac{1}{A}\frac{dV}{d\theta} = \frac{\Delta P_C}{32\mu L/g_c D_e^2} \quad \text{or} \quad \frac{dV}{d\theta} = \frac{A}{\mu}\frac{\Delta P_C}{g_c R_C} \quad \text{(English)}$$

where
A = total filtering area, ft^2 [m^2]
θ = time of filtration, s
m = viscosity of filtrate, $lb/ft \cdot s$ [$Pa \cdot s$]
L = thickness of cake, ft [m]
ΔP_C = pressure drop across cake, lb_f/ft^2 [N/m^2]
R_C = resistance of cake = $32 L/D^2$, 1/ft [1/m]
D_e = equivalent diameter, ft; [m] for the cake of flow area A, ft^2 [m^2]
g_c = Newton's law proportionality factor = 32.174 $ft \cdot lb/lb_f \cdot s^2$.

In actual practice, the ΔP_C across the cake is not easily measurable and, therefore, the resistance of the whole filter (not only of the cake) with a pressure drop ΔP is considered.

Then
$$R_T = R_1 + R_C + R_2 \tag{7.3a}$$

where
R_T = total resistance
R_1 = resistance of the filter medium
R_2 = resistance of the slurry leads and channels

R_2 is generally neglected. This reduces Equation 7.3a to

$$R_T = R_1 + R_c \tag{7.3b}$$

Assume R_1 to be the resistance of a fictitious cake of mass W_1 lb [kg] and to have been deposited by the filtrate of volume V_1 [m³] ft³. Then

$$R_T = R_1 + R_C = \frac{(W + W_1)\alpha}{A} \quad (7.4)$$

also $$R_C = \frac{W\alpha}{A} \quad \text{and} \quad R_1 = \frac{W_1\alpha}{A} \quad (7.4a)$$

where α is the average specific resistance of cake, ft/lb$_m$ [m/kg].

Then $$W + W_1 = \frac{x}{1 - mx}(V + V_1) \quad \text{and} \quad \frac{dV}{d\theta} = \frac{A}{\mu} \frac{\Delta P}{R_T} \quad \text{(SI)} \quad (7.5)$$

$$W + W_1 = \frac{x}{1 - mx}(V + V_1) \quad \text{and} \quad \frac{dV}{d\theta} = \frac{A}{\mu g_c} \frac{\Delta P}{R_T} \quad \text{(English)} \quad (7.5a)$$

After eliminating R_T and $W + W_1$ from Equations 7.4 and 7.5, one obtains

$$(V + V_1)dV = (V + V_1)d(V + V_1) = \frac{A^2 \Delta P(1 - mx)}{\mu \rho x \alpha} d\theta \quad (7.6)$$

Constant-Pressure Filtration

Integration of Equation 7.6 yields (assuming ΔP constant)

$$\int_0^{V+V_1} (V + V_1)d(V + V_1) = \frac{A^2 \Delta P(1 - mx)}{\mu \rho x \alpha} \int_{-\theta_1}^{\theta} d\theta \quad (7.6a)$$

where θ_1 is the imaginary time to lay down the fictitious cake. From Equation 7.6a the completion of integration gives

$$(V + V_1)^2 = \frac{2A^2 \Delta P(1 - mx)}{\mu \rho x \alpha}(\theta + \theta_1) \quad \text{at const. } \Delta P \quad (7.7)$$

Equation 7.7 can be written as

$$(V + V_1)^2 = C(\theta + \theta_1) \quad (7.8)$$

where $$C = \frac{2A^2 \Delta P(1 - mx)}{\mu \rho x \alpha} \text{ m}^6/\text{s (ft}^6/\text{s)} \quad (7.8a)$$

α will be constant for the noncompressible cakes. The constants C, θ_1, and V_1 are obtained by the use of the equation obtained by differentiating Equation 7.8. Thus

$$\frac{d\theta}{dV} = \frac{2V}{C} + \frac{2V_1}{C} \quad (7.8b)$$

A plot of $\Delta\theta/\Delta V$ versus V gives the slope = $2/C$ and the intercept $2V_1/C$, from which the constants V_1 and C can be calculated.

Constant-Rate Filtration

When the filtration is conducted at a constant filtration rate, the ΔP changes during the filtration and is given by

$$\Delta P = \frac{\rho x \mu \alpha}{A^2(1-mx)} \left(\frac{dV}{d\theta}\right)_C^2 (\theta + \theta_1) \quad (7.9)$$

A plot of ΔP versus θ is a straight line, which gives

$$\text{Slope} = \frac{\rho x \mu \alpha}{A^2(1-mx)} \left(\frac{dV}{d\theta}\right)_C^2 \quad (7.10)$$

and

$$\text{Intercept} = \frac{\rho \mu x \alpha \theta_1 \left(\frac{dV}{d\theta}\right)_C^2}{A^2(1-mx)} \quad (7.10a)$$

from which the constants α and θ_1 can be calculated.

Rate of Washing

If displacement washing is assumed, the rate of washing is constant and equal to the rate of filtration at the end of the filtration cycle. Thus, the rate of washing is given by

$$\frac{dV}{d\theta} = \frac{C}{2(V+V_1)} \quad (7.10b)$$

where $dV/d\theta$ is the rate of the filtration at $\theta = \theta_f$ or at the end of the filtration period.

Kozeny-Carman Equation

By mass balance,

$$(1-X)LA\rho_s = V\rho m = \frac{(V+XLA)\rho x}{1-x} \quad (7.11)$$

x = mass fraction of solids in slurry

where
- L = thickness of the cake, [m] ft
- A = area of filter medium, [m^2] ft^2
- X = porosity of the cake = volume of void space/total volume of the cake
- ρ_s = density of solids in the cake, [kg/m^3] lb/ft^3
- ρ = density of the filtrate, [kg/m^3] lb/ft^3
- m = mass ratio of dry cake/filtrate
- V = volume of filtrate, [m^3] ft^3.

The velocity through a bed (assuming laminar flow) is given by

$$u = \frac{1}{A}\frac{dV}{d\theta} = \frac{\overline{K}\, w_f \rho}{L\mu} = \frac{K\Delta P_C}{L\mu} \quad (7.12)$$

where
  w_f = frictional losses/unit mass of fluid, [N · m/kg] ft · lb$_f$/lb
  u = velocity, [m/s] ft/s
  K = permeability of cake, [m^2] lb · ft^3/lb$_f$ · s^2
  ΔP_C = pressure drop through the cake, [N/m^2] lb$_f$/ft^2

Solving Equation 7.11 for L, substituting in Equation 7.12, and defining $\alpha = 1/K\rho_s(1 - X)$,

$$\frac{dV}{d\theta} = \frac{A^2 \Delta P_C}{Vm\alpha\rho\mu} = \frac{\Delta P_C}{V\alpha m\mu\rho/A^2} \quad (7.13)$$

$$= \frac{\Delta P_C A^2}{2VC_V} \quad (7.13a)$$

where
$$C_V = \frac{\alpha m \rho \mu}{2} \quad (7.14)$$

However, in actual practice, the total ΔP needs to be used since ΔP_C is not known. If it is assumed that the fictitious equivalent resistance of an equivalent cake thickness L_e, which forms an equivalent filtrate volume V_e, is R_e, then equivalent resistance is given by

$$R_e = \frac{V_e \alpha m \rho \mu}{A^2} \quad \text{or} \quad \frac{2C_V V_e}{A^2} \quad (7.15)$$

Therefore
$$\frac{dV}{d\theta} = \frac{A^2 \Delta P}{\alpha m \rho \mu (V + V_e)} = \frac{A^2 \Delta P}{2C_V(V + V_e)} \quad (7.16)$$

where ΔP is the total resistance. For a constant-pressure filtration, the integration of Equation 7.16 gives

$$\theta = \frac{C_V(V^2 + 2VV_e)}{A^2 \Delta P} \quad (7.17)$$

Solving Equation 7.11 for V, substituting for V in Equation 7.12, defining $C_L = \mu[s(1-x)(1-X) - xX]/2K\rho x$ and an equivalent cake thickness L_e,

$$\frac{dL}{d\theta} = \frac{\Delta P}{2C_L(L + L_e)} \quad (7.18)$$

At constant porosity,

$$\theta = \frac{C_L(L^2 + 2LL_e)}{\Delta P} \quad (7.19)$$

Constant-Pressure Filtration

If the rate data are at constant ΔP, the plot of $\Delta\theta/\Delta V$ versus V gives

$$\text{Slope} = \frac{2C_V}{A^2 \Delta P} \qquad \text{Intercept} = \frac{2C_V V_e}{A^2 \Delta P} \quad (7.20)$$

from which the constants C_V and V_e are calculated.

298 Chapter 7 Filtration

Constant-Rate Filtration

If the data are taken at constant rate of filtration, V_e is constant, and then

$$\Delta P = \frac{2C_V}{A^2}\left(\frac{dV}{d\theta}\right)_C V + \frac{2C_V}{A^2}\left(\frac{dV}{d\theta}\right)_C V_e \quad (7.21)$$

A plot of ΔP versus V enables one to find the constants in the filtration equation. Thus

$$\text{Slope} = 2\frac{C_V}{A^2}\left(\frac{dV}{d\theta}\right)_C \qquad \text{intercept} = 2\frac{C_V}{A^2}\left(\frac{dV}{d\theta}\right)V_e \quad (7.21a)$$

If the area of the filter and the rate of filtration are known, C_V and V_e can be calculated using Equation 7.21a.

The permeability K of the Kozeny-Carman Equation 7.12 and the specific resistance α of Ruth's Equation 7.4 are related by

$$K = \frac{AL}{\alpha W} \quad (7.22)$$

Equation 7.10 can be written in a simplified form as

$$\frac{d\theta}{dV} = \frac{V+V_1}{C'} \quad \text{where } C' = \frac{C}{2} \quad \text{hence} \quad \frac{dV}{d\theta} = \frac{C'}{V+V_1} \quad (7.23)$$

Example 7.1

Volume of the filtrate collected is 1 gal when the filtration rate is 1.5 gpm, and 5 gal when the filtration rate is 0.6 gpm. Calculate the volume collected when the filtration rate is 0.75 gpm. What is the total volume collected in 10 min? Assume constant-pressure filtration.

Solution

From Equation 7.23

$$d\theta = \frac{1}{C'}(V+V_1)dV$$

By integration,

$$\theta = \frac{1}{C'}\int(V+V_1)\,dV = \frac{1}{C'}\left(\frac{V^2}{2}+V_1V\right) \quad \text{(a)}$$

Calculate constants C' and V_1. When $dV/d\theta = 1.5$ gpm, $V = 1$ gal.

Therefore

$$\frac{3}{2} = \frac{C'}{1+V_1} \quad \text{or} \quad 3+3V_1 = 2C' \quad \text{(b)}$$

When $dV/d\theta = 0.6$ gpm, $V = 5$ gal, and

$$\frac{3}{5} = \frac{C'}{5+V_1} \quad \text{or} \quad 15+3V_1 = 5C' \quad \text{(c)}$$

Solving (b) and (c) for C' and V_1, $C' = 4$ and $V_1 = \frac{5}{3}$.

When the filtration rate is $\frac{3}{4}$ gpm, the filtrate volume collected is obtained by the substitution of $dV/d\theta = \frac{3}{4}$, $V_1 = \frac{5}{3}$, and $C' = 4$ in Equation 7.23 and is given by

$$\frac{3}{4} = \frac{4}{V + \frac{5}{3}} \quad \text{or} \quad V = \frac{11}{3} = 3.67 \text{ gal}$$

The time required to collect 3.67 gal is

$$\frac{1}{4}\left[\frac{11^2}{2(9)} + \frac{5}{3}\left(\frac{11}{3}\right)\right] = 3.21 \text{ min}$$

Total Volume Collected at the End of 10 min. Substituting $\theta = 10$ min in the integrated Equation (a) gives

$$10 = \frac{1}{4}\left(\frac{1}{2}V^2 + \frac{5}{3}V\right)$$

By simplification, $240 = 3V^2 + 10V$ or $3V^2 + 10V - 240 = 0$; i.e., $V^2 + \frac{10}{3}V - 80 = 0$. From which

$$V = \frac{1}{2}[-\frac{10}{3} + \sqrt{\left(\frac{10}{3}\right)^2 + 320}] = 7.43 \text{ gal}$$

Example 7.2

E. L. McMillen and H. A. Webber obtained the following data in the constant-pressure filtration of slurry of $CaCO_3$ in water. The mass fraction of solids in the slurry to the press was 0.139. The filter area was 1 ft². Mass ratio of wet cake to dry cake was 1.47 and dry-cake density was 73.5 lb/ft³. Rate data obtained at 50 psi were as follows:

Time, s	0	19	68	142	241	368	702
Filtrate, lb	0	5	10	15	20	25	35

$\rho_{solids} = 164$ lb/ft³ $\mu_{filtrate} = 0.86$ cP

Calculate the specific resistance of the cake α, the filter medium resistance R_1, and the porosity of the cake.

Solution

Equation 7.8, which is in terms of volume, can be converted to units of pounds with the use of the density ρ as follows:

$$\frac{\rho d\theta}{3600 \rho dV} = \frac{2\rho V}{C\rho} + \frac{2V_1\rho}{\rho C}$$

in which θ is in seconds. Since $\rho dV = d(\rho V) = dW'$ and $\rho V = W'$, the preceding reduces to

$$\frac{d\theta}{dW'} = \frac{7200W'}{\rho^2 C} + \frac{7200W'_1}{\rho^2 C}$$

In the preceding equation, W' is the filtrate, lb, and W'_1 is the filtrate, lb, corresponding to the fictitious cake. Thus at constant ΔP, a plot of $d\theta/dW'$ versus W' will be a straight line. To make this plot for the example, the calculations of Table 7.1 are first made.

Table 7.1 Calculations of the filtration rates and average filtrate flows

q (Given) s	W' (Given) lb	Δq s	$\Delta W'$ lb	$\dfrac{\Delta q}{\Delta W'}$ Calculated lb/s	$\overline{W'} = \dfrac{(W')_1 + (W')_2}{2}$ lb
0	0	—	—	—	—
19	5	19	5	3.8	2.5
68	10	49	5	9.8	7.5
142	15	74	5	14.8	12.5
241	20	99	5	19.8	17.5
368	25	127	5	25.4	22.5
524	30	156	5	31.2	27.5
702	35	178	5	35.6	32.5

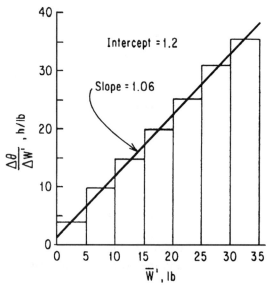

Exhibit 1 A plot of $\Delta\theta/\Delta W'$ versus W' (Example 7.2)

A plot of $\Delta\theta/\Delta W'$ versus $\overline{W'} = \tfrac{1}{2}(W'_1 + W'_2)$ is made in Exhibit 1. From this exhibit, intercept = 1.2 and

$$\text{Slope} = \frac{35.6 - 3.8}{32.5 - 2.5} = \frac{31.8}{30} = 1.06 = \frac{7200}{\rho^2 C}$$

Therefore

$$C = \frac{7200}{\rho^2(1.06)} = \frac{7200}{62.4^2(1.06)} = 1.744 \text{ ft}^6/\text{h}$$

$$\text{Intercept} = 1.2 = \frac{7200 W'_1}{C\rho^2}$$

Therefore
$$W_1' = \frac{1.2C\rho^2}{7200} = \frac{1.2(1.744)(62.4)^2}{7200} = 1.13 \text{ lb}$$

$$V_1 = \frac{1.13}{\rho} = \frac{1.13}{62.4} = 0.018 \text{ ft}^3$$

Calculation of θ_1.

$$\int_{-\theta}^{\theta} d\theta = \frac{7200}{\rho^2 C} \int_0^{(W'+W_1')} (W' + W_1') dW'$$

or

$$\theta + \theta_1 = \frac{3600}{\rho^2 C}(W' + W_1')^2$$

Substitution of the values at $\theta = 19$ s, and $W' = 5$ lb in the preceding equation gives

$$19 + \theta_1 = \frac{3600}{(62.4)^2(1.744)}(5 + 1.13)^2$$

from which

$$\theta_1 = 0.92 \text{ s} = 2.56 \times 10^{-4} \text{ h}$$

Specific Resistance. From Equation 7.8a,

$$\frac{2A^2 \Delta P(1-mx)}{\mu \rho x \alpha} = C$$

Therefore
$$\alpha = \frac{2A^2 \Delta P(1-mx)}{\mu \rho x C}$$

$$= \frac{2(1)^2(144)(50)[1-1.47(0.139)]}{2.07(62.4)(0.139)(1.744)}$$

$$= 365. \text{h}^2 \cdot \text{lb}_f/\text{lb}^2$$

Filter-Medium Resistance R_1. By definition, Equation 7.4, the filter medium resistance is given by

$$R_1 = \frac{W_1' \alpha}{A}$$

$$= \frac{(1.13 \text{ lb})(365.9 \text{ h}^2 \cdot \text{lb}_f/\text{lb}^2)}{1 \text{ ft}^2}$$

$$= 413.5 \text{ h}^2 \cdot \text{lb}_f/\text{ft}^2 \cdot \text{lb}$$

Porosity of Cake. $\quad(1 - X)\rho_S = 73.5$
Therefore $\quad X = 0.552 \quad$ or $\quad 55.2$ percent

Instead of deriving the relations of filtration theory in terms of mass ratio m (mass of wet cake/mass of dry cake) and mass fraction of solids, it is also possible to apply Kozeny-Carman relation using the initial concentration of slurry expressed as kg/m³.

The Kozeny-Carman relation is as follows:

$$-\frac{\Delta P_c}{L} = \frac{k_1 \mu u (1-\varepsilon)^2 S_0^2}{\varepsilon^3} \tag{7.24}$$

where
k_1 = a constant, which in most cases equals 4.17 for random particles of definite size and shape
μ = viscosity of filtrate, Pa · s (lb$_m$/ft · s)
u = linear velocity based on filter area, m/s (ft/s)
ε = void fraction or porosity of cake
L = thickness of cake, m (ft)
S_o = specific surface area of particle, m² (ft²) per m³ (ft³) volume of solid particles
ΔP_C = pressure drop in the cake, N/m² (lb$_f$/ft²)

For English units, the righthand side of Equation 7.24 is divided by g_c. The linear velocity based on empty cross-sectional area is

$$u = \frac{dV/dt}{A} \tag{7.25}$$

where
A = filter area, m²(ft²)
V = total volume of filtrate, m³(ft³) up to time (t) in seconds.

A material balance permits relation of the cake thickness L and the filtrate volume V as

$$LA(1-\varepsilon)\rho_P = c_s(V + \varepsilon LA) \tag{7.26}$$

where c_s = solids in kg/m³ (lb$_m$/ft³) of filtrate.

Substituting Equation 7.25 in Equation 7.24 and using Equation 7.26 to eliminate L the final relation is obtained.

$$\frac{dV}{Adt} = \frac{-\Delta P_c}{\frac{k_1(1-\varepsilon)S_0^2}{\rho_P \varepsilon^3} \frac{\mu c_s V}{A}} = \frac{-\Delta P_c}{\alpha \frac{\mu c_s V}{A}} \tag{7.27}$$

where α = specific cake resistance, m/kg (ft/lb$_m$) = $\dfrac{k_1(1-\varepsilon)S_0}{\rho_P \varepsilon^3}$ \qquad (7.28)

The filter medium resistance can be taken into account by writing the similar relation as in Equation 7.27 as follows:

$$\frac{dV}{Adt} = \frac{-\Delta P_f}{\mu R_m} \quad (7.29)$$

where
R_m = resistance of filter medium to filtrate flow, m^{-1} (ft^{-1})
ΔP_f = pressure drop, N/m^2 (lb$_f$/ft^2)

Resistances of the cake and the filter medium are in series and therefore can be combined. Thus

$$\frac{dV}{Adt} = \frac{-\Delta P}{\mu \left(\frac{\alpha c_s V}{A} + R_m\right)} \quad (7.30)$$

where total pressure drop = $\Delta P = \Delta P_c + \Delta P_f$ (7.30a)

Sometimes R_m is replaced by $\frac{\mu \alpha c_s V_e}{A}$, which results in the equation

$$\frac{dV}{Adt} = \frac{-\Delta P}{\frac{\mu \alpha c_s}{A}(V + V_e)} \quad (7.31)$$

Equation 7.31 is derived in the earlier section in a different manner.

Filtration Rate in Batch Process

Many times filtration is carried out at constant pressure.
Equation 7.30 is rearranged to give (Geankoplis)

$$\frac{dt}{dV} = \frac{\mu \alpha c_s}{A^2(-\Delta P)}V + \frac{\mu}{A(-\Delta P)}R_m = K_P V + B \quad (7.32)$$

where K_P is in s/m^6 (s/ft^6) and B in s/m^3 (s/ft^3)

$$K_P = \frac{\mu \alpha c_s}{A^2(-\Delta P)} \qquad B = \frac{\mu R_m}{A(-\Delta P)} \quad \text{(SI units)} \quad (7.32a)$$

$$K_P = \frac{\mu \alpha c_s}{A^2(-\Delta P)g_c} \qquad B = \frac{\mu R_m}{A(-\Delta P)g_c} \quad \text{(English units)} \quad (7.32b)$$

Integration of Equation 7.32 gives

$$t = \frac{K_P}{2}V^2 + BV \quad (7.33)$$

Dividing by V gives

$$\frac{t}{V} = \frac{K_P}{2}V + B \quad (7.34)$$

Equation 7.34 is used to plot batch filtration data to get the constants in the equation and from these constants specific resistance and filter medium resistance are determined.

If α is shown to be constant experimentally, then the cake is incompressible. If it is not constant, the following form is used to correlate α:

$$\alpha = \alpha_0(-\Delta P)^z \qquad (7.35)$$

Taking logarithms of both sides, $\ln \alpha = \ln \alpha_0 + z \ln(-\Delta P)$. Thus if a series of runs are made at different pressures and αs calculated at each pressure, a plot of $\ln \alpha$ vs $\ln(-\Delta P)$ will give a straight line whose slope is z, the compressibility of the cake.

Example 7.3

Data for laboratory filtration of $CaCO_3$ slurry are reported by Ruth and by McCabe and Smith. These data are also used for illustration by Geankoplis. In a series of three filtration runs at three different pressures, filtrate volumes were collected as functions of time. The runs were carried out at pressures of 338, 194.4, and 46.2 kPa respectively. Data for the three runs are given in Table 7.2. Other data are as follows:

$$\text{Slurry concentration } c_s = 23.47 \text{ kg/m}^3$$
$$\text{Viscosity of filtrate} = 0.9 \times 10^{-3} \text{ Pa}$$
$$\text{The filter area, } A = 0.0439 \text{ m}^2$$

Calculate the constant α for each run. Determine the compressibility of the cake.

Table 7.2 Experimental filtration data for Example 7.3

	Pressure 338 kPa			Pressure 46.2 kPa			Pressure 194.4 kPa		
ts	V × 10³	(t/V) × 10⁻³	ts	t/V × 10³	(t/V) × 10⁻³	ts	V × 10³	(t/V) × 10⁻³
0	0		17.3		34.6	6.3	0.5	12.6
4.4	0.498	8.84	41.3	1.0	41.3	14	1	14
9.5	1	9.5	72	1.5	48.0	24.2	1.5	16.13
16.3	1.501	10.86	108.30	2.0	54.15	37	2	18.5
24.6	2.0	12.3	152.00	2.5	60.8	51.7	2.5	20.68
34.7	2.498	13.89	201.70	3.0	67.23	69	3	23
46.1	3.002	15.36				88.8	3.5	25.37
59.0	3.506	16.83				110	4	27.5
73.69	4.004	18.38				134	4.5	29.78
89.4	4.502	19.86				160	5.0	32.0
107.3	5.009	21.42						

(t in seconds, volume in m³)

From the given data t/V was calculated. These values are given in column 3 of each run.

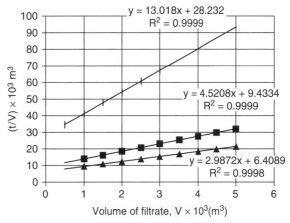

Exhibit 2 Plots of filtration data (Example 7.3)

The values of t/V are plotted in Exhibit 7.2. The slope of the line is $K_P/2$ and the intercept is $B = \frac{\mu R_m}{A(-\Delta P)}$. From the slope and intercept, α and R_m are calculated as follows:

From slope,

$$K_P = 2 \times 2.9872 \times 10^6 = \frac{\mu \alpha c_s}{A^2(-\Delta P)} = \frac{0.9 \times 10^{-3}(\alpha)(23.47)}{0.0439^2(338 \times 10^3)}$$

$$\alpha = 1.842 \times 10^{11} \text{ m/kg}$$

From intercept,

$$B = \frac{\mu R_m}{A(-\Delta P)} = \frac{0.9 \times 10^{-3} R_m}{0.0439 \times 338 \times 10^3} = 6408.9$$

$$R_m = 1.057 \times 10^{11} \text{ m}^{-1}$$

Similar calculations are done for other two pressures also. The results are tabulated in Table 7.3.

Table 7.3 Calculations of α and R_m

Pressure kPa	$\alpha \times 10^{-11}$	$R_M \times 10^{-11}$
338.0	1.842	1.057
194.4	1.604	0.895
46.2	1.097	0.64

The values $\ln \alpha$ versus $\ln(-\Delta P)$ are plotted in Exhibit 7.3.

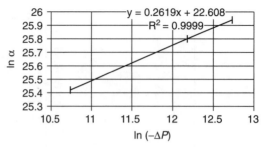

Exhibit 3 Plot of lnα vs ln($-\Delta$P)(Example 7.3)

The slope of the line is $z = 0.2619$, the compressibility of the cake.

RECOMMENDED REFERENCES

1. Geankoplis, *Transport Processes and Separation Process Principles,* 4th ed., Prentice-Hall, 2004.
2. Badger and Banchero, *Introduction to Chemical Engineering,* McGraw-Hill, 2004.

CHAPTER 8

Membrane Separation

OUTLINE

WHAT IS A MEMBRANE? 307
Classification of Membranes and Membrane Technology ■ Application of Membrane Technology

REVERSE OSMOSIS (RO) 308
Percent Recovery

ULTRAFILTRATION (UF) 314

ELECTRODIALYSIS (ED) 314

PERVAPORATION 314

NOMENCLATURE 315

RECOMMENDED REFERENCES 315

Although membranes play vital roles in the transport of matter and energy in the cells of all living organisms, industrial application of membranes started rather recently in the 1960s. Since then membranes have found applications in chemical industry and health science alike. The technology of membrane separation is used wherever a separation between molecules is required, mostly without a change of phase.

WHAT IS A MEMBRANE?

A membrane is a thin micro-porous sheet of material that controls the passage of molecules across it based on the size, shape, or charge of the molecule, and its affinity for the molecules, thereby selectively separating specific materials from a mixture. A membrane may be a bio-membrane, which may be a pliable sheet or lining of a connective tissue from an animal or vegetable body; it may also be an artificial or synthetic membrane.

Classification of Membranes and Membrane Technology

Membrane technology may be classified broadly into two groups depending on the method of separation.

1. The pore method

    In the pore method, the difference of pressure and concentration across the pores or the intermolecular spaces of the membrane causes the separation. The technologies that fall in this group are reverse osmosis, nanofiltration, ultra-filtration, microfiltration, dialysis, gas separation, pervaporation, etc.

2. The carrier method

    In the carrier method, the difference of electrical potential, pressure, concentration, and conjugated energy across the membrane causes the separation. The technologies that fall in this group are electrodialysis, electrolysis, etc.

Application of Membrane Technology

During the last forty years, substantial progress has been made to apply the membrane technology in the following major fields:

- Purification of water, abatement of pollution
- Concentration or recovery of an active ingredient from a solution
- Separation of gases and the components of azeotropic mixture
- Biomedical use such as blood purification in artificial kidney, drug delivery system.

REVERSE OSMOSIS (RO)

Reverse osmosis is a unit operation in which a certain component of a feed solution, usually the solvent, passes through a semipermeable membrane, while others, usually the dissolved solids, are retained primarily in the concentrate.

The separation of the solvent from a solution through reverse osmosis requires an application of pressure on the solution contained by the semipermeable membrane to overcome the osmotic pressure of the solution. It is therefore imperative to understand osmotic pressure.

When a solution is separated from the solvent by a semipermeable membrane with both phases at the same temperature, the excess pressure that must be applied on the solution in order to check the inflow of solvent and establish thermodynamic equilibrium is called the osmotic pressure of the solution. When pressure is applied on the solution side so that the chemical potential of the solvent in the solution side is increased to exactly match the chemical potential of the solvent in the solvent side to establish thermodynamic equilibrium, the inflow of solvent stops.

Osmosis is a natural phenomenon of establishing thermodynamic equilibrium between two solutions of different concentrations. Osmotic pressure is a solution property, which is independent of the membrane. This property always manifests in a solution irrespective of whether the solution encounters a membrane or not. When two solutions are in contact with each other without an intervening semipermeable membrane, solvent flow still proceeds from the lower concentration side to the higher concentration side, and diffusion of solute proceeds from the higher concentration to the lower concentration side until a thermodynamic equilibrium is established. Mother Nature, as though in sharp contrast with human nature, always tries to minimize the difference between the haves and the have-nots.

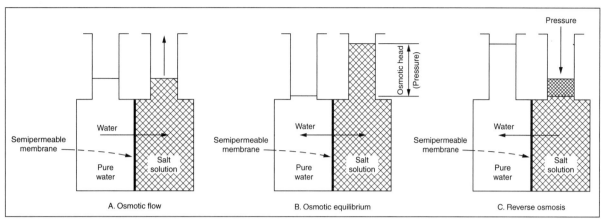

Figure 8.1 Chemical potential of water in salt solution is raised, osmotic flow is reversed

Osmotic pressure of solution is represented by

$$\pi = -\left[\frac{RT}{V_w^0}\right]\ln(\gamma_w x_w) \tag{8.1}$$

From known osmotic pressure and solute concentration, the activity coefficient of solvent may determined:

$$\gamma_w = \left[\frac{1}{1-x_s}\right]\left[e^{-\frac{V_w^0 \pi}{RT}}\right] \tag{8.2}$$

For an ideal solution, we can assign $\gamma_w = 1$, and Equation 8.2 reduces to:

$$\pi_{\text{ideal}} = -\left[\frac{RT}{V_w^0}\right]\ln(x_w) \tag{8.3}$$

A parameter called van't Hoff's coefficient is defined by

$$\phi = \frac{\pi}{\pi_{\text{ideal}}} \tag{8.4}$$

In Equation 8.3, if we approximate the logarithm term $\ln(x_w) = \ln(1-\sum x_s) \approx -\sum x_s$

$$\pi_{\text{ideal}} = -\left[\frac{RT}{V_w^0}\right][-\sum x_s] = \frac{(\sum x_s)RT}{V_w^0}$$

For dilute solution, $c_s = \frac{x_s}{V_w^0}$. Therefore, the above equation reduces to:

$$\pi_{\text{ideal}} = (\sum c_s)RT \tag{8.5a}$$

Substituting the value of R

$$\pi_{ideal} = 1.206 c_s T \text{ psi} \tag{8.5b}$$

Equation 8.5 represents the van't Hoff equation of osmotic pressure. According to this equation, the osmotic pressure of a very dilute ideal solution is independent of the nature of the solvent. Although there is a similarity between this equation and the equation of ideal gas ($p = cRT$), there is a fundamental difference between the van't Hoff equation and the ideal gas equation. The ideal gas equation applies to any composition of a mixture of ideal gases, whereas the van't Hoff equation applies to very dilute ideal solutions. The existence of deviations from the van't Hoff equation does not necessarily imply a non-ideal solution. Measurement of osmotic pressure at various concentrations and extrapolation of the pressure to zero concentration may be used to determine the molecular weight of a polymer. Osmotic pressure measurement can also help to determine the activity coefficient of a component. Two solutions are isotonic when their osmotic pressures are identical at the same temperature.

In the computation of concentration of solute, some basic terms should be understood.

Molal Concentration or Molality
Molality is the mols of solute per kg of solvent. Thus, if 'a' gms of solute of molecular weight M is dissolved in 'b' gms of solvent, then molal concentration, c_m, is given by

$$c_{molal} = \frac{a(1000)}{b(M)} \tag{8.6}$$

Molar Concentration or Molarity
Molarity is the mols of solute per liter of solution. Thus, if 'a' gms of solute of molecular weight M is dissolved in 'b' gms of solvent and the resulting solution has a density of 'd' gm/cc,

$$c_{molar} = \left[\frac{(a)(d)(1000)}{(a+b)(M)} \right] \tag{8.7}$$

Normality
Normality is the number of gram-equivalents per liter of solution. Thus, if 'a' gms of solute of equivalent weight E_w are dissolved in 'b' gms of solvent and the resulting solution has a density of 'd' gm/cc,

$$N = \left[\frac{(a)(d)(1000)}{(a+b)(E_w)} \right] \tag{8.8}$$

It is obvious from Equations 8.7 and 8.8 that molarity and normality become identical when molecular weight and equivalent weights are the same.

Parts per Million (ppm)
For liquids and solids, parts per million means a mass fraction. For gases, it is generally taken as mol or volume fraction.

Mathematical models to predict performance of reverse osmosis unit are

$$Q_w = \left[\frac{K_w A}{\tau}\right](\Delta P - \Delta \pi) \tag{8.9}$$

$$= K_{wo}(\Delta P - \Delta \pi) \tag{8.10}$$

$$Q_S = \left[\frac{K_s A}{\tau}\right](\Delta C) = K_{so}(\Delta C) \tag{8.11}$$

where
- Q_w = solvent permeation rate through membrane
- Q_s = solute flow rate through membrane
- K_w = solvent permeability coefficient
- K_s = solute permeability coefficient
- K_{wo} = overall solvent permeability constant of a given unit
- K_{so} = overall solute permeability coefficient of a given unit
- A = membrane surface area
- τ = membrane thickness
- ΔP = Hydraulic pressure difference across membrane
- ΔC = Solute concentration difference across membrane
- $\Delta \pi$ = osmotic pressure difference across the membrane

In a perfect semipermeable membrane, which does not exist, the solute permeability coefficient is zero.

In a batch operation, where the concentration of the solute in the feed increases with time, the solvent permeation falls even with constant hydraulic pressure difference across the membrane. This is due to the increase in osmotic pressure since the concentration of solute is increased, and due to concentration polarization. When solvent molecules travel toward the membrane, they accompany the solute molecules with them. As a result, the concentration of solute at the membrane surface increases dramatically compared with the bulk. This increases the osmotic pressure, which reduces and finally stops the permeation. In a well-designed system, concentration polarization can be essentially eliminated. In some cases, for a fixed hydraulic pressure difference, the permeation flux for a given system may be represented by

$$Q_w = K_{wo} c_s^{-n} \tag{8.12}$$

The two constants K_{wo} and n may be determined for a given system by collecting data of Q_w versus c_s under the desired hydraulic pressure and temperature, and plotting the data on a log-log plot.

For a batch process

$$-\frac{dV}{d\theta} = Q_w = K_{wo} c_s^{-n} \tag{8.13}$$

and,

$$V_0 c_{so} = V c_s \tag{8.14}$$

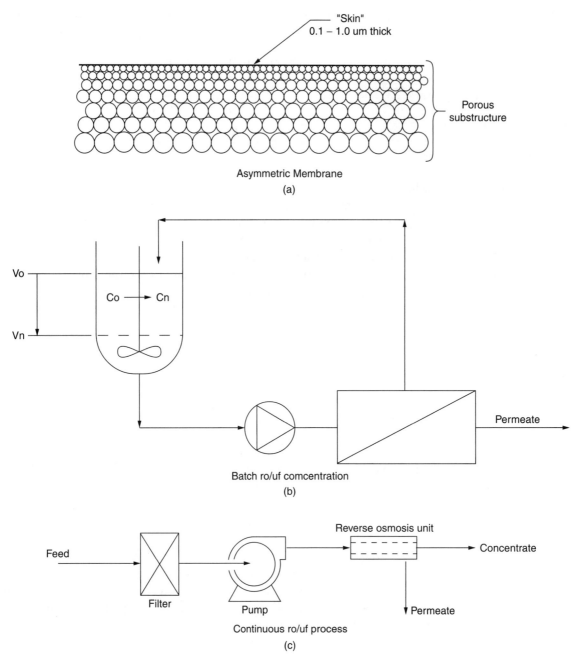

Figure 8.2 (a) Asymmetric membrane, (b) Batch RO/UF concentration, and (c) Continuous RO/UF process

From Equations 8.13 and 8.14, the time to concentrate from an initial volume, V_0, of a batch to a final volume, V_{final}, may be calculated by integration when n is not 1.

$$\theta = \left[\frac{(V_0 c_{so})^n}{K_{wo}(1-n)}\right]\left[V_0^{(1-n)} - V_{\text{final}}^{(1-n)}\right] \tag{8.15}$$

Equation 8.15 is based on the assumption that no solute permeates through the membrane. In actual practice there is loss of solute through the membrane. Here is some terminology used in reverse osmosis technology.

Feed
The solution that is fed at the inlet of the membrane system is called the feed.

Permeate
The material, mostly solvent, that passes through the membrane wall is called the permeate.

Concentrate
The material that comes out of the process side of the membrane after removal of the permeate is called the concentrate. Concentrate = feed − permeate.

One has to be careful in applying the terminology. Depending on the application, the concentrate or the permeate may be termed as the product. Thus, in the desalination of seawater, the permeate is the drinking water, and is called the product. On the other hand, in the concentration of dyestuff, the concentrate is called the product.

Percent Recovery

The recovery or percent conversion in an RO process is the ratio of the specific outlet stream flow to feed stream flow, both measured in the identical unit. Thus

$$\text{Permeate recovery} = \frac{\text{permeate flow rate}}{\text{feed flow rate}}, \text{ for a continuous or batch process}$$

$$= \frac{\text{permeate volume collected}}{\text{feed volume used}}, \text{ for batch process} \quad (8.16)$$

It can be shown by a simple material balance of a batch process:

$$\frac{\text{Permeate volume}}{\text{Feed volume}} = \frac{C_{cd} - C_{fd}}{C_{cd} - C_{pd}}$$

$$= 1 - \frac{C_{fd}}{C_{cd}}, \quad \text{when} \quad C_{pd} \approx 0 \quad (8.17)$$

This is used to define the volumetric efficiency of an RO unit in the desalination industry or to calculate approximately the amount of permeate to collect in order to achieve the desired concentration of an active ingredient in a batch process. On the other hand, in the concentration of dyestuff, the percent recovery refers to the concentrate side.

Another term, *solute rejection, R*, is used to define the concentration efficiency in the dyestuff and food industry in batch concentration. The solute rejection factor is used to measure the efficiency of a membrane in the concentration of active ingredient. The dyestuff, for example, is synthesized in an aqueous solution. The ability of the membrane to selectively reject the dyestuff molecules when the solution approaches the membrane to permeate across it is a measure of the efficiency of the membrane.

$$R_d = 1 - \frac{C_{pd}}{C_{cd}} \quad (8.18)$$

where

C_{pd} = concentration of dyestuff in permeate.
C_{cd} = concentration of dyestuff in concentrate.
C_{cf} = concentration of dystuff in feed.
R_d = dyestuff recovery or dyestuff rejected by membrane, time-independent

It can be shown by a differential analysis for a batch concentration

$$\text{Fractional product loss} = 1 - \left[\frac{V_{final}}{V_0}\right]^{(1-R_d)} \quad (8.19)$$

The RO process separates the solvent from the solution without a change of phase and does not require steam except to maintain an optimum temperature for the process. There are limitations of maximum allowable temperature and maximum allowable pressure, and silica content on the feed. Generally when the hydraulic driving force is very high compared with the osmotic pressure, higher operating temperature increases the permeation even though the osmotic pressure increases with temperature.

ULTRAFILTRATION (UF)

Like RO, UF makes the separation between the solute and the solvent without a change of phase. In the process, the membrane rejects the species of high molecular weights (1000–80,000) at the surface of the membrane. The pore sizes of UF membranes are 0.001 to 0.02 micrometer. The hydraulic pressure requirement is 10 to 100 psig.

Reverse osmosis and ultrafiltration have applications in the concentration of dissolved solids both in production of consumer materials and waste management. The concentrated material in the waste treatment is generally incinerated thereby saving additional cost of burning dilute waste.

ELECTRODIALYSIS (ED)

In electrodialysis, a direct electrical current is used to transport ions across the membrane, which is impervious to solvent. In the desalting process, for example, the ED unit has anode and cathode at opposite ends. An anion-selective membrane is placed next to anode, followed by a cation-selective membrane; the alternating placement of membranes is continued until the membrane next to the cathode is a cation-selective membrane.

Like RO, ED also removes solutes without a phase change. The system requires pretreatment of feed if colloids, silica, etc. are present. The optimum application of ED is when the total dissolved solids are between 1000–5000 mg/L range.

PERVAPORATION

Pervaporation is a membrane separation process in which a liquid is placed on one side of the membrane and the permeate is removed as a vapor from the other side through a vacuum pump. The permeate is then condensed. Thus pervaporation is a membrane separation in which the material undergoes a change of phase. Pervaporation can be used as a fractionation stage in a distillation column in series with a reflux condenser.

NOMENCLATURE

c_{sf} = concentration of solute in feed, gms/liter solution
c_{sp} = concentration of solute in permeate, gms/liter solution
c_s = concentration gmol/liter for non-electrolyte solute, for electrolytes it represents g-ion/liter
p = pressure, atm
Q_w = permeation rate, volume/time
R = Gas-law constant, 0.08205 (liter)(atm)/(mol)(K)
R_d = membrane rejection or solute recovery
T = Temperature, K
V_w^0 = partial molar volume of solvent, liter/mol. For water at 293 K it is 0.018 L/mol
x_w = mol fraction of solvent
x_s = mol fraction of solute
π = osmotic pressure of solution, atm, unless otherwise specified
γ_w = activity coefficient of the solvent
θ = time

RECOMMENDED REFERENCES

1. Applegate, "Membrane Separation Process," *Chemical Engineering*, McGraw-Hill, June 1984.
2. Sourirajan, *Reverse Osmosis,* Ottawa: National Research Council of Canada, 1977.

CHAPTER 9

Mass Transfer Fundamentals

OUTLINE

MOLECULAR DIFFUSION 318
Fick's Law ■ Steady State Equimolar Counter Diffusion of A and B (Gas) ■ Molecular Diffusion of Gases A and B plus Convection ■ Steady-State Diffusion of A into Stagnant Non-Diffusing B (Gas)

MOLECULAR DIFFUSION IN LIQUIDS 323
Equations for Diffusion in Liquids ■ Determination of Diffusion Coefficients (gases) ■ Diffusivity in a Multi-Component Mixture ■ Estimation of Diffusivities for Liquids

MOLECULAR DIFFUSION IN SOLIDS 329
Diffusion of Liquids in Porous Solids ■ Diffusion of Gases in Porous Solids

CONVECTIVE MASS TRANSFER 333
Turbulent Diffusion Equation for Mass Transfer (Convective Mass Transfer)

TURBULENT MASS TRANSFER 336
Film Model ■ Higbie Penetration Model ■ Surface Renewal Theory of Danckwerts ■ Turbulent Boundary Layer Theory

DIMENSIONLESS NUMBERS IN MASS TRANSFER 338
Mass Transfer for Flow Inside Pipes ■ Mass Transfer for Turbulent Flow Inside Pipes ■ Mass Transfer from a Gas into a Falling Liquid Film

INTERPHASE MASS TRANSFER 341
Overall Mass Transfer Coefficients

MASS TRANSFER IN PACKED BEDS 343

PHASE EQUILIBRIA 346

VAPOR-LIQUID EQUILIBRIA 346
Binary and Multicomponent Systems ∎ The Phase Rule and Duhem's Theorem ∎ Ideal Mixtures (Raoult's Law) ∎ Relative Volatility α_{ij} ∎ Henry's Law ∎ The Chemical Potential and Phase Equilibria ∎ Nonideal Systems ∎ Calculation of Activity Coefficients ∎ VLE by Modified Raoult's Law ∎ Equilibrium Vaporization Ratios (Equilibrium Constants) K ∎ Dew Point ∎ Bubble Point ∎ Flash Calculations ∎ Dew-Point Calculation Using K Values

ENTHALPY-COMPOSITION DIAGRAMS (BINARY SYSTEM) 365

RECOMMENDED REFERENCES 367

Separation processes (formerly called "unit operations") are common to all chemical and physical process industries. These processes have certain fundamental mechanisms in common. The mechanism of diffusion or mass transfer occurs in distillation, absorption, membrane separation, drying, adsorption, crystallization, liquid-liquid extraction, leaching, and ion exchange. Heat transfer is involved in evaporation, distillation, drying, and crystallization. Momentum transfer is involved in filtration, mixing, and sedimentation. In this chapter, we cover the fundamental principles of mass transfer operations.

MOLECULAR DIFFUSION

Mass transfer involves movement of molecules under some driving force such as concentration, temperature (thermal diffusion), or voltage differences (electro-diffusion). However, mass transfer discussed here will be dealing primarily with concentration difference. Mass transfer takes place by two mechanisms: (1) molecular diffusion and (2) convective diffusion. Mass transfer may occur in the same phase from one region to the other or from one distinct phase to another.

Molecular diffusion is the movement of molecules through a fluid by their random motion. The motion is random (not in a straight path) because the moving molecules collide with other molecules and change direction after collision. In *convective mass transfer*, analogous to heat or momentum convective transfer, the molecular movement is aided by agitation or fluid flow.

Fick's Law

The basic relation for mass transfer by *molecular diffusion* in a binary mixture of A and B is given by

$$J_A = -D_{AB} \frac{dc_A}{dz} \qquad (9.1)$$

where
J_A = molar flux of component A along axis z due to molecular diffusion, kg mol of A/s · m^2
D_{AB} = molecular diffusivity of A in the solution containing A and B, m^2/s
C_A = concentration of A, kg mol/m^3
z = distance coordinate in the direction of molecular diffusion, m.

Consistent English and cgs system units may also be used.

Another expression for the same case is

$$J_A = -c_m D_{AB} \frac{dy_A}{dz} \qquad (9.2)$$

where

y_A = mol fraction of A
c_m = average molar density of the mixture, kg mol/m^3

Example 9.1

A mixture of H_2 and NH_3 is contained in a pipe at 298 K and one atm, which is maintained constant. At one end of the pipe at point 1, the partial pressure p_{A1} of NH_3 is 0.65 atm and at another end 0.3 m away from point 1, p_{A2} of NH_3 is 0.15 atm. Calculate the flux of NH_3 at steady state if the diffusivity of the NH_3-H_2 mixture at 298 K is 0.783×10^{-4} m^2/s. What is the flux in lb mol/h · ft^2?

Solution

Total pressure is constant. At one atm the mixture may be assumed to exhibit ideal gas behavior. Therefore

$$PV = nRT$$

$$\frac{n}{V} = \frac{P}{RT} = c$$

where

n = kg mol of $NH_3 + H_2$
$R = 8314.3$ m^3 · Pa/kg mol · K
$T = 298$ K

c in kg mol of NH_3 ($NH_3 + H_2$)/m^3 $\quad c_{A1} = \dfrac{P_{A1}}{RT} = \dfrac{0.65 \times 1.0133 \times 10^5}{RT}$

$$c_{A2} = \frac{P_{A2}}{RT} = \frac{0.15 \times 1.033 \times 10^5}{RT}$$

In steady state, the flux J_A is constant and D_{AB} is given as constant. Therefore

$$J_{Az} \int_z^{z_2} dz = -D_{AB} \int_{A1}^{c_{A2}} dc_A$$

$$J_{Az} = \frac{D_{AB}(c_{A1} - c_{A2})}{z_2 - z_1} = \frac{0.783 \times 10^{-4}(1.0133 \times 10^5)(0.65 - 0.15)}{RT(z_2 - z_1)}$$

$$= \frac{0.783 \times 10^{-4}(1.0133 \times 10^5)(0.65 - 0.15)}{8314.3 \times 298 \times 0.3}$$

$$= 5.337 \times 10^{-6} \frac{\text{kg mol A}}{\text{s} \cdot \text{m}^2}$$

$$J_{Az} = 5.337 \times 10^{-6} \frac{\text{kg mol A}}{\text{s}} \left| \frac{2.202 \text{ lb mol}}{\text{kg mol}} \right| \frac{3600 \text{ s}}{\text{h}} \left| \frac{}{\text{m}^2 \times 10.76 \text{ ft}^2/\text{m}^2} \right.$$

$$= 3.932 \times 10^{-3} \frac{\text{lb mol A}}{\text{h} \cdot \text{ft}^2}$$

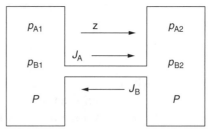

Figure 9.1 Equimolar counter diffusion of gases A and B

Steady State Equimolar Counter Diffusion of *A* and *B* (Gas)

The two components *A* and *B* diffuse in opposite direction at steady rate and in equimolar proportion. This process is shown in Figure 9.1 schematically.

Partial pressure $P_{A1} > P_{A2}$ and $P_{B2} > P_{B1}$. Total pressure *P* is constant and the gases exhibit ideal gas behavior. With these assumptions, the flux of component *A*, J_A is given by

$$J_{Az} = \frac{D_{AB}}{z_2 - z_1}(C_{A1} - C_{A2}) = \frac{D_{AB}}{RT(z_2 - z_1)}(p_{A1} - p_{A2}) \qquad (9.3)$$

Also $D_{AB} = D_{BA}$. The flux of B is equal but opposite in sign.

Molecular Diffusion of Gases *A* and *B* plus Convection

This case of diffusion takes into account the convective flow or net movement of the entire phase consisting of the mixture of *A* and *B*. The flux given by Equations 9.1 to 9.3 is based on the assumption that the fluid is stationary and there was no convective flow of the entire phase. The movement of molecules was entirely under the influence of concentration difference. This flux can be expressed in terms of a velocity of diffusion of *A* past a fixed point as follows:

$$J_A = v_{Ad}C_A \left(\frac{m}{s}\frac{kg\,mol\,A}{m^3}\right) \qquad (9.4)$$

where v_{Ad} = diffusion velocity of *A* in m/s. To get flux due to molecular diffusion as well as bulk phase movement, the bulk phase velocity is added to the diffusion velocity to give net velocity of *A* as

$$v_A = v_{Ad} + v_m \qquad (9.5)$$

where
  v_A = velocity of *A* relative to a stationary point, m/s
  v_m = velocity of bulk fluid comprising *A* and *B*, m/s

If Equation 9.5 is multiplied by c_A, one obtains

$$c_A v_A = c_A v_{Ad} + c_A v_m \qquad (9.6)$$

where
$c_A v_A = N_A$ = total flux of component A due to diffusion and convection
$c_A v_{Ad} = J_A$ = flux of component A due to molecular diffusion.
$c_A v_m$ = convective flux of A relative to the reference stationary point

Then Equation 9.6 can be written as

$$N_A = J_A + c_A v_m \tag{9.7}$$

Total convective flux $N = cv_m = N_A + N_B$ (9.7a)

Solving for v_m, $v_m = \dfrac{N_A + N_B}{c}$ (9.7b)

Replacing J_A with Fick's law relation and v_m by Equation 9.7b, a relation for N_A is obtained.

$$N_A = -cD_{AB} \frac{dx_A}{dz} + \frac{c_A}{c}(N_A + N_B) \tag{9.8}$$

In a similar way, $N_B = -cD_{BA} \dfrac{dx_B}{dz} + \dfrac{c_B}{c}(N_A + N_B)$ (9.9)

In order to solve Equations 9.8 and 9.9 the relation between fluxes N_A and N_B is required.

For equimolar counterdiffusion of A and B, $N_A = -N_B$ and the convection term in 9.8 or 9.9 becomes zero. Equations 9.8 and 9.9 then reduce to simple Fick's law relation for molecular diffusion in stagnant fluid.

Steady-State Diffusion of A into Stagnant Non-Diffusing B (Gas)

The case of component A diffusing through a stagnant layer of B at steady state happens very often. In this case, there is a boundary at the end of the diffusion path that is impermeable to component B. For example, the boundary at liquid surface in the tube (Figure 9.2) is impermeable to diffusion of air into acetone as air is very insoluble in acetone. Only acetone vapor can diffuse through the stagnant layer of air in the tube. This vapor is carried away by the stream of air flowing by the open mouth of the tube.

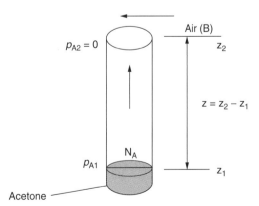

Figure 9.2 Acetone diffusing through stagnant air

The flux of diffusing component A through stagnant B can be derived from Equation 9.8 by putting $N_B = 0$ as B cannot diffuse. The flux is then given by

$$N_{Az} = \frac{D_{AB}P}{RT(z_2 - z_1)} \frac{1}{(p_B)_{lm}} (p_{A1} - p_{A2}) = \frac{cD_{AB}}{(z_2 - z_1)} \ln \frac{x_{B2}}{x_{B1}} \quad (9.10)$$

where

$$p_{lm} = \frac{p_{B2} - p_{B1}}{\ln(p_{B2}/p_{B1})}$$

with p_{B1} and p_{B2} partial pressures of non-diffusing B.

In deriving Equation 9.10, it is implicitly assumed that the level of the diffusing component A in the tube is maintained constant. In case this is not done by continuously making up for the loss of diffusing A, the level will drop. However, the variation of the path length z with time can be taken into consideration by assuming a pseudo-steady-state condition as the level drops very slowly. The resulting expression (Geankoplis) is as follows:

$$t_F = \frac{\rho_A (z_F^2 - z_0^2) RT p_{Bm}}{2 M_A D_{AB} P (p_{A1} - p_{A2})} \quad (9.11)$$

Example 9.2

NH$_3$ gas (A) and N$_2$ gas are counter-diffusing through a glass tube connected to two large chambers at the two ends of the tube at 298 K and 101.3 kPa. The length of the tube is two feet and its inside diameter is 0.0834 ft. The partial pressure of ammonia is constant at 30 kPa in one chamber and 6.7 kPa in the other. The diffusivity of NH$_3$ at 298 K and 101.3 kPa is 2.3×10^{-5} m^2/s. (a) Calculate the diffusion flux of NH$_3$ in lb mol/h at steady state. (b) Calculate the partial pressure of NH$_3$ at mid point of the tube.

Solution

Data: Temperature = 298 K = 536.4°R
30 kPa = (30/101.3)(14.7) = 4.353 psia
6.7 kPa = (6.7/101.3)(14.7) = 0.972 psia
T = 298(1.8) = 536.4°R

$$R = 10.73 \frac{\text{psia} \cdot \text{ft}^3}{\text{lb mol} \cdot {}^\circ R}$$

Cross sectional area of pipe $= \frac{\pi D^2}{4} = \frac{\pi (0.0834)^2}{4} = 0.005463 \text{ ft}^2$

Diffusivity of NH$_3$ = $2.3 \times 10^{-5} \frac{\text{m}^2}{\text{s}} \times 3.8736 \times 10^4 = 0.891 \frac{\text{ft}^2}{\text{h}}$

(a) $$J_A = \frac{0.891 \times (4.353 - 0.972)}{10.73 \times 536.4 \times 2} = 2.617 \times 10^{-4} \frac{\text{lb mol}}{\text{ft}^2 \cdot \text{h}}$$

Flux of NH$_3$ = $0.005463 \times (2.617 \times 10^{-4}) = 1.43 \times 10^{-6} \frac{\text{lb mol}}{\text{h}}$

(b) At steady state the flux is the same passing past any point along the length of the tube.

Therefore $\quad 2.617\times 10^{-4} = \dfrac{0.891\times (4.353 - p_{Am})}{10.73\times 536.4 \times 1}$

Then $\quad 4.353 - p_{Am} = \dfrac{2.617\times 10^{-4} \times 10.73 \times 536.4 \times 1}{0.891} = 1.6905$

Therefore, partial pressure of NH_3 at mid point of the tube is

$$p_{Am} = 4.353 - 1.6905 = 2.6625 \text{ psia} = 18.35 \text{ kPa}$$

Example 9.3

Ammonia gas is diffusing through air under steady-state conditions in nondiffusing air. Total pressure is 101.3 kPa and the temperature of the ammonia liquid is maintained at 273 K. The partial pressure of ammonia at one point is 0.1333×10^5 Pa and at another point 2 cm away it is 0.067×10^5 Pa. Diffusivity D_{AB} for the mixture is 0.198×10^{-4} m²/s at 101.3 kPa and 273 K. What is the flux of ammonia in kg mol/h.m²?

Solution

This is a direct application of Equation 9.10

$$p_{B1} = 1.013 \times 10^5 - 0.1333 \times 10^5 = 0.8797 \times 10^5 \text{ Pa}$$

$$p_{B2} = 1.013 \times 10^5 - 0.067 \times 10^5 = 0.946 \times 10^5 \text{ Pa}$$

$$p_{Blm} = \left(\dfrac{(0.946 - 0.8797)10^5}{\ln\dfrac{0.9460}{0.8797}} \right) = 0.9124 \times 10^5 \text{ Pa}$$

Flux by Equation 9.10

$$N_{Az} = \dfrac{D_{AB} P}{RT(z_2 - z_1)} \dfrac{1}{(p_B)_{lm}} (p_{A1} - p_{A2})$$

Substitute the values in the preceding equation and calculate the flux.

$$N_A = \dfrac{(0.198\times 10^{-4})\times (1.013\times 10^5)(0.1333 - 0.067)10^5}{8314.3 \times 273 \times 0.02 \times 0.9124 \times 10^5} = 3.21\times 10^{-6} \dfrac{\text{kg mol}}{\text{s}\cdot\text{m}^2}$$

MOLECULAR DIFFUSION IN LIQUIDS

Diffusion in liquids is important in separation processes such as distillation, absorption, and liquid-liquid extraction. It also plays an important role in oxygenation of river waters. The rate of molecular diffusion in liquids is much slower than in the gaseous phase. This is because the molecules in a liquid are closer compared to gas. Therefore more collisions of the diffusing molecule occur in liquid phase, thereby decreasing the diffusion rate. However, the concentration differences are higher in liquids, which results in greater driving force.

Equations for Diffusion in Liquids

Equimolar Counterdiffusion

Because in liquids the molecules are closer, both attractive forces between molecules and greater resistance to diffusion are important. Starting with Fisk's law, one obtains the following equation for equimolar counterdiffusion ($N_A = -N_B$), similar to Equation 9.3 for gases:

$$N_{Az} = \frac{D_{AB}(c_{A1} - c_{A2})}{z_2 - z_1} = \frac{D_{AB}}{(z_2 - z_1)}\left(\frac{\rho}{M}\right)_{av}(x_{A1} - x_{A2}) \qquad (9.12)$$

and

$$c_{AV} = \left(\frac{\rho_m}{M}\right)_{AV} = \frac{\left(\frac{\rho_1}{M_1} + \frac{\rho_2}{M_2}\right)}{2}$$

where
- c_{AV} = average total concentration of $A + B$, kg mol/m^3
- M_1, M_2 = average molecular weights at points 1 and 2 kg mass/kg mol
- q_1, q_2 = densities at points 1 and 2, q = average density, all in kg/m^3
- D_{AB} = average diffusivity, m^2/s.

Steady State Diffusion of A into Stagnant Liquid Film of B

This is a frequently occurring case of diffusion in liquids. $N_B = 0$. Equation 9.10 is rewritten in terms of concentrations with $c_{AV} = P/RT$, $c_{A1} = p_{A1}/RT$, and $x_{BM} = p_{BM}/P$. The equation for liquids is obtained.

$$N_{Az} = \frac{D_{AB}c_{AV}}{(z_2 - z_1)}\frac{1}{(x_B)_{lm}}(x_{A1} - x_{A2}) \quad \text{where} \quad x_{BM} = \frac{x_{B2} - x_{B1}}{\ln\frac{x_{B2}}{x_{B1}}} \qquad (9.13)$$

For dilute solutions, $x_{BM} \approx 1.0$ and c is essentially constant. Then Equation 9.13 reduces to

$$N_A = \frac{D_{AB}(c_{A1} - c_{A2})}{z_2 - z_1} \qquad (9.13a)$$

Example 9.4

Calculate the rate of diffusion of NaCl across a film of water (non-diffusing) solution 1.5 mm thick at 18°C when the concentrations on opposite sides of the film are 24 and 4 wt % NaCl, respectively. The diffusivity of NaCl is 1.3×10^{-5} cm^2/s at 18°C. Densities of 24 and 4 wt % NaCl at 18°C are 1181 and 1027 kg/m^3.

Solution

$$z = 0.0015 \text{ m}, \quad M_A = 58.5 \quad M_B = 18.02,$$
$$D_{AB} = 1.3 \times 10^{-5} \text{ cm}^2/\text{s} = 1.3 \times 10^{-9} \text{ m}^2/\text{s}$$

Density of 24 % solution is 1081 kg/m^3

Therefore $$x_{A1} = \frac{0.24/58.5}{0.24/58.5 + 0.76/18.02} = \frac{0.0041}{0.0463} = 0.0886$$

$$x_{B1} = 1 - x_{A1} = 1.0 - 0.0886 = 0.9114$$

$$M = \frac{1}{0.0463} = 21.6 \text{ kg/kg mol}$$

$$\frac{\rho}{M} = \frac{1081}{21.6} = 50.05 \text{ kg mol/m}^3$$

Similarly, for 4% solution

$$x_{A2} = \frac{0.04/58.5}{0.04/58.5 + 0.96/18.02} = \frac{0.000684}{0.054} = 0.0127$$

$$x_{B2} = 1 - 0.0127 = 0.9873$$

$$M = \frac{1}{0.054} = 18.52 \text{ kg/kg mol}$$

$$\frac{\rho}{M} = \frac{1027}{18.52} = 55.45 \text{ kg mol/m}^3$$

$$\left(\frac{\rho}{M}\right)_{AV} = \frac{50.05 + 55.45}{2} = 52.75 \frac{\text{kg mol}}{\text{m}^3}$$

$$x_{BM} = \frac{0.9873 - 0.9114}{\ln\frac{0.9873}{0.9114}} = 0.9488$$

$$N_A = \frac{D_{AB}}{Z x_{BM}} \left(\frac{\rho}{M}\right)_{AV} (x_{A1} - x_{A2})$$

$$= \frac{1.3 \times 10^{-9}}{0.0015 \times 0.9488} \times (52.75)(0.0886 - 0.0127)$$

$$= 3.657 \times 10^{-6} \frac{\text{kg mol}}{\text{m}^2 \cdot \text{s}}$$

Determination of Diffusion Coefficients (gases)

Experimental determination of diffusion coefficients for binary gas mixtures is made by several methods, such as,

1. evaporation of liquid in a narrow tube (Stefan's method, Equation 9.11)

2. two bulb method

3. rate of evaporation in case of solids such as naphthalene.

Experimental diffusivity data are tabulated in Perry's handbook (Perry and Green).
If experimental data are not available, estimation of diffusivities can be made using theoretically derived equations. One such relation for a pair of nonpolar

molecules is based on the Lennard-Jones function and is as follows:

$$D_{AB} = \frac{1.8583 \times 10^{-7} T^{3/2}}{P \sigma_{AB}^2 \Omega_{D,AB}} \left(\frac{1}{M_A} + \frac{1}{M_B} \right)^{1/2} \quad (9.14)$$

where
D_{AB} = diffusivity m²/s
T = temperature K
M_A, M_B = molecular weights of the binary components kg/kg mol
P = absolute pressure in atm
σ = average collision diameter Å
$\Omega_{D,AB}$ = collision integral based on Lennard-Jones potential

where
T = temperature, K
M_A and M_B = molecular weights
P = Pressure, atm
$\Omega_D = f(kT/\varepsilon_{AB})$, collision integral
k = Boltzmann's constant
$\varepsilon_{AB}, \sigma_{AB}$ Lennard-Jones force constants for the binary

The values of ε_{AB} and σ_{AB} are obtained from the values for the pure components by the following relations:

$$\frac{\varepsilon_{AB}}{k} = \left(\frac{\varepsilon_A}{k} \frac{\varepsilon_B}{k} \right)^{1/2} \quad \text{and} \quad \sigma_{AB} = \frac{1}{2}(\sigma_A + \sigma_B)$$

Values of ε/k and σ are available in the literature. If not, they may be estimated by the following relations:

$$\frac{\varepsilon}{k} = 0.75 T_c \quad \text{and} \quad \sigma = \frac{5}{6} V_c^{1/3}$$

where
T_c = critical temperature, K
V_c = critical volume, cm³/g mol
σ = Lennard-Jones potential parameter, A°

A semiempirical but more convenient method using atomic diffusion volumes that is preferable is given below.

$$D_{AB} = \frac{1.00 \times 10^{-7} T^{1.75} (1/M_A + 1/M_B)^{1/2}}{P \left[(\Sigma v_A)^{1/3} + (\Sigma v_B)^{1/3} \right]^2} \, m^2/s \quad (9.15)$$

where
$\Sigma v_A, \Sigma v_B$ = sum of the structural volume increments
v = the atomic diffusion volume of a simple molecule m³/1000 atoms × 10³
P = pressure in atm.

Σv is obtained by summation of the atomic diffusion volumes.

Diffusivity in a Multi-Component Mixture

The diffusivity of a component in a mixture is expressed by the Stefan-Maxwell equation. A simplified version of the equation is

$$D_{1-mix} = \frac{1}{\sum_{i=2}^{n} y'_i / D_{1-i}} \quad \text{where} \quad y'_i = y_i / \sum_{i=2}^{n} y_j \tag{9.16}$$

y'_i = mol fraction of component i, calculated by excluding component i.

Estimation of Diffusivities for Liquids

For liquids, the Wilke and Chang relation is given by

$$D_{AB}^O = 1.173 \times 10^{-16} \left[(\phi M_B)^{1/2} \frac{T}{\mu_B V_A^{0.6}} \right] \tag{9.17}$$

where
- D_{AB}^O = diffusivity of component A in a mixture of A and B m²/s
- ϕ = association parameter of solvent B
- ϕ = 2.26 for water as solvent, 1.9 for methanol, 1.5 for ethanol, and 1 for unassociated solvents such as benzene and ethyl ether
- μ_B = viscosity of B, Pa · s (centipoises) or kg/m · s
- V_A = molar volume of solute A at its normal boiling point, m³/kg mol
- M_B = molecular weight of component B, kg/kg mol

When water is the solute, values obtained from Equation 9.16 should be multiplied by a factor of (1/2.3).

When values of V_A are greater than 0.5 m³/kg, the following equation should be used:

$$D_{AB} = \frac{9.96 \times 10^{-16} T}{\mu V_A^{1/3}} \tag{9.18}$$

where
- D_{AB} = diffusivity m²/s
- μ = viscosity of solution in Pa · s or kg/m · s
- T = temperature in K

Note: To convert values in m²/s into ft²/h, use the conversion factor 3.875×10^4.

Example 9.5

(a) Estimate the diffusivity of n-butane in nitrogen at 101.32 kPa pressure and 298 K temperature by Fuller et al. method. Compare with the reported experimental value of 0.096×10^{-4} m²/s.

(b) What will be the diffusivity at 50°C and 2 atm?

Solution

Data: Molecular weight of n-butane = 58.12
Molecular weight of nitrogen = 28.02

From table of atomic diffusion volumes (Perry's handbook)

Σv_B = Diffusion volume for nitrogen = 17.9
Σv_A = Diffusion volume for n-butane (C_4H_{10}) = 4(16.5) + 10(1.98) = 85.8
$\underline{T}$ = 298 K   P = 101.32 kPa = 1 atm

(a) Substituting in Equation 9.15

$$D_{AB} = \frac{1.0 \times 10^{-7}(298)^{1.75}(1/58.12 + 1/28.02)^{0.5}}{1.0\left[(85.8)^{1/3} + (17.9)^{1/3}\right]^2} = 0.0996 \times 10^{-4} \, m^2/s.$$

This value is greater than experimental value of 0.096×10^{-4} by

$$\frac{(0.0996 - 0.096) \times 10^{-4}}{0.096 \times 10^{-4}} \times 100 = 3.75 \text{ percent}$$

(b) From Equation 9.15, $D_{AB} \propto \dfrac{T^{1.75}}{P}$

$T = 273.15 + 50 = 323.15$ K $P = 2$ atm

$$(D_{AB})_{T_2} = \frac{T_2^{1.75}/P_2}{T_1^{1.75}/P_1}(D_{AB})_{T_1} = \left(\frac{323.15}{298}\right)^{1.75} \times \frac{1}{2} \times 0.0996 \times 10^{-4}$$

$$= 0.05739 \times 10^{-4} \, m^2/s$$

Example 9.6

Estimate diffusion coefficient for acetic acid in water at 25 °C by Wilke-Chang correlation. The reported experimental value is 1.26×10^{-9} m²/s.

Solution

Data: T = 273 + 25 = 298 K

ϕ = 2.6 Association parameter for water

$*V_A = 2(0.0148) + 4(0.0037) + 1(0.012) + 1(0.0074) = 0.0638 \dfrac{m^3}{kg\,mol}$

$\mu = 0.8937 \times 10^{-3}$ Pa·s

$$D_{AB} = \frac{1.173 \times 10^{-16}(2.6 \times 18.02)^{1/2}(298)}{(0.8937 \times 10^{-3})(0.0638)^{0.6}} = 1.395 \times 10^{-9} \, m^2/s$$

Calculated value is greater by $\dfrac{10^{-9}(1.395 - 1.26)}{1.26 \times 10^{-9}} \times 100 = 10.71$ percent.

---

* 2 carbon, 4 hydrogen, 1 oxygen double-bonded as carbonyl, and 1 oxygen in acid as OH

MOLECULAR DIFFUSION IN SOLIDS

Although rates of diffusion of gases and liquids in solids are lower, diffusion mass transfer in solids is very important. Catalytic reactions, leaching of solids, separation of fluids using membranes, diffusion of gases through polymer films, drying, adsorption, and reverse osmosis are some examples of operations where diffusion plays an important role. Two types of diffusion occur in solids. In one type, diffusion follows Fick's law and does not depend very much on the structure of the solid. However, diffusion in porous solids does depend a great deal on the actual structure and void channels.

Fick's law can be applied for diffusion in solids under the following conditions: (1) concentration gradient is independent of time, (2) diffusivity is constant and independent of concentration, and (3) there is no bulk flow. Then, the flux is given by

$$N_A = -D_A \frac{dc}{dz} \tag{9.19}$$

where D_A is the diffusivity of A through the solid. For constant D_A, the following relations are obtained:

Diffusion through a solid slab: $N_A = \dfrac{D_A(c_{A1} - c_{A2})}{z_2 - z_1}$ (9.20)

Other solid shapes: $w = N_A S_{av} = D_A S_{av}(c_{A1} - c_{A2})/z$ (9.21)

where
S_{av} = average cross section for diffusion.

Radial diffusion through a solid cylinder of inner and outer radii r_1 and r_2 and of length L

$$S_{av} = \frac{2\pi L(r_2 - r_1)}{\ln(r_2/r_1)} \quad \text{and} \quad z = r_2 - r_1 \tag{9.22}$$

$$N_A = D_{AB}(c_{A1} - c_{A2}) \frac{2\pi L}{\ln(r_2/r_1)} \tag{9.23}$$

Radial diffusion through a spherical shell: with inner and outer radii r_1 and r_2

$$S_{av} = 4\pi r_1 r_2 \quad \text{and} \quad z = r_2 - r_1 \tag{9.24}$$

Example 9.7

Hydrogen gas at 15.0 bar A and 30°C is transported through a steel pipe with ID and OD of 58.5 and 89 mm, respectively, to a reaction system. Henry's law constant for the solubility of hydrogen in steel is reported to be 1.67 bar · m³/kg mol. The diffusion coefficient of hydrogen in steel is 0.3×10^{-12} m²/s.

Calculate the mass flux of loss hydrogen by diffusion per hour per 100 meter length of pipe in units of kg/h.

Assume Henry's law applies at 15 barA pressure. Henry's law: $p = Hc$.

Solution

The concentration at the inner surface of pipe $c_{A1} = p/H = 15/1.67 = 8.982$ kg mol/m³.
The concentration at the outer surface of the pipe can be taken as zero.

Since hydrogen is diffusing through radially, the effective average area for diffusion must be estimated. For a cylinder, the average diffusional area is given by

$$A_{av} = \frac{2\pi L(r_2 - r_1)}{\ln(r_2/r_1)}$$

where r_2 and r_1 are outer and inner radii respectively.

$$r_2 = 80 \times 10^{-3} \text{ m} \quad \text{and} \quad r_1 = 58.5 \times 10^{-3} \text{ m}$$

Therefore $A_{av} = \dfrac{2\pi(100)(89 \times 10^{-3} - 58 \times 10^{-3})/2}{\ln(89 \times 10^{-3}/58.5 \times 10^{-3})} = 22.835 \text{ m}^2/100 \text{ m length}$

$$z = (1/2)(89 - 58.5) = 15.25 \text{ mm} = 15.25 \times 10^{-3} \text{ m}$$

Molar flux (assuming diffusivity to be constant) is given by

$$N_A = \frac{0.3 \times 10^{-12} \times 22.835(8.982 - 0)}{15.25 \times 10^{-3}} = 4.035 \times 10^{-9} \text{ kg mol/s}$$

Mass flux = $4.035 \times 2 \times 10^{-9} = 8.07 \times 10^{-9}$ kg/s H_2 per 100 m length.

Therefore, mass flux per hour = $8.07 \times 10^{-9} \times 3600$
$= 2.905 \times 10^{-5}$ kg/h per 100 m length.

Solubility of a Solute Gas in a Solid

Solubility of a solute gas (A) in a solid is usually expressed as S in m³ of solute (at STP of 0°C and 1 atm) per m³ solid per atm partial pressure of solute (A) or

$$S = \frac{\text{m}^3(\text{STP})}{\text{atm} \cdot \text{m}^3 \text{solid}} \quad \text{(SI units)}$$

$$S = \frac{\text{cm}^3(\text{STP})}{\text{atm} \cdot \text{cm}^3 \text{solid}} \quad \text{(cgs units)}$$

These can be converted to kg mol/m³solid by using the fact that one kg mol occupies 22.414 m³ at STP as follows:

$$c_A = \frac{S \text{ m}^3(\text{STP})/\text{m}^3 \text{solid} \cdot \text{atm}}{22.414 \text{ m}^3(\text{STP})/\text{kg mol A}} p_A \text{ atm} = \frac{Sp_A}{22.414} \frac{\text{kg mol A}}{\text{m}^3 \text{solid}} = \text{(SI units)}. \quad (9.25)$$

Also $\quad c_A = \dfrac{Sp_A}{22414} \dfrac{\text{g mol A}}{\text{cm}^3 \text{solid}}$ (cgs units) (9.25a)

Diffusion Through Polymers

In many cases data for gas diffusion in solids are reported in terms of permeability P_m.

$$N_A = \frac{D_{AB} S_A (p_{A1} - p_{A2})}{22.414(z_2 - z_1)} \quad (9.26)$$

The product $D_{AB} S_A$ is called permeability P_m. Its units are

$$P_m = D_A S_A \frac{\text{m}^3(\text{STP})}{\text{s} \cdot \text{m}^2 \text{C.S} \cdot \text{atm/m}} \quad (9.27)$$

Equation 9-26 then becomes

$$N_A = \frac{P_m(p_{A1} - p_{A2})}{22.414(z_2 - z_1)} = \frac{P_m(p_{A1} - p_{A2})}{22.414 z} \text{ kg mol/s} \cdot \text{m}^2 \quad (9.28)$$

where
- N_A = diffusional flux of A kg mol/s · m²
- D_A = diffusivity of A, m²/s
- S_A = solubility coefficient m³ gas (STP)/(m³ solid) · atm
- z = thickness of polymeric membrane, m
- $P = D_A S_A$ where P = permeability, m³ gas (STP)/(m² · s)(atm/m)

For several solids in series having thickness $L_1, L_2, L_3 \cdots$ Equation 9.27 becomes

$$N_A = \frac{p_{A1} - p_{A2}}{22.414} \frac{1}{L_1/P_{M1} + L_2/P_{M2} + L_3/P_{M3} + \cdots} \quad (9.29)$$

Example 9.8

A nylon film 0.15 mm thick is used for packaging a product. Its permeability for oxygen is 0.029×10^{-12} m³(O₂) (STP)/s · m² · atm/m. If the partial pressure of O₂ is 0.21 atm on the outside of the package and 0.01 atm inside the package, calculate diffusion flux of O₂ at steady state.

Solution

Thickness of film = 0.00015 m = 0.15 mm

$$N_A = \frac{P_m(p_{A1} - p_{A2})}{22.414 z} = \frac{0.029 \times 10^{-12}(0.21 - 0.01)}{22.414 \times 0.00015} = 1.725 \times 10^{-12} \text{ kg mol/s} \cdot \text{m}^2$$

Diffusion of Liquids in Porous Solids

Porous solids have pores or interconnected voids in the solid, which affect the diffusion. If the voids are completely filled with the solvent, the solute diffuses through the solvent taking a tortuous path. This tortuous path length is greater than $z_2 - z_1$ by a factor called tortuosity τ. Diffusion does not occur in the inert solid. Instead the solute traverses through the solvent in the voids of the solid. In such a case the following equation can be written for the diffusion of the solute at steady state:

$$N_A = \frac{\varepsilon D_{AB}(c_{A1} - c_{A2})}{\tau(z_2 - z_1)} \quad (9.30)$$

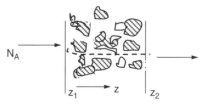

Figure 9.3 Diffusion in porous solid

where

ε = void fraction
τ = a correction factor that accounts for the effect of the longer path of diffusion in the porous solid.

The three quantities ε, τ, and D_{AB} are combined together and the combined quantity is called effective diffusivity. Thus

$$\text{Effective diffusivity} = \frac{\varepsilon D_{AB}}{\tau} \quad (9.30a)$$

The tortuosity factor varies from 1.5 to 5 for inert solids.

Diffusion of Gases in Porous Solids

If the void space in a porous solid is filled with a gas, and if the pores are large enough, the Fick's law diffusion occurs and the flux is then given by

$$N_A = \frac{\varepsilon D_{AB}}{\tau} \frac{(c_{A1} - c_{A2})}{z_2 - z_1} = \frac{\varepsilon D_{AB}}{\tau RT} \frac{p_{A1} - p_{A2}}{z_2 - z_1} \quad (9.31)$$

The approximate values of τ at different values of ε for diffusion of gases are

ε	0.2	0.4	0.6
τ	2.0	1.75	1.65

When the pores are quite small, and of the order of mean free path of gas, other types of diffusion such as Knudsen diffusion occur.

Example 9.9

A sintered silica solid plate 2 mm thick is porous. Its void fraction is 0.3 and tortuosity is 4.0. The pores are filled with water at 298 K. At one face of the plate the concentration of acetic acid is 0.12 g mol/liter and fresh water flows rapidly past the other face. Considering only resistance to diffusion of acetic acid is in the porous solid, calculate the diffusion of acetic acid at steady state. The diffusivity of acetic acid is 1.26×10^{-9} m²/s. What is the effective diffusivity?

Solution

$C_{A1} = 0.12/1000 = 1.2 \times 10^{-4}$ g mol/cm³ $= 0.12$ kg mol/m³
$C_{A2} = 0$ (Water is running fast by the side of the other face.)
Thickness of plate = 2 mm = 0.002 m

Therefore

$$N_A = \frac{\varepsilon \times D_{AB}(c_{A1} - c_{A2})}{\tau(z_2 - z_1)} = \frac{0.3 \times 1.26 \times 10^{-9}(0.12 - 0)}{4(0.002 - 0)}$$

$$= 6.0 \times 10^{-9} \text{ kg mol CH}_3\text{COOH/s} \cdot \text{m}^2$$

$$\text{Effective diffusivity} = \frac{\varepsilon D_{AB}}{\tau} = \frac{0.3 \times 1.26 \times 10^{-9}}{4} = 9.45 \times 10^{-11} \frac{\text{m}^2}{\text{s}}$$

CONVECTIVE MASS TRANSFER

In the preceding sections we reviewed molecular diffusion in stagnant fluids or fluids in laminar flow where the rates of diffusion are slow. In order to get rapid mass transfer, fluid velocity is increased and turbulent flow is resorted to. When a solute is dissolving from a surface into a flowing turbulent fluid, its concentration is very high at the surface and decreases as the distance from the surface increases (Figure 9.4). Three zones of mass transfer can be visualized. Adjacent to the surface there is a thin viscous film. In this film most of the mass transfer occurs by molecular diffusion. A larger portion of the concentration drop occurs in this zone. A transition or buffer zone is next to the film region and in this zone mass transfer takes place by both molecular and turbulent diffusion. In the turbulent region next to the transition zone the mass transfer is mainly due to turbulent diffusion. In the turbulent zone, concentration drop is lowest as the eddies tend to keep the concentration uniform. The average concentration $\bar{c}_A$ is slightly greater than the minimum but far below that near the surface.

Turbulent Diffusion Equation for Mass Transfer (Convective Mass Transfer)

On the basis of similarity between momentum, heat, and mass transfer, the following equation for turbulent mass transfer can be written:

$$J_{Az} = -(D_{AB} + \varepsilon_M)\frac{dc_A}{dz} \tag{9.32}$$

where ε_M = turbulent or masseddy diffusivity, m²/s.

ε_M is near zero at the surface and increases with increasing distance from the wall. Since variation of ε_M is not generally known, an average $\bar{\varepsilon}_M$ is used. With this assumption, Equation 9.32 is integrated between points 1 and 2 to give

$$J_A = \frac{D_{AB} + \bar{\varepsilon}_M}{z_2 - z_1}(c_{A1} - c_{A2}) \tag{9.33}$$

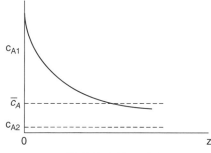

Figure 9.4 Concentration profile in turbulent mass transfer from a surface to a turbulent fluid

Since $z_2 - z_1$ is often not known, a convective mass transfer coefficient is defined by the relation

$$\text{Mass flux} = (\text{mass transfer coefficient})(\text{driving force})$$

or
$$N_A = k'_c(c_{A1} - c_{A2}) \tag{9.34}$$

where k'_c = convective mass transfer coefficient $= \dfrac{D_{AB} + \bar{\varepsilon}_M}{z_2 - z_1} \quad \dfrac{\text{kg mol}}{\text{s} \cdot \text{m}^2 \cdot (\text{kg mol/m}^3)}$ or m/s.

It is to be obtained experimentally. c_{A2} is the concentration at point 2 in kg mol/m³, which is usually the average bulk concentration $\bar{c}_{A2}$.

Mass-Transfer Coefficient for Equimolar Counterdiffusion

Starting with the equation,

$$N_A = -c(D_{AB} + \varepsilon_M)\frac{dx_A}{dz} + x_A(N_A + N_B) \tag{9.35}$$

Putting $N_A = -N_B$ for the case of equimolar counterdiffusion, integrating at steady state, and substituting k'_c for $\frac{D_{AB} + \bar{\varepsilon}_M}{z_2 - z_1}$ the following relation is obtained.

$$N_A = k'_c(c_{A1} - c_{A2}) \tag{9.36}$$

Equation 9.36 defines the mass transfer coefficient. Usually the concentrations are expressed in terms of mol fractions in the case of liquids and in terms of partial pressures in the case of gases. Hence the mass transfer coefficient for the case of equimolar counter diffusion can be defined in many ways as follows:

For gases: $\quad N_A = k'_c(c_{A1} - c_{A2}) = k'_G(p_{A1} - p_{A2}) = k'_y(y_{A1} - y_{A2}) \tag{9.37}$

For liquids: $\quad N_A = k'_c(c_{A1} - c_{A2}) = k'_L(c_{A1} - c_{A2}) = k'_x(x_{A1} - x_{A2}) \tag{9.38}$

All these mass transfer coefficients are related to each other. For example, from Equation 9.37

$$N_A = k'_c(c_{A1} - c_{A2}) = k'_y(y_{A1} - y_{A2}) = k'_y\left(\frac{c_{A1}}{c} - \frac{c_{A2}}{c}\right) = \frac{k'_y}{c}(c_{A1} - c_{A2}) \tag{9.39}$$

Therefore
$$k'_c = \frac{k'_y}{c} \tag{9.40}$$

The relations of mass transfer coefficients are given in Table 9.1.

Mass-Transfer Coefficients for A Diffusing Through Stagnant Nondiffusing B

In this case $N_B = 0$ and Equation 9.35 gives at steady state

$$N_A = \frac{k'_c}{x_{BM}}(c_{A1} - c_{A2}) = k_c(c_{A1} - c_{A2}) \quad (9.41)$$

$$= \frac{k'_x}{x_{BM}}(c_{A1} - c_{A2}) = k_x(x_{A1} - x_{A2}) \quad (9.41a)$$

$$x_{BM} = \frac{x_{B2} - x_{B1}}{\ln(x_{B2} - x_{B1})} \quad \text{and} \quad y_{BM} = \frac{y_{B2} - y_{B1}}{\ln(y_{B2} - y_{B1})} \quad (9.42)$$

We can rewrite the preceding equations using other units as follows:

For gases: $N_A = k_c(c_{A1} - c_{A2}) = k_G(p_{A1} - p_{A2}) = k_y(y_{A1} - y_{A2}) \quad (9.43)$

For liquids: $N_A = k_c(c_{A1} - c_{A2}) = k_x(x_{A1} - x_{A2}) = k_L(c_{A1} - c_{A2}) \quad (9.44)$

Values of mass transfer coefficients will depend on the units chosen for the driving force, viz. the concentration differences, partial pressures, or mole fractions. Units of mass transfer coefficients in different units are given in Table 9.2.

For gases,

$$k_g = (k_y/P_t) = (k_c/RT) \quad \text{because} \quad p_A = y_A P_t \quad \text{and} \quad c_A = (p_A/RT) \quad (9.45)$$

Table 9.1 Conversions between mass transfer coefficients

Gases:

$$k'_c c = k_c \frac{P}{RT} = k_c \frac{p_{BM}}{RT} = k'_G P = k_G p_{BM} = k_y y_{BM} = k'_y = k_c y_{BM} c = k_G y_{BM} P$$

Liquids:

$$k'_c c = k'_L c = k_L x_{BM} c = k'_L \rho/M = k'_x = k_x x_{BM}$$

(where ρ = density of liquid and M = molecular weight)

Table 9.2 Units of mass-transfer coefficients

	SI Units	cgs units	English Units
k_c, k_L, k'_c, k'_L	m/s	cm/s	ft/h
k_x, k_y, k'_x, k'_y	$\dfrac{\text{kg mol}}{\text{s} \cdot \text{m}^2 \cdot \text{mol frac}}$	$\dfrac{\text{g mol}}{\text{s} \cdot \text{cm}^2 \cdot \text{mol frac}}$	$\dfrac{\text{lb mol}}{\text{h} \cdot \text{ft}^2 \cdot \text{mol frac}}$
k_G, k'_G	$\dfrac{\text{kg mol}}{(\text{s} \cdot \text{m}^2 \cdot \text{Pa})}$ or $\dfrac{\text{kg mol}}{(\text{s} \cdot \text{m}^2 \cdot \text{atm})}$	$\dfrac{\text{g mol}}{\text{s} \cdot \text{cm}^2 \cdot \text{atm}}$	$\dfrac{\text{lb mol}}{\text{h} \cdot \text{ft}^2 \cdot \text{atm}}$

Example 9.10 (Use of Relations in Table 9.1)

A pure component A is diffusing in stagnant nondiffusing B. The value of k_G was determined experimentally and found to be 7.11×10^{-3} kg mol/s · m² · atm. The total pressure was 1 atm and temperature was 298 K. Partial pressures of A were $p_{A1} = 0.2$ atm and $p_{A2} = 0.03$ atm. k'_G and flux for A are to be estimated for the case of counterdiffusion of A and B under the same conditions of flow and concentrations.

(a) Calculate the flux of A for the case of A diffusing in nondiffusing stagnant B.

(b) Estimate k'_G and flux of A if A and B are counterdiffusing under the same conditions of flow and concentrations.

Solution

(a) The flux of A = $k'_G (p_{A1} - p_{A2})$

$$= 7.11 \times 10^{-3}(0.2 - 0.03) = 1.21 \times 10^{-3} \frac{\text{kg mol}}{\text{s} \cdot \text{m}^2}$$

(b) From Table 9.1a, $k'_G P = k_G p_{BM}$, therefore $k'_G = \frac{k_G p_{BM}}{P}$

$p_{B1} = 1 - 0.2 = 0.8$ atm and $p_{B2} = 1 - 0.03 = 0.97$ atm

$$p_{BM} = \frac{0.97 - 0.8}{\ln(0.97/0.8)} = 0.8823$$

$$k'_G = \frac{7.11 \times 10^{-3}(0.8823)}{1.0} = 6.273 \times 10^{-3} \frac{\text{kg mol}}{\text{s} \cdot \text{m}^2 \cdot \text{atm}}$$

Flux of A = $k'_G (p_{A1} - p_{A2}) = 6.273 \times 10^{-3}(0.2 - 0.03)$

$$= 1.066 \times 10^{-3} \frac{\text{kg mol}}{\text{s} \cdot \text{m}^2}$$

TURBULENT MASS TRANSFER

Most practical situations of mass transfer involve turbulent flow. In these situations, it is not possible to describe the flow conditions adequately in mathematical terms. Therefore, one relies mainly on experimental data. But these are limited in scope with respect to fluid conditions and the range of its properties. In order to extend the applicability of experimental data, many theories or models have been developed. Other methods involve the use of analogy between mass, heat, and momentum transfer to develop correlations for mass transfer in turbulent flow. Some of these models will be very briefly reviewed next.

Film Model

The film model assumes that the resistance to mass transfer resides entirely in a stagnant thin film and the concentration difference lies entirely in this film. It also assumes that the concentration difference is only due to molecular diffusion.

This model is based on the film model used for convective heat transfer. The mass flux through this film is given by

$$N_A = k'_c(c_{A1} - c_{A2}) = \frac{D_{AB}}{\delta_f}(c_{A1} - c_{A2}) \tag{9.46}$$

Higbie Penetration Model

The film model assumes no accumulation of diffusing species in the film and a steady state is assumed for transfer across the film. Higbie pointed out that in industrial gas-liquid contacting equipment contact times are very small, which prevents attainment of steady state. He developed his penetration model to take into account the transient nature of solute diffusion. Based on his derivation, the average flux and time average mass-transfer coefficient are given by

$$N_{A,av} = 2(c_{Ai} - c_o)\left(\frac{D_{AB}}{\pi t}\right)^{1/2} \quad \text{and} \quad k_L = 2(D_{AB'}/\pi t)^{1/2} \tag{9.47}$$

Also $\quad Sh = 1.13 \, Re^{0.5} Sc^{0.5} \quad k'_c = \sqrt{\frac{4 D_{AB}}{\pi t_L}} \tag{9.48}$

Surface Renewal Theory of Danckwerts

An important improvement in the Higbie penetration model was proposed by Danckwerts. Higbie assumed equal exposure time for repeated contacts of the fluid with the interface. Danckwerts used a wide range of contact times and averaged the various degrees of penetration. He derived the following expressions for mass flux and the mass transfer coefficient:

$$N_{Aav} = (c_{Ai} - c_{Ao})\sqrt{D_{AB} \cdot s} \quad \text{and} \quad k_L av = \sqrt{D_{AB} \cdot s} \tag{9.49}$$

Many more penetration theory models have been developed, which give the exponent on D_{AB} from 0.5 to 1.0. Dobbins' combination film-surface-renewal theory, Lightfoot's surface-stretch theory, and Toor-Marchello's modification of surface renewal theory are also available.

Turbulent Boundary Layer Theory

This theory explains the mass transfer between a fixed surface and a turbulent stream of fluid. There is no slip at the wall and the turbulence is suppressed in the fluid in contact with the wall. At the wall, the transfer is by molecular diffusion whereas eddy diffusion prevails in the turbulent stream far from the wall. In between these two positions both molecular and eddy diffusion contribute to mass transfer. As in heat and momentum transfer, turbulent mass transfer flux can be written as the sum of two fluxes, one due to turbulence and the other due to laminar boundary as follows:

$$(N_A)_{\text{total}} = -(D_{AB} + \varepsilon_D)\frac{dc_A}{dy} \tag{9.50}$$

where
ε_D = eddy mass diffusivity

With the use of the turbulent boundary layer theory, an expression for the mass transfer coefficient has been developed for turbulent mass transfer. The expressions derived are

Local convective mass transfer coefficient for a flat plate

$$\frac{k_c' x}{D_{AB}} = N_{Sh,x} = 0.332 N_{Re,x}^{1/2} N_{Sc}^{1/3} \quad (9.51)$$

Mean mass transfer coefficient for a flat plate,

$$\frac{k_c' L}{D_{AB}} = 0.664 Re^{0.5} Sc^{1/3} \quad (9.52)$$

Analogy correlations (heat transfer and momentum transfer) and *J* factor methods are also applied to determine mass transfer coefficients. Mass transfer coefficients for general case of *A* and *B* and convective flow using film theory: Two film theory assumes that mass transfer takes place through a thin film next to the wall or surface by molecular diffusion. The thickness of the film is determined using experimental value of k_c' for dilute solutions by (from Equation 9.35)

$$k_c' = \frac{D_{AB}}{\delta_f} \quad (9.53)$$

Then using Equation 9.35 we can write

$$N_A = -c \frac{D_{AB} dx_A}{dz} + x_A (N_A + N_B) \quad (9.53\text{a})$$

$x_A(N_A + N_B)$ is the convective term. Rearranging and integrating from $z = 0$ to $z = \delta_f$ = the following equation for N_A is obtained:

$$N_A = \frac{N_A}{N_A + N_B} (k_c' c) \ln \frac{N_A/(N_A + N_B) - x_{A2}}{N_A/(N_A + N_B) - x_{A1}} \quad (9.53\text{b})$$

DIMENSIONLESS NUMBERS IN MASS TRANSFER

A number of dimensionless groups are used in correlating mass transfer data. These are

$$Sh = \text{Sherwood number} = \left(\frac{kD}{D_{AB}}\right) \quad Re = \text{Reynolds' number} = \left(\frac{Du\rho}{\mu}\right)$$

$$Sc = \text{Schmidt number} = \left(\frac{\mu}{\rho D_{AB}}\right) \quad Gr = \text{Grashof number} = \left(\frac{gD^3 \rho \Delta\rho}{\mu^2}\right)$$

(9.54)

The product of *Re* and *Sc* is called the Peclet number *Pe*, and the Stanton number *St* is the ratio *Sh/Pe* (Sherwood number/Peclet number).

Mass Transfer for Flow Inside Pipes

Laminar flow: For laminar flow (Re < 2100) in pipes, a plot of

$$\frac{c_A - c_{Ao}}{c_{Ai} - c_{Ao}} \text{ vs } \frac{W}{D_{AB} \rho L}$$

is available (Geankoplis) where W = flow in kg/s and L = length of mass transfer section in m. c_{Ai} = concentration at the interface. For liquids that have small values of D_{AB}, data are correlated by the equation (for $\frac{W}{D_{AB}\rho L} > 400$)

$$\frac{c_A - c_{Ao}}{c_{Ai} - c_{Ao}} = 5.5 \left(\frac{W}{D_{AB} \cdot \rho \cdot L} \right)^{-2/3} \tag{9.55}$$

Mass Transfer for Turbulent Flow Inside Pipes

When $\quad \frac{Du\rho}{\mu} > 2100 \quad$ for gases or liquids

$$N_{sh} = k'_c \frac{D}{D_{AB}} = \frac{k_c p_{BM}}{P} \frac{D}{D_{AB}} = 0.023 \left(\frac{Du\rho}{\mu} \right)^{0.83} \left(\frac{\mu}{\rho D_{AB}} \right)^{1/3} \tag{9.56}$$

Depending on the nature of the mass transfer coefficient used, expressions for the Sherwood number will change. Thus the Sherwood number is

$$Sh = \frac{Fl}{cD_{AB}} = \frac{k_G p_B, \bar{M}^{RTl}}{D_{AB} P_t} = \frac{k_c p_B, \bar{M}^l}{p_t D_{AB}} \tag{9.57}$$

Mass Transfer from a Gas into a Falling Liquid Film

For $\quad Re < 100, k_{L,av} = 3.41 \frac{D_{AB}}{\delta} \quad$ or $\quad Sh_{av} = \frac{k_{l,av}\delta}{D_{AB}} = 3.41 \tag{9.58}$

where δ = film thickness

For $\quad Re > 100 \; k_{L,av} = \left(\frac{6D_{AB}\Gamma}{\pi\rho\delta L} \right)^{1/2} \quad$ and $\quad Sh_{av} = \left(\frac{3}{2\pi} \frac{\delta}{L} Re \, Sc \right)^{1/2} \tag{9.58a}$

where
δ = film thickness
Γ = mass rate of liquid flow per unit width of film in the *x* direction
L = length in the *z* direction.

For a falling film, the film thickness is given by

$$\delta = \left(\frac{3\mu\Gamma}{\rho^2 g}\right)^{1/3} = \left(\frac{3\bar{u}_y\mu}{\rho g}\right)^{1/2} \quad (9.59)$$

$k_{L,av}$ is given by: $\quad k_{L,av} = \dfrac{u_{\bar{y}}\delta}{L}\ln\dfrac{c_{Ai}-\bar{c}_A}{c_{A,i}-\bar{c}_{A,L}} \quad (9.60)$

Then flux, $N_{A,av} = k_{L,av}(c_{A,i} - \bar{c}_{A,L})_M$ where $(c_{A,i} - \bar{c}_{A,L})$ is the logarithmic average concentration difference over length L.

Mass Transfer for Flow Past Spheres

For very low Reynolds' numbers, Sherwood number

$$\frac{k_c'D_P}{D_{AB}} = N_{sh} = 2 \quad \text{and} \quad k_c' = \frac{2D_{AB}}{D_P} \quad (9.61)$$

where N_{sh} = Sherwood number for stationary fluid.

For other situations, the following equations are applicable:

Gases: Sc 0.6 to 2.7 and Re no. from 2 to 2000, the following equation is applicable:

$$Sh = 2 + 0.552\,Re^{0.53}\,Sc^{1/3} \quad \text{[Frossling Equation]} \quad (9.62)$$

Liquids: Equations for convective mass transfer are

For Reynolds' number, $2 < N_{Re} < 2000$, $N_{Sh} = 2 + 0.95 N_{Re}^{0.5} N_{Sc}^{1/3} \quad (9.62\text{a})$

For Reynolds' number, $2000 < N_{Re} < 17000$, $N_{sh} = 0.347 N_{Re}^{0.62} N_{Sc}^{1/3} \quad (9.62\text{b})$

Mass transfer in flow parallel to flat plates

Unconfined flow parallel to plate: $\dfrac{Lu\rho}{\mu} < 15000\ J_D = 0.664 N_{Re,L}^{-0.5} \quad (9.63)$

In terms of Sherwood number, $\dfrac{\bar{k}_c L}{D_{AB}} = N_{Sc} = 0.664 N_{Re,L}^{0.5} N_{Sc}^{1/3} \quad (9.64)$

where L is the length of the plate in the direction of flow.

Example 9.11

Water is flowing through a tube of 3 m length of which the last 2 m at one end are coated with benzoic acid. The inside diameter of the tube is 0.635 cm. Other data are as follows:

Velocity of water in tube = 3 m/s
Solubility of benzoic acid in water = 2.948×10^{-2} kg mol/m^3
Viscosity of water = 0.8718×10^{-3} Pa·s
$\rho = 1000$ kg/m^3
$D_{AB} = 1.245 \times 10^{-9}$ m^2/s

Assume flow is stabilized before water enters the zone of benzoic acid coating.

Calculate the outlet concentration of benzoic acid and the benzoic acid dissolved in water in kg mol/s.

Solution

$$D = 0.635 \text{ cm} = 0.00635 \text{ m}$$

$$\text{Area of mass transfer} = A = \pi DL = (\pi)(0.00635)(2) = 0.0399 \text{ m}^2$$

$$\text{Volumetric flow} = \frac{\pi D^2}{4} u = \frac{\pi (0.00635^2)(3)}{0.8718 \times 10^{-3}} = 9.5 \times 10^{-5} \frac{\text{m}^3}{\text{s}}$$

$$\text{Reynolds' number:} \frac{D u \rho}{\mu} = \frac{0.00635(3)(1000)}{0.8718 \times 10^{-3}}$$

$$= 21851 > 2000 \quad \therefore \text{Turbulent region}$$

$$\text{Schmidt number} = \frac{\mu}{\rho D_{AB}} = \frac{0.8718 \times 10^{-3}}{1000 \times 1.245 \times 10^{-9}} = 700.3$$

$$\text{Sherwood number} \frac{\bar{k}_c D}{D_{AB}} = 0.023 \, (21851)^{0.83} \, (700.3)^{0.33} = 814.7$$

$$\text{Then } k'_c = 814.7 \times \frac{1.245 \times 10^{-9}}{0.00635} = 1.6 \times 10^{-4} \frac{\text{m}}{\text{s}}$$

Because the solution is very dilute, $\bar{k}_c \approx k_c = 1.6 \times 10^{-4} \frac{\text{m}}{\text{s}}$

$$\text{Now } N_A A = A k_c \frac{(c_{Ai} - c_{A1}) - (c_{Ai} - c_{A2})}{\ln \frac{c_{Ai} - c_{A1}}{c_{Ai} - c_{A2}}} \quad \text{and} \quad N_A A = V \cdot c_{A2}$$

$$c_{Ai} = 2.948 \times 10^{-2} \text{ kg mol/m}^3, \quad c_{A1} = 0.0$$

Substituting the value $c_{A=0}$, equating the two relations, and simplifying gives

$$\ln \frac{c_{Ai}}{c_{Ai} - c_{A2}} = \frac{A k_c}{V} = \frac{0.0399(1.6 \times 10^{-4})}{9.5 \times 10^{-5}} = 0.0672$$

Solving for c_{A2}, gives $c_{A2} = 1.792 \times 10^{-3}$ kg mol/m^3

$$\text{Benzoic acid dissolved/s} = N_A A = V(c_{A2}) = 9.5 \times 10^{-5} \times 1.792 \times 10^{-3}$$

$$= 1.7 \times 10^{-7} \frac{\text{kg mol}}{\text{s}}$$

INTERPHASE MASS TRANSFER

Lewis and Whitman assumed that in mass transfer between two fluid phases, resistances to mass transfer occur only in the fluids and there is no resistance to solute transfer across the interface between the two phases. This is called the principle of two resistances (Figure 9.5).

At the interface, $y_{A,il}$, and x_{Ail} are in equilibrium. If k_x and k_y are local mass transfer coefficients, $N_A = k_y(y_{AG} - y_{Ai}) = k_x(x_{Ai} - x_{AL})$, where $(y_{AG} - y_{Al})$ and $(x_{Ai} - x_{Al})$

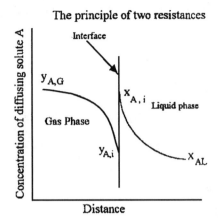
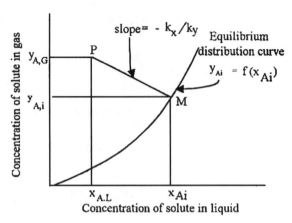

Figure 9.5 Mass transfer between two phases

are the driving forces for mass transfer in the gas and the liquid phases. A rearrangement gives the relation

$$\left(\frac{y_{Ag} - y_{Ai}}{x_{AL} - x_{AI}}\right) = -\frac{k_x}{k_y}$$

Thus if the mass transfer coefficients are known, the interfacial concentrations and the flux N_A can be determined graphically or analytically by solving the preceding equation in conjunction with the equilibrium curve.

Overall Mass Transfer Coefficients

As in heat transfer and mass transfer, one can write overall mass transfer coefficients in terms of the individual mass transfer coefficients in terms of the bulk concentrations. With liquid phase as basis, it is possible to derive the following relation:

$$\frac{1}{K_x} = \frac{1}{mk_y} + \frac{1}{k_x} \qquad (9.65)$$

If the gas phase is chosen as the basis to derive the overall mass transfer coefficients, the relation is

$$\frac{1}{K_y} = \frac{1}{k_y} + \frac{m}{k_x} \qquad (9.66)$$

where m is the slope of the equilibrium curve assumed to be a straight line.

It should be noted that if the gas phase resistance controls $\dfrac{1}{K_y} \approx \dfrac{1}{k_y}$ (9.66a)

and if the liquid phase resistance controls $\dfrac{1}{K_x} \approx \dfrac{1}{k_x}$. (9.66b)

These relations imply that

$$y_{Ag} - y^*_{Ai} \approx y_{Ag} - y_{Ai} \text{ for gas phase controlling.} \qquad (9.66c)$$

$$x_A^* - x_{AL} \approx x_{Ai} - x_A \text{ for liquid phase controlling.} \qquad (9.66d)$$

In practice, average mass transfer coefficients are used, and these are correlated by dimensionless numbers.

For general cases where (1) mass transfer rates are high, (2) diffusion of more than one substance is involved, or (3) equimolar counterdiffusion is not involved, F_G and F_L should be used. Thus mass transfer flux is given by

$$N_A = \frac{N_A}{\sum N} F_G \ln \frac{N_{A'} \sum N - y_{Ai}}{N_{A'} \sum N - y_{Ag}} = \frac{N_A}{\sum N} F_L \ln \frac{N_{A'} \sum N - x_{AL}}{N_{A'} \sum N - x_{Ai}} \quad (9.67)$$

where F_G and F_L are generalized gas and liquid phase mass transfer coefficients for component A. One can also define generalized overall mass transfer coefficients F_{OG} and F_{OL}. The following relations can be derived for two simple cases.

1. Diffusion of one component:

$$eN_{A'}F_{OG} = eN_{A'}F_G = m' \frac{1-x_{A,L}}{1-y_{A,G}}(1-e^{-N_{A'}F_L})$$

and $\quad e^{-N_{A'}F_{OL}} = \frac{1}{m''} \frac{1-y_{AG}}{1-x_{Al}}(1-e^{N_{A'}F_G}) + e^{-N_{A'}F} \quad (9.68)$

2. Equimolar counterdiffusion:

$$\frac{1}{F_{OG}} = \frac{1}{F_G} + \frac{m'}{F_L} \quad \text{and} \quad \frac{1}{F_{OL}} = \frac{1}{m''F_G} + \frac{1}{F_G} \quad (9.69)$$

where m' and m'' are slopes of chords of the equilibrium curve and are dimensionless.

MASS TRANSFER IN PACKED BEDS

Mass transfer in packed beds takes place between two flowing phases. The usual approach is to determine the product of the mass transfer coefficient and the interfacial area per unit volume of packed bed. If a is the interfacial area per unit volume of packed bed,

$$a = \frac{6(1-\varepsilon)}{D_P} \quad (9.70)$$

The preceding equation applies when the solids are spheres. Then

$$A_i = aV_t \quad (9.71)$$

where A_i = total interfacial area in a packed tower of packed volume V_t. Then molar flow rate $N_A A$ of the diffusing species A is given by

$$N_A A = Ak_c = \frac{(c_{Ai} - c_{A1}) - (c_{Ai} - c_{A2})}{\ln \frac{c_{Ai}-c_{A1}}{c_{Ai}-c_{A2}}} \quad (9.72)$$

where

c_{Ai} = concentration at the surface of solid kg mol/m^3
c_{A1}, c_{A2} = bulk fluid concentrations at inlet and outlet, kg mol/m^3

Material balance gives another relation.

$$N_A A = V_t (c_{A2} - c_{A1}) \qquad (9.73)$$

However, mass transfer coefficients in packed towers are correlated in terms of a height of a transfer unit, H. The height of the transfer unit is defined as follows:

$$H_G = \frac{G}{k_y a} \qquad (9.74)$$

where G = moles of gas/(unit time · unit cross section of empty tower)

Similarly,
$$H_L = \frac{L}{k_x a} \qquad (9.75)$$

where L = moles of liquid/(unit time · unit cross section of empty tower). H can also be defined to include the overall mass transfer coefficient as follows:

$$H_{OG} = \frac{G}{K_y a} \quad \text{(Overall height of transfer unit based on gas phase)} \qquad (9.76)$$

$$H_{OL} = \frac{L}{K_x a} \quad \text{(Overall height of transfer unit based on liquid phase)} \qquad (9.77)$$

Analytical expressions for H_{OG} and H_{OL}

$$H_{OG} = \frac{G}{K_y a (1-y)_{lm}} = \frac{G}{K_{Ga} P_t (1-y)_{lm}} \quad \text{where} \quad (1-y)_{lm} = \frac{(1-y^*)-(1-y)}{\ln \frac{(1-y^*)}{(1-y)}} \qquad (9.78)$$

$$H_{OL} = \frac{L}{K_x a (1-x)_{lm}} = \frac{L}{K_L a C (1-x)_{lm}} = \frac{L M_L}{K_L a \rho_L (1-x)_{lm}} \qquad (9.79)$$

where
$$(1-x)_{lm} = \frac{(1-x)-(1-x^*)}{\ln \frac{1-x}{1-x^*}} \qquad (9.79a)$$

where M_L = molecular weight of liquid solvent

The overall mass transfer coefficients $K_G a$ and $K_y a$ can be calculated from the individual gas and liquid phase mass transfer coefficients and equilibrium relationship. Table 9.3 will be useful to identify correct units to be used in a particular case.

Liquid-phase mass transfer coefficients may also be expressed in different units as follows:

(a) $\dfrac{1}{K_L a} = \dfrac{1}{k_L a} + \dfrac{m'_c}{k_G a}$ (b) $\dfrac{1}{K_L a} = \dfrac{1}{k_L a} + \dfrac{1}{m_c k_G a}$ (c) $\dfrac{1}{K_L a} = \dfrac{1}{k_L a} + \dfrac{\rho_m}{m k_G a p_t}$

(d) $\dfrac{1}{K_L a} = \dfrac{1}{k_l a} + \dfrac{\rho m}{m_x k_G a}$ (e) $K_L a = K_x a/c$ (f) $k_L a = k_x a/c$ $\qquad (9.80)$

Mass Transfer in Packed Beds 345

Table 9.3 Units of mass transfer coefficients

(a) $\dfrac{1}{K_G a} = \dfrac{1}{k_G a} + \dfrac{m_c}{k_L a}$

Here m_c is given by the equilibrium relationship $p^* = m_c c$ where p^* is in atm, and c in kg mol/m^3.
In this equation, m'_c is given by $p^* = c/m'_c$
where m is given by $y^* = mx$

(b) $\dfrac{1}{K_G a} = \dfrac{1}{k_G a} + \dfrac{1}{m'_c k_L a}$

and $\rho_m = \dfrac{\text{density of solution, kg/m}^3}{\text{molecular weight of solution}}$

(c) $\dfrac{1}{K_G a} = \dfrac{1}{k_G a} + \dfrac{m p_t}{k_L a \rho_m}$

where m_x is given by $\rho_m p^* = m_x x$
and m is given by $y^* = mx$

(d) $\dfrac{1}{K_G a} = \dfrac{1}{k_G a} + \dfrac{mx}{k_L a \rho_m}$

(e) $\dfrac{1}{K_G a} = \dfrac{1}{k_G a} + \dfrac{m}{k_x a}$

(f) $K_G a = K_y a/p_t$

(g) $K_G a = K_y a/p_t$

By analogy with H_{OL} and H_{OG}, the gas film and liquid film transfer units can also be defined as follows:

(a) $H_{tG} = \dfrac{G}{k_y a(1-y)_{lm}} = \dfrac{G}{k_y a}$ (b) $H_{tG} = \dfrac{G}{k_G a P_t (1-y)_{lm}} \approx \dfrac{G}{k_G a P_t}$

(9.81)

(c) $H_{tL} = \dfrac{L}{k_x a(1-x)_{lm}} \approx \dfrac{L}{k_x a}$ (d) $H_{tL} = \dfrac{L}{k_L a C(1-x)_{lm}} \approx \dfrac{L}{k_L a C}$

Example 9.12

Air is flowing in a tube whose inside surface is wetted with water. The temperature is 21°C. Calculate the flux of water evaporated from the pipe wall at a point where y_A, the bulk concentration (in mole fraction) of H$_2$O, is 0.0015. Additional data are as follows: ID of tube = 0.102 m, air velocity = 15.3 m/s, pressure = 1 bar A., $D_{AB} = 0.072$ m^2/h, μ of air at 21°C = 0.018 mPa·s, density of air = 1.2 kg/m^3.

$$Sh = 0.023 Re^{0.83} Sc^{0.33} \qquad N_A = K_G(P_{A1} - P_{A2}), \quad K_G = K_C/(RT)$$

Solution

$$Re = \dfrac{du\rho}{\mu} = \dfrac{0.102(15.3)(1.2)}{0.018 \times 10^{-3}} = 104040 \qquad D_{AB} = 0.072 \text{ m}^2/\text{h}$$

$$Sc = \dfrac{\mu}{\rho D_{AB}} = \dfrac{0.018 \times 10^{-3}(3600)}{1.2(0.072)} = 0.75$$

$$Sh = 0.023 Re^{0.83} Sc^{1/3} = 0.023(104040)^{0.83}(0.75)^{1/3} = 305$$

$$D_i = 0.102 \text{ m}$$

or $\quad Sh = \dfrac{k_c D}{D_{AB}} = 305$

or $\quad Sh = \dfrac{k_c D}{D_{AB}} = 305$

Therefore $\quad k_c = \dfrac{0.072}{0.102} \times 305 = 215.3$ m/h

Vapor pressure of water at $21°C = 0.025$ bar

$p_{B1} = 1 - 0.025 = 0.975$ bar $\quad p_{B2} = 1 - 0.0015 = 0.9985$ bar

$p_{BM} = \dfrac{p_{B1} - p_{B2}}{\ln(p_{B1}/p_{B2})} = \dfrac{0.9985 - 0.975}{\ln(0.9985/0.975)} = 0.986$ bar

$k_G = \dfrac{k_c}{RT} = \dfrac{215.3}{0.08314 \times 294} = 8.81$ kg mol/(h·m²·bar kg mol)

$N_A = k_G(p_{A1} - p_{A2}) = 8.81 \left[\dfrac{\text{kg mol}}{\text{h·m}^2 \cdot \text{bar}}\right] \times (0.025 - 0.0015)$ (bar)

$= 0.207$ kg mol/h·m²

PHASE EQUILIBRIA

So far we have reviewed mostly properties of pure substances or mixtures of constant compositions. Mass transfer operations such as distillation, absorption, and extraction involve contact of different phases but also changes in composition. When two or more phases come in contact, their compositions change because of mass transfer between them. The degree of change and the rate of transfer depend on how far the system is away from its equilibrium state. Thus phase equilibria are important in quantitative treatment of mass transfer operations. In industrial practice, chemical engineers have to deal with vapor-liquid, liquid-liquid, solid-liquid, and vapor-solid systems, although vapor-liquid systems are encountered more often than the others.

VAPOR-LIQUID EQUILIBRIA

In this section we review vapor-liquid equilibria and the calculation of phase compositions at equlibrium in distillation. We will recapitulate Raoult's and Henry's laws that allow the calculation of temperature, pressure, and phase compositions in vapor-liquid equilibrium. We will also review modifications of Raoult's law, which allows dealing with mixtures of species of chemically dissimilar nature. Bubble and dew points, and flash calculations will also be covered in this section.

Binary and Multicomponent Systems

A binary system consists of two components. A multicomponent system consists of more than two components. Both multicomponent and binary systems can be either ideal or nonideal. Ideal behavior of systems is found at low pressures and normal distilling temperatures. Nonideal behavior is exhibited by systems consisting of substances of dissimilar nature or where temperature and pressure conditions are severe.

The Phase Rule and Duhem's Theorem

According to phase rule for nonreacting systems, the number of variables that may be independently fixed for a system at equilibrium is the difference between the total number of variables that describe the intensive state of the system and the number of independent equations that can be written relating the variables. If the system contains N chemical components at given T and P, the phase rule variables are $N-1$ mole fractions and T and P. The degrees of freedom F are given by (Smith et al.)

$$F = 2 - \pi + N \tag{9.82}$$

Duhem's rule pertains to a closed system at equilibrium for which both extensive and intensive states of the system are fixed. In this case, there are $2 + (N-1)\pi$ phase rule intensive variables and π extensive variables expressing the masses or moles of the phases. Therefore total number of variables is $(\pi-1)N + N = \pi N$. We can write N additional independent material balance equations (one each for each component) giving the total number of independent equations $= (\pi-1)N + N = \pi N$. The difference between the number of variables and the number of independent equations is then $2 + N\pi - \pi N = 2$. Thus for a closed system of constant component masses, the equilibrium state is completely determined when two independent variables are fixed. The phase rule, however, fixes the number of independent intensive variables. Therefore depending on the degrees of freedom F for a system, the number of extensive variables that can be chosen will also be determined.

For a binary system at equilibrium comprising two phases (vapor and liquid), $N = 2$, and $\pi = 2$. Therefore, $F = 2$. Thus, fixing two variables determines binary system equilibrium. A ternary system with two phases in equilibrium has $F = 3$ and will be defined by three variables. For an n component system, and two phases in equilibrium, there are n degrees of freedom.

Boiling Point (T-xy) and Equilibrium (x-y) Diagrams (Binary Systems).

For binary systems, the differences in compositions of the liquid and vapor phases are graphically represented by either boiling point (temperature-composition, $T-xy$) or equilibrium (composition, y vs x only) diagrams at constant pressure. In the former, boiling point temperature is plotted against liquid and vapor compositions, whereas in the latter, the equilibrium vapor phase compositions of more volatile components are plotted against the corresponding liquid phase compositions as in Figure 9.6 a and b.

Systems that obey Raoult's law are called ideal systems. Those systems that do not obey Raoult's law are nonideal systems. Boiling point and equilibrium diagrams such as shown in Figure 9.6a and b are typical of ideal binary systems. Boiling point diagrams and equilibrium diagrams for nonideal systems exhibit different behavior such as shown in Figure 9.7a and b. For example, an acetone-chloroform system is a maximum boiling point system, whereas the ethyl acetate-ethanol system shows a minimum boiling point. At the azeotropic point, vapor and liquid composition are the same ($x_{Az} = y_{Az}$). For maximum boiling point systems, this happens at T_{max} and for minimum boiling point systems at $T_{min.}$

Such diagrams are typical of binary systems that obey Raoult's law.

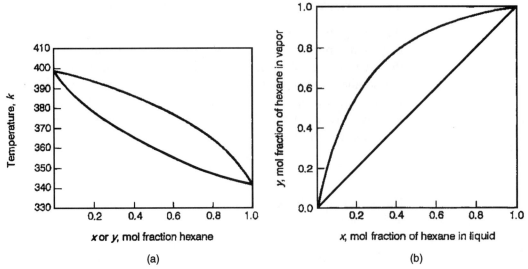

Figure 9.6 (a) T-xy diagram or boiling point diagram (b) Equilibrium or x-y diagram

For a liquid and vapor mixture at equilibrium, the ratio of the moles of the liquid phase to moles of the vapor phase is given by the inverse lever rule

$$\frac{L}{V} = \frac{y_{AV} - x_{AF}}{z_{AF} - x_{AL}} \tag{9.83}$$

L = mols of liquid phase
V = mols of vapor phase in equilibrium with the liquid
y_{AV} = mol fraction of component A in vapor phase
z_{AF} = mol fraction of component A in the feed before vaporization
x_{AL} = mol fraction of component A in the liquid phase at equilibrium

The boiling point and equilibrium diagrams of Figure 9.6a and b are typical of ideal binary systems that follow Raoult's law. Nonideal systems that do not follow Raoult's law show a different behavior, such as a maximum or a minimum boiling point.

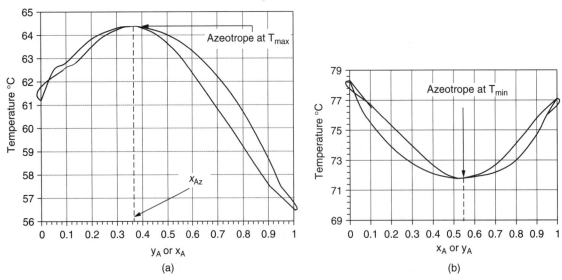

Figure 9.7 (a) System acetone chloroform, maximum boiling point azeotrope and (b) System ethyl acetate-ethanol, minimum boiling point azeotrope

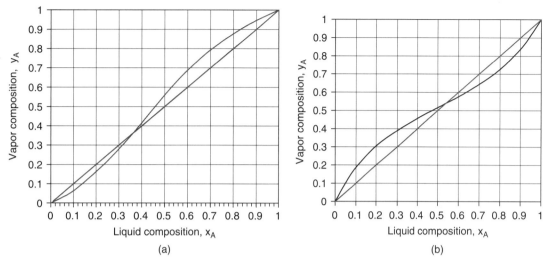

Figure 9.8 (a) System acetone-chloroform *x-y* diagram and (b) System ethyl acetate-ethanol x-y diagram

On the *x-y* diagram, the equilibrium curves cross the 45° line at the maximum or minimum boiling point (Figure 9.8a and b).

Azeotropic mixtures are very common. They cannot be completely separated by ordinary distillation methods since at the azeotropic composition $y_A = x_A$. One of the most important is the ethanol-water azeotrope, which occurs at 78.4°C and 1 atm. Azeotropism disappears in this system at pressures below 70 mm Hg.

Some substances exhibit very large positive deviations from ideality in that they are not completely soluble in the liquid state and form a pair of partially immiscible components. Isobutanol-water is such a system. The boiling point and equilibrium diagrams for this system are shown in Figure 9.9a and b.

Immiscible Liquids

If the liquids are completely insoluble in each other, each exerts its own vapor pressure at the prevailing temperature. When the sum of the two vapor pressures

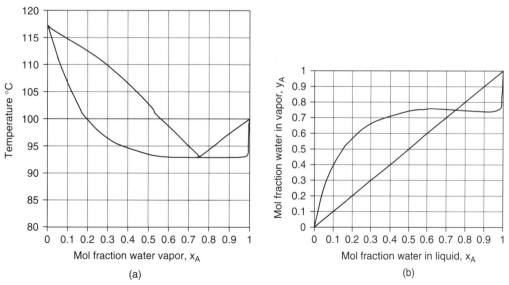

Figure 9.9 (a) System water-I-butanol and (b) Heterogeneous azeotrope

equals the total pressure, the mixture boils. The vapor composition is then easily calculated by application of the ideal gas law.

$$P_A + P_B = P \quad \text{and} \quad y = \frac{P_A}{P} \tag{9.84}$$

Ideal Mixtures (Raoult's Law)

Ideal gas, ideal liquid, and ideal gas and liquid mixtures are the basis of many quantitative equilibrium calculations. Raoult's law relates the partial pressure of a component in the vapor phase of a gaseous mixture to its concentration (mol fraction) in the liquid phase and its vapor pressure at the temperature of the mixture in contact with the liquid at equilibrium. Mathematically it can be stated as follows:

$$p_i = x_i P_i \tag{9.85}$$

where
- p_i = partial pressure of the component i in the vapor phase
- x_i = mol fraction of component i in the liquid phase
- P_i = vapor pressure of pure component i at the temperature of the liquid-vapor mixture

By combining Raoult's law with Dalton's law of partial pressures, we obtain an expression relating mixtures of ideal vapors and liquids in equilibrium.

$$P = \sum_i^n p_i = y_i \quad P = \sum_i^n x_i P_i \tag{9.86}$$

For a single component, $y_i P = x_i P_i$ (9.86a)

where P = total pressure

The vapor pressure of a pure component is a unique property of the pure component and is a function of temperature. It increases with an increase in temperature. A component that has a higher vapor pressure at a given temperature than another component is called more volatile. Vapor pressures as a function of temperature, including the Antoine equation, are discussed in Chapter 2.

For systems that obey Raoult's law it is possible to calculate the boiling-point diagram from the vapor pressures of the pure components and the following relations that can be obtained by application of Raoult's and Dalton's laws to ideal binary systems.

$$p_A + p_B = P$$

$$P_A x_A + P_B (1 - x_A) = P \tag{9.87}$$

$$y_A = \frac{p_A}{P} = \frac{P_A x_A}{P}$$

$$y_A = \frac{P_A x_A}{P_A x_A + P_B (1 - x_A)} = \frac{P_A x_A}{P} \quad x_A = \frac{P - P_B}{P_A - P_B} \tag{9.88}$$

where P_A and P_B are vapor pressures of the components A and B, respectively.

Example 9.13

(Use of Raoult's Law)

The vapor pressure data for the system hexane-octane are given in the following table. The system can be assumed to be ideal and obeying Raoult's law. Plot the $T\text{-}xy$ and $x\text{-}y$ diagrams for hexane-octane system at 101.3 kPa pressure.

	Vapor Pressure	
$T°C$	n-Hexane kPa	n-Octane kPa
68.7	101.3	16.1
79.4	136.7	23.1
93.3	197.3	39.2
107.2	284.0	57.9
125.7	456.0	101.3

Solution

First liquid compositions are calculated using Equation 9.87. The relation

$$x_A = \frac{P - P_B}{P_A - P_B}$$

is used for this purpose. Then vapor composition is calculated by

$$y_A = \frac{p_A}{P} = \frac{P_A x_A}{P}$$

Calculations done on a spreadsheet are given in the following table. The boiling point and equilibrium diagrams are plotted in Exhibit 1a and b respectively.

$T°C$	$P - P_A$	$P_A - P_B$	$x_A = \dfrac{P - P_B}{P_A - P_B}$	$y_A = \dfrac{P_A x_A}{P}$
68.7	85.2	85.2	1	1
79.4	78.2	113.6	0.68838	0.928940
93.3	64.2	160.2	0.40075	0.780531
107.2	43.4	226.1	0.19195	0.538143
125.7	0	354.7	0	0

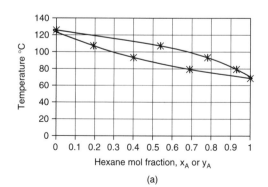 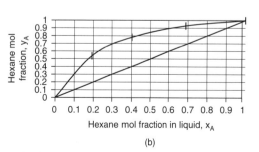

(a) (b)

Exhibit 1 $T\text{-}x$, $x\text{-}y$ or equilibrium diagrams for Example 9.13

Relative Volatility α_{ij}

For a vapor phase in equilibrium with its liquid phase, the relative volatility of a component i with respect to component j is given by

$$\alpha_{ij} = \frac{y_i/x_i}{y_j/x_j} \qquad (9.89)$$

where y is the mole fraction of a component in the vapor phase and x is the mole fraction of the same component in the liquid phase. The relative volatility is the measure of separability of two components by distillation. The larger the value of α_{ij} above unity, the greater the degree of separability.

For a binary system, $y_j = 1 - y_i$ and $x_j = 1 - x_i$; therefore

$$\alpha = \left(\frac{y_i}{1-y_i}\right)\left(\frac{1-x_i}{x_i}\right) \qquad (9.90)$$

In general

$$\alpha = \left(\frac{y}{1-y}\right)\left(\frac{1-x}{x}\right) \qquad (9.91)$$

from which

$$y = \frac{\alpha x}{1+(\alpha-1)x} \qquad x = \frac{y}{\alpha + y(1-\alpha)} \qquad (9.91a)$$

For ideal mixtures, α is the ratio of the vapor pressures or

$$\alpha_{ij} = \frac{P_i}{P_j} \qquad (9.92)$$

and for a binary system, Equation 9.88 applies.

Example 9.14

(Calculation of Relative Volatility)

From the data of Example 9.13, calculate the relative volatilities of n-hexane-n-octane system at various temperatures.

Solution

For n-hexane-n-octane system, which follows Raoult's law, relative volatility

$$\alpha = \frac{P_A}{P_B}$$

The calculations of α are presented in the following table.

Temperature °C	Vapor Pressure of Hexane kPa	Vapor Pressure of Octane kPa	$\alpha = P_A/P_B$
68.7	101.3	16.1	6.3
79.4	136.7	23.1	5.92
93.3	197.3	39.2	5.32
107.2	284.0	57.9	4.91
125.7	456.0	101.3	4.50

Henry's Law

If the pressure is low enough so that the vapor phase can be assumed to be an ideal gas, Henry's law can be applied. For a component present in liquid phase at very low concentrations, Henry's law states that the partial pressure of the component in the vapor phase is directly proportional to its mol fraction in the liquid phase.

Thus
$$p_A = y_i P = H_i x_i \tag{9.93}$$

where H = Henry's law constant characteristic of the system. Values of H_i are to be obtained experimentally. *Henry's law* applies to the vapor pressure of a solute in a dilute solution (Raoult's law applies to the vapor pressure of the solvent).

The Chemical Potential and Phase Equilibria

The chemical potential of a species in a mixture is related to free energy G and is defined by

$$\mu_i \equiv \left[\frac{\delta(nG)}{\delta n_i} \right]_{P,T,n_j} \tag{9.94}$$

With this definition the following equation is written for n moles in a single fluid system.

$$d(nG) = (nV)dP - (nS)dT + \sum \mu_i dn_i \tag{9.95}$$

Equation 9.95 is a fundamental property relation for single fluid systems of constant or variable mass or composition. For the case of one mole, Equation 9.95 can be written as follows:

$$dG = VdP - SdT + \sum \mu_i dx_i \tag{9.96}$$

For a closed system consisting of two phases in equilibrium it can be shown that for a component i

$$\mu_i^\alpha = \mu_i^\beta = \cdots = \mu_i^\pi \tag{9.97}$$

The relationship of Equation 9.97 indicates that multiple phases are in equilibrium with each other at the same temperature and pressure when the chemical potential of each component species is the same in all the phases.

Absolute values of chemical potentials do not exist. While this does not preclude the use of chemical potentials in phase equilibrium analysis, application of equlibrium criteria is simplified by introduction of a new quantity called fugacity that takes the place of μ_i.

Fugacity, and Fugacity Coefficient and Phase Equilibria

The concepts of fugacity and the fugacity coefficient of a component in gas and liquid mixtures are already reviewed in Chapter 3. In a multicomponent system at equilibrium, the fugacity of each component is the same in all phases. Thus, in a two-phase system of a multicomponent liquid mixture in equilibrium with its vapor

$$f_i^V(T,P,y) = f_i^L(T,P,x) = f_i^{sat} \tag{9.98}$$

Another formulation of vapor-liquid equilibrium is based on fugacity coefficients and is equally valid.

$$\phi_i^V = \phi_i^L = f_i^{sat} \tag{9.99}$$

Fugacity of a component i in a liquid mixture is usually calculated by the relation

$$\hat{f}_i^L = x_i \gamma_i f_i^L \tag{9.100}$$

where
- f_i^L = fugacity of pure liquid component i
- x_i = mol fraction of component i in liquid mixture
- γ_i = activity coefficient of component i
- $\hat{f}_i^L$ = fugacity of a component i in a liquid mixture

The fugacity of a pure component can be calculated by using the relation

$$f_i^l = \phi_i^{sat} P_i^{sat} \exp\left\{v_i^L\left(P - P_i^{sat}\right)/RT\right\} \tag{9.101}$$

The exponential in the preceding equation is called the Poynting factor

where
- ϕ_i^{sat} = fugacity coefficient of pure saturated i
- P_i^{sat} = saturation pressure of pure component i
- V_i^L = molar specific volume of pure liquid component i

Very often the system pressure is low and close to atmospheric. In this situation,

$$f_i^L = P_i^{sat} \tag{9.102}$$

For the case of vapor in equilibrum with the liquid mixture, the fugacity can be calculated by the relation

$$f_i^V = y_i \hat{\phi}_i^V P \tag{9.103}$$

where
- y_i = mol fraction of component i in vapor
- $\hat{\phi}_i^V$ = fugacity coefficient of component i in vapor
- P = system pressure.

The following relation allows estimation of the fugacity coefficient from PVT data.

$$\ln \hat{\phi}_i^V = \int_0^P \left(\frac{v_i}{RT} - \frac{1}{P} \right) dP \tag{9.104}$$

Once $\hat{\phi}_i^V$ is determined it is easy to calculate f_i^V with the use of Equation 9.103.

Nonideal Systems

If the components show strong interactions (physical or chemical) then the boiling point and equilibrium diagrams for binary systems are quite different from those of the ideal systems (Van Winkle, Henley, Perry, Treybal, Smith and Van Ness). Azeotropic mixtures have a critical liquid composition at which the liquid and vapor compositions are the same. In all systems (both multicomponent and binary), the partial pressures of components of a real solution will deviate from those calculated by Raoult's law. Nonideal behavior may exist in either or both phases. For system pressures close to atmospheric, the real behavior may be accounted for by including a correction factor γ_i, called the activity coefficient, in Raoult's law. This results in the following relation:

$$p_i = \gamma_i x_i P_i \tag{9.105}$$

The product $\gamma_i x_i$ is termed the activity a_i of the component i. The activity coefficient γ_i varies with temperature and composition and can be >1 or <1. By incorporating activity and fugacity coefficients γ_i and $\hat{\phi}_i$ in Raoult's law, the following relation results:

$$\hat{\phi}_i y_i P = \gamma_i x_i P_i \tag{9.106}$$

With the use of this relation, the relative volatility in a nonideal system is shown to be

$$a_{ij} = \frac{y_i/x_i}{y_j/x_j} = \frac{\gamma_i P_i \hat{\phi}_j}{\gamma_j P_j \hat{\phi}_i} \tag{9.107}$$

Calculation of Activity Coefficients

Many empirical and semi-empirical equations have been proposed to predict the variation of the activity coefficients with temperature and composition. For nonideal binary systems, van Laar and Margules equations are widely used. They are described briefly next.

van Laar Equations
These are as follows:

$$T \ln \gamma_1 = \frac{B}{[1 + A(x_1/x_2)]^2} \quad \text{and} \quad T \ln \gamma_2 = \frac{AB}{(A + x_2/x_1)^2} \tag{9.108}$$

where A and B are constants that are assumed to be independent of temperature. Another simple form of Equation 9.108 is

$$\ln \gamma_1 = \frac{B/T}{\left[A(x_1/x_2)+1\right]^2} \quad \text{and} \quad \ln \gamma_2 = \frac{(AB/T)\left(\frac{x_1}{x_2}\right)}{[A(x_1/x_2)]^2} \quad (9.108a)$$

At the azeotropic composition

$$x_i = y_i \quad \text{and} \quad \gamma_i = P/P_i, \quad \text{Also} \quad \gamma_i = y_i P / x_i P_i \quad (9.109)$$

Therefore, knowing the azeotropic composition, the constants A and B can be calculated.

If the temperature variable is not included, another more frequently used van Laar equation model results. These equations are

$$\ln \gamma_1 = A_{12}\left(1 + \frac{A_{12} x_1}{A_{21} x_2}\right)^{-2} \quad \text{and} \quad \ln \gamma_2 = A_{21}\left(1 + \frac{A_{21} x_2}{A_{12} x_1}\right)^{-2} \quad (9.110)$$

where A_{12} and A_{21} are van Laar constants.

Carlson and Colburn expressed Equation 9.110 in a different format as follows:

$$\log \gamma_1 = \frac{A' x_2^2}{[(A'/B')x_1 + x_2]^2} \quad \text{and} \quad \log \gamma_2 = \frac{B' x_1^2}{[(B'/A')x_2 + x_1]^2} \quad (9.110a)$$

The constants A', B' of equation 9.110a are related to van Laar constants by the following equations:

$$A' = \frac{B}{2.303T} \quad \text{and} \quad B' = \frac{B}{2.303AT} \quad (9.110b)$$

Margules Equations

These are given by

$$\ln \gamma_1 = x_2^2[A + 2x_1(B-A)] \quad \text{and} \quad \ln \gamma_2 = x_1^2[B + 2x_2(A-B)] \quad (9.111)$$

The Wilson equation for binary systems is based on excess free energy and contains only two parameters. It is written as follows:

$$\frac{G^E}{RT} = -x_1 \ln(x_1 + x_2 \Lambda_{12}) - x_2 \ln(x_2 + x_1 \Lambda_{21})$$

$$\ln \gamma_1 = -\ln(x_1 + x_2 \Lambda_{12}) + x_2 \left(\frac{\Lambda_{12}}{x_1 + x_2 \Lambda_{12}} - \frac{\Lambda_{21}}{x_2 + x_1 \Lambda_{21}} \right) \quad (9.112)$$

$$\ln \gamma_2 = -\ln(x_2 + x_1 \Lambda_{21}) + x_1 \left(\frac{\Lambda_{12}}{x_1 + x_2 \Lambda_{12}} - \frac{\Lambda_{21}}{x_2 + x_1 \Lambda_{21}} \right)$$

For infinite dilution Wilson equations are

$$\ln \gamma_1^\infty = -\ln \Lambda_{12} + 1 - \Lambda_{21} \quad (9.113)$$

$$\ln \gamma_2^\infty = -\ln \Lambda_{21} + 1 - \Lambda_{12}$$

Other models such as NRTL equation, UNIQUAC equation, and the UNIFAC method are of greater complexity. The Wilson and other models are extendable to multicomponent systems. However, these are more suitable for computer calculations.

For a detailed treatment of activity coefficients and their estimation, the student is referred to other thermodynamic and distillation texts (Van Winkle, Perry, Treybal, Smith et al.).

VLE by Modified Raoult's Law

To take into account the deviations from ideality in the liquid phase γ, the *activity coefficient* is incorporated in Raoult's law as follows:

$$y_i P = x_i \gamma_i P_i^{sat} \quad (9.114)$$

Activity coefficients are functions of temperature and liquid phase composition. They are to be obtained experimentally. If activity coefficients are known, bubble point and dew point can be made.

Since $\sum y_i = 1$, Equation 9.114 can be written as

$$P = \sum_i x_i \gamma_i P_i^{sat} \quad (9.115)$$

Also we can write from Equation 9.114,

$$P = \frac{1}{\sum_i y_i / (\gamma_i P_i^{sat})} \quad (9.116)$$

And these equations can be used to make VLE calculations for nonideal systems.

Equilibrium Vaporization Ratios (Equilibrium Constants) K

The K value is extremely useful in hydrocarbon distillation, absorption, and stripping calculations. It is defined by the following relation:

$$K_i = \frac{y_i}{x_i} \quad (9.117)$$

In terms of vapor pressure, activity coefficient, and the fugacity coefficient, K_i is given by

$$K_i = \frac{\gamma_i P_i}{\hat{\phi}_i P} \quad (9.118)$$

For an ideal system, both γ_i and $\hat{\phi}_i$ are unity and K_i reduces to $K_i = P_i/P$.

Dew Point

The dew point of a mixture is the temperature and pressure at which the first drop of the liquid is formed when a vapor mixture is cooled. This may be expressed in terms of the equilibrium relationship; thus

$$\sum \frac{y_i}{K_i} = \sum x_i = 1.0 \qquad (9.119)$$

For ideal systems,

$$\sum x_i = \sum \frac{y_i}{P_i} P_t = 1.0 \qquad (9.120)$$

The temperature at which the first drop forms is called the *dew-point temperature*, and the corresponding pressure is the *dew-point pressure*.

Bubble Point

The bubble point of a liquid mixture is the temperature and pressure at which the liquid begins to boil such that the vapor is in equilibrium with the liquid. The condition can be expressed by the relation

$$\sum K_i x_i = \sum y_i = 1.0 \qquad (9.121)$$

For ideal systems,

$$\sum y_i = \sum \frac{P_i x_i}{P_t} = 1.0 \qquad (9.122)$$

In the preceding equations, K_i is the equilibrium constant of a component defined by Equation 9.117.

Flash Calculations

Flash calculation is an important application of vapor-liquid equilibria. A flash means a partial evaporation or vaporization of a liquid, which is at its bubble point pressure or greater when the pressure is reduced. This partial vaporization results in two coexisting phases, one vapor and the other liquid in equilibrium. A flash calculation involves the determination of relative quantities of vapor and liquid phases, their compositions at known T, P, and also the overall composition. Since two independent variables T and P are specified for a system of fixed overall composition, the problem is solvable on the basis of Duhem's theorem.

There are four types of flash calculations that are of interest to chemical engineers.

These are

Bubble Pressure	(1) Calculate y_i and P given x_i and T
Dew Pressure	(2) Calculate x_i and P given y_i and T
Bubble temperature	(3) Calculate y_i and T given x_i and P
Dew temperature	(4) Calculate x_i and T given y_i and P

The most important flash is the *PT* flash.

For a multicomponent mixture, the following equations apply:

$$x_i = \frac{z_i/L}{K_i V/L + 1} \quad \text{or} \quad x_i = \frac{z_i/V}{K_i + L/V} \tag{9.123}$$

where
L = moles of undistilled liquid
V = moles of vapor (distilled liquid)
z_i = mol fraction of component i in the feed
K_i = ratio of equilibrium vapor to equilibrium liquid
 $= y_i/x_i$
x_i = mol fraction of component i in undistilled liquid
y_i = mol fraction of component i in distilled vapor

For an ideal system,

$$K_i = \frac{P_i}{P}$$

The solution of the flash-vaporization problem involves trial and error. Since $y_1 = K_i x_i$,

$$y_i = \frac{K z_i/V}{K_i + L/V}$$

and either $\Sigma x_i = 1$ or $\Sigma y_i = 1.0$.

If 1 mol of the liquid is considered and f is the fraction vaporized, then $(1-f)$ is the liquid left behind and by material balance,

$$z_i = f y_i + (1-f) x_i \tag{9.124}$$

Therefore,

$$y_i = -\frac{1-f}{f} x_i + \frac{z_i}{f} \tag{9.125}$$

Equation 9.121 is easier to use in solving the flash-vaporization problems. In the case of a multicomponent system (more than two components), a trial-error solution is required. In the case of the binary systems, the fraction vaporized is given by

$$f = \frac{z_i(K_1 - K_2)/(1 - K_2) - 1}{K_1 - 1} \tag{9.126}$$

Example 9.15

A liquid-phase mixture of hydrocarbons has the following composition at 125 psia: C_2 = 2.0 mol %, $^=C_3$ = 29 mol %, C_3 = 29.0 mol %, n-C_4 = 40.0 mol %. Calculate the bubble-point temperature of the mixture at 125 psia. K values are given in Exhibit 2.

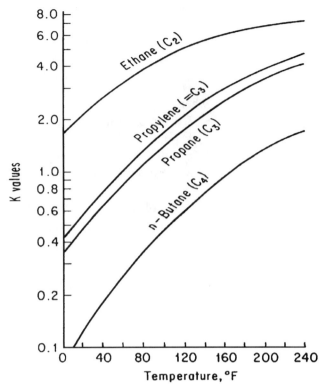

Exhibit 2 Approximate K values for light hydrocarbons

Solution

At the bubble point
$$y_i = Kx_i$$
$$\Sigma y_i = \Sigma K_i x_i = 1.0$$

If noncondensables are present at the bubble point,

$$\Sigma y_i = \Sigma K_i x_i = 1 - y_{nc} \qquad i \neq nc$$

The bubble point is to be calculated by trial and error. A temperature is assumed, K values at this temperature are obtained, and $\Sigma K_i x_i$ is calculated. y_{nc} is the mole fraction of the noncondensables. A useful relation to obtain the next temperature estimate is

$$t_{n+1} = t_n - 100(\Sigma K_i x_i - 1.0)$$

Assumed temperature $t_1 = 100°F$. See Table 9.4.

Table 9.4

Component	x_i	$t_1 = 100°F$		$t_2 = 86.6°F$		$t_3 = 86.5°F$	
		K_i	$y_i = K_i x_i$	K_i	$y_i = K_i x_i$	K_i	$y_i = K_i x_i$
C_2	0.02	4.3	0.086	3.90	0.078	3.89	0.0778
C_3	0.29	1.65	0.4785	1.43	0.4147	1.43	0.4148
$={C_3}$	0.29	1.44	0.4176	1.22	0.3538	1.21	0.3509
$n\text{-}C_4$	0.40	0.46	0.1840	0.39	0.1548	0.39	0.1564
			1.1661		1.0013		0.9999

Therefore, the bubble-point temperature is 86.5°F. The vapor phase composition is

$$C_2 = 7.78 \text{ mol\%} \quad\quad C_3 = 41.48 \text{ mol\%}$$
$$^=C_3 = 35.09 \text{ mol\%} \quad\quad n\text{-}C_4 = 15.64 \text{ mol\%}$$

Dew-Point Calculation Using K Values

Knowing the composition of a vapor phase, the composition of the liquid phase in equilibrium with this vapor can be obtained. At the dew point,

$$\sum \frac{y_i}{K_i} = \sum x_i = 1.0 \quad\quad (9.127)$$

If nonvolatiles are present,

$$\sum \frac{y_i}{K_i} = 1.0 - x_{nv} \quad i \ne nv \quad\quad (9.128)$$

x_{nv} is the mole fraction of the nonvolatile component in the liquid phase.

To calculate the dew point, assume a temperature, get K values at the assumed temperature, and then calculate $\sum y_i/k_i$. If $\sum y_i/k_i \ne 1.0$, assume another temperature and repeat the calculation. A useful equation to obtain the next temperature for an estimate is

$$t_{n+1} = t_n + 100\left(\sum \frac{y_i}{K_i} - 1.0\right)$$

Example 9.16

A vapor phase mixture at 125 psia has the following composition: $C_2 = 2$ mol %, $^=C_3 = 29$ mol %, and $n\text{-}C_4 = 40$ mol %. Calculate the dew-point temperature and the composition of the condensed phase at the dew-point temperature. K values can be obtained from Figure 9.6.

Solution

Assume a temperature and estimate the composition. The results are shown in Table 9.5.

Table 9.5

Component	y_i	Assume $t = 125°F$		Assume $t = 119°F$		Mol %
		K_i	$x_i = y_i/K_i$	K_i	$x_i = y_i/K_i$	
C_2	0.02	5.10	0.00392	4.6	0.00435	0.435
$^=C_3$	0.29	2.18	0.13302	2.05	0.14146	14.146
C_3	0.29	1.85	0.15676	1.75	0.16571	16.571
$n\text{-}C_4$	0.40	0.625	0.6400	0.58	0.68966	68.966
			$\sum x_i = 0.9337$		$\sum x_i = 1.00118$	

362 Chapter 9 Mass Transfer Fundamentals

Table 9.6

Component	Mol %
C_2	0.435
$^=C_3$	14.146
C_3	16.571
$n\text{-}C_4$	68.966
	100.118

The dew-point temperature is 119°F. The composition of condensed liquid is shown in Table 9.6.

Example 9.17

A liquid mixture at 125 psia has the composition given in Example 9.16. Calculate the fraction vaporized and the compositions of the liquid and vapor phases if it is flashed isothermally at 100°F and 125 psia.

Solution

The dew point of the mixture is 119°F assuming the vapor composition is identical with the liquid composition given in the problem. The bubble point of the mixture is 86.5°F. The flash temperature is between the dew and bubble points.
  Feed Composition

$$y_i = \frac{z_{Fi} K_i / V}{K_i + L/V}$$

Assume 1 mol of feed. Then fraction vaporized $= f = V$ and the fraction of feed remaining as liquid $= L = 1 - V$. Hence

$$\frac{L}{V} = \frac{1-V}{V}$$

See Table 9.7.
  From the last trial, when $f = 0.45$, $\sum y_i \cong 1.0$. Hence the fraction vaporized is 0.45 mol/mol of feed, and the fraction of feed remaining as liquid is 0.55 mol/mol of feed. The liquid compositions are then calculated using the relation $x_i = y_i/K_i$ and are given in Table 9.8.

Table 9.7 Calculation of y_i at Different Values of $f = V$

	x_F	K_{100}	$f = 0.5$	$f = 0.41$	$f = 0.45$
C_2	0.02	4.30	0.0325	0.0366	0.0346
$^=C_3$	0.29	1.69	0.3693	0.3820	0.3740
C_3	0.29	1.44	0.3413	0.3525	0.3475
C_4	0.40	0.46	0.2520	0.2363	0.2431
	1.00		$\sum x_i = 0.9951$	$\sum x_i = 1.0074$	$\sum x_i = 0.9992 \cong 1$

Table 9.8

Component	Vapor Composition, y_i	Liquid Composition, $x_i = y_i/K_i$
C_2	0.0346	0.00805
$^=C_3$	0.3740	0.2213
C_3	0.3475	0.2430
C_4	0.2431	0.5285
	$\Sigma y_i = 0.9992$	$\Sigma x_i = 1.00085$

Example 9.18

A mixture of 60 mol % n-hexane, 10 mol % n-heptane, and the remainder steam is cooled at constant pressure 1 atm from an initial temperature of 205°C. Hydrocarbons and water are immiscible. Calculate (a) the temperature at which condensation first occurs, (b) the composition of the first liquid phase, (c) the temperature at which the second liquid phase appears, and (d) the composition of the second liquid phase as it appears.

Vapor pressures of n-hexane and in n-heptane are given by

$$\ln P = A + \frac{B}{T}$$

where T is in K, P is in mm Hg, and constants A and B are as given below.

	A	B
Hexane	17.7109	−787
Heptane	17.9184	−193

Solution

Check for the condensation of steam. For the vapor pressure of steam, refer to steam tables. The vapor pressure of steam at the point of condensation is

$$P_s = 0.3(760) = 228 \text{ mm} \quad \text{Hg} = 4.41 \text{ psi}$$

From steam tables at 4.41 psi,

$$t = 155 + \frac{4.41 - 4.203}{4.741 - 4.203}(5)$$
$$= 156.92°F = 69.4°C = 342.4 \text{ K}$$

This is lower than 478 K. Hence, the steam will condense when cooled to this temperature. Check for condensation of n-heptane and n-hexane. Vapor pressure of n-hexane at 342.4 K = 773.4 mm Hg. Vapor pressure of n-heptane at 342.4 K = 290.80 mm Hg. Assume Raoult's law applies.

Then $\quad y_i P_t = x_i P_i \quad x_i = \dfrac{y_i P_t}{P_i}$

Hence, for hexane $\quad x_{hex} = \dfrac{0.6(760)}{773.4} = 0.5896$

For heptane $\quad x_{hep} = \dfrac{0.1(760)}{290.8} = 0.2613$

and $\quad \Sigma x_i = 0.5896 + 0.2613 = 0.8509 < 1.0$

Read just x_{hex} and x_{hep}

$$x_{hex} = \dfrac{0.5896}{0.8509} = 0.693$$

$$x_{hep} = \dfrac{0.2613}{0.8509} = \dfrac{0.307}{1.00}$$

Example 9.19

A liquid mixture of cyclohexane (1) and phenol (2) is in equilibrium with its vapor at 417.15 K. Liquid phase composition of cyclohexane is 0.6 mol fraction. Calculate the equilibrium pressure P and vapor composition y_i. Additional data are as follows:

(a) Activity coefficients are given by $\ln \gamma_1 = Ax_2^2$ and $\ln \gamma_2 = Ax_1^2$

(b) At 417.15 K, $P_1^{sat} = 75.2 \text{ kPa}$ and $P_2^{sat} = 31.66 \text{ kPa}$

(c) The binary system forms an azeotrope at 417.15 K. The azeotropic composition is $x_{1,az} = y_{1,az} = 0.294$

Solution

Relations for activity coefficients are suggested but the value of A is not given. From azeotropic data,

$$\gamma_1 = \dfrac{y_i P}{x_i P_1} = \dfrac{P}{P_1} \text{ since } x_1 = y_1 \text{ for an azeotrope}$$

Similarly $\quad \gamma_2 = \dfrac{P}{P_2}$

Then $\quad \dfrac{\gamma_1}{\gamma_2} = \dfrac{P/P_1}{P/P_2} = \dfrac{P_2}{P_1} = \dfrac{31.66}{75.2} = 0.421 \quad \therefore \gamma_1 = 0.421 \gamma_2$

Also from the given expressions for activity coefficients,

$$\dfrac{\ln \gamma_1}{\ln \gamma_2} = \dfrac{Ax_2^2}{Ax_1^2} = \dfrac{x_2^2}{x_1^2} = \dfrac{0.706^2}{0.294^2} = 5.767$$

Substituting γ_1 for and solving for ln γ_1, In $\gamma_2 = -0.1815$ which gives $\gamma_2 = 0.834$.

Then
$$\gamma_1 = 0.421 \times 0.834 = 0.351$$

This allows us to calculate A in the equations for activity coefficients as follows:

$$\ln \gamma_1 = Ax_2^2 = A(0.706)^2 = \ln(0.351) \quad \text{which gives } A = -2.1.$$
Now at 417.15 K, $x_1 = 0.6 \quad y_1 = ? \quad A = -2.1$
$$\ln \gamma_1 = Ax_2^2 = -2.1(0.4)^2 = -0.336 \quad \text{and} \quad \gamma_1 = 0.7146$$
$$\ln \gamma_2 = Ax_1^2 = -2.1(0.6)^2 = -0.756 \quad \text{and} \quad \gamma_2 = 0.4695.$$

Calculation of equilibrium pressure

$$y_1 P + y_2 P = P = (0.6)(0.7146)(75.2) + (0.4)(0.4695)(31.66) = 38.19 \text{ kPa}$$

Calculation of vapor composition

$$y_1 P = x_1 \gamma_1 P_1^{sat} = 0.6 \times 0.7146 \times 75.2$$

$$y_1 = \frac{0.6 \times 0.7146 \times 75.2}{38.19} = 0.844 \quad \therefore y_2 = 1 - 0.844 = 0.156$$

ENTHALPY-COMPOSITION DIAGRAMS (BINARY SYSTEM)

The enthalpy-composition diagram is a plot of vapor and liquid enthalpies versus the vapor and liquid compositions. This diagram must be constructed for a temperature range covering the two-phase region at the pressure of distillation. Since enthalpy has no absolute value, a reference state is needed to compute enthalpy changes. This is usually taken as 77°F or 25°C because most heats of mixing data are available at this temperature. In case one of the components is water, it is convenient to use 32°F or 25°C to calculate relative enthalpies. Once the basis is chosen, one can compute the relative enthalpies of liquid and vapor mixtures at a given pressure in the following manner:

$$\text{Enthalpy of liquid mixture}, \hat{h}_{\text{mixture}} = x_A \hat{h}_A + x_B \hat{h}_B + \Delta \hat{H}_{\text{mixing}} \quad (9.129)$$

where
$\hat{h}_{\text{mixture}}$ = relative enthalpy of liquid mixture per unit mass
$\hat{h}_A, \hat{h}_B$ = enthalpies of pure components (liquid state) per unit mass relative to reference temperature
$\Delta \hat{H}_{\text{mixing}}$ = heat of mixing (solution) per unit mass at the given temperature.

The heat of mixing is usually available at 77°F (25°C), therefore if the reference temperature is different from this, both $\hat{h}_A$ and $\hat{h}_B$ need to be calculated at 77°F (25°C) by the following equation:

$$(\hat{h}_A)_t - (\hat{h}_A)_{\text{ref}} = \int_{t_{\text{ref}}}^{t=77°F} C_{PA} dt \tag{9.130}$$

which reduces to $(\hat{h}_A)_t = \int_{t_{\text{ref}}}^{t=77°F} C_{PA} dt \quad \text{if} \quad (\hat{h}_A)_{\text{ref}} = 0$ (9.131)

In a similar manner, $(\hat{h}_B)_t = \int_{t_{\text{ref}}}^{t=77°F} C_{PB} dt \quad \text{if} \quad (\hat{h}_B)_{\text{ref}} = 0$ (9.132)

Once $\hat{h}_A$ and $\hat{h}_B$ at 77°F (25°C) are calculated, they are used to calculate $\hat{h}_{\text{mixture}}$ at any composition of liquid at 77°F (25°C). When this is done, relative enthalpies of liquid mixtures at any other temperature can be calculated by

$$\hat{h}_{\text{mixture at any } t} = \hat{h}_{\text{mixture at 77°F= 25°C}} + \int_{77°F= 25°C}^{T} C_{Px} dt \tag{9.133}$$

where C_{Px} is average heat capacity of the liquid mixture at concentration x.

After computing relative enthalpies of liquid mixtures at the bubble point temperatures, the next step is to calculate relative enthalpies of vapors in equilibrium with saturated liquid mixtures at their respective bubble point temperatures. The following data are required to make these calculations:

- Latent heats of vaporization of the two pure components
- Heat of mixing data for vapors
- Heat capacities of pure liquids and vapors
- Bubble point temperatures as a function of composition
- Dew point temperatures of the vapors as a function of composition
- Vapor-liquid equilibrium data

The heats of mixing of vapors are usually negligible and ignored. Alternatively, latent heats of vaporization of a given composition can be used to construct the enthalpy curve of saturated vapor. For practical purposes the latent heat of a liquid mixture can be calculated by the equation

$$\lambda_{\text{mixture}} = x_A \lambda_A + x_B \lambda_B \tag{9.134}$$

Then enthalpy of the saturated vapor of the same composition as the liquid is given by

$$\hat{H}_V = \hat{h}_{\text{mixture at bubble point temperature}} + \lambda_{\text{mixture}} \tag{9.135}$$

Enthalpy of a saturated vapor in equilibrium with the saturated liquid can be calculated using the composition of the vapor in equilibrium with the liquid. Thus if y_A and y_B are equilibrium compositions of components A and B, then the enthalpy of equilibrium vapor will be given by the equation

$$\hat{H}_V = y_A[C_{PA}(t_V - t_{\text{ref}}) + \lambda_A] + y_B[C_{PB}(t_V - t_{\text{ref}}) + \lambda_B] \tag{9.136}$$

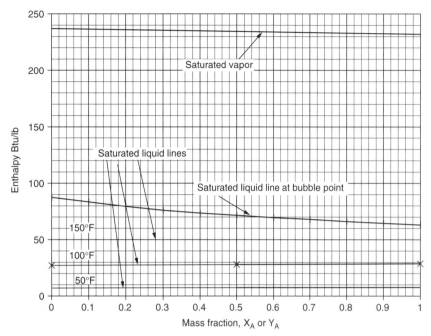

Figure 9.10 Enthalpy composition diagram

In making enthalpy calculations, it should be remembered that the *enthalpy is a state property* and any convenient path may be chosen to arrive at that state.

When no heat of solution data are available and only pure component data are used to construct the enthalpy-composition diagram, the saturated vapor and saturated liquid enthalpy lines will be straight. An example of an enthalpy composition diagram is shown in Figure 9.10.

RECOMMENDED REFERENCES

1. Geankoplis, *Transport Processes and Separation Principles,* 4th ed., Prentice-Hall, 2003.
2. AIChE Modular Instruction Series, *Mass Transfer,* vol. 1–4, 1985.
3. Greencorn and Kessler, *Transfer Operations,* McGraw-Hill, 1972.
4. Perry and Green, *Perry's Handbook for Chemical Engineers,* 50th ed., McGraw-Hill, 1999.
5. Sherwood, Pigford, and Wilke, *Mass Transfer,* McGraw-Hill, 1974.
6. Treybal, *Mass Transfer Operations,* 4th ed., McGraw-Hill, 1994.
7. Smith et al., *Chemical Engineering Thermodynamics,* 6th ed., McGraw-Hill, 2004.
8. Treybal, *Mass-Transfer Operations,* 4th ed., McGraw-Hill, 1994

CHAPTER 10

Distillation

OUTLINE

DISTILLATION METHODS 370
Equilibrium or Flash Distillation (Continuous Operation) ■ Differential Distillation ■ Simple Steam Distillation ■ Single-stage Equilibrium Contact and Constant Molal Overflow ■ Distillation with Reflux

PLATE-TO-PLATE CALCULATIONS 379
Sorel's Method ■ Lewis Method ■ Graphical Methods

BINARY SYSTEMS: McCABE-THIELE METHOD 380
Overall Material Balance: Binary Systems

FEED-PLATE LOCATION 384
Overall Energy Balance ■ Heating and Cooling Requirements ■ Minimum Reflux

COLD REFLUX 388
Partial Condensation ■ Total Reflux ■ Multifeed Column

SIDE-STREAM WITHDRAWAL 391
Special Cases of Rectification Using McCabe-Thiele Method

USE OF OPEN STEAM 393
Case 1 ■ Case 2 ■ Optimum Reflux Ratio ■ Fenske Equation

GRAPHICAL STAGES AT LOW CONCENTRATIONS 395

SEPARATION OF BINARY AZEOTROPIC MIXTURES 409
Extractive Distillation ■ Azeotropic Distillation ■ Multicomponent Distillation

SHORTCUT METHODS 413
Total Reflux in Multicomponent Distillation ■ Minimum Reflux Ratio for Multicomponent Distillation ■ Shortcut Method for Number of Theoretical Stages at Operating Reflux Ratios ■ Feed Plate Location

DESIGN CALCULATIONS FOR PACKED TOWERS 419
Tray Efficiencies ■ Efficiencies in Packed Towers

SIZING OF TRAY AND PACKED TOWERS 426
General Features of Tray Columns

RECOMMENDED REFERENCES 432

Distillation is an operation to separate various components from a liquid solution by vaporizing a part of the liquid mixture to form two phases, vapor and liquid. The separation of the components depends on the distribution of the components between the two phases. For separation to take place, the composition of the vapor phase must be different from the composition of the liquid phase. Distillation design calculations involve material and energy balances, and vapor-liquid equilibrium relationships. We have already reviewed vapor-liquid equilibria in Chapter 9. In this chapter, we review various methods of distillation and designing equipment for the same.

DISTILLATION METHODS

Distillation is carried out by two main methods. The first involves the generation of vapor by boiling the liquid mixture to be separated in a single stage and recovering the vapor, which is condensed. No part of the condensed liquid is returned to the stage as a reflux to contact the vapors rising from the still. In the second method, a portion of the condensed liquid is returned to the column consisting of a series of trays or stages through which it flows down to the still, in this case called the reboiler. Vapors from the reboiler rise through the trays and thus are brought in intimate contact with the liquid on the trays. This method is called by various names: *fractional distillation*, *rectification*, or *distillation with reflux*. Fractional distillation is used when sharp separations are desired and continuous large-scale operations are to be carried out.

Single-still distillation operations are of three types: (1) equilibrium or flash distillation, (2) simple batch or differential distillation, and (3) simple steam distillation.

Equilibrium or Flash Distillation (Continuous Operation)

In continuous flash distillation (Figure 10.1), the mixture to be separated is first heated and then fed continuously to a flash chamber. The separated vapor and liquid are withdrawn from the flash chamber continuously.

For a binary system, the process calculations can be done as follows:

Overall material balance: $\quad F = V + L \quad$ (10.1)

Component material balance: $\quad Fz_F = Vy + Lx = Vy + (F - V)x \quad$ (10.2)

The fraction vaporized is $\quad \dfrac{V}{F} = \dfrac{z_F - x}{y - x} \quad$ (10.3)

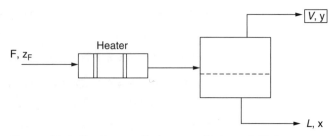

Figure 10.1 Continuous flash or equilibrium distillation

Equations 10.1 and 10.2 in conjunction with the equilibrium diagram allow us to determine the relative amounts of vapor and liquid produced for a binary system. A convenient way to determine the amount of vapor is to plot Equation 10.2 on the equilibrium diagram and locate the intersection point of Equation 10.2 and the equilibrium line. The intersection point gives the vapor and liquid compositions in equilibrium.

The energy balance for the process can be written as follows:

$$Fh_F = VH_v + Lh_L \quad (10.3a)$$

where
F = moles of feed
V = moles of flashed vapor
L = moles of undistilled liquid
h_F, H_v, and h_L = molar enthalpies of feed, vapor, and the condensed liquid

For a *multicomponent* mixture, the following equations apply:

$$x_i = \frac{z_{Fi}/L}{(K_i V/L)+1} \quad \text{or} \quad x_i = \frac{z_{Fi}/V}{K_i + L/V} \quad (10.4)$$

since
$$y_i = K_i x_i, \quad y_i = \frac{K_i z_{Fi}/V}{K_i + L/V} \quad (10.5)$$

and either $\Sigma x_i = 1.0$ or $\Sigma y_i = 1.0$. If one mole of the feed liquid is considered and f is the fraction vaporized, then $(1 - f)$ is the liquid left behind. Then by material balance,

$$z_{Fi} = f y_i + (1-f) x_i \quad (10.6)$$

From which

$$y_i = -\frac{1-f}{f} x_i + \frac{z_{Fi}}{f} \quad (10.7)$$

For multicomponent (>2 components) mixtures, a trial-and-error solution is required.

For equilibrium, $\quad y_i = K_i \alpha_i x_i = K_B \alpha_i x_i \quad (10.7a)$

where
$\alpha_i = K_i/K_B$
K_B = equilibrium constant of the base component.

Solving for x_i and adding for all components

$$\sum x_i = \sum \frac{z_{iF}}{f(K_B \alpha_i - 1) + 1} = 1.0 \quad (10.7b)$$

In the case of binary systems, the fraction vaporized is given by

$$f = \frac{z_{Fi}(K_1 - K_2)/(1 - K_2) - 1}{K_1 - 1} \quad (10.8)$$

Equilibrium flash calculations have been covered in Chapter 9.

Differential Distillation

A liquid is charged to a heated kettle or still and slowly boiled; the vapor formed is continuously withdrawn as rapidly as possible as it forms to a condenser and is condensed out. The condensed vapor (distillate) is collected. The portion of the condensate first collected will be richer in more volatile component. The Rayleigh equation relating final moles L_2 in the still to the initial charge L_1 moles of the liquid is

$$\ln \frac{L_1}{L_2} = \int_{x_{i2}}^{x_{i1}} \frac{dx_i}{y_i - x_i} \quad (10.9)$$

In theory, the vapor formed is in equilibrium with the liquid. The solution of this equation requires graphical integration. The integral is evaluated by plotting $1/(y_i - x_i)$ versus x_i and using Simpson's rule or other method to find the integral.

If the relative volatility α is constant, the solution of the equation is

$$\ln \frac{L_1}{L_2} = \frac{1}{\alpha - 1} \left(\ln \frac{x_{i1}}{x_{i2}} + \alpha \ln \frac{1 - x_{i2}}{1 - x_{i1}} \right) \quad (10.10)$$

For nonideal systems,

$$y = \frac{\gamma x P}{v P_t} = Kx \quad (10.11)$$

and

$$\ln \frac{L_1}{L_2} = \int_{x_{i2}}^{x_{i1}} \frac{dx_i}{x_i(K_i - 1)} = \int_{x_{i2}}^{x_{i1}} \frac{dx_i}{x_i(\gamma_i P_i / v_i P_t - 1)} \quad (10.12)$$

The average composition of total material distilled, y_{av} can be calculated by the equation

$$L_1 x_1 = L_2 x_2 + (L_1 - L_2) y_{av} \quad (10.13)$$

Example 10.1

A liquid mixture of 100 kg mol containing 50 mol% heptane and 50 mol% ethyl benzene is differentially distilled at 101.3 kPa until 40 mol is distilled. What is the average composition of the total vapor distilled and the composition of the liquid left in the still?

The vapor-liquid equilibrium data are given in Table 10.1.

Table 10.1 Equilibrium data for heptane-ethyl benzene system

x	y	1/(y − x)	x	y	1/(y − x)
0	0		0.3	0.575	3.60
0.08	0.23	6.67	**0.45**	**0.705**	**3.92**
0.1	**0.27**	**5.9**	0.485	0.73	4.08
0.2	**0.45**	**4.0**	0.79	0.94	8.77
0.25	0.514	3.79	1.0	1.0	

Solution

The given data were plotted to get the equilibrium diagram (see Exhibit 1),

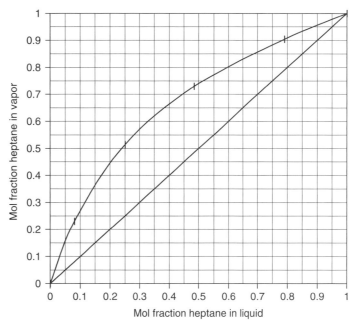

Exhibit 1 Vapor-liquid equilibrium diagram heptane-ethyl benzene system

and some intermediate values of y at different values of x were obtained to calculate $1/(y - x)$ at closer range. The interpolated data is given in bold in Table 10.1

A plot of $1/(y - x)$ vs x was prepared and is given in Exhibit 2.

Data given are: $L_1 = 100$ kg mol, $x_1 = 0.5$ mol fraction, $L_2 = 60$ kg mol, and moles distilled, $V = 40$ kg mol. Substituting the values in Equation 10.9, we obtain

$$\ln \frac{100}{60} = \int_{x_2}^{x_1 = 0.5} \frac{dx_i}{y_i - x_i} = 0.51$$

Since x_2 is not known, it has to be obtained by trial and error. A value of x_2 is assumed and the area under the curve from 0.5 to x_2 is obtained by trapezoidal rule (calculations are not shown here). A value of $x_2 = 0.304$ gave area under the curve $= 0.5094 \doteq 0.51$ the required value. Therefore $x_2 = 0.304$.

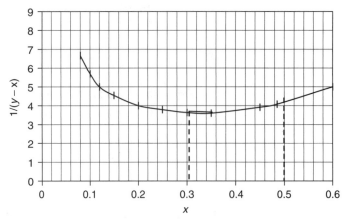

Exhibit 2 Plot of 1/(y −x) versus x for Example 10.1

Then average composition of 40 kg moles distilled can be calculated by Equation 10.13 as

$$100(0.5) = 60(0.304) + 40 y_{av}$$

and

$$y_{av} = 0.794$$

Simple Steam Distillation

High boiling liquids cannot be purified by distillation at atmospheric pressure. It is also possible that the components may decompose at the high temperatures. Simple steam distillation permits a separation at lower temperature and pressure. Steam distillation is usually used to separate high boiling materials from nonvolatile impurities.

If a component B is high boiling and immiscible with water (component A) they will form two separate phases when put together. On boiling this mixture at 101.3 kPa (1 atm), both water and component B will exert their own vapor pressures. Also, there will be three phases present: two liquid phases and a vapor phase (non-volatile inerts in small amounts are ignored). By application of the phase rule for 3 phases and 2 components, the degrees of freedom F is given by

$$F = 2 - 3 + 2 = 1$$

Therefore if the total pressure is fixed, the system state is fixed. When the sum of the vapor pressures equals the total pressure, the mixture will boil. Then the vapor composition is

$$y_A = \frac{P_A}{P} \quad \text{and} \quad y_B = \frac{P_B}{P} \tag{10.14}$$

As long as two phases are present, the mixture will boil at a constant temperature. So long as liquid water is present, the high boiling component also boils at a lower temperature, which is far below its normal boiling point. The vaporized mixture is condensed and the resulting two phases separated. A major disadvantage of this method is that a large amount of heat has to be used to vaporize water, which has a high latent heat of vaporization.

Figure 10.2 Concept of equilibrium stage in distillation

The ratio of moles of B distilled to moles of steam used A is given by

$$\frac{n_B}{n_A} = \frac{P_B}{P_A} \tag{10.15}$$

Other more complex cases of steam distillation are discussed by Van Winkle.

Single-stage Equilibrium Contact and Constant Molal Overflow

A vapor stream V_2 and a liquid stream L_0 are brought into contact in a stage as in Figure 10.2 and allowed to come to equilibrium. We consider only a binary system here meaning streams L_0 and V_2 consist of the same two components A and B but are of different compositions.

Overall and component material balances can be written as follows:

$$\text{Overall balance:} \quad V_2 + L_0 = V_1 + L_1 \tag{10.16}$$

$$\text{Component balance:} \quad V_2 y_2 + L_0 x_0 = V_1 y_1 + L_1 x_1 \tag{10.16a}$$

where

x_0, x_1 = mol fractions of component A (more volatile) in liquid
x_0 = mol fraction in liquid feed = mol fraction in liquid leaving the stage y_1
and y_2 are similarly defined for the vapor phase streams V_1 and V_2.

If the sensible heat effects are small and negligible and heat of mixing can be ignored, and the latent heats of vaporization of the two components are the same, then when one mol of A vaporizes, 1 mol of B must condense. This means mols of vapor V_1 leaving the stage will equal the mols entering V_2.

Since V_1 and V_2 are equal, L_0 and L_1 also are equal. This is termed *constant molal overflow*. Knowing the amounts and compositions of the liquid and vapor feeds to the stage, the compositions of the outlet streams can be determined in conjunction with an equilibrium diagram. An example will make the concept clear.

Example 10.2

100 kg mol of a vapor at its dew point and 101.32 kPa containing 0.4 mol fraction heptane (A) and 0.6 mol fraction ethyl benzene (B) is brought into contact with 120 kg mol of a liquid at the boiling point and containing 0.3 mol fraction of heptane and the rest ethyl benzene. The outlet streams leave the stage in equilibrium with each other. Calculate the amounts and compositions of the exit streams. Assume constant molal overflow.

Solution

The process diagram is the same as in Figure 10.3. The equilibrium diagram is shown in Exhibit 3.

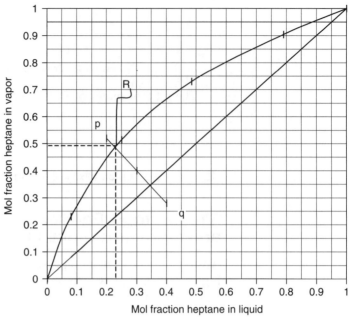

Exhibit 3 Equilibrium diagram for system heptane-ethyl benzene. Solution of Example 10.2

With the notation of Equation 10.16, given data are as follows:

$V_2 = 100$ kg mol, $y_{A2} = 0.4$, $L_0 = 120$ kg mol and $x_{A0} = 0.3$ mol fraction.

For constant molal overflow, $L_0 = L_1$ and $V_2 = V_1$.
Then making a material balance on component A gives

$$L_0 x_{A0} + V_2 y_{A2} = L_1 x_{A1} + V_1 y_A$$

By substituting the given values in the above relation,

$$120(0.3) + 100(0.4) = 120(x_{A1}) + 100(y_{A1})$$

To solve this equation, we need to know the relationship between y_{A1} and x_A. The relationship is provided by the equilibrium relationship. Since no analytical relationship is available between y_{A1} and x_{A1}, a trial-and-error solution is required. Best way is to plot the above relationship on the equilibrium diagram and locate its intersection with the equilibrium curve. This intersection point will yield the correct solution. For various values of x_{A1}, calculate values of y_{A1}.

x_{A1}	0.2	0.3	0.4
y_{A1}	0.52	0.4	0.28

The intersection point is $x_{A1} = 0.225$ $y_{A1} = 0.49$.

Distillation with Reflux

Rectification or continuous distillation with reflux can be considered as a series of equilibrium flash vaporization stages arranged one above the other in such a manner that the vapor rising from the reboiler passes countercurrently to the flow of liquid

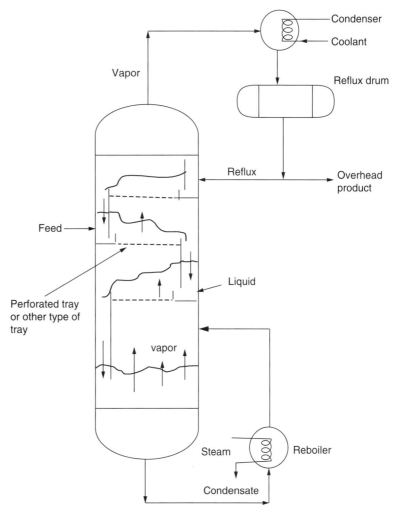

Figure 10.3 Schematics of a rectification tray tower (only 3 trays are shown)

descending downward through the stages in a column. The vapors come in contact with the liquid and exchange heat energy and material with the effect that more volatile components from the liquid are vaporized while less volatile components from the vapors are condensed. At each stage the vapor leaving the stage is in equilibrium with the liquid leaving the stage. The principle of equilibrium stage is explained by considering a single-stage contact earlier. Example 10.2 illustrates the calculation procedure for the case of constant overflow. The trays are arranged in a tower vertically. A schematic diagram of a distillation column is shown in Figure 10.3.

A distillation unit usually consists of (a) reboiler where the vapor is generated, (b) a fractionating column having a number of plates or trays of some type (bubble cap, sieve) through which vapors rise in countercurrent contact with the descending liquid forming a pool on each tray and (c) a condenser, which condenses vapors from the column. A part of the condensed liquid is returned to the column as reflux. Since heat and material exchange is brought about by intimate contact, different types of internals have been devised to bring about intimate contact of rising vapor and the descending liquid. The more common internals are bubble caps and sieve or perforated plates. Another relatively new development is turbogrid plate that consists of a number of horizontal bars with narrow slots in between. Turbogrid plate has much more separation effectiveness and much less pressure drop.

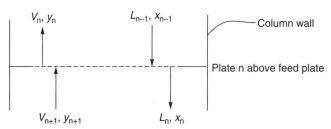

Figure 10.4 Theoretical plate in rectification section above feed plate

Another piece of equipment that is frequently used to bring vapor and liquid in effective contact is the *packed tower*. For moderate throughputs, a packed tower is much more cost effective.

The design of a tower is based on the principle of the equilibrium stage and includes the calculation of number of equilibrium stages required. In practice, the stage is not as perfect as assumed, and therefore an efficiency factor has to be used to obtain the actual number of trays. The next step is to size the tower, specify its internals, material of construction, etc. We will review these aspects of distillation with reflux next.

Theoretical Stage

A theoretical or ideal stage in a column is akin to an equilibrium stage. It may be defined as a plate holding a pool of liquid from which a vapor rises and a liquid descends such that the composition of rising vapor is in equilibrium with the composition of the liquid. This assumes that the mixing of vapor and the liquid is perfect. In Figure 10.4, plate n above the feed plate in a column is shown where the plates are numbered beginning from the top of the column. Notice that the subscript to a stream refers to the plate from which it originates.

A material and heat balance on the plate can be written in the following manner:

$$\text{Material balance:} \quad V_{n+1} + L_{n-1} = V_n + L_n \quad (10.17)$$

$$\text{Component balance:} \quad V_{n+1} y_{n+1} + L_{n-1} x_{n-1} = V_n y_n + L_n x_n \quad (10.17a)$$

Similarly, an energy balance can be written for the plate using enthalpies of the stream as follows:

$$V_{n+1} \hat{H}_{n+1} + L_{n-1} \hat{h}_{n-1} = V_n \hat{H}_n + L_n \hat{h}_n \quad (10.18)$$

In general the latent heats of vaporization of the components to be separated are not equal and although sensible heat effects may be small and heat losses from the column may be ignored, the streams are $V_{n+1} \neq V_n$ and $L_{n-1} \neq L_n$. This means the molal overflows from plate to plate will not be constant. The calculations in that case will be more complex.

If, however, latent heats of vaporization of the components are equal; this fact can be combined with Trouton's rule to make it more useful. Trouton's empirical approximation states that the molal latent heat of vaporization divided by the absolute boiling point temperature is constant for large groups of chemically similar compounds. This means that a mole of one component requires as much heat to vaporize it as a mole of another component. The heat of vaporization of any mixture on the plate is independent of the composition of the mixture provided the temperature change is negligible. In this case, $V_{n+1} = V_n$ and $L_{n-1} = L_n$. This result is general in that it applies to all plates between the top plate and the feed plate. Thus we have *constant molal vaporization* and *constant molal underflow* (liquid). If this assumption can be made, the design calculation of the number of theoretical plates is greatly simplified.

A portion of the condensed distillate called *reflux* is returned to the top plate in the column. This causes vaporization of the more volatile (light) component at each stage and condensation of the less volatile (heavy) component in the top or above the feed portion of the column. This enrichment of the more volatile component is called *rectification* and the portion of the column where it occurs is termed the *rectification section*.

In the section below the feed plate, hot vapor from the reboiler passes through each stage, causing vaporization of the more volatile or light component and condensation of the less volatile or heavy component. Thus the more volatile component is stripped from the liquid. This process is called *stripping* and the portion of the column where it is done is termed the *stripping section* of the column.

In multicomponent distillation, the light (more volatile) and the heavy (less volatile) components between which the required separation is specified are termed the *key components* (*light* key and *heavy* key). For the design of a column, the number of theoretical stages required for a given separation is first computed. The actual number of stages is then determined by using a *plate efficiency factor*.

PLATE-TO-PLATE CALCULATIONS

One method of determining the number of equilibrium stages is to make plate-to-plate calculations. Conditions at the two ends (top and bottom) of the column are known. Hence either the top or bottom can be selected as the starting point for the calculations.

Sorel's Method

This method does plate-to-plate calculations without the assumption of constant molal overflow. It uses material and enthalpy balances together with equilibrium calculations to determine the flows of liquid and vapor from the plate, temperature of the plate, and the composition of each stream from the plate. The method assumes the operating pressure, the reflux ratio, temperature or enthalpy of the reflux stream, and the use of a total condenser. These calculations are rigorous, tedious, and time consuming for hand calculations. If relevant data are available, this method is suitable for high-speed computer calculation.

Lewis Method

This method also uses plate-to-plate calculations but simplifies them by assuming constant molal overflow from stage to stage in both the rectifying and stripping sections. This is equivalent to assuming equimolal latent heats and heat capacities and no heat of mixing. He also assumed the reflux L_0 to be a saturated liquid. Even then, like Sorel's method, it is tedious and time consuming.

Graphical Methods

Graphical methods are developed to determine the equilibrium stages because of the tedious trial-and-error calculations for each plate involved in plate-to-plate calculations. Two important methods for binary systems are (a) McCabe-Thiele and (b) Poncho-Savarit.

Poncho-Savarit Graphical Method

The Poncho-Savarit graphical method uses an enthalpy-composition diagram for solving material and energy balances for binary systems. For the construction of ideal stages on this diagram, no simplifications, such as assumption of constant molal overflow, are needed. However, almost all schools in the U.S. have stopped teaching the Poncho-Savarit method and therefore we will not review it in this book.

McCabe-Thiele Method

The McCabe-Thiele method is based on the Lewis modification of the Sorel method. To simplify calculations, it makes a number of assumptions, as follows:

1. Constant molal overflow in the rectifying section as well as in the stripping section, which means equimolal latent heats and heat capacities and no heat of mixing.

2. The reflux L_0 to the top plate in the column is a saturated liquid.

3. The column pressure is constant.

4. The reflux ratio is also assumed.

We shall now review the McCabe-Thiele graphical method at length.

BINARY SYSTEMS: McCABE-THIELE METHOD

Analysis of plate columns is based on the overall material and energy balances and phase equilibria. The important factors in the design calculations of the plate columns are (1) number of plates required, theoretical and actual, for a desired separation; (2) heat input to the reboiler; (3) heat output from the condenser; (4) plate spacing and diameter of column; and (5) type and construction of plates.

Overall Material Balance: Binary Systems

Referring to Figure 10.5, the total material balance gives

$$F = D + B \tag{10.19}$$

and component A balance gives

$$Fz_F = Dx_D + Bx_B \tag{10.20}$$

where
F = feed, lb mol/h [kg mol/h]
D = distillate, lb mol/h [kg mol/h]
B = bottoms product, lb mol/h [kg mol/h]

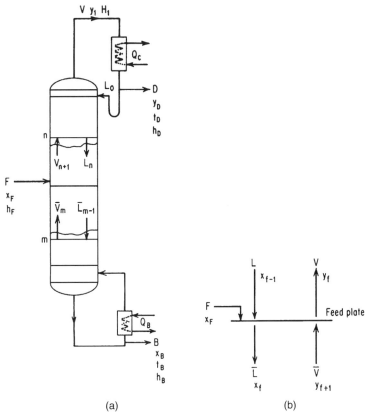

Figure 10.5 (a) Material and energy balances for a binary system around distillation column; (b) Material balance over the feed plate

From Equations 10.19 and 10.20, one obtains

$$\frac{D}{F} = \frac{x_F - x_B}{x_D - x_B} \quad \text{and} \quad \frac{B}{F} = \frac{x_D - x_F}{x_D - x_B} \tag{10.21}$$

The operating line equations are different for the rectification and stripping sections.

Reflux at It's Bubble Point, Total Condenser

The equation of the operating line is

$$y_{n+1} = \frac{L}{V} x_n + \frac{D}{V} x_D \tag{10.22}$$

where L is the liquid flow, lb mol/h [kg mol/h], from plate to plate in the rectifying section and V is the vapor flow, lb mol/h [kg mol/h], from plate to plate in the rectifying section. Also,

$$y_{n+1} = \frac{R}{R+1} x_n + \frac{x_D}{R+1} \tag{10.23}$$

where the external reflux ratio is

$$R = \frac{L_0}{D} = \frac{L}{D} \qquad (10.24)$$

The slope of the operating line is

$$\frac{L}{V} = \frac{R}{R+1} \qquad (10.25)$$

and the intercept of the line is $x_D/(R + 1)$.

Stripping Section
The equation of the operating line is

$$y_{m+1} = \frac{\bar{L}}{\bar{V}} x_m - \frac{B}{\bar{V}} x_B \qquad (10.26)$$

where $\bar{L}$ is the liquid overflow in stripping section, lb mol/h [kg mol/h], and $\bar{V}$ is the vapor flow in stripping section, lb mol/h [kg mol/h], or

$$y_{m+1} = \frac{\bar{L}}{\bar{L} - B} x_m - \frac{B}{\bar{L} - B} x_B \qquad (10.27)$$

Using the material balance at the feed plate (Figure 10.5b),

$$F + L + \bar{V} = V + \bar{L} \qquad (10.28)$$

$$\bar{L} = L + qF \quad \text{or} \quad V = \bar{V} + (1-q)F \qquad (10.28a)$$

$$q = \frac{\bar{L} - L}{F} = \frac{\text{heat to convert 1 mol of feed to saturated vapor}}{\text{latent heat of vaporization per mol of feed}} \qquad (10.29)$$

The equation of the operating line (stripping section) in terms of q is

$$y_{m+1} = \frac{L + qF}{L + qF - B} x_m - \frac{Bx_B}{L + qF - B} \qquad (10.30)$$

The equation of the q line is

$$y = \frac{q}{q-1} x - \frac{x_F}{q-1} \qquad (10.31)$$

The method of the McCabe-Thiele graphical solution is shown in Figure 10.6.

Values of q
The values of q are determined as follows:

Cold feed: $q > 1$ $\qquad q = 1 + (C_p)_L \dfrac{t_b - t_F}{\lambda} \qquad (10.32)$

Feed at boiling point: $\qquad q = 1 \qquad (10.32a)$

Partially vaporized feed: $0 < q < 1 \qquad q = \dfrac{f_\ell \lambda}{\lambda} = f_\ell \qquad (10.32b)$

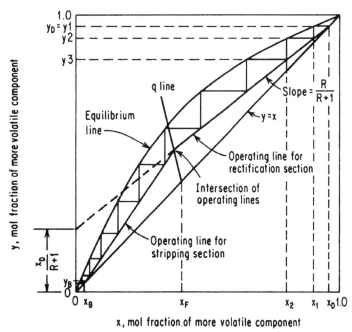

Figure 10.6 McCabe-Thiele solution for the number of theoretical plates for a binary system

where f_l is the liquid fraction in the feed.

Feed at dew point (saturated vapor): $\quad q = 0 \quad$ (**10.32c**)

Feed, superheated vapor: $\quad q < 0 \quad q = \dfrac{C_{pv}(t_F - t_d)}{\lambda} \quad$ (**10.32d**)

where
  t_d = dew-point temperature, °F [K]
  $(C_p)_L$ = specific heat of liquid feed, Btu/lb mol · °F [kJ/kg · K]
  C_{pv} = specific heat of superheated vapor feed, Btu/lb mol · °F [kJ/kg · K]
  t_b = boiling point, °F [K]
  t_F = feed temperature, °F [K]
  λ = latent heat, Btu/lb mol of feed [kJ/kg mol]

Note that the q or feed line has a slope of ∞ (i.e., it is vertical) when $q = 1$ or the feed is at its boiling point. It is parallel to the x axis when the feed is a saturated vapor.

pa: q line when the feed is cold liquid.

pb: q line when the feed is liquid at its boiling point. The line is vertical.

pc: q line when the feed is a mixture of liquid and vapor (partially vaporized feed).

pd: q line when the feed is a saturated vapor. The line is parallel to the x axis.

pe: q line when the feed is a superheated vapor.

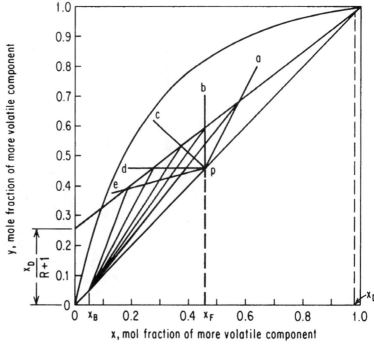

Figure 10.7 Effect of feed condition on q line

The q line also passes through the intersection point of the 45° line and the vertical line through x_F. All the intersections of the operating lines fall on the q line. The effect of the feed condition on the q line is shown in Figure 10.7.

FEED-PLATE LOCATION

In order to obtain the number of theoretical plates by graphical method, the operating lines are drawn that intersect each other on the q line. Starting at the top point given by x_D, y_D (see Figure 10.5), stages are stepped off using the equilibrium curve and the rectification section operating line. As shown in Figure 10.5, shift to the use of the stripping section operating line was made immediately after the intersection point was passed. This results in a minimum number of theoretical stages. The shift could be made earlier (before passing the intersection point) or later (after passing the intersection point). However, this results in more theoretical plates as is shown in Figure 10.8. In both the cases early or later entry of the feed to the column is found to require a larger number of theoretical plates. Thus the least total number of trays will result if the transition is made immediately after crossing the intersection point of the operating lines. This is the basis to be used in the design of new columns.

When an existing column is to be adapted to a new separation, the choice of feed location is restricted to existing nozzles. Slope of the operating lines or reflux ratio have to be determined by trial and error.

The method of introducing the feed to the column will depend on the condition of feed. If it is all vapor, it is introduced below the feed plate. If it is all liquid, it is added to the liquid flowing down from the tray above the feed plate. If it is a mixture of part vapor and part liquid, they are first separated in a separator. The vapor portion is admitted below the feed tray, whereas the liquid is added above the tray.

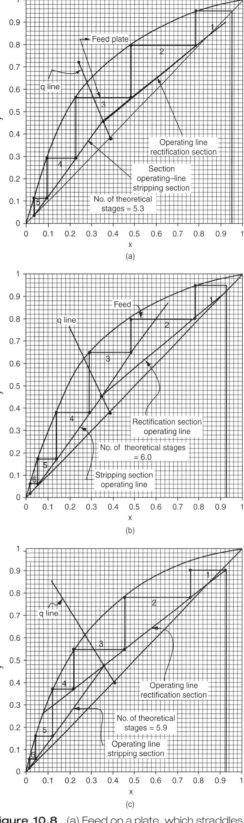

Figure 10.8 (a) Feed on a plate, which straddles the two operating lines, (b) Early entry of feed into column, and (c) Delayed entry of feed

Overall Energy Balance

Assume heat loss to the surroundings of $Q_L = 0$. Energy balance around the whole column gives

$$Fh_F + Q_B = Dh_D + Bh_B + Q_c \qquad (10.33)$$

Condenser duty is given by

$$Q_c = D(R+1)(H_1 - h_D) \qquad (10.34)$$

where h and H represent enthalpy, Btu/lb mol [kJ/kg mol]. Subscripts B, D, F, and 1 indicate bottom, distillate, feed, and tray 1, respectively. If no subcooling of the condensate is done,

$$Q_c = D(R+1)\lambda \qquad (10.34a)$$

where λ is the latent heat of condensation of overhead vapor, Btu/lb mol [kJ/kg mol].

If t_F is taken as the datum temperature, the feed is 100 percent liquid, and t_D and t_B are distillate and bottoms temperatures respectively,

$$h_F = 0 \qquad h_D = C_{PD}(t_D - t_F) \qquad h_B = C_{PB}(t_B - t_F) \qquad (10.34b)$$

where C_{PD} and C_{PB} are the molar specific heats of the distillate and bottoms, respectively. Hence

$$Q_B = DC_{PD}(t_D - t_F) + BC_{PB}(t_B - t_F) + Q_C \qquad (10.35)$$

If heat loss Q_L is not negligible, the reboiler duty would be $Q_B + Q_L$.

Heating and Cooling Requirements

Steam Required at the Reboiler

Assuming radiation from the column to be small and negligible and the column to be operating adiabatically, the amount of the steam required at the reboiler is given by

$$S = \frac{\overline{V}\lambda}{\lambda_s} = \frac{Q_B}{\lambda_s} \text{ lb/h [kg/h]} \qquad (10.36)$$

where
- λ = latent heat of mixture, Btu/lb mol [kJ/kg mol]
- λ_s = latent heat of steam, Btu/lb [kJ/kg]
- $\overline{V}$ = vapor flow in stripping section, lb mol/h [kg mol/h]

Cooling Water Required at the Condenser

With no subcooling and total condensation, the cooling water required at the condenser is given by

$$\frac{V\lambda}{t_2 - t_1} \text{ lb/h [kg/h]} \qquad (10.37)$$

and with subcooling of the condensate, the cooling water needed is

$$\frac{V[\lambda + C_{PD}(t_b - t_R)]}{t_2 - t_1} \text{ lb/h [kg/h]} \qquad (10.38)$$

where

t_b = bubble-point temperature of the overhead vapor, °F [K]
t_R = actual temperature of condensed distillate, °F [K]
C_{PD} = specific heat of the distillate, Btu/lb mol · °F [K]
V = vapor flow in the rectification section, lb mol/h = $(R + 1)D$
$t_2 - t_1$ = temperature rise of cooling water, °F [K]
λ = latent heat of overhead vapor mixture, Btu/lb mol [kJ/kg mol]

Minimum Reflux

When the equilibrium curve shows concavity downward, the minimum reflux ratio (Figure 10.9) can be calculated by

$$R_m = \frac{x_D - y'}{y' - x'} \tag{10.39}$$

where R_m is the minimum reflux ratio. At the minimum reflux ratio, the number of plates required is infinite for a given separation. When $q = 1$, i.e., the feed is at the bubble point, the minimum reflux ratio for a binary system can be calculated by

$$R_m = \frac{1}{\lambda - 1} \left[\frac{x_D}{x_F} - \frac{\alpha(1 - x_D)}{1 - x_F} \right] \tag{10.40}$$

When $q = 0$, i.e., the feed is saturated vapor at the dew point, the minimum reflux ratio can also be calculated by

$$R_m = \frac{1}{\alpha - 1} \left(\frac{\alpha x_D}{y_F} - \frac{1 - x_D}{1 - y_F} \right) - 1 \tag{10.41}$$

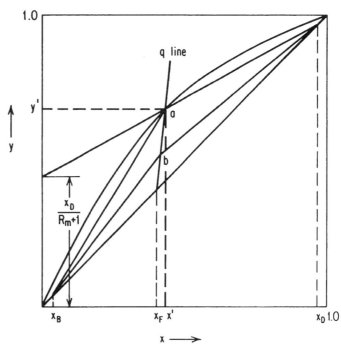

Figure 10.9 Minimum reflux ratio when the equilibrium curve is concave downward (*Note*: $x' = x_F$ when the feed is at it's boiling point)

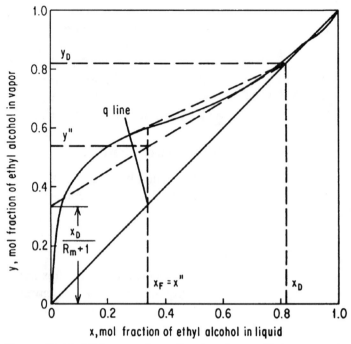

Figure 10.10 Minimum reflux ratio when the equilibrium curve is concave upward (ethyl alcohol–water system)

In these cases, α is estimated at the conditions represented by the intersection of the q line and the equilibrium curve. Equation 10.39 cannot be applied in all cases. If the equilibrium curve has an unusual curvature, the minimum reflux R_m is to be found by drawing an operating line through $y_D = x_D$ that is tangent to the equilibrium curve. The slope of this tangent line is to be used to find R_m (Figure 10.10):

$$R_m = \frac{x_D - y''}{y'' - x''} \qquad (10.42)$$

Alternatively, R_m can be calculated from the intercept of the operating line drawn through the point (x_D, y_D) and tangential to the equilibrium line. Note that R_m is independent of the q line in this case.

COLD REFLUX

If the reflux is cooled below its bubble point, the amount of vapor from the top tray is less because of condensation caused by heating the reflux from t_R to t_b. In this case, internal reflux L_n is different from L_0, the reflux fed to the top tray, and is given by

$$L_n = L_0 + \frac{L_0 C_{PD}(t_b - t_R)}{\lambda_{av}} \qquad (10.43)$$

Therefore, apparent reflux ratio is

$$R' = \frac{L_n}{D} = \frac{L}{V - L} = R\left[1 + \frac{C_{PD}(t_b - t_R)}{\lambda_{av}}\right] \qquad (10.44)$$

where C_{PD} is the specific heat, Btu/lb mol · °F [kJ/kg mol · K], and λ_{av} is the average latent heat of overhead vapor mixture, Btu/lb mol [kJ/kg mol].

The equation of the operating line in the rectification or enriching section is given by

$$y_{n+1} = \frac{R'}{R'+1} x_n + \frac{x_D}{R'+1} \qquad (10.45)$$

Partial Condensation

A partial condenser may produce one of the following results: (1) if the time of contact of the liquid and vapor is adequate, equilibrium condensation occurs; (2) if the condensate is removed rapidly, differential condensation occurs; and (3) if cooling is rapid, no mass transfer occurs and the compositions of the vapor and liquid are the same. In case 1, the condenser rates as one theoretical plate. In the design of new equipment, it is safer to ignore the enrichment by the partial condensation and include an additional theoretical tray in the column itself, since it is difficult to ensure that equilibrium condensation will occur.

Total Reflux

At total reflux, the operating lines coincide with the 45° line. Then the number of plates is minimum, but there is no product (Figure 10.11).

Multifeed Column

Equations for the operating lines and q lines are as follows (Figure 10.12). For enriching section,

$$y_{n+1} = \frac{L}{V} x_n + \frac{D}{V} x_D = \frac{L}{L+D} x_n + \frac{D}{L+D} x_D = \frac{R}{R+1} x_n + \frac{1}{R+1} x_D \qquad (10.46)$$

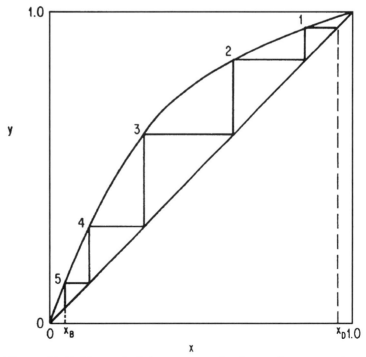

Figure 10.11 Theoretical plate construction for a column operating at total reflux (*Note*: The operating line coincides with the 45° line.)

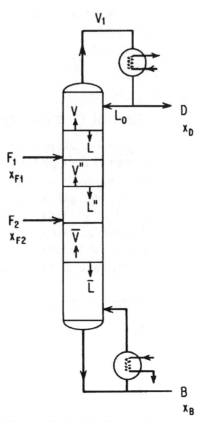

Figure 10.12 Multifeed column

For the intermediate section (column section in between two feed points),

$$y_{p+1} = \frac{L''}{V''} x_p + Dx_D - Fx_{F1} \tag{10.47}$$

For stripping section,

$$y_{m+1} = \frac{\bar{L}}{\bar{V}} x_m - \frac{\bar{B}}{\bar{V}} x_B \tag{10.48}$$

where
$$L'' = L + L_{F1} \quad \text{and} \quad V'' = L'' + D - F_1$$
$$\text{Also,} \quad \bar{L} = L'' + L_{F2} \quad \text{and} \quad \bar{V} = L' - B$$

Equations of the q lines for feeds 1 and 2 are

$$y = \frac{q_1}{q_1 - 1} x - \frac{x_{F1}}{q_1 - 1} \quad \text{for feed 1} \tag{10.49}$$

$$y = \frac{q_2}{q_2 - 1} x - \frac{x_{F2}}{q_2 - 1} \quad \text{for feed 2} \tag{10.50}$$

The preceding equations then become, for the rectification section,

$$y_{n+1} = \frac{R}{R+1} x_n + \frac{x_D}{R+1} \tag{10.51}$$

and for the intermediate section, the q-line equation is given by

$$y_{p+1} = \frac{L + q_1 F_1}{L + q_1 F_1 + D - F_1} x_p + \frac{Dx_D - F_1 x_{F1}}{L + q_1 F_1 + D - F_1} \qquad (10.52)$$

The bottom section operating line is given by

$$y_{m+1} = \frac{L + q_1 F_1 + q_2 F_2}{L + q_1 F_1 + q_2 F_2 + D - F_1 - F_2} x_m - \frac{Dx_D - F_1 x_{F1} - F_2 x_{F2}}{L + q_1 F_1 + q_2 F_2 + D - F_1 - F_2} \qquad (10.53)$$

SIDE-STREAM WITHDRAWAL

When the side stream is in the enriching section, the operating line above side stream is given by (refer to Figure 10.13)

$$y_{n+1} = \frac{L}{V} x_n + \frac{Dx_D}{V} = \frac{R}{R+1} x_n + \frac{1}{R+1} x_D \qquad (10.54)$$

For the section between the feed point and side stream, the operating-line equation is

$$y_{p+1} = \frac{L''}{V} x_p + \frac{Dx_D + L_s x_s}{V} \quad \text{since} \quad V'' = V \qquad (10.55)$$

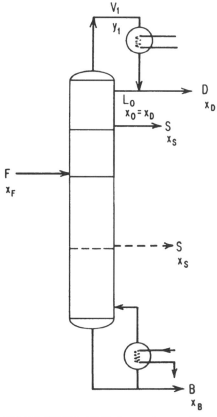

Figure 10.13 Distillation column with side-stream withdrawal

The locus of the intersection of the two operating lines is given by

$$(V'' - V)y = (L - L'')x - L_s x_s \tag{10.56}$$

Since $V'' = V$ and $V'' - V = 0$, the locus of the intersection of the operating line passes through the 45° line at x_s and has a slope of ∞; therefore it is a vertical line. Since $L_s = L - L''$ or $L'' = L - L_s$, so

$$y_{p+1} = \frac{L - L_s}{V} x_p + \frac{Dx_D + L_s x_s}{V} \tag{10.57}$$

At the feed point, the equation of the q line is

$$y = \frac{q}{q-1} x - \frac{x_F}{q-1} \tag{10.58}$$

The bottom-section operating-line equation is

$$y_{m+1} = \frac{(L - L_s) + qF}{(L - L_s) + qF - B} x_m - \frac{Bx_B}{(L - L_s) + qF - B} \tag{10.59}$$

If the side stream is withdrawn from the bottom section, the following equations apply:

$$y_{m+1} = \frac{\bar{L}_m}{\bar{V}_{m+1}} x_m - \frac{L'_s x'_s + Bx_B}{\bar{V}_{m+1}} \quad \text{above side stream} \tag{10.60}$$

$$\left.\begin{array}{l} V''_{r+1} y_{r+1} = L''_s x''_s + Bx_B \\[6pt] y_{r+1} = \dfrac{L'' x_r}{V''_{r+1}} x_r - \dfrac{Bx_B}{V''_{r+1}} \end{array}\right\} \quad \text{below side stream} \tag{10.61}$$

where $L''_r = L_m - L_s$ and $V''_{r+1} = V$. In this case also, the side-draw line is vertical because $V''_{r+1} = \bar{V}$.

Special Cases of Rectification Using McCabe-Thiele Method

Stripping Column Distillation

Sometimes in a particular case, complete column consisting of both rectification and stripping sections is not used. Instead, either a stripping tower or a rectification column is used. In stripping-column distillation, the feed is added to the top plate. The feed is usually saturated liquid at its boiling point. The overhead product is the vapor rising from the top plate. No reflux is returned to the tower. The column operates as a stripping tower and the vapor removes the more volatile component from the descending liquid. Making usual assumptions of constant molal overflows, a material balance for a plate together with the section of the column below it yields

$$y_{m+1} = \frac{L_m}{V_{m+1}} x_m - \frac{Bx_B}{V_{m+1}} \tag{10.62}$$

This stripping line equation intersects the $y = x$ line at $x = x_B$. The slope of the line L_m/V_{m+1} is constant. Using q line and stripping operating line, the theoretical steps can be stepped off. If the tower consists of only rectifying section, then also theoretical stages could be stepped off in a similar manner.

USE OF OPEN STEAM

Open steam is useful only where water is one of the components and is removed as a bottom product (Figure 10.14).

Case 1

Saturated Steam at Tower Pressure
The overall balance is

$$F + S = D + B \tag{10.63}$$

where S is flow rate of steam used.

$$Fx_F = Dx_D + Bx_B \tag{10.63a}$$

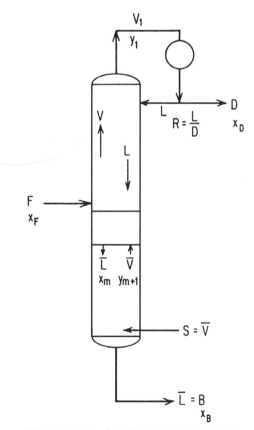

Figure 10.14 Stripping column with use of open steam

The operating line for the rectifying section is given by

$$y_n = \frac{L}{V}x_n + \frac{Dx_D}{V} = \frac{R}{R+1}x_n + \frac{x_D}{R+1} \tag{10.64}$$

q line

$$y = \frac{q}{q-1}x - \frac{x_F}{q-1} \tag{10.65}$$

For saturated steam at tower pressure, the material balance for the lower section is given by

$$\bar{L}x_m + S(0) = \bar{V}y_{m+1} + Bx_B \tag{10.66}$$

and since $\bar{L} = B$ and $\bar{V} = S$,

$$\frac{\bar{L}}{\bar{V}} = \frac{y_m + 1}{x_m - x_B} \quad \text{or} \quad y_{m+1} = \frac{\bar{L}}{\bar{V}}x_m - \frac{\bar{L}}{\bar{V}}x_B \tag{10.67}$$

when $x_m = x_B$, $y_{m+1} = 0$ at the bottom of the column. Therefore the stripping-section operating line passes through $y = 0$, and $x = x_B$.

Case 2

Superheated Steam

If steam is superheated, the liquid on the bottom tray vaporizes. The amount of vaporization is given by

$$\frac{S(\hat{H}_s - H_s)}{\lambda_m} \quad \text{lb mol/h [kg mol/h]} \tag{10.68}$$

where
  S = lb mol/h of superheated steam [kg mol/h]
  $\hat{H}_s$ = enthalpy of superheated steam, Btu/lb mol [kJ/kg mol]
  H_s = enthalpy of saturated steam at tower pressure, Btu/lb mol [kJ/kg mol]
  λ_m = molar latent heat of bottom liquid mixture at tower pressure, Btu/lb mol [kJ/kg mol]

Therefore, $\quad \bar{V} = S + \dfrac{S(\hat{H}_s - H_s)}{\lambda_m} = S\left(1 + \dfrac{\hat{H}_s - H_s}{\lambda_m}\right) \tag{10.69}$

from which the internal $\frac{\bar{L}}{\bar{V}}$ is calculated.

Optimum Reflux Ratio

As the reflux ratio is increased from the theoretical minimum, the total cost of running a distillation system comprising the depreciation of the installed equipment and the operating cost of coolant, heating medium, and electricity first decreases, reaches the minimum value, and then increases. The reflux ratio at which the total cost is minimum is called the optimum reflux ratio. In practice, a reflux ratio equal

Fenske Equation

The minimum number of plates at total reflux can be obtained graphically on the xy diagram between composition x_D and x_B using the 45° line as the operating line for both the rectifying and stripping sections.

For ideal mixtures and constant relative volatility, the minimum number of plates N_{min} can be calculated by the Fenske equation

$$N_{min} = \frac{\log[x_D(1-x_B)/x_B(1-x_D)]}{\log \alpha_{AB}} - 1 \tag{10.70}$$

If α changes moderately over the column,

$$\alpha_{av} = \sqrt{\alpha_{top} \alpha_{bottom}} \tag{10.71}$$

Some engineers include the relative volatility α_{am} obtained using the arithmetic mean of the conditions at the top and bottom of the column and generate a three point geometric-average-relative volatility α_{av} by the following relation

$$\alpha_{av} = \sqrt[3]{\alpha_{top} \alpha_{bottom} \alpha_{am}} \tag{10.71a}$$

GRAPHICAL STAGES AT LOW CONCENTRATIONS

At low concentrations, McCabe-Thiele graphical construction is difficult to carry out even when scales are enlarged. In such cases the equilibrium data and the operating line are plotted on log-log paper in the low concentration region. The construction of graphical stages is then relatively more accurate because the equilibrium and the operating lines do not converge rapidly.

In the low concentration region, Henry's law is obeyed and the relation $y = kx$ is applied to construct the equilibrium line if experimental data are not available.

Smoker's equation is useful to calculate analytically the number of theoretical stages. It is not limited to the range of low concentrations but can be used over the entire range of concentrations. However, the equation must be applied separately to the rectifying and stripping sections of the tower and average α's for each section are to be determined. The equation is as follows:

$$n = \frac{\log \dfrac{x'_D \left[1 - \dfrac{mc(\alpha-1)}{-mc^2} x'_n \right]}{x'_n \left[1 - \dfrac{mc(\alpha-1)}{-mc^2} x'_D \right]}}{\log(\alpha/mc^2)} \tag{10.72}$$

where
  n = number of theoretical stages required to give separation between x'_D and x'_n,

  $x'_n = x_n - k \quad x'_D = x_D - k$

  k = composition of the liquid where the operating line intersects the equilibrium line.

The operating line equation is

$$y = mx + b \tag{10.73}$$

$$c = 1 + (\alpha - 1)k \tag{10.74}$$

$$a = \text{average relative volatility (assumed constant)} = \frac{\alpha_{top} + \alpha_{bottom}}{2} \tag{10.75}$$

Example 10.3

(English Units)

A fractionating column is to be designed to separate 10,000 lb/h of a mixture of 40 mol% benzene and 60 mol% toluene into an overhead product containing 97 mol% benzene and a bottoms product containing 98 mol% toluene.

(a) Establish an overall material balance on the column.

(b) Calculate the minimum number of plates at total reflux graphically and by the use of the Fenske equation.

(c) Using a reflux ratio of 3.7 mol per mol of distillate product, calculate the number of theoretical plates and the minimum reflux ratio, and indicate the position of the feed plate for the following cases:

1. if the feed is liquid at its boiling point,

2. if the feed is saturated vapor,

3. if the feed is liquid at 20°C (specific heat of feed = 0.44 Btu/lb · °F),

4. if the feed is a mixture of 67 mol% vapor and 33 mol% liquid.

Data: In all cases, assume total condenser is employed. Consider the reboiler as one theoretical plate. Molal latent heat of benzene = 13,248 Btu/lb mol. Molal latent heat of toluene = 14,328 Btu/lb mol. The equilibrium data for the benzene-toluene system are given below.

x	1	0.78	0.581	0.411	0.258	0.130	0.0
y	1	0.90	0.777	0.632	0.456	0.261	0.0

Solution

(a) *Overall Material Balance*. For total feed, the weight of 1 mol of feed is

$$0.4(78) + 0.6(92) = 86.4 \text{ lb/lb mol}$$

so the total feed is

$$\frac{10,000}{86.4} = 115.74 \text{ lb mol/h}$$

Overall balance gives

$$F = B + D$$

or

$$115.74 = B + D \tag{1}$$

Benzene balance gives

$$0.4F = 0.02B + 0.97D$$

or

$$0.02B + 0.97D = 0.4(115.74) = 46.296$$

or

$$B + 48.5D = 2314.8 \tag{2}$$

Solving (1) and (2) simultaneously gives

$$D = 46.296 \text{ lb mol/h} \quad B = 69.444 \text{ lb mol/h}$$

The benzene in the distillate product is

$$0.97(46.296) = 44.91 \text{ lb mol/h}$$

The toluene in the distillate product is

$$0.03(46.296) = 1.39 \text{ lb mol/h}$$

The benzene in the bottoms product is

$$0.02(69.44) = 1.39 \text{ lb mol/h}$$

The toluene in the bottoms product is

$$0.98(69.44) = 68.05 \text{ lb mol/h}$$

(b) This part solves for the minimum number of theoretical plates at total reflux.

Graphical Method

Plot the equilibrium diagram. In this case the operating lines coincide with the 45° line. Starting from the distillate composition, step off the theoretical plates by the McCabe-Thiele method (Exhibit 4). When this is done, the number of theoretical plates is found to be 8, which includes the reboiler. The minimum number of theoretical plates = 8 − 1 = 7 at total reflux.

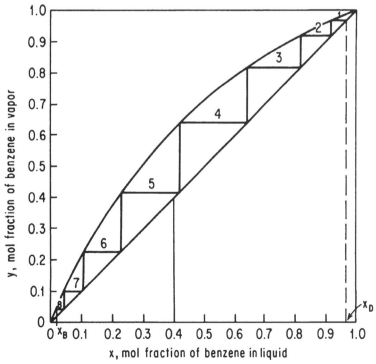

Exhibit 4 Construction of the minimum number of theoretical plates at total reflux

Fenske Equation

The Fenske equation is

$$N_{min} + 1 = \frac{\log[x_D(1-x_B)/x_B(1-x_D)]}{\log \alpha_{AB}}$$

$$x_D = 0.97 \quad x_B = 0.02$$

For α_{top}, $y = 0.97$ and $x = 0.918$ (from equilibrium diagram), so

$$\alpha_{top} = \left(\frac{y}{1-y}\right)_t \left(\frac{1-x}{x}\right)_t = \frac{0.97}{0.03} \times \frac{1-0.918}{0.918} = 2.89$$

Similarly,

$$\alpha_{bottom} = \frac{0.041}{0.959} \times \frac{0.98}{0.02} = 2.095$$

$$\alpha_{av} = \sqrt{\alpha_{top} \alpha_{bottom}} = \sqrt{2.89(2.095)} = 2.461$$

$$N_m + 1 = \frac{\log[0.97(1-0.02)/0.02(1-0.97)]}{\log 2.461}$$

$$N_m = 8.2 - 1 = 7.2 \text{ theoretical plates}$$

(c) *Calculations of Theoretical Stages for Different Conditions*

1. The feed is liquid at its boiling point. The q line is vertical. Draw a vertical line through $x_F = 0.4$ meeting the equilibrium line at $y' = 0.63$. In this case, $x' = 0.40$.

$$\text{Intercept of rectification operating line} = \frac{0.97}{3.7+1} = 0.2064$$

Draw the operating line on the equilibrium diagram by joining x_D on the 45° line and the point $x = 0$, $y = 0.2064$. The operating line for the stripping section passes through the intersection point of the q line and the operating line for the rectification section and also through a point at which the vertical at $x_B = 0.02$ cuts the 45° line. When the theoretical stages are stepped off, it is found that the number of theoretical plates is 11.7, out of which one corresponds to the reboiler (Exhibit 5). The number of theoretical plates in the column = 11.7 − 1 = 10.7. The minimum reflux ratio in this case is

$$R_m = \frac{x_D - y'}{y' - x'} = \frac{0.97 - 0.63}{0.63 - 0.40} = 1.48$$

Alternatively, R_m can be calculated by drawing the straight line through $x_D = y_D = 0.97$ and passing through the intersection of the q and equilibrium lines and making an intercept of $x_D/(R + 1) = 0.39$; hence $R_m = 1.49$. The feed location is at the sixth theoretical plate.

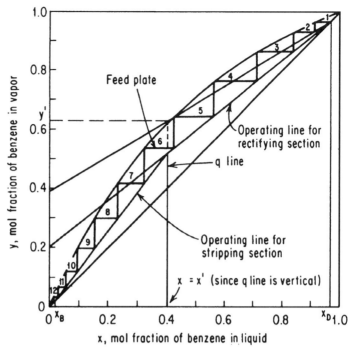

Exhibit 5 Determination of theoretical plates, Example 10.3c(1). Feed: liquid at its boiling point; q line vertical

2. In this case, the q line is horizontal. In Exhibit 4, the operating lines are plotted in the same manner as in Case 1, except that the intersection point of the two operating lines now lies on the horizontal line.

    The number of theoretical plates from Exhibit 6 is $15.4 - 1 = 14.4$. The feed location in this case is at the eighth theoretical plate. The minimum reflux ratio,

$$R_m = \frac{x_D - y'}{y' - x'} = \frac{0.97 - 0.4}{0.4 - 0.212} = 3.032$$

3. The equation of the q line is obtained in the following manner:

    Molecular weight of feed = 86.4
    Molar specific heat of feed = 86.4(0.44)
    $\qquad\qquad\qquad\qquad$ = 38.016 Btu/lb mol · °F
    Latent heat of feed = 0.4(13,248) + 0.6(14,328)
    $\qquad\qquad\qquad\qquad$ = 13,896 Btu/lb mol

From the boiling-point diagram of benzene and toluene, the bubble point of feed is 95°C = 203°F. The feed temperature is 20°C = 68°F.

$$q = \frac{\lambda + C_p(t_b - t_F)}{\lambda}$$

$$= \frac{13,896 + 38.016(203 - 68)}{13,896} = 1.37$$

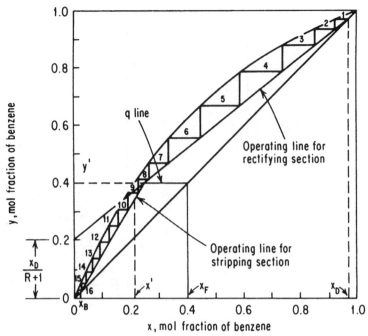

Exhibit 6 Determination of the theoretical plates, Example 10.3c(2). Feed: Saturated vapor; q line horizontal

and the slope of the q line is

$$\frac{q}{q-1} = \frac{1.37}{1.37-1} = 3.703$$

The q line thus has a slope of 3.703. Draw the q line in the same manner as in Case 1, except that the q line in this case will have a slope of 3.703. The operating lines are drawn in Exhibit 7 in the same manner as in Case 1. From

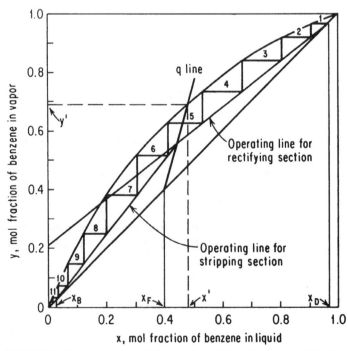

Exhibit 7 Determination of theoretical plates, Example 10.3c(3)

Exhibit 7 the number of theoretical plates including the reboiler is 10.5. The number of theoretical plates in the column is $10.5 - 1 = 9.5$. The minimum reflux ratio in this case is

$$R_m = \frac{x_D - y'}{y' - x'} = \frac{0.97 - 0.695}{0.695 - 0.48} = 1.28$$

The feed plate is the fifth plate.

4. In this case, the feed is a mixture of vapor and liquid, which means the feed is at its boiling point. The fraction of the feed to be vaporized is $q = 0.33$. Then, the equation of the q line is given by

$$y = \frac{0.33}{0.33 - 1}x - \frac{0.4}{0.33 - 1} = -0.493x + 0.597$$

The q line is drawn as before but with a slope of -0.493. The operating lines and the theoretical number of stages are stepped off in the usual manner as shown in Exhibit 8. From the figure, the number of theoretical stages is $12 - 1 = 11$ stages in the column. The minimum reflux ratio is

$$R_m = \frac{x_D - y'}{y' - x'} = \frac{0.97 - 0.468}{0.468 - 0.265} = 2.473$$

The feed plate location is on the seventh plate.

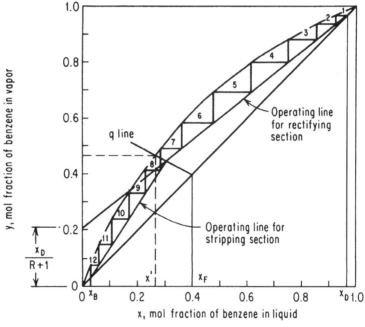

Exhibit 8 Determination of theoretical plates, Example 10.3c(4)

Example 10.3-SI

(SI System)

A fractionating column is to be designed to separate 5000 kg/h of a mixture of 40 mol% benzene and 60 mol% toluene to produce a distillate containing 97 mol% benzene and a bottoms product, containing 98 mol% toluene. The column is to be operated at one atmosphere and total condenser is to be used. The other data are as follows:

Bubble point of feed liquid = 368 K

Feed temperature = 293 K, feed specific heat = 1.6×10^2 kJ/(kg mol · K)

Molal latent heat of vaporization for benzene = 3.08×10^4 kJ/kg mol

Molal latent heat of vaporization for toluene = 3.33×10^4 kJ/kg mol.

Equilibrium diagram for benzene-toluene system is given in Exhibit 9.

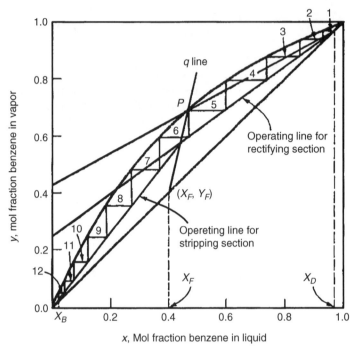

Exhibit 9 Equilibrium diagram for benzene-toluene system (Solution to Example 10.3-SI)

Calculate or obtain the following:

(a) the slope of q line

(b) minimum reflux ratio

(c) number of theoretical stages required if reflux ratio two times the minimum is used

(d) feed plate location

(e) material balance, condenser duty, and reboiler duty (neglecting heat losses)

(f) liquid and vapor molal overflows in the rectifying and stripping section.

Solution

(a) Heat required to raise 1 mol of feed to 368 K = $1 \times 1.6 \times 10^2 (368 - 293)$
$$= 12000 \text{ kJ/kg mol.}$$

λ of feed mixture = $0.4(3.08 \times 10^4) + 0.6(3.33 \times 10^4) = 3.23 \times 10^4$ kJ/kg mol

Therefore, $$q = \frac{12000 + 3.23 \times 10^4}{3.23 \times 10^4} = 1.37$$

$$\text{Slope of } q \text{ line} = \frac{q}{q-1} = \frac{1.37}{1.37 - 1.0} = 3.7$$

(b) Refer to Exhibit 9 for solution.

First draw the q line with slope of 3.7 and passing through the intersection point (x_F, y_F) of vertical from x_F and 45° line to intersect the equilibrium line at pinch point P. Join (x_D, y_D) and P and produce the line joining the two points to meet y axis at intercept = 0.425.

$$\text{Thus } \frac{x_D}{R_m + 1} = 0.425 \quad \text{from which } R_m = 1.28$$

Minimum reflux ratio is 1.28.

(c) Actual reflux ratio = 2 × 1.28 = 2.56

$$\text{Intercept of rectifying section operating line on } y \text{ axis } \frac{x_D}{R+1} = \frac{0.97}{2.56 + 1} = 0.27$$

Operating line for rectifying section is drawn from (x_D, y_D) to intersect y axis at $y = 0.27$.

In Exhibit 9, draw the operating line for stripping section. Step off the stages in the usual manner.

Number of theoretical plates = 12.6 − 1(Reboiler) = 11.6

(d) Crossover from one operating line to the other occurs at 6th plate. Feed plate = 6.

(e) Material balance (overall): $F = D + B$

Component balance: $Fx_F = Dx_D + Bx_B$

Molecular weight of feed = 0.4(78) + 0.6(92) = 86.4

Feed = 5000/86.4 = 57.87 kg mol/h, $x_D = 0.97$ $x_B = 0.02$

Substitute these values in the material balance equations and get the following:

$$D + B = 57.87$$
$$D(0.97) + B(0.02) = 57.9(0.4)$$

Solving, $D = 23.16$ $B = 34.71$ kg mol/h respectively.

Condenser Duty
Assume no subcooling.

Vapor flow from the column to the condenser = 1.92 × 23.16 + 23.16 = 67.63 kg mol/h

$$\lambda = 0.97(3.08 \times 10^4) + 0.03(3.33 \times 10^4) = 3.09 \times 10^4 \text{ kJ/kg mol of distillate.}$$

Then $Q_C = 67.63 \times (3.09 \times 10^4) = 2.09 \times 10^6$ kJ/h.

Reboiler Duty

Assume specific heat of liquid is independent of composition.

Heat required to raise distillate to bubble point (355 K) = $23.16(355 - 293)(1.6 \times 10^2)$
$$= 2.3 \times 10^5 \text{ kJ/h}$$

Heat required to raise bottoms product to 382.8 K = $34.71(382.8 - 293)(1.6 \times 10^2)$
$$= 4.99 \times 10^5 \text{ kJ/h}$$

Reboiler duty = $Q_C + Q_{SD} + Q_{SB} = 2.09 \times 10^6 + 2.3 \times 10^5 + 4.99 \times 10^5$
$$= 2.82 \times 10^6 \text{ kJ/h}$$

(f) Vapor and liquid flows in column:

Flows from plate to plate in rectifying section:

Liquid flow: $L = L_0 = 2.56 \, D = 2.56(23.15) = 59.26$ kg mol/h
Vapor flow: $V = L_0 + D = D(R + 1) = 23.16(2.56 + 1) = 82.41$ kg mol/h

Flows in stripping section:

Liquid flow: $\bar{L} = L + qF = 59.26 + 1.37(57.87) = 123.75$ kg mol/h

Vapor flow: $\bar{V} = V - (1-q)F = 82.41 - (1-1.37)57.87 = 103.83$ kg mol/h

---

Example 10.4

In Example 10.3c3

(a) Obtain the plate-to plate liquid and vapor molar flows in each section of the column.

(b) Obtain the condenser duty and number of gallons per minute of cooling water required if the water enters the condenser at 86°F and leaves at 106°F. The distillate leaves at the bubble point of 180°F, and the bottoms leaves at 229°F.

(c) Obtain reboiler duty and amount of steam in pounds per hour if steam at 25 psia is used. Assume the heat loss from the column to be negligible. Assume also C_p of liquid independent of composition.

(d) Establish the operating-line equation for the stripping section.

Solution

(a) *Molar Flows*

$q = 1.37$ calculated in Example 10.3
$R = 3.7$ by the problem statement
$V = L_0 + D = RD + D = D(R + 1) = 46.296(3.7 + 1)$
$\quad\quad\quad\quad\quad\quad\quad\quad\quad\quad\quad\quad\quad = 217.6$ lb mol/h
$L = L_0 = 3.7(46.296) = 171.3$ lb mol/h
$\bar{L} = L + qF = 171.3 + 1.37(115.74) = 329.9$ lb mol/h
$\bar{V} = V - (1 - q)F = 217.6 - (1 - 1.37)\,115.74$
$\quad = 215.6 + 0.37(115.74) = 260.4$ lb mol/h

(b) *Condenser Duty*

$$Q_c = V\lambda$$

$$\lambda = 0.97(13{,}248) + 0.03(14{,}328) = 13{,}280 \text{ Btu/lb mol}$$

$$Q_c = 217.6(13{,}280) = 2{,}889{,}728 \text{ Btu/h}$$

$$\text{Flow of water} = \frac{2{,}889{,}728}{(106-86)500} = 288.98 \ [289 \text{ gpm}]$$

(c) C_p independent of composition. With this assumption the molar specific heat of the feed as calculated in Example 10.3c3 is 38.02 Btu/lb mol·°F. Using the feed temperature of 20°C (= 68°F) as the datum temperature one obtains Q, the heat required to heat distillate D to 180°F, as

$$Q = 46.296(38.02)(180 - 68) = 197140 \text{ Btu/h}$$

Q_{SB}, the heat required to heat bottoms product to 229°F, is

$$Q_{SB} = 69.444(38.02)(229 - 68) = 425{,}082 \text{ Btu/h}$$

Reboiler Duty

$$Q_B = Q_c + Q_{SD} + Q_{SB} = 2{,}889{,}728 + 197{,}140 + 425{,}082$$
$$= 3{,}511{,}950 \text{ Btu/h}$$

Latent heat of steam $\lambda_s = 959.4$ Btu/lb

$$\text{Steam} = \frac{3{,}511{,}950}{959.4} = 3660 \text{ lb/h}$$

Alternatively,

$$\text{Steam required at the reboiler} = \frac{\overline{V}\lambda}{\lambda_s}$$

Now $\lambda = 0.98(14{,}328) + 0.02(13{,}248) = 14{,}306.4$ Btu/lb·mol

$$S = \frac{260.4(14{,}306)}{959.4} = 3883 \text{ lb/h}$$

(d) *Operating-Line Equation for the Stripping Section*

$$y_{m+1} = \frac{L+qF}{L+qF-B}x_m - \frac{Bx_B}{L+qF-B}$$

$$= \frac{329.9}{329.9-69.44}x_m - \frac{69.44(0.02)}{329.9-69.44}$$

or $\quad y_{m+1} = 1.27x_m - 0.005332$

Example 10.5

An acetone-water solution containing 20 mol% acetone is to be fractionated at 1 atm pressure. Ninety-eight percent of the acetone is to be received in the distillate at a concentration of 97 mol %. The feed is available at the bubble point at 80°F.

The reflux will be returned at the bubble point. Open steam at 20 psig will be used at the base of the tower. Assume the heat loss from the tower to be negligible.

Calculate:
(a) the minimum reflux ratio
(b) a practical reflux ratio
(c) steam required per mole of feed
(d) the number of theoretical trays for a practical reflux ratio, and
(e) suggest actual number of trays based on the efficiency value obtained from the O'Connell plot.

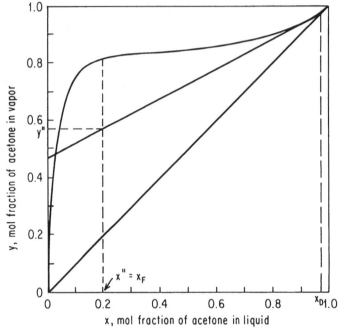

Exhibit 10 Minimum reflux ratio determination (Example 10.5)

Solution

(a) The equilibrium diagram is plotted in Exhibit 10, with vertical line at a tangent to the equilibrium curve from x_D. The minimum reflux ratio is

$$\frac{x_D - y''}{y'' - x_F} = \frac{0.97 - 0.573}{0.573 - 0.2} = \frac{0.397}{0.373} = 1.064$$

The minimum reflux ratio can more easily be found from the intercept as

$$R_m = \frac{x_D}{\text{intercept}} - 1 = \frac{0.97}{0.47} - 1 = 1.064$$

(b) The optimum reflux ratio ranges from 1.2 to 2 times the minimum reflux ratio. Choose 2 times the minimum reflux ratio as the practical reflux ratio. The practical reflux ratio is 2.128. For the material balance, the basis is 100 lb mol/h feed. From the statement of the problem,

$$D = \frac{100(0.2)(0.98)}{0.97} = 20.20 \text{ lb mol/h}$$

R = actual reflux ratio = $1.064(2.0) = 2.128$

$$L_0 = L = RD = 2.128(20.20) = 43.0 \text{ lb mol/h}$$
$$V = L_0 + D = 43.0 + 20.20 = 63.2 \text{ lb mol/h}$$

Since the feed is at the bubble point, $q = 1$ and the q line is vertical.

$$\overline{L} = L + qF = 43 + 1(100) = 143 \text{ lb mol/h}$$
$$\overline{V} = V + F(q-1) = 63.2 + 100(1-1) = 63.2 \text{ lb mol/h}$$

(c) *Calculation of Steam Requirements.* For steam at 20 psig,

$$H_s = \text{enthalpy of steam at 1 atm} = 1150.5 \text{ Btu/lb}$$
$$\lambda = 970.3 \text{ Btu/lb neglecting contribution of acetone}$$

$\overline{H}_s$ of steam at 20 psig = 1166.9 Btu/lb

$$\overline{V} = S\left[1 + \frac{1166.9 - 1150.3}{970.3}\left(\frac{18}{18}\right)\right]$$

$$S = \frac{63.2}{1.01711} = 62.1 \text{ lb mol/h}$$

Steam to tower = S = 62.1 lb mol/h
Steam required per mole of feed = 0.621 lb mol

(d) *Balance on Bottom Plate*

$$B = S + \overline{L} - \overline{V} = 62.1 + 143.0 - 63.2 = 141.9 \text{ lb mol/h}$$

Bottoms Composition

Acetone in distillate = 0.97(20.2) = 19.594 lb mol/h
Acetone in bottoms = 20 − 19.594 = 0.406 lb mol/h

$$\text{Mol\% acetone in bottoms} = \frac{0.406}{141.9}(100) = 0.286 \text{ mol\%}$$
$$= 0.003 \text{ mol fraction}$$

Calculation of Theoretical Plates. The intercept of the operating line on the y axis is

$$\frac{x_D}{R+1} = \frac{0.97}{2.128+1} = 0.31$$

Draw this line on the equilibrium diagram in Exhibit 11. It passes through $x_D = y_D = 0.97$ and its intercept on the y axis is 0.31.

The operating line in the stripping section passes through ($y = 0$, $x_B = 0.003$) and the point of intersection of the q line with the operating line of the rectification section. This is drawn in Exhibit 11.

The number of the theoretical plates from the graph is 10 since there is no reboiler.

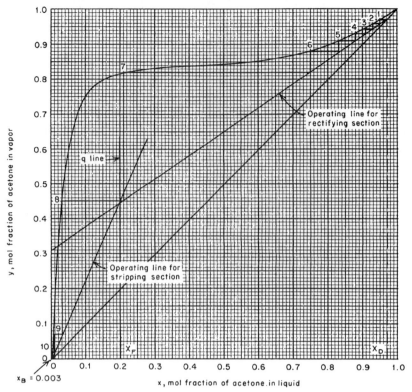

Exhibit 11 Graphical determination of theoretical stages (Example 10.5)

(e) *Calculation of Actual Number of Plates.* From the composition of vapor at the top of the column and the equilibrium data,

$$\text{Column top temperature} = 135°F$$

$$\text{Column bottom temperature} = 212°F$$

$$\text{Average temperature} = \frac{212+135}{2} = 173.5°F$$

Average Viscosity

Acetone: $\mu = 0.2$ cP
H$_2$O: $\mu = 0.35$ cP
Average: $\mu_m^{1/3} = 0.2(0.2)^{1/3} + (0.35)^{1/3}(0.8)$
$\mu_m = 0.32$ cP

Average α

$$\alpha_{\text{top}} = \frac{0.97/(1-0.97)}{0.96/(1-0.96)} = 1.35$$

$$\alpha_{\text{bottom}} = \frac{0.076/(1-0.076)}{0.003/(1-0.003)} = 27.33$$

$$\alpha_{av} = \sqrt{1.35(27.33)} = 6.07$$
$$\alpha_{av}\mu_l = 6.07(0.32) = 1.94$$

From O'Connell plot,

$$\text{Efficiency} = 0.41$$

$$\text{Actual number} = \frac{10}{0.41} = 24.4$$

The number of actual plates must be an integer.

SEPARATION OF BINARY AZEOTROPIC MIXTURES

A large number of binary systems form azeotropic mixtures. It is frequently necessary to separate them into their components. At azeotropic composition the relative volatility is one and therefore rectification is not possible. In such cases some technique must be used to go past the azeotropic composition. Two methods are used in practice to separate azeotropic mixtures.

1. Distillation in combination with some other separation process such as decantation, extraction, and crystallization. Near azeotropic composition can be produced by distillation and then one of these separation processes can be used to produce two fractions having compositions on each side of the azeotrope. The two fractions can then be subjected to fractional distillation separately to yield essentially pure components, and fractions of azeotropic composition, which are recycled to the distillation process.

2. Modification of relative volatility, changing system pressure or adding a third component to the mixture. A two-column system and change in distillation conditions, azeotropes of the following type can be separated by straight fractional distillation.

    a. *Compounds with close boiling points.* Compounds whose vapor pressure curves cross or which have somewhat different slopes fall into this group. Referring to Figure 10.15, it is seen that at pressure indicated by line 1 the two vapor pressure curves intersect each other. At the pressure and temperature indicated by the intersection point, vapor pressures are the same and the boiling points of the two compounds are very close or equal. The relative volatility is 1. At pressures indicated by the line 2 and by the line 3, the boiling points of the two compounds are different and relative volatility >1. Therefore they can be separated at these pressures.

    b. *Homogeneous azeotropes, whose compositions are very highly pressure sensitive.* If there is a substantial change of azeotropic composition with small change in pressure, these mixtures can be separated by using a two-column fractionation scheme as shown in Figure 10.16a.

    c. Heterogeneous azeotropes whose components form two liquid phases and a vapor phase at the azeotropic temperature as shown in Figure 10.17b can be separated by using two-column system and a separator vessel to allow the two phases to separate and settle. The scheme is shown in Figure 10.17a. If the feed composition is such that it forms two liquid phases at the temperature of the separator, the feed is directly introduced into the separator vessel. If it consists of a single phase, it is fed to the column, which is distilling the material that has the composition nearest to that of feed. The scheme of two-column separation is shown in Figure 10.17a.

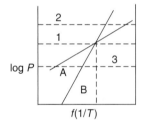

Figure 10.15 Effect of pressure on boiling points and relative volatility

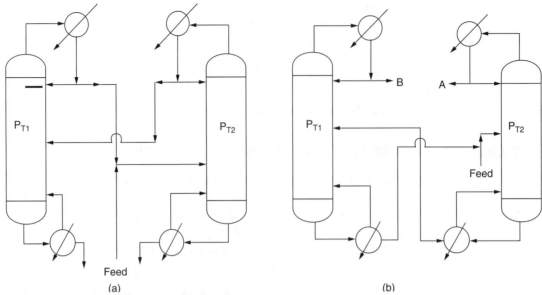

Figure 10.16 Separation of minimum and maximum boiling azeotropes by two column scheme of fractionation (a) Separation scheme for minimum boiling azeotropes (b) Separation scheme for maximum boiling azeotropes (Feed composition is azeotropic composition)

Extractive Distillation

A binary mixture, which is difficult to separate by ordinary fractional distillation, has a third component called solvent added, which influences the relative volatility of the two components and thus permits the separation of the two components. The added solvent is of low volatility and therefore does not vaporize itself appreciably. Examples of extractive distillation are toluene-isooctane separation with phenol, use of sulfuric acid in nitric acid concentration, butene-butane separation with either acetone or furfural, and acetone-methanol separation with water.

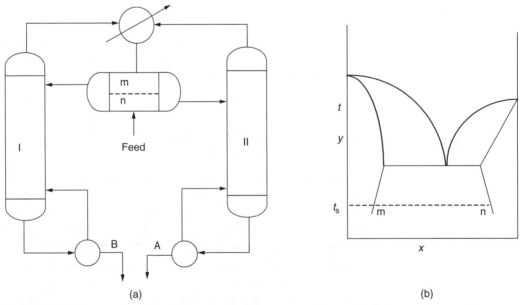

Figure 10.17 Separation scheme for a heterogeneous azeotrope (a) Scheme of separation (b) Heterogeneous minimum boiling azeotrope

Extractive column design calculations to determine the number of theoretical stages can be made by pseudobinary method, pseudoternary method, or multicomponent methods. Use of the McCabe-Thiele diagram has been described by Chambers. A detailed review of these methods is beyond the scope of this book. The reader is referred to Van Winkle for detailed discussion of these methods.

Azeotropic Distillation

This method involves addition of a third component, often called entrainer, to the mixture. The entrainer forms a low boiling azeotrope with one of the components of the mixture and this azeotrope distills over. The volatility of the entrainer is such that it can be easily separated from the original constituent. Examples of separations by the azeotropic method are acetic acid-water separation with butyl acetate, ethanol and isopropanol separations from water using pentane as entrainer. General methods of design for multicomponent distillation can be applied.

Multicomponent Distillation

All systems that contain more than two components are termed multicomponent systems. The principles of design of columns for multicomponent systems are the same in many respects as those for binary systems. However, the calculations are much more complex and tedious. There is one mass balance equation for each component. Heat balance is also similar to that in binary systems. We have already covered use of equilibrium data, calculations of bubble and dew point temperatures, and flash calculations in Chapter 9.

Number of Distillation Towers Needed

For a two-component system, one tower was required to separate the two components. However, in a multicomponent mixture of n components, $n - 1$ towers will be needed if each component is to be obtained in pure or enriched form.

Design Calculations

General principles of column design calculations are applicable. The equilibrium stages can be determined by stage-to-stage calculations, which are rather tedious and presently are done using high-speed computers. Computer software programs such as Aspen or Process are used to provide rigorous solutions to the multicomponent design problems.

Shortcut Calculation Methods

Shortcut methods are useful to get approximate solutions to multicomponent distillation rapidly. We will now review these methods.

Key Components

If the components of a multicomponent distillation are arranged in order of their relative volatilities, there will be some components that will be more volatile compared to others. A single fractionation tower can allow separation between two components only. The two specific adjacent components that are separated are called the light key, which is more volatile, and the heavy key, which is less volatile. The components more volatile than the light key are called *light components*. The components that are less volatile than the heavy key are termed *heavy components*. Light components will be present in the bottoms product in small amounts while the heavy components will be present in the distillate in small amounts. Both the light and heavy key components are present in distillate and in bottoms in significant amounts.

In some instances, the keys selected are not adjacent but have an intermediate boiling component between them. The three components are then called light key, heavy key, and intermediate (boiling) or distributed key.

The relative volatilities are computed with respect to the heavy key (base component) and are given by

$$\alpha_i = \frac{K_i}{K_{hk}} \tag{10.76}$$

For a bubble point calculation,

$$\sum y_i = \sum K_i x_i = K_{hk} \sum \alpha_i x_i = 1.0 \tag{10.77}$$

As discussed in Chapter 9, the calculations are by trial and error. A temperature is assumed, and then α_is are calculated using values of K_is at this temperature. Then K_{hk} is calculated from $K_{hk} = 1/\sum \alpha_i x_i$. After the bubble temperature is fixed by trial and error, the vapor composition is calculated by

$$y_i = \frac{\alpha_i x_i}{\sum(\alpha_i x_i)} \tag{10.78}$$

For dew point calculation, $\sum x_i = \sum \left(\frac{y_i}{K_i}\right) = \frac{1}{K_{hk}} \sum \frac{y_i}{\alpha_i} = 1.0.$ (10.79)

$K_{hk} = 1/\sum \alpha_i x_i$. When the dew point temperature is fixed, the liquid composition is calculated by

$$x_i = \frac{y_i/\alpha_i}{\sum y_i/\alpha_i} \tag{10.80}$$

For flash distillation of multicomponent mixture, as shown in Chapter 9, the following equation applies:

$$y_i = -\frac{1-f}{f}x_i + \frac{z_{iF}}{f} \tag{10.81}$$

At equilibrium, the preceding becomes

$$y_i = K_{hk}\alpha_i x_i = -\frac{1-f}{f} + \frac{z_{iF}}{f} \tag{10.82}$$

Solving for x_i and summing up for all components,

$$\sum x_i = \sum \frac{z_{iF}}{f(K_{hk}\alpha_i - 1) + 1} = 1.0 \tag{10.82a}$$

This is solved by trial and error.

SHORTCUT METHODS

Total Reflux in Multicomponent Distillation

The Fenske equation is used to calculate minimum number of stages at total reflux. It applies to any two components in a multicomponent mixture. When applied to heavy key and light key components, the Fenske equation becomes

$$N_m = \frac{\log\left[\left(\frac{x_{LD}}{x_{HD}}\right)\left(\frac{x_{HB}}{x_{LB}}\right)\right]}{\log(\alpha_{L,AV})} \qquad (10.83)$$

where
- N_m = minimum number of theoretical plates in the column
- x_{LD} = mol fraction of light key in distillate
- x_{HD} = mol fraction of heavy key in distillate
- x_{LB} = mol fraction of light key in bottoms
- x_{HB} = mol fraction of heavy key in bottoms

Average value of α_L is calculated from values of α_L at the top temperature and at bottoms temperature and is given by

$$\alpha_{L,AV} = \sqrt{\alpha_{LD}\alpha_{LB}} \qquad (10.84)$$

The distillate dew point and bottoms boiling point temperatures are to be calculated by trial and error because the distribution of other components in both distillate and bottoms is not known. This equation should be used if temperature difference between the top and bottoms temperatures is not more than 80°C. If it exceeds this value, Winn's method should be used. To locate the feed plate, the Fenske equation may be applied between the feed and bottoms composition for the light and heavy keys.

If the reboiler and partial condenser (if used) are taken as theoretical stages, the number of theoretical stages in the column will be $N - 2$ stages.

To estimate the distribution of other than light and heavy key components in distillate and bottoms at total reflux, the following equation obtained by rearrangement of Equation 10.83 should be used:

$$\frac{x_{iD}D}{x_{iB}B} = (\alpha_{i,AV})^{N_m} \frac{x_{HD}D}{x_{HW}B} \qquad (10.85)$$

Minimum Reflux Ratio for Multicomponent Distillation

As in binary distillation, in multicomponent distillation also, at minimum reflux, R_m an infinite number of trays is required for the given separation of the key components. Step-by-step procedure of calculating R_m is extremely tedious by hand calculations. Underwood's shortcut method to calculate R_m uses constant average α values and assumes constant molal overflows in both sections of the tower. The two equations to be solved are as follows:

$$\sum_{1}^{n} \frac{\alpha_i z_{iF}}{\alpha_i - \theta} = 1 - q \qquad (10.86)$$

and

$$R_m + 1 = \sum_{1}^{n} \frac{\alpha_i x_{iD}}{\alpha_i - \theta} \qquad (10.87)$$

where
- α_i = relative volatility of component i in the mixture
- z_{iF} = mol fraction of i in the feed
- q = the number of moles of saturated liquid formed from 1 mol of feed on the feed plate. For a bubble point feed, $q = 1$, for a dew point feed, $q = 0$.
- θ = a value that lies between the α values of light and heavy keys

θ is obtained by trial and error from Equation 10.86 and then substituted in Equation 10.87 to calculate R_m.

If there is a distributed key or a component, which has a relative volatility between that of the light key and that of the heavy key, then two values of θ need to be obtained. One of the values lies between α_{LK} and α_{DK} and the other lies between α_{DK} and α_{HK}. Then solving Equation 10.86, get those two values of θ and substitute in the following equation:

$$R_m + 1 = \sum_{1}^{n} \frac{\alpha_i x_D}{(\alpha_i - \theta)} + \frac{\alpha_i x_{dD}}{(\alpha_i - \theta)} \qquad (10.88)$$

This process of substitution of two values of θ in Equation 10.88 yields two equations and two unknowns R_m and x_{dD}. They can be solved simultaneously to get R_m and x_{dD} (Van Winkle).

Shortcut Method for Number of Theoretical Stages at Operating Reflux Ratios

A number of correlations in graphical form are available to determine the number of theoretical stages for an operating or actual reflux ratio. These are Brown and Martin correlation, Gilliland correlation and that of Erbar and Maddox. Erbar and Maddox correlation is the latest. It is shown in Figure 10.18.

In this correlation, the operating reflux ratio R based on the flow rates at the top of the column is plotted against N_m/N with $R_m/(R_m + 1)$ as a parameter. N is the number of theoretical stages for a given R. R_m is the minimum reflux ratio obtained by Underwood method. N_m is the minimum number of theoretical stages obtained by Fenske method.

Feed Plate Location

Kirkbride (see M. Van Winkle) devised an empirical correlation to estimate the ratio of the numbers of theoretical plates above and below the feed plate. This can be used to locate the feed stage. The correlation is as follows:

$$\log \frac{N_R}{N_s} = 0.206 \log \left[\left(\frac{x_{HF}}{x_{LF}} \right) \frac{B}{D} \left(\frac{x_{LB}}{x_{HD}} \right)^2 \right] \qquad (10.89)$$

Shortcut Methods   415

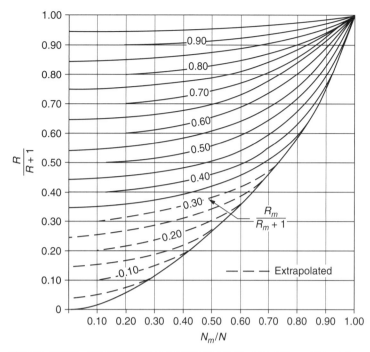

Figure 10.18 Erbar-Maddox correlation between reflux ratio and number of stages (R_m based on Underwood method). [From J. H. Erbar and R. N. Maddox, *Petrol. Refiner*, 40(5), 183 (1961). With permission.]

Example 10.6

100 kg mol of a mixture of *n*-butane (A), *n*-pentane (B), *n*-hexane (C), and *n*-heptane (D) at its boiling point and at 405.3 kPa pressure is fed to a fractionating column. The composition of the mixture is butane 40 mol percent, pentane 25 percent, hexane 20 percent, and heptane 15 percent. Ninety-five percent of pentane is recovered in distillate and 95 percent of hexane in bottoms. Calculate the following:

(a) Moles per hour and composition of distillate and bottoms

(b) Top and bottoms temperatures of column

(c) Minimum stages at total reflux and distribution of other components in distillate and bottoms

(d) Minimum reflux ratio using Underwood's method

(e) Number of theoretical stages at an operating reflux ratio of 1.3 times the minimum using the Erbar-Maddox correlation

(f) Location of feed tray using the Kirkbride method.

Solution

Since separation is specified between pentane and hexane, they are the light (pentane) and heavy (hexane) key components. For a first trial to calculate the top and bottom column temperatures we assume that all butane (lighter than the light key) is in the distillate and all heptane (heavier than heavy key) is in the bottoms. The material balance with these assumptions becomes

(a)

	Feed F		Distillate D		Bottoms B	
Comp	z_{iF}	$z_{iF}F$	$y_{iD}=x_{iD}$	$y_D D$	x_B	$x_B B$
butane	0.40	40	0.618	40.0	0.00	0.000
pentane	0.25	25	0.367	23.75	0.0355	1.25
hexane	0.20	20	0.015	1.0	0.539	19.0
heptane	0.15	15	0.00	0.00	0.4255	15.00
	1.0000	100.0	1.0000	64.75	1.0000	35.25

For calculating the dew point or temperature at the top of the column, a temperature is assumed, K values are read from chart, and relative volatilities are calculated with heavy key as base component Σx_i. For the present problem, trials were done at temperatures of 65, 67, and 70 and final trial was made at 66°C. Details of unsuccessful trials are not given here to save space. The results of the final trial at 66°C are given below.

Comp	y_{iD}	K at 66°C	α_i	y_i/α_i	x_i
C_4	0.618	1.76	6.769231	0.091295	0.352204
C_5	0.367	0.625	2.40	0.152917	0.589929
C_6	0.015	0.26	1.0	0.015	0.057868
C_7	0.000	0.094	0.3577	0.0000	0
				$K_H = 0.259212$	1.0000

The dew point temperature is 66°C.

Bubble point temperature at the bottom of the column is also calculated by trial and error.

The results of the final trial are as given in the following table.

Results of final trial for bubble point temperature at the bottom of the column. Temperature 134°C

Component	y_{iD}	K_i	α_i	$\alpha_i x_i$	y_i
C_4	0.000	5.3	4.4534	0.000	0.000
C_5	0.0355	2.6	2.185	0.077568	0.092669
C_6	0.539	1.19	1.00	0.539	0.644
C_7	0.4255	0.62	0.52	0.221154	0.264209
				0.837042	1.0009
				$K_H = 1.193715$	

Therefore bubble point temperature at the bottom of the column is 134°C.

(b) *Minimum Theoretical Plates (Stages) at Total Reflux.* The α_i values to be used for the light key are as follows:

$\alpha_{LD} = 2.4$ at 66°C (at column top) and 2.185 at 134°C (at column bottom).

$$\alpha_{L,AV} = \sqrt{2.4 \times 2.185} = 2.29$$

$$N_m = \frac{\log\left[\frac{0.367}{0.015}\left(\frac{0.539}{0.0355}\right)\right]}{\log 2.29} = 7.14$$

Number of theoretical trays in the column = 7.14 − 1(Reboiler) = 6.14

(c) *Distribution of Other Components.* Distribution of the other components can be calculated using Equation 10.85.

For component butane (A)

$$\alpha_{A,AV} = \sqrt{6.769231 \times 4.4534} = 5.491$$

$$\frac{x_{AD}D}{x_{AB}B} = (\alpha_{AV})^{7.14} \frac{x_{HD}D}{x_{HB}B} = (5.491)^{7.14} \frac{0.015 \times 64.75}{0.539 \times 35.25} = 9765.5$$

By material balance, $x_{AD}D = 40 = 9765.5 x_{AB}B + x_{AB}B$

Therefore $x_{AB}B = \dfrac{40}{9766.5} = 0.00409$, $x_{AB} = \dfrac{0.00409}{35.25} = 0.00012 = 1.2 \times 10^{-4}$

Then $x_{AD}D = 40 - 0.00409 = 39.996$ kg mol

For component heptane (D)

Average α_{AV} for component heptane, $\alpha_{AV} = \sqrt{0.52 \times 0.3577} = 0.4313$

$$\frac{x_{DD}D}{x_{DB}B} = (\alpha_{D,AV})^{N_m} \frac{x_{HD}D}{x_{HB}B} = (0.4313)^{7.14} \frac{0.015 \times 64.75}{0.539 \times 35.25} = 0.000126$$

Again by material balance, $x_{DD}D + x_{DB}B = 15$

$$0.000126\, x_{DB}B + x_{DB}B = 15$$

Therefore, $x_{DB}B = \dfrac{15}{1.00026} = 14.998$ kg mol

$x_{DD}D = 15 - 14.998 = 0.002$ kg mol

A summary of the results of the preceding calculations is given in the following table.

Component	$x_D D$	$y_D = x_D$	$x_B B$	x_B
C_4	39.996	0.6177	0.004	0.0001135
C_5	23.75	0.3668	1.25	0.0355
C_6	1.000	0.01544	19.00	0.53900
C_7	0.002	0.00003	14.998	0.4255
	64.748	0.99997	35.252	1.0001135

Using the new composition, recalculations of dew point and bubble point do not change their values by very much. Therefore these values will be accepted.

(d) *Calculation of Minimum Reflux by Underwood's Method*

Two equations to be solved are

$$1 - q = \sum \frac{\alpha_i z_{iF}}{\alpha_i - \theta}$$

and

$$R_m + 1 = \sum \frac{\alpha_i x_{iD}}{\alpha_i - \theta}$$

Use first equation to get θ by trial and error. Substitute θ so obtained in the second equation to get R_m. Since the feed is at its boiling point, all is liquid and hence $q = 1$. Value of θ was found by spreadsheet calculation so that $\sum \alpha_i z_{iF}/\alpha_i - \theta = 0$. A value of $\theta = 1.208545$ gave $\sum \alpha_i z_{iF}/\alpha_i - \theta = -1.8 \times 10^{-6}$, which is very close to 0. This value is substituted in the second equation to find R_m. Calculations are given in following table.

Component	$\alpha_i x_{iD}$	x_{iD}	$\alpha_i - u$	$\alpha_i x_{iD}/(\alpha_i - u)$
C_4	3.3916	0.6177	4.2824	0.79203
C_5	0.83997	0.3668	1.0814	0.77675
C_6	0.01544	0.01544	-0.2086	-0.07402
C_7	0.000013	0.00003	-0.7773	-1.7×10^{-5}
				1.5

Therefore $R_m + 1 = 1.5$ and minimum reflux ratio is $R_m = 0.5$

(e) *Theoretical plates from Erbar-Maddox correlation*

$$\frac{R}{R+1} = \frac{1.3 \times 0.5}{1.3 \times 0.5 + 1} = 0.394$$

$$\frac{R_m}{R_m + 1} = \frac{0.5}{0.5 + 1} = 0.333$$

From Erbar-Maddox chart, N_m/N is approximately 0.42.

Therefore $N = (7.14)/0.42 = 17$ theoretical stages. Assuming reboiler is a theoretical stage, theoretical trays in column $= 17 - 1 = 16$ theoretical stages.

(f) *Feed Plate Location using Kirkbride Equation*

Kirkbride equation is $\log \dfrac{N_R}{N_s} = 0.206 \log \left[\left(\dfrac{x_{HF}}{x_{LF}}\right) \dfrac{B}{D} \left(\dfrac{x_{LB}}{x_{HD}}\right)^2 \right]$

$$= \log \frac{N_R}{N_s} = 0.206 \log \left[\frac{(0.2)}{0.25} \times \frac{35.252}{64.748} \left(\frac{0.0355}{0.01544}\right)^2 \right] = 0.07462$$

Hence $\dfrac{N_R}{N_s} = 1.1875$, also $N_R + N_s = 16$ calculated in (e).

Solving $N_s = 7.3$ and $N_R = 16 - 7.3 = 8.7$

Therefore feed plate is 8.7, $\doteq$ 9th plate from the top.

DESIGN CALCULATIONS FOR PACKED TOWERS

In packed towers the contact of vapor rising in the column and liquid flowing downward is continuous, unlike in tray towers where the contact is stagewise. Formerly packed columns were used for small capacities. However, with the development of very efficient packings such as pall rings and the need to increase the capacity, to reduce the pressure drop and where the hold up must be small as in heat sensitive materials, packed columns in larger sizes are now extensively used in distillation also. Dumped or random packings such as ceramic Raschig rings, Lessing rings, ceramic, metallic, and plastic intalox saddles, metal and plastic structured packings pall rings, berl saddles and tellerettes are used. Structured packings such as Koch-sulzer, Hyperfil, and Goodloe are also available. Design of packed towers is based on the concept of transfer units or NTUs.

Vapor flow rates and their compositions are indicated in Figure 10.19a and operating equilibrium diagram in Figure 10.19b. Making the assumption that over a differential height dz of packing, constant overflow exists, then the mass transfer

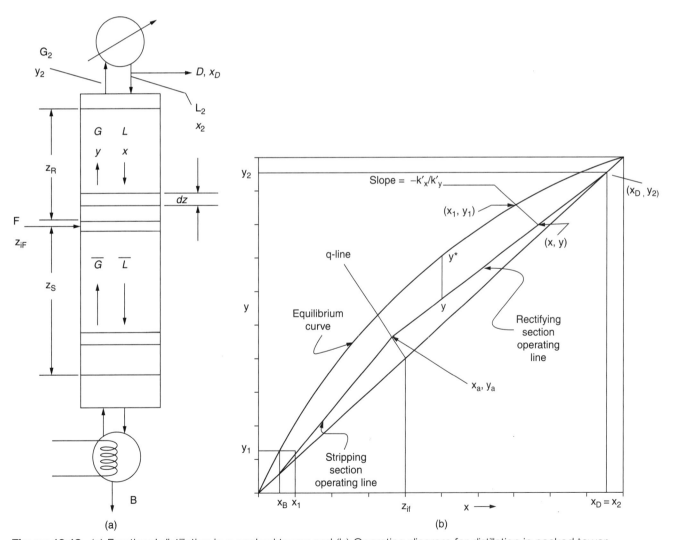

Figure 10.19 (a) Fractional distillation in a packed tower and (b) Operating diagram for distillation in packed tower

between phases is by equimolar counterdiffusion and hence $N_A = -N_B$. Therefore mass flux of A is given by

$$N_A = \frac{d(Gy)}{a dz} = k'_y(y_i - y) = \frac{d(Lx)}{a dz} = k'_x(x - x_i) \tag{10.90}$$

Hence $\quad z_R = \int_0^{z_R} dz = \int_{(Gy)_a}^{(Gy)_2} \frac{d(Gy)}{k'_y a(y_i - y)} = \int_{(Lx)_a}^{(Lx)_2} \frac{d(Lx)}{k'_x a(x - x_i)} \tag{10.91}$

where z_R is the total packing height in the rectifying section above feed plate. A similar equation with appropriate integration limits applies to the stripping section.

For any point (x, y) on the operating line, the corresponding point (x_i, y_i) on the equilibrium curve is obtained by the intersection of a line of slope $= -k'_x/k'_y = -k'_x a/k'_y a$ and drawn from (x, y) as shown in Figure 10.19b.

When the main resistance to mass transfer is in the vapor, it is appropriate to use the first integral of Equation 10.91 involving y. When the resistance is mainly in liquid phase, it is better to use the integral involving x.

When simplifying assumptions can be made that V and L are constant in any section of the tower, the heights of transfer units are given by

$$H_{tG} = \frac{G}{k'_y a} \quad \text{and} \quad H_{tL} = \frac{L}{k'_x a} \tag{10.92}$$

when they are also more or less constant or some average values can be used for them, then Equation 10.91 can be written as

$$z_R = H_{tG} \int_{y_a}^{y_2} \frac{dy}{y_i - y} = H_{tG} N_{tG} \tag{10.93}$$

where z_R = packing height in rectification section.

Similar equations can be written for the stripping section. The difficulty in using this procedure is in calculating $k'_y a$ and k'_x. If the equilibrium curve is essentially straight, we can use overall mass transfer coefficients and overall gas phase and liquid phase transfer units according to the following equations:

$$z_R = H_{OG} \int_{y_a}^{y_2} \frac{dy}{y^* - y} = H_{OG} N_{OG} \tag{10.94}$$

$$z_R = H_{OL} \int_{x_a}^{x_2} \frac{dx}{x - x^*} = H_{OL} N_{OL} \tag{10.95}$$

where $\quad H_{OG} = \frac{G}{K'_y a} \quad \text{and} \quad H_{OL} = \frac{L}{K'_x a} \tag{10.96}$

In such cases, the simplified equations given below could also be used under certain conditions to avoid graphical integration.

$$N_{OG} = \frac{y_1 - y_2}{(y - y^*)_M} \tag{10.97}$$

where $(y - y^*)_M$ is the logarithmic mean of the concentration differences at the two ends of the tower.

If the solutions are dilute and if Henry's law applies, two more equations can be obtained as follows:

$$N_{OG} = \frac{\ln\left[\frac{y_1 - mx_2}{y_2 - mx_2}\left(1 - \frac{1}{A}\right) + \frac{1}{A}\right]}{1 - 1/A} \quad (10.98)$$

where $A = L/mG$. For stripper, the corresponding equation in terms of N_{OL} is

$$N_{OL} = \frac{\ln\left[\frac{x_2 - y_1/m}{x_1 - y_1/m}(1 - A) + A\right]}{1 - A} \quad (10.99)$$

Graphical plots for number of transfer units or for the number of theoretical plates versus $\frac{y_1 - mx_2}{y_2 - mx_2}$ or $\frac{x_2 - y_1/m}{x_1 - y_1/m}$ are available. (Refer to M. Van Winkle, Treybal.)

The following equations relate the overall coefficients with the individual phase transfer coefficients:

$$\frac{1}{K'_y} = \frac{1}{k'_y} + \frac{m}{k'_x} \quad (10.100)$$

$$\frac{1}{K'_x} = \frac{1}{mk'_y} + \frac{1}{k'_x} \quad (10.101)$$

$$H_{OG} = H_G + \frac{mG}{L} H_L \quad (10.102)$$

$$H_{OL} = H_L + \frac{L}{mG} H_L \quad (10.103)$$

where m is the slope of the straight equilibrium curve.

Tray Efficiencies

In design calculations of theoretical stages required for a desired separation, it is assumed that the stage is an equilibrium stage, that is, the vapor leaving the stage is in equilibrium with the liquid leaving. This implicitly assumes that on a tray, there is perfect mixing and there is sufficient time for vapor and liquid to contact. In practice this is not the case and a theoretical stage is not 100 percent efficient. As a result actual number of trays to be put into a column for a given separation will be larger than the calculated number of trays. This requires the introduction of an efficiency factor to arrive at the actual number of trays. In distillation, three types of efficiencies are defined. They are (1) overall tray efficiency, (2) Murphree plate efficiency, and (3) point efficiency.

1. Overall Tray Efficiency, E_o

This has a reference to the entire column and is defined as the ratio of the number of theoretical trays to the number of actual trays used. Thus

$$E_o = \frac{\text{Number of theoretical trays}}{\text{Number of actual trays}} = \frac{N_{\text{theoretical}}}{N_{\text{actual}}} \quad (10.104)$$

When the calculated number of theoretical trays is 9, that includes a reboiler, and if the reboiler is taken as one theoretical plate, the number of theoretical trays in the column will be 8. Then if the overall efficiency is 50 percent, the number of actual trays to be used will be 8/0.5 = 16 actual trays in the column. If the reboiler cannot be treated as one theoretical plate, the actual number of trays in the column will be 9/0.5 = 18.

The overall plate efficiency is dependent on several factors. The interested reader is referred to Chapter 18 in this book and Chapter 18 of Perry's handbook for details. An empirical correlation is given by O'Connell in graphical form (refer to Figure 18.23a, Perry's Handbook) for determining overall efficiency. Lockett developed the following equation to represent the data of the O'Connell plot.

$$E_O = 0.492(\mu_L \alpha)^{-0.245} \qquad (10.105)$$

Equation 10.105 can also be used for sieve and valve trays.

By making a number of assumptions Backowski (Perry's handbook, Chapter 18) developed a semitheoretical equation for overall efficiency in a binary distillation

$$E_O = \frac{1}{1 + 3.7 \times 10^4 \frac{KM}{h' \rho_l T}} \qquad (10.106)$$

where
  y^* = vapor phase concentration at equilibrium with the liquid, mol fraction
  $K = y^*/x$ = vapor-liquid equilibrium ratio
  x = liquid phase composition at equilibrium, mol fraction
  M = molecular weight
  h' = effective liquid depth, mm
  ρ_l = liquid density, kg/m^3
  T = temperature, K

For sieve or valve plates, $h' = h_w$, outlet weir height, mm. For bubble caps h' = height of static seal. AIChE method is more comprehensive and rigorous.

2. Murphree Efficiency

The Murphree tray efficiency for the vapor phase is given by

$$E_M = \frac{y_n - y_{n+1}}{y_n^* - y_{n+1}} \qquad (10.107)$$

where
  E_M = Murphree plate efficiency
  y_n^* = composition of vapor in equilibrium with the liquid leaving stage n
  y_n, y_{n+1} = actual nonequilibrium values for the vapor streams leaving nth and $(n + 1)$th plate

y_n^* is to be read from the equilibrium curve. If the Murphree efficiency is assumed constant for all plates in the column, it is possible to use it to obtain a pseudoequilibrium curve, which in turn can be used to obtain the actual number of stages for the column. For this, the pseudoequilibrium curve is drawn below the true equilibrium curve. To obtain a point on this curve, the vertical line is drawn between the true equilibrium curve and the operating line. The point of the pseudoequilibrium curve will be marked on this vertical line as indicated by the percent efficiency. For example, if the Murphree plate efficiency is 50 percent, the point will lie midway on the vertical.

3. Point Efficiency

The point or local efficiency on a tray is defined by the following equation:

$$E_P = \frac{y'_n - y'_{n+1}}{y^*_n - y'_{n+1}} \tag{10.108}$$

where
- y'_n = mol fraction of vapor at a specific point n in the plate
- y'_{n+1} = mol fraction of the vapor entering the plate n at the same point
- y^*_n = mol fraction of the vapor that would be in equilibrium with x'_n at the same point

In small-diameter towers the vapor agitates the liquid quite vigorously so that the liquid has uniform composition on the tray. In that case, $E_M = E_P$.

When the slopes of equilibrium and operating lines are constant. Murphree tray efficiency is related to overall efficiency by the following equation:

$$E_O = \frac{\log[1 + E_M(mG/L - 1)]}{\log(mG/L)} \tag{10.109}$$

If equilibrium and operating lines are curved and not straight, Murphree efficiency can be used to construct a pseudoequilibrium curve. This curve and the curved operating lines can then be employed to determine actual nonideal plates. The reader is referred to reference 1 listed at the end of this chapter for details and illustration.

Efficiencies in Packed Towers

In packed towers the vapor flow and the liquid flow are truly countercurrent. HETP in m (ft) or height of packing equivalent to a theoretical plate is defined as the height of packing that is required to effect the same separation as a theoretical plate. Thus vapor leaving one HETP of packing will be in equilibrium with the liquid that leaves the same HETP. Total height of packing required to effect a given separation is then

$$H = n(\text{HETP}) \tag{10.110}$$

where n is the number of theoretical stages required.

As in absorption, in distillation also, packed towers are designed on the basis of transfer units. Using the transfer units (N_{OG}) concept, the height of a packed tower is given by

$$H = H_{OG} N_{OG} \tag{10.111}$$

where H_{OG} is the overall height of a transfer unit in m(ft).

HETP and H_{OG} can be related by the following equation:

$$\text{HETP} = H_{OG} \frac{\ln[m/(LV)]}{m/LV - 1} \tag{10.112}$$

where m is the slope of the equilibrium line. L and V are molar flows. For a tower HETP can be defined as

$$\text{HETP} = T/E_O$$

where T = tray spacing.

Efficiency of Random Packings
HETP for random packings can be estimated by the following equation:

$$\text{HETP} = 0.018 d_P \quad \text{(SI units)}$$
$$\text{HETP} = 1.5 d_P \quad \text{(English units)} \tag{10.113}$$

where HETP is in m and d_P is in mm with SI units and in ft and inches in English units.

Equation 10.113 is valid for low viscosity liquids.

For vacuum service,

$$\text{HETP} = 0.018 d_P + 0.15 \quad \text{(SI units)}$$
$$\text{HETP} = 1.5 d_P + 0.5 \quad \text{(English units)} \tag{10.114}$$

Efficiency of Structured Packings
Efficiencies of structured packings can be calculated using the following relations:

$$\text{HETP} = 100/a + 0.1 \text{ (SI units)}$$
$$\text{HETP} = 100/a + 0.33 \text{ (SI units)} \tag{10.115}$$

where HETP is in m and a is packing surface area in m^2/m^3 in SI units. In English units, HETP is in ft and a is in ft^2/ft^3. Practical values of HETP range from 0.3 m to 0.6 m (1 ft to 2 ft). Manufacturers of packings are also a good source for getting packing efficiencies.

Example 10.7

An equimolal mixture of benzene and toluene is to be fractionated in a packed tower at 101.3 kPa pressure to produce a distillate containing 96 percent benzene and a bottoms product containing 4 percent benzene. The feed is at its boiling point. A reflux ratio of 3 will be used. The equilibrium data is plotted in Exhibit 12. How many transfer units are needed to effect the desired separation?

Solution

First the equilibrium diagram was prepared. A vertical q line is drawn as the feed is at its boiling point. The operating line for rectifying section is then drawn from $x, y_D = 0.96$ to meet the y ordinate at intercept = 0.24. From the intersection point of q line and rectification operating line, another operating line is drawn to meet the diagonal at $x_B = 0.04$. The equilibrium diagram is given in Exhibit 12.

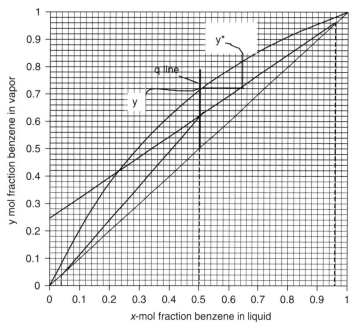

Exhibit 12 Equilibrium diagram, q line, and operating lines, Example 10.7

For a given value of y on the operating line, y^* was read and the corresponding value of $1/(y^* - y)$ calculated. Calculations are done for both rectifying and stripping sections of the column. The results are tabulated in the following table.

	Rectifying Section				Stripper Section			
y	y^*	$y^* - y$	$1/(y^* - y)$	y	y^*	$y^* - y$	$1/(y^* - y)$
0.615	0.713	0.098	10.2	0.04	0.08	0.04	25.0
0.70	0.80	0.10	10.0	0.10	0.16	0.06	16.7
0.80	0.88	0.08	12.5	0.20	0.31	0.11	9.1
0.90	0.94	0.04	25.0	0.30	0.45	0.15	6.7
0.96	0.986	0.026	49.9	0.40	0.543	0.143	7.4
				0.50	0.63	0.13	7.69
				0.6	0.705	0.105	9.52
				0.615	0.713	0.098	10.2

These results are plotted in Exhibit 13a and b to get the number of transfer units for both the sections of the tower.

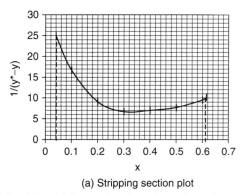
(a) Stripping section plot

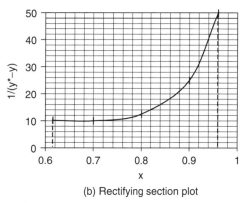
(b) Rectifying section plot

Exhibit 13 Determination of number of transfer units for Example 10.7

Areas under the curves were determined by Simpson's rule. For the rectifying section, the number of NTUs is 5.34; for the stripping section, 5.2. Thus total number of transfer units for the packed column is 5.34 + 5.2 = 10.54.

SIZING OF TRAY AND PACKED TOWERS

One of the factors that affects the efficiency of trays or packed towers is the entrainment of liquid by vapors rising through the column. Tray columns of various kinds or packed towers are designed for the purpose of improving contact between vapor and liquid so that the interfacial area increases and a better approach to equilibrium is obtained on each plate. In tray towers the maximum vapor velocity that can be used is limited either by entrainment or by the liquid handling capacity of the tray downcomer. Excessive vapor velocity may cause the liquid in the downcomer to back up to the tray above, and entrainment reduces the efficiency of fractionation. The large pressure drop leads to a condition of *flooding*. The design of column is therefore based on an allowable vapor velocity, for which there are different criteria for tray towers and for packed towers.

Tray Towers

In sizing the required diameter of a tray column such as sieve, bubble cap, or valve trays, the allowable vapor velocity is obtained using Fair's correlation. According to this correlation the allowable vapor velocity is given by the equation

$$u_{max} = K_v \left(\frac{\sigma}{20}\right)^{0.2} \sqrt{\frac{\rho_L - \rho_v}{\rho_v}} \text{ ft/s} \qquad (10.116)$$

where
- u_{max} = allowable vapor velocity, ft/s (velocity is based on net cross-sectional area, and is equal to total cross-sectional area minus the area allowed for downspouts)
- ρ_L, ρ_V = densities of liquid and vapor, kg/m³ or lb$_m$/ft³
- σ = surface tension of the liquid, dyn/cm [mN/m]
- K_v = allowable velocity factor, ft/s. This is to be obtained from Fair's correlation given in Figure 10.20
- L, G = Flow rates of liquid and vapor, kg/h or lb$_m$/h

Value of K_v from Equation 10.116 has to be modified for several reasons. First it should be multiplied by a factor of 0.9 to account for the downcomer or downspout area of 10 percent. Equation 10.116 is valid for nonfoaming liquids. So for foaming liquids, K_v should be multiplied again by 0.9. Another design factor called flood factor of 0.80 should be used in order to operate 20 percent below flooding point. For columns of diameter under 1 m or 3 ft, the flood factor of 0.65 to 0.75 should be used.

General Features of Tray Columns

Shell and Trays

Towers are constructed of several materials depending on the corrosion characteristics of the materials to be handled. Metals and alloys, glass, glass lined metals impervious carbon, and plastics are examples of materials used. Small-diameter

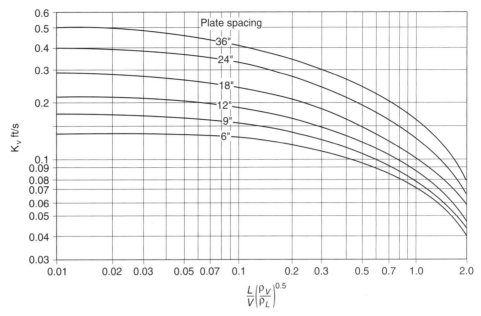

Figure 10.20 Estimation of K_v values for allowable vapor velocity. (From Fair, J.R., Petro/Chem. Eng., 33 (10), 45 (1961). (With permission)

towers are provided with hand holes and larger towers are provided with manways to about every 10th tray to facilitate cleaning.

Tray Spacing

In order to allow vapor and entrained liquid droplets to separate, the trays are placed in the tower at certain spacings. Spacing is an important parameter in determining the allowable vapor velocity in the tower. Spacings used vary generally from 12 in. to 48 in. but when cost is an important factor, spacings as small as 6 in. are also used. Recommended tray spacings are as follows (Treybal):

Tower Diameter		Tray Spacing	
m	ft	m	in
		0.15	6 min.
1 or less	4 or less	0.50	20
1–3	4–10	0.60	24
3–4	10–12	0.75	30
4–8	12–24	0.90	36

Downspouts or Downcomers

The liquid is made to flow from one tray to the next by means of downspouts or downcomers. Depending on the diameter of the column, these may be circular pipes or segmental tower cross sections set aside for liquid flow as in Figure 10.21. In segmental weirs, the vertical plates used to confine liquid flow are usually tapered to have a larger cross section at tray exit to allow the froth to disengage into liquid and vapor. Adequate seal is required to prevent vapor rising through the downcomer to the tray above. If there is a tendency to accumulate dirt, weep holes should be provided in the seal pot to facilitate drainage during shutdown.

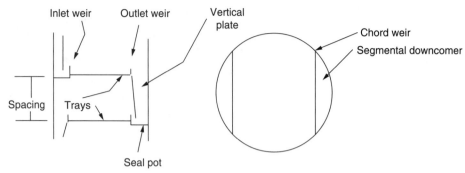

Figure 10.21 Trays and associated components

Weirs

The depth of liquid on the tray that is required for vapor-liquid contacting is maintained by overflow or outlet weir. Straight weirs are common but multiple V-notch weirs are also used if variation in liquid depth is to be made less sensitive to liquid flow.

Liquid flow over the tray may be directed in several ways (Figure 10.22): (a) Cross flow, (b) reverse flow, (c) split flow, (d) radial flow, and (e) cascade trays. Cross flow is more common. However, when liquid flow and column diameter are large, split flow is used to reduce flow path length of the liquid.

For a more detailed review including calculations of pressure drop, liquid back up in downcomers, aeration factor and foam densities, a reference is made to Perry's handbook, Van winkle, Geankoplis, Badger and Banchero, and Smith.

Packed Tower Design and Packed Tower Internals

The design of packed towers in distillation is similar to that in absorption. Vapor-liquid contacting in packed towers is countercurrent. Characteristics of various packings are given in Chapter 11. For a given type and size of packing (Raschig rings, pall rings, saddles, etc.) there is an upper limit to the rate of vapor flow, called the *flooding* velocity. Above this velocity the packed tower cannot operate. At low vapor velocities, the downward liquid flow is not essentially influenced by vapor velocity. However, at a vapor flow rate termed the *loading point*, the vapor flow

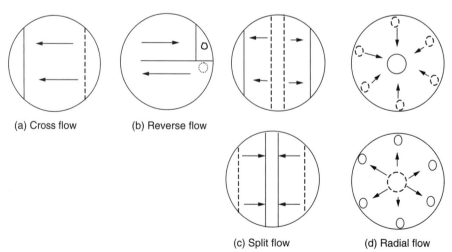

Figure 10.22 Liquid flow types in tray towers

starts to hinder the liquid flow downward. As the vapor rate is increased, hold up of liquid in the packing increases and at the flooding point it becomes impossible for the liquid to descend down through the packing. The loading in a packed tower starts at approximately at 65 to 70 percent flooding.

Pressure Drop in Packings

Pressure drop in packings is a very important consideration in the design of packed towers. Latest plot of Strigle correlating pressure drop in random packings is shown in Figure 10.23a and that of Kister and Gill for structured packings in Figure 10.23b.

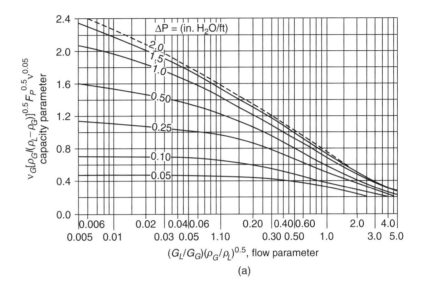

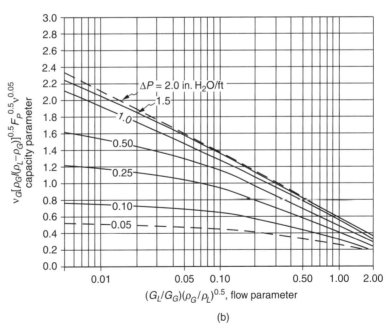

Figure 10.23 (a) Pressure-drop correlation for random packings by Strigle. (Reprinted from *Random Packings and Packed Towers*, R. F. Strigle, © 1987. With permission from Elsevier Science.) and (b) Pressure-drop correlation for structured packings by Kister and Gill. (From H.Z. Kister, *Distillation Design*, Petrochemical Engineers, Chem. Engr. Progress 87 (Feb) 32 (1991). Reproduced with permission. © 1991 AIChE.)

In these figures the flow parameter (abscissa value) is $(G_L/G_G)(\rho_G/\rho_L)^{0.5}$ and the capacity parameter (ordinate value) is

$$v_G \left[\frac{\rho_G}{(\rho_L - \rho_G)}\right]^{0.5} F_p^{0.5} v^{0.05} \qquad (10.117)$$

where

ρ_G, ρ_L = vapor and liquid densities, lb_m/ft^3
$v_G = G_G/\rho_G$
G_G, G_L = vapor and liquid mass velocities, $lb_m/s \cdot ft^2$
F_p = packing factor, ft^{-1}
v = kinematic viscosity $\mu_L/(\rho_L/62.4)$ = in centistokes
μ_L = liquid viscosity in cp

Note the capacity factor is not dimensionless and therefore only the units given above should be used.

Flooding Pressure Drop in Random and Structured Packings

Figures 10.23a and 10.23b do not include a flooding curve. Kister and Gill empirical equation to predict limiting pressure drop at flooding is

$$\Delta P_{\text{flood}} = 0.115 F_p^{0.7} \qquad (10.118)$$

where ΔP_{flood} = pressure drop in inches of water per foot of packing height. To convert from English to SI units, the conversion factor is 1.00 in water/ft of packing = 83.34 mm H$_2$O per m height of packing (3.281 in./m = 3.281 × 25.4 mm/in. = 83.34 mm H$_2$O/m packing).

Example 10.8

Obtain a preliminary diameter for a tray tower separating a mixture of benzene and toluene. For the design, the conditions at the bottom plate, where fluid properties can be taken essentially as those of toluene, will be used. The data for the conditions at bottom plate are as follows:

Temperature = 386 K

Pressure = 119 kPa

Vapor flow rate = 200 kg mol/h

Liquid flow rate = 225 kg mol/h

Molecular weight = 92

Density of liquid = 846 kg/m^3

surface tension of fluid = 18 dyn/cm

Tray spacing = 24 in. or 0.61 m

Solution

Calculate density of vapor.

$$\rho_V = \frac{92}{22.414(386/273.15)(101.3/119)} = 3.41 \text{ kg/m}^3$$

$\rho_L = 846 \text{ kg/m}^3$ (given in data)

V = Vapor flow rate = 200 kg mol/h = 200 × 92 = 18400 kg/h
L = Liquid flow rate = 220 kg mol/h = 220 × 92 = 20240 kg/h

Calculate flow parameter (abscissa) value.

$$\frac{L}{V}\left(\frac{\rho_V}{\rho_L}\right)^{0.5} = \frac{20240}{18400}\left(\frac{3.41}{846.2}\right)^{0.5} = 0.0698 \approx 0.07$$

Spacing is given as 24 in. However some space is taken by liquid level and plate thickness. Allow about 4 in for this. Then 20-in. clearance spacing is available for disengagement of vapor and liquid. So get K_V (allowable velocity factor) from Fair's correlation (Figure 10.20) with 20 in. for useful spacing as

$K_V = 0.29$ ft/s as read from Figure 10.20.

This factor should be modified for some design factors as follows:

Use 0.9 for allowing 10 percent of tower cross section as downcomer area

Use a factor of 0.95 to account for foaming (assume moderate foaming for benzene-toluene system)

Use a factor of 0.8-in. order to operate the tower 20 percent below flooding.

Then maximum allowable velocity is calculated as follows:

$$v_{max} = 0.29(0.9)(0.95)(0.8)\left(\frac{18}{20}\right)^{0.2}\left(\frac{846.2-3.41}{3.41}\right)^{0.5} = 3.05 \text{ ft·s} = 0.93 \text{ m/s}$$

Note that in the preceding equation, the densities can be used either in kg/m^3 or lb$_m$/ft^3 because we are dealing with a ratio of densities. However the resulting v_{max} is in ft/s according to Fair's correlation.

Now calculate the vapor flow rate.

$$\text{Vapor flow rate} = \frac{18400}{3.41 \times 3600} = 1.499 \text{ m}^3/\text{s}$$

$$\text{Total cross-sectional area of tower} = \frac{1.499}{0.93} = 1.612 \text{ m}^2$$

If D is inside diameter of the tower, its cross-sectional area is $\frac{\pi D^2}{4}$. Then

$$\frac{\pi D^2}{4} = 1.612 \text{ m}^2$$

Hence $\quad D = \sqrt{\dfrac{4 \times 1.612}{\pi}} = 1.466 \cong 1.47 \text{ m} = 4.82 \text{ ft}$

Towers are generally constructed in diameters with 6 in. increments. Therefore, the above diameter will be increased to 5 ft.

Example 10.9

For the same flow rates and conditions a packed column is to be designed using 2 in. Pall rings. What will be the diameter of the packed column? The kinematic viscosity of toluene can be taken as 0.31 centistokes. $F_p = 27$ ft^{-1}

Solution

$$\text{Flow parameter} = \frac{L}{V}\left(\frac{\rho_V}{\rho_L}\right)^{0.5} = \frac{20240}{18400}\left(\frac{3.41}{846.2}\right)^{0.5} = 0.0698 \approx 0.07$$

Calculate the flooding pressure drop.

$$\Delta P_{\text{flood}} = 0.115(F_p)^{0.7} = 0.115(27)^{0.7} = 1.155 \text{ in. H}_2\text{O/ft of packing}$$

At an abscissa value of 0.07 and pressure drop of 1.16 in. H$_2$O/ft of packing, the ordinate from Figure 10.23a is about 1.58. Then

$$v_G\left[\frac{\rho_G}{\rho_L - \rho_G}\right]^{0.5} F_p^{0.5} v^{0.05} = 1.58$$

$$v_G\left[\frac{3.41}{846.2 - 3.41}\right]^{0.5} 27^{0.5} 0.31^{0.05} = 1.58$$

$$\text{Flood velocity, } v_G = \frac{1.58}{0.0636 \times 5.196 \times 0.9431} = 5.07 \text{ ft/s}$$

Design for 50 percent of flood. Then actual velocity = 5.07/2 = 2.535 ft/s = 0.7726 m/s. Cross-sectional area of tower = 1.612/0.7726 = 2.09 m^2.

$$D = \sqrt{\frac{4 \times 2.09}{\pi}} = 1.63 \text{ m}$$

RECOMMENDED REFERENCES

1. Badger and Banchero, *Introduction to Chemical Engineering,* McGraw-Hill, NY, 2004.
2. McCabe and Smith, *Unit Operations of Chemical Engineering,* 6th ed., McGraw-Hill, NY, 2000.
3. Geankoplis, *Transport Processes and Separation Process Principles,* 4th ed., Prentice-Hall, NJ, 2004.
4. Van Winkle, *Distillation,* McGraw-Hill, NY, 1967.
5. Treybal, *Mass Transfer Operations,* 4th ed., McGraw-Hill, NY, 1994.

CHAPTER 11

Absorption

OUTLINE

EQUILIBRIUM SOLUBILITY CURVES 433
Determination of Solvent Flow Rate ■ Determination of Packing Height Z ■ Estimation of $K_G a$ for a System from the $K_G a$ of Another System ■ Graphical Determination of N_{OG} ■ Determination of Diameter of Column

RECOMMENDED REFERENCES 457

Absorption is an important unit operation used in industry. Problems involving absorption are frequently given in the P.E. examination. Candidates are required to become familiar with equilibrium solubility relationships; absorption theory, including the height of transfer unit; the numbers of transfer units, mass-transfer coefficients, etc.; and the determination of the diameter and height of the absorption column.

Generally, the solution of problems on absorption requires establishing a material balance equation, an equation of the operating line, and an equation or graphical representation of the equilibrium solubility relationship. Occasionally the solvent flow rate is not specified, and the candidate is advised to use a multiple of the minimum solvent flow rate. One must carefully examine the units of various parameters given in the problem statement and use the appropriate equation.

EQUILIBRIUM SOLUBILITY CURVES

Equilibrium solubility relationships are available in many forms. Some of them may be converted into equations of the type $Y^* = f(X)$ or $y^* = f(x)$ which can be readily used for the calculation of the transfer units by integration.

Figure 11.1 shows the operations of the countercurrent and cocurrent gas-absorption towers, which may be any of the following types: packed, spray and bubble-cap, or sieve tray. The design of the packed absorption towers will be dealt with here. The solvent gas and solvent liquid flows are essentially unchanged in quantity as they pass through the tower and it is convenient to express the material balances in terms of these. The equations of the material balance and operating line can be easily derived for various conditions of the absorption operation and are as follows:

1. **Countercurrent Absorption (All Concentrations)**

    Material balance: $G_S(Y_1 - Y_2) = L_S(X_1 - X_2)$ (11.1)
    Operating line: $G_S(Y - Y_2) = L_S(X - X_2)$

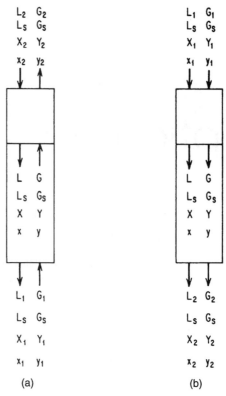

Figure 11.1 Absorption in packed column
(a) Countercurrent,
(b) Cocurrent

2. **Countercurrent Absorption (Dilute Solution)**

   Material balance: $\quad G_S(y_1 - y_2) = L_S(x_1 - x_2) \quad$ (11.2)
   Operating line: $\quad G_S(y - y_2) = L_S(x - x_2)$

3. **Cocurrent Absorption (All Concentrations)**

   Material balance: $\quad G_S(Y_1 - Y_2) = L_S(X_2 - X_1) \quad$ (11.3)
   Operating line: $\quad G_S(Y_1 - Y) = L_S(X - X_1)$

4. **Cocurrent Absorption (Dilute Solutions)**

   Material balance: $\quad G_S(y_1 - y_2) = L_S(x_2 - x_1) \quad$ (11.4)
   Operating line: $\quad G_S(y_1 - y) = L_S(x - x_1)$

The gas stream at any point in the tower is the total gas flow G, which includes the diffusing solute of mole fraction y, partial pressure p, or mole ratio Y, and a nondiffusing (insoluble) gas of flow rate G_S. Similarly, the liquid stream consists of total flow L containing the soluble gas of mole fraction x or mole ratio X and the essentially nonvolatile solvent of flow rate L_S. The auxiliary equations relating these quantities are

$$Y = \frac{y}{1-y} = \frac{p}{P_t - p} \quad (11.5a)$$

$$y = \frac{Y}{1+Y} = \frac{p}{P_t} \quad (11.5b)$$

$$X = \frac{x}{1-x} \tag{11.5c}$$

$$x = \frac{X}{1+X} \tag{11.5d}$$

$$G_S = G(1-y) \tag{11.6a}$$

$$L_S = L(1-x) \tag{11.6b}$$

where
- G_S = solvent gas flow rate, lb mol/h · ft^2
- L_S = solvent liquid flow rate, lb mol/h · ft^2
- L = total liquid flow rate, lb mol/h · ft^2
- Y = concentration of solute in the gas, lb mol solute/lb mol solvent gas
- X = concentration of solute in liquid, lb mol solute/lb mol solvent liquid
- y = concentration of solute in gas, mole fraction
- x = concentration of solute in liquid, mole fraction
- p = partial pressure of solute gas
- P_t = total pressure
- G = total gas flow rate, lb mol/h · ft^2

In the auxiliary Equation 11.5a and 11.5b, p and P_t must have the same units.

Determination of Solvent Flow Rate

The following steps may be followed to determine the liquid solvent flow rates required to affect a given absorption:

1. The average gas flow rate is first calculated by

$$\text{Average gas flow rate} = \frac{G_{\text{in}} + G_{\text{out}}}{2} = G_{av}$$

where the subscripts refer to the inlet, outlet, and average flow rates. G_{in} is calculated from the given data and G_{out} is calculated as follows:

$$G_{\text{out}} = G_{\text{in}} - \text{solute absorbed} + \text{solvent carried over as vapor}$$

$$\text{Solvent carried over} = \frac{p_s G_S}{P_t - p_s}$$

where p_s is the vapor pressure of the solvent at the operating temperature.

2. If the equilibrium solubility relationship is available in the form $y^* = mx$, estimate the solvent flow rate by

$$L_{av} = 1.6 m G_{av}$$

where the star indicates the equilibrium composition and m is a constant.

3. If the equilibrium solubility relationship is available in the form $Y^* = f(X)$, then proceed as follows. (Refer to Fig. 11.2a and b.)

   a. Plot the equilibrium solubility line Y versus X.
   b. Locate the point A (X_2, Y_2). Note that for a pure solvent, $X_2 = 0$ and Y_2 is obtained from Y_1 and the specified percent recovery.
   c. Locate Y_1 on the ordinate and draw a horizontal line CB through Y_1 intersecting the equilibrium curve at B.
   d. Join AB without crossing the equilibrium curve, as in Figure 11.2b.

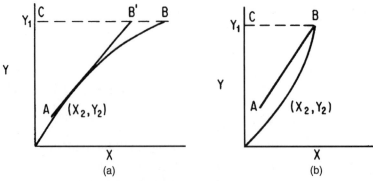

Figure 11.2 Estimation of minimum solvent flow: (a) Equilibrium curve convex toward operating line; (b) Equilibrium curve concave toward operating line

If line AB tends to cross the equilibrium curve as in Figure 11.2a, draw AB' tangential to the equilibrium curve. The slope of line AB or AB' is L_S/G_S. Minimum solvent flow rate is the slope of AB or AB' times G_S. The operating solvent flow rate is 1.5 times the minimum solvent flow rate.

The multiplier 1.5 in the preceding relation is not a constant factor. It is determined by an economic analysis.

Note: In the above, symbols, L, L_S, G, or G_S may be in lb mol/h · ft^2 or lb mol/h.

The method of graphical determination of the minimum solvent flow rate is illustrated in Example 11.3 and Exhibit 1a.

Determination of Packing Height Z

There are three most commonly used equations for determining the height of a packed column. These are

1. Principal resistance is in the gas phase. (This may be used when neither the gas nor the liquid phase resistance dominates.)

$$Z = N_{OG}H_{OG} \text{ ft} \tag{11.7}$$

2. Principal resistance is in the liquid phase. (This may be used when neither gas nor liquid phase resistance dominates.)

$$Z = N_{OL}H_{OL} \text{ ft} \tag{11.8}$$

3. Resistance may be either in the gas phase or liquid phase or both phases.

$$Z = N_P(\text{HETP}) \text{ ft} \tag{11.9}$$

where HETP is the height equivalent of a theoretical plate.

In the preceding equations, the number of transfer units, N_{OG} and N_{OL}, and the number of theoretical stages, N_P, can be determined as follows:

Determination of N_{OG}

1. For countercurrent absorption, and for any concentration of solute, the number of transfer units is given by

$$N_{OG} = \int_{y_2}^{y_1} \frac{(1-y)_{lm}\, dy}{(1-y)(y-y^*)} = \int_{y_2}^{y_1} k\, dy \tag{11.10}$$

where

$$(1-y)_{lm} = \frac{(1-y^*)-(1-y)}{\ln[(1-y^*)/(1-y)]} \quad (11.11)$$

and

$$k = \frac{(1-y)_{lm}}{(1-y)(y-y^*)} \quad (11.12)$$

N_{OG} is determined by graphical integration. A value of x is assumed between the known or specified limits (i.e., x_1 to x_2) and the corresponding value of y^* is obtained from the equilibrium relationship, which is expressed as $y^* = f(x)$. For the same value of x, the corresponding value of y is obtained from the material-balance equation. From the values of y and y^* thus obtained, the value of k can be determined from Equation 11.12. A plot of k versus y is made. The area under the curve gives N_{OG}. The equation can be simplified by replacing the logarithmic average by the arithmetic average as follows:

$$N_{OG} = \int_{y_2}^{y_1} \frac{dy}{y-y^*} + \frac{1}{2}\ln\frac{1-y_2}{1-y_1} = \int_{y_2}^{y_1} \frac{dY}{Y-Y^*} - \frac{1}{2}\ln\frac{1+Y_1}{1+Y_2} \quad (11.13)$$

2. In some countercurrent absorptions involving dilute solutions, both the equilibrium and operating lines may be straight. Then the equilibrium line is of the form $y^* = mx + c$ where m and c are constants. In these cases, the number of transfer units is given by

$$N_{OG} = \frac{y_1 - y_2}{(y-y^*)_{lm}} \quad (11.14)$$

where

$$(y-y^*)_{lm} = \frac{(y-y^*)_{bottom} - (y-y^*)_{top}}{\ln[(y-y^*)_{bottom}/(y-y^*)_{top}]} \quad (11.15)$$

In the preceding equation, the subscripts "bottom" and "top" denote the conditions at the bottom and top of packing.

3. In some countercurrent absorptions in dilute solutions, the equilibrium and operating lines are straight but not parallel. If the equilibrium line follows Henry's law, $y^* = mx$, and N_{OG} is given by

$$N_{OG} = \frac{1}{1-mG/L}\ln\left[\frac{(1-mG/L)(y_1-mx_2)}{y_2-mx_2} + \frac{mG}{L}\right] \quad (11.16)$$

In Equation 11.16, m, G, and L must be in consistent units. When y^* and x are in mole fraction, G and L are molar flows.

4. If the conditions are the same as in (3) and if the equilibrium and operating lines are straight and parallel, i.e., $y^* = mx = (L/G)\, x$, N_{OG} is given by

$$N_{OG} = \frac{y_1 - y_2}{y_2 - mx_2} \tag{11.17}$$

5. For the countercurrent absorption when dissolved solute reacts with the solvent so that there is no equilibrium pressure or concentration of the dissolved solute, i.e., $m = 0$ (e.g., absorption of CO_2 in a caustic solution), N_{OG} is given by

$$N_{OG} = \ln\left[\frac{\ln(1-y_1)}{\ln(1-y_2)}\right] \tag{11.18}$$

6. When the conditions are the same as in (5) but the solution is dilute ($y < 0.05$), N_{OG} can be calculated by the relation

$$N_{OG} = \ln\frac{y_1}{y_2} \tag{11.19}$$

7. For cocurrent absorption, when other conditions are the same as in (3) (see Figure 11.1b for an explanation of the symbols), N_{OG} is given by

$$N_{OG} = \frac{1}{1 + mG/L} \ln\frac{y_1 - mx_1}{y_2 - mx_2} \tag{11.20}$$

N_{OG} can also be estimated by the graphical method.

Determination of N_{OL}

For countercurrent absorption, and for any concentration of the solute, N_{OL} is given by

$$N_{OL} = \int_{x_2}^{x_1} \frac{(1-x)_{lm}\, dx}{(1-x)(x^*-x)} = \int_{x_2}^{x_1} k\, dx \tag{11.21}$$

where

$$k = \frac{(1-x)_{lm}}{(1-x)(x^*-x)} \quad \text{and} \quad (1-x)_{lm} = \frac{(1-x)-(1-x^*)}{\ln[(1-x)/(1-x^*)]}$$

N_{OL} is determined by a graphical integration in the same manner as the determination of N_{OG}. A value of y is assumed between the known or specified limits (i.e., y_1 and y_2), and the corresponding value of x^* is obtained from the equilibrium line expressed in the form $y = f(x^*)$. For the same value of y, the corresponding value of x is obtained from the material-balance equation. Hence, the value of k is determined. The equation can be simplified by replacing the logarithmic average by the arithmetic average as follows:

$$N_{OL} = \int_{x_2}^{x_1} \frac{dx}{x^*-x} + \frac{1}{2}\ln\frac{1-x_2}{1-x_1} = \int_{x_2}^{x_1} \frac{dX}{X^*-X} - \frac{1}{2}\ln\frac{1+X_1}{1+X_2} \tag{11.22}$$

If the equilibrium relationship is linear, the equation may be used to estimate N_{OL} without resorting to the graphical integration.

Determination of N_P

Determination of N_p is by the analytical method or the graphical method.

1. **Analytical Method.** This is applicable when the operating line is straight, the solution is dilute, and Henry's law is applicable, i.e., $y^* = mx$. The number of theoretical plates is then given by

$$N_P = \frac{\ln[(1 - mG/L)(y_1 - mx_2)/(y_2 - mx_2) + mG/L]}{\ln(L/mG)} \quad (11.23)$$

2. **Graphical Method.** (See the following pages.)

Relation Between N_{OG} and N_P

N_{OG} and N_P are related by the following equation:

$$N_{OG} = N_P \frac{\ln A}{A - 1} \quad (11.24)$$

where $A = mG/L$.

Determination of Height of Transfer Units H_{OG} and H_{OL}

These are experimentally determined and have the dimension of length. The analytical expression for H_{OG} is

$$H_{OG} = \frac{G}{K_y a(1-y)_{lm}} = \frac{G}{K_G a P_t (1-y)_{lm}} \quad (11.25)$$

where

$$(1-y)_{lm} = \frac{(1-y^*) - (1-y)}{\ln[(1-y^*)/(1-y)]}$$

If the soluble gas concentration is low, $(1 - y)_{lm}$ may be assumed equal to unity.

In the preceding equation $K_G a$ and $K_y a$ are the overall mass-transfer coefficients. These overall coefficients may be calculated from the individual gas and liquid-film mass-transfer coefficients and the equilibrium solubility relationship. However, care must be exercised to identify the units of these film coefficients and the units of the parameters in the equilibrium relationship.

The analytical expression for H_{OL} is

$$H_{OL} = \frac{L}{K_x a(1-x)_{lm}} = \frac{L}{K_L a C(1-x)_{lm}} = \frac{LM_L}{K_L a \rho_L (1-x)_{lm}} \quad (11.26)$$

where

$$(1-x)_{lm} = \frac{(1-x) - (1-x^*)}{\ln[(1-x)/(1-x^*)]}$$

and M_L = molecular weight of the liquid solvent.

Estimation of $K_G a$ for a System from the $K_G a$ of Another System

If $K_G a$ is known for one system, the $K_G a$ of another system can be determined, if the diffusivities for the two systems are known, by using

$$K_G a(\text{unknown}) = K_G a(\text{known}) \left[\frac{D_v(\text{unknown})}{D_v(\text{known})}\right]^{0.56}$$

In the preceding pages, $K_G a$, $K_L a$, $K_x a$, $K_y a$, $k_G a$, $k_L a$, $k_x a$, and $k_y a$ indicate the mass-transfer coefficients. The capital letters indicate the overall mass-transfer coefficients; the small letters indicate the individual mass-transfer coefficients. The units of these mass-transfer coefficients are denoted by their subscripts as

$$K_G a, k_G a = \text{lb mol/h} \cdot \text{ft}^3 \cdot \text{atm}$$
$$K_L a, k_L a = \text{lb mol/h} \cdot \text{ft}^3 \cdot (\text{lb mol/ft}^3)$$
$$K_x a, k_x a, K_y a, k_y a = \text{lb mol/h} \cdot \text{ft}^3 \cdot \text{mol fraction}$$

where
- M_L = molecular weight of liquid
- ρ_L = density of liquid, lb/ft^3
- C = concentration of solute in liquid, lb mol/ft^3
- D_v = diffusivity, ft^2/s

Graphical Determination of N_{OG}

The following steps (see Fig. 11.3) are used: (1) Plot the equilibrium and operating line Y versus X; (2) draw the vertical lines, such as AB and DF, and draw a line bisecting these vertical lines; (3) starting from X_2, Y_2, draw triangle CMN such that $CO = OM$. This completes one unit of N_{OG}. Repeat constructing triangles like this until point $A(X_1, Y_1)$ is covered. For practical purposes, the last triangle covering (X_1, Y_1) may be taken as one unit of N_{OG}.

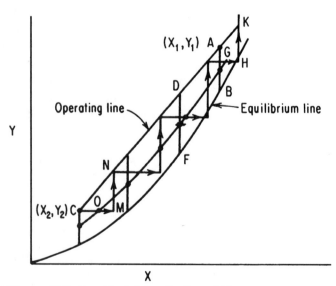

Figure 11.3 Graphical determination of N_{OG}

Table 11.1 Selection of packing size

Process Condition	Packing Size
Gas flow rate < 500 ft³/min (0.236 m³/s)	$\frac{3}{4}$-in. (19 mm)
Gas flow rate = 500 – 2000 ft³/min (0.236 – 0.944 m³/s)	1-in. (25 mm)
Gas flow rate > 2000 ft³/min or vacuum (0.944 m³/s) operation	$1\frac{1}{2}$–2-in. (38–50 mm) or bigger

The total number of triangles to cover the end composition is N_{OG}. Note that the triangles of N_{OG} may not touch or may cross the equilibrium curve.

Determination of Diameter of Column

The steps to be followed to estimate the diameter of an absorption column are

1. Draw a sketch as in Figure 11.1a or b, and show the gas and liquid flows and the compositions.

2. If not given, select a suitable type of packing. For operating pressure ≥ 100 mm Hg, the ring- or saddle-type packings may be selected. For the lower operating pressures, the systematic packings may be considered.

3. If not otherwise dictated, select the packing size[2] from Table 11.1.

4. Select the unit pressure drop, in. H₂O/ft (or Pa/m), of the packing from Table 11.2.

5. Calculate the numerical value of $(L/G)\sqrt{\rho_G/\rho_L}$. If L is not known, for estimation purposes one may choose a value such that $(L/G)\sqrt{\rho_G/\rho_L}$ lies between 0.02 and 1. From $(L/G)\sqrt{\rho_G/\rho_L}$ and the chosen unit pressure drop, find the ordinate from Figure 11.4 and calculate the design mass velocity.

$$G' = \left[\frac{(\text{ordinate})(\rho_G)(\rho_L - \rho_G)}{(C)(F_p)(v^{0.1})}\right]^{0.5}$$

where G' = lb/s · ft² or kg/s · m² and consistent units are to be used. For F_p, see Table 11.3.

Table 11.2 Selection of unit pressure drop* for nonfoaming systems

Type of Operation	Unit Pressure Drop	
	In. H₂O/ft of packing height	Pa/m of packing height
Atmospheric and high pressure	0.25 to 1.0 maximum 1.5	204 to 816 maximum 1224
Vacuum operation	0.05 to 0.2 maximum 1.0	41 to 164 maximum 816

*The pressure drop constraints given here are general guidelines and are not absolute recommendations. For vacuum operation, for example, a range of 0.1 to 1.0 inH₂O/ft of packing has also been recommended[4] depending upon the system. Foaming limits the capacity and changes the ΔP characteristics of packing. In such cases, the design value of the unit pressure drop needs to be modified. A consideration of the effect of foaming is beyond the scope of this text.

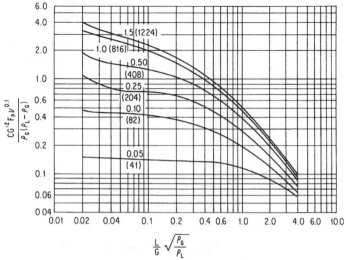

Figure 11.4 Generalized pressure-drop correlation. Parameter of curves is pressure drop in inches of water per foot of packing. Figures in parentheses are pascals per meter of packing height. In the British units, the symbols in the ordinate are G' (gas rate) = lb/ft^2 · s, $C = 1$, v (viscosity of liquid in centistokes) = cP/sp. gr., ρ_L, ρ_G (liquid and gas densities) = lb/ft^3. In SI units, G' = kg/m^2 · s, $C = 42.84$, v = kinematic viscosity in centistrokes, ρ_L and ρ_G = kg/m^3. The symbols in the abscissa must have consistent units: G and L must be in the same unit and ρ_G, ρ_L must be in the same unit (*By permission, Koch-Glitsch, LP. Wichita, Kansas.*)

6. The column diameter is to be calculated by

$$\text{Column diameter} = \begin{cases} 0.0188 \left(\dfrac{G_T}{G'}\right)^{0.5} \text{ ft} & G_T = \text{gas flow rate, lb/h} \\ 1.128 \left(\dfrac{G_T}{G'}\right)^{0.5} \text{ m} & G_T = \text{gas flow rate, kg/s} \end{cases}$$

7. This step involves checking the packing size and wetting rate. Maximum allowable packing sizes are given in Table 11.4.

The *wetting rate* is obtained by dividing the liquid flow in gpm or m^3/s by the cross section in ft^2 (or m^2). This rate should be greater than or equal to the minimum wetting rate, which can be calculated by using the relationships given in Table 11.5.

The *minimum wetting rate* thus obtained is not the absolute minimum below which flow conditions are poor and above which the packing becomes thoroughly wet. It is desirable that the column operates above the minimum wetting rate estimated by the criteria given in Table 11.5. If the actual wetting rate is below the minimum wetting rate, one of the following should be used: higher packing size, an increased pressure drop, a combination of both, a systematic packing, or recirculation of liquid. The turbulence created by a lower packing size would improve the mass transfer and may compensate for the loss of wetting.

Note: Distillation in a packed column can be run at a wetting rate lower than what is predicted by the above formulas without adversely affecting the packing efficiency.

Table 11.3 Properties of various packings for determination of column diameter

Type of * Packing	Material	Nominal Packing Size, inch or #								
		1/4	3/8	1/2	5/8	3/4	1	1-1/2	2	3 or 3 1/2
Super Intalalox Saddles	Ceramic	—	—	—	—	—	60	—	30	—
Super Intalalox Saddles	Plastic	—	—	—	—	—	40	—	28	18
Intalox Saddles	Ceramic	725 (300) [75]	330	200 (190) [78]	—	145 (102) [77]	92 (102) [77]	52 (59) [80]	40 (36) [79]	22 (28) [80]
Intalox Snow Flake	Plastic						13 (28)	13 (28)	13 (28)	
IMTP	metal				51		41	24	18	12
Hy-Pak Rings	metal	—	—	—	—	—	45	29	26	16
Pall rings	Plastic	—	—	—	95 (104) [87]	—	55 (63) [90]	40 (39) [91]	26 (31) [92]	17
Type of * Packing	Material	1/4	3/8	1/2	5/8	3/4	1	1-1/2	2	3 or 3 1/2
Pall rings	Metal	—	—	—	81 (104) [93]	—	56 (63) [94]	40 (39) [95]	27 (31) [96]	18
Raschig rings	Ceramic	b,x 1600 (217) [62]	b,x 1000	c 580 (112) [64]	c 380	c 255 (74) [72]	d 179 (58) [74]	e 93 (37) [73]	f 65 (28) [74]	g,x 37 (19) [75]
Raschig rings [a]	Metal	x 700	x 390	x 300 (122) [85]	170 (103) [87]	155 (81) [89]	x 155	—	—	—
Raschig rings [b]	Metal			410 (111) [73]	300	220 (75) [80]	144 (56) [86]	83 (39) [90]	57 (29) [92]	x 32 (20) [95]
Berl Saddles	Ceramic	x 900 (274) [60]	—	x 240 (142) [62]	—	170 (87) [66]	110 (76) [68]	65 (46) [71]	x 45 (32) [72]	—
Tellerettes	Plastic	—	—	—	—	—	36 (55) [87]	—	h 11 (28) [95]	16 (30) [92]

Packing factor (F_p), specific surface (ft^2/ft^3), and % free gas space of wet and dump-packed packings. Numbers without any bracket indicates F_p, numbers within ( ) indicates specific surface and numbers within [ ] indicates % free gas space. (Ref. 1,3,5)

Notes: a = 1/32" wall   c = 3/32" wall   e = 3/16" wall   g = 3/8" wall   h = type K
         b = 1/16" wall   d = 1/8" wall   f = 1/4" wall   x = Extrapolated

*All packings except Tellerettes are products of NORTON Company, Akron, OH. Tellerettes is a product of Ceilcote, Berea, OH.

Table 11.4 Maximum allowable packing sizes

Packing Type	Maximum Allowable Packing Size
Raschig and partition ring	1/30th of column diameter
Saddle type	1/15th of column diameter
Pall-type rings	1/10th of column diameter

Table 11.5 Minimum wetting rate[2]

Packing Size	Minimum Wetting Rate*	
	gpm/ft^2	m^3/s · m^2
<3 in (75 mm)	0.106A_p	2.2 × 10$^{-5}A_p$
≥3 in (75 mm)	0.164A_p	3.4 × 10$^{-5}A_p$

*A_p is the specific surface, ft^2/ft^3 or m^2/m^3. The specific surfaces of the packings are given elsewhere.[3d]

Example 11.1

A gas stream containing 10 percent by volume CO_2 at a pressure of 16 in. H_2O, a temperature of 176°F, and a flow rate of 70,000 lb/h is to be countercurrently scrubbed with 195,000 lb/h of 8 percent (by weight) caustic solution at 176°F. The entering gas has the following composition (by volume): CO_2 = 10%, N_2 = 80%, and O_2 = 10%. The pressure at the top of the column is 0.1 psig. A pressure drop of 0.38 in. H_2O per foot of packing will be used. Determine the diameter of the column for 70 percent recovery of CO_2. Choose a suitable plastic packing. The caustic solution has a viscosity of 0.9 cP and a specific gravity of 1.06 at the operating temperature. Check the wetting of the packing.

Solution

Step 1. Establish the gas and liquid flows and estimate $(L/G)\sqrt{\rho_G/\rho_L}$. At the bottom of the column, the gas flow is 70,000 lb/h.

$$\text{Average molecular weight of gas} = \Sigma M_i x_i$$
$$= 44(0.1) + 28(0.8) + 32(0.1)$$
$$= 30$$

$$CO_2 \text{ in the gas stream} = \frac{70,000}{30}(0.1) = 233.3 \text{ lb mol/h}$$

For 70 percent recovery,

$$CO_2 \text{ absorbed} = 233.3(0.7)(44) = 7185 \text{ lb/h}$$

$$\text{Inert gas in feed} = \frac{70,000}{30}(1 - 0.1) = 2100 \text{ lb mol/h}$$

$$\text{NaOH conc. in feed solvent} = \frac{\frac{8}{40}}{\frac{8}{40} + \frac{92}{18}}$$
$$= 0.0377 \text{ mol fraction.}$$

The vapor pressure of water at 176°F is 6.87 psia. The water vapor pressure of solution is 6.87(1 − 0.0377) = 6.61 psia. Because the pressure at the top of the column

Table 11.6

Component	mol/h	Mole Fraction
CO_2	233.3(0.3) = 70	0.018
Water vapor	1751.2	0.447
N_2	2100(8/9) = 1866.66	0.476
O_2	2100(1/9) = 233.33	0.060
Total	3921.19	1.001 ≈ 1.00

0.1 psig, assuming saturation by water vapor, water vapor in inert gas at the top of the column is

$$\frac{6.61}{14.8 - 6.61} = 0.807 \text{ lb mol/lb mol inert gas}$$

Gas remaining after CO_2 absorption = 2100 + 233.3(0.3)
= 2170 lb mol/h

Water vapor carried by the gas = 2170(0.807) = 1751.2 lb mol/h
Top gas composition is listed in Table 11.6.
The molecular weight of gas at the top of the column is

$$0.018(44) + 18(0.447) + 28(0.476) + 32(0.06) = 24.09$$

The gas density at the top is

$$\frac{PM_W}{RT} = \frac{14.8(24.09)}{10.731(460+176)} = 0.052 \text{ lb/ft}^3$$

The gas flow at the top is

$$3921.19(24.09) = 94{,}461.5 \text{ lb/h}$$

$$\frac{L}{G}\sqrt{\frac{\rho_G}{\rho_L}} \text{ at top} = \frac{195{,}000}{94{,}461.5}\sqrt{\frac{0.052}{1.06(62.4)}} = 0.058$$

The liquid flow at the bottom of the column equals caustic flow in plus gas flow in minus gas flow out.

$$195{,}000 + 70{,}000 - 94{,}462 = 170{,}538 \text{ lb/h}$$

The gas density at the bottom of the column is

$$\frac{15.28(30)}{10.72(460+176)} = 0.067 \text{ lb/ft}^3$$

$$\frac{L}{G}\sqrt{\frac{\rho_G}{\rho_L}} \text{ at bottom of column} = \frac{170{,}538}{70{,}000}\sqrt{\frac{0.067}{1.06(62.4)}} = 0.078$$

$$\frac{L}{G}\sqrt{\frac{\rho_G}{\rho_L}} \text{ average} = \frac{0.058 + 0.078}{2} = 0.07$$

Step 2. Select a $1\frac{1}{2}$-in plastic pall ring as packing,[3c] because

$$\text{Gas flow} = \frac{70{,}000}{0.067(60)} = 17{,}413 \text{ ft}^3/\text{min} > 2000 \text{ ft}^3/\text{min}$$

$$F_p = 32$$

Step 3. The specified pressure drop is 0.38 inH$_2$O per foot of packing.

Step 4. From Figure 11.4 the ordinate is 1.0 at 0.38 in. H$_2$O per foot. Then

$$G' = \left[\frac{(\text{ordinate})(\rho_G)(\rho_L - \rho_G)}{CF_p(v)^{0.1}}\right]^{1/2} = \left[\frac{1.0(0.063)(66.14 - 0.063)}{32(0.9/1.06)^{0.1}}\right]^{1/2}$$

$$= 0.3636 \text{ lb/ft}^2 \cdot \text{s}$$

$$G_T = \text{average vapor flow rate} = \frac{94{,}462 + 70{,}000}{2} = 82{,}231 \text{ lb/h}$$

$$\text{Column diameter} = 0.0188 \left(\frac{G_T}{G'}\right)^{1/2} = 0.0188 \left(\frac{82{,}231}{0.3636}\right)^{1/2}$$

$$= 8.94 \text{ ft} \quad \text{say 9 ft}$$

Packing size is well below $\frac{1}{10}$ of the column diameter.

$$\text{Minimum wetting rate} \approx 0.1 A_p$$

$$= 0.1(39) = 3.9 \text{ gpm/ft}^2$$

$$\text{Actual wetting rate} = \frac{182{,}769}{\left(\frac{\pi}{4}\right)(9)^2 (1.06)(500.4)} = 5.42 \text{ gpm/ft}^2$$

[*Note*: 1 gpm (s) 500.4 lb/h, where s = specific gravity]

The selected packing will have sufficient wetting.

Example 11.2

Estimate the packing height for the problem in Example 11.1. The overall mass-transfer coefficient $K_G a$ may be obtained from Figure 18.86 of Perry.[3a] $K_G a$ is proportional to $L^{0.28}$. The temperature correction factor for $K_G a$ may be obtained from the following data:

At 52°C and $L = 3000$, $K_G a = 18$ lb mol/h · ft^3 · atm.
At 55°C and $L = 3000$, $K_G a = 20$ lb mol/h · ft^3 · atm.

L is liquid mass velocity in lb/h · ft^2.

Solution

Step 1. Estimate the percent conversion of NaOH to Na$_2$CO$_3$. From Example 11.1, CO$_2$ absorbed = 7185.6 lb/h and

$$2\text{NaOH} + \text{CO}_2 \rightarrow \text{NaCO}_3 + \text{H}_2\text{O}$$
$$44 \text{ lb CO}_2 \rightarrow 2(40) \text{ lb NaOH}$$

$$7185.6 \text{ lb CO}_2 = \frac{2(40)(7185.6)}{44} = 13{,}064.7 \text{ lb NaOH}$$

$$\text{NaOH in caustic} = 195{,}000(0.08) = 15{,}600 \text{ lb/h}$$

$$\text{Conversion} = \frac{13{,}064.7}{15{,}600} = 83.7 \text{ percent.}$$

Step 2. Estimation of $K_G a$. It is to be noted that $K_G a$ for CO_2 absorption depends on the temperature, percent conversion, caustic normality, and superficial liquid mass velocity; it is independent of gas mass velocity.

$$K_G a \text{ of the system} = C_1 C_2 C_3 (K_G a)_R$$

where

$(K_G a)_R = K_G a$ at the reference point
C_1 = correction for normality and conversion
C_2 = correction for temperature
C_3 = correction for superficial mass velocity

a. Estimate $K_G a$ at the top of the tower. From Exhibit 1(a), the overall mass transfer coefficient value or $K_G a$ value for the $1\frac{1}{2}$-in. plastic pall ring may be obtained as follows:

$$(K_G a)_R = 2.2 \text{ lb mol/h} \cdot \text{ft}^3 \cdot \text{atm}$$

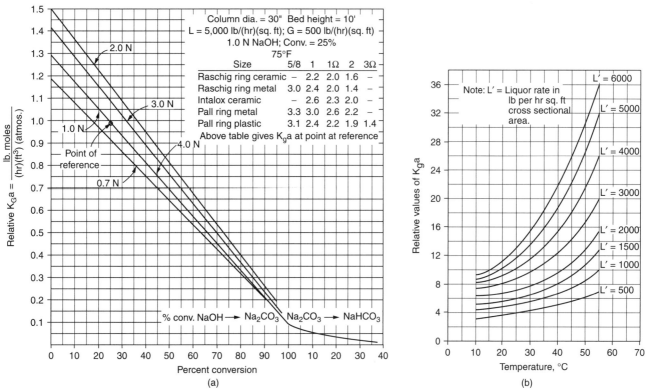

Exhibit 1 (a) Relative $K_G a$ versus percent conversion. (b) Relative $K_G a$ versus temperature Example 11.2

at the reference point. Conditions at the reference point are

Caustic solution	1 N NaOH
Conversion	25 percent
Temperature	75°F
Superficial liquid mass velocity	5000 lb/h · ft²

In this problem, normality is the number of gram moles in 1000 ml of solution. The basis is 100 g solution.

$$\text{Volume} = (100 \text{ g})\left(\frac{1}{1.06 \text{ g/ml}}\right) = 94.34 \text{ ml}$$

$$\text{Number of g} \cdot \text{mol} = \frac{8}{40} = 0.2$$

$$\text{Normality} = \frac{0.2(1000)}{94.34} = 2.1$$

The conversion at the top is 0 percent and the temperature is 176°F. At 0 percent conversion and 176°F (80°C), the correction factor for normality and conversion C_1 is 1.5 (from Exhibit 1(a)). The correction for the temperature from Exhibit 1(b) is found as follows:

$$\text{Superficial liquid mass velocity} = \frac{195,000}{\frac{1}{4}\pi(9^2)}$$

$$= 3065 \text{ lb/h} \cdot \text{ft}^2$$

At 52°C and $L' = 3000$ lb/h·ft², $K_Ga = 18$. At 55°C and $L' = 3000$ lb/h·ft², $K_Ga = 20$. Rate of increase of K_Ga is

$$\frac{20-18}{55-52} = 0.67/°C$$

C_2 is the correction for temperature (176°F = 80°C).

$$C_2 = \frac{18 + 0.67(80-52)}{18} = 2.04$$

C_3 is the correction for superficial mass velocity.

$$C_3 = \left(\frac{3065}{5000}\right)^{0.28} = 0.87$$

Hence, at the top of the tower

$$K_Ga = (2.2)(1.5)(2.04)(0.87)$$
$$= 5.86 \text{ lb mol/h} \cdot \text{ft}^3 \cdot \text{atm}$$

b. K_Ga at bottom of the tower. Correction factor for 2.1 N and 83.7 percent conversion is 0.33. At the bottom

$$K_Ga = 2.2(0.33)(2.04)(0.87)$$
$$= 1.29 \text{ lb mol/h} \cdot \text{ft}^3 \cdot \text{atm}$$

Taking the logarithmic mean,

$$K_G a = \frac{5.86 - 1.29}{\ln(5.86/1.29)} = 3.02 \text{ lb mol/h} \cdot \text{ft}^3 \cdot \text{atm}$$

Step 3. Estimation of G, y_1, y_2 and $(1-y)_{lm}$ follows.
At bottom of tower,

$$\text{Total gas flow} = \frac{70,000}{30} = 2333.33 \text{ lb mol/h}$$

$$(1-y)_{lm} = -\frac{y_1}{\ln(1-y_1)} = -\frac{0.1}{\ln 0.9} = 0.949$$

At top of tower, the inert gas in feed is

$$2333.33(1-0.1) \text{ B } 2100 \text{ lb mol/h}$$

NaOH concentration in feed solvent is

$$\frac{\frac{8}{40}}{\frac{8}{40} + \frac{92}{18}} = 0.0377$$

The vapor pressure of water at 176°F is 6.87 psia. The vapor pressure of the solution is $6.87(1-0.0377) = 6.61$ psia. The water vapor at the top of the tower is

$$\frac{6.61}{14.8 - 6.61} = 0.807 \text{ lb mol/lb mol inert gas}$$

The inert gas in feed is 2100 lb mol/h, and the gas remaining after absorption of CO_2 is

$$2100 + 2333.33(0.1)(0.3) = 2170 \text{ lb mol/h}$$

The water vapor carried by the gas is

$$0.807(2170) = 1751.20 \text{ lb mol/h}$$

The total gas flow at top of tower is

$$2170 + 1751.2 = 3921.2 \text{ lb mol/h}$$

Average gas flow is

$$\frac{1}{2}(3921.2 + 2333.33) = 3127.27 \text{ lb mol/h}$$

$$G = \frac{3127.27}{\frac{1}{4}\pi(9)^2} = 49.16 \text{ lb mol/h} \cdot \text{ft}^2$$

The amount of CO_2 in the exit gas is

$$2333.33(0.1)(0.3) = 70 \text{ lb mol/h}$$

$$y_2 = \frac{70}{3921.2} = 0.0179$$

$$(1-y)_{lm} = -\frac{0.0179}{\ln(1-0.0179)} = 0.991$$

Average $(1-y)_{lm} = \frac{1}{2}(0.949 + 0.991) = 0.97$

$$N_{OG} = \ln\frac{\ln(1-y_1)}{\ln(1-y_2)} = \ln\frac{\ln(1-0.1)}{\ln(1-0.0179)} = 1.764$$

$$H_{OG} = \frac{G}{K_G a P_t (1-y)_{lm}} = \frac{49.16}{3.02(1)(0.97)} = 16.78 \text{ ft}$$

Total packed height $= N_{OG} H_{OG} = 1.764(16.78) = 29.6$ say 30 ft.

Example 11.3

The vent gas from a reactor containing nitrogen and a solute gas of molecular weight of 80 is to be freed of the solute gas to meet the environmental standard by countercurrent absorption with a solvent in a packed column. The overhead gas from the packed column is vented to the atmosphere. Available data are as follows: vent gas flow is 0.2 m³/s at 0.11 MPa and 300 K. The concentration of the solute gas in the inert from the reactor is 4.5 percent by volume.

The concentration of the solute gas in the vent from the absorption column should not be more than 0.2 percent by volume. The equilibrium curve of the solute gas in the absorbing solvent is shown in Exhibit 1a. Pure solvent of molecular weight 240 and viscosity 0.9 mPa·s is used, and the design solvent flow rate will be 1.5 times the minimum.

The packing is 25 mm polypropylene Pall ring. The specific gravity of the solvent is 1.1. The overall mass-transfer coefficient is given by

$$K_G a = 25 \ G'^{\,0.7} L'^{\,0.07} \text{ kg/s} \cdot \text{m}^3 \cdot \text{MPa}$$

where $G' = \text{kg/s} \cdot \text{m}^2$ and $L' = \text{kg/s} \cdot \text{m}^2$. The diameter of the column is 387 mm and has a packing height of 3 m.

Determine whether the column is suitable for the service.

Solution

Step 1. First determine whether the column meets the pressure-drop requirement. Base calculations on the inlet conditions. The molecular weight of the inlet gas is

$$M_w = 0.045(80) + 0.955(28) = 30.34 \text{ kg/kg mol}$$

$$\text{Gas density} = \frac{PM_w}{RT} = \frac{0.11 \text{ MPa} \times 10^6 (30.34 \text{ kg/kg mol})}{(8314 \text{ Pa} \cdot \text{m}^3/\text{kg mol} \cdot \text{K}) 300 \text{ K}}$$

$$= 1.338 \text{ kg/m}^3$$

Liquid density $= 1.1 \times 10^3 \text{ kg/m}^3$

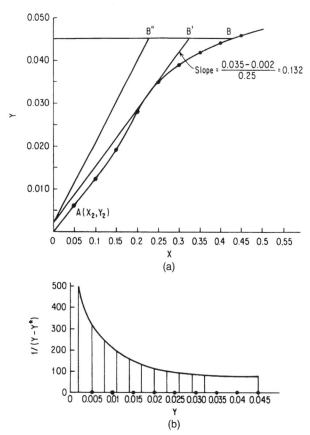

Exhibit 2 (a) Minimum and design solvent flow rates (Example 11.4); (b) Integration for Example 11.4

$$\text{Gas flow rate} = (0.2 \text{ m}^3/\text{s})(1.338 \text{ kg/m}^3) = 0.2676 \text{ kg/s}$$
$$= 0.00882 \text{ kg mol/s}$$

$$\text{Gas mass velocity} = \frac{0.2676}{0.785(0.387)^2} = 2.276 \text{ kg/m}^2 \cdot \text{s}$$

Following the procedure outlined in connection with Figure 11.2a, the solvent flow rate is determined as follows. Because a pure solvent is used, $X_2 = 0$ and

$$y_1 = 0.045 \quad Y_1 = \frac{0.045}{1 - 0.045} = 0.0471$$

$$G_2 = 0.00882(1 - 0.045) = 0.00842 \text{ kg mol/s}$$

$$y_2 = 0.002 \text{ maximum} \quad \text{or} \quad Y_2 = \frac{y_2}{1 - y_2} = \frac{0.002}{1 - 0.002} \doteq 0.002$$

(X_2, Y_2) is located in Exhibit 2a.

$$\text{Minimum solvent flow rate} = (\text{slope of the line } AB)G_S$$
$$= 0.132(0.00842) = 0.001111 \text{ kg mol/s}$$
$$= (0.001111 \text{ kg mol/s})(240 \text{ kg/kg mol})$$
$$= 0.27 \text{ kg/s}$$
$$\text{Actual solvent flow rate} = 0.27(1.5) = 0.405 \text{ kg/s}$$

$$\text{Solvent mass velocity} = \frac{0.405}{0.785(0.387)^2} = 3.445 \text{ kg/m}^2 \cdot \text{s}$$

$$\text{Liquid viscosity} = 0.9 \text{ mPa} \cdot \text{s} = 0.9 \times 10^{-3} \text{ (kg/m} \cdot \text{s}^2\text{)(s)}$$
$$= 9 \times 10^{-4} \text{ kg/m} \cdot \text{s}$$

$$\text{Kinematic liquid viscosity} = \frac{9 \times 10^{-4} \text{ kg/m} \cdot \text{s}}{1.1 \times 10^3 \text{ kg/m}^3} = 8.182 \times 10^{-7} \text{ m}^2/\text{s}$$

Referring to Figure 11.4,

$$\frac{L}{G}\left(\frac{\rho_G}{\rho_L}\right)^{0.5} = \frac{3.445}{2.276}\left(\frac{1.338}{1.1 \times 10^3}\right)^{0.5} = 0.053$$

$$\frac{CG'^2 F_p v^{0.1}}{\rho_G(\rho_L - \rho_G)} = \frac{42.84(2.276)^2(52)(8.182)^{0.1} \times (10^{-7})^{0.1}}{1.338(1100 - 1.338)} = 1.93$$

Locating 0.053 (abscissa) and 1.93 (ordinate) on Figure 11.4, the pressure drop is read as

$$\Delta P = 612 \text{ Pa/m of packing}$$

$$\text{Total pressure drop} = (612 \text{ Pa/m})(3 \text{ m}) = 1836 \text{ Pa} = 0.001836 \text{ MPa}$$

The allowable pressure drop including the line loss is 0.01 MPa. Hence, the column would work as far as the pressure drop and flooding are concerned.

Step 2. Determine the number of transfer units using Equation 11.13.

$$N_{OG} = \int_{Y_2}^{Y_1} \frac{dY}{Y - Y^*} - \frac{1}{2}\ln\frac{1 + Y_1}{1 + Y_2}$$

The first half of the righthand side requires graphical integration. Locate the operating line with a slope of $0.132(1.5) = 0.198$ and draw the operating line AB'' (Exhibit 2a). From the operating line, it may be seen that X varies from $X_2 = 0$ to $X = 0.228$ mol/mol. So between $X_2 = 0$ and $X_1 = 0.228$, assume values of X, read the values of Y from the operating line and Y^* from the equilibrium line, and tabulate as in Table 11.7. Plot $1/(Y - Y^*)$ versus Y to find the integral of the preceding equation (Exhibit 2b). The two righthand columns of the table will be

Table 11.7 Calculations for evaluation of integral

			$\int_{Y_2}^{Y_1} \frac{dY}{Y - Y^*}$		
X	Y	Y^*	$1/(Y - Y^*)$	$\frac{1}{2}(Y + Y^*)$	$\frac{1}{2}(Y + Y^*)_{\text{average}}$
$X_2 = 0$	0.0020	0	500	0.001	
0.05	0.01140	0.006	185.2	0.0087	
0.1	0.0210	0.012	111.1	0.0165	
0.15	0.0304	0.019	87.7	0.0247	0.0206
0.2	0.0398	0.028	84.7	0.0339	
$X_1 = 0.228$	0.045	0.0324	79.4	0.0387	

used to estimate H_{OG}. The area under the curve, partly by Simpson's rule (Exhibit 2b), is 6.162.

$$N_{OG} = 6.162 - \frac{1}{2} \ln \frac{1+0.0471}{1+0.002} = 6.14$$

Step 3. Determine the height of transfer unit from Equation 11.25.

$$H_{OG} = \frac{G}{K_G a P_t (1-y)_{lm}}$$

$$(1-y)_{lm} \approx (1-Y)_{lm} = \frac{(1-Y)+(1-Y^*)}{2} = 1 - \frac{Y+Y^*}{2}$$

$$= 1 - \frac{0.0206}{2} = 0.990$$

$$K_G a = 25 G'^{0.7} L'^{0.07}$$
$$G' = 2.276 \text{ kg/m}^2 \cdot \text{s}$$
$$L = 3.445 \text{ kg/m}^2 \cdot \text{s}$$

$$K_G a = 25(2.276)^{0.7}(3.443)^{0.07} = 48.48 \text{ kg/mg}^3 \cdot \text{s} \cdot \text{MPa}$$

$$H_{OG} = \frac{G'}{K_G a P_t (1-y)_{lm}} = \frac{2.276}{48.48(0.11)(0.99)} = 0.431$$

Height of packing required = $N_{OG} H_{OG}$ = 6.14(0.431) = 2.65 m

Because the available packed height is 3 m, the column would be adequate for the service specified in the problem.

Example 11.4

A gas stream containing 0.05 mol fraction of the solute gas is to be countercurrently scrubbed in a packed column to reduce the solute content to 0.0001 mol fraction. Estimate the packing height. Use these data: gas mass velocity = 10 lb mol/h·ft^2; liquid mass velocity = 400 lb mol/h·ft^2; overall mass-transfer coefficient based on overall driving force in liquid phase = 120 lb mol/ft^3·h·mol fraction. The equilibrium solubility relationship is given by $y = 20x$ where y and x are mole fractions of the solute in the vapor phase and liquid phase, respectively. The solvent entering the column is pure.

Solution

Because the principal resistance is in the liquid phase,

$$N_{OL} = \int_{x_2}^{x_1} \frac{dx}{x^* - x} + \frac{1}{2} \ln \frac{1-x_2}{1-x_1}$$

Equilibrium line: $\quad y = 20x^* \quad$ or $\quad x^* = 0.05y$

Operating line: $\quad G(y - y_2) = L(x - x_2)$

or
$$y = \frac{L}{G}(x - x_2) + y_2$$
$$= \frac{L}{G}x + y_2 \quad \text{since} \quad x_2 = 0$$
$$= \frac{400}{10}x + 0.0001$$
$$= 40x + 0.0001$$
$$x^* = 0.05y = 0.05(40x + 0.0001) = 2x + 0.000005$$
$$x^* - x = x + 0.000005$$

Overall material balance:

$$G(y_1 - y_2) = L(x_1 - x_2) = 10(0.05 - 0.0001) = 400x_1$$
$$x_1 = 0.00125$$

$$N_{OL} = \int_{x_2=0}^{x_1=0.00125} \frac{dx}{x + 0.000005} + \frac{1}{2}\ln\frac{1-x_2}{1-x_1}$$
$$= \ln\frac{x_1 + 0.000005}{x_2 + 0.000005} + \frac{1}{2}\ln\frac{1-x_2}{1-x_1}$$
$$= 5.33$$

where $\quad x_1 = 0.00125 \quad$ and $\quad x_2 = 0$

$$H_{OL} = \frac{L}{K_x a(1-x)_{lm}} \doteq \frac{L}{K_x a} = \frac{400}{120} = 3.33$$

Height of packing = 5.53(3.33) = 18.4 ft.

Example 11.5

The capacity of a column is to be increased by 55 percent by replacing the existing 1-in. metal Raschig rings ($F_P = 115$) with some other packing without significantly changing the percent flooding. The end compositions, L/G ratio, pressure, temperature, etc., remain unchanged. Select the packing.

Solution

The new packing factor required is

$$F_{P2} = \left(\frac{G_1}{G_2}\right)^2 F_{P1} = \left(\frac{G_1}{1.55G_1}\right)^2 (115) = 47.9$$

From Table 18.5 of Ref. 3, select 1-in. metallic Pall rings.

Example 11.6

During the unloading of a tank car into a storage tank, air containing 0.02 mol fraction of a water-soluble gas comes out of the storage tank. This air is to be scrubbed with water in a countercurrent packed column to reduce the concentration of the gas to 0.0001 mol fraction. The following data are available: gas flow rate = 1000 scfm/ft² tower cross section; pure water rate = 1500 lb/h·ft² tower cross section; equilibrium relationship $y^* = 1.8x$. y^* and x are mole fractions of the solute in vapor and liquid phase, respectively. $K_y a = 2$ lb mol/ft³·h·mol fraction. Determine the packing height.

Solution

By Equation 11.16,

$$N_{OG} = \frac{1}{1 - mG/L} \ln\left[\frac{(1 - mG/L)(y_1 - mx_2)}{y_2 - mx_2} + \frac{mG}{L} \right]$$

$$m = 1.8 \quad G = \frac{1000}{359} = 2.79 \text{ lb mol/h} \cdot \text{ft}^2$$

$$y_1 = 0.02 \quad y_2 = 0.0001 \quad x_2 = 0$$

$$L = \frac{1500}{18} = 83.33 \text{ lb mol/h} \cdot \text{ft}^2$$

$$\frac{mG}{L} = \frac{1.8(2.79)}{83.33} = 0.0603 \qquad 1 - \frac{mG}{L} = 1 - 0.0603 = 0.9397$$

$$N_{OG} = \frac{1}{0.9397} \ln\left[\frac{0.9397(0.02)}{0.0001} + 0.0603 \right]$$

$$= 1.0642(5.236) = 5.572$$

$$H_{OG} = \frac{G}{K_y a} = \frac{2.79}{2} = 1.395$$

Height = $N_{OG} H_{OG}$ = 5.572(1.395) = 7.77 ft

Example 11.7

A gas mixture containing 0.01 mol fraction of a solute is to be scrubbed at a pressure of 1 atm in a packed column. The gas flow rate is 200 lb mol/h and the pure solvent flow rate is 200 lb mol/h. The tower is 5 ft in diameter and 10 ft high. The overall height of the transfer unit for the operating condition and packing used is given by

$$H_{OG} = 5 \frac{G^{0.1}}{L^{0.4}}$$

where G and L are gas and liquid flows, lb/h·ft². The molecular weights of the gas and liquid are 29 and 18, respectively. The equilibrium relationship is given by $p = x$, where p is the partial pressure of the solute in atmosphere and x is mol fraction of the solute in the solution.

Estimate: (a) Percent recovery for countercurrent absorption, (b) percent recovery for cocurrent absorption.

Solution

Step 1. Estimate H_{OG}, $H_{OG} = 5G^{0.1}/L^{0.4}$.

$$G = \frac{200(29)}{\frac{1}{4}\pi(5^2)} = 295.39 \text{ lb/h} \cdot \text{ft}^2$$

$$L = \frac{200(18)}{\frac{1}{4}\pi(5)^2} = 183.35 \text{ lb/h} \cdot \text{ft}^2$$

$$H_{OG} = 5(295.39)^{0.1}(183.35)^{-0.4} = 1.098$$

Step 2. Estimate N_{OG}.

$$N_{OG} = \frac{\text{height}}{H_{OG}} = \frac{10}{1.098} = 9.1$$

Step 3. Estimate N_{OG} using analytical expression. Express the equilibrium line in the form $y^* = mx$.

Equilibrium line: $\quad p = x$

$$y^* = \frac{p}{P_t} = \frac{x}{P_t} = x \quad \text{since} \quad P_t = 1 \text{ atm}$$

Slope of equilibrium line $= m = 1$. So the operating and equilibrium lines are parallel.

a. *Countercurrent Flow*

$$N_{OG} = \frac{y_1 - y_2}{y_2 - mx_2} = \frac{y_1 - y_2}{y_2} \quad \text{because} \quad x_2 = 0$$

$$= \frac{y_1}{y_2} - 1$$

From Step 2,

$$\frac{y_1}{y_2} - 1 = 9.1 \qquad y_2 = 0.099 y_1$$

$$\text{Recovery} = \frac{G(y_1 - y_2)}{Gy_1} 100 = \frac{0.01 - 0.01(0.099)}{0.01} 100$$

$$= 90.1\%$$

b. *Cocurrent Flow.*

$$N_{OG} = \frac{1}{1 + mG/L} \ln \frac{y_1 - mx_1}{y_2 - mx_2} = 9.1$$

$$m = \frac{L}{G} = 1 \qquad y_1 = 0.01 \qquad x_1 = 0$$

$$\frac{1}{1+1} \ln \frac{0.01}{y_2 - x_2} = 9.1$$

$$\frac{0.01}{y_2 - x_2} = e^{2(9.1)}$$

$$y_2 = x_2$$

Equation of the operating line:

$$x_2 = x_1 + \frac{G}{L}(y_1 - y_2) = y_1 - y_2 = 0.01 - y_2$$

From these two equations,

$$y_2 = 0.01 - y_2 \qquad y_2 = 0.005$$

$$\text{Recovery} = \frac{G(y_1 - y_2)}{Gy_1}100 = \frac{0.01 - 0.005}{0.01}100 = 50\%$$

RECOMMENDED REFERENCES

1. Eckert, *Chemical Engineering,* vol. 82, April 14, 1975, pp. 70–76.
2. Morris and Jackson, *Absorption Towers,* Butterworth, London, 1953.
3. Perry (ed.), *Chemical Engineers' Handbook,* 5th ed., McGraw-Hill, New York, 1973: (a) p. 18–45; (b) p. 18–47, (c) p. 18–22. (d) p. 18–24.
4. Dolan, Norton Co., Akron, Ohio, private communication, Sept. 28, 1982.
5. Strigle, *Random Packings and Packed Towers, Design and Applications,* Gulf Publishing Company, Houston, 1994.

CHAPTER 12

Liquid-Liquid Extraction and Leaching

OUTLINE

LIQUID-LIQUID EXTRACTION 459
Liquid-Liquid Equilibria ■ Ternary Systems

CONJUGATE PHASES 464
Construction of Conjugate Line ■ Extract and Raffinate Phases ■ Selectivity of Solvent ■ Ponchon-Savarit Diagram for Solvent Extraction ■ Liquid-Liquid Extraction Calculations ■ Application to Multistage Countercurrent System

EXTRACTION EQUIPMENT 472
Design of Packed Towers for Extraction Using Mass Transfer Coefficients

LEACHING 476
Leaching Calculations ■ Equilibrium Relationship in Leaching ■ Representation of Equilibrium Data ■ Calculation of Equilibrium Stages ■ Material Balance Equations

RECOMMENDED REFERENCES 491

In this chapter, we review two operations: LL-Extraction and Leaching. In these operations a soluble component, either a solid or liquid, is removed from a solid or a liquid with the use of a solvent. During the contact of the solvent with the solute, liquid, or solid, two phases exist and the solute or solutes can diffuse from one phase to the other. During the contact of the two phases, the components of the original mixture redistribute between the phases. The phases are then separated by physical means. One phase is enriched and the other depleted in one or more solutes.

LIQUID-LIQUID EXTRACTION

Liquid-liquid or solvent extraction is an operation in which a solute dissolved in one liquid phase is transferred to a second liquid phase. The two phases (both liquid) are chemically quite different. This results in the separation of the components according to physical and chemical properties. In the review of liquid-liquid equilibria in this chapter, the solvent is assumed to be insoluble in the solution from which solute is to be extracted.

Liquid-Liquid Equilibria

The criteria of liquid-liquid equilibrium are the same as for vapor-liquid equilibrium reviewed in Chapter 9. In both phases, at the same temperature and pressure, the fugacity of each component must be the same in all phases.

Ternary Systems

Generally there are three components and two phases in equilibrium in a liquid-liquid extraction process. Application of the phase rule gives $F = 3 - 2 + 2 = 3$. The variables are temperature, pressure, and four concentrations. Only two concentrations can be specified for each phase because the third is automatically fixed, as the total of the three mass fractions must be equal to one.

Equilibrium in a three-component system can be represented on (1) equilateral triangle, (2) right angle triangle, or (3) Ponchon-Savarit diagrams or X-Y plots. The right triangle and Ponchon-Savarit diagrams are more convenient to use. Compositions may be expressed in terms of the mass fractions or mole fractions. Methods of the construction of the equilibrium diagrams are illustrated by example later in the chapter. We will first review some terms encountered in liquid-liquid extraction.

Extract Solvent   The solvent that is used to recover the solute from another solution is termed the *extract solvent*.

Extract Phase   The liquid phase, which is rich in extracted solute, is termed the *extract phase*.

Raffinate Solvent   The *raffinate solvent* is the solvent from which the desired solute is to be separated.

Raffinate Phase   The *raffinate phase* is the liquid phase, which is lean in solute to be extracted.

Conjugate Phases   Two phases, which exist in equilibrium such that their compositions are independent of the total two-phase mixture, are called *conjugate solutions*.

Distribution of Solute between Two Immiscible Liquids   In dilute solutions, the equilibrium distribution of a solute between two phases is expressed in terms of a *distribution coefficient K*

$$K = \frac{C_E}{C_R} \qquad (12.1)$$

where C_E and C_R are concentrations of solute in extract and raffinate phases respectively. The applicability of the distribution law for practical calculations is limited.

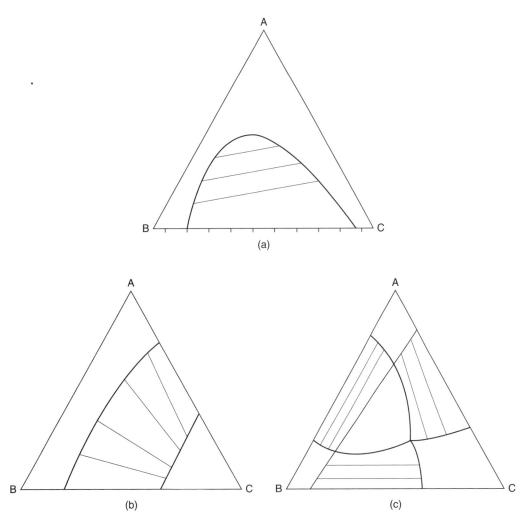

Figure 12.1 (a) Type I system, (b) Type II system, and (c) Type III system

Types of Ternary Liquid Systems

Three types of ternary liquid systems exist with respect to the partial miscibility of the three components comprising the system. These are shown in Figure 12.1.

Type I: Only one pair of components is partly or wholly immiscible. The other two pairs are completely miscible in all proportions. The equilibrium diagram for such a system is shown in Figure 12.1a. Chloroform-water acetic acid is an example of this type system.

Type II: In type II, component pairs *CA* and *CB* are partly miscible but the pair *AB* is miscible in all proportions. The ternary diagram for such a system is schematically shown in Figure 12.1b. n-heptane-methyl cyclohexane-water is an example of this system.

Type III: In this type all three pairs are partly miscible as shown in Figure 12.1c.

Triangular Coordinates and Equilibrium Data   Equilateral triangular coordinates are often used to represent the ternary system equilibrium data. In equilateral triangle there are three axes and the sum of distances from any point within the triangle is equal to the altitude of the triangle. An equilateral triangular diagram is shown in Figure 12.2. In this diagram the three corners represent the three components *A*, *B*,

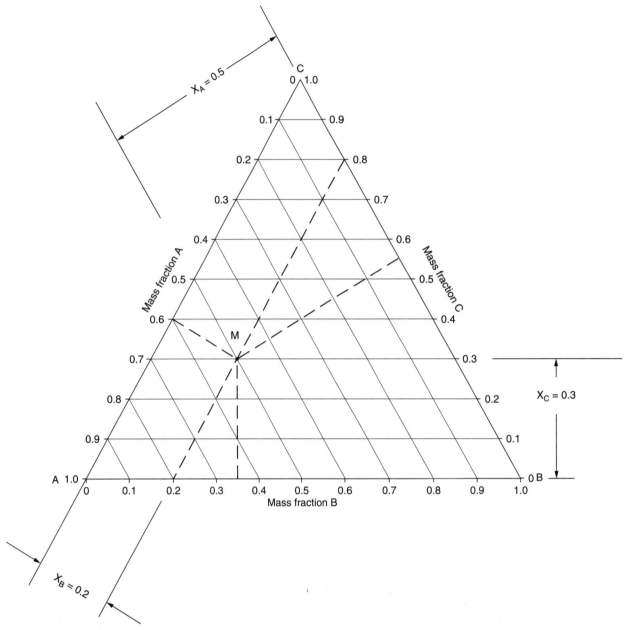

Figure 12.2 Equilateral triangle coordinates to represent liquid-liquid equilibrium data

and C. A point such as M within the triangle represents a mixture of the three components A, B, and C.

The perpendicular distances from point M to the three bases (sides of the triangle) represent the component mass fractions, X_A, X_B, and X_C. For example, the distance from M to base AB represents the mass fraction of the component C, X_C at the mixture point M. In a similar manner, the mass fractions X_A, X_B are given by the perpendicular distances from M to BC and CA respectively.

$$X_A + X_B + X_C = 1 \tag{12.2}$$

A typical phase diagram for a system in which two components A and B are partially miscible is shown in Figure 12.3.

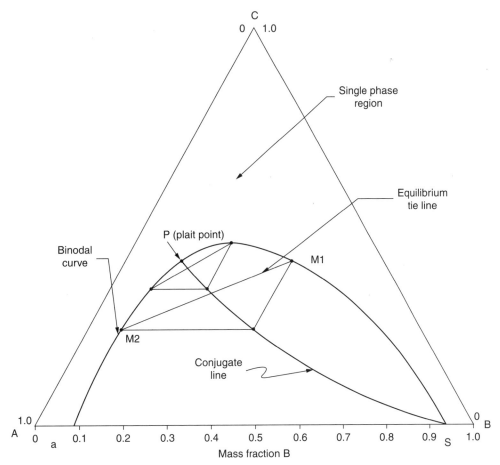

Figure 12.3 Schematics of liquid-liquid equilibrium data for a ternary system on equilateral triangle coordinates

Referring to Figure 12.3, component C is completely soluble in component A as well as component B. The components A and B are slightly soluble in each other. The region enclosed by the curved line R, P, S is a two-phase region. A mixture at point M in this region will separate into two phases whose compositions are given by points M_1 and M_2. The curve RPS is called *binodal curve*. A line connecting two equilibrium compositions on the binodal curve is called a tie line. The point P on the binodal curve where the two phases disappear is called the *plait point*. The plait point may not necessarily be at the apex of the binodal curve. A single phase exists in the region not enclosed by the binodal curve.

Instead of constructing a number of tie lines and cluttering the diagram, a single line (curve) called the conjugate line may be constructed to represent the equilibrium data. The conjugate line can then be used to read the equilibrium compositions.

Rectangular Coordinates An equilateral triangle diagram has some disadvantages, for example, it doesn't scale up or down easily. Also the construction to find the ideal number of stages gets very crowded. More convenient and useful is to plot the data on rectangular coordinates and use an isosceles right angle triangle to represent the phase equilibrium data. The concept of conjugate phases is explained next.

CONJUGATE PHASES

The equilibrium compositions of conjugate phases are plotted in Figure 12.4. A line such as AA' connecting two conjugate phases is called a *tie line*. At the point P, the plait point, the tie line disappears. At this point, the two conjugate phases are mutually soluble. The solubility curve EPF is obtained experimentally. Instead of showing many tie lines, a conjugate line or curve is drawn. This can be used to draw any tie line and get the compositions of the two phases.

Construction of Conjugate Line

From the experimental data, the compositions of a few conjugate phases, such as AA', BB', and CC', are first plotted. Then straight lines parallel to two sides of the triangle are drawn from each composition of a conjugate phase to intersect each other as at Q, R, and S. A curve through points P, Q, R, S, and N is drawn, and this is the conjugate line. To draw a desired tie line, a reverse procedure is followed; e.g., if the tie line for the conjugate phase D' is required, a straight line $D'T$ is drawn parallel to $0A$ to cut the conjugate curve at T. From T, a straight line TD is drawn, parallel to $0S$ to cut the solubility curve in D. Then DD' is the required tie line.

Extract and Raffinate Phases

The solutions lying within the boundaries of the solubility curve will separate into two phases, which are in equilibrium. The mass ratio of the two phases resulting in a mixture (e.g., as represented by a point K in the immiscibility region), is given by the ratio of opposite line segments. Thus

$$\frac{\text{Mass of phase } E}{\text{Mass of phase } R} = \frac{KB}{KB'}$$

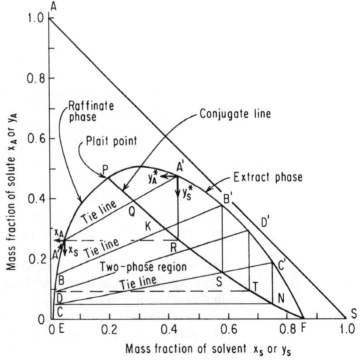

Figure 12.4 Construction of phase-equilibrium diagram on right triangle diagram

In the preceding equation, E represents the extract phase and R represents the raffinate phase. If A and S are the solute to be recovered and the solvent, respectively, the separation is increased as the plait point approaches zero and the line increases in slope.

Selectivity of Solvent

A solvent is said to be selective for a given solute when the mass fraction of the solute in the extract layer is greater than that in the raffinate layer. Selectivity of the solvent S for a solute component A is defined as follows:

$$\text{Selectivity} = \frac{y_A/y_B}{x_A/x_B} = \frac{y_A x_B}{x_A y_B} \quad (12.3)$$

where
    y = mass fraction of the component in the extract phase
    x = mass fraction in the raffinate phase
    A = desired solute
    B = raffinate solvent

A selectivity diagram can be prepared from the conjugate-phase equilibrium data or from the binodal solubility curve, as shown in Figure 12.5.

Ponchon-Savarit Diagram for Solvent Extraction

This diagram is especially useful for calculations involving the continuous countercurrent extraction with reflux. The X and Y coordinates are given by

$$X_A \text{ or } Y_A = \frac{\text{mass of solute component}}{\text{mass of solute + mass of raffinate solvent}} \quad (12.4)$$

$$X_s \text{ or } Y_s = \frac{\text{mass of extract solvent}}{\text{mass of solute + mass of raffinate solvent}} \quad (12.5)$$

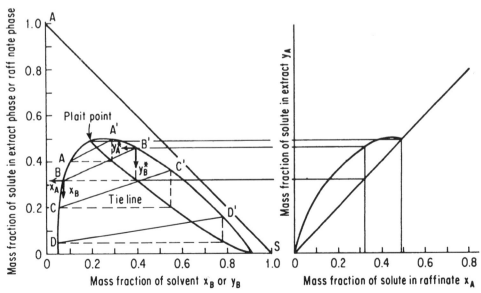

Figure 12.5 Construction of selectivity diagram from phase-equilibrium diagram

Liquid-Liquid Extraction Calculations

Calculations in liquid-liquid extraction involve material balances, computation of relative amounts of phases separated, and number of stages if countercurrent stagewise operation is to be used. These methods are reviewed next.

Single-Stage Equilibrium Extraction

Single-stage extraction may be conducted batchwise or continuously. The streams entering or leaving a stage are shown in Figure 12.6. An overall material balance over the stage gives

$$E_0 + R_0 = E_1 + R_1 = M_1 \tag{12.6}$$

The solute balance gives

$$E_0(y_A)_0 + R_0(x_A)_0 = E_1(y_A)_1 + R_1(x_A)_1 \tag{12.6a}$$

The point M_1 is located on the line joining E_0 and R_0 by the mixture rule or more easily by a calculation using the relations

$$R_0(x_A)_0 + E_0(y_A)_0 = M_1(x_A)_{M_1} \quad \text{solute} \tag{12.6b}$$

Thus

$$(x_A)_{M_1} = \frac{R_0(x_A)_0 + E_0(y_A)_0}{M_1} \tag{12.6c}$$

and

$$(x_s)_{M_1} = \frac{R_0(x_s)_0 + E_0(y_s)_0}{M_1} \tag{12.6d}$$

The point M_1 lies on line $R_0 E_0$, and the tie line passing through M_1 (Figure 12.7) will give the compositions of the two phases R_1 and E_1 into which the mixture separates.

Also

$$\frac{\text{Mass of } E_1}{\text{Mass of } R_1} = \frac{\overline{R_1 M_1}}{\overline{M_1 E_1}} \tag{12.6e}$$

If E_0 is pure solvent, the point E_0 will coincide with apex S.

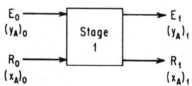

Figure 12.6 Single-stage extraction: material balance

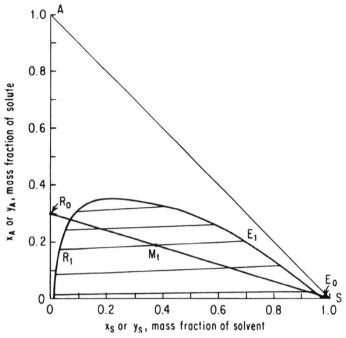

Figure 12.7 Single-stage extraction: representation on right-triangular diagram

Stagewise Extraction

For the multistage extraction with fresh solvent in each stage, the equilibrium relationship is

$$(y_A)_n = f(x_A)_n \tag{12.7a}$$

when

$$(y_A)_n = K(x_A)_n \quad K = \text{const} \tag{12.7b}$$

$$(x_A)_n = \left(\frac{b}{b + sK}\right)^n (x_A)_0 \tag{12.7c}$$

where
b = mass of raffinate solvent
s = mass of extract solvent
n = number of stages

In Equations 12.7a, b, and c, the masses b and s are assumed to be the same for each stage.

Example 12.1

One hundred kilograms of a solution of acetic acid (solute) and water (raffinate solvent) containing 40 wt% acetic acid is to be extracted two times using methylisobutylketone as the solvent. The extraction is isothermal at 25°C. Seventy-five kgs of ketone is to be used in each extraction.

(a) Determine the quantities and compositions of the various streams.

(b) Find how much solvent would be required if the same final raffinate concentration were to be obtained with a single extraction.

The tie line data are plotted in Exhibit 1a as the binodal curve.

Solution

(a) Prepare the equilibrium curve by plotting y_A versus x_A (Exhibit 1b).

Stage 1:

Overall balance: $R_0 + S_1 = E_1 + R_1 = M_1$

or $100 + 75 = E_1 + R_1 = 175 = M_1$

Solute balance: $R_0(x_A)_0 + S_1(y_A)_{S_1} = M_1(x_A)_{M_1}$

Hence $100(0.40) + 75(0) = 175(x_A)_{M_1}$

Thus $(x_A)_{M_1} = \dfrac{40}{175} = 0.229$

Solvent balance: $R_0(x_S)_0 + S_1(y_S)_1 = M_1(x_S)_{M_1}$

Hence $100(0) + 75(1) = 175(x_S)_{M_1}$

or $(x_S)_{M_1} = \dfrac{75}{175} = 0.429$

With the help of these coordinates, locate M_1 on $\overline{R_0 S}$ as shown in Exhibit 1a. [*Note*: The location of M_1 needs $\overline{R_0 S}$ and one of the coordinates $(x_A)_{M_1}$ and $(x_S)_{M_1}$.]

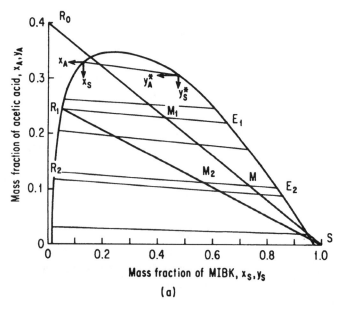

(a)

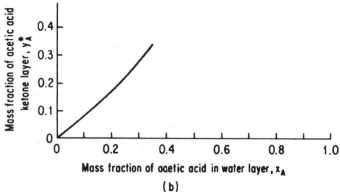

(b)

Exhibit 1 Multistage extraction on right-triangular diagram: (a) Solution of Example 12.1 (b) Auxiliary equilibrium diagram for locating tie lines

Now with the help of the distribution and equilibrium curves, locate the line $\overline{R_1E_1}$ passing through M_1 cutting the binodal curve in R_1 and E_1.
First calculate E_1 and R_1 by the mixture rule.

$$\frac{R_1}{E_1} = \frac{M_1E_1}{M_1R_1} = \frac{0.65 - 0.429}{0.429 - 0.05} = 0.58$$

$$R_1 = 0.58E_1 \qquad R_1 + E_1 = 175 \text{ kg}$$
$$1.58E_1 = 175 \qquad E_1 = 111 \text{ kg}$$
$$R_1 = 175 - 111 = 64 \text{ kg}$$

Stage 2:

Overall material balance gives

$$R_1 + S_2 = R_2 + E_2 = M_2 = 64 + 75 = 139 \text{ kg}$$

By the solute balance

$$R_1(x_A)_1 + S_2(y_A)_{S_2} = M_2(x_A)_{M_2}$$

or $\qquad 64(0.245) + 0 = 139(x_A)_{M_2}$

Therefore $\qquad (x_A)_{M_2} = \dfrac{64(0.245)}{139} = 0.113$

Join R_1 and S and locate M_2 on the line (Exhibit 1a). Locate R_2 and E_2 on the binodal curve by drawing the tie line through M_2. (*Note:* This involves locating the tie line by interpolation.)

$$E_2 = \frac{M_2[(x_A)_{M_2} - (x_A)_2]}{(y_A)_2 - (x_A)_2}$$

$$= \frac{139(0.113 - 0.131)}{0.104 - 0.131} = 93 \text{ kg}$$

Therefore $\qquad R_2 = 139 - 93 = 46 \text{ kg}$

(b) The amount of the solvent for the same raffinate concentration with a single extraction is calculated in the following manner. The final raffinate concentration is $(x_A)_2 = 0.131$. The mixture point will be the intersection point of the tie line $\overline{R_2E_2}$ and $\overline{R_0S}$. By material balance,

$$R_0 + S = R + E = M$$

and by solute balance,

$$R_0(0.4) + S(0) = M(x_A)_M$$
$$(x_A)_M = 0.107 \quad \text{from Exhibit 1a}$$
$$M = R_0 + S = 100 + S$$

Therefore $\qquad 100(0.4) + S(0) = 100(0.107) + S(0.107)$

or $\qquad S = \dfrac{100(0.4 - 0.107)}{0.107 - 0} = 273.8 \text{ kg}$

Thus, for the same raffinate concentration, a single-stage extraction requires 273.8 kg compared with 150 kg of solvent for the two-stage extraction.

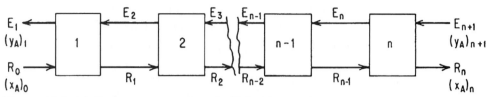

Figure 12.8 Multistage countercurrent liquid-liquid extraction system

Application to Multistage Countercurrent System

Either stepwise calculations or graphical methods are used. Figure 12.8 illustrates a continuous countercurrent multistage extraction system. Overall material balance gives

$$E_{n+1} + R_0 = E_1 + R_n \qquad (12.8)$$

Solute balance gives

$$E_{n+1}(y_A)_{n+1} + R_0(x_A)_0 = E_1(y_A)_1 + R_n(x_A)_n \qquad (12.9)$$

Extract solvent balance gives

$$E_{n+1}(y_S)_{n+1} + R_0(x_S)_0 = E_1(y_S)_1 + R_n(x_S)_n \qquad (12.10)$$

The procedure used to obtain the number of the equilibrium stages is illustrated by Example 12.2.

Example 12.2

Two thousand kg per hour of an acetic acid solution in water containing 30 percent acid is countercurrently extracted with isopropyl ether, so as to reduce the acid concentration to 2 percent in the solvent-free raffinate. Determine (a) minimum amount of the solvent per kg of feed solution and (b) the number of theoretical stages if the solvent-to-feed-solution ratio used is 1.6 times the minimum. The tie-line data are plotted in Exhibit 2 on a triangular diagram.

Solution

(a) Composition of R_0 is $(x_A)_0 = 0.3$. Composition of R'_n (free of solvent) equals 0.02. Composition $(y_s)_{n+1}$ equals 1.0.

These points are plotted as R_0, R'_n and S. Joint R'_n to S. The point of intersection of $\overline{R'_n S}$ with the solubility curve at R_n gives the composition of the raffinate stream R_n. The tie line passing through R_0, which cuts $\overline{R_n S}$ to the right nearest* to S, gives the Δ for the minimum solvent-to-feed-solution ratio. Let V_m be the intersection point of the binodal curve and $\overline{R_0 \Delta_m}$. Join V_m and R_n. From M_m, the intersection point of $\overline{V_m R_n}$ and $\overline{R_0 S}$, the minimum solvent-to-feed-solution ratio is

$$\frac{\overline{R_0 M_m}}{\overline{M_m S}} = \frac{8.2}{5.0} = 1.64$$

Thus, 1.64 kg solvent per kg feed solution is the minimum solvent.

---

*It would have been farthest from S if the point of intersection were on the left side of the diagram.

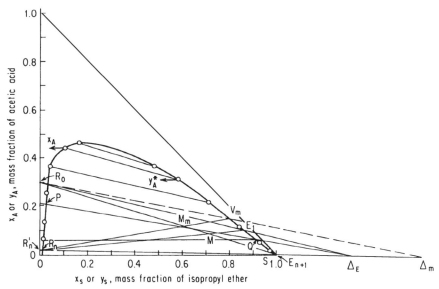

Exhibit 2 Determination of minimum reflux ratio (Example 12.2)

(b) If 1.6 times the minimum solvent is used,

Actual solvent = 1.64(1.6)
= 2.624 kg per kg of feed solution

The composition of the actual addition point is calculated as

$$(y_A)_M = \text{mass fraction solute A} = \frac{0.3}{1+2.624} = 0.083$$

$$(x_S)_M = \text{mass fraction solvent B} = \frac{2.624}{3.624} = 0.724$$

This point is located on $\overline{R_0 S}$. $\overline{R_n M}$ is extended to intersect the solubility curve in E_1. This intersection point gives the composition of the strong solution leaving the system. Join $R_0 E_1$ and extend to cut $R_n S$ extended in Δ_E. Using the Δ_E point and the solubility curve, it is possible to graphically obtain the equilibrium stages by following the difference point and the tie lines, but when the number of the stages is somewhat large, the lines get crowded. It is therefore more convenient to prepare an auxiliary diagram, as in Figure 12.9. To get the operating line on this diagram, few lines are drawn from Δ_E to cut the solubility curve. For example, a line from Δ_E cuts the solubility curve in P and Q. y_A and x_A are read. Various corresponding values of y_A and x_A are obtained in this manner and tabulated below.

x_A	0.02	0.055	0.09	0.150	0.205	0.250	0.3 = $(x_A)_0$
y_A	0	0.010	0.02	0.04	0.06	0.08	0.1 = $(y_A)_1$

These are plotted as y_A versus x_A to give the operating line (Figure 12.9). The equilibrium curve is plotted using the tie-line data, and the stages are then stepped as in the McCabe-Thiele diagram. When this is done, the number of theoretical stages from the graph is 7.9.

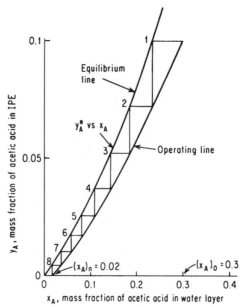

Figure 12.9 Determination of equilibrium stages on the distribution diagram (Example 12.2)

EXTRACTION EQUIPMENT

A variety of equipment is used to conduct liquid-liquid extractions. Processing is carried out batchwise or continuously depending on the nature of the system and production requirements. The equipment is primarily of two types: (1) equipment such as mixer-settler combination where mechanical agitation is provided for intimate contact and (2) equipment such as towers where mixing is effected by the flow of the fluids themselves.

Types of equipment employed are mixer-settler combination, spray towers, packed towers, sieve tray columns, pulsed packed and sieve tray towers, and mechanically agitated towers. We will review the design of packed towers here because it is specifically singled out in NCEE's recommended requirements for the PE Exam.

A more effective extraction tower is one packed with random packing such as Raschig rings, pall rings, Berl saddles, Pall rings, etc. The height equivalent to a theoretical stage (HETS) is generally greater for a packed tower compared to mechanically agitated or pulsed towers. In order to arrive at the tower size for a particular requirement, one must know the number of stages or the number of transfer units. The determination of equilibrium stages has been reviewed earlier in this chapter. Design of packed towers is described next.

Design of Packed Towers for Extraction Using Mass Transfer Coefficients

Mass transfer coefficients are used in the design of packed towers for liquid-liquid extraction in the same manner as in stripping in absorption. For extraction, the tower packed height is given by

$$z = H_{OL} N_{OL} = H_{OL} \int_{x_1}^{x_2} \frac{dx}{x^* - x} \qquad (12.11)$$

and

$$H_{OL} = \frac{L}{K'_x a S} \qquad (12.12)$$

If both the operating and equilibrium lines are straight and solutions are dilute, the integral of Equation 12.11 is evaluated to give

$$N_{OL} = \frac{x_1 - x_2}{(x^* - x)_M} \quad (12.13)$$

where $(x^* - x)_M$ is logarithmic mean given by

$$(x^* - x)_M = \frac{(x_1^* - x_1) - (x_2^* - x_2)}{\ln[(x_1^* - x_1)/(x_2^* - x_2)]} \quad (12.14)$$

and in terms of the extract phase,

$$N_{OV} = \frac{y_1 - y_2}{(y - y^*)_M} \quad (12.15)$$

The overall transfer units can also be determined by the following equation:

$$N_{Ol} = \frac{1}{(1-A)} \ln\left[(1-A)\left(\frac{x_2 - y_1/m}{x_1 - y_1/m}\right) + A\right] \quad (12.16)$$

Analytical equations for number of trays.

The analytical equation to calculate the number of theoretical trays N for extraction is the same as for stripping in gas-liquid separations and is given as

$$N = \frac{\ln\left[\frac{(x_2 - y_1/m)}{(x_1 - y_1/m)}(1-A) + A\right]}{\ln(1/A)} \quad (12.17)$$

where $A = L/mV$.

The value of A may vary in the tower. In that case an average A may be calculated as follows

$$A = \sqrt{A_1 A_2} \quad (12.18)$$

where A_1 and A_2 are the values at top and bottom.

An equation that allows us to evaluate the performance of an existing tower when the number of theoretical stages N is known is given by

$$\frac{x_2 - x_1}{x_2 - y_1/m} = \frac{(1/A)^{N+1} - (1/A)}{(1/A)^{N+1} - 1} \quad (12.19)$$

Flooding in Packed Towers for Liquid-Liquid Extraction

In packed towers for liquid-liquid extraction random packing is generally used. Flooding occurs in packed extraction towers when increasing the flow rate of the

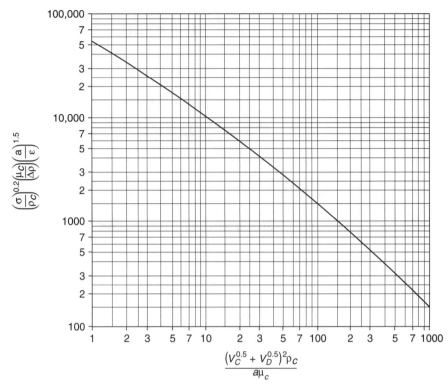

Figure 12.10 Flooding correlation for packed extraction towers.
From J.W. Crawford and C.R. Wilke, *Chem. Eng. Prog.*, 47, 423 (1951).
(Reproduced with permission)

continuous phase or the dispersed phase causes both phases to exit the tower at the outlet of the continuous phase.

A flooding correlation for packed extraction towers developed by Crawford and C. R. Wilke is readily available (Geankoplis or Perry's Handbook) and is given in Figure 12.10.

This correlation requires the use of U.S. customary units. In this correlation chart, the ordinate is

$$\left(\frac{\sigma}{\rho_C}\right)^{0.2} \left(\frac{\mu_C}{\Delta\rho}\right) \left(\frac{a}{g}\right)^{1.5} \text{ and abscissa is } \frac{\left(V_C^{0.5} + V_D^{0.5}\right)^2 \rho_C}{a\mu_C} \quad (12.20)$$

where the variables are defined as follows:

V_C, V_D = superficial velocities of the continuous and dispersed phases, ft/h
ρ_C, ρ_D = densities of continuous and dispersed phases, lb_m/ft^3
$\Delta\rho = |\rho_C - \rho_D|$
μ_C = viscosity of continuous phase, $lb_m/ft \cdot h$
a = specific surface area of packing, ft^2/ft^3
ε = void fraction of packed section
σ = interfacial tension between phases, lb_m/h^2 (*note:* 1.0 dyn/cm = 28572 lb_m/h^2).

Example 12.3

In Example 12.2, calculate

(a) the number of trays by analytical equation.
(b) the number of transfer units by analytical equation.

Solution

Make a sketch first. (Exhibit 3)

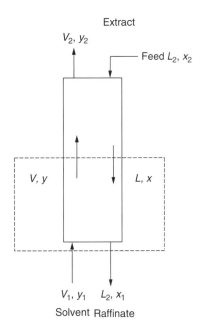

Exhibit 3 Entering and leaving streams from an extraction tower

$$L_2 = 2000 \text{ kg/h} \qquad L_1 = \frac{2000(1-0.3)}{1-0.02} = 1428.6 \text{ kg/h}$$

$$V_1 = 2000 \times 2.624 = 5248 \text{ kg/h}$$

acctic acid transferred to solvent $= 2000(0.3) - \frac{2000(1-0.3)}{} \left| \frac{0.02}{0.98} \right| = 571.4 \text{ kg/h}$

$$V_2 = 5248 + 571.4 = 5819.4 \text{ kg/h}$$

Slope of equilibrium line at lean end (bottom of column) $m_1 = 0.18/0.69 = 0.261$

Slope of equilibrium line at point (y_2, x_2) using Figure 12.9 = $m_2 = 0.448$

$$A_1 = \frac{1429}{0.261 \times 5248} = 1.043, \qquad A_2 = \frac{2000}{0.448(5819)} = 0.7672$$

$$A = \sqrt{A_1 A_2} = \sqrt{0.7672 \times 1.0433} = 0.8947$$

(a) Using Equation 12.17, because $y_1 = 0$ (pure solvent)

$$N = \frac{\ln\left[\frac{0.3-0.0}{0.02-0.0}(1-0.8947)+0.8947\right]}{\ln(1/0.8947)} = 8.1$$

(b) Number of transfer units by Equation 12.16, because $y_1 = 0$,

$$N_{OL} = \frac{1}{(1-0.8947)}\ln\left[(1-0.8947)\left(\frac{0.3}{0.02}\right)+0.8947\right] = 8.6.$$

LEACHING

Leaching involves extraction of a soluble component or components from a solid by contacting the solid with a solvent. It consists of two steps: (1) contacting the solid with a liquid phase and (2) separating the solution from the inert solid.

In actual practice, it is not possible to separate the solution from the solid completely. Therefore two streams, a liquid phase stream, which usually does not contain the insoluble solid, and a stream of inert solids containing the solution retained by them, are obtained on separation of the two phases. The solution phase is called the overflow while the insoluble solids containing the retained solution are called the underflow.

Leaching Calculations

Leaching calculations are based on the concept of an ideal stage as in liquid-liquid extraction, fractionation, or absorption. Ideal stage is defined as a stage in which the solution leaving the stage in overflow is of the same composition as that of the solution retained by the solid in the underflow. An efficiency factor has to be used to convert the ideal stages into actual stages to be used.

For purposes of calculations, most solid-liquid extraction systems are considered to consist of three components: (1) solute, A (2) insoluble inert solid, B, and (3) solvent component S. Solute may be existing on the surface of the inert solid or held in its structure.

Extraction calculations can be done by (1) stage-to-stage algebraic calculations using the material balance and equilibrium relationship and (2) graphical methods such as the use of the right triangular, Ponchon-Savarit, or McCabe-Thiele diagram. The multistage countercurrent leaching is the most important and should be reviewed in detail.

Equilibrium Relationship in Leaching

The assumptions generally made to simplify the leaching calculations are as follows: (1) The system consists of three components, a solute, a solvent, and an inert solid; (2) the inert solid (solute-free) is insoluble in the solvent and the flow rate of the inerts from stage to stage is constant; and (3) in the absence of adsorption of the solute by the inerts, an equilibrium is attained on complete solution of the solute in the solvent. The nonequilibrium condition is taken into consideration by an efficiency factor.

If the conditions of equilibrium are met, the concentration of the solution leaving a stage is the same as the concentration of the solution adhering to the inerts. The equilibrium relationship is, therefore, $x_e = y_e$.

Representation of Equilibrium Data

The published data on a particular system are generally given in terms of the amount of the solution retained by the inerts and the extract composition. These can be used to construct the triangular or Ponchon-Savarit diagram. Example 12.4 illustrates the method of the calculation of the equilibrium data for the triangular diagram.

Example 12.4

The experimental data on the retention of oil by livers are given in the first two columns of Table 12.1. Construct the right triangular diagram for the system.

Table 12.1 Retention of oil by livers

Experimental Data		Calculated Data		
Lb[kg] of Liver Oil in 1 lb[kg] of Solution (Overflow Composition) y_A	Lb[kg] of Solution Retained by 1 lb[kg] of Oil-Free Livers (K)	Underflow Compositions		
		$x_A = y_A K/(K+1)$	$x_S = (1-y_A)K/(K+1)$	$x_I = 1/(K+1) = 1-(x_A+x_S)$
0.00	0.205	0	0.170	0.830
0.10	0.242	0.020	0.175	0.805
0.20	0.286	0.044	0.178	0.778
0.30	0.339	0.076	0.177	0.747
0.40	0.405	0.115	0.173	0.712
0.50	0.489	0.164	0.164	0.672
0.60	0.600	0.225	0.150	0.625
0.65	0.672	0.261	0.141	0.598
0.70	0.765	0.3034	0.130	0.567
0.72	0.810	0.3222	0.125	0.553

The calculated mass fractions are given in the last three columns of Table 12.1. A right-triangular diagram is prepared as in Exhibit 5.

Solution

For smoothing out, the data can be plotted as in Exhibit 4.

Using the overflow and underflow compositions in the table, the mass fractions of the solute, the solvent, and the inerts are calculated from these data by the following relations:

$$x_A = \frac{y_A K}{K+1}, \qquad x_S = \frac{(1-y_A)K}{K+1}, \qquad x_I = \frac{1}{K+1} = 1-(x_A+x_S)$$

where
x_A = mass fraction of the solute
x_S = mass fraction of the solvent
x_I = mass fraction of the inert solid
K = lb[kg] of the solution retained by 1 lb[kg] of oil-free livers
A, S, I = solute, solvent, and inert solid, respectively

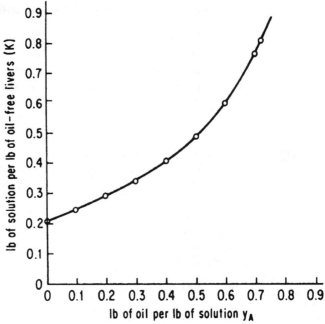

Exhibit 4 Retention of solution by livers as a function of solution concentration

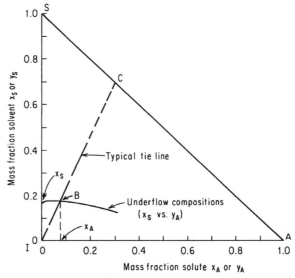

Exhibit 5 Oil extraction data on right-triangular diagram (Example 12.4)

Example 12.5

Prepare a Ponchon-Savarit diagram for the system in Example 12.4.

Solution

To prepare the Ponchon-Savarit diagram, the data given in Table 12.2 are first calculated. The Y coordinate is the ratio of the inerts to the solution retained by the inerts; the X coordinate is the solute fraction in the solution, or $y_A = x_A$.

In the right-triangular diagram, the tie lines pass through $0(x_I = 1.0)$, and in the Ponchon-Savarit diagram they are vertical for the ideal systems where $x_e = y_e$.

Table 12.2 Calculation of coordinates for the Ponchon-Savarit diagram A lb [kg] solute per lb [kg] solution

$x_A = \dfrac{A}{A+S} = y_A$	$Y_I = \dfrac{I}{A+S}$
0.00	4.88
0.10	4.13
0.20	3.50
0.30	2.95
0.40	2.47
0.50	2.05
0.60	1.67
0.65	1.49
0.70	1.31
0.72	1.24

Note that $Y_I = 1/K$, where K is obtained from Table 12.1. The data are plotted in Exhibit 6.

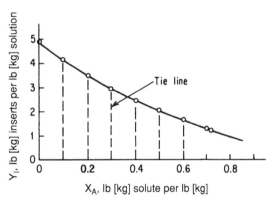

Exhibit 6 Oil extraction data on Ponchon-Savarit diagram (Example 12.5)

The equilibrium data can be represented on an equilateral triangle also. However, because of the inconvenience of the equilateral triangle, the right-triangular or Ponchon-Savarit diagram is preferred.

(*Note:* As an example, to read the composition of the underflow represented by point B, lines are drawn perpendicular to X and Y axes to give $x_S = 0.176$ and $x_A = 0.076$; then $x_I = 1 - 0.176 - 0.076 = 0.748$. To determine the composition of the overflow in equilibrium with the underflow represented by point B, the tie line is drawn through point B to intersect the hypotenuse SA in C. The point C gives the required overflow composition, which is $y_A = 0.3$, $y_S = 0.7$.)

$$X_A = A/(A+S); \quad Y_I = I/(A+S)$$

Calculation of Equilibrium Stages

Consider a countercurrent leaching battery as shown in Figure 12.11. The notation used is as follows:

V = overflow solution (no inerts in overflow)
L = underflow solution exclusive of inerts

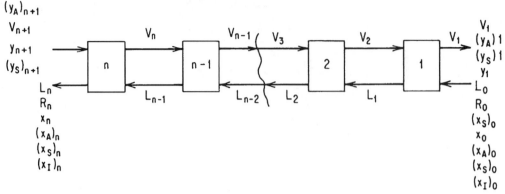

Figure 12.11 Multistage countercurrent extraction system

R = total underflow including the solution and inerts
x = solute concentration in the solution retained by the inerts, mass fraction
y = solute concentration in the extract phase, mass fraction
y_S = mass fraction of the solvent in the extract solution
y_A = mass fraction of the solute in the extract solution.

Because no inerts are present in the extract phase, $y = y_A$. The number of equilibrium stages can be calculated by the algebraic method by making stage-to-stage calculations, using mathematical equations in some cases, or using graphical methods. The inerts are assumed to be constant from stage to stage and insoluble in the solvent. Also, no inerts are present in the extract (usual assumption) or overflow solution.

Material Balance Equations

In terms of the overflow and underflow solutions (exclusive of inerts), the overall material balance can be written as

$$V_{n+1} + L_0 = V_1 + L_n \tag{12.21}$$

and the solute balance as

$$V_{n+1} y_{n+1} + L_0 x_0 = V_1 y_1 + L_n x_n \tag{12.22}$$

In establishing the preceding equations, the inerts are assumed constant from stage to stage. In terms of the total overflow and underflow compositions, the overall material balance is

$$V_{n+1} + R_0 = V_1 + R_n \tag{12.23}$$

$$V_{n+1}(y_A)_{n+1} + R_0(x_A)_0 = V_1(y_A)_1 + R_n(x_A)_n \tag{12.24}$$

and the operating line equation is given by

$$y_{n+1} = \frac{1}{1+(V_1 - L_0)/L_n} x_n + \frac{V_1 y_1 - L_0 x_0}{L_n + L_1 - L_0} \tag{12.25}$$

When the solution retained by the inerts is constant, both the underflow L_n and overflow V_n are constant and the equation of the operating line is a straight line. Because the equilibrium line is also straight, the number of the stages is given by

$$N = \frac{\log[(y_{n+1} - y_n^*)/(y_1 - y_1^*)]}{\log[(y_{n+1} - y_1)/(y_n^* - y_1^*)]} \quad (12.26)$$

In leaching, $y_1^* = x_0$ and $y_n^* = x_n$, so Equation 12.26 becomes

$$N = \frac{\log[(y_{n+1} - x_n)/(y_1 - x_0)]}{\log[(y_{n+1} - y_1)/(x_n - x_0)]} \quad (12.27)$$

where y_1^* and y_n^* are the equilibrium compositions of clear solutions at points indicated by subscripts.

Equation 12.27 cannot be used for the entire extraction battery if L_0 differs from $L_1, L_2, \ldots, L_n$ (underflows within the system). In this case, the compositions of all the streams entering and leaving the first stage are separately calculated by a material balance, and then Equation 12.27 is applied to the remaining cascade.

Example 12.6 illustrates the method of solution of the problems when the solution retention by the inerts is constant.

Example 12.6

A countercurrent extraction system is to treat 100 tons/h of sliced sugar beets with fresh water as solvent. Analysis of the beets is as follows: water 48 percent, sugar 12 percent, with the balance being pulp. If 97 percent sugar is to be recovered and the extract phase leaving the system is to contain 15 percent sugar, determine the number of cells required if each ton of the dry pulp retains 3 tons of solution.

Solution

Basis of calculation: 100 tons of fresh sliced beets

Sugar in beets	12 tons
Water in beets	48 tons
Pulp in beets	40 tons

Because 97 percent of the sugar is to be recovered, the sugar in the final extracted solution is

$$0.97(12) = 11.64 \text{ tons}$$

Water in the final extract solution = $(0.85/0.15)(11.64) = 65.96$ tons, and

$$V_1 = 11.64 + 65.96 = 77.6 \text{ tons}$$

Each ton of pulp retains 3 tons of solution. Hence the solution underflow = $3(40) = 120$ tons.

$$L_1 = L_2 \ldots L_n = 120 \text{ tons} \quad \text{but} \quad L_0 = 60 \text{ tons}$$

(Note that inert dry pulp is excluded.) Therefore, first complete the calculations on the first stage. By overall balance,

$$L_0 + V_{n+1} = V_1 + L_n \quad \text{or} \quad 60 + V_{n+1} = 77.6 + 120$$
$$V_{n+1} = 137.6 \text{ tons}$$

For material balance over first stage

$$L_1 = 120 \text{ tons solution} \quad x_1 = y_1 = 0.15$$
$$L_1 + V_1 = L_0 + V_2$$
or
$$120 + 77.6 = 60 + V_2$$
$$V_2 = 137.6 \text{ tons}$$

By solute balance over first stage

$$137.6(y_2) + 12 = 11.64 + 120(0.15)$$
$$y_2 = 0.1282$$

Now, because both the underflow and overflow solutions are constant for the remaining $(N-1)$ stages, Equation 12.31 can be applied as follows:

$$N - 1 = \frac{\log[(y_{n+1} - x_n)/(y_2 - x_1)]}{\log[(y_{n+1} - y_2)/(x_n - x_1)]}$$

Note the replacement of y_2 and x_1 for y_1 and x_0, respectively, because the first stage is dealt with separately.

The substitution of the values in the preceding equation gives

$$N - 1 = \frac{\log\left[(0 - \frac{0.36}{120})/(0.1282 - 0.15)\right]}{\log\left[(0 - 0.1282)/(\frac{0.36}{120} - 0.15)\right]}$$

$$N - 1 = 14.5 \quad \text{and} \quad N = 15.5 \text{ cells}$$

This is a fractional number of stages and in actual practice 16 cells will be used. This would give a slightly better recovery than is assumed in this example.

When the amount of the solvent retained by the inerts is constant, the number of the stages is given by the equation

$$N = \frac{\log[(y'_{n+1} - x')/(y'_1 - x'_0)]}{\log[(y'_{n+1} - y'_1)/(x'_n - x'_0)]} \tag{12.28}$$

where x' is the mass ratio of the solute to solvent in the underflow and y' is the mass ratio of the solute to solvent in the overflow.

Equation 12.28 also needs the calculation of the first stage separately, because the solvent content of the fresh solids to the first stage is different from that in the other stages.

Example 12.7

In Example 12.6, determine the number of cells if each ton of dry pulp retains 3.5 tons of water. (This is left as an exercise to the reader.)

Solution

$N = 19$ stages

When solution retention data by the inerts are available, the number of stages for a given extraction can be obtained by stage-to-stage calculations. An example is given next.

Example 12.8

Oil is to be extracted from meal with the use of benzene as solvent. The unit is to treat 1000 kg of meal (oil-free solid)/h. The feed meal contains 400 kg oil and 25 kg benzene. The wash solution contains 10 kg oil dissolved in 655 kg benzene. The discharged solids are to contain 60 kg unextracted oil. Solution retention data are given below.

Concentration Kg of oil/kg solution	0	0.1	0.2	0.3	0.4	0.5	0.6	0.7
Solution retained kg/kg of solid	0.5	0.505	0.515	0.53	0.55	0.571	0.595	0.620

Assuming countercurrent extraction, compute by stagewise calculations the following: (a) the composition of the strong solution, (b) the weight of solution leaving with the extracted meal, (c) the amount of strong solution, (d) the number of stages required.

Solution

Step 1: To simplify interpolation, plot the solution retention data vs. the mass fraction of the solute in the solution. This is done in Exhibit 7 (curve A).

Step 2: Establish overall material balance. Assume there is no inert meal in the overflow. V_1 and L_n are not known. However, the composition and amount of L_n can be established from the solution retention data. Because L_n has an equilibrium composition, the solvent in L_n is found from the solution retention data as follows:

y_A in solution	0	0.1	0.2	0.3	0.4	0.5	0.6	0.7
kg oil/kg inerts	0	0.0505	0.103	0.159	0.22	0.2855	0.357	0.434

These data are plotted in Exhibit 7 as the curve B. The ratio of oil to inerts in $L_n = \frac{60}{1000} = 0.06$. From the curve B, for the oil/inerts ratio of 0.06, the solution concentration $(y_A)_n = 0.12$. Therefore, the solvent amount in the underflow in L_n is

$$\frac{(1-0.12)(60)}{0.12} = 440 \text{ kg/h}$$

Therefore $L_n = 440 + 60 = 500$ kg/h (inert solid free).

Step 3: Establish overall material balance.

$$L_0 + V_{n+1} = V_1 + L_n$$
$$V_{n+1} = 655 + 10 = 665 \text{ kg/h}$$
$$L_0 = 400 + 25 = 425 \text{ kg/h}$$
$$L_n = 500 \text{ kg/h}$$

Therefore

$$V_1 = 425 + 665 - 500 = 590 \text{ kg/h}$$

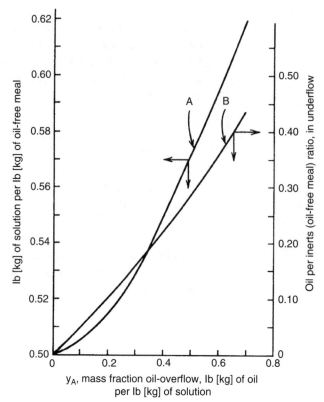

Exhibit 7 Solution retention data and oil/oil-free meal ratios as a function of solution concentration (Example 12.8)

The amount of oil in the strong solution is

$$400 + 10 - 60 = 350 \text{ kg/h}$$

The oil concentration of the strong solution is

$$(y_A)_1 = \frac{350}{590} = 0.5932$$

Step 4: Material balance over first stage

$$L_0 + V_2 = V_1 + L_1$$
$$L_0 = 425 \text{ kg/h}$$
$$V_1 = 590 \text{ kg/h}$$

From curve A of Exhibit 7, corresponding to $y_A = 0.5932$, ordinate = 0.593 and $L_1 = 0.593(1000) = 593$ kg/h. Hence

$$V_2 = 590 + 593 - 425 = 758 \text{ kg/h}$$
$$x_1 = (y_A)_1 = 0.5932$$

Solute balance on first stage

$$L_0 x_0 + V_2(y_A)_2 = L_1 x_1 + V_1(y_A)_1$$

or
$$400 + 758(y_A)_2 = 593(0.5932) + 590(0.5932)$$

Hence

$$(y_A)_2 = 0.398$$
$$\text{Solute in } V_2 = 0.398(758) = 301.7 \text{ kg/h}$$

Step 5: Material balances on stages 2 to n: The calculations are continued for second stage onward in a similar manner to that used in calculating material balance on the first stage. For example, the concentration of V_2 enables us to obtain L_2 from Exhibit 8 and also $(y_A)_2 = x_2$. These values enable us to establish the overall material and solute balances on stage 2 and so on. A summary of the results is given in Table 12.3.

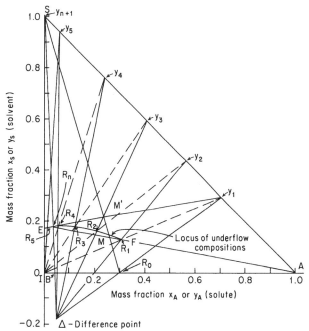

Exhibit 8 Graphical solution for the number of theoretical stages (Example 12.8)

Table 12.3 Summary of stage-to-stage calculations of example 12.8

	Solution Underflow Leaving Stage $n = L_n$					Solution Entering Stage $n = V_{n+1}$			
	Quantities kg/h		Composition L_n			Quantities, kg/h		Composition V_{n+1}	
	Solution L_n	**Solution and Inert** R_n^*	**Oil** $L_n x_n$	**Solvent** $L_n(x_S)_n$	$(x_A)_n^\dagger$	V_{n+1}	**Oil** $V_{n+1}(y_A)_{n+1}$	**Solvent** $V_{n+1}(y_S)_{n+1}$	**Mass Fraction Oil** $(y_A)_{n+1}$
n									
0	425	1425	400	25.0	0.9411	590	350.0	240	0.5932
1	593	1593	351.8	241.2	0.5932	758	301.8	456.2	0.3982
2	550	1550	219.0	331.0	0.3982	715	169.0	546.0	0.2364
3	520	1520	122.9	397.1	0.2364	685	72.9	612.1	0.1064
4	505.5	1505.5	53.8	451.7	0.1064	670.5	3.8	666.7	0.0057
			60	←(Given)→		665	10	655	0.015

*Includes 1000 kg of oil-free meal. $R_n = 1000 + L_n$.
†Meal-free basis, or composition of L_n (not R_n).

Solution

(a) Mass fraction oil in strong solution = 0.5932

(b) Amount of solution leaving with extracted meal ≐ 500 kg/h

(c) Amount of strong solution = 590 kg/h

(d) Number of stages = 4

Note: Calculations as shown in Table 12.3 indicate that fewer than four stages are actually required.

Use of Triangular Diagram

The compositions in terms of the mass fractions are useful in the calculation of the number of the equilibrium stages with the use of a triangular diagram. The material-balance equations are given as

Total material balance:
$$R_0 + V_{n+1} = R_n + V_1 \qquad (12.29)$$

(Note here that R_0 is the total underflow including the inerts, which move from stage to stage unchanged.) The solute balance gives

$$R_0(x_A)_0 + V_{n+1}(y_A)_{n+1} = R_n(x_A)_n + V_1(y_A)_1 \qquad (12.30)$$

By a rearrangement of the material balances on the intermediate stages $m - 1$, m, $m + 1$, etc., to express differences in the flows of the streams between the stages, the net flows toward the nth stage are obtained. The net total flow toward the nth stage is given by

$$R_{m-1} - V_m = R_m - V_{m+1} = R_{m+1} - V_{m+2} \qquad (12.31)$$

and the net flow of the solute towards the nth stage is given by

$$R_{m-1}(x_A)_{m-1} - V_m(y_A)_m = R_m(x_A)_m - V_{m+1}(y_A)_{m+1} \qquad (12.32)$$

Example 12.9

Oil is to be extracted from halibut livers by continuous countercurrent multistage extraction with ether. The solution retention data are given in Table 12.1. Halibut livers contain 0.3 mass fraction oil. The fresh solids charge is 100 kg/h, and 95 percent of the oil is to be recovered.

Assuming the fresh solvent to be pure, calculate (a) the composition of the solids discharged from the last stage; (b) minimum solvent-to-livers ratio; (c) the composition of the solution leaving the first stage using 1.6 times the minimum ratio; (d) the number of equilibrium stages; and (e) the number of actual stages required if stage efficiency is 73.3 percent.

Solution

Step 1: Data (Table 12.4) of Example 12.4 are plotted on the triangular diagram in Figure 12.13.

Table 12.4 Composition of feed solids

Mass fraction oil	$(x_A)_0 = 0.3$
Oil in fresh solids	$0.3(100) = 30$ kg
Inerts in fresh solids	$0.7(100) = 70$ kg
Oil in discharged solids	3 kg
Mass fraction of oil in discharged solids (excluding solvent)	$3/73 = 0.0411$

Locate this point on $\overline{IA}$ and call it B. Join $\overline{By_{n+1}}$ to intersect EF, the locus of the underflow in R_n. The composition of stream R_n is

$$\text{Oil} = 0.033 \quad \text{ether} = 0.179 \quad \text{inerts} = 0.788$$

These are read directly from the graph.

Step 2: Calculation of minimum solvent/solids ratio
Join R_n and A and $\overline{y_{n+1}R_0}$.

$$\text{Minimum } \frac{\text{solvent}}{\text{solids}} = \frac{V_{n+1}}{R_0} = \frac{(x_A)_0 M}{M y_{n+1}} = \frac{1.9}{11.4} = 0.17$$

Step 3: Actual solvent/solids ratio = $0.170(1.6) = 0.272$.

Step 4: Location of addition point
Total solvent required = $0.272(100) = 27.2$ lb/100 kg solids

$$\text{Solvent + solids} = 127.2 \text{ kg}$$

$$(x_A)_{M'} = \frac{30}{127.2} = 0.2358$$

$$(x_S)_{M'} = \frac{27.2}{127.2} = 0.2138$$

$$(x_I)_{M'} = \frac{70}{127.2} = 0.5504$$

Locate this point on $\overline{y_{n+1}R_0}$ as M'.

Step 5: Join R_n and M' and extend the line to meet the hypotenuse in y_1 [that is $(y_A)_1$ also]. The composition of the solution leaving stage 1 (from graph) is

$$(y_A)_1 = 0.705 \quad (y_S)_1 = 0.295 \quad (y_I)_1 = 0$$

Step 6: Join $\overline{y_1(x_A)_0}$ and extend to intersect $\overline{y_{n+1}R_n}$. The intersection point is the difference point. Coordinates of difference point are

$$(x_A)_\Delta = +0.05, \ (x_S)_\Delta = -0.18, \quad (x_I)_\Delta = 1 - (x_A)_\Delta - (x_S)_\Delta = +1.13$$

Join the tie line $\overline{y_1 I}$ to intersect the underflow locus EF in R_1. Join $\overline{\Delta R_1}$ meeting the hypotenuse in y_2. Join $y_2 I$ to intersect the locus of the underflow compositions, EF in R_2. Join $\overline{\Delta R_2}$ meeting the hypotenuse in y_3. Proceed in the above manner to obtain other stages till R_n is passed. From the graph, about 4.4 equilibrium stages are required.

Step 7:
$$\text{Actual stages} = \frac{\text{theoretical stages}}{\text{efficiency}} = \frac{4.4}{0.733} \doteq 6$$

Step 8: The compositions of the streams can be read directly from the graph for any stage.

The use of the Ponchon-Savarit method to solve multistage extraction problems is illustrated by Example 12.10.

Example 12.10

Oil is to be extracted from halibut livers by means of ether. The solution retention data are given in Table 12.2. Halibut livers contain 0.257 mass fraction oil. If 95 percent of the oil is to be extracted and the strong solution from the system is to contain 0.7 mass fraction oil, determine

(a) quantity and composition of the discharged solids

(b) kilograms of ether (oil free) required to treat 1000 kg charge

(c) number of ideal stages required

(d) number of actual stages if the stage efficiency is 70 percent.

Solution

Step 1: Overall Material Balance A, the oil in feed solids, is 257 kg; I, the inerts in feed solids, is 743 kg. The coordinates of points V_{n+1} and R_0 are

Point R_0: $\quad X = \dfrac{257}{257+0} = 1.0 \quad Y = \dfrac{743}{257} = 2.891$

Point V_{n+1}: $\quad\quad X = 0 \quad Y = 0$

Plot these points on the Ponchon-Savarit diagram (Exhibit 9).

Step 2: The amount of oil in strong solution is

$$0.95(257) = 244.15 \text{ kg}$$

The amount of solvent in strong solution is

$$\frac{0.3}{0.7}(244.15) = 104.64 \text{ kg}$$

The Ponchon-Savarit coordinates of this point are

$$X = \frac{244.15}{104.64 + 244.15} = 0.7 \quad Y = 0$$

Plot this point V_1 on the diagram of Exhibit 9.

Step 3: Location of point R_n: The ratio of Y/X for point R_n is

$$\frac{\text{Inerts}}{\text{Solute in } R_n} = \frac{743}{12.85} = 57.82$$

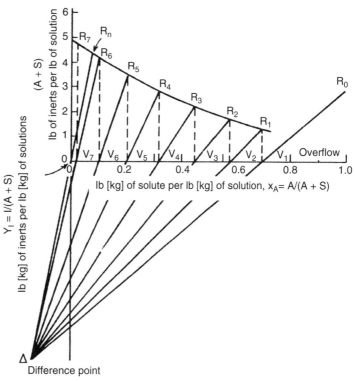

Exhibit 9 Theoretical stages by Ponchon-Savarit diagram (Example 12.14)

Draw a line through the origin with a slope Y/X to cut the underflow line in R_n (because R_n lies on the underflow curve). From the graph, $X = 0.075$, $Y = 4.34$, and

$$\frac{A}{S+A} = 0.075 \quad \text{and} \quad \frac{I}{S+A} = 4.34$$

or

$$S + A = \frac{743}{4.34} = 171.2 \text{ kg}$$

but $\quad A$ in $R_n = 12.85$ kg
$\quad\quad S$ in $R_n = 171.2 - 12.85 = 158.35$ kg

Therefore $\quad R_n = 12.85 + 158.35 + 743 = 914.2$ kg

Step 4: Overall solvent balance gives

$$\frac{\text{Solvent}}{1000 \text{ kg solids}} = \frac{158.35 + 104.65}{1000} = 263 \text{ kg}/1000 \text{ kg}$$

or solvent = 0.263 kg per kg of fresh solids. The composition of discharged solids is

Mass fraction solute: $\quad \dfrac{12.85}{914.2} = 0.014$

Mass fraction solvent: $\quad \dfrac{158.35}{914.2} = 0.173$

Mass fraction inerts: $\quad \dfrac{743}{914.2} = 0.813$

Step 5: Step the stages as follows. Extend $\overline{R_0V_1}$ and $\overline{R_nV_{n+1}}$ to find the difference point Δ. From V_1 draw a perpendicular (tie line) to intersect the underflow curve in R_1. Join ΔR_1 to intersect the overflow line in V_2. Draw a perpendicular from V_2 to cut the underflow curve in R_2. Proceed in this manner to obtain the remaining stages till a tie line falls on the other side of R_n. In Exhibit 9, the desired separation requires between 6 and 7 stages, or roughly 6.3 stages.

Step 6: Actual number of stages equals

$$\frac{\text{Number of equilibrium stages}}{\text{Efficiency}} = \frac{6.3}{0.7} = 9$$

Caution: In using the Ponchon-Savarit diagram, note the difference in Y and X scales when plotting the line of slope Y/X as in Step 3 in Example 12.14.

Mass Transfer in Leaching

When soluble components are extracted from inside of porous particles by means of a solvent, various mass transfer steps occur in the extraction process. The first essential step is the transfer of the solvent to the surface of the particle, followed by its diffusion into the interior of the particle to dissolve the solute. After dissolution the solute diffuses to the surface of the particle through the solute-solvent mixture existing in the void space of the porous particle. Finally the solute is transferred to the bulk of the solution. Usually the solvent transfer to solid is not the limiting step compared to diffusion into the interior of the particle and diffusion of solute to the surface after dissolution.

The process of dissolution may be either physical or by chemical reaction that converts an insoluble solid into a soluble component. In general, most resistance to mass transfer lies within the solid. Solids therefore need to be prepared to facilitate the penetration of solvent into the solid and diffusion of the dissolved solute to the surface.

Rate of Leaching When Dissolving a Solid   When a solid is dissolved and transferred to solvent solution, the rate of mass transfer from solid surface to solvent solution is the controlling step. The rate of mass transfer of solute A into the solution of volume V m³ is given by

$$\frac{\overline{N}_A}{A} = k_L(c_{AS} - c_A) \tag{12.33}$$

where
  $\overline{N}_A$ = kg of solute A dissolving in solvent /s
  A = surface area of particle, m²
  k_L = mass transfer coefficient, m/s
  c_{AS} = saturation solubility of solute A in the solution, kg mol/m³
  c_A = concentration of solute A in the solution at time t seconds, kg mol/m³.

By material balance, the rate of accumulation in the solution is given by

$$\frac{V dc_A}{dt} = \overline{N}_A = k_L(c_{AS} - c_A) \tag{12.34}$$

Integration of Equation 12.34 with boundary conditions at $t = 0$, $c_A = c_{AO}$ and at $t = t$, $c_A = c_A$ yields the solution

$$\frac{c_{As} - c_A}{c_{As} - c_{AO}} = e^{-(k_L A/V)t} \tag{12.35}$$

As the solid dissolution progresses, the interfacial area also increases. Mass transfer coefficient may also change. If the particles are small (<0.6 mm), the mass transfer coefficient k'_L can be calculated by the equation

$$k'_L = \frac{2D_{AB}}{D_P} + 0.31 N_{Sc}^{-2/3} \left(\frac{\Delta\rho \mu_c g}{\rho_c^2} \right)^{1/3} \tag{12.36}$$

Where
D_{AB} = diffusivity of solute A in solution, m²/s
D_P = diameter of solid particle, m
μ_c = viscosity of solution, kg/m·s
g = 9.80665 m/s²
$\Delta\rho = |\rho_P - \rho_c|$ difference in densities, always positive
ρ_C = density of the continuous phase, kg/m³
ρ_P = density of particle, kg/m³.

Mass transfer equations for larger diameter particles or particles in highly turbulent mixers are also available (Geankoplis).

RECOMMENDED REFERENCES

1. Geankoplis, *Transport Processes and Separation Process Principles,* 4th ed., Prentice-hall, NJ, 2004.
2. McCabe and Smith, *Unit Operations of Chemical Engineering,* 6th ed., McGraw-Hill, NY, 2000.
3. Badger and Banchero, *Introduction to Chemical Engineering,* McGraw-Hill, NY, 2004.

CHAPTER 13

Adsorption

OUTLINE

ADSORBENTS AND THEIR PHYSICAL PROPERTIES 493

TYPES OF ADSORPTION 494

EQUILIBRIUM RELATIONSHIPS IN ADSORPTION 495

ADSORPTION FROM LIQUIDS 497

STAGEWISE OPERATIONS 497
Single Stage Adsorption ■ Design of Fixed Bed Adsorption Column

RECOMMENDED REFERENCES 510

In adsorption processes, one or more components of a liquid or gas stream are adsorbed on the surface of a solid adsorbent and thus the separation of adsorbed species from the liquid or gas stream is effectively accomplished. Adsorption has been used to remove trace quantities of water from gas or liquid streams and to remove undesirable impurities such as hydrogen sulfide and mercaptans from natural gas. It has also been used to remove organic pollutants from water. These processes are classified as purification processes because the components to be removed are present in very low concentrations. Application of adsorption as a separation process has gained importance recently. Separation of paraffins from aromatics and fructose from glucose are two good examples of adsorption as a separation process.

ADSORBENTS AND THEIR PHYSICAL PROPERTIES

The earliest commonly used adsorbents were activated carbon and silica gel. Development of molecular sieve adsorbents, especially of synthetic zeolites, has greatly enhanced the potential of adsorption as a separation process. Commercially available adsorbents are materials that are characterized by their highly porous structure and relatively large pore surface. The pore volume may be up to 50 percent of total particle volume and the pore surface area may be as large as 2000 m^3/g. They are available in sizes from 0.1 mm to 12 mm.

Activated Carbon
Activated carbon is made by thermal decomposition of carbonaceous material followed by activation with steam or carbon dioxide at temperatures ranging from 700°C to1100°C.

Silica Gel
Silica is a partially dehydrated form of polymeric colloidal silicic acid. It is prepared by acid treatment of sodium silicate and then drying. Its chemical composition may be expressed as $SiO_2 \cdot nH_2O$.

Activated Alumina
Activated alumina is prepared either from Bauxite ($Al_2O_3 \cdot 3H_2O$) or from monohydrate by dehydration and recrystallization at elevated temperatures.

Molecular Sieve Zeolites
Zeolites are porous crystalline aluminosilicates. Micropore structure of zeolites is determined by the crystal lattice, which is precisely uniform. Si/Al ratio in a zeolite is never less than one but there is no upper limit, which makes it possible to produce a wide variety of zeolites having different properties. Adsorptive properties vary from Al-rich sieves to microporous silicas. The former have affinity for water and other polar molecules, the latter are hydrophobic (no affinity for water) and adsorb n-paraffins in preference to water. The transition from hydrophilic to hydrophobic occurs at a Si/Al ratio of 8 to 10. A, X and Y and ZSM-5 type zeolites are used for drying, CO_2 removal from natural gas, paraffin separation, and air separation (type A), H_2 purification, removal of mercaptans from natural gas, and Xylene separation (type X). Type Y and ZSM-5 are mainly used for xylene separation.

Synthetic Polymers or Resins
These are porous spherical beads. They are made from polymerizable monomers. They are two types. Those made from unsaturated aromatics such as styrene and divinylbenzene are used for adsorption of nonpolar organics from aqueous solutions. The other type is made from acrylic esters and is used to remove polar solutes.

Other adsorbents in use are Fuller's earths, which are natural clays, activated clays, decolorizing carbons, etc. Physical properties of some adsorbents are listed in Table 13.1.

Table 13.1 Physical properties of some adsorbents

Type	Densities g/cm³		Pore Volume (cm³/g)	Diameter Particle (mm)	Pore (Å)	Surface Area (m²/g)
	Bulk	Particle				
Activated carbon	0.44–0.48	0.75–0.85	0.15–0.5	1–5	15–20	300–1250
Activated alumina	0.60–0.85	1.20–1.40		2–12	25–30	200–500
Silica gel	0.40–0.75	1.22	0.43–1.15	1–7	20–140	340–800
Zeolites	0.60–0.70	1.00–1.70		1–5	3–10	
Synthetic polymers	0.42–0.7			0.5	0.08–0.7	

TYPES OF ADSORPTION

There are two types of adsorption:

1. Physical
2. Chemisorption.

Physical or van der Waal's Adsorption

Physical adsorption is the result of intermolecular forces of attraction between molecules of the solid and the substance adsorbed. When the force of attraction between the molecules of adsorbent and the molecules of adsorbate is greater than the attractive forces between the adsorbate molecules themselves, the adsorbate gas will condense on the surface of the solid even if its pressure is lower than the vapor pressure at the prevailing temperature. This is a reversible process. The adsorbed component can be desorbed, that is, removed from the adsorbent, either by lowering the pressure of the gas phase or by increasing the temperature. Heat of adsorption released in physical adsorption is of the order of heat of sublimation of the gas.

Chemisorption

Chemisorption is the result of chemical interaction between the solid and the adsorbed substance. The adhesive force is much greater than that observed in physical adsorption. The heat liberated in chemisorption is large and nearly equal to the heat of reaction. The process in most of the cases is irreversible and on desorption will be found to have undergone a chemical change. Chemisorption is of importance in catalytic reactions.

EQUILIBRIUM RELATIONSHIPS IN ADSORPTION

Figure 13.1 shows the various types of adsorption isotherms that are encountered in sorption operations. Some exhibit convex curves and some exhibit concave curves. There are also isotherms with inflections. The concentration in the solid phase is usually expressed as q, kg adsorbate per kg adsorbent, and in the fluid phase (gas or liquid) as c, kg adsorbate/m^3 fluid.

Data that follow a linear relationship can be expressed in terms of Henry's law relationship as follows:

$$q = Kc \quad \text{or} \quad q = K'p \tag{13.1}$$

where q and c are in moles/unit volume in both adsorbed and fluid phases. By application of ideal gas law, $K = K'RT$. In terms of surface concentration,

$$n_s = \frac{K}{A_s} = \frac{K'}{A_s}P \tag{13.2}$$

where
 A_s = specific surface area per unit volume of the adsorbent
 n_s = surface concentration.

In adsorption where the concentrations are very low and Henry's law is followed, the molecules adsorbed on the surface are isolated from their neighbors.

Langmuir developed the theoretical model for monolayer adsorption. This model represents the type I equilibrium curve. He made the following assumptions:

1. Molecules are adsorbed at a fixed number of well-defined localized sites.

2. Each site holds one adsorbate molecule.

3. All sites are equivalent in regard to energy.

4. There is no interaction between molecules adsorbed on the neighboring sites.

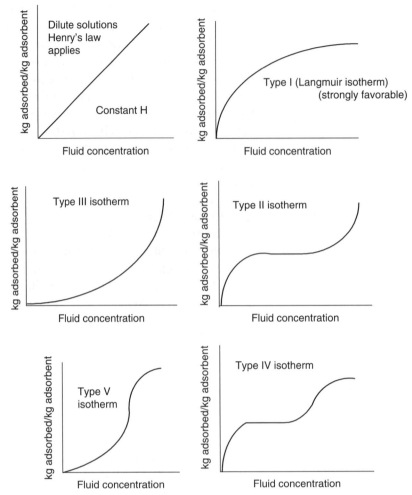

Figure 13.1 Types of adsorption isotherms (gas phase operations)

With these assumptions and considering the molecule exchange between adsorbed and gaseous phases, and because at equilibrium the rates of adsorption and desorption are equal, the following relation was derived:

$$\frac{q}{q_s} = \frac{bp}{1+bp} \qquad (13.3)$$

Or in terms of concentration c, a more useful relation is obtained as

$$q = \frac{q_0 c}{K+c} \qquad (13.4)$$

where q, q_0 are in kg adsorbate/kg solid adsorbent and K is in kg/m³.

Most of the adsorption systems show that the degree of adsorption decreases as the temperature is raised. This is very useful from a practical point of view because adsorption can be carried out at low temperature and desorption at higher temperature.

BET model for multilayer adsorption was developed by Brunauer, Emmett, and Teller and is as follows:

$$\frac{q}{q_s} = \frac{b(p/p_s)}{(1-p/p_s)(1-p/p_s+bp/p_s)} \qquad (13.5)$$

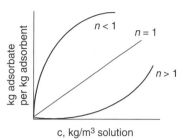

Figure 13.2 Types of Freundlich isotherms

ADSORPTION FROM LIQUIDS

The adsorption in liquid phase is represented by Freundlich isotherm equation. This relation is given by

$$q = Kc^n \qquad (13.6)$$

where K and n are constants. A log-log plot of q versus c is used to determine the constants n and K. Depending on the value of n, the shape of Freundlich isotherm changes. These are shown qualitatively in Figure 13.2.

STAGEWISE OPERATIONS

Adsorption, like extraction and leaching, can be carried out in a single stage or multiple stages. Multiple stage operations can be crosscurrent or countercurrent.

Single Stage Adsorption

Single-stage batch operation (Figure 13.3) is used to adsorb solutes from liquids when the quantities to be treated are in small amounts. As in liquid-liquid extraction, liquid phase adsorption can be carried out in a single stage batchwise or in a continuous manner in multiple stages using mixer-settler type equipment. Crosscurrent stage contact can also be used. To make design calculations, a material balance and an equilibrium relationship such as Freundlich or Langmuir isotherm are required.

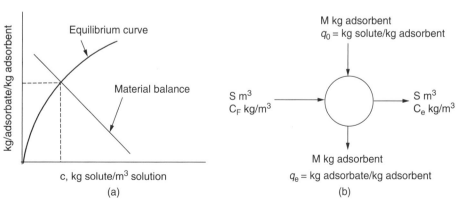

Figure 13.3 Single-stage adsorption

A material balance on the adsorbate gives the following equation:

$$q_0 M + c_F S = q_e M + c_e S \qquad (13.7)$$

where
- q_0 = initial amount of solute adsorbed by the adsorbent
- q_e = equilibrium concentration of solute adsorbed, kg/kg adsorbent
- M = mass of adsorbent charged initially
- S = volume of feed solution used, m³
- c_e = concentration of solute in solution at equilibrium

To solve the problem, the equilibrium curve and the material balance equation (q versus c) are plotted on the same diagram. The intersection point of the two gives the solution.

The derivation of Equation 13.7 assumes that the amount of solution retained (not adsorbed) by the solid after filtering or settling is very small and can be neglected. The quantity of adsorbent usually used is very small in comparison to the amount of liquid treated, therefore this assumption does not introduce serious error in the material balance.

The Freundlich equation can be used to describe adsorption of this type. For the concentration units used in Equation 13.6, the equation for the final equilibrium condition can be written as

$$q_e = K c_e^n \qquad (13.8)$$

Usually, in batch operation, fresh adsorbent is used and $q_0 = 0$. Therefore, the material balance Equation 13.7 reduces to

$$c_F S = q_e M + c_e S \qquad (13.9)$$

By rearranging equation 13.9, we get

$$\frac{M}{S} = \frac{c_F - c_e}{q_e} = \frac{c_F - c_e}{K c_e^n} \qquad (13.9a)$$

Equation 13.9a permits calculation of adsorbent/solution ratio analytically for a given change in solution concentration, c_F to c_e.

Multistage Crosscurrent Operation

In order to obtain economy in the usage of adsorbent for the removal of a given amount of solute, the solution is treated with separate smaller batches of adsorbent rather than a single batch. Thus if M kg of an absorbent are required to effect a certain separation in a single batch and correspondingly M_1 and M_2 kgs are the amounts used to effect the same separation in two-stage contact, then it is found that $(M_1 + M_2) < M$. This split method is mostly applied in batch operations although continuous operation is possible. Adsorbent economy is especially important when it is very expensive. The savings increase with number of batches but the savings are offset because of increased costs of filtration and attendant increased handling costs. Therefore, more than two stages will rarely be used.

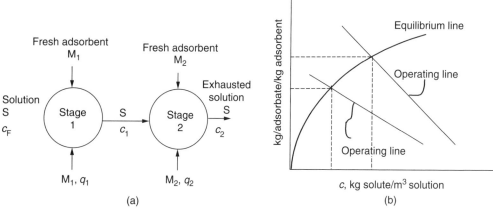

Figure 13.4 Two-stage crosscurrent adsorption

Although the two contacts can be done in the same vessel one after the other, in the interest of greater production rate two-stage crosscurrent systems are usually used. Such a two-stage system is shown in Figure 13.4.

Material balance on the adsorbate can be written for each stage as follows:

$$\left. \begin{array}{ll} \text{Stage 1:} & q_0 M_1 + c_F S = c_1 S + q_1 M_1 \\ \text{Stage 2:} & q_0 M_2 + c_1 S = c_2 S + q_2 M_2 \end{array} \right\} \quad (13.10)$$

Equations 13.10 provide two equations whose slopes are different. If the adsorbent amounts used are equal, the operating lines will be parallel. The least amount of adsorbent that can be used requires usage of unequal amounts in each stage except when the equilibrium isotherm is linear. In the general case, the relative amounts to be used in each stage will have to be established by trial-and-error calculation.

When the Freundlich equation can be applied and the fresh absorbent contains no solute, $q_0 = 0$, then Equations 13.10 can be written after rearrangement as follows:

$$\frac{c_F - c_1}{q_1} = \frac{M_1}{S} \quad \text{and} \quad \frac{c_1 - c_2}{q_2} = \frac{M_2}{S} \quad (13.11)$$

Substituting for q_1 and q_2 using the Freundlich equation, and adding the two equations gives the following:

$$\frac{c_F - c_1}{Kc_1^n} + \frac{c_1 - c_2}{Kc_2^n} = \frac{M_1}{S} + \frac{M_2}{S} \quad (13.12)$$

or

$$\frac{K}{S}(M_1 + M_2) = \frac{c_F - c_1}{Kc_1^n} + \frac{c_1 - c_2}{Kc_2^n} \quad (13.12a)$$

For a minimum of total adsorbent that is to minimize $(M_1 + M_2)$, differentiate righthand side of Equation 13.12a with respect to c_1 and equate to zero. Thus

$$\frac{d}{dc_1}\left[\frac{c_F - c_1}{c_1^n} + \frac{c_1 - c_2}{c_2^n}\right] = 0 \quad (13.13)$$

On evaluating the derivative, one obtains the following relation:

$$\left(\frac{c_1}{c_2}\right)^n - nc_F c_1 + n - 1 = 0 \quad (13.14)$$

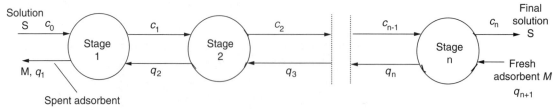

Figure 13.5 Multistage countercurrent system

Equation 13.14 can be solved by trial and error for c_1 and then M_1 and M_2 can be calculated by Equation 13.11.

Multistage Countercurrent Operation

To obtain greater economy in adsorbent usage and larger scale operation, a countercurrent system is more suitable. When stagewise methods are used for treating liquids, the countercurrent operation can only be simulated. There is complete analogy between continuous countercurrent adsorption and other continuous countercurrent operations such as distillation and gas absorption. Figure 13.5 shows a multistage countercurrent system.

A steady-state mass balance gives

$$S(c_0 - c_n) = M(q_1 - q_{n+1}) \quad \text{for adsorption} \tag{13.15}$$

This provides the operating line, which passes through the coordinates of terminal conditions (q_1, c_0) and (c_n, q_{n+1}) and has a slope M/S. The method of constructing the McCabe-Thiele diagram (Figure 13.6) is the same as in distillation or absorption. As the number of stages is increased, the amount of adsorbent at first decreases rapidly but approaches the minimum value very slowly. In practice, when filtration must be used to separate solids from the liquid between the stages, most often it is not economical to use more than two stages in a countercurrent cascade.

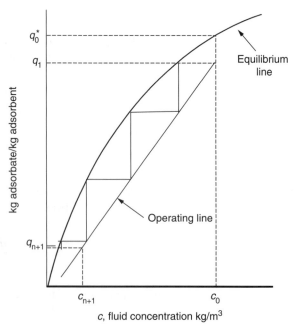

Figure 13.6 McCabe-Thiele construction in adsorption

Design of Fixed Bed Adsorption Column

Fixed bed adsorption is widely used to adsorb solutes from both liquid and gaseous systems. Fluid to be treated is usually passed downward through the bed. Design of the adsorber and time of adsorption and desorption cycles are the two main considerations. Mass transfer resistance is also very important in fixed bed adsorption.

The concentrations of the solute in the fluid phase and solid adsorbent phase change with time and with position in the fixed bed as the adsorption continues. At the start of the adsorption cycle, it is assumed that the adsorbent (fresh or desorbed) contains no solute. At the start of the adsorption cycle, when the fluid comes in contact with the inlet of the bed, most of mass transfer and adsorption takes place in a narrow width of the bed. This narrow zone is called the mass transfer zone. After a short time, the solid at the entrance of the fixed bed gets saturated, and the position of the mass transfer zone moves next to the saturated zone. The mass transfer zone is S-shaped. As the fluid passes through the bed, its solute concentration drops rapidly with distance and reaches zero well before the end of the bed.

The profile of the concentration change is shown in Figure 13.7.

The adsorbent at the entrance would be nearly saturated. This concentration will remain constant up to the mass transfer zone, where it starts dropping off very rapidly to almost zero. The dashed line at t_3 shows the concentration in the fluid phase in equilibrium with the solid. Difference in concentrations is the driving force for mass transfer.

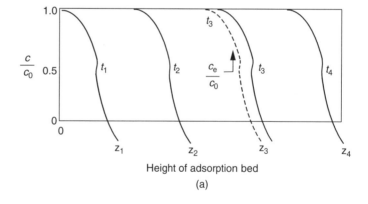

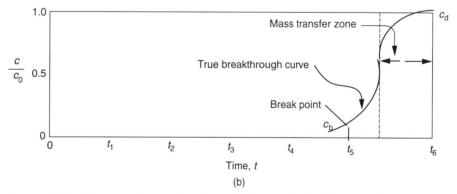

Figure 13.7 Concentration profiles for a fixed bed: (a) Profiles at various positions and times and (b) Breakthrough concentration profile in the fluid at outlet of bed

As described above and seen in Figure 13.7a, most of the adsorption at any time takes place in the prevalent mass transfer zone. As the fluid continues to flow, the position of the mass transfer zone moves down the column. However, the outlet concentration continues to remain zero until the mass transfer zone starts to reach the tower outlet at time t_4. At this point and time, the outlet concentration begins to rise and at t_5, it has risen to c_b. This concentration, which is decided to be the maximum that can be discarded, is called breakpoint concentration. It is frequently set between $c_b/c_0 = 0.01$ and 0.05. After the break point, the concentration rises very rapidly to c_d. The value of c_d/c_0 is taken as the point where $c_d \cong c_0$. For a narrow mass transfer zone, the breakthrough curve is very steep and most of the bed capacity is used at the break point.

There are a number of theoretical methods proposed to predict the mass transfer zone and concentration profiles in the bed. However, because of the uncertainties in the results obtained, experimental breakthrough data are required for a reliable scale up.

The total capacity of a fixed bed if the entire bed comes to equilibrium with the feed is proportional to the area between the curve and a line at $c/c_0 = 1.0$ as is shown in Figure 13.8.

The total shaded area represents the total or stoichiometric capacity of the bed as given by the following equation:

$$t_t = \int_0^\infty \left(1 - \frac{c}{c_0}\right) dt \qquad (13.16)$$

where t_t = time equivalent to the total capacity. Usable capacity of the bed up to the breakpoint time t_b is given by the cross-hatched area and is expressed as follows:

$$t_u = \int_0^{t_b} \left(1 - \frac{c}{c_0}\right) dt \qquad (13.17)$$

In Equation 13.17, t_u is time equivalent to usable capacity or the time at which the exit stream concentration reaches the arbitrarily set maximum allowable value of the breakpoint concentration. Numerical integration of Equations 13.16 and 13.17 can be done by spreadsheet. t_u/t_t is the fraction of the total fixed bed capacity or length that is utilized up to the break point. Hence fraction of total bed height that is used up to break point is

$$Z_B = \frac{t_u}{t_t} Z_T \text{ in m} \qquad (13.18)$$

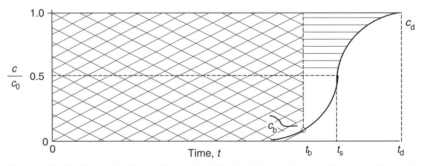

Figure 13.8 Determination of capacity of a fixed bed column using breakthrough curve

The length of the unused bed Z_{UNB} in m is given by

$$Z_{UNB} = \left(1 - \frac{t_u}{t_t}\right) Z_T \qquad (13.19)$$

where Z_T is the total length or height of the fixed bed. Z_{UNB} is the height or length of the mass transfer zone. It depends on the fluid velocity. It may be determined experimentally under conditions that are to be used for large-scale operation. The length required for the usable capacity up to break point can be calculated. Then total length of the fixed bed is given by

$$Z_T = Z_{UNB} + Z_B \qquad (13.20)$$

Instead of following the above procedure, which requires determination of areas by integration, it may be assumed that the breakthrough curve is symmetrical at $c/c_0 = 0.5$ and t_s. Then the value of $t_t = t_s$.

The following guidelines will be useful in scaling up experimental data in the design of a fixed bed adsorber:

1. Typical heights of fixed beds in gas adsorption – 0.3 to 1.5 m (1 to 5 ft) with downflow of gas

2. Superficial gas velocities – 0.15 to 0.5 m/s (0.5 to 1.7 ft/s)

3. Adsorbent particle size – 4 to 50 mesh (0.3 to 5 mm)

4. Adsorption time – 0.5 to up to 8 h

5. For liquids, the superficial velocity – 0.3 to 0.7 cm/s.

Fixed bed operation is carried out in a batch cyclic manner to keep flow continuity. There are basically four methods that are used. Two or three adsorbers are used in parallel. They differ from each other in the manner of regeneration of the bed in the desorption cycle.

After the adsorption cycle is completed, the fluid flow is diverted to one of the regenerated beds and the adsorber, which has completed the adsorption cycle, is placed under regeneration cycle. The regeneration of a bed is carried out by one of the following methods, whichever is suitable for the particular system:

1. temperature swing cycle

2. pressure swing cycle

3. inert purge gas stripping cycle

4. displacement purge cycle.

Steam stripping is frequently used in the regeneration of solvent recovery systems using activated carbon.

Temperature Swing Cycle

This is the most common regeneration method. In this method the bed is regenerated by heating with a stream of hot gas (sometimes hot liquids) to a temperature at which desorption of the adsorbed component takes place. Desorped components are removed from the bed through the hot fluid stream. Thermal swing has the following advantages: It is good for removing strongly adsorbed species. A small change in temperature causes a large change in equilibrium concentration of the adsorbate; desorped species could be recovered at high concentration. Disadvantages include thermal aging of the adsorbent, heat loss and accompanying inefficiency in energy usage, and the impossibility of rapid cycling, resulting in inefficient use of the adsorbent.

Pressure Cycle
In this process, desorption is achieved by reducing the pressure at constant temperature and then purging the bed at low pressure. This mode of operation can be used only in case of gaseous adsorbates. It is advantageous when the adsorbed component is required in high concentration. Rapid cycling allows efficient use of the adsorbent. Very low pressures may be required. Removal of adsorbate from the adsorbent may not be complete.

Purge Gas Stripping
The bed is regenerated by purging with nonadsorbing inert gas. This method is applicable only when the adsorbate species is weakly attached to adsorbent. It cannot normally be used when the adsorbate is to be recovered as the Concentration of the recovered material will be very low.

Displacement Desorption
The adsorbed species are displaced by a stream containing a competitive adsorbed species. It is good for strongly attached species, and avoids thermal deterioration of the adsorbent. However, product separation and recovery are needed.

Example 13.1

Geankoplis et al. give equilibrium isotherm data for adsorption of glucose from an aqueous solution by activated alumina. These data are as follows:

$C(g/cm^3)$	0.0040	0.0087	0.019	0.027	0.094	0.195
q(g solute/g alumina)	0.026	0.053	0.075	0.082	0.123	0.129

Determine the isotherm expression that fits the data and the constants in the equation.

Solution

We can try Freundlich or Langmuir equation and see which one fits the data better.

Freundlich equation is $\quad q = Kc^n$
then $\quad \ln q = \ln K + n \ln c$

and a plot of $\ln q$ versus $\ln c$ should be a straight line.
On the other hand if the data fit the Langmuir equation given by

$$q = \frac{q_0 c}{K + c}$$

taking the inverse, $\quad \dfrac{1}{q} = \dfrac{K}{q_0}\dfrac{1}{c} + \dfrac{1}{q_0}$

and the plot of $1/q$ versus $1/c$ will be a straight line. The data are plotted in Exhibit 1a and b.

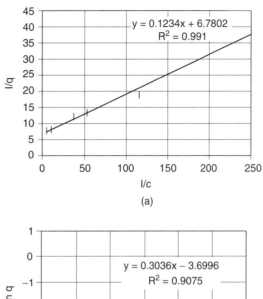

(a)

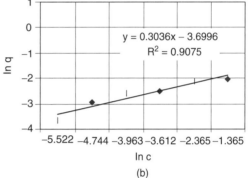

(b)

Exhibit 1 Fitting adsorption data (a) Langmuir isotherm and (b) Freundlich isotherm

The fit to Langmuir equation is clearly better (R^2 for Langmuir model = 0.991 while that for Freundlich model = 0.9075). Therefore we accept Langmuir equation to represent the data and calculate the constants as follows:

Intercept = $6.7802 = 1/q_0$
Therefore $q_0 = 1/6.7802 = 0.1475$ g solute/g alumina
Slope = $0.1234 = K/q_0$.
Therefore $K = 0.1234 \times 0.1475 = 0.0182$ g/cm^3

The Langmuir equation that fits the data is $q = \frac{0.1475\,c}{0.0182+c}$ g solute/g alumina and c is in g/cm^3.

Example 13.2

A sugar solution containing 50 percent sucrose by weight is colored due to the presence of small quantities of impurities. It is to be decolorized at 350 K by treatment with activated carbon. The data for equilibrium adsorption isotherm are as follows:

kg carbon/kg dry sucrose	0	0.005	0.01	0.015	0.02	0.03
Color removed %	0	47	70	83	90	95

The initial color of the sucrose solution is 20 color units/kg dry sucrose.

(a) Convert the equilibrium data to color units/kg sucrose and to color units/kg carbon. Check whether the data follow Freundlich equation. If they do, calculate the constants of the equation.

(b) Calculate the amount of activated carbon that needs to be used to reduce the color concentration to 2.5 percent of the original in a one-stage contact for treating 1000 kg of sucrose solution.

Solution

First convert the data to q = color units/kg carbon and color units/kg sucrose so that a plot can be made to check whether the data obey Freundlich equation.

Taking the second data point we see 47 percent color is removed when carbon added is 0.005 kg per kg dry sucrose.

The equivalent color units removed = $20 \times 0.47 = 9.4$ color units.

Color units that can be removed from the solution = $9.4/0.005 = 1800$ units/kg carbon.

Color units not removed or remaining with sugar = $20 - 9.4 = 10.6$ color units/kg sucrose.

Similar calculations are done for the other data points and the results are given in the following table.

kg carbon/kg sucrose	0	0.005	0.01	0.015	0.02	0.03
Color removed, color units/kg sucrose	0	9.4	14.0	16.6	18.0	19.0
Color units remaining/kg dry sucrose	20	10.6	6.0	3.4	2.0	1.0
Color adsorbed, color units/kg carbon		1800	1400	1106.7	900.0	633.0
ln c		2.361	1.792	1.224	0.693	00.0
ln q		7.5	7.244	7.009	6.8024	6.41

(a) To check whether the data fit Freundlich equation a plot of ln q versus ln c is done and shown in Exhibit 2. Here q = color units adsorbed/kg carbon and c = color units remaining in solution/kg of dry sucrose. Taking the logarithm of two sides of the equation, $q = Kc^n$, we have ln q = ln K + $n$ln c.

The plot of ln q vs ln c is made in Exhibit 2.

The plot gives a straight line and the R^2 value is 0.9945. This shows a good fit to Freundlich equation.

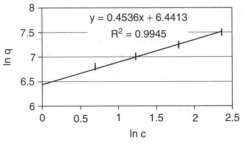

Exhibit 2 Plot of lg q vs. lg c for Example 13.2

The slope of the line = $n = 0.4536$.

The intercept of the line is 6.4413. Therefore $\ln K = 6.4413$ and $K = e^{6.4413} = 627.22$.

The material balance for the solute in terms of color units on single-stage contact is

$$q_F M_C + S c_F = M_C q + S c$$

where

M_c = mass of carbon, kg
S = mass of sucrose, kg
c = final color of sucrose solution, color units/kg sucrose
c_F = initial color of solution, color units/kg sucrose = 20 units
q_F = initial color of carbon, color units/kg carbon

The above is the operating line and is plotted on the equilibrium diagram in Exhibit 3.

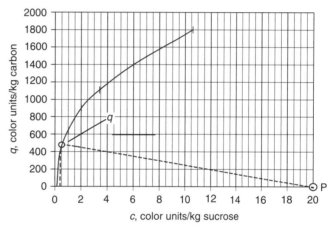

Exhibit 3 Equilibrium diagram and operating line, Example 13.2

Because initial concentration of color of solution is 20 color units and $q_F = 0$, the point P(20, 0) is located at $c = 20$ as shown. The final condition is located on the equilibrium line vertically at the final color concentration of 0.5 color units/kg sucrose. Then

$$\frac{M_c}{S} = \frac{20 - 0.5}{460 - 0} = 0.0424 \text{ kg carbon per kg sucrose}$$

Total carbon to be used = 500(0.0424) = 21.2 kg carbon.

Alternatively, instead of using equilibrium diagram, final equilibrium color concentration can be calculated from the Freundlich equation as follows:

$$q = 627.22 c^{0.4536} = 627.22 c^{0.4536} = 458.0$$

$$\frac{M_c}{S} = \frac{20 - 0.5}{458} = 0.0426 \text{ kg carbon/kg sucrose}$$

Therefore carbon to be used = 500(0.0426) = 21.3 kg carbon.

Example 13.3

J.J. Collins [Chem. Eng. Progress. symp. Ser., 63(74), 31(1967)] obtained data for the removal of water vapor from nitrogen gas in a bed packed with molecular sieves at 28.3°C. The bed height was 0.268 m. The breakthrough data are given below.

Time in hours, t	0	9	9.2	9.6	10	10.4
c (kg H$_2$O/kg N$_2$ ×10^6)	<0.6	0.6	2.6	21	91	235
Time in hours, t	10.8	11.25	11.5	12.0	12.5	12.8
c (kg H$_2$O/kg N$_2$ ×10^6)	418	630	717	855	906	926

The other data are as follows:

Bulk density of adsorbent solids in column = 712.8 kg/m^3

Initial water concentration in N$_2$ = c_0 = 926 × 10^{-6} kg water/kg N$_2$

Initial water concentration in adsorbent solid = 0.01 kg water/kg solid adsorbent

Mass velocity of N$_2$ gas through bed = 4052 kg/m$^2 \cdot$ h

Breakpoint value of c/c_0 = 0.02 (set arbitrarily).

(a) Determine the breakpoint time, the fraction of total capacity used up to break point, the length of the unused bed, and the saturation capacity of the adsorbent.

(b) For a proposed column length, Z_T = 0.5 m, compute the breakpoint time and fraction of total capacity used.

Solution

The given concentration data is first converted into c/c_0 ratios and plotted as a function of time. The values of c/c_0 were obtained by dividing c by c_0 = 926 × 10^{-6} and are given below.

Time in hours, t	0	9	9.2	9.6	10	10.4
c/c_0	<0.00065	0.00065	0.00281	0.0227	0.0983	0.254
Time in hours, t	10.8	11.25	11.5	12	12.5	12.8
c/c_0	0.4514	0.0680	0.7743	0.9233	0.9784	1.00

The values of c/c_0 are plotted versus time in hours in Exhibit 4.
Graphical integration gives the areas A_1 = 9.58 h and A_2 = 1.33 h

Therefore
$$t_t = \int_0^\infty (1 - c/c_0)dt = 9.58 + 1.33 = 10.91 \text{ h}$$

Time equivalent to usable capacity up to break point

$$t_u = \int_0^{t_b=9.58} (1 - c/c_0)dt = A_1 = 9.58 \text{ h}$$

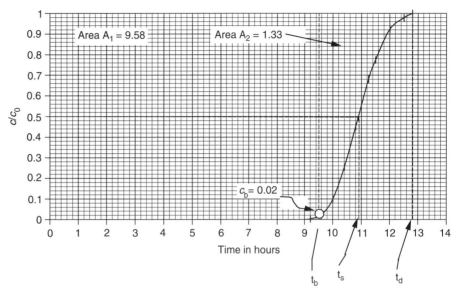

Exhibit 4 Determination of capacity of a fixed bed using breakthrough curve

Hence the fraction of total capacity used up to break point is

$$\frac{t_u}{t_t} = \frac{9.58}{9.58 + 1.33} = 0.878$$

Therefore length of the bed used up to the break point = $0.878 \times 0.268 = 0.2353$ m. Thus $Z_B = 23.53$ cm.

Length of unused bed $Z_{UNB} = (1 - \frac{t_u}{t_t})Z_T = (1 - 0.878)(26.8 \text{ cm}) = 3.27$ cm

Saturation capacity of the adsorbent:

$$\text{Water vapor adsorbed} = 926 \times 10^{-6} \frac{\text{kg water}}{\text{kg N}_2} \left| \frac{4052 \text{ kg}}{\text{m}^2 \cdot \text{h}} \right| \frac{10.91 \text{h}}{} \left| \frac{1 \text{m}^2}{} \right| = 40.94 \text{ kg}$$

The N_2 flowing through a cross section of 1 m² flows through a bed of 0.268 m height.

Therefore it flows through

$$0.268 \text{ m}^3 \times 712.8 \text{ kg/m}^3 = 191.03 \text{ kg of adsorbent}$$

Hence water vapor adsorbed by adsorbent = 40.94/191.3 = 0.214 kg water/kg N_2

Initial concentration of water in adsorbent = 0.01 kg water/kg adsorbent

Total saturation capacity of bed = 0.214 + 0.01 = 0.224 kg water/kg adsorbent.

(b) Here $Z_T = 40$ cm. From part a, unused equivalent height $Z_{UNB} = 3.27$ cm.

Therefore useful height of column = 40 − 3.27 = 36.73 cm.

Total time is obtained by direct comparison with useful length and time in part (a). thus

$$\text{New break time} = \frac{Z_2}{Z_1} \times t_{t1} = \frac{36.73}{23.53} \times 10.91 = 14.95 \text{ h}$$

Useful equivalent time up to break point = 14.95 − 1.33 (from part (a))
= 13.62 h

Therefore, fraction of used bed = 13.62/14.95 = 0.911.

RECOMMENDED REFERENCES

1. Geankoplis, *Transport Processes and Separation Process Principles,* 4th ed., Prentice-Hall, NJ, 2004.
2. Treybal, *Mass Transfer Operations,* 4th ed, McGraw-Hill, NY, 1994.

CHAPTER 14

Psychrometry, Humidification, and Drying

OUTLINE

PSYCHROMETRY AND HUMIDIFICATION 511
Definitions in Psychrometry ■ Wet-Bulb Temperature t_w ■ Adiabatic Saturation Temperature t_s ■ The Psychrometric or Humidity Chart

COOLING TOWERS 520

DRYING 525
Equilibria ■ Rate-of-Drying Curve ■ Rate of Drying

RECOMMENDED REFERENCES 533

Humidification involves transfer of vapor (particularly water vapor) from the liquid phase to a gaseous mixture of air and vapor. Dehumidification is the reverse transfer of vapor from its gaseous mixture to a liquid state. Although humidification and dehumidification apply to other mixtures of vapors such as benzene, most applications are related to air–water vapor mixtures. Drying is an operation where small amounts of water are removed from wet materials, as opposed to large amounts removed in evaporation.

From the aspect of the PE examination, these topics are relatively unimportant as the NCEES has allotted only 1 percent of the total examination to these topics (including adsorption). Nevertheless, some of the basics of these subjects include principles that are useful or required in solving problems in other areas. We will therefore review these subjects very briefly.

PSYCHROMETRY AND HUMIDIFICATION

In dealing with processes involving exchange of mass and energy between a liquid and an essentially insoluble gas, the principles of psychrometry are useful. Important operations to consider are the cooling, humidification, and dehumidification of a gas. A familiarity with the use of the psychrometric chart for the air–water vapor system at atmospheric pressure is also essential.

Definitions in Psychrometry

The following is a review of the important definitions used in psychrometry.

Humidity Y

Sometimes this is also called *specific humidity* or humidity ratio. When P is the total pressure, which is low enough (near atmospheric) for the gas mixture to be considered ideal, the humidity is given by

$$Y = \frac{M_A p_A}{M_B (P - p_A)} \qquad (14.1)$$

where

M_A = molecular weight of vapor
M_B = molecular weight of dry gas
p_A = partial pressure of vapor in gas, atm or [kPa]
P = total pressure, atm [101.3 kPa]
Y = lb [kg] of vapor per lb[kg] of vapor-free (dry) gas

Saturation Humidity Y_s

When $p_A = P_A$, the vapor pressure of the liquid at the prevailing temperature and pressure, the gas is saturated with the vapor and the saturation humidity is given by

$$Y_s = \frac{M_A P_A}{M_B (P - P_A)} \qquad (14.2)$$

Molal Absolute Humidity Y′

This is defined as the number of moles of vapor per mole of the dry gas or

$$Y' = \frac{M_A p_A}{M_B (P - p_A)} \qquad (14.3)$$

Thus from Equations 14.1 and 14.3,

$$Y = Y' \frac{M_A}{M_B} \qquad (14.4)$$

For the case of air-water vapor mixtures, $M_A = 18.02$ for water, and molecular weight of dry air is $M_B = 28.97 \cong 29$.

Percentage Saturation or Percent Absolute Humidity

This is equal to the ratio of the actual humidity to the saturation humidity multiplied by one hundred. Thus

$$\text{Percent absolute humidity} = 100 \frac{Y}{Y_s} \quad \text{mass basis} \qquad (14.5)$$

$$\text{Percent absolute molar humidity} = 100 \frac{Y'}{Y'_s} \quad \text{molar basis} \qquad (14.6)$$

---

* In this chapter, to avoid confusion, the symbol Y, not H, is used for humidity because H is also used for enthalpy.

Percent Relative Humidity or Relative Humidity

This is a term that expresses the actual humidity of a gas at a given temperature as a percentage of its saturation humidity at the same dry-bulb temperature in terms of partial pressures or

$$\text{Percent relative humidity, } Y_R = 100 \frac{p_A}{P_A} \qquad (14.7)$$

Example 14.1

The air at a certain place was at 30°C and 101.33 kPa. The partial pressure of water vapor in air was found to be 2.84 kPa. Using vapor pressure data, calculate the following:

(a) humidity of air Y

(b) saturation humidity, Y_S, and percentage humidity, Y_P

(c) percentage relative humidity, Y_R

Solution

From steam tables the vapor pressure of water at 30°C is $P_A = 4.246$ kPa.

$$p_A = 2.84 \text{ kPa}, \quad P = 101.33 \text{ kPa}$$

(a) Using Equation 14.1,

$$Y = \frac{18.02}{29} \frac{p_A}{P - p_A} = \frac{18.02}{29} \times \frac{2.84}{101.3 - 2.84} = 0.01792 \text{ kg water vapor/kg air}$$

(b) Using Equation 14.2, the saturation humidity is calculated as

$$Y_S = \frac{18.02}{29} \frac{P_A}{P - P_A} = \frac{18.02}{29} \times \frac{4.246}{101.3 - 4.246} = 0.02718 \text{ kg water vapor/kg air}$$

Percentage absolute humidity from Equation 14.5 is

$$Y_P = 100 \frac{Y}{Y_S} = 100 \frac{0.01792}{0.02718} = 36.79 \text{ percent}$$

(c) Percentage relative humidity by equation 14.7 is

$$Y_R = 100 \frac{p_A}{P_A} = 100 \frac{2.84}{4.246} = 66.9 \text{ percent}$$

Dew-Point Temperature

This is the temperature at which the vapor-gas mixture becomes saturated when the gas mixture is cooled at constant total pressure. This means the partial pressure of water vapor equals the vapor pressure of water at the saturation temperature.

Humid Volume v_H

This is the total volume of a unit mass of dry (vapor-free) air (gas) plus the water vapor (vapor) it contains at 101.3 kPa or 1 atm. By the ideal gas law, v_H is related to the humidity and temperature by the equation

$$v_H = 22.414\left(\frac{1}{M_B} + \frac{Y}{M_A}\right)\left(\frac{T(K)}{273}\right) \text{ m}^3/\text{kg dry air (SI)} \tag{14.8}$$

$$= (2.834 \times 10^{-3} + 4.556 \times 10^{-3}Y)T \text{ m}^3/\text{kg dry air} \tag{14.8a}$$

where T is in K

$$v_H = 359\left(\frac{1}{M_B} + \frac{Y}{M_A}\right)\left(\frac{t_G + 460}{492}\right) \text{ ft}^3/\text{lb}_m \text{ dry air (English)} \tag{14.9}$$

$$= (0.02519 + 0.0405Y)T \text{ °R ft}^3/\text{lb}_m \text{ dry air} \tag{14.9a}$$

where T is in degrees Rankine.

For a saturated gas, if $Y = Y_s$, the volume is the saturated volume v_s.

Humid Heat c_s

This is the heat required J (or kJ) to raise the temperature of 1 kg dry gas plus the vapor it contains by 1 K or 1°C at constant pressure. In the English system, c_s is the heat required to raise the temperature of 1 lb dry gas plus the vapor it contains by 1°F at constant pressure. Thus

$$c_S = c_{PB} + c_{PA}Y \tag{14.10}$$

where c_{PB} and c_{PA} are specific heats of the gas and vapor, respectively, at constant pressure, kJ/kg · K dry air (Btu/lb · °F dry air).

The heat capacities of both air and water vapor may be considered constant over the temperature range usually encountered. Given this assumption, c_{PB} and c_{PA} are constant and have the values 1.005 kJ/kg dry air · K and 1.882 kJ/kg water vapor · K respectively and Equation 14.10 can be written as

$$c_s = 1.005 + 1.88Y \text{ kJ/kg dry air} \cdot \text{K (SI)}$$
$$c_s = 0.24 + 0.45Y \text{ Btu/lbm dry air} \cdot \text{°F (English)} \tag{14.11}$$

Relative Enthalpy or Humid Enthalpy

This is the enthalpy of 1 kg (1 lb) of the dry carrier gas together with the amount of vapor it contains. By choosing t_0 as the base or reference temperature for both the components and basing the enthalpy of the component A or the liquid A at t_0, the enthalpy of the gas-vapor mixture at t_G (the temperature of the gas) is given by

$$H = c_{PB}(t_G - t_0) + Y\lambda_0 + c_{PA}Y(t_G - t_0) \text{ J or kJ/kg dry gas}, t \text{ in K or °C}$$
$$H = c_{PB}(t_G - t_0) + Y\lambda_0 + c_{PA}Y(t_G - t_0) \text{ Btu/lb dry gas}, t \text{ in °F} \tag{14.12}$$

where λ_0 is the latent heat of evaporation, kJ/kg (Btu/lb), of the liquid A at t_0. From the definition of humid heat, the preceding expression can be written as

$$H = c_s(t_G - t_0) + Y\lambda_0 \tag{14.13}$$

Note that the enthalpy of air-water mixtures is with reference to liquid water at 0°C (273.15 K) (SI system) or 32°F in the English system. We can then write the following relations for air-water vapor mixtures:

$$H_v = c_s(T - T_0) + Y\lambda_0 = (1.005 + 1.88Y)(T - T_0) + Y\lambda_0 \text{ kJ/kg dry air} \quad (14.14)$$

where T is in °C

$$H_v = c_s(T - T_0) + Y\lambda_0 = (0.24 + 0.45Y)(T - T_0) + Y\lambda_0 \text{ Btu/lb dry air}$$

where T is in °F.

In all the equations, consistent units must be used when calculating values in SI or English units.

The following values for water and air are worth remembering:

λ_0 = 1075.1 Btu/lb latent heat of vaporization of water at 32°F
= 2501.7 kJ/kg at 0°C

C_p of water vapor = 0.45 Btu/lb · °F = 1.882 kJ/kg · °C
C_p of dry air = 0.24 Btu/lb°F = 1.005 kJ/kg dry air

Wet-Bulb Temperature t_w

This is the steady-state dynamic equilibrium temperature reached by a small mass of a liquid evaporating under adiabatic conditions into a continuous large stream of gas. At the wet-bulb temperature, the vapor is transferred to the surrounding gas by a process of diffusion through the gas film surrounding the small mass of the liquid while the sensible heat is transferred from the air to the liquid. This sensible heat is used up in the evaporation of the liquid, which escapes to the gas-vapor mixture through the film. When the rate of heat transfer from the gas to the liquid equals the rate of heat required for the evaporation, the liquid attains the wet-bulb temperature. For this condition, the following equation can be established:

$$t_G - t_w = \frac{\lambda_w k_G M_B P}{h_G}(Y_w - Y) \quad (14.15)$$

But $M_B P k_G = k_Y$, where k_Y is the mass transferred, lb/h · ft² · ΔY, and k_G is the mass transferred, lbmol/h · ft² · atm. Therefore,

$$Y_w - Y = \frac{(h_G/k_Y)(t_G - t_w)}{\lambda_w} \quad (14.16)$$

where
h_G/k_Y = the *psychrometric ratio*
$t_G - t_w$ = the wet-bulb depression, °C or K (°F)
λ_w = the latent heat of vaporization at t_w, kJ/kg H₂O (Btu/lb$_m$)
h_G = the gas film coefficient of the heat transfer, kW/m² · K (Btu/h · ft² · °F)

The ratio h_G/k_Y is given by

$$\frac{h_G}{k_Y} = \begin{cases} 0.294 \, Sc^{0.56} & \text{for air–water–vapor mixtures} \\ c_s \left(\dfrac{Sc}{Pr}\right)^{0.56} & \text{for other gases and liquids} \end{cases} \quad (14.17)$$

where

$$Sc = \text{Schmidt number} = \frac{\mu}{\rho D} \quad (14.18)$$

and

$$Pr = \text{Prandtl number} = \frac{c_p \mu}{k} \quad (14.19)$$

The ratio Sc/Pr is called the *Lewis number*. For the air–water-vapor mixtures

$$Sc \doteq Pr \quad \text{and} \quad \frac{h_G}{k_Y} = c_S = 1.005 \text{ (SI units) or } 0.24 \text{ (English units)} \quad (14.20)$$

Adiabatic Saturation Temperature t_s

When a limited amount of a gas is in contact with a large amount of a liquid adiabatically, the temperature of the liquid will remain constant at t_s, the adiabatic saturation temperature, and the gas will be humidified and cooled. With sufficient intimate contact, the gas will leave at t_s in a condition of saturation. Because t_G and p_A are not constant, the steady-state equations cannot be written. However, an enthalpy balance may be written with t_s as datum and using the inlet values t_G and Y as follows:

$$(c_{PB} + Yc_{PA})(t_G - t_s) + Y\lambda_{as} = Y_{t_s} \lambda_{as} \quad (14.21)$$

which can be rewritten as

$$\frac{Y_{t_s} - Y}{t_G - t_s} = \left(\frac{c_S}{\lambda_{as}}\right) = \frac{1.005 + 1.882Y}{\lambda_{as}} \quad \text{(SI)} \quad (14.22)$$

$$\frac{Y_{t_s} - Y}{t_G - t_s} = \left(\frac{c_S}{\lambda_{as}}\right) = \frac{0.24 + 0.45Y}{\lambda_{as}} \quad \text{(English)} \quad (14.23)$$

Equations 14.16 and 14.22 are identical if $h_G/k_Y = c_S$. This relation is approximately true in the case of a water vapor–air system, and hence the wet-bulb and adiabatic saturation temperatures are effectively equal for the air–water vapor mixtures below a temperature of 65.6°C (150°F).

The Psychrometric or Humidity Chart

The properties for any given gas-vapor mixture as defined before, can be conveniently represented on a chart called the *psychrometric chart*. One has to deal with the air-water system more frequently. Because of the importance of air–water vapor systems, complete charts are available in both the British engineering and SI units. A version of the chart in SI units is given in Figure 14.1 and a version in fps units is shown in in Figure 14.2 for ready reference.

Psychrometry and Humidification 517

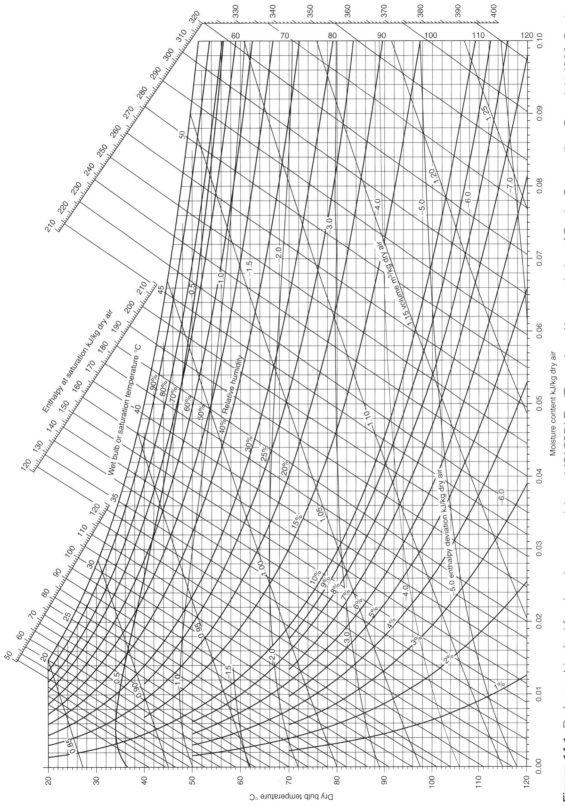

Figure 14.1 Pychrometric chart for air–water vapor mixture at 101.325 kPa. (Reproduced by permission of Carrier Corporation, Copyright 1980, Carrier Corporation)

518 Chapter 14 Psychrometry, Humidification, and Drying

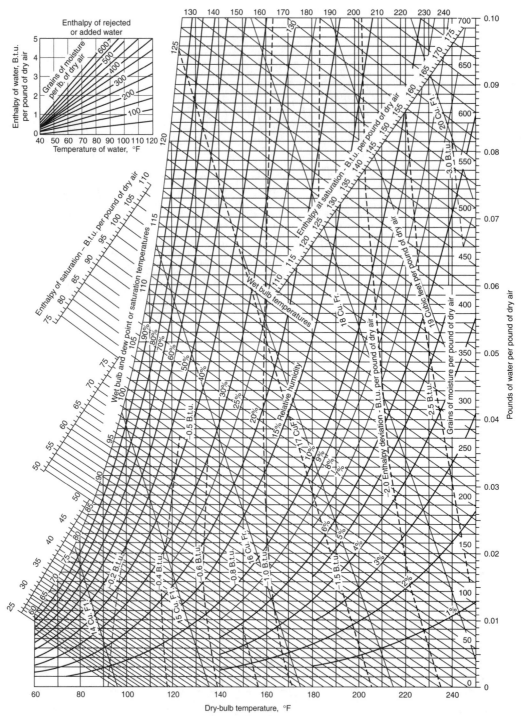

Figure 14.2 Psychrometric chart for air-water-vapor system at 1 atm barometric pressure: 29.92 in Hg. (Reproduced by permission of Carrier Corporation, Copyright 1980, Carrier Corporation)

Example 14.2

(Use of Humidity Chart)

Moist air at 50°C has a wet-bulb temperature of 40°C. Get various properties of the wet air from the humidity chart.

Solution

First locate the intersection point of a vertical line from 50°C dry bulb and the wet-bulb temperature 40°C. To find the wet-bulb line, move along the saturation curve. Alternatively, follow the vertical line at 40°C until it intersects the 100 percent humidity line. From the wet-bulb temperature of 40°C, proceed along the wet-bulb line till it intersects the 50°C dry bulb line. Once this intersection point is established it is easy to read various properties of air.

(a) *Dew Point*. From the intersection point, proceed horizontally to connect to the saturation curve. From the saturation curve, proceed down to abscissa and read dew point = 34.2°C.

(b) *Relative Humidity*. By interpolation between 50 and 60 percent relative humidity curves, the relative humidity is 55.5 percent.

(c) From the intersection point, proceed horizontally to read *humidity* = 0.043 kg water vapor per kg of dry air.

(d) *Humid Volume*. The point lies between humid volume curves for 0.95 and 1.

Again by interpolation, the humid volume is 0.978 m³/kg dry air.

(e) Using the saturated enthalpy curve and the enthalpy deviation lines, the enthalpy of saturated air at 50°C dry bulb and 40°C wet bulb is 166.5 kJ/kg dry air.

Interpolation between the enthalpy deviation lines gives a deviation of −0.3 kJ.

Therefore, humid enthalpy of air is 166.5 − 0.3 = 1.662 kJ.

Example 14.3

Air at dry-bulb temperature of 75°F and wet-bulb temperature of 65°F is heated to 250°F and fed to a dryer. If the air leaves the dryer 90 percent saturated and if the dryer operation is adiabatic, use the relative-humidity chart to obtain the following: (a) dew point, humidity, percent relative humidity, and humid volume of fresh air; (b) heat needed to heat 5000 ft³ of fresh air from 75 to 250°F; (c) temperature and humidity at which air leaves the dryer; and (d) water evaporated in the dryer per 100 lb dry air.

Solution

(a) For the air-water mixtures, the adiabatic and wet-bulb temperatures may be considered the same. To find the humidity of the fresh air, start at the intersection point of ordinate at 65°F, move to the right along the adiabatic line, and locate the intersection point of the adiabatic cooling line and the ordinate at 75°F. This point gives the initial condition of the air. At this point, from the humidity chart,

Absolute humidity of air = 0.0109 lb/lb dry air.

From the above point, move leftward along the constant-humidity line and locate the intersection point of the saturation curve and the constant-humidity line. Read the saturation temperature or dew point as 59.6°F. The point for the initial condition lies between the 50 and 60 percent relative humidities. By a rough interpolation, relative humidity is 58 percent. The humid volume of the fresh air lies between 13.5 and 14 ft³/lb. Again by a rough interpolation, humid volume is 13.7 ft³/lb.

(b)
$$\text{Mass of 5000 ft}^3 \text{ of air} = \frac{5000}{13.7} = 365 \text{ lb}$$

From Equation 14.11,

$$\text{humid heat of initial air} = 0.24 + 0.45(0.0109)$$
$$= 0.245 \text{ Btu/lb dry air} \cdot °F$$

The heat needed to raise the temperature of air from 75 to 250°F is 365(0.245)(250 − 75) = 15,649 Btu.

(c) Starting at 250°F and absolute humidity equal to 0.0109 lb/lb dry air, follow an adiabatic line toward the saturation curve. The intersection of the adiabatic line passing through the startup point and the saturation curve gives the following values:

$$\text{Saturation temperature} = 102°F$$
$$\text{Saturation humidity} = 0.046 \text{ lb/lb dry air}$$

Air leaving the dryer is 90 percent saturated. Therefore, the actual humidity of the air exiting from the dryer is

$$Y = 0.9(0.046) = 0.0414 \text{ lb/lb dry air}$$

Again from the humidity chart, the intersection point of constant humidity (0.0414 lb/lb dry air) and the adiabatic cooling line gives

$$\text{Dry bulb temperature of exit air} = 122°F$$

(d) Water evaporated in the dryer/100 lb dry air is

$$100(0.0414 - 0.0109) = 3.05 \text{ lb}$$

COOLING TOWERS

Mass balance over a differential volume $S\, dZ$ of the cooling tower (Figure 14.3a) gives

$$dL' = G'dY \tag{14.24}$$

and enthalpy balance gives

$$d(L'h) = L'C_L dt_L = G'dH \tag{14.25}$$

Integration over the tower from $z = 0$ to $z = z$ gives

$$H_2 - H_1 = \frac{L'C_L}{G'}(t_{L2} - t_{L1}) \tag{14.26}$$

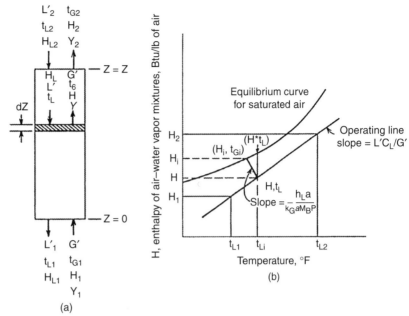

Figure 14.3 Cooling tower: (a) Stream terminology (b) Operating diagram

This equation defines the operating line for the tower in the H-versus-t_L coordinate system. By a consideration of the simultaneous mass and heat transfer from the interface to the bulk of a gas-vapor mixture and assuming the Lewis relationship, it can be shown that the number of transfer units is given by

$$(\text{NTU})_G = \int_{H_1}^{H_2} \frac{dH}{H_i - H} = \frac{k_G a M_B P S z}{G'} \quad (14.27)$$

Then the height of the transfer unit is given by

$$(\text{HTU})_G = \frac{z}{(\text{NTU})_G} = \frac{G'}{k_G a M_B P S} \quad (14.28)$$

where
  L' = superficial mass velocity of liquid, lb/h · ft² [kg/s.m²]
  G' = superficial mass velocity of dry gas, lb/h · ft² [kg/s.m²]
  c_L = specific heat of liquid at constant pressure, Btu/lb · °F [J or kJ/kg · K]
  t_{L1} = temperature of water leaving the tower, °F [K or °C]
  t_{L2} = temperature of water entering the tower, °F [K or °C]
  z = height of packing, ft [m]
  S = cross section of tower, ft² [m²]
  $k_G a$ = mass transfer coefficient, lb · mol/h · ft³ · atm [kg mol/s · m³ · Pa]
  a = interfacial surface per unit volume of packing, ft²/ft³ [m²/m³]

Determination of the $(\text{NTU})_G$ by Equation 14.27 requires the value of the interface gas enthalpy H_i corresponding to the enthalpy H of the bulk of the gas at any given tower cross section. The expression relating these quantities can be derived by a consideration of the rate of heat transfer to the interface from the liquid and is given by

$$\frac{H_i - H}{t_{Gi} - t_L} = -\frac{h_L a}{k_G a M_B P} = -\frac{h_L a}{k_Y a} \quad (14.29)$$

If $h_L a$ and $k_G a$ values are available, it is possible to determine H_i values. For this, the saturation curve and the operating line are plotted first. At a series of points between the top and bottom, lines are drawn with a slope of $-h_L a/k_G a M_B P$ to cut the saturation curve and the operating line (Figure 14.3b). The corresponding values of H_i and H are read and tabulated. The $(NTU)_G$ can then be found by graphical integration from a plot of $1/(H_i - H)$ versus H. Knowing the $(NTU)_G$, z can be calculated.

$$\int_{H_1}^{H_2} \frac{dH}{H_i - H} = \frac{H_2 - H_1}{(H_i - H)_{av}} = N_{tG} \tag{14.30}$$

The term $(H_2 - H_1)/(H_i - H)_{av}$ denotes the number of gas enthalpy transfer units N_{tG}. The height of the column is then given by

$$z = H_{tG} N_{tG} \tag{14.31}$$

where the height of gas enthalpy transfer unit is

$$H_{tG} = \frac{G'}{k_Y a} \tag{14.32}$$

The overall number and overall height of transfer units are given by

$$N_{tOG} = \int_{H_1}^{H_2} \frac{dH}{H^* - H} = \frac{k_G a z}{G'} = \frac{z}{H_{tOG}} \tag{14.33}$$

In this case, the resistance in liquid phase is assumed negligible; therefore, H^* can be used in place of H_i, the enthalpy at the interface.

Equation 14.33 can also be used in the form

$$\frac{K_G a z}{L'} = \int_{t_{L1}}^{t_{L2}} \frac{dt_L}{H^* - H} \tag{14.34}$$

which is the frequently used form in the cooling-tower industry.

Example 14.4

Water is cooled from 110 to 85°F by countercurrent contact with air in an induced-draft cooling tower at 1 atm. The air enters at dry-bulb temperature of 85°F, and the design wet-bulb temperature for the location is 75°F. The water rate is 1250 lb/h · ft² of the tower cross section and the dry air rate is 1.5 times the minimum. Calculate the height of transfer unit $(NTU)_G$ and the packed height Z. The tower is packed with wooden slats.[†]

Solution

Step 1: Plot saturation enthalpy vs. temperature (Exhibit 1).

Step 2: Enthalpy of air at dry-bulb temperature of 85°F, wet-bulb temperature of 75°F = 38.6 Btu/lb (from the humidity chart). This point is located on the graph as point A, whose coordinates are

$$t_{L1} = 85°F \qquad H_1 = 38.6 \text{ Btu/lb}$$

[†] The transfer coefficients are given by $h_L a = 0.03(L')^{0.51}(G')$, $k_G a = 0.04(L')^{0.26}(G')^{0.72}$.

Step 3: From point A, draw a tangent AB' to the enthalpy curve.

$$\text{Slope of tangent} = \frac{89 - 38.6}{110 - 85} = 2.016$$

$$\frac{L'C_L}{G'_{min}} = 2.016 \quad \text{or} \quad G'_{min} = \frac{1250}{2.016} = 620 \text{ lb/h} \cdot \text{ft}^2$$

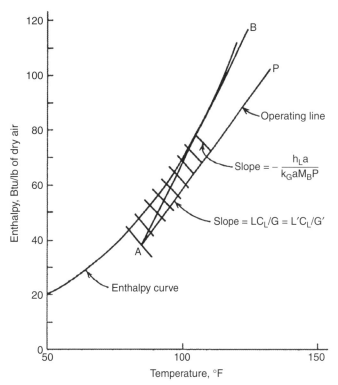

Exhibit 1 Determination of minimum L/G ratio (Example 14.4)

$C_L = 1$ for water. Actual $G' = 620(1.5) = 930$ lb/h · ft². Then the slope of the operating line = $1250(1)/930 = 1.344$. The operating line AP with slope 1.344 is drawn through A. Then

$$H_2 - H_1 = \frac{L'C_L}{G'}(t_{L2} - t_{L1})$$

$$H_2 = H_1 + \frac{LC_L}{G}(t_{L2} - t_{L1})$$

$$= 38.6 + \frac{1250(1)}{930}(110 - 85)$$

$$= 38.6 + 33.6 = 72.2 \text{ Btu/lb}$$

Step 4: Calculation of transfer coefficients

$$-\frac{h_L a}{k_G a M_B P} = -\frac{0.03(1250)^{0.51}(930)}{0.04(1250)^{0.26}(930)^{0.72}(29)(1)} = -1.042$$

Table 14.1 Calculation of $1/(H_i - H)$

H	H_i	$H_i - H$	$1/(H_i - H)$
72.2	78.4	6.2	0.161
68.0	73.5	5.5	0.181
64.0	69.1	5.1	0.196
56.0	60.5	4.5	0.222
52.0	56.4	4.4	0.227
48.0	52.5	4.5	0.222
42.6	47.6	5.0	0.200
38.6	43.8	5.2	0.192

Draw a series of lines, each of slope -1.042 to cut the operating line and the equilibrium curve, read from the graph the corresponding values of H_i and H, and prepare Table 14.1.

Make a plot of $1/(H_i - H)$ versus H and determine the area under the curve between H_2 and H_1 or by Simpson's rule,

$$(NTU)_G = \int_{H_1}^{H_2} \frac{dH}{H_i - H} = \frac{16.8}{3}[0.192 + 4(0.236) + 0.161]$$

$$= 7.3 \text{ transfer units}$$

Step 5: Calculation of $(HTU)_G$ and z

$$k_G a = 0.04(1250)^{0.26}(930)^{0.72} = 35 \text{ mol/h} \cdot \text{ft}^3 \cdot \text{atm}$$

$$(HTU)_G = \frac{930}{35(29)(1)} = 0.92 \text{ ft}$$

$$z = (NTU)_G (HTU)_G = 7.3(0.92) = 6.72 \text{ ft}$$

Example 14.5

Using the data from Example 14.4, evaluate the following: (a) number of gas enthalpy transfer units; (b) height of gas enthalpy transfer unit; and (c) overall packing height on the basis of the gas-phase transfer units.

Solution

The number of gas enthalpy transfer units is

$$\frac{H_2 - H_1}{(H_i - H)_{av}} = N_{tG}$$

$$H_2 - H_1 = 72.2 - 38.6 = 33.6 \text{ Btu}$$

$$(H_i - H)_1 = 5.2 \text{ Btu} \quad (H_i - H)_2 = 6.2 \text{ Btu}$$

$$(H_i - H)_{av} = \frac{6.2 - 5.2}{\ln(6.2/5.2)} = 5.69$$

$$N_{tG} = \frac{H_2 - H_1}{(H_i - H)_{av}} = \frac{33.6}{5.69} = 5.9 \text{ units}$$

$$H_{tG} = \frac{G'}{k_y a} = \frac{930}{k_G a M_B P} = \frac{930}{35(29)} = 0.92 \text{ ft}$$

$$z = 5.90(0.92) = 5.43 \text{ ft}$$

DRYING

Drying generally refers to the removal of small amounts of volatile liquid, usually water, from solids by evaporation into a gas (usually air) stream. Drying operations are carried out either batchwise or on continuous basis. Many types of drying equipment are in use, for example tray, turbo, flash, fluid bed, tunnel, spray, rotary, thin film, vacuum shelf indirect dryers. We will briefly review this operation in this section.

Equilibria

When a wet solid is brought in contract with air, the solid tends to lose moisture if the humidity of the air is lower than the humidity corresponding to the moisture content of the solid. If the air is more humid than the solid in contact with it, the solid will gain moisture until equilibrium is attained.

Equilibrium Moisture
The portion of water in the wet solid that cannot be removed by the air in its contact is called the *equilibrium moisture*. Since equilibrium is reached between the air and the solid, there is no humidity driving force between them.

Bound and Unbound Water in Solids
Bound water in a solid exerts an equilibrium vapor pressure lower than that of pure water at the same temperature. Unbound water exerts the equilibrium vapor pressure of pure water at the prevailing conditions.

Free Moisture
The free moisture is the amount of water in a wet solid in excess of the equilibrium moisture. Thus

$$X = X_T - X^*$$

where
 X = free moisture content of wet solid, lb/lb[kg/kg] dry solid
 X_T = total moisture content of wet solid, lb/lb [kg/kg] dry solid
 X^* = equilibrium moisture content of solid, lb/lb [kg/kg] dry solid

Many times the moisture content is given as percent moisture on a dry basis, which is

$$\text{Percent moisture on dry basis} = \frac{\text{kg H}_2\text{O}}{100 \text{ kg dry air}} \times 100$$

If free moisture content is plotted against time, a drying curve is obtained as shown in Figure 14.4.

The drying curve is drawn on the basis of experimental data and can be used to obtain **time of drying** during constant rate of evaporation of free moisture from the surface of the solid. Thus if a solid represented by the drying curve of Figure 14.4 is to be dried from a free moisture content X_1 = 0.35 kg H$_2$O/kg dry solid to say X_2 = 0.2 kg H$_2$O/kg dry solid, we read from Figure 14.2-1a for X_1 = 0.35, t_1 = 1 h and for X_2 = 0.2, t_2 = 3.7 h. Therefore time of drying is 3.7 − 1.0 = 2.7 hours.

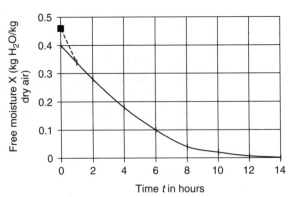

Figure 14.4 Typical drying curve. (Free moisture X vs time, t in hours)

Rate-of-Drying Curve

When the drying rates are determined and plotted against the moisture content of the solid, a curve such as in Figure 14.5 is obtained. There are usually two major parts to the rate curve, a portion over which the rate of drying is constant and another during which the drying rate is falling. Different solids and different drying conditions, however, may result in curves of different shapes in the falling-rate period.

In the constant-rate period, the evaporation rate is constant because of the presence of unbound moisture. The evaporation takes place from the surface. The constant drying rate can be expressed in terms of a constant mass-transfer coefficient as

$$N_c = k_Y(Y_s - Y) \tag{14.35}$$

where
Nc = drying rate, lb/h · ft^2
k_Y = water evaporated, lb/h · ft^2 · ΔY [kg/h · m^2 · ΔY]
Y_s = saturation humidity at liquid-surface temperature, lb moisture/lb dry gas [kg moisture/kg dry gas]
Y = humidity of drying air or gas, lb moisture/lb dry air or gas [kg moisture/kg dry air or gas]

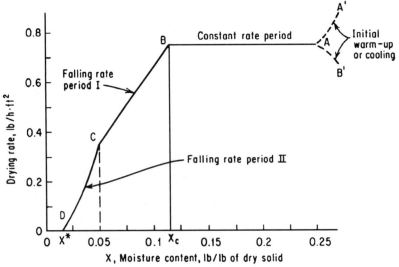

Figure 14.5 Rate of drying curve for solid

Under constant drying conditions k_Y is constant and, therefore, the rate of evaporation N_c is constant. If the solid is initially cold, its temperature slowly rises to t_s, the ultimate surface temperature (part *AB* of curve).

When the average value of the moisture content reaches a value of X_c equal to the critical moisture content, the drying rate begins to fall. The part *BC* represents the first falling-rate period in which unsaturated surface drying takes place. On further drying, diffusion from the interior of the solid controls the drying process, and the second falling-rate period is obtained (part *CD*). This part of the curve may not exist entirely in many cases, depending upon the solid and the drying conditions.

When the moisture content reaches the equilibrium moisture content value X^*, there is no more evaporation of water and the drying stops.

Rate of Drying

This can be expressed by the following equation

$$N = \frac{L_s dX}{A dt} \tag{14.36}$$

The time of drying t is given by

$$t = \int dt = \frac{L_s}{A} \int_{X_2}^{X_1} \frac{dX}{N} \tag{14.37}$$

where
  L_s = dry solid, lb
  M_D = dry solid lb/h·ft² for continuous drying
     = L_s/A
  A = drying surface, ft²[m²], or the cross section of bed for through circulation Drying
  t = time of drying, s

Constant Rate Period. If X_1 and $X_2 \geq X_c$, the critical moisture content, then

$$t_c = \frac{L_s(X_1 - X_2)}{AN_c} \tag{14.38}$$

where
  N_c = constant rate of drying, lb/h·ft² [kg/h·m²]
  A = drying surface or cross-sectional area perpendicular to gas flow, ft²[m²]
  t_c = time of drying in constant rate period, h

Falling Rate Period. If X_1 and X_2 are both less than X_c, the integral in Equation 14.37 has to be determined graphically in the general case when the integral cannot be obtained mathematically, and by integration, in the special case when rate of the drying curve is a straight-line function of X. Thus, if

$$N = mX + b \tag{14.39}$$

where m is the slope of the falling-rate curve and b is its intercept. Then

$$t_f = \frac{L_s}{A} \int_{X_2}^{X_1} \frac{dX}{mX + b} = \frac{L_s}{mA} \ln \frac{mX_1 + b}{mX_2 + b} \tag{14.40}$$

but

$$N_1 = mX_1 + b \quad N_2 = mX_2 + b \quad m = \frac{N_1 - N_2}{X_1 - X_2} \tag{14.41}$$

Therefore

$$t_f = \frac{L_s(X_1 - X_2)}{A(N_1 - N_2)} \ln \frac{N_1}{N_2} = \frac{L_s(X_1 - X_2)}{AN_m} \tag{14.42}$$

where the log mean drying rate is given by

$$N_m = \frac{N_1 - N_2}{\ln(N_1/N_2)} \tag{14.43}$$

Very often the entire falling rate can be represented by a straight line or can be so approximated; then

$$t_f = \frac{L_s(X_c - X^*)}{N_c A} \ln \frac{X_1 - X^*}{X_2 - X^*} \tag{14.44}$$

Example 14.6

A batch of solid for which the drying rate curve of Figure 14.5 applies is to be dried from a moisture content of 20 percent to a moisture content of 2 percent. Initial weight of the solids is 600 lb and the drying surface is 1 ft²/10 lb dry weight. Estimate the total time of drying.

Solution

Initial moisture content $X_1 = 0.2/0.8 = 0.25$ lb/lb dry solid. Final moisture content $X_3 = 0.02/0.98 = 0.020$ lb/lb dry solid. From Figure 14.5, critical moisture content, $X_c = 0.115$ lb/lb dry solid. The moisture content at the end of the first falling rate period is

$$X_2 = 0.05 \text{ lb/lb dry solid}$$

The rate of drying in constant-rate period from Figure 14.5 is

$$N_c = 0.75 \text{ lb/h} \cdot \text{ft}^2$$

Constant-rate period

$$\frac{L_s}{A} = \frac{\text{lb dry solid}}{\text{area}} = \frac{10}{1} = 10$$

$$t_c = \frac{L_s(X_1 - X_c)}{AN_c} = \frac{10(0.25 - 0.115)}{0.75} = 1.8 \text{ h}$$

Falling-Rate Period I.

$$N_c = 0.75 \text{ lb/h} \cdot \text{ft}^2 \quad N_2 = 0.35 \text{ lb/h} \cdot \text{ft}^2$$

Table 14.2 Calculation of 1/N at various values of X

X	N	$\frac{1}{N}$	X	N	$\frac{1}{N}$
0.05	0.35	2.94	0.045	0.28	3.571
0.04	0.22	4.55	0.035	0.16	6.25
0.03	0.116	8.62	0.025	0.078	12.821
0.02	0.035	28.57	0.0225	0.058	17.24

Applying Equation (14.42) to the falling rate period gives

$$t_2 = \frac{L_s(X_1 - X_2)}{A(N_1 - N_2)} \ln \frac{N_1}{N_2}$$

$$= \frac{10(0.115 - 0.05)}{1(0.75 - 0.35)} \ln \frac{0.75}{0.35} = 1.24 \text{ h}$$

Falling-Rate Period II. Since the rate curve is not a straight line, graphical integration is required. Table 14.2 is prepared using the values of N read from Figure 14.5.

A plot of $1/N$ versus moisture content X is made (not shown here), and the area under the curve is determined to be 0.283. Therefore, the time of drying from $X_2 = 0.05$ to $X_3 = 0.02$ lb/lb dry solid is

$$t_f = \frac{10}{1} 0.283 = 2.83 \text{ h}$$

Then, total time of drying is 1.8 + 1.24 + 2.83 = 5.87 h.

Estimation of Drying Rate and Time of Drying During Constant Rate Period

Use of drying curve and the rate of drying curve during constant rate period has been described earlier. Both these methods require experimental data. Whenever possible experimental data should be obtained for the design of drying equipment. However for the constant rate period it is possible to estimate the drying rate and drying time with the use of predicted transfer coefficients.

Equation to predict constant rate drying is derived as follows:

Drying of material occurs by mass transfer of water vapor from the saturated surface through an air film to the air or gas phase. Since the moisture content is above the critical moisture value, there is sufficient liquid present in the solid, which moves to the surface and keeps it saturated. Assuming conduction and radiation are negligible, the only heat transfer to the solid surface is by convection from air and is given by

$$q = h_c(T - T_W)A \qquad (14.45)$$

where
q = rate of heat transfer, J/s (Btu/h)
h_c = convection heat transfer coefficient, W/m² · K (Btu/h · ft² · °F)

Flux of water vapor from the surface of the solid is given by equation and is

$$N = \frac{k'_y}{x_{BM}}(y_s - y) = k_y(y_w - y) \qquad (14.46)$$

since $t_s \cong t_w$ in case of air-water vapor mixture,

where
- N = kg mol of water vapor evaporating/s · m²
- A = surface area of evaporation, m²
- x_{BM} = log mean mol fraction of air (inert)
- $x_{BM} \cong 1.0$ and $k'_y \cong k_y$ since the mixture is dilute
- $y_s = y_w$ in case of air-water vapor mixture.
- y_w = mol fraction water vapor in the air at the surface
- y = mol fraction of water vapor in bulk of air

Now
$$y = \frac{Y/M_A}{1/M_B + Y/M_A} \qquad (14.47)$$

where M_A = molecular weight of water vapor M_B = molecular weight of air

since Y is small, $Y/M_A \ll 1/M_B$ and we can take $1/M_B + Y/M_A \approx 1/M_B$.

Hence
$$y = \frac{YM_B}{M_A} \qquad (14.48)$$

Also the heat required to vaporize N mols of water vapor is given by

$$q = M_A N \lambda_w A \qquad (14.49)$$

where λ_w = the latent heat of vaporization of water at t_w, J/kg (Btu/lb)

From equations 14.45, 14.46 and 14.49, the rate of drying in the constant-rate period is obtained as follows

$$R_C = \frac{q}{A\lambda_w} = \frac{h_c(T - T_w)}{\lambda_w} = k_y M_B(Y_W - Y) \qquad (14.50)$$

This equation is exactly similar to the equation for the wet bulb temperature. However, rate of drying calculated using heat transfer coefficient is more reliable as the error in the determination of T_W affects the difference $(Y_w - Y)$ more than the difference $(T - T_w)$. The rate of drying in constant rate period can then be written as follows

$$R_C = \frac{h_c}{\lambda_w}(T - T_w\,°C)(3600) \quad \text{(SI units)}$$

$$\qquad (14.51)$$

$$R_C = \frac{h_c}{\lambda_w}(T - T_w\,°F) \quad \text{(English units)}$$

The use of equations 14.51 to estimate rates of drying in constant rate period requires value of h_c, the heat transfer coefficient. This can be calculated by the following equations

Air-flow *parallel* to evaporating surface:

For mass velocity $G = 2450 - 29300$ kg/h·m² (500-6000 lb$_m$/h·ft²) or velocity of 0.61–7.6 m/s (2–25 ft/s), the convective heat transfer coefficient is given by

$$h_c = 0.0204 G^{0.8} \quad \text{(SI units)} \tag{14.52}$$

$$h_c = 0.0128 G^{0.8} \quad \text{(English units)} \tag{14.52a}$$

where
$G = u\rho$ kg/h·m²(lb$_m$/h·ft²)
h_c = heat transfer coefficient, W/m²·K(Btu/h·ft²·°F)

Air-flow *perpendicular* to evaporating surface:

$$h_c = 1.17 G^{0.37} \quad \text{(SI units)}$$

$$h_c = 0.37 G^{0.37} \quad \text{(English units)} \tag{14.53}$$

These equations are applicable for $3900 < G < 19500$ kg/h·m² or a velocity of 0.9 to 4.6 m/s in SI units and for $800 < G < 4000$ lb$_m$/h·ft² or a velocity of 3–15 ft/s in English units.

The time of drying during the constant-rate period can be estimated by the following relation:

$$t = \frac{L_s \lambda_w (X_1 - X_2)}{A h_c (T - T_w)} = \frac{L_s (X_1 - X_2)}{A k_y M_B (Y_W - Y)} \tag{14.54}$$

Example 14.7

(Estimation of Constant-Rate Drying Rate and Time)

An insoluble granular material is dried in a tray dryer of 0.5m × 0.5m exposed surface. The tray is insulated on all sides and the bottom so that heating and evaporation take place from the exposed surface only. Air at 50°C dry-bulb and 27°C wet-bulb temperature flows parallel to the exposed surface at a velocity of 4 m/s. Initial moisture content of the dry solid is 0.3 kg water/kg dry solid. The critical moisture content of the material is 0.1 kg/kg dry solid. The density of dry solid is 1927 kg/m³. Using SI unit humidity chart, answer the following questions:

(a) What is the dew point of fresh air to the dryer?

(b) What is percent absolute humidity of the fresh air?

(c) What is the density of humid air?

(d) What is the initial free moisture content of the solid?

(e) What is the enthalpy of fresh air?

(f) Estimate the drying rate for constant-rate drying.

(g) Calculate total drying rate for the given surface

(h) Estimate the time of drying the material from its initial moisture content of 0.3 kg/kg dry solid to its critical moisture content.

Solution

Use humidity chart in SI units to get air properties.

(a) Proceeding vertically at 50°C, locate the intersection point with 27°C wet-bulb temperature line. From the intersection point, proceed horizontally toward the saturation curve and locate the intersection of the saturation curve and the wet bulb line. From this intersection point read vertically down to get the dew point temperature. It is 18°C.

(b) Again from the intersection point of the saturated line and the wet-bulb line read to the right the absolute humidity = 0.013 kg moisture/kg dry air. Proceed along wet-bulb line to the saturation line and from the intersection point read horizontally to the saturation humidity Y_s = 0.0227 kg/kg dry air. Then percent absolute humidity is

$$\text{Percent absolute humidity} = 100 \times (Y/Y_s)$$
$$= 100 \times (0.013/0.0227)$$
$$= 57.3 \text{ percent}$$

(c) First obtain the humid volume. You can get this from the chart by interpolating or by calculation. By calculation, the humid volume is

$$v_H = (2.83 \times 10^{-3} + 4.56 \times 10^{-3} \times 0.013)(323.15) = 0.934 \text{ m}^3/\text{kg dry air}$$

Then the density of humid air to the dryer is

$$\rho = \frac{1.0 + 0.013}{0.934} = 1.085 \text{ kg/m}^3$$

(d) Initial moisture content (total) = 0.3 kg/kg dry solid

Critical moisture content = 0.1 kg/kg dry solid.

Therefore initial free (unbound) moisture = 0.3 − 0.1 = 0.2 kg/kg dry solid.

(e) Enthalpy of fresh air to dryer:

From the intersection point found in part a of this problem, proceed along the wet-bulb line, go beyond saturation line, and read enthalpy value on the enthalpy line as 85.2 kJ/kg dry air. This is the enthalpy of saturated air. To get the enthalpy of fresh air, interpolate between enthalpy deviation lines represented by −1.0 and −1.2 (the intersection point lies in between these two curves) as −0.11 kJ/kg dry air.

Then enthalpy of fresh air = 85.2 − 0.11 = 85.09 kJ/kg dry air.

(f) Drying rate R_C

$$\text{Mass velocity } G = u\rho \times 3600 = 4 \times (1.085)(3600) = 15624 \text{ kg/m}^2 \cdot \text{h}$$

Flow is parallel to the surface, therefore

$$h_c = 0.0204 G^{0.8} = 0.0204(15624)^{0.8} = 46.2 \frac{\text{W}}{\text{m}^2 \cdot \text{K}}$$

Constant-rate period drying rate R_C.

$$\lambda_w = 2550.8 - 113.5 = 2437.55 \text{ kJ/kg from steam tables (SI units)}$$

$$R_C = \frac{h_c}{\lambda_w}(T - T_w)(3600) = \frac{46.2}{2437.6 \times 1000}(50 - 27)(3600) = 1.57 \text{ kg/h} \cdot \text{m}^2$$

(g) Total drying rate = $R_C A$ = 1.57(0.25 = 0.3925 kg H_2O/h

(h) Volume of cake = $0.5 \times 0.5 \times 0.0254 = 0.0035 \text{ m}^3$

Because the solid is insoluble in water, the water is in the voids of the solid. So the calculated volume is also the volume of the solid as if there were no water in it. Density of solids = 1926.5 kg/m^3, hence L_s = 0.00635(1926.5 = 12.23 kg.

Then time to dry the cake from 0.3 to 0.1 kg/kg solids moisture content is given by

$$t = \frac{L_s(X_1 - X_2)}{Ah_c(T - T_w)} = \frac{12.23 \times 2437.55 \times 1000(0.3 - 0.1)}{0.25 \times 46.2(50 - 27)(3600)} = 6.23 \text{ h}$$

RECOMMENDED REFERENCES

1. Geankoplis, *Transport Processes and Separation Process Principles,* 4th ed., Prentice-Hall, NJ, 2004.
2. Perry and Green, *Chemical Engineers' Handbook,* platinum ed., McGraw-Hill, NY, 1999.
3. Treybal, *Mass Transfer Operations,* 4th ed., McGraw-Hill, NY, 1994.

CHAPTER 15

Chemical Reaction Engineering

OUTLINE

CHEMICAL REACTION ENGINEERING 536
Classification of Reactions ■ Rate of Chemical Reaction and Rate Constant ■ Reactions at Constant Volume and Temperature ■ Elementary and Nonelementary Reactions ■ Molecularity and Order of Reaction ■ Law of Mass Action ■ Chemical Equilibria ■ Effect of Temperature on the Rate of Reaction ■ Temperature Dependency of Reaction Rate from Collision and Transition Theories

INTERPRETATION OF KINETIC DATA AND THE CONSTANTS OF THE RATE EQUATION 550
Simple Irreversible Reactions ■ Irreversible Reactions ■ Empirical Rate Equations of the nth Order for Irreversible Reactions

REACTOR DESIGN 566
Batch Reactors ■ Steady-State Flow Reactor ■ Ideal Stirred-Tank Reactor ■ Plug-Flow Reactor ■ Packed Bed Reactors (PBR) ■ Fluidized Bed Reactors (FBR)

MASS AND ENERGY BALANCES 567
Ideal Batch Reactor ■ Flow Reactors ■ Steady-State Mixed-Flow Reactor or Continuous-Flow Stirred-Tank Reactor ■ Steady-State Plug-Flow Reactor ■ Plug-Flow Reactor ■ Volume of Back-Mix Reactor ■ Volume of Plug-Flow Reactor ■ Multiple Reactors in Series

PRODUCT DISTRIBUTION AND TEMPERATURE 579
Series Reactions: Maximizing the Desired Product ■ Plug-Flow or Batch Reactor ■ Mixed Flow Reactor ■ Recycle Reactors ■ Batch Recycle Reactor ■ Packed-Bed Reactor (PBR) ■ Catalysts and Catalytic Reactors ■ Design of Reactors for Gas-Solid Reactions

STOICHIOMETRIC TABLES 593
Stoichiometric Table for Flow Systems ■ Steady State Nonisothermal Reactor Design

OPTIMAL TEMPERATURE PROGRESSION 597

Heat Addition or Removal ■ Heat Transfer to Tubular Reactors (PFR and PBR) ■ Nonisothermal Continuous-Flow Reactors ■ Adiabatic Plug-Flow Tubular Reactor ■ Steady-State Plug-Flow Reactor with Heat Transfer ■ Multiple Steady States

RECOMMENDED REFERENCES 612

From the point of view of the P.E. examination, the important areas of knowledge in the field of reaction engineering are reaction parameters, reaction rates, reactor design and evaluation, and heterogeneous reaction systems. In this chapter we will briefly review these topics. Eleven percent of the P.E. examination is allotted to chemical reaction engineering.

CHEMICAL REACTION ENGINEERING

A chemical process is designed to produce economically a desired useful and marketable product by chemical reaction and physical treatment (separation, purification, etc.) from suitable raw materials. The design and operation of chemical reactors is one of the important tasks facing a chemical engineer. To accomplish the desired objective, it is necessary to specify the type and size of the reactor system, the desired extent of reaction (i.e., approach to equilibrium or maximum conversion), operating conditions such as pressure and temperature, and auxiliary equipment.

Two important questions are (1) how far the reaction proceeds toward equilibrium condition and (2) how fast is the desirable approach to equilibrium. The problems of maximum yield and the heat transfer required from or to the reactor system are adequately addressed by thermodynamics. Chemical kinetics deals with the mechanism and the rate of chemical reaction.

The principal functions of chemical kinetics from the point of view of chemical engineers are the following:

1. Establishing mechanism of the chemical reaction

2. Obtaining experimental rate data to find how the reaction proceeds

3. Correlating the data

4. Designing a suitable reactor system

5. Specifying operating conditions such as temperature, pressure, methods of control, and auxiliary equipment.

Classification of Reactions

Reactions are classified in many ways. On the basis of mechanism, they are classified as

1. *Irreversible*. Reactions that proceed in one direction. For example,

$$A \rightarrow B \quad \text{or} \quad A + B \rightarrow R + S$$

2. *Reversible*. Reactions in which reactants react to give products and products in turn react to convert back to reactants

$$A \rightleftharpoons B \quad \text{or} \quad A + B \rightleftharpoons R + S$$

3. *Simultaneous or Parallel.* Reactants undergoing two different reactions at the same time

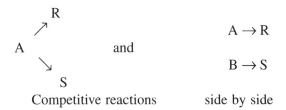

        Competitive reactions   side by side

4. *Consecutive.* In consecutive reactions, a reactant converts to a product, which in turn reacts to another product.

$$A \rightarrow R \rightarrow S$$
Consecutive or series reactions

Another important classification is based on the number of phases present. If the reaction takes place in a single phase, it is called homogenous. If two or more phases are present, the reaction is called heterogeneous. Reactions are also classified as exothermic or endothermic. In exothermic reactions, heat is generated and heat needs to be removed from the reaction system. In endothermic reactions, heat is absorbed and therefore heat needs to be added to the system.

Other classifications are based on order of reaction, thermal operating condition, molecularity, catalyst or no catalyst, and type of equipment used. Lastly, one more classification is based on the mode of operation, namely batch, semibatch, or continuous.

Rate of Chemical Reaction and Rate Constant

In quantitative terms the rate of reaction is expressed as the number of units of mass of some reactant that is converted to a product per unit time per unit volume V of the system. This can be expressed in terms of equations as follows:

$$r = -\frac{1}{V}\frac{dn}{dt} \text{ in general} \tag{15.1}$$

$$r = -\frac{d(n/V)}{dt} = -\frac{dC}{dt} \text{ at constant volume only} \tag{15.1a}$$

where
    n = number of moles of reactant present at time t
    C = concentration kg mol/m^3 [lb mol/ft^3]
    r = reaction rate

If we write the rate equation in terms of the reactant that is converted in time t, such that $x = n_0 - n$, the rate equation can be written as

$$r = +\frac{1}{V}\left(\frac{dx}{dt}\right) \tag{15.2}$$

In case of some reactions, it is possible to separate the effects of the quantities n_i from the effects of other variables. The rate equation can then be written as follows:

$$r = -\frac{1}{V}\frac{dn}{dt} = kf(n_a, n_b, \ldots) \tag{15.3}$$

In Equation 15.3, the proportionality constant k is called the *rate constant*. It is also called the *specific reaction rate* (SRR), or the *rate coefficient*. It is not truly a constant. By definition, it is independent of the masses of the reactants but is affected by other variables, which influence the chemical reaction. When the gases or solutions are not ideal, the rate constant is dependent on the concentrations. In this case activities should be used in place of the concentrations. It should be noted here that these equations apply to only simple or elementary reactions. To describe complex reactions, more than one reaction constant is required.

Reactions at Constant Volume and Temperature

Rate equations can be written for elementary reactions using stoichiometric relationships.

A list of such rate equations for some simple reactions at constant volume and temperature is given in Table 15.1.

Reaction: $\quad aA + bB + cC \rightarrow$ products

Rate equation: $\quad \dfrac{dx}{dt} = k(n_{A0} - x)^p \left(n_{B0} - \dfrac{bx}{a}\right)^q \left(n_{C0} - \dfrac{cx}{a}\right)^r$

Table 15.1 Reactions at constant volume and temperature*

Order	Reaction	Rate Equation	Integral
0	A → products	$\dfrac{dx}{dt} = k$	$x - x_0 = k(t - t_0)$
$\tfrac{1}{2}$	A → products	$\dfrac{dx}{dt} = k(n_{A0} - x)^{1/2}$	$(n_{A0} - x_0)^{1/2} - (n_{A0} - x)^{1/2} = \dfrac{k(t - t_0)}{2}$
1	A → products	$\dfrac{dx}{dt} = k(n_{A0} - x)$	$\ln \dfrac{n_{A0} - x_0}{n_{A0} - x} = k(t - t_0)$
2	2A → products	$\dfrac{dx}{dt} = k(n_{A0} - x)^2$	$\dfrac{1}{n_{A0} - x} - \dfrac{1}{n_{A0} - x_0} = k(t - t_0)$
2	A + B → products	$\dfrac{dx}{dt} = k(n_{A0} - x)(n_{B0} - x)\,(n_{A0} \neq n_{B0})$	$\dfrac{1}{n_{B0} - n_{A0}} \ln \dfrac{(n_{A0} - x_0)(n_{B0} - x)}{(n_{B0} - x_0)(n_{A0} - x)} = k(t - t_0)$
3	3A → products	$\dfrac{dx}{dt} = k(n_{A0} - x)^3$	$\left(\dfrac{1}{n_{A0} - x}\right)^2 - \left(\dfrac{1}{n_{A0} - x_0}\right)^2 = 2k(t - t_0)$

*n_{A0}, n_{B0}, and n_{C0} are initial moles of A, B, and C, respectively.

Elementary and Nonelementary Reactions

A reaction in which the rate equation corresponds to a stoichiometric equation is called an elementary reaction. For example, the rate equation for the reaction $A + B \rightarrow R$ is

$$-r_A = k C_A C_B \tag{15.4}$$

Here the rate is simply proportional to the concentrations of A and B. When the rate equation does not correspond to the stoichiometry of the reaction, it is called nonelementary. For example, the rate equation for the reaction between H_2 and Br_2 to produce HBr is

$$r_{HBr} = \frac{k_1 C_{H2} C_{Br2}^{1/2}}{k_2 + C_{HBr}/C_{Br2}} \tag{15.5}$$

Molecularity and Order of Reaction

Molecularity of an elementary reaction is the number of the molecules involved in the reaction. This applies only to the elementary reactions, in which case the molecularity can be one, two, or occasionally three.

In the case of many complex reactions, the rate equation is found empirically. Hence the exponents of the concentration terms in the rate equation are not related to the stoichiometric coefficients and are different. The rate equation then can be written as follows:

$$r_A = k C_A^p C_B^q \cdots C_D^s \tag{15.6}$$

The sum of these empirically determined exponents is called the *overall order of the reaction*. Then the order of the reaction is given by

$$n = p + q \cdots + s \tag{15.6a}$$

where n is the overall order of reaction (it may be a fractional number). p is the order of the reaction with respect to the reactant A, q with respect to B, and r with respect to C, respectively. Orders with respect to other components are defined the same way.

Dimensions of Rate Constant k

When the rate expression for a homogeneous reaction is written in terms of Equation 15.6, the dimensions of the rate constant k for the nth order reaction are

$$(\text{time})^{-1}(\text{concentration})^{1-n} \tag{15.6b}$$

For a first-order reaction, Equation 15.6a reduces to a very simple form

$$(\text{time})^{-1} \tag{15.6c}$$

Law of Mass Action

The law of mass action states that the rate of chemical reaction is proportional to the active masses of the participants in the reaction (Guldberg and Waage law).

Thermodynamic activity is generally taken as the active mass.

Chemical Equilibria

In case of elementary reversible reactions such as $A + B \leftrightarrows R + S$, the rate of formation of, say, R and that of its disappearance by reverse reaction are respectively

$$r_{R,\text{forward}} = k_1 C_A C_B \quad \text{and} \quad -r_{R,\text{reverse}} = k_2 C_R C_S \tag{15.7}$$

At equilibrium the two rates are equal and there is no net formation of R. Hence we can write

$$r_{R,\text{forward}} + r_{R,\text{reverse}} = 0 \tag{15.7a}$$

$$\text{or} \quad \frac{k_1}{k_2} = \frac{C_R C_S}{C_A C_B} \tag{15.7b}$$

If K is defined as equilibrium constant, Equations 15.7a and b can be combined to give

$$K = \frac{k_1}{k_2} = \frac{C_R C_S}{C_A C_B} \tag{15.8}$$

Thus in terms of kinetics, the equilibrium state is a dynamic steady state in which there is a constant interaction between reactants and products.

From a thermodynamic point of view, a system is in equilibrium with its surroundings at a given temperature and pressure if the free energy of the system is minimum. For any departure from the equilibrium

$$(\Delta G)_{P,T} > 0 \tag{15.9}$$

From a chemical kinetics point of view, the system is in equilibrium if the rates of forward and reverse elementary reactions are equal.

The standard free-energy change between the free energies of the products and reactants at standard state, ΔG^o, is given by

$$\Delta G^o = -RT \ln K \tag{15.10}$$

where
 K = equilibrium constant
 T = absolute temperature, K
 R = gas law constant

For a reaction of the type $aA + bB \rightarrow cC + dD$, the equilibrium constant K is defined by the relation

$$K = \frac{a_C^c a_D^d}{a_A^a a_B^b} \tag{15.11}$$

where a's are the equilibrium activities. Activity a_i in terms of fugacity is given by

$$a_i = \frac{f_i}{f_i^o} \tag{15.12}$$

If the standard state chosen is unit fugacity or $f_i^o = 1$,

$$K = \frac{f_C^c f_D^d}{f_A^a f_B^b} \tag{15.13}$$

For ideal gas behavior,

$$K_P = \frac{p_C^c p_D^d}{p_A^a p_B^b} \tag{15.14}$$

where the p's are the partial pressures. Since $p_i = y_i P_t$ where P_t is the total pressure and y_i is the mole fraction of component i, K_P in the case of the gaseous reactions is given by

$$K_P = \frac{(y_c P_t)^c (y_d P_t)^d}{(y_a P_t)^a (y_b P_t)^b} \tag{15.15}$$

$$= K_y P_t^{[(c+d)-(a+b)]} \tag{15.16}$$

If the reaction takes place at 1 atm the pressure term in the preceding equation is 1, then

$$K_p = K_y \tag{15.17}$$

The Van't Hoff equation relates K to the heat of reaction by

$$\frac{d \ln K}{dT} = \frac{\Delta H_T^o}{RT^2} \tag{15.18}$$

where ΔH_T^o is the standard-state enthalpy change for the reaction. If ΔH_T^o is approximately independent of temperature, then Equation 15.18 can be integrated to give

$$\ln \frac{K_2}{K_1} = \frac{-\Delta H_T^o}{R} \left(\frac{1}{T_2} - \frac{1}{T_1} \right) \tag{15.19}$$

where K_2 is the equilibrium constant at temperature T_2 and K_1 is the equilibrium constant at temperature T_1. If ΔH^o varies with temperature and if the heat capacities of the reactants and products are expressed by equations of the type

$$C_p^o = \alpha + \beta T + \gamma T^2 + \cdots \tag{15.20}$$

$$\Delta C_p^o = \Delta \alpha + \Delta \beta T + \Delta \gamma T^2 + \cdots \tag{15.21}$$

and the reaction is of the type

$$n_a A + n_b B + \cdots \to n_r R + n_s S + \cdots \tag{15.22}$$

where

$$\left. \begin{array}{l} \Delta \alpha = (\Sigma n \alpha)_{\text{products}} - (\Sigma n \alpha)_{\text{reactants}} \\ \Delta \beta = (\Sigma n \beta)_{\text{products}} - (\Sigma n \beta)_{\text{reactants}} \\ \Delta \gamma = (\Sigma n \gamma)_{\text{products}} - (\Sigma n \gamma)_{\text{reactants}} \end{array} \right\} \tag{15.23}$$

then substituting ΔC_p^o in Equation 3.23 and integrating yields

$$\Delta H_T^o = I_H + \Delta\alpha T + \Delta\beta \tfrac{1}{2}T^2 + \Delta\gamma \tfrac{1}{3}T^3 + \cdots \qquad (15.24)$$

where I_H is the constant of integration.

When use is made of Equations 15.18 and 15.24, K as a function of the temperature is given by integrating Equation 15.18 as

$$\ln K = -\frac{I_H}{RT} + \frac{\Delta\alpha}{R}\ln T + \frac{\Delta\beta}{R}\left(\frac{1}{2}\right)T + \frac{\Delta\gamma}{R}\left(\frac{1}{6}\right)T^2 + \cdots + I \qquad (15.25)$$

Knowing the K at one temperature, it is possible to evaluate I in the preceding equation, and therefore the evaluation of the K is possible at any other temperature. Also since $\Delta G_T^o = -RT \ln K$, one can get

$$\Delta G_T^o = I_H - \Delta\alpha T \ln T - \tfrac{1}{2}\Delta\beta T^2 - \tfrac{1}{6}\Delta\gamma T^3 - IRT \qquad (15.26)$$

or
$$\Delta G_T^o = I_H + I_G T - \Delta\alpha T \ln T - \tfrac{1}{2}\Delta\beta T^2 - \tfrac{1}{6}\Delta\gamma T^3 \qquad (15.27)$$

where $I_G = -(IR)$. $\qquad (15.27a)$

Knowing ΔH_{298}, the heat of the reaction at 25°C, and the specific heats of the components, it is possible to evaluate the constants in the preceding equations. Note the following conditions:

If $G^o < 0$, i.e., negative, spontaneous reaction takes place in the standard states.

If $G^o = 0$, there is equilibrium in standard states.

If $G^o > 0$, i.e., positive, then $K < 1$, and therefore the reactants in the standard state will not react to produce the products in the standard states.

Example 15.1

Calculate the equilibrium constant K and equilibrium conversion for the following reaction:

$$C_2H_6(g) \rightarrow C_2H_4(g) + H_2(g)$$

at a temperature of 1000 K, if the reaction takes place at 1 atm. pressure. Data on ΔG^o, ΔH^o, and C_p are given in Table 15.2.

Table 15.2 Data on ΔG_{298}^o, ΔH_{298}^o, and C_p values

Component	ΔG_{298}^o kJ/mol	ΔH_{298}^o kJ/mol	C_p kJ/mol·K
C_2H_6	−32.886	−84.667	$0.0096 + 8.37 \times 10^{-5}T$
C_2H_4	68.124	52.3	$0.0117 + 12.55 \times 10^{-5}T$
H_2	0	0	$0.0289 + 1.67 \times 10^{-5}T$

Solution

Calculate ΔG^o_{298} and ΔH^o_{298} for the reaction

$$\Delta G^o_{298} = \Sigma(\Delta G^o_{298})_{products} - \Sigma(\Delta G^o_{298})_{reactants}$$

where *P* and *R* denote the products and reactants respectively.

$$\Delta G^o_{298} = 68.124 + 0 - (-32.886)$$
$$= 101.01 \text{ kJ/mol}$$
$$\Delta H^o_{298} = 52.3 + 0 - (-84.667)$$
$$= 136.967 \text{ kJ/mol}$$

$$C_p = \begin{cases} 0.0289 + 1.67 \times 10^{-5}T & \text{for } H_2 \\ 0.0117 + 12.55 \times 10^{-5}T & \text{for } C_2H_4 \\ -0.0096 - 8.37 \times 10^{-5}T & \text{for } -C_2H_6 \end{cases}$$

By addition, $C_p = 0.031 + 5.85 \times 10^{-5}T$

Calculation of I_H and ΔG^o

From the preceding ΔC_p equation, $\Delta\alpha = 0.031$, $\Delta\beta = 5.85 \times 10^{-5}$. $\Delta\gamma = 0$. From Equation 15.24,

$$\Delta H^o_T = I_H + (\Delta\alpha T + \Delta\beta \tfrac{1}{2}T^2 + \cdots)$$

or
$$I_H = \Delta H^o_T - (\Delta\alpha T + \Delta\beta \tfrac{1}{2}T^2 + \cdots).$$

At 298 K, $\Delta H^o_T = 136.967$ kJ/mol

$$I_H = 136.967 - [0.031(298) + 5.85 \times 10^{-5}\left(\tfrac{1}{2}\right)(298)^2]$$
$$= 125.31 \text{ kJ/mol}$$

At $T = 298$ K, using Equation 15.27,

$$101.01 = 125.131 - 0.031(298)\ln 298 - 2.925 \times 10^{-5}(298)^2 + I_G(298)$$

which gives

$$I_G = 0.1044$$

$$\Delta G^o_T = 125.131 - 0.031T \ln T - 2.925 \times 10^{-5}T^2 + 0.1044T$$

At $T = 1000$ K,

$$\Delta G^o_{1000} = 125.131 - 0.031(1000)\ln 1000 - 2.925 \times 10^{-5}(1000)^2 + 0.1044(1000)$$
$$= -13.859 \text{ kJ/g mol}$$

Then

$$\ln K = \frac{-\Delta G^o}{RT} = -\frac{-13.859 \times 10^3}{8.314(1000)} = 1.667$$

and solving for K gives $K = 5.296$.

Calculation of Conversion

Assuming components behave as ideal gases ($P_T = 1$ atm),

$$K = \frac{f_{H_2} f_{C_2H_4}}{f_{C_2H_6}} = \frac{y_{H_2} y_{C_2H_4}}{y_{C_2H_6}} \frac{P_T P_T}{P_T}$$

$$= \frac{y_{H_2} y_{C_2H_4}}{y_{C_2H_6}} \quad \text{because} \quad P_T = 1 \text{ atm}$$

At equilibrium, let X be the number of moles of C_2H_6 converted. The moles of each component in the reaction mixture are

	Initial	Equilibrium
C_2H_6	1	$1-X$
C_2H_4	0	X
H_2	0	X

Therefore, total moles = $1 + X$ at equilibrium. The mole fractions of the components are then calculated as

$$y_{C_2H_6} = \frac{1-X}{1+X} \quad y_{H_2} = \frac{X}{1+X} \quad y_{C_2H_4} = \frac{X}{1+X}$$

Therefore

$$K = \frac{[X/(1+X)][X/(1+X)]}{(1-X)/(1+X)} = \frac{X^2}{1-X^2} = 5.296$$

$$X^2 = \frac{5.296}{6.296}$$

and $X = 0.917$ (positive acceptable root of the quadratic)

Therefore, the conversion of C_2H_6 to C_2H_4 is 91.7 percent.

Effect of Temperature on the Rate of Reaction

It has been noted earlier that the rate constant k is not truly a constant. Although by definition it is independent of masses in ideal systems, it is strongly dependent on temperature. Arrhenius suggested that the temperature dependence of the reaction rate constant k can be correlated by the following equation:

$$k = \alpha e^{-E/RT} \tag{15.28}$$

where
- α = frequency factor or preexponential factor
- E = the activation energy, J/mol or cal/mol
- R = gas constant = 8.314 J/mol·K = 1.987 cal/mol·K
- T = absolute temperature, K

Equation 15.28 has been verified empirically over a wide range of temperatures. It is determined by carrying out the reaction at different temperatures. By taking the logarithm of both sides of Equation 15.28,

$$\ln k = \ln \alpha - \frac{E}{R}\left(\frac{1}{T}\right) \tag{15.29}$$

From Equation 15.29 it will be seen that a plot of $\ln k$ versus $1/T$ will be a straight line with a slope of $-E/R$ and in intercept of $\ln \alpha$, which makes it possible to determine both E and α.

Example 15.2

Determine the activation energy and frequency factor given the following data for the bimolecular formation of methyl ether in an ethyl alcohol solution.

T,°C	0	6	12	18	24	30
$k \times 10^5$ L/g·mol·s	5.6	11.8	24.5	48.8	100	208

Solution

$$k = \alpha e^{-E/RT}$$

$$\ln k = \ln \alpha - \frac{E}{RT}$$

or $\log k = \log \alpha - \dfrac{E}{2.3RT}$ (converts natural logarithm to logarithm to base 10)

Thus if $\log k$ is plotted against $1/T$ (T is in Kelvin), a straight line is obtained. The slope of this line is $-E/(2.3R)$, and the intercept is $\log \alpha$. Thus E and α can be determined from a plot of $\log k$ versus $1/T$. For the given data, the following table is prepared:

T,°K	273	279	285	291	297	303
$\frac{1}{T} \times 10^3$	3.663	3.584	3.509	3.436	3.367	3.300
$K \times 10^5$	5.6	11.8	24.5	48.8	100	208

A plot of k versus $1/T$ is made on a semilogarithmic paper as shown in Exhibit 1.

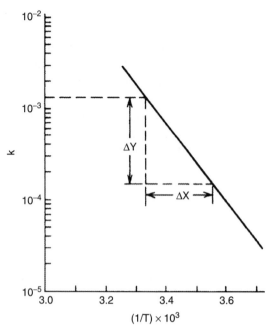

Exhibit 1 Plot of k versus $1/T$ (Example 15.2)

From the graph, the slope is

$$-\frac{E}{2.3R} = \frac{\log(1.3\times 10^{-3}) - \log(1.5\times 10^{-4})}{(3.34 - 3.56)\times 10^{-3}}$$

$$= \frac{-\log 8.67}{0.22\times 10^{-3}} = -4263$$

Therefore, $E = -2.3(R)(-4263) = 2.3(1.987)(4263) = 19482$ cal/g mol.

Now
$$\log k = -\frac{E}{2.3R}\frac{1}{T} + \log \alpha$$

At $1/T = 3.34 \times 10^{-3}$, $\quad \log k = \log(1.3\times 10^{-3}) = -2.88606$

$$-2.88606 = -\frac{19{,}482(3.34\times 10^{-3})}{2.3(1.987)} + \log \alpha$$

from which $\quad \alpha = 2.25 \times 10^{11}$

By calculating the values at other readings, an average α can be determined.

Another way to calculate the activation energy is to use the decade method. For this, we write two equations for k as follows:

$$\log k_1 = \log A - \frac{E}{2.3R}\left(\frac{1}{T_1}\right) \quad (p)$$

$$\log k_2 = \log A - \frac{E}{2.3R}\left(\frac{1}{T_2}\right) \quad (q)$$

(15.30)

Subtraction of p from q gives $\log \dfrac{k_2}{k_1} = -\dfrac{E}{2.3R}\left(\dfrac{1}{T_2} - \dfrac{1}{T_1}\right)$ (15.31)

Now choose $1/T_1$ and $1/T_2$ such that $k_2 = 0.1k_1$. Therefore log $(k_1/k_2) = 1$.
From exhibit 1,

$$\text{When} \quad k_1 = 0.001 \quad \frac{1}{T_1} = 0.003367$$

$$\text{When} \quad k_2 = 0.0001 \quad \frac{1}{T_2} = 0.003602$$

Then $E = -\dfrac{(2.3)(R)\log(k_2/k_1)}{1/T_2 - 1/T_1} = \dfrac{2.3R}{1/T_2 - 1/T_1} = \dfrac{(2.3)\left(8.314 \frac{J}{mol.K}\right)}{(0.003602 - 0.003367/K)}$

$$= 81.37 \text{ kJ/mol} = 19447 \text{ cal/g mol}$$

A rule of thumb says that the rate of a reaction doubles for every 10°C increase in temperature. However, this is not true in all situations. It is valid only for a specific combination of activation energy and the temperature. Arrhenius relation involves activation energy and temperature, and both affect the reaction rate. For a given activation energy, the temperature range of 10°C over which the rate will exactly double can be found as follows:

If k_1 and k_2 are the rate constants at two temperatures T_1 and T_2 such that $T_2 = T_1 + 10$, K and the reaction rate is doubled from T_1 to T_2, we can write as follows using the Arrhenius relation:

$$\ln k_1 = \ln \alpha - \frac{E}{RT_1} \quad \text{at temperature } T_1 \text{ K} \tag{15.32a}$$

and $\quad \ln k_2 = \ln \alpha - \dfrac{E}{RT_2} \quad \text{at temperature } T_2 = T_1 + 10 \text{ K} \tag{15.32b}$

By subtracting Equation 15.32a from Equation 15.32b,

$$\ln \frac{k_2}{k_1} = -\frac{E}{R}\left(\frac{1}{T_2} - \frac{1}{T_1}\right) \tag{15.33}$$

After substitution, $T_2 = T_1 + 10$ and $\frac{k_2}{k_1} = 2$. This then results in

$$\log 2 = \frac{E}{2.3(R)}\left[\frac{10}{T_1^2 + 10T_1}\right] \tag{15.34}$$

which can be simplified to a quadratic as follows:

$$T_1^2 + 10T_1 - \frac{10E}{(2.3)R\log 2} = 0 \tag{15.35}$$

This is a quadratic in T_1 and for a given E, there is a unique value of T_1. For $E = 53600$ J/g mol, the equation becomes ($R = 8.314$ J/g mol · K)

$$T_1^2 + 10T_1 - 93029.6 = 0 \tag{15.35a}$$

The positive and acceptable root of the preceding equation is

$$T_1 = \frac{-10 + \sqrt{100 + 4 \times 93029.6}}{2} = 300 \text{ K} \qquad (15.35b)$$

Therefore, the rate of reaction doubles from 300 to 310 K.

At some other temperatures, the reaction rate increase may be closer to 2 times but not exactly so. For example, if we consider the range 330–340 K, calculations with $E = 53600$ kJ/mol and using Arrhenius relation yields $k_2/k_1 = 1.7764$, which is less than 2 times for 10°C increase. With $E = 147$ kJ/mol, Equation 15.35 gives a solution, $T_1 = 500.3$ K. Thus for $E = 147$ KJ/g mol, the reaction rate doubles in the temperature range of 500–510 K.

Temperature Dependency of Reaction Rate from Collision and Transition Theories

Collision theory yields the following relation:

$$k = \alpha T^{1/2} e^{-E/RT} \qquad (15.36)$$

whereas the transition theory predicts

$$k = \alpha T e^{-E/RT} \qquad (15.37)$$

or in general,

$$k = \alpha T^m e^{-E/RT} \qquad (15.38)$$

Differentiating the logarithms of both sides of Equation 15.38 gives

$$\frac{d \ln k}{dT} = \frac{m}{T} + \frac{E}{RT^2} = \frac{mRT + E}{RT^2} \qquad (15.39)$$

Since $mRT < E$ for most of the reactions, the preceding gives

$$\frac{d \ln k}{dT} = \frac{E}{RT^2} \qquad (15.40)$$

which shows Arrhenius theory is a good approximation to both the collision and transition-state theories.

Example 15.3

The rate of a bimolecular reaction at 500 K is 10 times the rate at 400 K. Calculate the activation energy of this reaction by

(a) the Arrhenius equation

(b) the collision theory

Solution

(a) Let k_1 be the reaction rate at 400 K and k_2 at 500 K.

$$k_1 = \alpha \exp\left[-\frac{E}{R(400)}\right] \quad \text{and} \quad k_2 = \alpha \exp\left[-\frac{E}{R(500)}\right]$$

$$\frac{k_2}{k_1} = 10 = \exp\left[-\frac{E}{R}\left(\frac{1}{500} - \frac{1}{400}\right)\right] = \exp\left(\frac{E}{2000R}\right)$$

$$\ln 10 = \frac{E}{2000R} \quad \text{and} \quad E = 2000(1.987) \ln 10 = 9150 \text{ cal/g mol}$$

(b) *Activation Energy by Collision Theory*

$$\frac{k_2}{k_1} = \left(\frac{500}{400}\right)^{1/2} \exp\left[-\frac{E}{R}\left(\frac{1}{500} - \frac{1}{400}\right)\right] = 10$$

from which
$$\exp\left(\frac{E}{2000R}\right) = 10\left(\frac{400}{500}\right)^{1/2} = 8.94427$$

and
$$\frac{E}{2000R} = \ln 8.94427$$

Hence $E = 2000(1.987)(\ln 8.94427) = 8707$ cal/g mol

Example 15.4

The third-order gas-phase reaction $2NO + O_2 \rightarrow 2NO_2$ has a specific reaction rate of

$$k_c = 2.65 \times 10^4 \text{ L}^2/(g \cdot \text{mol})^2 \cdot s$$

at 30°C and 1 atm. Find k_p and k_n. Show clearly the conversion of units.

Solution

$$k_c = (RT)^n k_p = \left(\frac{RT}{P_T}\right)^n k_n$$

Here $n = 3$, $P_T = 1$ atm, and $T = 273 + 30 = 303$ K.

$$k_p = \frac{k_c}{(RT)^n} = \frac{2.65 \times 10^4 \text{ L}^2/(g \cdot \text{mol})^2 \cdot s}{[(0.08206 \text{ L} \cdot \text{atm/g} \cdot \text{mol} \cdot \text{K})(303 \text{ K})]^3}$$

$$= \frac{2.65 \times 10^4}{0.08206^3} \frac{\text{L}^2}{(g \cdot \text{mol})^2 \cdot s} \left[\frac{(g \cdot \text{mol})^3}{\text{L}^3 \cdot \text{atm}^3 \cdot 303^3}\right]$$

$$= 1.7239 \text{ g} \cdot \text{mol/L} \cdot \text{atm}^3 \cdot s$$

and
$$k_n = P_T^n k_p = 1^3 \text{atm}^3 (1.7239) \text{ g} \cdot \text{mol/L} \cdot \text{atm}^3 \cdot s$$

$$= 1.7239 \text{ g} \cdot \text{mol/L} \cdot s$$

INTERPRETATION OF KINETIC DATA AND THE CONSTANTS OF THE RATE EQUATION

Kinetic data are generally correlated by trial. Batch reactors are used to determine rate law parameters. Concentrations are measured as a function of time and then one of the methods such as differential, integral or least squares of data analysis is used to determine reaction order, frequency factor α, and the rate constant. However, in most cases, the stoichiometry of the reaction suggests the form of the rate equation, which should be tried first. When the mathematical equation is written, the next step is to find the constants in the rate equation. For this the following methods are used: (1) method of differential, (2) method of integrated equation, (3) method of halftime, (4) method of reference curves, and (5) method of k calculation.

Simple Irreversible Reactions

Consider the general equation

$$a\,A + b\,B + \cdots \rightarrow c\,C + d\,D + \cdots$$

At constant volume and temperature, the rate equation in terms of x (mols converted), i.e., $x = (n_{A0} - n_A)$ will be

$$\frac{dx}{dt} = k(n_{A0} - x)^a \left(n_{B0} - \frac{bx}{a}\right)^b \cdots$$

$$= k'(n_{A0} - x)^{a+b+\cdots} = k'(n_{A0} - x)^n$$

Method of Differentiation

In this method, the plot of log (dx/dt) versus log $(n_{A0} - x)$ is made. If the plot is a straight line, the slope of the curve is n and the intercept is log k'. If the curve obtained is not a straight line, the reaction is probably complex.

When a reaction is irreversible, the reaction constants can be determined by numerically differentiating concentration versus time data. This method is applicable when the rate of reaction is mainly a function of the concentration of one reactant as in the reaction

$$A \rightarrow \text{Products and the rate law is } -r_A = kC_A^n \quad (15.41)$$

By using the method of excess reactant, it is possible to determine the relationship between $-r_A$ and the concentration for other reactants. For example if the reaction is

$$A + B \rightarrow \text{Products and the rate law is } -r_A = k_A C_A^\alpha C_B^\beta \quad (15.42)$$

In one set of experiments, the reaction is carried out in excess of B to determine α and in another in excess of A to determine β. When α and β are determined, the specific reaction rate k_A can be calculated from the measurement of $-r_A$ at known concentrations of A and B.

The differentials dx/dt can be obtained from the concentration versus time data by three different methods:

(a) graphical differentiation

(b) numerically by using differentiation formulas

(c) differentiating a polynomial, which fits the data.

For more details and examples of the differentiation method, a reference is made to Folger.

Method of Integration

Integration of the rate equation yields ($n \neq 1$):

$$\left(\frac{1}{n_{A0}-x}\right)^{n-1} - \left(\frac{1}{n_{A0}}\right)^{n-1} = (n-1)kt \tag{15.43}$$

When $n = 1$, the solution of the rate equation is

$$\ln\left(\frac{n_{A0}}{n_{A0}-x}\right) = kt \tag{15.44}$$

Therefore, for a first-order irreversible reaction, a plot of $\ln(n_{A0} - x)$ versus t will give a straight line from which the reaction rate constant k can be calculated.

The integral method involves a trial-and-error procedure to obtain the reaction order.

Zero-Order Reaction. For a zero order reaction,

$$r_A = -k \quad \text{or} \quad \frac{dC_A}{dt} = -k \tag{15.45}$$

Integration with $C_A = C_{A0}$ at $t = 0$, gives

$$C_A = C_{A0} - kt \tag{15.45a}$$

In this case a plot of C_A versus t is a straight line with slope $-k$.

First-Order Reaction: If the reaction is of first order the rate equation is

$$-\frac{dC_A}{dt} = kC_A \tag{15.46}$$

In this case, with $C_A = C_{A0}$ at $t = 0$, the integrated solution is

$$\ln\frac{C_{A0}}{C_A} = kt \tag{15.46a}$$

and a plot of $\ln\frac{C_{A0}}{C_A}$ versus t is a straight line with slope of k.

Second-Order Reaction: If the reaction is second order, the rate equation is

$$-\frac{dC_A}{dt} = kC_A^2 \tag{15.47}$$

and with initial condition $C_A = C_{A0}$, solution is

$$\frac{1}{C_A} - \frac{1}{C_{A0}} = kt \tag{15.48}$$

In this case, the plot of $1/C_A$ versus t is a straight line with slope k.

If in all cases tried a linear plot is not obtained, the assumed reaction orders do not fit the data and the reaction is not zero, first or second order.

Method of Halftimes
At 50 percent conversion, the integrated equations are

$$\frac{dx}{dt} = k(n_{A0} - x) \quad \text{First order:} \quad t_{1/2} = \frac{\ln 2}{k} \tag{15.49}$$

$$\frac{dx}{dt} = k(n_{A0} - x)^2 \quad \text{Second order:} \quad t_{1/2} = \frac{1}{kn_{A0}} \tag{15.50}$$

$$\frac{dx}{dt} = k(n_{A0} - x)^n \quad n\text{th order:} \quad t_{1/2} = \frac{2^{n-1} - 1}{k(n-1)n_{A0}^{n-1}} \tag{15.51}$$

Thus
$$\log t_{1/2} = \log \frac{2^{n-1} - 1}{k(n-1)} - (n-1)\log n_{A0} \tag{15.52}$$

Therefore, a plot of $\log t_{1/2}$ versus $\log n_{A0}$ will give a straight line whose slope is $1 - n$, which determines n. k is then calculated from

$$k = \frac{2^{n-1} - 1}{t_{1/2}(n-1)n_{A0}^{n-1}} \tag{15.53}$$

However, data from several experiments as a function of the initial quantity n_{A0} must be available.

Method of Initial Rates
In this method a series of runs is carried out at different initial concentrations, C_{A0}, and the initial rate of reaction, $-r_{A0}$, is determined for each run. The initial rate, $-r_{A0}$, can be found by differentiating the data and extrapolating to zero time. By plotting in various ways or by numerical analysis for relating $-r_{A0}$ to C_{A0}, the proper rate law can be obtained.

Method of Reference Curves
A reference plot of the percent conversion x/n_{A0} versus $t/t_{0.9}$, the fraction of the time required for 90 percent conversion, can be used to determine the order of a reaction. Each curve of this plot is determined by a unique value of n. The data for a reaction under investigation are plotted on the same scale as this plot, and the order of the reaction is found by superimposition.

Method of k Calculation

In this method a value for the order of the reaction is assumed and k values are calculated at various experimental data points. If the k values calculated are nearly constant, the assumed n value is correct.

Example 15.5

For the irreversible thermal dissociation of paraldehyde at 260°C and constant volume the following data (Table 15.3) were obtained:

Table 15.3 Total pressure data as a function of time for Example 15.5

Time, h	0	1	2	3	4	∞
P total, mmHg	100	173	218	248	266	300

Determine the order of the reaction and the rate constant.

Solution

Paraldehyde decomposes according to the equation

$$(CH_3CHO)_3 \rightarrow 3CH_3CHO$$

If the reaction is first order, the rate equation is given by

$$r_A = -\frac{dC_A}{dt} = k_1 C_A$$

Assuming the reactant and product obey the ideal-gas law, the concentration is given by

$$C_A = \frac{n_A}{V} = \frac{p_A}{RT}$$

Therefore, in terms of the partial pressure p_A,

$$-\frac{dp_A}{dt} = k_1 p_A$$

$$\frac{dp_A}{p_A} = -k_1 dt$$

$$\ln p_A = -k_1 t + C \quad \text{(a constant)}$$

At $t = 0$, $p_A = p_{A0}$, $C = \ln p_{A0}$. The solution of the differential equation is

$$\ln \frac{p_A}{p_{A0}} = -k_1 t$$

Thus, the graph of $\ln (p_A/p_{A0})$ versus t should be a straight line. Now

$$(CH_3CHO)_3 \rightarrow 3CH_3CHO$$

Table 15.4 Calculations of p_A/p_{A0}

Time, h	P_T, mmHg at 0°C	p_A	p_A/p_{A0}
0	100 = p_{A0}	100	1.0
1	173	63.5	0.635
2	218	41.0	0.410
3	248	26	0.26
4	266	17	0.17
	300	0	0.000

If n_A is the number of moles of paraldehyde at time t and n_{A0} the moles of paraldehyde at $t = 0$, then moles of acetaldehyde = $3(n_{A0} - n_A)$ at time t. Thus total moles at time $t = n_T = n_A + 3(n_{A0} - n_A)$. Assuming the gas law applies, we can get (because $n = pV/RT$, and V and T are constant)

$$P_T = p_A + 3(p_{A0} - p_A)$$

where
P_T = total pressure
p_A = partial pressure of A (paraldehyde) at time t
p_{A0} = partial pressure of A at time $t = 0$
$p_A = \frac{1}{2}(3p_{A0} - P_T)$

Using the above relation for P_T, Table 15.4 is prepared.

A plot of p_A/p_{A0} versus t on semilog paper (Exhibit 2) gives a straight line. Therefore the assumed first order for the reaction is correct. Slope of the straight line is $-k_1/2.3$.

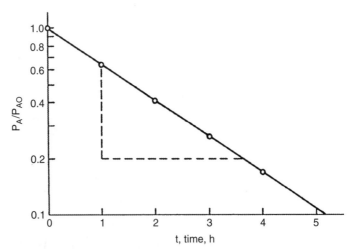

Exhibit 2 Plot of p_A/p_{A0} versus t (Example 15.5)

Thus
$$\frac{k_1}{2.3} = -(\text{slope})$$

and
$$k_1 = -\frac{2.3(0+1)}{0-5.2} = 0.442 \, \text{h}^{-1}$$

Example 15.6

The reaction $2NOCl \rightarrow 2NO + Cl_2$ is studied at 200°C. The concentration of NOCl initially consisting of NOCl only changes according to data in Table 15.5:

Table 15.5 Reaction rate data for Example 15.6

t, s	0	200	300	500
N_{NOCl}, g mol/L	0.02	0.0159	0.0144	0.0121

Find the order of the reaction and the rate constant.

Solution

If the reaction is second order, the rate equation is

$$-\frac{dC_A}{dt} = k_1 C_A^2$$

which on integration and with boundary condition $C_A = C_{A0}$ at $t = 0$ yields the solution.

$$\frac{1}{C_A} - \frac{1}{C_{A0}} = k_1 t$$

Thus a plot of $1/C_A$ versus t will be a straight line of slope k_1. Calculated values of $1/C_A$ as a function of t are given in Table 15.6.

Table 15.6 Calculations of $1/C_A$ as a function of time

t	0	200	300	500
$1/C_A$	50	62.3	69.44	82.64

$1/C_A$ versus t is plotted in Exhibit 3. The graph is a straight line. Therefore, the assumed order of the reaction is correct. The slope of the line is

$$k_1 = \frac{86-50}{550-0} = \frac{36}{550} = 0.0655 \text{ L/g mol} \cdot \text{s}$$

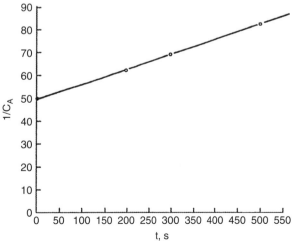

Exhibit 3 Plot of $1/C_A$ versus t (Example 15.6)

Irreversible Reactions

Unimolecular First-Order Reactions

These are of the type

$$A \rightarrow products$$

Rate equation
$$-r_A = -\frac{dC_A}{dt} = kC_A$$

Boundary condition
$$C_A = C_{A0} \quad \text{at} \quad t = 0$$

Solution
$$\ln \frac{C_A}{C_{A0}} = -kt \quad \text{or} \quad \frac{C_A}{C_{A0}} = e^{-kt} \tag{15.54}$$

In terms of fractional conversion X_A, the rate equation is

$$\frac{dX_A}{dt} = k(1 - X_A) \tag{15.55}$$

With boundary condition $X_A = 0$ at $t = 0$, the solution of Equation 15.55 is

$$-\ln(1 - X_A) = kt \tag{15.56}$$

Only the simple first-order equations can be treated this way. Complex first-order reactions cannot be treated by the above method.

Bimolecular Second-Order Reactions

These are of the type

$$A + B \rightarrow products \tag{15.57}$$

Rate equation
$$-r_A = -\frac{dC_A}{dt} = -\frac{dC_B}{dt} = kC_A C_B \tag{15.58}$$

The amounts of A and B that have reacted are equal at any time t. Let these be

$$C_{A0} X_A = C_{B0} X_B \tag{15.58a}$$

The rate equations can be written in terms of X_A as

$$-r_A = C_{A0} \frac{dX_A}{dt} = k(C_{A0} - C_{A0} X_A)(C_{B0} - C_{A0} X_A) \tag{15.58b}$$

Let $M = C_{B0}/C_{A0}$; then

$$-r_A = C_{A0} \frac{dX_A}{dt} = kC_{A0}^2 (1 - X_A)(M - X_A) \tag{15.58c}$$

from which

$$\int_0^{X_A} \frac{dX_A}{(1 - X_A)(M - X_A)} = C_{A0} k \, dt \tag{15.58d}$$

The solution if $M \neq 1$ is

$$\ln \frac{M - X_A}{M(1 - X_A)} = C_{A0}(M - 1)kt \qquad M \neq 1 \qquad (15.58e)$$

or
$$\ln \frac{C_A}{MC_A} = (C_{B0} - C_{A0})kt \qquad (15.59)$$

The plots of $\ln(C_B/C_A)$ versus t or $\ln(M - X_A)/M(1 - X_A)$ versus t are straight lines.

Special Case When M = 1
If $M = 1$, i.e., if at $t = 0$ the reactants are equal in molar concentration, the preceding equations are indeterminate. In this case, the rate equations become

$$-r = \frac{dC_A}{dt} = -\frac{dC_B}{dt} = kC_A^2 = kC \qquad (15.60)$$

The boundary conditions are, at $t = 0$, $C_A = C_{A0} = C_{B0}$. The solution is

$$\frac{1}{C_A} - \frac{1}{C_{A0}} = kt \qquad (15.60a)$$

In terms of X_A- fractional conversion,

$$-r_A = C_{A0}\frac{dX_A}{dt} = C_{B0}\frac{dX_B}{dt} = kC_A^0(1 - X_A)^2$$

Therefore
$$\frac{dX_A}{dt} = kC_{A0}(1 - X_A)^2$$

With the boundary condition, $X_A = 0$ at $t = 0$, the solution is

$$\frac{1}{C_{A0}}\frac{X_A}{1-X_A} = kt \qquad (15.60b)$$

The preceding rate equations and solutions will also apply to the reactions of the type

$$2A \rightarrow \text{products}$$

Empirical Rate Equations of the nth Order for Irreversible Reactions

Rate equation: $\qquad -r_A = -\dfrac{dC_A}{dt} = kC_A^n \qquad (15.61)$

Boundary conditions: $\qquad C_A = C_{A0} \quad \text{at} \quad t = 0$

Solution: $\qquad C_A^{1-n} - C_{A0}^{1-n} = (n-1)kt \quad n \neq 1 \qquad (15.61a)$

The value of n must be found by trial and error. In terms of X_A,

Rate equation: $-r_A = C_{A0} \dfrac{dX_A}{dt} = k C_{A0}^n (1-X_A)^n$ (15.61b)

Boundary condition: $X_A = 0 \quad t = 0$

Solution: $(1-X_A)^{1-n} - 1 = (n-1) C_{A0}^{n-1} k t$ (15.61c)

Zero-Order Reaction

Rate equation: $-r_A = -\dfrac{dC_A}{dt} = k$ (15.62)

Boundary condition: $C_A = C_{A0}$

Solution: $C_{A0} - C_A = kt$ (15.62a)

or $C_{A0} X_A = kt$ (15.62b)

In this case, the conversion is proportional to time.

Reactions in Parallel

Consider elementary reactions of the following type:

$$A \xrightarrow{k_1} B \qquad A \xrightarrow{k_2} C$$

The rate equations for these reactions are

$$-r_A = -\dfrac{dC_A}{dt} = k_1 C_A + k_2 C_A = (k_1 + k_2) C_A \qquad (15.63)$$

$$r_B = \dfrac{dC_B}{dt} = k_1 C_A \qquad (15.63a)$$

$$r_C = \dfrac{dC_C}{dt} = k_2 C_A \qquad (15.63b)$$

Boundary condition: $C_A = C_{A0}$ at $t = 0$

$C_B = C_{B0} \quad C_C = C_{C0}$ at $t = 0$

Solution: $-\ln \dfrac{C_A}{C_{A0}} = (k_1 + k_2) t$ (15.63c)

and in terms of X_A,

$$-\ln(1 - X_A) = (k_1 + k_2) t \qquad (15.63d)$$

If $-\ln(C_A/C_{A0})$ or $-\ln(1-X_A)$ is plotted versus t, the slope of the resulting straight line is $k_1 + k_2$. Also

$$\frac{r_B}{r_C} = \frac{dC_B}{dC_C} = \frac{k_1}{k_2}$$

from which
$$\frac{C_B - C_{B0}}{C_C - C_{C0}} = \frac{k_1}{k_2} \quad (15.63e)$$

or
$$C_B - C_{B0} = \frac{k_1}{k_2}C_C - \frac{k_1}{k_2}C_{C0} \quad (15.63f)$$

Thus the slope of the straight-line plot of C_B versus C_C gives the ratio k_1/k_2. Knowing $k_1 + k_2$ and k_1/k_2 gives the individual values of k_1 and k_2.

Homogeneous Catalyzed Reactions

A catalyzed reaction can be represented as follows:

$$A \xrightarrow{k_1} P$$

$$A + C \xrightarrow{k_2} P + C$$

where C and P represent the catalyst and product, respectively. The reaction rates are

$$-\left(\frac{dC_A}{dt}\right)_1 = k_1 C_A \quad (15.64)$$

$$-\left(\frac{dC_A}{dt}\right)_2 = k_2 C_A C_C \quad (15.64a)$$

The overall rate of disappearance of A then is

$$-\frac{dC_A}{dt} = k_1 C_A + k_2 C_A C_C \quad (15.64b)$$

The catalyst concentration remains unchanged.

Boundary condition: $\quad C_A = C_{A0} \quad \text{at} \quad t = 0 \quad (15.64c)$

Solution: $\quad -\ln\dfrac{C_A}{C_{A0}} = -\ln(1 - X_A) = (k_1 + k_2 C_C)t = kt \quad (15.64d)$

where $k = k_1 + k_2 C_C$. In this case, a series of runs is made with various concentrations of the catalyst and a plot is made of $k = (k_1 + k_2 C_C)$ versus C_C. The slope of this straight line is k_2, and the intercept is k_1.

The Autocatalytic Reaction

When one of the products of a reaction acts as a catalyst, it is called an autocatalytic reaction. The simplest reaction is given by

$$A + B \xrightarrow{k} B + B$$

for which the rate equation is

$$-r_A = -\frac{dC_A}{dt} = kC_A C_B \qquad (15.65)$$

Because, when A is consumed, the total moles of A and B remain unchanged, at any time t, the following relation holds:

$$C_0 = C_A + C_B = C_{A0} + C_{B0} = \text{const}$$

Then the rate equation becomes

$$-r_A = -\frac{dC_A}{dt} = kC_A(C_0 - C_A) \qquad (15.65a)$$

Integration with the use of partial fractions yields the solution

$$\ln \frac{C_{A0}(C_0 - C_A)}{C_A(C_0 - C_{A0})} = \ln \frac{C_B/C_{B0}}{C_A/C_{A0}} = C_0 kt \qquad (15.65b)$$

In terms of the initial reaction ratio $M = C_{B0}/C_{A0}$ and the fractional conversion X_A of A, the solution is

$$\ln \frac{M + X_A}{M(1 - X_A)} = C_{A0}(M + 1)kt = (C_{A0} + C_{B0})kt \qquad (15.65c)$$

Reactions in Series

A typical example of reactions in series is

$$A \xrightarrow{k_1} B \xrightarrow{k_2} C$$

The rate equations are

$$-r_A = -\frac{dC_A}{dt} = k_1 C_A \qquad (15.66)$$

$$r_B = \frac{dC_B}{dt} = k_1 C_A - k_2 C_B \qquad (15.66a)$$

$$r_C = \frac{dC_C}{dt} = k_2 C_B \qquad (15.66b)$$

and the initial conditions are

$$C_A = C_{A0} \quad C_B = 0 \quad C_C = 0 \quad \text{at} \quad t = 0$$

Concentration of A by integration is

$$-\ln \frac{C_A}{C_{A0}} = k_1 t \quad \text{or} \quad C_A = C_{A0} e^{-k_1 t} \tag{15.66c}$$

Using the relation of Equation 15.66c in Equation 15.66a, one obtains

$$\frac{dC_B}{dt} + k_2 C_B = k_1 C_{A0} e^{-k_1 t} \tag{15.66d}$$

which can be integrated by the method of *integrating factor* and using initial condition $C_{B0} = 0$ at $t = 0$, find the constant of integration to yield

$$C_B = C_{A0} k_1 \left(\frac{e^{-k_1 t}}{k_2 - k_1} + \frac{e^{-k_2 t}}{k_1 - k_2} \right) \tag{15.66e}$$

Because the total number of moles does not change, $C_{A0} = C_A + C_B + C_C$. Using this, one obtains

$$C_C = C_{A0} \left(1 + \frac{k_2}{k_1 - k_2} e^{-k_1 t} + \frac{k_1}{k_2 - k_1} e^{-k_2 t} \right) \tag{15.66f}$$

The maximum concentration of B occurs at

$$t_{\max} = \ln \frac{k_2 / k_1}{k_2 - k_1} \tag{15.66g}$$

and the maximum concentration of B is given by

$$\frac{C_{B\max}}{C_{A0}} = \left(\frac{k_1}{k_2} \right)^{k_2 / (k_2 - k_1)} \tag{15.66h}$$

First-Order Reversible Reactions

Reaction:
$$A \underset{k_2}{\overset{k_1}{\rightleftarrows}} B \tag{15.67}$$

Rate equations:
$$\frac{dC_B}{dt} = -\frac{dC_A}{dt} = k_1 C_A - k_2 C_B \tag{15.67a}$$

or in terms of X_A (if $M = C_{B0}/C_{A0}$),

$$C_{A0} \left(\frac{dX_A}{dt} \right) = k_1 (C_{A0} - C_{A0} X_A) - k_2 (C_{A0} M + C_{A0} X_A) \tag{15.67b}$$

The boundary conditions are: at $t = 0$, $M = C_{B0}/C_{A0}$, $X_A = 0$. At equilibrium,

$$\frac{dC_A}{dt} = 0 \quad \text{or} \quad \frac{dX_A}{dt} = 0 \tag{15.67c}$$

Therefore, at equilibrium condition, K_e is given by

$$K_e = \frac{C_{Be}}{C_{Ae}} = \frac{M + X_{Ae}}{1 - X_{Ae}} \tag{15.67d}$$

where equilibrium constant

$$K_e = \frac{k_1}{k_2} \tag{15.67e}$$

If the preceding equations are combined, the rate equation in terms of the equilibrium conversion is

$$\frac{dX_A}{dt} = \frac{k_1(M+1)}{M + X_{Ae}}(X_{Ae} - X_A) \tag{15.67f}$$

If the concentrations are measured in terms of X_{Ae}, the equilibrium conversion, integration gives

$$-\ln\left(1 - \frac{X_A}{X_{Ae}}\right) = -\ln\frac{C_A - C_{Ae}}{C_{A0} - C_{Ae}} = \frac{M+1}{M + X_{Ae}}k_1 t \tag{15.68}$$

and a plot of $-\ln(1 - X_A/X_{Ae})$ versus t will be a straight line.

Second-Order Reversible Reactions

Bimolecular-type second-order reactions are as follows:

$$A + B \underset{k_2}{\overset{k_1}{\rightleftharpoons}} R + S \tag{15.69}$$

$$2A \underset{k_2}{\overset{k_1}{\rightleftharpoons}} R + S \tag{15.69a}$$

$$2A \underset{k_2}{\overset{k_1}{\rightleftharpoons}} 2R \tag{15.69b}$$

$$A + B \underset{k_2}{\overset{k_1}{\rightleftharpoons}} 2R \tag{15.69c}$$

With the restrictions that $C_{A0} = C_{B0}$, $C_{R0} = C_{S0} = 0$ in Equation 15.69, the rate equation becomes

$$-r = -\frac{dC_A}{dt} = C_{A0}\frac{dX_A}{dt} = k_1 C_A^2 - k_2 C_R^2 \tag{15.69d}$$

$$= k_1\left[C_{A0}^2(1 - X_A)^2 - \frac{k_2}{k_1}X_A^2\right] \tag{15.69e}$$

where the fractional conversion $X_A = 0$ at $t = 0$ and $dX_A/dt = 0$ at equilibrium, and therefore the final solution is

$$\ln\frac{X_{Ae} - 2X_{Ae} - 1)X_A}{X_{Ae} - X_A} = 2k_1 C_{A0}\left(\frac{1}{X_{Ae}} - 1\right)t \tag{15.70}$$

All other reversible second-order rate equations (15.69) have the same solution with the boundary conditions assumed in the above solution.

Reversible Reactions in General

For orders other than 1 or 2, the integration of the rate equation is difficult. Therefore, the differential method of analysis should be used to search the form of the rate equation. Sometimes a complex equation of the type below fits the data well

$$-r_A = -\frac{dC_A}{dt} = k_1 \frac{C_A}{1+k_2 C_A} \quad (15.71)$$

Taking reciprocals, one obtains

$$\frac{1}{-r_A} = \frac{1+k_2 C_A}{k_1 C_A} = \frac{1}{k_1}\frac{1}{C_A} + \frac{k_2}{k_1} \quad (15.71a)$$

and a plot of $1/-r_A$ versus $1/C_A$ would be a straight line with a slope equal to $1/k_1$ and an intercept of k_2/k_1.

Another method of analysis can be obtained in the above case by multiplying each side of Equation 15.71a by k_1/k_2 and solving for $-r_A$ to yield another form as

$$-r_A = \frac{k_1}{k_2} - \frac{1}{k_2}\frac{-r_A}{C_A} \quad (15.72)$$

on the basis of which, $-r_A$ versus $-r_A/C_A$ will be a linear plot, and values of k_1 and k_2 can be determined from the slope $-1/k_2$ and intercept k_1/k_2.

Example 15.7

The hydrolysis of methyl acetate is an autocatalytic reaction and is first order with respect to methyl acetate and first order with respect to acetic acid. The reaction is elementary, bimolecular, and can be considered irreversible at constant volume for design purposes. The following data are given:

Initial concentration of methyl acetate = 0.5 g mol/L

Initial concentration of acetic acid = 0.05 g mol/L.

The conversion in 1 h is 60 percent in a batch reactor. Calculate

(a) the rate constant and indicate the rate equation,

(b) the time at which the rate passes through a maximum, and

(c) based on the above information, the type of optimum reactor system you would specify for the plant to process 200 ft³/h. What would be the reactor volume in this system?

Solution

$$CH_3COOCH_3 + H_2O \rightarrow CH_3COOH + CH_3OH$$
$$A \rightarrow B$$

(a) For autocatalytic reaction, $A + B \rightarrow B + B$ and

$$-r_A = -\frac{dC_A}{dt} = k_1 C_A C_B$$

Using the solution for the autocatalytic reaction, the value of the initial reactant ratio M can be calculated.

$$\ln\frac{M+X_A}{M(1-X_A)} = (C_{A0}+C_{B0})k_1 t$$

$$M = \frac{C_{B0}}{C_{A0}} = \frac{0.05}{0.5} = 0.1$$

At $t = 1$ h, $X_A = 0.6$, $\ln\frac{0.1+0.6}{0.1(1-0.6)} = (0.5+0.05)k_1(1)$

$$k_1 = \frac{\ln(0.7/0.04)}{0.05} = 5.204 \text{ L/g mol} \cdot \text{h}$$

The rate equation is

$$-r_A = -\frac{dC_A}{dt} = 5.2 C_A C_B \quad \text{L/g mol} \cdot \text{h}$$

(b) Rate is maximum when $C_A = C_B$. But

$$C_A + C_B = C_{A0} + C_{B0} = 0.5 + 0.05 = 0.55$$

Therefore $\quad C_A = C_B = \frac{1}{2}(0.55) = 0.275$

and then $\quad X_A = \frac{C_{A0}-C_A}{C_{A0}} = \frac{0.5-0.275}{0.5} = 0.45$

Find t by substituting $X_A = 0.45$ and $k_1 = 5.204$ found in part (a) in the integrated equation. Thus

$$\ln\frac{0.1+0.45}{0.1(1-0.45)} = 0.55(5.204)t$$

from which, $t = 0.8045$ h.

(c) For solution of this part, refer to Example 15.12.

Example 15.8

The gas-phase decomposition $A \to B + 2C$ is conducted in a constant-volume reactor. Runs 1 to 5 were conducted at 100°C; run 6 was carried out at 110°C.

(a) From the data given in Table 15.7 below, determine the reaction order and specify the reaction rate.

(b) What are the activation energy and frequency factor for this reaction?

Table 15.7 Half-life $t_{1/2}$ as function of initial concentration C_{A0}

No.	C_{A0}, g mol/L	Half-Life $t_{1/2}$, min
1	0.025	4.1
2	0.0133	7.7
3	0.0100	9.8
4	0.050	1.96
5	0.075	1.30
6	0.025	2.0

Solution

(a)
$$t_{1/2} = \frac{2^{n-1}-1}{k(n-1)} \frac{1}{C_{A0}^{n-1}}$$

$$\ln t_{1/2} = \ln \frac{2^{n-1}-1}{k(n-1)} + (1-n) \ln C_{A0}$$

The plot of C_{A0} versus $t_{1/2}$ on log-log paper is shown in Exhibit 4. From the graph the slope of the line $= -1.0$. Therefore, the order of the reaction is given by $1 - n = -1.0$. Thus the order of the reaction is 2, and specific reaction rate, when $t_{1/2} = 5$ min and $C_{A0} = 0.02$ g·mol/L at 100°C, is

$$k_{100} = \frac{2^{n-1}-1}{n-1} \frac{1}{C_{A0}^{n-1} t_{1/2}}$$

$$= \frac{2-1}{1} \frac{1}{0.02(5)} = \frac{1}{0.10} = 10 \text{ L/g mol} \cdot \text{min}$$

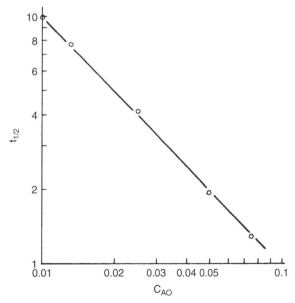

Exhibit 4 Plot of $t_{1/2}$ versus C_{A0} (Example 15.8)

At 110°C, $C_{A0} = 0.025$ g mol/L, $t_{1/2} = 2$ min, and

$$k_{110} = \frac{2-1}{1}\frac{1}{0.025(2)} = 20 \text{ L/g mol} \cdot \text{min}$$

(b) *Activation Energy*

$$k_{100} = \alpha e^{-E/R(373)} \quad \text{and} \quad k_{110} = \alpha e^{-E/R(383)}$$

$$\ln \frac{k_{110}}{k_{100}} = -\frac{E}{R(383)} + \frac{E}{R(373)}$$

$$\ln \frac{20}{10} = \ln 2 = \frac{E}{R}\left(\frac{1}{373} - \frac{1}{383}\right)$$

or

$$E = \frac{\ln 2(1.987)}{\frac{1}{373} - \frac{1}{383}} = 19676 \text{ cal/g mol}$$

Frequency Factor
At 100°C,

$$\alpha = \frac{10}{\exp[-19676/1.987(373)]} = 3.39 \times 10^{12} \text{ L/g mol} \cdot \text{min}$$

At 110°C,

$$\alpha = \frac{20}{\exp[-19676/1.987(383)]} = 3.39 \times 10^{12} \text{ L/g mol} \cdot \text{min}$$

REACTOR DESIGN

Reaction equipment in which homogeneous reactions are carried out are of three types: (1) batch reactors, (2) steady-state flow reactors, and (3) unsteady-state flow or semibatch reactors.

Batch Reactors

In a batch reactor, neither the reactants nor the products flow into or leave the system when the reaction is carried out. They are either the constant-volume or constant-pressure (variable volume) reactors. The operation of a batch reactor is unsteady state. Although the composition throughout the reactor is ideally uniform at a given instant, it changes with time.

Steady-State Flow Reactor

Steady-state flow reactors are of two types: (1) ideal stirred tank or CSTR, also known as mixed reactor or mixed flow reactor; (2) plug-flow reactor, also known as piston flow, ideal tubular, or unmixed flow reactor.

Ideal Stirred-Tank Reactor

In this reactor, the contents are well mixed and uniform in concentration throughout. The composition of the exit stream from this reactor is the same as the composition of the fluid in the reactor.

Plug-Flow Reactor

The flow of fluid in a plug-flow reactor is orderly with no backward or forward mixing or diffusion of the fluid elements in the direction of the flow path. There may be lateral mixing of the fluid in a plug-flow reactor. The plug flow is characterized by the fact that the residence time in the reactor is the same for all the elements of the fluid.

In the following treatment, the reactor volume V is the volume of the reaction space or the volume of the fluid.

Packed Bed Reactors (PBR)

A packed bed reactor, also called a fixed bed reactor, is mostly a tubular reactor that is packed with solid catalyst particles. It is a heterogeneous reaction system that is used to carry out catalyzed gas phase reactions. Temperature control in packed beds is not easy. Replacement of catalyst is difficult. Sometimes channeling of gas flow occurs that results in nonutilization or ineffective use of parts of the packed bed. The advantage of the packed bed reactor is that it gives the highest conversion per unit weight of catalyst compared to other catalytic reactors.

Fluidized Bed Reactors (FBR)

This is also a catalytic reactor and is similar to CSTR in that its heterogeneous contents are well mixed. As a result there is even temperature distribution throughout the moving catalyst column. It can handle large amounts of feed and catalyst solids and has a good temperature control. Catalyst replacement and regeneration are easier. This reactor and the catalyst regeneration equipment unit have high costs.

MASS AND ENERGY BALANCES

Ideal Batch Reactor

Assuming the reaction A $\rightarrow$ products, the mole balance on species A in a batch reactor of volume V, where the composition is uniform throughout, results in

$$\frac{dn_A}{dt} = r_A V \tag{15.73}$$

The equation is true for both constant and variable volume. Since the reactant A is disappearing, we write the equation in the form

$$-\frac{dn_A}{dt} = -r_A V \tag{15.74}$$

Now in terms of the conversion X_A, the reaction rate is

$$-\frac{dn_A}{dt} = -\frac{dn_{A0}(1-X_A)}{dt} = n_{A0}\frac{dX_A}{dt} \tag{15.75}$$

$$n_{A0}\frac{dX_A}{dt} = -r_A V \tag{15.75a}$$

The solution of this equation with the initial condition $t = 0$, $X_A = 0$, is given by

$$t = n_{A0}\int_0^{X_A} \frac{dX_A}{-r_A V} \tag{15.76}$$

which gives a relation showing the time required to obtain a conversion X_A. If the density of the fluid remains constant, one obtains

$$t = C_{A0}\int_0^{X_A} \frac{dX_A}{-r_A} = -\int_{C_{A0}}^{C_A} \frac{dC_A}{-r_A} \tag{15.77}$$

If the volume of the reaction mixture changes proportionately with the conversion, the equation becomes

$$t = n_{A0}\int_0^{X_A} \frac{dX_A}{-r_A V_0(1+\varepsilon_A X_A)} = C_{A0}\int_0^{X_A} \frac{dX_A}{-r_A(1+\varepsilon_A X_A)} \tag{15.78}$$

where ε_A is the fractional change in volume of the system between complete conversion and no conversion of reactant A.

The reaction time t is the measure of the processing rate in a batch reactor.

Flow Reactors

The performance of flow reactors is evaluated in terms of space time and space velocity. These terms are defined as follows:

Space Time. This is the reciprocal of space velocity and is given by

$$\tau = \frac{1}{S} = \text{time required to process one reactor volume of feed measured to specified conditions}$$
$$= \text{Time}$$

Space Velocity. The space velocity is related to space time as follows:

$$S = \frac{1}{\tau} = \text{number of reactor volumes of feed at specified conditions which can be treated in unit time}$$
$$= (\text{time})^{-1} \tag{15.79}$$

Note that the values of the space time and space velocity will depend on the conditions of the stream, viz., temperature, pressure, and state. If the conditions are those of the feed stream, the following relations can be established:

$$\tau = \frac{1}{S} = \frac{C_{A0}V}{F_{A0}} \tag{15.80}$$

$$\tau = \frac{(\text{moles A feed/volume of feed})(\text{volume of reactor})}{\text{moles A feed/time}}$$

$$= \frac{V}{V_0} = \frac{\text{reactor volume}}{\text{volumetric feed rate of A}} \tag{15.81}$$

If a standard condition is chosen to express the volumetric feed rate, the following relation holds:

$$\tau' = \frac{1}{S} = \frac{C'_{A0}V}{F_{A0}} = \frac{C'_{A0}}{C_{A0}} \tag{15.82}$$

where the prime denotes the values at standard conditions chosen.

Steady-State Mixed-Flow Reactor or Continuous-Flow Stirred-Tank Reactor

The design equations for a stirred-tank reactor (Figure 15.1) are established. Mole balance on component A, with the reactor as the system, gives

$$n_E - n_B = \bar{n}_B = \bar{n}_I \Delta t - \bar{n}_O \Delta t + n_P \Delta t \tag{15.83}$$

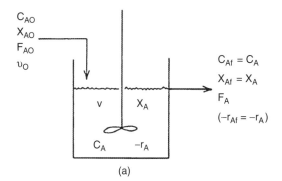

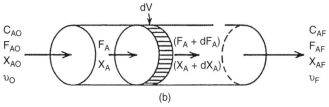

Figure 15.1 (a) Steady-state mixed-flow reactor (b) Steady-state plug-flow reactor

where n_B and n_E are moles in the beginning and end of the accounting period Δt. $n_E - n_B = 0$ from steady-state operation. Therefore

$$\bar{n}_I - \bar{n}_O = -\bar{n}_P$$

$\bar{n}_I$ = input of A moles/time = $F_{A0}(1 - X_{A0}) = F_{A0}$ if $X_{A0} = 0$

$\bar{n}_O$ = output of A moles/time = $F_A = F_{A0}(1 - X_A)$

and
$$-n_P = -r_A V$$

which is the disappearance of A in moles per unit time

$$= \frac{\text{moles A reacting}}{\text{time(volume of fluid)}} (\text{volume of reactor})$$

With no conversion in the inlet stream, one obtains

$$F_{A0} - F_{A0}(1 - X_A) = -r_A V \tag{15.83a}$$

or
$$F_{A0} X_A = -r_A V \tag{15.83b}$$

Then volume of back-mix reactor is

$$V = \frac{F_{A0} X_A}{-r_A} \tag{15.83c}$$

Also
$$\frac{V}{F_{A0}} = \frac{X_A}{-r_A} = \frac{\tau}{C_{A0}}$$

or
$$\tau = \frac{1}{S} = \frac{V}{v_0} = \frac{V C_{A0}}{F_{A0}} = \frac{C_{A0} X_A}{-r_A} \tag{15.83d}$$

where X_A and r_A are evaluated at the exit stream conditions. If $X_{A0} \neq 0$, i.e., the feed is partially converted, then

$$\frac{V}{F_{A0}} = \frac{X_{Af} - X_{Ai}}{(-r_A)_f} \tag{15.83e}$$

where f and i denote the exit and inlet conditions, respectively,

or
$$\tau = \frac{V C_{A0}}{F_{A0}} = \frac{C_{A0}(X_{Af} - X_{Ai})}{(-r_A)_f} \tag{15.83f}$$

For the special case, when the density is constant,

$$\frac{V}{F_{A0}} = \frac{X_A}{-r_A} = \frac{C_{A0} - C_A}{C_{A0}(-r_A)} \tag{15.84}$$

or
$$\tau = \frac{V}{v} = \frac{C_{A0} - C_A}{-r_A} \tag{15.84a}$$

By combining rate law and mole balance $\tau = \dfrac{C_{A0} - C_A}{kC_A}$ (15.84b)

from which the effluent concentration of A, $C_A = \dfrac{C_{A0}}{1 + \tau k}$ (15.84c)

Since there is no volume change (liquid phase reaction),

$$C_A = C_{A0}(1 - X) \tag{15.84d}$$

By combining Equations 15.84c and d, $X = X = \dfrac{\tau k}{1 + \tau k}$ (15.84e)

For a first-order reaction, the product $k\tau$ is called the reaction Damkohler number. It is a dimensionless number, which permits quick estimate of the degree of conversion that can be obtained in continuous flow reactors. For the first- and second-order irreversible reactions the Damkohler numbers are

$$Da = \dfrac{-r_{A0}V}{F_{A0}} = \dfrac{kC_{A0}V}{vC_{A0}} = k\tau \quad \text{and} \quad Da = \dfrac{kC_{A0}^2 V}{vC_{A0}} = C_{A0}k\tau \text{ respectively} \tag{15.84f}$$

A value of $Da \leq 0.1$, will usually give less than 10 percent conversion while a value of Da 10 or greater will usually give greater than 90 percent conversion.

Steady-State Plug-Flow Reactor

Referring to Figure 15.1b, the mass balance on a difference element gives

$$M_E - M_B = \bar{M}_I \Delta t - \bar{M}_O \Delta t + \sum M_P \Delta t \tag{15.85}$$

$$M_E - M_B = 0 \quad \text{because steady state}$$

$$-(M_I - M_O) = +\sum M_P$$

$$M_I = F_A \quad M_O = F_A + dF_A \quad -\sum M_P = -r_A dV$$

and $\quad dF_A = +r_A dV$ (15.85a)

also $\quad dF_A = d[F_{A0}(1 - X_A)] = -F_{A0} dX_A$ (15.85b)

From Equations 15.85a and b,

$$F_{A0} dX_A = -r_A dV \tag{15.85c}$$

from which, after separation of variables and integration,

$$V = F_{A0} \int_0^{X_A} \dfrac{dX_A}{-r_A}, \tag{15.86}$$

which gives the plug-flow reactor volume for conversion X and then

$$\tau = \frac{V}{v_0} = C_{A0} \int_0^{X_A} \frac{dX_A}{-r_A} \tag{15.87}$$

If the feed on which the conversion is based is partially converted at the entrance to the reactor,

$$\frac{V}{F_{A0}} = \frac{V}{C_{A0}v_0} = \int_{X_{Ai}}^{X_{Af}} \frac{dX_A}{-r_A} \tag{15.88}$$

or

$$\tau = \frac{V}{v_0} = C_{A0} \int_{X_{Ai}}^{X_{Af}} \frac{dX_A}{-r_A} \tag{15.89}$$

For the special case when density is constant,

$$\frac{V}{F_{A0}} = \frac{\tau}{C_{A0}} = \int_0^{X_{Af}} \frac{dX_A}{-r_A} = -\frac{1}{C_{A0}} \int_{X_{A0}}^{C_{Af}} \frac{dC_A}{-r_A} \tag{15.90}$$

or

$$\tau = \frac{V}{v_0} = C_{A0} \int_0^{X_{Af}} \frac{dX_A}{-r_A} = -\int_{C_{A0}}^{C_{Af}} \frac{dC_A}{-r_A} \tag{15.91}$$

Example 15.9

A reaction A $\to$ R is to be carried out in a batch reactor. The rates of the reaction as a function of C_A are given in Table 15.8.

How long must a batch be reacted to reach a concentration of 0.3 mol/L from initial concentration of $C_{A0} = 1.3$ mol/L? The reaction is liquid phase.

Solution

Since the reaction is liquid phase, the density can be considered constant, and then t is given by

$$t = -\int_{C_{A0}}^{C_A} \frac{dC_A}{-r_A}$$

Table 15.8 Rates of reaction as a function of C_A

C_A, mol/L	$-r_A$, mol/L · min	$\frac{1}{-r_A}$
0.1	0.100	10.000
0.2	0.300	3.34
0.3	0.500	2.00
0.4	0.600	1.67
0.5	0.500	2.00
0.6	0.250	4.00
0.7	0.100	10.00
0.8	0.060	16.67
1.0	0.050	20.00
1.3	0.045	22.20
2.0	0.042	23.81

We plot $1/-r_A$ versus C_A and obtain the value of the integral Exhibit 5 ($-dC_A/-r_A$) between $C_A = 0.3$ and $C_{A0} = 1.3$ mol/L. The area under the

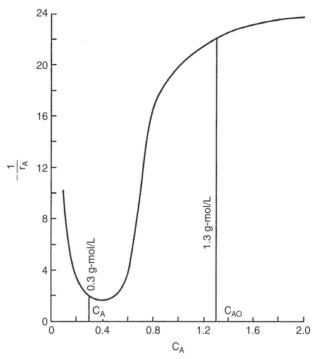

Exhibit 5 Plot of $1/-r_A$ versus C_A (Example 15.9)

curve is found by Simpson's rule as

$$\text{Area} = -\frac{0.1}{3}[2 + 4(1.67) + 2(2) + 4(4) + 2(10) + 4(16.7) + 2(18.5)$$
$$+ 4(20) + 2(20.9) + 4(21.7) + 22.22]$$
$$= -12.8$$
$$t = -(-12.8) = 12.8 \text{ min}$$

Example 15.10

Hydrolysis of CH_3COOCH_3 is an autocatalytic reaction and has a specific reaction rate constant $k = 5.2$ L/g mol · h. The initial concentration of CH_3COOCH_3 is 0.5 g mol/L, and the initial concentration of CH_3COOH is 0.05 g mol/L. The reaction is first order with respect to both methyl acetate and acetic acid. On the basis of the above information, what type of optimum reactor system would you specify for the plant to process 200 L/h of feed? What would be the reactor volume in this system?

Solution

Since the reaction is liquid phase, the density changes are negligible.

$$CH_3COOCH_3 + H_2O \rightarrow CH_3COOH + CH_3OH$$
$$A \rightarrow B$$

Table 15.9 Calculation of rates and conversions (example 15.10)

C_A	C_B	$-r_A = 5.2C_AC_B$	$\dfrac{1}{-r_A}$	$X_A = \dfrac{C_{A0}-C_A}{C_{A0}}$
$0.5(C_{A0})$	0.05	0.13	7.69	0
0.4	0.15	0.312	3.20	0.2
0.3	0.25	0.39	2.56	0.4
0.2	0.35	0.364	2.75	0.6
0.1	0.45	0.234	4.27	0.8
0.05	0.50	0.130	7.69	0.9

For a continuous back-mix reactor, the reactor volume is given by

$$V = \frac{F_{A0}X_A}{-r_A} = \frac{C_{A0}-C_A}{C_{A0}(-r_A)}F_{A0} = \frac{v_0 C_{A0}X_A}{-r_A}$$

$$F_{A0} = 200(0.5) = 100 \text{ g mol/h}$$

Assume that the conversion required is 90 percent.

For a back-mix reactor, the reactor volume is the total area bounded by the rectangle covering the conversions X_A and 0 multiplied by $v_0 C_{A0}$ or F_{A0} (Figure 15.2a), whereas for a plug-flow reactor, the volume is area under the curve only multiplied by $v_0 C_{A0}$ or F_{A0} (Figure 15.2b). Calculate points from Table 15.9 to plot the curve. For minimum reactor volume, use back mix for the first portion and then plug flow to complete (Figure 15.3a). The rate is maximum when $X_A = 0.45$ or $C_A = 0.275$. Assuming $X_{Af} = 0.9$,

$$\text{Volume of back-mix reactor} = F_{A0}\left(\frac{X_A}{-r_A}\right)$$

$$= 100(2.54)(0.45)$$

$$= 114.3 \text{ L}$$

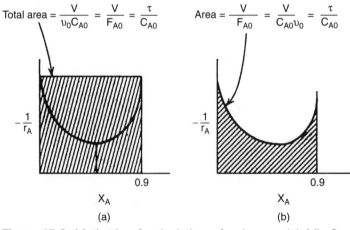

Figure 15.2 Methods of calculation of volumes: (a) Mix-flow reactor; (b) Plug-flow reactor

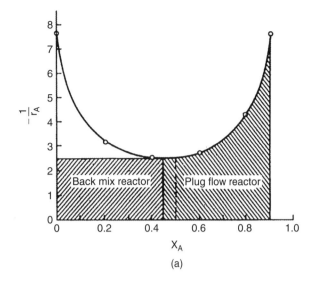

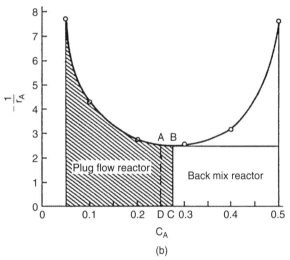

Figure 15.3 Solution of Example 15.10 (a) In terms of conversion, (b) In terms of concentration

Plug-Flow Reactor

Use trapezoidal rule from 0.45 to 0.5 and Simpson's rule from 0.50 to 0.9.

$$A = \frac{1}{2}(a+b)h = \frac{1}{2}(0.05)(2.54 + 2.5) = 0.126$$

$$= \frac{0.05}{3}[10.23 + 4(2.64 + 2.95 + 3.75 + 5.25) + 2(2.75 + 3.3 + 4.27)]$$

$$= \frac{0.05}{3}(89.23) = 1.4872$$

Total $\quad A = 0.126 + 1.4872 = 1.6132$

$V = F_{A0}(1.6132) = 100(1.6132) = 161.3$

Volume of back—mix reactor = 114.3 L

Volume of plug—flow reactor = 161.3 L

The problem can be solved directly in terms of concentrations instead of conversions as done above (Figure 15.3b). For $X_{Af} = 0.9$, $C_{Af} = 0.05$ mol/L.

Volume of Back-Mix Reactor

$$\text{Space time } \tau = \frac{C_{A0}X_A}{-r_A} = \frac{C_{A0} - C_A}{-r_A} = \frac{V}{v}$$

$$= \frac{C_{A0} - C_A}{-r_A}$$

$$= (0.5 - 0.275)(2.54) = 0.5715 \text{ h}$$

$$V = v\tau = 0.5715(200) = 114.6 \text{ L}$$

Volume of Plug-Flow Reactor

This requires the determination of area under the curve from $C_A = 0.05$ to $C_A = 0.275$. Determine the area $ABCD$ by trapezoidal rule and the area $ADEF$ by Simpson's rule.

$$\text{Area } ABCD = \frac{1}{2}(0.025)(2.54 + 2.5) = 0.063 = \tau_{ABCD}$$

$$\text{Area } ADEF = \frac{0.025}{3}[7.69 + 2.54 + 4(5.15 + 3.8 + 3.04 + 2.62)$$

$$+ 2(4.27 + 3.4 + 2.75)]$$

$$= 0.746 \text{h} = \tau$$

$$\text{Plug flow} = 0.063 + 0.746 = 0.809 \text{h}$$

Volume of plug-flow reactor = $0.809(200) = 161.8$ L (dm^3)

Multiple Reactors in Series

In some cases it is desirable to use a series of stirred-tank reactors in which the exit stream from one reactor serves as the feed stream to the next. Analysis of such reactors is considered next.

Equal Size Mixed-Flow Reactors in Series

Consider n reactors of equal volume in series. With steady state and no density change, a material balance on the reactor i for component A gives

$$\tau = \frac{C_0 V_i}{F_0} = \frac{V_i}{v} = \frac{C_0(X_i - X_{i-1})}{-r_A} \tag{15.92}$$

Since $\Delta \rho = 0.0$, i.e., $\varepsilon = 0$, the preceding can be written in terms of the concentrations as

$$\tau = \frac{C_{Ai-1} - C_{Ai}}{kC_{Ai}} \tag{15.93}$$

where $-r_A = kC_{Ai}$ is the specific reaction rate for first-order system, or

$$\frac{C_{Ai-1}}{C_{Ai}} = 1 + k\tau_i \qquad (15.94)$$

Since the space time (or mean residence time t) is the same for all the reactors,

$$\frac{C_{A0}}{C_{An}} = \frac{1}{1-X_{An}} = \frac{C_{A0}}{C_{A1}} \frac{C_{A1}}{C_{A2}} \cdots \frac{C_{An-1}}{C_{An}} = (1+k\tau)^n \qquad (15.95)$$

or

$$C_{An} = \frac{C_{A0}}{(1+kt)^n} \qquad (15.96)$$

For the system as a whole by rearrangement

$$\tau_n = n\tau = \frac{n}{k}\left[\left(\frac{C_{A0}}{C_{An}}\right)^{1/n} - 1\right] \qquad (15.97)$$

Notice that the preceding equation reduces to the plug-flow equation in the limit as $n \to \infty$

$$\tau = \frac{1}{k} \ln \frac{C_{A0}}{C_A} \qquad (15.98)$$

For reactions other than first order, explicit solution for C_{Ai} in terms of C_{A0} is quite complicated. For a small number of stages, numerical solutions are available. For plug flow (second-order rate equation),

$$\frac{C_{A0}}{C_A} = 1 + C_{A0}k\tau_p \qquad (15.99)$$

where τ_p is the space time for plug flow.

Comparisons of the performance of a series of n equal-size mixed reactors with a plug-flow reactor for first- and second-order reactions are available in the form of graphs.

Multiple Reactions and Maximization of Desired Product

We have reviewed the concept of yield and rate selectivity in chemical reactions in Chapter 2. In many situations, various reactions are taking place simultaneously in the reactor and as a result both desired and undesired products are produced. Economic success of chemical manufacture requires that the production of the desired product should be maximized while that of the undesired product is minimized. Increased yield of the desired product combined with subsequent reduced costs of purification because of low production and therefore lower content of undesired impurity in the desired product should form the basis for selection of reaction system and reaction conditions. In this section, we review multiple reactions and the reactor selection to improve the yield and selectivity of the desired product.

Multiple reactions are mainly (a) reactions in series, (b) reactions in parallel (competing reactions), and (c) independent reactions.

Series reactions, also called consecutive reactions, involve the formation of an intermediate product, which reacts further to form a second product. For example,

$$A \to B \to C$$

In *parallel reactions,* a reactant reacts to form two products simultaneously by two different pathways, for example

In *multiple reactions,* both parallel and series reactions take place together simultaneously.

Independent reactions involve, as the name implies, reactions that take place in the same reaction volume but have different reactants and products.

Parallel Reactions: Maximizing the Desired Product

Effect of Reaction Orders   For competing reactions, consider the reaction of A reacting to D, a desired product and to U, an undesirable product. The reactions are as follows:

$$\text{Reaction 1:} \quad A \xrightarrow{k_1} D$$

$$\text{Reaction 2:} \quad A \xrightarrow{k_2} U$$

The rate equations for the two reactions are

$$r_D = k_D C_A^{n_1} \tag{15.100}$$

$$r_U = k_U C_A^{n_2} \tag{15.101}$$

The rate of disappearance of A by the two reactions is given by

$$-r_A = r_D + r_U = r_D = k_D C_A^{n_1} + k_U C_A^{n_2} \tag{15.102}$$

The rate selectivity S_{DU} for the desired product D is given by

$$S_{DU} = \frac{r_D}{r_U} = \frac{k_D}{k_U} C_A^{n_1 - n_2} \tag{15.103}$$

In the preceding equations, n_1 is the order of the desired reaction and n_2 is the order of the undesired reaction. The method to follow to maximize the rate selectivity parameter will depend on the relative values of n_1 and n_2.

Case 1: $n_1 > n_2$. The order of desired reaction is greater than that of the undesired reaction. Then if $a = n_1 - n_2$, a is positive number. Equation 15.103 then becomes

$$S_{DU} = \frac{r_D}{r_U} = \frac{k_D}{k_U} C_A^{a} \tag{15.104}$$

To make S_{DU} as large as possible, C_A should be as large as possible during the reaction. This means if the reaction is liquid phase, use of diluents should be minimal. If the reaction is gas phase, the reaction should be run without adding inerts, and at high pressures to keep volume low so that molar concentration is higher. A batch or plug-flow reactor should be used in this case, because in these reactors the concentration starts at a high value and progressively decreases during the course of reaction. A CSTR is not a good choice of reactor in this case because in a perfectly mixed CSTR the concentration within the reactor is always at it's lowest value and equals the exit concentration.

Case 2: $n_2 > n_1$. The order of undesired reaction is greater than the order of the desired reaction. We let $n_2 - n_1 = b$, a positive number. Then

$$S_{DU} = \frac{r_D}{r_U} = \frac{k_D C_A^{n_1}}{k_U C_A^{n_2}} = \frac{k_D}{k_U C_A^{n_2-n_1}} = \frac{k_D}{k_U C_A^b} \quad (15.105)$$

Inspection of Equation 15.105 reveals that the concentration term is in the denominator. Hence to make S_{DU} large, $\frac{k_D}{k_U C_A^b}$ must be large. This means C_A should be as small as possible during the course of the reaction. Low concentration of A during the reaction could be maintained by using a CSTR and by diluting the feed with inerts. A recycle reactor will also serve the purpose because the part of product recycled to the reactor will act as a diluent and maintain the concentration low.

Case 3: $n_1 = n_2$. The desired and undesired reactions are of the same or equal order.

$$S_{DU} = \frac{r_D}{r_U} = \frac{k_D}{k_U} C_A^{n_1-n_2} = \frac{k_D}{k_U} \text{ because } n_1 - n_2 = 0 \quad (15.105a)$$

Hence the product distribution is fixed by k_D/k_U and is independent of the concentration C_A. In this case, the distribution is unaffected by the type of reactor used. Reactor volume requirement will dictate the design.

PRODUCT DISTRIBUTION AND TEMPERATURE

Besides controlling the concentration, product distribution may be controlled by changing the temperature at which the reaction is carried out and by using a catalyst. Main advantage of catalyst is it's selectivity in promoting a particular reaction while suppressing the other.

If two competing steps in multiple reactions have specific rate constants k_1 and k_2, then

$$\frac{k_1}{k_2} = \frac{\alpha_1 e^{-E_1/RT}}{\alpha_2 e^{-E_2/RT}} = \frac{\alpha_1}{\alpha_2} e^{(E_2-E_1)/RT} \propto e^{(E_2-E_1)/RT} \quad (15.106)$$

Depending on whether E_1 is greater or smaller than E_2, k_1/k_2 changes. When T rises, k_1/k_2 increases if $E_1 > E_2$. The ratio decreases if $E_1 < E_2$. A high temperature favors the reaction of higher activation energy; a low temperature favors the reaction of lower activation energy.

Effect of Activation Energies The favorable temperature (high or low) required for a reaction to maximize the production of a desired product can be determined if the activation energies for the desired and undesired reactions are known. Let these be E_D and E_U respectively. The sensitivity of the rate selectivity parameter to temperature is then obtained by taking the ratio of the specific reaction rates in terms of the Arrhenius equation as follows

$$\frac{k_D}{k_U} = \left(\frac{\alpha_D}{\alpha_U}\right) e^{-[(E_D - E_U)/RT]} \tag{15.107}$$

Two cases to be considered are:

Case 1: $E_D > E_U$. In this case, as the temperature increases, the specific rate of the desired reaction k_D (and correspondingly the overall rate r_D) increases more rapidly compared to the specification rate k_U of the undesired reaction. In such a case, the reaction should be carried out at the highest possible temperature that is practical in order to maximize the rate selectivity parameter.

Case 2: $E_U > E_D$. In this case the reaction needs to be carried out at low temperature to maximize S_{DU}. However, if the temperature is too low, the reaction may not proceed to any significant extent.

Series Reactions: Maximizing the Desired Product

For consecutive reactions, the most important variable is time: space time for a flow reactor and real time for a batch reactor.

$$A \xrightarrow{k_1} B \xrightarrow{k_2} C$$

If the specific reaction rate $k_1 < k_2$, that is if the reaction B $\rightarrow$ C is faster than the reaction A $\rightarrow$ B, then B disappears into C faster than it is produced. Consequently it will be very difficult to produce B. If, however, the rate of reaction B $\rightarrow$ C is slower than the reaction A $\rightarrow$ B, then significant amounts of B will be produced and accumulate because of slow disappearance of B into C. Therefore time needed to carry out the reaction to produce maximum of B has to be calculated exactly.

For a detailed treatment of finding proper temperature to use, see Levenspiel.

Plug-Flow or Batch Reactor

Equation 15.66 applies to this case. These equations relate the concentration with time for all components of the unimolecular-type reactions. If reaction time is replaced by the space time, the equations apply to plug-flow reactors also. Thus

$$\frac{C_A}{C_{A0}} = e^{-k_1 \tau} \tag{15.108}$$

$$\frac{C_B}{C_{A0}} = \frac{k_1}{k_2 - k_1}\left(e^{-k_1\tau} - e^{-k_2\tau}\right) \tag{15.109}$$

$$C_C = C_{A0} - C_A - C_B$$

The maximum concentration, the intermediate, and the time at which it occurs are given by

$$\frac{C_{B,max}}{C_{A0}} = \left(\frac{k_1}{k_2}\right)^{k_2/(k_2-k_1)} \quad (15.110)$$

$$\tau_{p,opt} = \frac{1}{k_{\log mean}} = \frac{\ln(k_2/k_1)}{k_2 - k_1} \quad (15.111)$$

The time at which B is maximum is also the point at which the rate of formation of C is most rapid.

Mixed Flow Reactor

By taking material balance for steady state over the reactor (Figure 15.4), the following relations can be obtained:

$$\frac{C_A}{C_{A0}} = \frac{1}{1+k_1\tau_m} \quad (15.112)$$

$$\frac{C_B}{C_{A0}} = \frac{k_1\tau_m}{(1+k_1\tau_m)(1+k_2\tau_m)} \quad (15.113)$$

$$\frac{C_C}{C_{A0}} = \frac{k_1 k_2 \tau_m^2}{(1+k_1\tau_m)(1+k_2\tau_m)} \quad (15.114)$$

The space time at which maximum concentration of B occurs is found by equating the derivative of Equation 15.112 to zero, and is given by

$$\tau_{m,opt} = \frac{1}{\sqrt{k_1 k_2}} \quad (15.115)$$

Maximum concentration of B is given by

$$\frac{C_{B,max}}{C_{A0}} = \frac{1}{\left[(k_2/k_1)^{1/2} + 1\right]^2} \quad (15.116)$$

Similar analyses can be made for other reactions such as bimolecular or complex reactions.

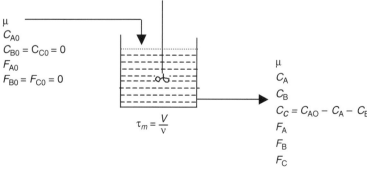

Figure 15.4 Variables for reactions in series occurring in a CSTR

Recycle Reactors

A recycle reactor consists of a plug-flow reactor through which fresh feed and a portion of the product as a recycle are fed. They are of two types: (1) continuous flow recycle reactor (Figure 15.5a) and (2) batch-recycle reactor (Figure 15.5b). Recycle reactors are used when the reaction is autocatalytic. Other situations where a recycle reactor is useful are

(a) to maintain near isothermal operation

(b) to favor a certain selectivity.

They are also useful as laboratory reactors to obtain chemical reaction rate data.

The recycle ratio in the continuous-flow recycle reactor (Figure 15.5a) can be varied from zero to infinity. As the recycle ratio is increased, the behavior of the recycle reactor changes from plug flow ($R = 0$, no back mixing) to mixed flow ($R \to \infty$ and complete mixing). Thus recycling provides a method of simulating various degrees of back mixing with a plug-flow reactor. If there is no product or fresh feed but a large reservoir of the fluid is provided as shown in Figure 15.5b, then the system behaves as a tank-type batch reactor.

The procedure of designing recycle reactors is similar to the one followed for reactors without recycle. However, additional material balances are required at points M and S.

For the plug-flow reactor, the design equation is

$$\frac{V}{F'_{A0}} = \int_{X_{A1}}^{X_{A2}=X_{Af}} \frac{dX_A}{-r_A} \quad (15.117)$$

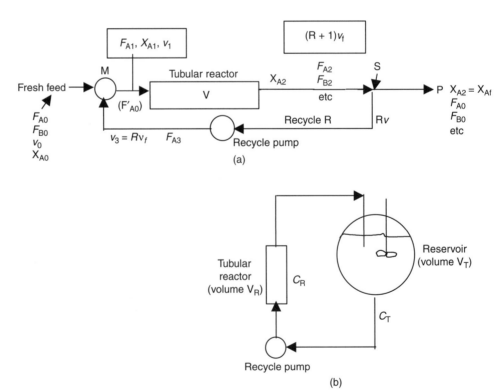

Figure 15.5 (a) Tubular-recycle reactor (b) Batch-recycle reactor

where F'_{A0} = feed rate of A if both the fresh feed and the recycle were unconverted. Also both F'_{A0} and X_{A1} are not known; they need to be written in terms of known quantities. If R is the recycle ratio at the split point, $F'_{A0} = RF_{A0}$. Then total feed to the reactor is

$$F'_{A0} = RF_{A0} + F_{A0} = (R+1)(F_{A0}) \quad (15.118)$$

Conversion X_{A1} can be written in terms of concentrations as follows:

$$X_{A1} = \frac{1 - C_{A1}/C_{A2}}{1 + \varepsilon_A C_{A1}/C_{A2}} \quad (15.119)$$

We assume pressure is constant. Therefore fresh feed and the recycle can be added directly (details not shown) to give

$$C_{A1} = \frac{F_{A1}}{v_1} = \frac{F_{A0} + F_{A3}}{v_0 + Rv_f} = \frac{F_{A0} + RF_{A0}(1 - X_{Af})}{v_0 + Rv_0(1 + \varepsilon_A X_{Af})} = C_{A0}\left(\frac{1 + R - RX_{Af}}{1 + R + R\varepsilon_A X_{Af}}\right) \quad (15.120)$$

Combining equations 15.119 and 15.120 yields

$$X_{A1} = \left(\frac{R}{R+1}\right) X_{Af} \quad (15.121)$$

After combining Equations 15.118 and 15.121, the following performance equation for the recycle reactor results:

$$\frac{V}{F_{A0}} = \left(R + 1\right) \int_{\left(\frac{R}{R+1}\right) X_{Af}}^{X_{Af}} \frac{dX_A}{-r_A} \quad (15.122)$$

When the density is constant, the preceding equation can be written in terms of the concentration as follows:

$$\tau = \frac{C_{A0} V}{F_{A0}} = (R+1) \int_{\frac{C_{A0} + RC_{Af}}{R+1}}^{C_{Af}} \frac{dC_A}{-r_A} \quad (15.123)$$

For the first-order reaction and $\varepsilon_A = 0$, the integration of the performance equation for the recycle reactor gives

$$\frac{k\tau}{R+1} = \ln\left[\frac{C_{A0} + RC_{Af}}{(R+1)C_{Af}}\right] \quad (15.124)$$

For second-order reaction, 2A → Products, $-r_A = kC_A^2$, $\varepsilon_A = 0$ integration gives

$$\frac{kC_{A0}\tau}{R+1} = \frac{C_{A0}(C_{A0} - C_{Af})}{C_{Af}(C_{A0} + RC_{Af})} \quad (15.125)$$

For other more complex rate equations direct integration may not be possible. In that case, graphical or numerical integrations may be used.

Batch Recycle Reactor

In deriving material balance for the batch-recycle reactor shown in Figure 15.5b, the following assumptions are made:

1. The volume of the connecting lines and the pump are negligible.
2. The reaction occurs only in the tubular reactor.
3. Volumes of reservoir and the reactor are constant.
4. There is contribution to the accumulation term from both the reactor and the reservoir.
5. The contents of the reservoir are well mixed by agitation.

With these assumptions, a material balance around reactor and reservoir gives

$$\int_0^{V_R} r\,dV_R = \frac{d}{dt}\int C_R\,dV_R + \frac{dC_T}{dt} \qquad (15.126)$$

The concentration C_R varies with position in the tubular reactor. It can be expressed in terms of the recycle flow and the rate r_i. However, the major application of the batch recycle reactor is as a differential reactor. In this situation, the concentration change in the recycle reactor is very small and then $C_R \approx C_T$. This constraint reduces the equation to

$$r = \left(\frac{V_R + V_T}{V_R}\right)\frac{dC_T}{dt} \qquad (15.127)$$

This is the rate equation for a batch recycle reactor operating in differential mode. To obtain differential mode of operation, a high recycle rate and small reactor volume are required.

Example 15.11

The photochemical decomposition of acetone has been studied by Smith et al. in a batch recycle reactor. The reaction conditions were 370K and 116kPa. The following data were obtained:

t	0	1.25	2.25	3.25	4.25
$C_{C_2H_6} \times 10^3$ kg mol/m^3	0	2.0	6.5	8.5	11.8

The chemical equation is $CH_3COCH_3 \rightarrow CO + C_2H_6$

The other operating data were as follows:

$$Q = 3.33 \times 10^{-4} \text{ m}^3/\text{s (circulation rate)}$$
$$\text{Volume of reservoir } V_T = 5.5 \times 10^{-3} \text{ m}^3$$
$$\text{Volume of reactor} = V_R = 62.6 \times 10^{-6} \text{ m}^3$$
$$\text{Initial concentration of acetone} = 5.7 \times 10^{-3} \text{ kg mol/m}^3.$$

Calculate

(a) the rate of reaction as function of the acetone concentration

(b) the concentration difference $C_T - C_e$

(c) conversion per pass.

Solution

First calculate the concentrations of acetone as a function of time. We know that the concentration of ethane (C_2H_6) is zero at $t = 0$. On decomposition of acetone, one mol of ethane is produced per mole acetone decomposed. Therefore we can get the concentration of acetone at a given time by subtracting the concentration of ethane from the initial concentration of acetone. Thus the concentrations of acetone as a function of time are

t (h)	0	1.25	2.25	3.25	4.25
$C_{CH_3COCH_3} \times 10^3$	5.7	5.68	5.635	5.615	5.582

The data are plotted as concentration of acetone versus time in hours in Exhibit 6.

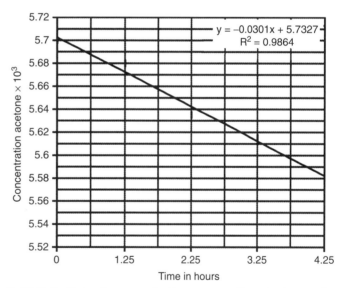

Exhibit 6 Plot of concentration versus time in hours for Example 15.11

Slope of the line is -0.0301×10^{-3}.

(a) $r = \left(\dfrac{62.6 \times 10^{-6} + 5.5 \times 10^{-3}}{62.6 \times 10^{-6}} \right)(-0.0301 \times 10^{-3}) = -2.675 \times 10^{-3}$ kg mol/h·m^3

(b) $C_T - C_e = -\dfrac{62.6 \times 10^{-6}}{3.33 \times 10^{-4}} \left(\dfrac{5.5 \times 10^{-3}}{62.6 \times 10^{-6} + 5.5 \times 10^{-3}} \right)(-2.695 \times 10^{-3})$

$= 5.01 \times 10^{-4}$ kg mol/m^3

(c) Highest conversion per pass is obtained when C_T is the lowest, that is, at 4.25 hours. At this point the conversion per pass is given by

$$(X)_{pass} = \frac{C_T - C_e}{C_T} = \frac{5.01 \times 10^{-4}}{5.582 \times 10^{-3}} = 0.09 \text{ or } 9 \text{ percent}$$

Example 15.12

The following reactions are carried out in the gas phase

$$A + B \rightarrow D \quad r_D = k_1 C_A^2 C_B^1$$
$$A + B \rightarrow R \quad r_R = k_2 C_A C_B$$
$$A + B \rightarrow S \quad r_S = k_3 C_A^3 C_B$$

D is the desired product. R and S are the undesired side products. To maximize the selectivity of D that is $S_{D,RS} = \frac{r_D}{r_R + r_S}$,

(a) What type reaction system would you suggest?

(b) Calculate C_A at which point selectivity will be maximum.

(c) Sketch $S_{D,Rs}$ as function of C_A

Solution

(a) $$S_{D,RS} = \frac{r_D}{r_R + r_S} = \frac{k_1 C_A^2 C_B}{k_2 C_A C_B + k_3 C_A^3 C_B} = \frac{k_1 C_A}{k_2 + k_3 C_A^2}$$

Expression for selectivity shows that to make S_D large, C_A should be as small as possible. Use a CSTR to maintain concentration low. The concentration will always be low in a CSTR and equal to the exit concentration.

(b) To obtain concentration at which point selectivity will be maximum, take derivative with respect to C_A and equate it to zero.

$$\frac{dS}{dC_A} = \frac{d}{dC_A}\left[\frac{k_1 C_A}{k_2 + k_3 C_A^2}\right] = 0$$

$$= k_1 \left(k_2 + k_3 C_A^2\right) - k_1 C_A^2 (2 k_3 C_A) = 0$$

From which the concentration at which S is maximum is obtained as

$$C_A = C_A^* = \sqrt{\frac{k_2}{k_3}}$$

where C_A^* is used to distinguish it from other concentrations. Numerical values of specific reaction constants are not given, therefore numerical values of C_A^* and S cannot be calculated. A conceptual sketch is given in Exhibit 7.

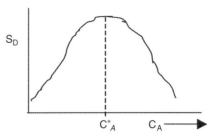

Exhibit 7 Sketch of selectivity versus C_A

Example 15.13

A decomposes to three different products B, D, and C according to the following:

$$\begin{array}{ll} B & r_B = 1 \\ \nearrow & \\ A \to D & r_D = 2C_A \text{ (desired)} \\ \downarrow & \\ C & r_C = C_A^2 \end{array}$$

Find the maximum concentration of D expected for isothermal operations in

(a) a mixed reactor

(b) a plug flow reactor.

If unconverted A can be separated from the product and recycled, which reactor should be chosen?

Assume $C_{A0} = 2$.

Solution

D is the desired product. Therefore writing fractional yield in terms of D,

$$\phi(D/A) = \frac{dC_D}{dC_B + dC_D + dC_C} = \frac{2C_A}{1 + 2C_A + C_A^2} = \frac{2C_A}{(1+C_A)^2}$$

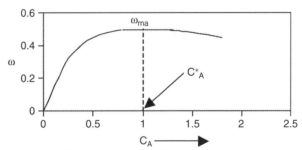

Exhibit 8 Fractional yield of desired product versus concentration

Differentiating fractional yield with respect to C_A, we get

$$\frac{d\varphi_{Df}}{dC_A} = \frac{d}{dC_A}\left[\frac{2C_A}{(1+C_A)^2}(2-C_A)\right] = 0$$

Equating the derivative to 0, $C_A = 1$ and $\varphi = 0.5$.

(a) *Mixed reactor:* Most D is formed when the rectangle under the φ versus C_A curve has the maximum area. However, simple expressions for k values in this example and the various quantities permit analytical solution, which is as follows:

$$C_{Df} = \varphi \cdot (-\Delta C_A) = \frac{2C_A}{(1+C_A)^2}(C_{A0} - C_A)$$

Differentiating and equating the derivative to zero to find the condition at which most D is formed gives

$$\frac{dC_{Df}}{dC_A} = \frac{d}{dC_A}\left[\frac{2C_A}{(1+C_A)^2}(2-C_A)\right] = 0$$

From this equality, the optimum operating conditions for a mixed reactor are

$$C_{Df} = 0.67 \quad \text{at} \quad C_{Af} = 0$$

Plug-flow reactor: The production of D is maximum when the area under the φ versus C_A curve is maximum. This is true for 100 percent conversion of A. Thus

$$C_{Df} = -\int_{C_{A0}}^{C_{Af}} \varphi \, dC_A = \int_0^2 \frac{2C_A}{(1+C_A)^2} dC_A$$

Value of this integral gives the optimum conditions for a plug-flow reactor.

$$C_{Df} = 0.867 \text{ at } C_{Af} = 0$$

(c) *Any reactor with separation and recycle of unconverted reactant.*

No reactant leaves the system unconverted. Therefore, it is necessary to operate at conditions that give highest fractional yield. This is at $C_A = 1$ and $\varphi = 0.5$. Therefore a mixed reactor should be used and it should be operated at $C_A = 1$. Then 50 percent of reactant A that is not converted is recycled to the system, forming product.

Our calculations were based on $C_{A0} = 2$. Product formed per mol of the reactant A in the above three cases is (moles of D formed per mole of A fed to the reactors)

(a) mixed reactor : $0.67/2 = 0.335$

(b) plug-flow reactor: $0.867/2 = 0.4335$

(c) mixed reactor with recycle: 0.5.

Thus the mixed reactor with separation and recycle of unconverted reactant gives the highest product distribution.

Example 15.14

Three products are formed from a reactant A. D is the desired product while B and C are two side products that are undesired. The reactions are gas phase and their rates are given by

Reaction A → D $\quad r_D = 0.0003 e^{\left[39000\left(\frac{1}{300} - \frac{1}{T}\right)\right]} C_A \quad$ D is desired product

Reaction A → B $\quad r_B = 0.002 e^{\left[27000\left(\frac{1}{300} - \frac{1}{T}\right)\right]} C_A^{1.5} \quad$ Undesired side product

Reaction A → C $\quad r_D = 0.005 e^{\left[6500\left(\frac{1}{300} - \frac{1}{T}\right)\right]} C_A^{0.5} \quad$ Second undesired product.

How should these reactions be carried out to maximize D and minimize the concentrations of B and C? Specify reactor type, pressure (high or low), temperature, etc.

Solution

Examination of the rate equations shows that the frequency factors in the three equations are rather small and therefore comparable. On the basis of these the rates should not differ by much. However, the activation energies of the first two reactions are much greater than that of the third reaction. Therefore the rate of formation of C will be much smaller and negligible compared to the rates of the first two equations at high temperatures. This can be expressed by the following equation:

$$S_{DC} = \frac{r_D}{r_c} \text{ will be very large}$$

and we may choose to ignore the formation of C in comparison with D. We need to compare only the rates of formation of D and B and see how D can be maximized relative to B. The selectivity of A with respect to B can be written as

$$S_{DB} = \frac{r_D}{r_B} = \frac{0.15 e^{12000\left[\frac{1}{300} - \frac{1}{T}\right]}}{C_A^{0.5}}$$

From the preceding equation we notice that S_{DB} can be increased by carrying out the reaction at low concentrations of A. This will increase production of D and decrease that of B. Also to maximize the conversion of A to D, it would be desirable to operate the reactor at higher temperature to minimize the formation of C and at low concentration of A to minimize the formation of B. Thus the conditions to be used are

1. high temperature (because $E_1 > E_2$)

2. low concentrations of A

    (a) add inerts to lower the concentration

    (b) use low pressures (due to gas phase reaction)

    (c) use a CSTR or a recycle reactor.

Packed-Bed Reactor (PBR)

Tubular packed-bed reactor to carry out heterogeneous reactions is like a tubular reactor but packed with catalyst. The reaction rate is based on mass of catalyst rather than reactor volume.

In solid-fluid heterogeneous system, the rate of reaction of a species A is defined as

$$-r'_A = \text{g mol of A reacted}/s \cdot g \text{ catalyst} \tag{15.128}$$

The usual assumptions—namely there are no radial gradients in concentration, temperature or reaction rate—are made. A mole balance over catalyst weight ΔW and a time interval $\Delta \theta$ gives the following equation:

$$(F_A)_I \Delta \theta - (F_A)_O \Delta \theta + \bar{r}'_A \Delta W \Delta \theta = 0 \tag{15.129}$$
$$\text{in} - \text{out} + \text{generation} = \text{accumulation}$$

where

$(F_A)_I$ = moles entering the catalyst bed per unit time during time interval $\Delta \eta$
$(F_A)_0$ = moles leaving the catalyst bed per unit time during time interval $\Delta \eta$
ΔW = mass of catalyst, g
$\Delta \theta$ = difference element of time interval over which the mole balance is made
$\bar{r}'_A$ = average rate of reaction of A over time interval $\Delta \theta$

Dividing by ΔW and $\Delta \theta$ and writing the first two terms as difference, we get

$$-\frac{\Delta F_A}{\Delta W} + \bar{r}'_A = 0 \tag{15.130}$$

By taking limit as $\Delta W \to 0$, and transposing, we have

$$\frac{dF_A}{dW} = r'_A \tag{15.131}$$

Neglecting the pressure drop through the reactor and the catalyst decay, the design equation for the packed bed reactor results. This allows calculating the amount of catalyst needed with the following integrated equation:

$$W = \int_{F_{A0}}^{F_A} \frac{dF_A}{r'_A} \tag{15.132}$$

In terms of conversion X, the flow F_A is given by

$$F_A = F_{A0} - F_{A0} X \tag{15.133}$$

Substituting Equation 15.133 into Equation 15.132 gives

$$F_{A0} \frac{dX}{dW} = -r'_A \tag{15.134}$$

This is the differential form of the design equation in terms of conversion and should be used when analyzing packed-bed reactors that have pressure drop along the length of the reactor. Integration with boundary condition $W = 0$ when $X = 0$ gives

$$W = F_{A0} \int_0^X \frac{dX_A}{-r'_A} \tag{15.135}$$

This equation is useful to determine the catalyst weight required to obtain a conversion X when total pressure remains constant and the reactor operation is isothermal.

Catalysts and Catalytic Reactors

Catalysts are substances that affect the rate of reaction but are unchanged themselves. The catalytic reaction takes place at the fluid-solid interface. The solid contains many pores and the surface of these pores supplies the area needed for the high rate of reaction. A typical silica-alumina catalyst has a pore volume of 0.2 to 0.7 cm^3/g and a surface area of 200 to 500 m^2/g. Calculation of catalyst properties such as d_p, surface area, etc. are dealt with at length by Smith.

In order to determine the catalyst weight needed to accomplish a given conversion, the expression for the rate of reaction $-r'_A$ in terms of conversion is needed. Adsorption plays an important role in catalytic reactions. Most of the catalytic rate equations are based on the Langmuir-Hinshelwood adsorption model (see Chapter 13, Adsorption). To simplify treatment, the process of catalytic reaction is divided into several steps: adsorption on the catalyst surface, surface reaction, and desorption. In addition, external effects of mass transfer by diffusion must be taken into account. A detailed review of all these is beyond the scope of this book and the reader is referred particularly to Fogel, Felder and Rousso, and Smith for derivations of rate equations for several mechanisms.

The rate limiting step. When heterogeneous reactions are carried out at steady state, the rates of the three reaction steps, adsorption, surface reaction, and desorption are equal. Thus

$$-r'_A = r_{AD} = r_s = r_D \tag{15.136}$$

Many times one step in the series is found to be rate limiting or rate controlling.

Design of Reactors for Gas-Solid Reactions

Kinetic rate equations for catalytic or fluid-solid reactors are more complex, and this fact can make it difficult or impossible to solve the reactor design equation analytically. In most cases the design equation must be solved by graphical or numerical methods. The design equation for the packed-bed tubular reactor has already been reviewed in this chapter. Design equations for some other reactors are as follows:

Batch Reactor

For an ideal batch reactor, the differential design equation for a heterogeneous reaction is

$$N_{A0} \frac{dX}{dt} = -r'_A W \tag{15.137}$$

For a packed-bed reactor, the differential form of the equation for a heterogeneous reaction is

$$F_{A0}\frac{dX}{dt} = -r'_A \qquad (15.138)$$

Fluidized CSTR

For a packed-bed CSTR, the design equation is based on catalyst weight rather than volume of reactor and is

$$W = \frac{F_{A0}X}{-r'_A} \qquad (15.139)$$

Fluidized Bed Reactor

The design equation for the fluidized CSTR can be used for a fluidized-bed reactor. Fluidized-bed reactor can process large volumes of fluid. Conditions in the reactor are more uniform. Bubbling bed model of Kunii and Levenspiel is mostly used to describe reactions in fluidized beds. In this model the reactant gas flows upward in the reactor in the form of bubbles. The rate of reaction depends on the velocity of bubbles through the reactor and the rate of transport of gases in and out of the bubbles. To determine the velocity of bubbles through the bed, we need calculate the following:

(a) Porosity at minimum fluidization, ε_{mf}

$$\varepsilon_{mf} = 0.586\psi^{-0.72}\left(\frac{\mu^2}{\rho_s \eta d_p^3}\right)^{0.021} \qquad (15.140)$$

Where Ψ is the sphericity of the particles

(b) Minimum fluidization velocity, u_{mf}

$$u_{mf} = \frac{(\psi d_p)^2}{150\mu}[g(\rho_c - \rho_g)]\frac{\varepsilon_{mf}^3}{1-\varepsilon_{mf}} \qquad (15.141)$$

(c) Bubble size, d

$$\frac{d_{bm} - d_b}{d_{bm}} = e^{-0.3h/D_t} \qquad (15.142)$$

$$d_{bm} = 0.652[A_c(u_o - u_{mf})]^{0.4} \qquad (15.143)$$

(d) Velocity of bubble rise

$$u_b = u_o - u_{mf} + (0.711)(gd_b)^{1/2} \qquad (15.144)$$

(e) The transport coefficients are given by the following equations:

$$K_{bc} = 4.5\left(\frac{u_{mf}}{d_b}\right) + 5.85\left(\frac{D_{AB}^{1/2}g^{1/4}}{d_b^{5/4}}\right) \qquad (15.145)$$

$$K_{ce} = 6.78\left(\frac{\varepsilon_{mf}D_{AB}u_b}{d_b^3}\right)^{1/2} \qquad (15.146)$$

Reaction Rates

The following parameters need to be calculated first before one can determine the reaction rate parameters.

$$\text{Fraction of total bed occupied by bubbles, } \delta = \frac{u_o - u_{mf}}{u_b - u_{mf}(1+\alpha)} \quad (15.147)$$

$$\text{Fraction of the bed consisting of wakes} = \alpha\delta \quad (15.148)$$

where α is a function of the particle size.

Also we need to calculate volume of catalyst in the bubbles γ_b, volume of clouds γ_c, and volume of emulsion γ_e. These parameters are given by

$$\gamma_c = (1-\varepsilon_{mf})\left[\frac{3(u_{mf}/\varepsilon_{mf})}{u_{br} - (u_{mf}/\varepsilon_{mf})} + \alpha\right] \quad (15.149)$$

$$\gamma_e = (1-\varepsilon_{mf})\left(\frac{1-\delta}{\delta}\right) - \gamma_c - \gamma_b \quad (15.149\text{a})$$

$$\gamma_b = 0.01 \text{ to } 0.001 \quad (15.149\text{b})$$

After all these parameters are obtained by calculation, the amount of catalyst required for a given conversion can be calculated using the following equation:

$$W = \frac{\rho_c A_c u_b (1-\varepsilon_{mf})(1-\delta)}{k_{cat} K_R} \ln\frac{1}{1-X} \quad (15.150)$$

where

$$K_R = \gamma_b + \cfrac{1}{\cfrac{k_{cat}}{K_{bc}} + \cfrac{1}{\gamma_c + \cfrac{1}{\cfrac{1}{\gamma_e} + \cfrac{k_{cat}}{K_{ce}}}}} \quad (15.150\text{a})$$

k_{cat} = specific reaction rate determined from laboratory experiments.

STOICHIOMETRIC TABLES

A stoichiometric table is a representation of stoichiometric relationships between reacting molecules for a single reaction. It gives an overall material balance for a chemical reaction in the form of a table. It gives the following information:

(a) the various components present in the reaction system

(b) the initial molar quantities of each component present

(c) the change in the number of moles caused by the reaction

(d) the number of moles of each component remaining in the system at time t.

As an example consider the general reaction

$$aA + bB \rightarrow cC + dD \quad (15.151)$$

If A is taken as the limiting reactant and basis of calculation, dividing by the stoichiometric coefficient of A, one obtains

$$A + \frac{b}{a}B \rightarrow \frac{c}{a}C + \frac{d}{a}D \qquad (15.152)$$

With this equation, we can use 1 mol of A as basis and relate to it the moles of other components. Now if X is the conversion of A, and its initial moles are N_{A0}, the moles of A remaining after reaction is $N_A = N_{A0}(1-X)$. The stoichiometric table is then constructed for a batch system as follows in Table 15.10.

The stoichiometric coefficient $(d/a + c/a - b/a - 1)$ in the expression for total moles after the reaction represents the increase in total number of moles per mole of a reacted. This increase term occurs in calculations frequently and is given the symbol δ. Thus total number of moles can be expressed by the equation

$$N_T = N_{T0} + \delta N_{A0} X \qquad (15.153)$$

The concentration of a component is its moles per unit volume. For example, the concentration of component A is N_A/Ve. From the stoichiometric table, we can get the moles of each species and write it's concentration in terms of conversion of A. For example, the concentration of B can be written as follows:

$$C_B = \frac{N_B}{V} = \frac{N_{B0} - (b/a)N_{A0}X}{V} \qquad (15.154)$$

The preceding expression is simplified by defining a parameter Θ_i given by

$$\Theta_i = \frac{N_{i0}}{N_{A0}} = \frac{C_{i0}}{C_{A0}} = \frac{y_{i0}}{y_{A0}} \qquad (15.155)$$

Table 15.10 Stoichiometric table for a batch system

Component	Initial (mol)	Moles (Converted)	Moles (Remaining)
A	N_{A0}	$-(N_{A0}X)$	$N_A = N_{A0} - N_{A0}X$
B	N_{B0}	$-\frac{b}{a}(N_{A0}X)$	$N_B = N_{B0} - \frac{b}{a}N_{A0}X$
C	N_{C0}	$-\frac{c}{a}(N_{A0}X)$	$N_C = N_{C0} + \frac{c}{a}N_{A0}X$
D	N_{D0}	$-\frac{c}{a}(N_{A0}X)$	$N_D = N_{D0} + \frac{d}{a}N_{A0}X$
I (inerts present)	N_{I0}	—	$N_I = N_{I0}$
Total	N_{T0}		$N_T = N_{T0} + \left(\frac{d}{a} + \frac{c}{a} - \frac{b}{a} - 1\right)N_{A0}X$

Thus C_B can be expressed as

$$C_B = \frac{N_{A0}[N_{B0}/N_{A0} - (b/a)X]}{V} = \frac{N_{A0}[\Theta_B - (b/a)X]}{V} \quad (15.156)$$

where $\Theta_B = \frac{N_{B0}}{N_{A0}}$ by definition.

For constant volume system, the preceding simplifies to

$$C_B = C_{A0}\left(\Theta_B - \frac{b}{a}X\right) \quad (15.157)$$

Similar equations can be written for other components.

Stoichiometric Table for Flow Systems

Stoichiometric table for continuous flow systems is similar to that for a batch system except that N_{i0} in batch system has to be replaced by F_{i0}. Again take component A as the basis for calculation and divide by the stoichiometric coefficient of A in the equation $aA + bB \rightarrow cC + dD$ and get

$$A + \frac{b}{a}B \rightarrow \frac{c}{a}C + \frac{d}{a}D \quad (15.158)$$

In a flow system the concentration at a given point is obtained from F_A and the volumetric flow rate μ at that point using the following

$$C_A = \frac{F_A}{v} = \frac{\text{moles/time}}{\text{dm}^3/\text{time}} = \frac{\text{moles}}{\text{dm}^3 (\text{or liters})} \quad (15.159)$$

Concentration of B for example can be written as

$$C_B = \frac{F_B}{v} = \frac{F_{B0} - (b/a)F_{A0}X}{v} \quad (15.160)$$

In flow system the parameter Θ is defined as given below

$$\Theta_B = \frac{F_{B0}}{F_{A0}} = \frac{C_{B0}v_0}{C_{A0}v_0} = \frac{C_{B0}}{C_{A0}} = \frac{y_{B0}}{y_{A0}} \quad (15.161)$$

Similar definitions can be written for other components. The stoichiometric table for a flow system is then prepared as in Table 15.11.

For liquids, volume does not change very much with reaction when no phase change occurs. Then we can assume $v = v_0$. Then the expressions for concentrations can be simplified as follows:

Table 15.11 Stoichiometric table for a flow system

Component	Feed Rate to Reactor (mol/time)	Change in the Reactor (mol/time)	Exit Rate from Reactor (mol/time)
A	F_{A0}	$-(F_{A0}X)$	$F_A = F_{A0}(1-X)$
B	$F_{B0} = \Theta_B F_{A0}$	$-\dfrac{b}{a}(F_{A0}X)$	$F_B = F_{A0}\left(\Theta_B - \dfrac{b}{a}X\right)$
C	$F_{C0} = \Theta_C F_{A0}$	$-\dfrac{c}{a}(F_{A0}X)$	$F_C = F_{A0}\left(\Theta_C + \dfrac{c}{a}X\right)$
D	$F_{D0} = \Theta_D F_{A0}$	$-\dfrac{c}{a}(F_{A0}X)$	$F_D = F_{A0}\left(\Theta_D + \dfrac{d}{a}X\right)$
I (inerts)	$F_{I0} = \Theta_I F_{A0}$	—	$F_I = F_{A0}\Theta_I$
Total	F_{T0}		$F_T = F_{T0} + \left(\dfrac{d}{a} + \dfrac{c}{a} - \dfrac{b}{a} - 1\right)F_{A0}X$
			$F_T = F_{T0} + \delta F_{A0}X$

$$\left.\begin{array}{l} C_A = \dfrac{F_{A0}}{v_0}(1-X) = C_{A0}(1-X) \\ \\ C_B = C_{A0}\left(\Theta_B - \dfrac{b}{a}X\right) \end{array}\right\} \quad (15.162)$$

and similar equations can be written for other concentrations.

When volume changes occur with reaction, one can establish more involved relations using equation of state $PV = zN_TRT$. The reader is referred to Fogler for further details.

Steady State Nonisothermal Reactor Design

Up to now we reviewed basics of chemical kinetics and how we use them to design isothermal chemical reactors. The basic assumption was that there were no heat effects involved and the reactor was operating at a given fixed temperature. There was no exchange of heat energy with the surroundings and the reaction had no significant heat of reaction. However, many reactions in practice are either endothermic or exothermic and therefore heat has to be supplied or removed from the reaction in order to carry out the reaction at a reasonable rate and desired conversion.

Chemical reaction rate is a strong function of temperature. When heat is evolved or absorbed in a reaction or when heat is added or removed from the reaction system, there is a necessity to account for the temperature effect on the reaction rate. Basic design equations, rate equations, and stoichiometric relationships are applicable to the design of nonisothermal reactors but their evaluation is different because temperature variation along the length in a tubular reactor or the variation of temperature in a CSTR affects the reaction rate and conversion. Hence an energy balance is required besides a mass balance. Basic principles of energy balances and heats of reaction are reviewed in Chapter 3. In this section we will review the application of energy balance to some simple reactions and will briefly touch on the topics of multiple steady states, and runaway reactions.

OPTIMAL TEMPERATURE PROGRESSION

Optimal temperature progression is defined as the temperature variation, which minimizes V/F_{A0} for a given conversion of the reactant. The optimum may be a singular temperature (isothermal operation) or may be a changing temperature over a certain range (non-isothermal operation). It may change in a batch reactor with time, along the length of the plug-flow reactor or from stage to stage for a series of mixed-flow reactors.

For endothermic reactions, rise in temperature increases both reaction rate and conversion. Therefore highest allowable temperature should be used.

For exothermic reversible reactions, the situation is different. With rise in temperature, the rate of forward reaction increases but the maximum obtainable conversion decreases. Therefore a high temperature should be used when the system is far from equilibrium and the reaction rate is high. As the reaction approaches equilibrium, temperature should be lowered to shift the equilibrium to a favorable value. To start with we consider adiabatic operation.

When we have to take into consideration the heat of reaction, heat transfer, and work effects a general energy balance for a reaction system can be written as follows:

$$\frac{dU}{dt} = \sum_{i=1}^{n} H_{i0} F_{i0} - \sum_{i=1}^{n} H_i F_i + Q + W^* \qquad (15.163)$$

where subscript $i0$ indicates stream entering the reactor. And i denotes at the outlet.

For steady state $\frac{dU}{dt} = 0$ and the energy balance reduces to

$$0 = \sum_{i=1}^{n} H_{i0} F_{i0} - \sum_{i=1}^{n} H_i F_i + Q + W \qquad (15.164)$$

We now apply the energy balance for a general reaction aA + bB → cC + dD. We first write the equation as follows:

$$A + \frac{b}{a} B \rightarrow \frac{c}{a} C + \frac{d}{A} D \qquad (15.165)$$

The inlet and exit streams in the equation are expanded in the following manner:

In: $\sum H_{i0} F_{i0} = H_{A0} F_{A0} + H_{B0} F_{B0} + H_{C0} + H_{D0} + H_{I0} F_{I0}$ (15.166)

Out: $\sum H_i F_i = H_A F_A + H_B F_B + H_C F_C + H_D F_D + H_I F_I$ (15.166a)

---

*The energy balance is written here with modern convention for signs. Both heat and work are considered positive when they enter the system. Since schools have switched to this convention in the past few years, the same convention is followed in this book.

Expressing the molar flow rates in terms of conversion we have

$$F_A = F_{A0}(1-X) \quad F_B = F_{A0}\left(\Theta_B - \frac{b}{a}X\right) \quad F_C = F_{A0}\left(\Theta_C + \frac{c}{a}X\right)$$

$$F_D = F_{A0}\left(\Theta_D + \frac{d}{a}X\right) \quad F_I = \Theta_I F_{A0} \quad (15.167)$$

Substituting these values in Equations 15.166 and 15.166a and then subtracting 15.166a from 15.166,

$$\sum_{i=1}^{n} H_{i0}F_{i0} - \sum_{i=1}^{n} F_i H_i = F_{A0}\,[(H_{A0} - H_A) + (H_{B0} - H_B)\Theta_B$$

$$+ [(H_{C0} - H_C)\Theta_C + (H_{D0} - H_D)\Theta_D + (H_{I0} - H_I)\Theta_I]]$$

$$- \underbrace{\left(\frac{d}{a}H_D + \frac{c}{a}H_C - \frac{b}{a}H_B - H_A\right)}_{\Delta H_{RX}} F_{A0} X \quad (15.168)$$

The term in Equation 15.168 that is multiplied by $F_{A0}X$ is the heat of reaction at temperature T and is shown as ΔH_{RX}. The heat of reaction is always given on the basis of the component that is chosen as the basis of calculation. The general energy balance can now be written as follows:

$$\sum_{i=1}^{n} F_{i0}H_{i0} - \sum_{i=1}^{n} F_i H_i = F_{A0}\sum_{i=1}^{n}\Theta_i(H_{i0} - H_i) - \Delta H_{RX}(T)F_{A0}X \quad (15.169)$$

Combining Equations 15.164 and 15.169, steady state energy balance is written as follows:

$$Q + W - F_{A0}\sum_{i=1}^{n}\Theta_i(H_{i0} - H_i) - \Delta H_{RX}(T)F_{A0}X = 0 \quad (15.169a)$$

Given the heat capacity data as functions of T, changes in enthalpies can be calculated at any desired temperature. A reference temperature, usually 298.15 K, at which the heats of reactions are available is chosen. This aspect has been reviewed in Chapter 3.

In some situations, if the temperature range is small, mean heat capacities can be used to calculate the enthalpy changes. If mean or constant heat capacities are used, the general steady state balance becomes

$$Q + W - F_{A0}\sum\Theta_i \hat{C}_{Pi}(T - T_R) - F_{A0}X\left[\Delta H_{RX}^{\circ}(T_R) + \Delta \hat{C}_P(T - T_R)\right] = 0 \quad (15.170)$$

Heat Addition or Removal

Heat added or removed from the reactor is specified in many applications in terms of overall coefficient, U, area of heat transfer, A, and the difference between reaction temperature and the ambient temperature.

Continuous Stirred Tank Reactors

For heat transfer in CSTRs, reactor is generally provided with half-pipe or conventional jackets. An energy balance on the heat exchanger yields the relation

$$\dot{Q} = \dot{m}_C C_{P_C} \left\{ (T_{a1} - T) \left[1 - \exp\left(\frac{-UA}{\dot{m}_C C_{P_C}} \right) \right] \right\} \quad (15.171)$$

$$\dot{Q} = \frac{UA(T_{a1} - T_{a2})}{\ln[(T - T_{a1})/(T - T_{a2})]} \quad (15.172)$$

Heat Transfer to Tubular Reactors (PFR and PBR)

In these reactors heat transfer varies along the length of the reactor, total heat transferred has to be obtained by integration of heat flux equation

$$\text{For a PFR,} \quad \dot{Q} = \int^A U(T_a - T)\, dA = \int^V Ua(T_a - T)\, dV \quad (15.173)$$

where a = heat exchange area per unit volume of the reactor.

In case of tubular reactor of diameter D, the volume of reactor of length L is $(\pi D^2/4)L$ and area is πDL, so

$$a = \frac{\pi DL}{(\pi D^2/4)L} = \frac{4}{D} \quad (15.174)$$

For PBR, Equation 15.173 is written in terms of catalyst weight by dividing through by the bulk density of the catalyst to get

$$\frac{1}{\rho_b}\frac{d\dot{Q}}{dV} = \frac{Ua}{\rho_b}(T_a - T) \quad (15.175)$$

Since $dW = \rho_b dV$, heat transfer to PBR is given by

$$\frac{d\dot{Q}}{dW} = \frac{Ua}{\rho_b}(T_a - T) \quad (15.176)$$

Let us now apply the general energy balance to some simple cases of reactor operation.

Nonisothermal Continuous-Flow Reactors

When we apply the general energy balance equation to CSTR, F, the steady-state energy balance becomes

$$Q + W - F_{A0}\sum_{i=1}^{n}\int_{T_{i0}}^{T}\Theta_i C_{Pi}\, dT - \left[\Delta H_{RX}^0(T_R) + \int_{T_R}^{T}\Delta C_P\, dT\right] F_{A0} X = 0 \quad (15.177)$$

For a CSTR, mole balance gives $F_{A0}X = -r_A V$. Therefore one may replace $F_{A0}X$ by $-r_A V$ in the preceding equation if required. If we use constant or mean heat capacities, the energy balance reduces to

$$Q + W - \left[\Delta H_{RX}^o(T_R) + \Delta \hat{C}_P(T - T_R)\right] F_{A0} X = F_{A0} \sum_{i=1}^{n} \Theta_i \hat{C}_{Pi}(T - T_{i0}) \quad (15.178)$$

In many cases except when the fluid viscosity is very high, $W = 0$ (energy input of agitator is negligible), and if the operation is adiabatic $Q = 0$. Then Equation 15.178 becomes

$$X = \frac{\sum \Theta_i \dot{C}_{Pi}(T - T_R)}{-\left[\Delta H_{RX}^0(T_R) + \Delta \hat{C}_P(T - T_R)\right]} \quad (15.179)$$

The sensible heat term in the denominator of Equation 15.179 is usually negligible in comparison with the heat of reaction; hence the plot of conversion versus T will be a straight line. The design equation for a CSTR is

$$V = \frac{F_{A0} X}{-r_A}$$

Rewrite the equation in a slightly different form to get

$$\frac{Q + W}{F_{A0}} - X\left[\Delta H_{Rx}^0(T_R) + \Delta \hat{C}_P(T - T_R)\right] = \sum B\Theta_i \hat{C}_P(T - T_{i0}) \quad (15.180)$$

Equations 15.179 and 15.180 are used to design CSTR with or without heat exchange.

Consider a first-order irreversible liquid phase reaction A → B, which is carried out in a CSTR.

Based on mole balance the design equation is $V = \dfrac{F_{A0} X}{-r_A}$ \hfill (cstr.1)

Rate equation: $\quad -r_A = k C_A$ \hfill (cstr.2)

Rate constant: $\quad k = \alpha e^{-E/RT}$ \hfill (cstr.3)

Stoichiometry: $\quad v = v_0$ constant volume (liquid phase)

$$C_A = C_{A0}(1 - X) \quad \text{(cstr.4)}$$

Combining cstr.1 to cstr.4, we get

$$V = \frac{v_0}{\alpha e^{-E/RT}}\left(\frac{X}{1 - X}\right) \quad \text{(cstr.5)}$$

Case 1: Volume V is to be determined when the variables X, v_0, C_{A0}, and F_{A0} are specified. The following steps are to be used:

(1) assume pure A enters the reactor

(2) $\Delta \hat{C}_P = 0$

For adiabatic case, we solve Equation 15.179 for T to give

$$T = T_0 + \frac{X\left(-\Delta H_{RX}^0\right)}{\hat{C}_{P_A}} \tag{cstr.6}$$

For nonadiabatic case, $Q = UA(T_a - T)$ and Equation 15.180 is used with $W = 0$ to get T as a function of X.

$$T = \frac{F_{A0}X\left(-\Delta H_{RX}^0\right) + F_{A0}\hat{C}_{P_A}T_0 + UAT_a}{F_{A0}\hat{C}_A + UA} \tag{cstr.6a}$$

After the temperature is known, calculate k from Arrhenius equation and then V-volume of reactor using cstr.5.

Case 2: The variables v_0, C_{A0}, V, and F_{A0} are specified and the conversion, X, and the exit temperature, T, are to be calculated. The following procedure should be followed to get X and T.

For the adiabatic case, solve the energy balance (adiabatic case) for X as a function of T.

$$X_{EB} = \frac{\hat{C}_{PA}(T - T_0)}{-\left(\Delta H_{RX}^0(T_R)\right)} \tag{cstr.7}$$

For the nonadiabatic case with $Q = UA(T_a - T)$ and $W = 0$,

$$X_{EB} = \frac{UA(T - T_a)/F_{A0} + C_{P_A}(T - T_0)}{-\Delta H_{RX}^0} \tag{cstr.8}$$

Next solve Equation cstr.5 for X as a function of T using the following:

$$X_{MB} = \frac{\tau \alpha e^{-E/RT}}{1 + \tau \alpha e^{-E/RT}} \quad \text{where} \quad \tau = \frac{V}{v_0} \tag{cstr.9}$$

Next find the values of X and T that satisfy both the mole balance and the energy balance. This result can be obtained by numerical calculations by trial and error or plotting X vs T calculated from the two equations on the same graph. The intersection point gives the values of X and T as shown in Figure 15.6.

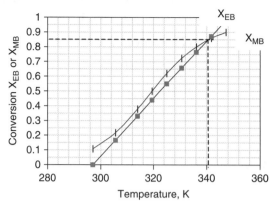

Figure 15.6 Conversions from energy balance and mole balance equations

Adiabatic Plug-Flow Tubular Reactor

Because the operation is adiabatic, $Q = 0$ and $W = 0$. Hence the general energy balance equation reduces to

$$X[-\Delta H_{Rx}(T)] = \int_{i0}^{T} \Sigma \Theta_i C_{Pi} dT \qquad (15.181)$$

The differential mole balance is given by

$$F_{A0} \frac{dX}{dV} = -r_A(X,T) \qquad (15.181a)$$

These two equations in combination allow calculation of the temperature, conversion, and concentration profiles along the length of the reactor. First one prepares a table of T as a function of X. Then $k(T)$ is calculated as a function of X and $-r_A$ as a function of X alone.

Usual procedure is then used to calculate the size of different reactors.

Steady-State Plug-Flow Reactor with Heat Transfer

Assuming $W = 0$ and then differentiating the resulting energy balance equation, we can obtain with some manipulation the following equation:

$$\frac{dT}{dV} = \frac{Ua(T_a = T) + (-r_A)[-\Delta H_{RX}(T)]}{F_{A0}(\Sigma \Theta_i C_{Pi} + X \Delta C_P)} = f(g,T) \qquad (15.182)$$

The mole balance equation is

$$\frac{dX}{dV} = \frac{-r_A}{F_{A0}} = f(X,T) \qquad (15.183)$$

Equations 15.182 and 15.183 are two differential equations and must be solved simultaneously and by numerical integration.

Solutions of most of the design problems involving energy balance have to be obtained by numerical or graphical integration. The solutions are very lengthy and take large space even for simple reactions. Programs such as Polymath and Matlab have to be used. To save space, example solutions are avoided here. The reader is requested to refer to Fogler for some good problems and their solutions involving energy balances.

Example 15.15

The liquid phase hydrolysis of dilute aqueous anhydride solution can be treated as first order and irreversible. A batch reactor for carrying out the hydrolysis is charged with 200 liters (200 dm^3) of anhydride solution at 25°C. The initial concentration of anhydride is 2.16×10^{-4} g·mol/cm^3, and other data are as follows:

density of solution $\rho = 1.05$ g/cm^3
specific heat of solution = 0.9 cal/g·°C   assume const.
heat of reaction = $-50{,}000$ cal/g·mol   assume const.

The rate of reaction is given by the equation

$$r_A = kC_A$$

where

$k = 18.11 \times 10^6 \exp(-11{,}000/RT)$
$C = $ g mol/cm^3
$T = $ K.

What time is required for a conversion of 70 percent when the reactor is operated adiabatically? What time is required for the same conversion if the reactor were operated isothermally?

Solution

For a batch reactor, time θ is given by

$$\theta = C_{A0} \int_0^{X_F} \frac{dX_A}{-r_A}$$

$$-r_A = 18.11 \times 10^6 [\exp(-11{,}000/1.987T)] C_{A0}(1 - X_A)$$

Establish energy balance. Since the reaction is adiabatic, $Q_t = 0$ and $W = 0$, and we can get

$$M_t C_p dT = F_{A0}(-\Delta H_R)\, dX_A$$

or $200(1000)(1.05)(0.9)dT = 200(1000)(2.16 \times 10^{-4})(50{,}000)\, dX_A$
or $dT = 11.43\, dX_A$

By integration, $T = 11.43 X_A + I$, where I is integration constant. When $X_A = 0$, $T = 280°$K, $I = 298$, and therefore

$$T = 11.43 X_A + 298$$

Using this expression for T, prepare Table 15.12 for various values of X_A from 0 to 0.7.

Table 15.12 Calculation of rates of reaction

X_A	T, K	$-r_A \times 10^4$	$\dfrac{1}{-r_A} C_{A0}$
0	298.0	0.3345	6.457
0.1	299.14	0.3591	6.684
0.2	300.29	0.3854	7.005
0.3	301.43	0.4133	7.467
0.4	302.57	0.4429	8.129
0.5	303.72	0.4746	9.101
0.6	304.86	0.5081	10.63
0.7	306.00	0.5437	13.24

Integration by the Simpson rule gives time for adiabatic reaction.

$$\frac{1}{3}h[6.457 + 4(7.76) + 13.24] = \frac{1}{3}(0.35)[6.457 + 4(7.76) + 13.24]$$
$$= 5.92 \text{ min}$$

For isothermal reaction at 25°C,

$$-r_A = 18.11 \times 10^6 [\exp(-11{,}000/298R)]C_0(1 - X_A)$$

Substitution in the expression for θ gives

$$\theta = \int_0^{X_{AF}} \frac{C_0 dX_A}{C_0(1 - X_A)(18.11 \times 10^6) \exp[-11{,}000/1.987(298)]}$$
$$= +6.456 \int_0^{X_{AF}} \frac{dX_A}{1 - X_A}$$
$$= -6.456 \ln[(1 - X_A)]_0^{X_{AF}}$$
$$= -6.456 [\ln(0.3) - \ln(1)] = 7.77 \text{ min}$$

In most cases, the solutions will not be as simple as in this problem. When the expression for T cannot be integrated analytically, numerical and graphical techniques will have to be employed.

Multiple Steady States

In this section we consider steady-state operation of a CSTR in which a first-order reaction is taking place. The following assumptions will be made in making the energy balance:

1. $W = 0$
2. $\Delta \dot{C}_P = 0$, therefore, $\Delta H_{RX}^0(T) = \Delta H_{RX}^0(T_R) \equiv \Delta H_{RX}^0$

With these assumptions for simplification, we can rearrange the energy balance Equation 15.178 and incorporate in it the mole balance and get the following:

$$(-r_A V)(-\Delta H_{Rx}) = F_{A0} C_{P0}(T - T_0) + UA(T - T_a) \tag{15.184}$$

where
$$C_{P0} = \Sigma \Theta_i \hat{C}_{Pi}$$

Let
$$\kappa = \frac{UA}{C_{P0} F_{A0}} \quad \text{and} \quad T_C = \frac{T_0 F_{A0} C_{P0} + UA T_a}{UA + C_{P0} F_{A0}} = \frac{\kappa T_a + T_0}{1 + \kappa} \tag{15.185}$$

Then Equation 15.184 can be simplified to

$$(-r_A V / F_{A0})\left(-\Delta H_{Rx}^0\right) = C_{P0}(1 + \kappa)(T - T_c) \tag{15.186}$$

Lefthand side of Equation 15.186 is termed the heat generated term.

$$G(T) = \left(-\Delta H_{Rx}^0\right)(-r_A V / F_{A0}) \tag{15.187}$$

The righthand side of Equation 15.186 is referred to as heat removed term. The heat is removed by flow and heat exchange.

$$R(T) = C_{P0}(1 + \kappa)(T - T_c) \tag{15.188}$$

Heat Removed Term R(T)

Equation 15.188 shows that $R(T)$ increases linearly with temperature with a slope $C_{P0}(1 + \kappa)$. As the feed temperature T_0 is increased, the new line has the same slope but shifts to the right as depicted in Figure 15.7.

Heat-generated term can be written in terms of conversion as follows:

$$G(T) = \left(-\Delta H_{Rx}^0\right) X \tag{15.189}$$

For a first-order liquid phase reaction, it can be shown using the CSTR mole balance, $V = \frac{F_{A0} X}{k C_A} = \frac{v_0 C_{A0} X}{k C_{A0}(1-X)}$ and solving for X

$$X = \frac{k\tau}{1 + k\tau} \tag{15.190}$$

Substitution of X in Equation 15.189 gives $G(T) = \dfrac{-\Delta H_{Rx}^0 k\tau}{1 + k\tau}$ (15.191)

Figure 15.7 Variation of heat removal line with T_0 and φ ($\kappa = UA/C_{P0}F_{A0}$)

Now replacing k in terms of Arrhenius equation, the following relation is obtained:

$$G(T) = \frac{-\Delta H^0_{Rx} \tau \alpha e^{-E/RT}}{1 + \tau \alpha e^{-E/RT}} \quad (15.192)$$

Solutions of this type for other reaction orders and for reversible reactions can be obtained by solving CSTR mole balance for X.

At very low temperatures, the second term in the denominator of Equation 15.192 can be neglected. Therefore

$$G(T) = -\Delta H^0_{Rx} \tau \alpha e^{-E/RT}$$

At high temperatures, the second term in the denominator dominates, and Equation 15.192 is reduced to

$$G(T) = -\Delta H^0_{Rx}$$

$G(T)$ curves are shown qualitatively in Figure 15.8 a and b.

Ignition-Extinction Curve

The points of intersection or the tangency points of $R(T)$ and $G(T)$ curves are the temperatures at which the reactor can operate at steady state. Depending on the feed (entering) temperature, there will be one or more intersections or tangency points of the $G(T)$ and $R(T)$ curves. (See Figure 15.9a and Figure 15.9b).

In Figure 15.9a the heat-generated curve intersects the heat-removed curve in three points but at two different feed temperatures T_{01} and T_{02}. When the entering or feed temperature is T_{01}, there is only one intersection point and therefore only one steady state given by T_{S1}. However, when the feed temperature is raised to T_{02}, there are two steady states given by T_{s2} and T_{s3}. At upper steady state, temperature and conversion are higher. However, to operate at the upper steady state, the feed temperature must be maintained above a certain level.

All multiple steady states are not, however, stable. This can be understood better from the ignition-extinction diagram presented in Figure 15.10.

From Figure 15.9 it is seen that there are twelve possible steady states at six different feed temperature values and consequently there are twelve corresponding steady state temperatures. By plotting these steady-state temperatures versus feed temperature, an extinction-ignition diagram is obtained as in Figure 15.10. The steady-state temperatures at different feed temperatures are given in Table 15.13.

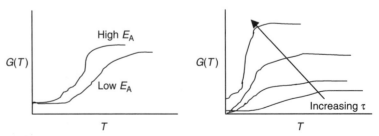

Figure 15.8 $G(T)$ Curves (schematics only) (a) Effect of activation energy (b) Effect of τ

Optimal Temperature Progression

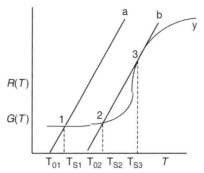

(a) Multiple steady states with variation of T_0

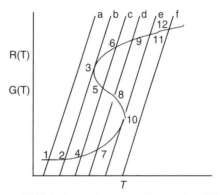

(b) Multiple steady states with variation of T_0

Figure 15.9 Finding multiple steady states when T_0 is varied

From Figure 15.10, it is seen that by increasing the feed temperature, the steady-state temperature increases along the bottom line (lower steady states) until T_{05} is reached. Any fractional degree increase in temperature T_{05} will cause the steady-state reactor temperature to jump to T_{S11} (see Figure 15.10). The temperature at which this jump occurs is called ignition temperature. A slight decrease below T_{02} would drop the steady-state reactor temperature to T_{S2}. Therefore T_{02} is called the extinction temperature. Middle points 5 and 8 indicate unstable steady-state conditions.

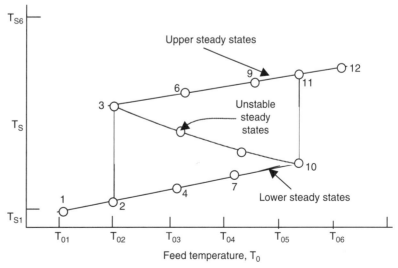

Figure 15.10 Feed temperature extinction-ignition curve

Table 15.13 Temperatures of multiple steady states

Entering Temperature		Reactor Temperature		
T_{01}		T_{S1}		
T_{02}	T_{S2}		T_{S3}	
T_{03}	T_{S4}	T_{S5}		T_{S6}
T_{04}	T_{S7}	T_{S8}		T_{S9}
T_{05}		T_{S10}	T_{S11}	
T_{06}		T_{S12}		

Runaway Reactions In many reaction systems, the temperature of the upper steady state may be high enough so that is undesirable or even dangerous to operate the reactor at that condition. Once the ignition temperature is exceeded, the reaction system will proceed to an upper state. Ignition temperature occurs at point 10 in Figure 15.9 at which the heat removed curve is tangent to the heat-generated curve. Therefore at this point the slopes of $R(T)$ and $G(T)$ curves are equal.

For heat removal curve,
$$\frac{dR(T)}{dT} = C_{P0}(1+\kappa) \tag{15.193}$$

Differentiating the heat–generated curve,

$$\frac{dG(T)}{dT} = \frac{d\left[-\Delta H_{Rx}\frac{-r_A V}{F_{A0}}\right]}{dT} = \frac{(-\Delta H_{RX})V}{F_{A0}}\frac{(-r_A)}{dT} \tag{15.194}$$

Now we make simplifying assumptions as follows:

1. The reaction is irreversible.
2. The reaction follows a power law model
3. Concentrations of reactants are weak functions of temperature.

Therefore the rate equation becomes

$$-r_A = (\alpha e^{-E/RT})f(C_i) \tag{15.195}$$

Differentiating with respect to temperature yields the following:

$$\frac{d(-r_A)}{dT} = \frac{E}{RT^2}\alpha e^{-E/RT}f(C_i) = \frac{E}{RT^2}(-r_A) \tag{15.196}$$

Equating 15.193 and 15.196,

$$C_{P0}(1+\kappa) = \frac{E}{RT^2}(-r_A)\frac{\Delta H_{Rx}}{F_{A0}}V \tag{15.197}$$

Now we divide Equation 15.184 by Equation 15.197 and obtain the following $<T$ value for a CSTR, which is operating at $T = T_r$.

$$\Delta T_{rc} = T_r - T_c = \frac{RT_r^2}{E} \qquad (15.198)$$

where ΔT_{rc} is the critical temperature difference. T_C is given by Equation 15.75. If this ΔT_{rc} is exceeded, the reaction system will move to the upper state. For many industrial reactions, this critical temperature difference will be about 15°C to 30°C.

Example 15.16

Obtain an expression for the heat-generated curve for a second-order liquid phase reaction with rate law

$$-r_A = kC_A^2$$

Solution

The heat-generated curve is given by the expression

$$G(T) = \left(-\Delta H_{Rx}^0\right)(-r_A V/F_{A0})$$

and the CSTR mole balance equation is

$$V = \frac{F_{A0}X}{kC_A^2} = \frac{v_0 C_{A0} X}{kC_{A0}^2(1-X)^2} \quad \text{since} \quad C_A = C_{A0}(1-X)$$

Now putting $V/v = \tau$, the previous can be written as

$$\tau = \frac{X}{kC_{A0}(1-2X+X^2)}$$

which simplifies to

$$k\tau C_{A0} X^2 - (1 + 2k\tau C_{A0})X + k\tau C_{A0} = 0$$

This is a quadratic and has the acceptable root given by

$$X = \frac{(1 + 2k\tau C_{A0}) - \sqrt{4k\tau C_{A0} + 1}}{2k\tau C_{A0}}$$

Substituting for k in terms of Arrhenius relation and subsequently in the general heat-generated curve, we get the final expression for $G(T)$ as follows:

$$G(T) = \frac{-\Delta H_{Rx}^0 \left[\left(2\tau C_{A0} \alpha e^{-E/RT} + 1\right) - \sqrt{4\tau C_{A0} + 1}\right]}{2\tau C_{A0} \alpha e^{-E/RT}}$$

where the heat of reaction ΔH_{Rx}^0 is calculated at the reference temperature.

Example 15.17

The first-order irreversible exothermic liquid phase reaction A → B is to be carried out in a CSTR. Species A and an inert are fed to the reactor in equimolar proportion. The feed rate of A to the reactor is 80 g mol/min. Data given are as follows:

$$\text{Heat of reaction} = -7500 \text{ cal/g mol}$$
$$\text{Heat capacity of A: 20 cal/g mol}$$
$$\text{Heat capacity of B: 20 cal/g mol}$$
$$\text{Heat capacity of Inerts: 30 cal/g mol}$$

T_a = ambient temperature = 300 K
UA = 8000 cal/g mol
τ = 100 min^{-1}
E = 40000 cal/g mol · K
k = 6.6 × 10^{-3} min^{-1} at 350 K

(a) Calculate and plot $G(T)$ and $R(T)$ curves.

(b) What is the reactor temperature if the feed is at 450 K?

(c) What is the conversion at the reactor temperature in b?

Solution

For a first-order reversible reaction, Equation 15.192 is to be used.

$$G(T) = \frac{-\Delta H_{Rx}^0 \tau \alpha e^{-E/RT}}{1 + \tau \alpha e^{-E/RT}}$$

To make this calculation we have all the data except the frequency factor α. This can be computed from the given data for k at 350 K using Arrhenius equation.

$$\alpha = \frac{6.6 \times 10^{-3}}{e^{-40000/(1.987 \times 350)}} = 6.3 \times 10^{22}$$

Other given data are heat of reaction $\Delta H_{Rx}^0 = -7500$ cal/g mol, $\tau = 100$ min. Substituting these values

$$G(T) = \frac{7500(100)(6.3 \times 10^{22})e^{-20231/T}}{1 + 100 \times 6.3 \times 10^{22} e^{-20231/T}}$$

Calculations are made at various temperatures on a spreadsheet. Some of these are given below in Table 15.14.

Table 15.14 Values of $G(T)$ at various temperatures

T(K)	300	310	315	320	325	340	350	355	360
G(T)	0.244	2.15	6.044	16.46	43.4	623.6	2488	3963.3	5340.4
T(K)	365	375	380	390	400	425	450	475	500
G(T)	6317	7192.8	7345.7	7460	7489	7499.4	7499.96	74.996	7500

These are plotted in Exhibit 9.

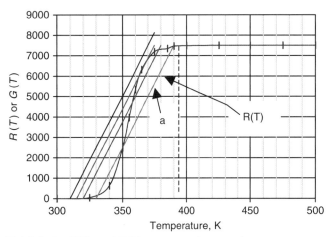

Exhibit 9 $G(T)$ and $R(T)$ curves vs temperature curves

Calculation of $R(T)$ curve:

$$\text{Feed rate of A} = 80 \text{ g mol/min}$$
$$\text{Feed rate of inerts} = 80 \text{ g mol/min}$$

Since one mol of B is produced per one mol of A converted, mols of A + B at any time is = 80 mol/min. and their heat capacities are the same. Hence heat capacity of mixture is

$$\text{Average } C_{P0} = (0.5 \times 20 + 0.5 \times 30)/2 = 25 \text{ cal/g mol} \cdot \text{K (A + B + inerts)}$$

$$\kappa = \frac{UA}{C_{P0} F_{A0}} = \frac{8000}{25(80)} = 4$$

$T_a = 300$ K (given)

$$T_C = \frac{T_0 + \kappa T_a}{1 + \kappa} = \frac{450 + 5(300)}{1 + 4} = 330 \text{ K}$$

$$R(T) = C_{P0}(1 + \kappa)(T - T_C)$$

To plot this, calculations are made as follows:

T(K)	330	350	375	380	390
R(T)	0	2500	5625	6250	7500

This is plotted in Exhibit 9 as line a. Reactor steady-state temperature is 390 K.

At 390 K $G(T) = 7460$ from the table.

Therefore conversion at 390°C = $\frac{G(T)}{\Delta H_{Rx}} = \frac{7460}{7500} = 0.995$. This indicates almost complete conversion.

At $T_0 = 450$ K.

With feed temperature at 450 K, there are 3 steady states for the reactor operation.

RECOMMENDED REFERENCES

1. Fogler, *Elements of Chemical Reaction Engineering,* 3rd ed., Prentice-Hall, NJ, 2004.
2. Levenspiel, *Chemical Reaction Engineering,* 3rd ed., John Wiley & Sons, NY, 1998.
3. Wallas, *Reaction Kinetics for Chemical Engineers,* 4th ed., McGraw-Hill, NY, 1959
4. Smith, *Chemical Engineering Kinetics,* McGraw-Hill, 4th ed. NY, 2004.

CHAPTER 16

Process Control

OUTLINE

CONTROL SYSTEMS 613

DEFINITIONS 615
Open Loop Control ■ Feedback Control ■ Feed-Forward Control

BLOCK DIAGRAMS 616

LAPLACE TRANSFORMS IN CONTROL SYSTEM ANALYSIS 617
Transfer Functions in Terms of Laplace Transforms ■ Overall System Transfer Functions

CONTROL ACTIONS 621

PROPORTIONAL BAND 622

DEVELOPMENT OF PROCESS MODELS 623
Dynamic Behavior of Basic Systems

FIRST-ORDER SYSTEMS 625

SECOND-ORDER SYSTEMS 628
Poles and Zeros of a Transfer Function

CONCEPT OF STABILITY AND STABILITY CRITERIA 634

RECOMMENDED REFERENCES 637

A number of process design aspects other than those discussed in the previous chapters have received some attention from time to time on the P.E. examination. These include process control, plant safety, explosion protection, waste treatment and disposal, water and energy conservation, and pollution control. While treatment of all of these subjects is beyond the limited space of this book, process control is very briefly dealt with in this chapter. Additionally, plant safety and environmental considerations are reviewed in Chapter 20.

CONTROL SYSTEMS

A control system or scheme is characterized by an output variable (e.g., temperature) that is automatically controlled through the manipulation of inputs (input variables). A typical control system is shown in Figure 16.1. In this scheme, the

614 Chapter 16 Process Control

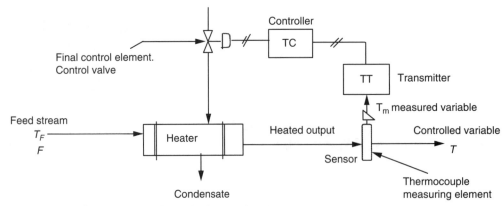

Figure 16.1 Components of a typical control system

temperature of the outlet stream from the heater is measured and used to control the steam flow to the heater.

It will be apparent that to automatically control a variable in a process requires the installation of a group of physical components as follows:

A Sensor or Measuring Element. The desired variable is measured and transmitted to the controller. Examples are thermocouples for measuring the temperature, venturimeter for measuring the flow rate, and gas cromatograph for measuring the composition. When the signals are continuous, such as electrical current or voltage, they are called analog. Digital signals are sent at discrete intervals of time.

Common measurement devices supply 4 to 20 mA or 0 to 5 V signal as a function of time. The pneumatic signal is usually 3 to 15 psig (English system) or 20 to 100 kPa (SI system). Usually the controller output of 3 to 15 psig would be sent to an actuator of a control valve. The pneumatic signal moves the valve stem. If the valve is too large, the signal of 3 to 15 psig can be amplified to supply enough pressure to move the valve stem. If the controller output is digital, there is usually either a D/I or an I/P converter needed. If sample times used are very small, the performance of digital controllers is nearly equal to that of analog controllers.

Transducers. Certain measurements cannot be used directly to control a process. They must be converted to physical entities such as electric voltage or current, or pneumatic signal, which can be transmitted with ease. Transducers are devices that convert measured signals into desired signals.

Transmission Lines. Transmission lines are used to carry the measured signals from the sensor to the controller. Many times the measured signals are very weak and it is not possible to transmit them over a long distance. In such cases, amplifiers or converters are used to raise the level or strength of the signal. As an example, the output of a thermocouple is a few millivolts. Hence it is amplified to a few volts.

Controller. This is a pneumatic, electronic, or digital device that compares the desired value of the variable (set point) with the actual measured value and sends out a signal to the final control element.

Final Control Element. This is generally a *control valve* but may be a variable transformer, depending on the application.

DEFINITIONS

At this point it may be useful to recall the definitions of some terms used in process control.

Dynamics. The variation in the behavior of the process with time is termed process dynamics. Process behavior without a controlling system to regulate it is called *open loop* response. The dynamic behavior of a process regulated by a controlling system is called its *closed loop* response.

Variables. Manipulated variables are typically the flow rates of streams that enter or leave the process. *Controlled variables* are those variables that are controlled to keep them constant as far as possible or keep them within certain bounds or limits. Typical variables usually controlled include flow rates, temperature, pressure, levels, composition, and pH. Uncontrolled variables are those variables that are not controlled.

Load Disturbances. The variations that occur in flow rates, temperature, or compositions entering the process (but sometimes leaving also) are *load disturbances*. There is no freedom to manipulate them, as these streams are set by upstream or downstream parts of the plant. The control system must be able to control the process even if load disturbances are present.

Open Loop Control

In an open loop control system, the inputs to the process are regulated independently without using the controlled output variable to adjust the inputs.

Feedback Control

In a closed loop or feedback control system, the controlled output variable is used to adjust the input to the process. The variable to be controlled is measured and the measured value is compared with the desired value or the set point. The difference or error is fed to the controller, which changes the manipulated variable in such a way that the controlled variable is returned back to the desired value. Two characteristics of feedback control are apparent: (1) Measurement of the error in the controlled variable is fed back to the manipulated variable and (2) action is taken after a change occurs in the process. In other words, output measurement is used to manipulate an input variable. For example, in Figure 16.2, change in level is measured and compared with the set point and this information is sent to

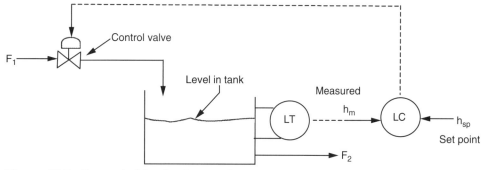

Figure 16.2 Concept of feedback control

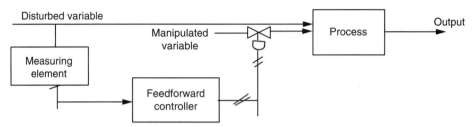

Figure 16.3 Concept of feed-forward control

the final control element, which in this case is a control valve. The control valve then either closes or opens more to reduce or increase the flow to maintain level in the tank.

In negative feedback, the difference between the set point and the measurement of the controlled variable is used to adjust the control element, which results in a tendency to reduce the error. If the signal to the comparator were obtained by adding the measurement to the set point, it is called positive feedback, which is inherently unstable. It is very rarely used.

Feed-Forward Control

In a feed-forward control, the measurement of one input variable is used to adjust another input variable. The idea is to take control action before the disturbance reaches the process. As shown in Figure 16.3, the disturbance in one input variable is detected before it enters the process, and corrective action is taken to readjust the manipulated input instead of waiting till the disturbance propagates all through the process.

The main drawback of FFC (feed-forward control) is its sensitivity to uncertainty. In practice, FFC is combined with FBC (feedback control) to account for the uncertainty (Bequette).

The method of *cascade control* is frequently used to decrease the process upsets. It involves the use of the output of a primary controller to adjust the set point of a secondary controller and is commonly used in feedback control systems. The ratio and selector controls are two other control modes, which require two or more interconnected instruments.

Servo and Regulatory Control Operations

In the servo operation mode, the control system is designed to follow the changes in the set point as closely as possible according to some prescribed function. On the other hand, in regulatory operation the control system is designed to keep output constant, that is, to maintain the controlled variable at a fixed value in spite of the changes in the load. The regulatory type of control systems is more common in chemical processes.

BLOCK DIAGRAMS

In control system analysis, block diagrams are used to show the functional relationship between the various parts of the system. Each part is represented by a rectangle or box with one input and one output. The "transfer function" (or its symbol), which is the mathematical relationship between output (response function)

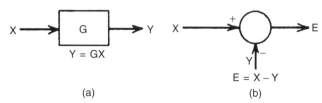

Figure 16.4 Components of a control system block diagram: (a) Block diagram of transfer function; (b) Comparator

and input (forcing function), is written inside the box (Figure 16.4a). The blocks are connected by arrows, which indicate the flow of the information in the system. The outputs and inputs are considered as signals.

The comparison of the signals is shown by circles with signs (+, −, ×, ÷) written outside them (Figure 16.4b). The symbols to denote addition (+), multiplication (×), or division (÷) may be written inside the circles. These circles are the comparators. Branching of a signal into more than one direction indicates the flow of the signal without any modification in it.

LAPLACE TRANSFORMS IN CONTROL SYSTEM ANALYSIS

The analysis of a control system begins with the determination of the transfer functions for the various parts of the system shown in the block diagram by the application of either the mass, energy, or force balance to each part of the system. The solutions of the developed linear differential equations are simplified with the use of the Laplace transforms, which make it possible to transform the differential equations into algebraic equations. The transformation replaces the independent time variable t by the complex variable s.

The Laplace transform of the time function $f(t)$ is defined by the operation (L is the Laplace operator)

$$L[f(t)] = \int_0^\infty f(t)e^{-st}dt = F(s) \qquad (16.1)$$

where $F(s)$ is a function of the complex variable s, a parameter that is constant with respect to the integration process. $F(s)$ is called the *Laplace transform of $f(t)$* as a function of the complex variable s. The following two useful theorems[1] concerning the Laplace transforms of functions can be easily established:

Theorem 1: $L[c_1 f_1(t) \pm c_2 f_2(t)] = c_1 L f_1(t) \pm c_2 L f_2(t)$ (16.1a)

Theorem 2: $Lf'(t) = sLf(t) - f^{(1)}(0) = sF(s) - f(0)$ (16.1b)

and in general, for the nth derivate,

$$LF^n(t) = s^n F(s) - s^{n-1}f(0) - s^{n-2}f^{(1)}(0) - \cdots - f^{n-1}(0) \qquad (16.1c)$$

where $f^{(i)}(0)$ is the ith derivate of $f(t)$ with respect to t and evaluated for $t = 0$.

In control system analysis, it is usually not necessary to perform the integration of Equation 16.1. Instead, the available Laplace transforms tables are used.

The initial- and final-value theorems are useful in checking the transfer functions for systems of known initial and final values. The final-value theorem is also useful in obtaining the steady-state error or offset of a control system. These theorems are

$$\text{Initial-value theorem:} \quad \lim_{t \to 0} f(t) = \lim_{s \to \infty} sF(s) \qquad (16.1\text{d})$$

$$\text{Final-value theorem:} \quad \lim_{t \to \infty} f(t) = \lim_{s \to 0} sF(s) \qquad (16.1\text{e})$$

Transfer Functions in Terms of Laplace Transforms

In terms of the Laplace transforms, the transfer function is the ratio of the Laplace transform of the output (response variable) to the Laplace transform of the input (the forcing or disturbing variable). The convention to represent the transfer function in the block diagram is $KG(s)$, where $G(s)$ represents the dynamic portion of the transfer function and K is the gain of the element. In the notation of the block diagram, (s) may be omitted from $G(s)$. The block diagram in Exhibit 1b uses this notation.

The evaluation of the transfer functions is illustrated by Example 16.1.

Example 16.1

Obtain the transfer function of the transient variation in the liquid height for the tank filling system in Exhibit 1a.

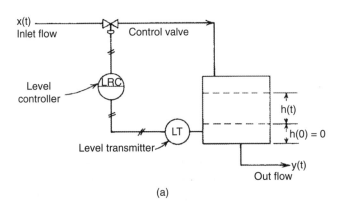

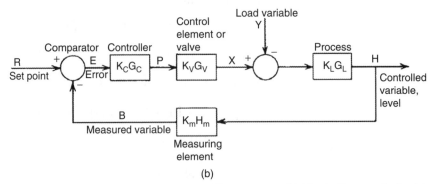

Exhibit 1 (a) Control system for liquid filling system; (b) Block diagram for liquid filling system (Example 16.1)

Solution

By making unsteady-state mass balance with the assumption of a constant density, one obtains

$$\rho A h_2 - \rho A h_1 = \overline{\rho x(t)} \Delta t - \overline{\rho y(t)} \Delta t$$

which, after canceling out ρ, can be written as

$$A \frac{\Delta h(t)}{\Delta t} = \overline{x(t)} - \overline{y(t)}$$

Taking limit as $t \to 0$ gives

$$A \frac{dh(t)}{dt} = x(t) - y(t)$$

which is rewritten by transferring $y(t)$ to the other side

$$x(t) = y(t) + A \frac{dh(t)}{dt}$$

where A is the cross-sectional area of the tank. Laplace transformation gives

$$X(s) = Y(s) + A[sH(s) - h(0)]$$

For simplification, let $h(0) = 0$. Therefore

$$X(s) = Y(s) + AsH(s)$$

Then the transfer function for the level system is given by

$$\frac{\text{Output}}{\text{Input}} = \frac{H(s)}{X(s) - Y(s)} = \frac{1}{As} = K_L G_L(s)$$

$1/As$ is the transfer function of the tank-filling system. It can be expressed as $K_L G_L(s)$ where $K_L = 1/A$ and $G_L(s)$ or $G_L = 1/s$. In the block diagram of Exhibit 1b, this transfer function is shown as $K_L G_L$.

Overall System Transfer Functions

By combining the transfer functions of the individual system elements, an overall system transfer function is obtained for the analysis of the transient response of the control system as a whole. The overall transfer function for a system can be obtained by the following two simple rules:

1. If there are several transfer functions in a series in a loop, the overall transfer function for the loop is the product of the individual transfer functions in the series. For example, in Exhibit 1b the overall transfer function $X(s)/E(s)$ is given by

$$\frac{X(s)}{E(s)} = (K_c G_c)(K_v G_v)$$

2. In a single-loop feedback system, the overall transfer function relating any two variables Y and X is given by the equation

$$\text{Overall transfer function} = \frac{Y}{X} = \frac{\pi_f}{1 \pm \pi_\ell} \qquad (16.2)$$

where π_f is the product of individual transfer functions between the locations of the signals Y and X and π_ℓ is the product of the individual transfer functions in the loop.

The plus sign is to be used in the denominator when the feedback is negative and the minus sign used when the feedback is positive. This rule can also be used to reduce a multiloop system to a single-loop system. Example 16.2 illustrates the method of obtaining an overall transfer function.

Example 16.2

Obtain the overall transfer functions $C(s)/U(s)$, and $C(s)/R(s)$ for the system shown in Exhibit 2.

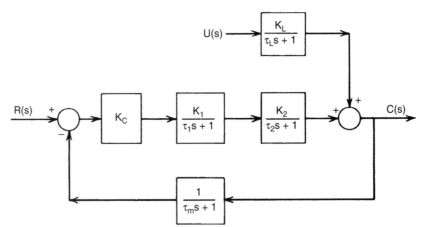

Exhibit 2 Control system (Example 16.2)

Solution

(a) For transfer function C/U,

$$\pi_f = \frac{K_L}{\tau_L s + 1} \qquad \pi_1 = \frac{K_c K_1 K_2}{(\tau_1 s + 1)(\tau_2 s + 1)(\tau_m s + 1)}$$

and

$$\frac{C}{U} = \frac{K_L/(\tau_L s + 1)}{1 + [K_c K_1 K_2/(\tau_1 s + 1)(\tau_2 s + 1)(\tau_m s + 1)]}$$

$$= \frac{K_L(\tau_1 s + 1)(\tau_2 s + 1)(\tau_m s + 1)}{(\tau_L s + 1)[(\tau_1 s + 1)(\tau_2 s + 1)(\tau_m s + 1) + K_c K_1 K_2]}$$

(b) For the transfer function between C and R,

$$\pi_f = K_c \frac{K_1}{\tau_1 s + 1} \frac{K_2}{\tau_2 s + 1}$$

and

$$\pi_\ell = \frac{K_c K_1 K_2}{(\tau_1 s + 1)(\tau_2 s + 1)(\tau_m s + 1)}$$

Then

$$\frac{C}{R} = \frac{K_c [K_1/(\tau_1 s + 1)][K_2/(\tau_2 s + 1)]}{1 + K_c K_1 K_2/(\tau_1 s + 1)(\tau_2 s + 1)(\tau_m s + 1)}$$

which simplifies to

$$\frac{C}{R} = \frac{K_c K_1 K_2 (\tau_m s + 1)}{(\tau_1 s + 1)(\tau_2 s + 1)(\tau_m s + 1) + K_c K_1 K_2}$$

CONTROL ACTIONS

Various types of controller actions are possible. These are described next.

On-Off or Bang-Bang Controller

In this case, when the measured variable is below the set point, the controller is on with a maximum output. However, the controller is off and the output is zero when the measured variable is above the set point. Because of the either-on-or-off mode of the operation, the on-off control action is inherently cyclic in nature. The on-off control is also called a *two-position control*. This type of control is seldom used in a continuous process because of the cycling nature of the response, surging flows, and wear on control valves.

Proportional and Floating Controllers

It is preferable to obtain non-cyclic steady operation when disturbances are absent. For this, the change in the controlled variable should be some continuous function of the error. Different types of the proportional and floating actions are given in Table 16.1.

In a proportional control, the controlled variable eventually becomes constant at a value, which differs from the desired set point resulting in a steady-state error or *offset*. There is no offset with the integral control because the controller output keeps changing until the error is reduced to zero. The integral action, however, has the disadvantage of the oscillatory response.

A derivative action is added to the proportional integral control to speed up the response so that the system returns to the original steady-state value rapidly with little or no oscillation.

Table 16.1 Proportional and floating control actions

Controller Type	Functional Relationship between Controller Output and Error	Transfer Function
Proportional	$P = K_c e$	$\dfrac{P}{E} = K_c$
Integral	$p = \dfrac{1}{\tau_i} \int_0^t e\, dt$	$\dfrac{P}{E} = \dfrac{1}{\tau_i s}$
Proportional + integral	$p = K_c \left(e + \dfrac{1}{\tau_i} \int_0^t e\, dt \right)$	$\dfrac{P}{E} = K_c \left(1 + \dfrac{1}{\tau_i s} \right)$
Proportional + derivative	$p = K_c \left(e + \tau_D \dfrac{de}{dt} \right)$	$\dfrac{P}{E} = K_c (1 + \tau_D s)$
Proportional + integral + derivative	$p = K_c \left(e + \dfrac{1}{\tau_i} \int_0^t e\, dt + \tau_D \dfrac{de}{dt} \right)$	$\dfrac{P}{E} = K_c \left(1 + \dfrac{1}{\tau_i s} + \tau_D s \right)$

where
- p = fractional change in controller output
- e = fractional change in error
- K_c = controller gain
- τ_i = integral time
- τ_D = derivative time

PROPORTIONAL BAND

The term *proportional band* is used to express the proportional control action and is defined as the percentage of the maximum range of the input variable required to cause 100 percent change in the controller output or

$$\% \text{ proportional band} = 100 \frac{\Delta e}{\Delta e_{max}} \tag{16.3}$$

where Δe_{max} is the maximum range of the error signal and Δe is the change in the error that gives a maximum change in the output of the controller. The proportional band is also called the proportional bandwidth and is often expressed as the percentage of the chart width.

For a proportional pneumatic controller, the maximum output is 3 to 15 psig or $\Delta P_{max} = 12$ psi, which is related to error signal by the equation

$$\Delta P_{max} = K_C\, \Delta e \tag{16.4}$$

Combining Equations 16.3 and 16.4 gives

$$\% \text{ proportional band} = \frac{100\, \Delta P_{max}}{K_C\, \Delta e_{max}}$$

where K_C is the gain of the controller.

Some pneumatic controllers are calibrated in the sensitivity units or psi/inch of the pen travel. With a 4-in. chart width and 3 to 15 psig controller output range, the sensitivity is given by the relation

$$S = \frac{15 - 3}{4} = 3 \text{ psi/in. travel} \tag{16.5}$$

Example 16.3

The liquid level at the bottom of a distillation column is controlled with a pneumatic proportional controller by throttling a control valve located at bottoms in the discharge line. The level may vary from 0.15 to 1 m from the bottom tangent line of the column. With the controller set point held constant, the output pressure of the controller varies from 100 (valve fully closed) to 20 kPa (valve fully open) as the level increases from 0.3 to 0.9 m from the bottom tangent line. (a) Find the percent proportional band and the gain of the controller. (b) If the proportional band is changed to 80 percent, find the gain and the change in level necessary to cause the valve to go from a fully open to a fully closed position.

Solution

(a) Proportional band

$$= 100 \frac{\Delta e}{\Delta e_{max}}$$

$$= \frac{\text{span of control variable for fully closed position of control valve}}{\text{full span or range}}$$

$$= \frac{0.9 - 0.3}{1 - 0.15} 100 = 70.6 \text{ percent}$$

$$\text{Gain of controller} = \frac{\Delta P}{\Delta e} = \frac{\text{change in controller output pressure}}{\text{corresponding change in control variable}}$$

$$= \frac{100 - 20}{0.9 - 0.3} = 133.3 \text{ kPa/m}$$

(b) If proportional band is 80 percent,

$$\frac{\Delta e}{\Delta e_{max}} = 0.8$$

$$\Delta e = 0.8 \Delta e_{max} = 0.8(1 - 0.15) = 0.68 \text{ m}$$

$$\text{Gain} = \frac{\Delta e}{\Delta e_{max}} = \frac{100 - 20}{0.68} = 117.65 \text{ kPa/m}$$

Hence, the change in the level necessary to cause the valve to go from a fully open to a fully closed position is 0.68 m, and the gain of the controller is 117.65 kPa/m.

DEVELOPMENT OF PROCESS MODELS

In order to devise a suitable control system for a process, the process behavior needs to be analyzed to derive differential equations describing the behavior of the process. To construct a block diagram of a control scheme for a process, mathematical modeling of the same is a prerequisite. Mathematical modeling helps to develop the control scheme. In addition, the controller can use the process model to anticipate the effect of a control action. There are also alternative controller design methods that make use of process models explicitly. Thus mathematical modeling of steady-state and dynamic behavior of a process is very important.

For process modeling, basic principles of chemical engineering such as material and energy balances, thermodynamics, kinetics, transport phenomena, etc. are required.

Differential equations are obtained for steady-state and transient operations of the process and these are used to obtain transfer functions and develop block diagrams for the control system. Depending on the complexity of the process, linear differential equations of the first order, second order, and of higher order are obtained. An example for a very simple case of level in a tank is given in Example 16.1. For more examples, reference is made to other texts (Bequette, Harriot, Coughanowr and Koppel, and Stephanopoulos).

Differential equations of linear models can be expressed in terms of *deviation variables,* that is, perturbations from a steady-state operating condition. Process models that describe the behavior of chemical processes in terms of material and energy balances are generally nonlinear. However, commonly used control systems are based on linear system models. Therefore nonlinear differential equations are linearized using Taylor expansion.

Dynamic Behavior of Basic Systems

The analysis of dynamic behavior of basic systems when their inputs change in some way is necessary as these more often form the component parts of a full control system. In order to study the transient response of a system, a transfer function is derived for the system and then its transient response is obtained by applying a disturbance or a forcing function to it to simulate dynamic behavior.

Forcing Functions

Step Function   A step disturbance is a sudden difference change of magnitude A from one level to the other, and thereafter remaining constant. Mathematical representation of this function is

$$X(t) = Au(t) \qquad (16.6)$$

If the size of step A is equal to 1, the function is called unit step function. The transform of this function is A/s and if it is a unit step function, it is simply $1/s$. This function is more frequently used in chemical process control analysis. The function is graphically shown in Figure 16.5.

Pulse Functions   A pulse is a function of arbitrary shape, usually rectangular or triangular. The function begins and ends at the same level. A sketch of rectangular pulse function is shown in Figure 16.6.

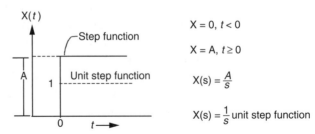

Figure 16.5 Step input and function

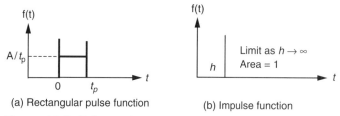

Figure 16.6 Pulse and impulse functions

Total integrated input of area magnitude A is applied over a time interval t_p. The function is $f(t) = A/t_p$ for $0 < t < t_p$ and $f(t) = 0$ for $t > t_p$. The Laplace transform of this function is

$$X(s) = \frac{A}{t_p}\left[\frac{1-e^{-t_p s}}{s}\right] \tag{16.7}$$

Impulse Function   This is known as Dirac delta function. This is an infinitely high pulse whose width is zero and area is unity. It is defined as

$$X(t) = A\delta t \tag{16.8}$$

The Laplace transform of this function is given by $X(s) = A$ and that of the unit impulse function is $X(s) = 1$.

Ramp Function   Ramp function varies proportionally to time.

$$F(t) = K(t) \quad \text{for} \quad t \geq 0 \tag{16.9}$$

where K is a constant. The Laplace transform of this function is K/s^2.

Sinusoidal Forcing Function   Response of a system to this forcing function is called frequency response, and is of great practical utility. In a linear system the input is sine wave with a frequency ω and the output is also a sine wave of the same frequency. The function is represented mathematically by the equations

$$X = 0 \quad \text{for} \quad t < 0 \quad \text{and} \quad X = A(\sin \omega t) \quad \text{for} \quad t \geq 0$$

Laplace transform of the sinusoidal forcing function is

$$X(s) = \frac{A\omega}{s^2 + \omega^2} \tag{16.10}$$

FIRST-ORDER SYSTEMS

If the transfer function of a system has a first-order denominator, the system is said to be first order. Only one parameter, called the time constant τ, is required to characterize the dynamic behavior of a first-order system. The differential equation for a linear first-order process is often written as

$$a_1 \frac{dy}{dt} + a_0 y = bf(t) \tag{16.11}$$

where $f(t)$ is the input and $y(t)$ is the output. If $a_0 = 0$, then dividing Equation 16.11 gives

$$\frac{a_1}{a_0}\frac{dy}{dt} + y = \frac{b}{a_0}f(t) \tag{16.12}$$

which can be written as

$$\tau_1 \frac{dy}{dt} + y = K_c f(t) \tag{16.13}$$

where
 τ_1 = time constant of the process and has the units of time
 K_c = gain of the process.

If $y(t)$ and $f(t)$ are in terms of deviation variables with a steady state as reference, the initial conditions will be

$$y(0) = 0 \quad \text{and} \quad f(0) = 0 \tag{16.14}$$

The transfer function of the first-order process is derived from Equation 16.13 and is given by

$$G(s) = \frac{y(s)}{f(s)} = \frac{K_c}{\tau_1 s + 1} \tag{16.15}$$

However, if $a_0 = 0$, we get from Equation 16.11

$$\frac{dy}{dt} = \frac{b}{a_1}f(t) = K'_c f(t) \tag{16.16}$$

The transfer function of Equation 16.16 is then

$$G(s) = \frac{y(s)}{f(s)} = \frac{K'_c}{s} \tag{16.17}$$

The response of first-order system described by the transfer function of Equation 16.15 can be obtained by applying a unit step change. We can write

$$y(s) = \frac{K_c}{s(\tau_1 s + 1)} = \frac{K_c}{s} - \frac{K_c \tau_1}{\tau_1 s + 1} \tag{16.18}$$

Inversion gives the solution of Equation 16.18 as follows:

$$y(t) = K_c(1 - e^{-t/\tau_1}) \tag{16.19}$$

If the forcing function were of magnitude A instead of unit step, the solution will be

$$y(t) = AK_c(1 - e^{-t/\tau_1}) \tag{16.19a}$$

The response can be plotted in dimensionless format as $y(t)/AK_c$ versus t/τ_1. This plot is shown in Figure 16.7.

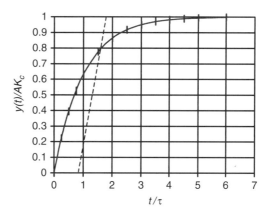

Figure 16.7 Response of first-order system to step input disturbance

Figure 16.7 is in dimensionless form and as such can be used to determine the response of any first-order system to a step disturbance regardless of the values of K_c, A, and τ_1.

Several features of the plot of Figure 16.7 are important enough to note here.

1. A first-order lag process is *self-regulating*. It reaches a new steady state after the disturbance is applied.

2. The slope of the response curve at $t = 0$ equals 1. This means that if the initial rate of change were to be maintained, the response would attain its final value in one time constant. The time constant τ_1 is the measure of the time required for the process to adjust to a change in its input.

3. The value of the response $y(t)/AK_c$ reaches 63.2 percent of its final value in time equal to one time constant. Ninety-eight percent of the final value (100%) is reached in a period of four time constants.

4. The ultimate value of the response is K_c for unit step change and AK_c for a step change of size A. This can also be inferred from Equations 16.19 and 16.19a by taking limits as $t \to \infty$. For a step change, the following relation is true

$$\Delta(\text{output}) = K_c \, \Delta(\text{input}) \tag{16.20}$$

From Equation 16.20, we can infer that if K_c is large, a small input will be required to effect a desired change in the output, whereas a large change in input would be required to produce the same change in output if K_c is small.

Dynamic response of a pure capacitive process given by the transfer function of Equation 16.17 is, however, quite different. For a unit step change, Equation 16.17 becomes

$$y(t) = \frac{K'_c}{s^2} \tag{16.21}$$

Inverting we obtain

$$y(t) = K'_c t \tag{16.21a}$$

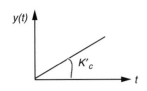

Figure 16.8
Unbounded response of pure capacitive process

We notice from Equation 16.21a that $y(t)$ increases linearly with an increase in time.
Thus

$$y(t) \to \infty \quad \text{as} \quad t \to \infty$$

The output is therefore unbounded. This response is plotted in Figure 16.8. The response of a number of first-order systems in series, whether interacting or not, and responses with other forcing functions can be derived in the same manner.

SECOND-ORDER SYSTEMS

A second-order system is characterized by a second-order differential equation. The differential equation can be written as follows:

$$a_2 \frac{d^2 y}{dt^2} + a_1 \frac{dy}{dt} + a_0 y = b f(t) \tag{16.22}$$

If $a_0 \neq 0$, Equation 16.22 can be rewritten in a standard form as follows:

$$\tau^2 \frac{d^2 y}{dt^2} + 2\xi\tau \frac{dy}{dt} + y = K_c f(t) \tag{16.23}$$

where
$\tau^2 = a_2/a_0$, $2\xi\tau = a_1/a_0$ and $K_c = b/a_0$
τ = natural period of oscillation of the system
ξ = defined as damping factor
K_c = process gain.

If Equation 16.23 is in terms of deviation variables based on steady-state reference state, the initial conditions are zero and then Laplace transform of Equation 16.23 is

$$G(s) = \frac{y(s)}{f(s)} = \frac{K_c}{\tau^2 s^2 + 2\xi\tau s + 1} \tag{16.24}$$

It requires two parameters, τ and ξ to describe the dynamic behavior of a second-order system,
where
ξ = damping coefficient
τ = period of oscillation.

If a unit step change is applied, Equation 16.24 becomes

$$y(s) = \frac{1}{s} \frac{1}{(\tau^2 s^2 + 2\xi\tau s + 1)} \tag{16.25}$$

The roots of the quadratic in Equation 16.25 will be real or complex depending on the value of the parameter ξ. The nature of the roots will in turn determine the

Table 16.2 Responses of a second-order system

ξ	Nature of Roots	Response
<1	Complex	Oscillatory or underdamped
=1	Real and equal	Critically damped
>1	Real	Nonoscillatory or overdamped

response function. The solution of Equation 16.25 can be written as follows:

$$y(s) = \frac{1/\tau^2}{s(s-s_1)(s-s_2)} \tag{16.26}$$

where s_1 and s_2 are the roots of the quadratic in Equation 16.25, which are given by

$$s_1 = -\frac{\xi}{\tau} + \frac{\sqrt{\xi^2-1}}{\tau} \quad \text{and} \quad s_2 = -\frac{\xi}{\tau} - \frac{\sqrt{\xi^2-1}}{\tau} \tag{16.27}$$

For three possible types of roots the system response will be as given in Table 16.2. The three cases are as follows:

Case 1: Step response when $\xi < 1$. In this case the inversion gives the following solution:

$$y(t) = 1 - \frac{1}{\sqrt{1-\xi^2}} e^{-\xi t/\tau} \sin\left(\sqrt{1-\xi^2}\,\frac{t}{\tau} + \tan^{-1}\frac{\sqrt{1-\xi^2}}{\xi}\right) \tag{16.28}$$

$y(t)$ in dimensionless form is plotted against t/τ in Figure 16.9 at various values of ξ as parameters.

For values of $\xi < 1.0$, the response curves are oscillatory. As ξ increases, the response becomes less oscillatory. This response is called *underdamped*.

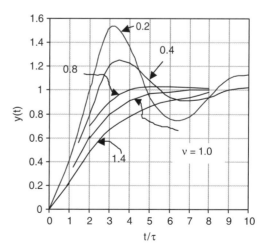

Figure 16.9 Response of second-order system to a step change

Case 2: Step response for $\xi = 1$. For this case the solution is

$$y(t) = 1 - \left(1 + \frac{t}{\tau}\right)e^{-t/\tau} \tag{16.29}$$

The response plotted in Figure 16.9 is nonoscillatory. This response is also called a *critically damped* response because for values of $\xi < 1$, the response begins to oscillate.

Case 3: Step response when $\xi > 1.0$. The solution for this case is

$$y(t) = 1 - e^{-\xi t/\tau}\left(\cosh\sqrt{\xi^2 - 1}\,\frac{t}{\tau} + \frac{\xi}{\sqrt{\xi^2 - 1}}\sinh\sqrt{\xi^2 - 1}\,\frac{t}{\tau}\right) \tag{16.30}$$

The response is plotted in Figure 16.9 for a value of $\xi = 1.4 > 1.0$. The response is not oscillatory but more sluggish. This is called an *overdamped* response.

Of the three responses of the second-order system listed in Table 16.2, the oscillatory or underdamped response is the most frequent in control systems. Hence, a number of special terms are in use to describe the underdamped response. These terms and their equations, which can be derived from the time response equation, are given below:

Overshoot for a unit step change is the amount by which the response exceeds the ultimate steady-state value. It is related to ξ by the equation

$$\text{Overshoot} = \exp(-\pi\xi/\sqrt{1-\xi^2}) \tag{16.31}$$

Decay ratio is defined as the ratio of successive peaks and is given by

$$\text{Decay ratio} = \exp(-2\pi\xi/\sqrt{1-\xi^2}) = (\text{overshoot})^2 \tag{16.31a}$$

Rise time, t_r: Time required for the response to first reach its ultimate value.

Response Time: Time required for the response to come within ± a certain percent (usually ±5%) of its ultimate value.

Period of Oscillation: The radian frequency (rad/time) of the oscillations of an underdamped response is the coefficient of t in the sine term in the solution of the characteristic equation for this case. Thus

$$\text{radian frequency, } \omega = \frac{\sqrt{1-\xi^2}}{\tau} \text{ radians/time} \tag{16.31b}$$

The radian frequency ω and cyclical frequency f are related by $\omega = 2\pi f$. Hence, the cyclical frequency is given by

$$Cyclical\ frequency = f = \frac{1}{T} = \frac{1}{2\pi}\frac{\sqrt{1-\xi^2}}{\tau} \quad \text{also } T = \text{period of oscillation}$$

$$= \text{time/cycle, from which, } T = \frac{2\pi\tau}{\sqrt{1-\xi^2}} \quad (16.31\text{c})$$

Natural Period of Oscillation: When damping is eliminated ($\xi = 0$), the system oscillates continuously without attenuation in amplitude. The radian frequency under these undamped conditions is $1/\tau$ and is referred to as natural frequency ω_n.

$$\omega_n = 1/\tau \quad (16.31\text{d})$$

$$Natural\ frequency = f_n = 1/T_n = 1/2\pi\tau \quad (16.31\text{e})$$

$$\frac{Actual\ frequency}{Natural\ frequency} = \frac{f}{f_n} = \sqrt{1-\xi^2} \quad (16.31\text{f})$$

Example 16.4

The block diagram of a control system having a first-order process, a measurement lag, and containing a two-mode (proportional-derivative) control action is shown in Exhibit 3. (a) Find an expression for the period of oscillation τ and the damping coefficient for the closed-loop response. Assume a regulatory operation. (b) Compute and compare the offset for a step change of the load when $\tau_1 = 2$ min, $\tau_m = 20$ s, $\xi = 0.8$, and when (i) $\tau_D = 0$, (ii) $\tau_D = 2$s.

Solution

For regular operation, $R(s) = 0$ and from Equation 16.2,

$$\frac{C(s)}{U(s)} = \frac{\text{product of transfer functions between } C \text{ and } U}{1+\text{product of all transfer functions in the closed loop}}$$

$$= \frac{1/(\tau_1 s + 1)}{1+[K(1+\tau_D s)/(\tau_m s + 1)(\tau_1 s + 1)]}$$

$$= \frac{\tau_m s + 1}{\tau_m \tau_1 s^2 + (\tau_m + \tau_1 + K\tau_D)s + (1+K)}$$

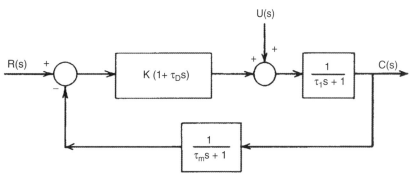

Exhibit 3 System block diagram (Example 16.4)

(a) The denominator (called the *characteristic equation*) of the preceding equation is expressed in the form of Equation 16.25.

$$\tau_m \tau_1 s^2 + (\tau_m + \tau_1 + K\tau_D)s + 1 + K = 0$$

or

$$s^2 + \frac{\tau_m + \tau_1 + K\tau_D}{\tau_m \tau_1}s + \frac{1+K}{\tau_m \tau_1} = 0$$

By a comparison of this equation with Equation 16.6, one obtains

$$2\xi\omega = \frac{\tau_m + \tau_1 + K\tau_D}{\tau_m \tau_1}$$

and

$$\omega^2 = \frac{1+K}{\tau_m \tau_1}$$

Hence

$$\omega = \frac{1}{\tau} = \sqrt{\frac{1+K}{\tau_m \tau_1}}$$

and

$$\xi = \frac{\tau_m + \tau_1 + K\tau_D}{2\omega \tau_m \tau_1} = \frac{\tau_m + \tau_1 + K\tau_D}{2\sqrt{(1+K)(\tau_m \tau_1)}}$$

Because the disturbance is a step of size a, $U(s) = a/s$ and

$$C(s) = \frac{\tau_m s + 1}{\tau_m \tau_1 s^2 + (\tau_m + \tau_1 + K\tau_D)s + (1+K)} \cdot \frac{a}{s}$$

By the final-value theorem Equation 16.1e, the offset is given by

$$\text{Offset} = sC(s)\Big|_{s \to 0} = \frac{(\tau_m s + 1)a}{\tau_m \tau_1 s^2 + (\tau_m + \tau_1 + K\tau_D)s + (1+K)} = \frac{a}{1+K}$$

(b) (i) For $\tau_1 = 2$ min $= 120$ s, $\tau_m = 20$ s, $\tau_D = 0$, $\xi = 0.8$,

$$\xi = \frac{\tau_m + \tau_1 + K\tau_D}{2\sqrt{(1+K)(\tau_m \tau_1)}}$$

$$0.8 = \frac{20 + 120 + 0}{2\sqrt{(1+K)(20)(120)}} = \frac{70}{\sqrt{(1+K)(2400)}}$$

Therefore, $(1 + K)(2400) = 4900/0.64 = 7656.25$ and $K = 2.19$.

$$\omega = \frac{1}{\tau} = \sqrt{\frac{1+K}{\tau_m \tau_1}} = \sqrt{\frac{1+2.19}{20(120)}} = 0.0365$$

$$\text{Offset} = \frac{a}{1+K} = \frac{a}{1+2.19} = \frac{a}{3.19} = 0.313a$$

(b) (ii) For $\tau_1 = 120$ s, $\tau_m = 20$ s, $\tau_D = 2$ s, $\xi = 0.8$

$$\xi = 0.8 = \frac{\tau_m + \tau_1 + K\tau_D}{2\sqrt{(1+K)(\tau_m\tau_1)}}$$

$$= \frac{20 + 120 + 2K}{2\sqrt{(1+K)(20)(120)}} = \frac{70+K}{\sqrt{(1+K)(2400)}}$$

Hence, squaring both sides and simplifying gives the quadratic equation

$$K^2 - 1396K + 3364 = 0$$

from which

$$K = \frac{1396 \pm \sqrt{1396^2 - 4(3364)}}{2}$$

which gives $K = 1393.6$ or 2.414. When $K = 1393.6$,

$$\omega = \frac{1}{\tau} = \sqrt{\frac{1+K}{\tau_m\tau_1}} = \sqrt{\frac{1+1393.6}{20(120)}} = 0.762$$

Then

$$\text{Offset} = \frac{a}{1+K} = \frac{a}{1393.6} = 0.0007a$$

When $K = 2.414$,

$$\omega = \frac{1}{\tau} = \sqrt{\frac{1+K}{\tau_m\tau_1}} = \sqrt{\frac{1+2.414}{20(120)}} = 0.0377$$

and

$$\text{Offset} = \frac{a}{1+K} = \frac{a}{1+2.414} = 0.293a$$

Poles and Zeros of a Transfer Function

The transfer function is defined as the ratio of Laplace transforms of output and input and is given by

$$G(s) = \frac{y(s)}{f(s)} \tag{16.32}$$

Generally, the transfer function is a ratio of two polynomials. The denominator is called the characteristic equation. The transfer function then can be expressed as

$$G(s) = \frac{b_m s^m + b_{m-1} s^{m-1} + \cdots + b_1 s + b_0}{a_n s^n + a_{n-1} s^{n-1} + \cdots + a_1 s + a_0} \tag{16.33}$$

The roots of the denominator are called the poles of the transfer function. Values of s that cause the numerator of Equation 16.33 to equal zero are called *zeros of the transfer function*. Factoring both numerator and denominator yields the *pole-zero form* of the transfer function.

$$G(s) = \left(\frac{b_m}{a_n}\right) \frac{(s-z_1)(s-z_2)\cdots(s-z_m)}{(s-p_1)(s-p_2)\cdots(s-p_n)} \qquad (16.34)$$

where
z_i = zeros of the transfer function
p_i = poles of the transfer function.

The *gain-time constant form* is commonly used for control system analysis.

CONCEPT OF STABILITY AND STABILITY CRITERIA

An important consideration in closed loop control systems is the stability of the system. A system is said to be stable if the output response is bounded for all bounded inputs. Unbounded response to a bounded input indicates that the system is unstable. This can be put in mathematical terms as follows:

If the response function in the time domain, $y(t)$ is bounded as $t \to \infty$ for all bounded inputs the system is said to be stable; if $y(t) \to \infty$ as $t \to \infty$, the system is unstable. It should be understood though that infinitely unbounded outputs exist only in theory and not in practice because all physical quantities have limitations. Therefore the term unbounded should be taken to mean very large.

The denominator of the closed-loop transfer function for a higher-order system is a polynomial larger than second-degree. For a control system to be stable, the roots of the characteristic equation, which are the poles of the transfer function, must be real or must occur as complex conjugate pairs. Moreover, the real parts of all poles must be negative for the system to be stable. In other words, a system is stable if all its poles lie in the left half of the s plane. The locations of the zeros have no effect on the stability of the system although they do influence the dynamic response.

Calculation of the roots of the characteristic equation is not required if the objective is only to find out whether the system is stable or not. Routh-Hurwitz criterion for stability allows one to test if any root is to the right of the imaginary axis. This test is limited to polynomial characteristic equations.

Routh's Procedure: The procedure consists of writing the characteristic equation in the form

$$a_0 s^n + a_1 s^{n-1} + a_2 s^{n-2} + \cdots + a_n = 0 \qquad (16.35)$$

where a_0 is positive. (If a_0 is originally negative multiply both sides by -1.) In this form it is necessary that all coefficients be positive if all the roots are to lie in the left half of the s plane. If any coefficient is negative, the system is definitely unstable and Routh's test is not required to answer the question of stability.

If all the coefficients are positive, the system may be stable or unstable. Then it is necessary to prepare Routh's array to determine the stability of the system. The Routh's array is prepared in the following manner:

Row				
1	a_0	a_2	a_4	a_6
2	a_1	a_3	a_5	a_7
3	b_1	b_2	b_3	
4	c_1	c_2	c_3	
5	d_1	d_2		
6	e_1	e_2		
7	f_1			
$n+1$	g_1			

The elements in the remaining rows are found by using the following formulas:

$$b_1 = \frac{a_1 a_2 - a_0 a_3}{a_1} \qquad b_2 = \frac{a_1 a_4 - a_0 a_5}{a_1} \ldots$$

$$c_1 = \frac{b_1 a_3 - a_1 b_2}{b_1} \qquad c_2 = \frac{b_1 a_5 - a_1 b_2}{b_1} \ldots$$

Notice that the elements in any row are obtained from the elements in the two preceding rows. Any row can be divided by a positive number. This will not affect the results of the test but will simplify computation.

Once Routh's array is obtained, the following rules should be followed to determine the stability of the system (Coughanowr and Koppel):

1. The necessary and sufficient condition for all the roots of the characteristic equation to have negative real parts is that all elements of the first column of Routh's array be positive and nonzero. Then the system is stable.

2. If some elements in the first column of Routh's array are negative (the system is unstable), the number of roots with positive real parts in the right half plane is equal to the number of sign changes in the first column.

Routh's array derived by algebraic method can only tell us whether the system is stable or not. It gives neither the actual roots of the characteristic equation nor any idea about the degree of stability.

Example 16.5

A control system with a negative feedback is shown in Exhibit 4.

(a) Write down the transfer function and the characteristic equation for the system.

(b) Determine how many zeros the transfer function has.

(c) Construct Routh's array for the system.

(d) Is the system stable for (1) $K_c = 9.5$ (2) $K_c = 11$ (3) $K_c = 6$?

(e) How many roots with positive real parts are in the right half of plane for $K_c = 11$?

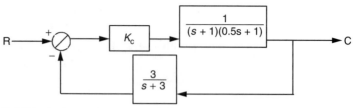

Exhibit 4 Control system with negative feedback

(f) Is the system stable for $K_c = 10$?

Solution

(a) For a feedback system transfer function

$$\frac{C(s)}{R(s)} = \frac{K_c\left(\frac{1}{(s+1)(0.5s+1)}\right)}{1+K_c\left(\frac{1}{(s+1)(0.5s+1)}\frac{3}{(s+3)}\right)}$$

$$= \frac{2K_c(s+3)}{s^3+6s^2+11s+6+6K_c}$$

The characteristic equation is

$$1+\frac{3K_c}{(s+1)(0.5s+1)(s+3)}=0$$

which on simplification gives

$$s^3+6s^2+11s+6(1+K_c)=0$$

(b) The transfer function has one zero $s = -3$.

(c) Construction of Routh's array: The coefficients of the characteristic polynomial are arranged in rows as follows:

Row		
1	$a_0 = 1$	$a_2 = 11$
2	$a_1 = 6$	$a_3 = 6(1+K_c)$
3	$b_1 =$	
4	c_1	

Divide the second row by 6, a positive constant. Then the array becomes

Row		
1	$a_0 = 1$	$a_2 = 11$
2	$a_1 = 1$	$a_3 = 1(1+K_c)$
3	$b_1 = 10 - K_c$	
4	$c_1 = 1 + K_c$	

$$b_1 = \frac{a_1 a_2 - a_0 a_3}{a_1} = \frac{1\times 11 - (1+K_c)}{1} = 10 - K_c$$

$$c_1 = \frac{b_1 a_3 - a_1 b_2}{b_1} = \frac{(10-K_c)(1+K_c)-0}{10-K_c}$$

$$= 1 + K_c \text{ if } (10 - K_c) = 0 \text{ otherwise } 0$$

(d) 1. For $K_c = 9.5$, third row element in first column $= 10 - 9.5 = 0.5$, a positive number. Then all the elements in the first column are positive. Hence the system is stable.

2. For $K_c = 11$, $(10 - K_c) = 10 - 11 = -1$ a negative number. Thus, the element in the third row and first column is negative. Since all elements in first column are not positive, the system is unstable.

3. For $K_c = 6$, third row first column element is positive. Hence all elements in the first column are positive and the system is stable.

(e) For $K_c = 11$, b_1 in the first column is negative. Hence there are two sign changes, one from a_1 to b_1 and second from b_1 to c_1. Therefore there are two roots with positive real parts in the right plane.

(f) When $K_c = 10$, b_1 and c_1 elements in the first column of Routh's array become zero and the system is unstable.

For additional review of advanced topics such as root locus methods, frequency analysis, Bode diagrams, Nyquist stability criteria, nonlinear methods such as phase-plane analysis, controller tuning and system analysis based on discrete time models including z-transforms, the references given at the end of this chapter should be consulted. Frankly, from the point of the P.E. examination, which assigns only 2 percent of the exam to Process Control, very rigorous study of advanced topics is not warranted.

RECOMMENDED REFERENCES

1. Coughanowr and Koppel, *Process Systems Analysis and Control,* McGraw-Hill, NY, 1991.
2. Harriott, *Process Control,* McGraw-Hill, NY, 1984.
3. Bequette, *Process Control, Modeling Design and Simulation,* Prentice-Hall, NJ, 2003.
4. Luyben and Luyben, *Essentials of Process Control,* McGraw-Hill, NY, 1997.
5. Stephanopoulos, *Chemical Process Control,* Prentice-Hall, NJ, 1984.

CHAPTER 17

Corrosion and Materials of Construction

OUTLINE

CORROSION TYPES AND CORROSION CONSIDERATION OF MATERIALS 639
Uniform Attack ■ Galvanic or Bimetal Corrosion ■ Crevice Corrosion ■ Pitting ■ Intergranular Corrosion ■ Selective Leaching ■ Erosion Corrosion ■ Stress Corrosion ■ Hydrogen Damage

MATERIALS OF CONSTRUCTION AND THEIR BEHAVIOR 645
Metals ■ Non-metals

LOW-TEMPERATURE APPLICATION OF METALS 652

MAXIMUM ALLOWABLE STRESS IN TENSION VERSUS TEMPERATURE 653

COST CONSIDERATIONS 653

RECOMMENDED REFERENCES 654

Proper design consideration is the foundation for the economic life of the equipment of a chemical plant. Selection of proper material of construction is essential to maintain product quality and gain competitive edge. An understanding of how to prevent and control corrosion is required to reduce the economic losses caused by corrosion. The appropriate design considerations are

■ Corrosion types and consideration of corrosion of metals and non-metals

■ Thermal effect on allowable stress of a material of construction

■ Cost consideration.

CORROSION TYPES AND CORROSION CONSIDERATION OF MATERIALS

There are eight forms of corrosion. (Refer to Figures 17.1 and 17.2 to avoid or minimize various forms of corrosion.

1. Uniform attack

2. Galvanic or bimetal corrosion

3. Crevice corrosion

4. Pitting

5. Intergranular corrosion

6. Selective leaching

7. Erosion corrosion

8. Stress corrosion

Uniform Attack

This form of corrosion takes place uniformly on the surface of the material. It is characterized by a chemical or an electrochemical reaction. Because the corrosion is uniform, equipment life is predictable when the corrosion rate is known. This type of corrosion is preventable by proper choice of materials dictated by corrosion test, inhibitors, or cathodic protection. Alternatively, a corrosion allowance, typically 0.125 in. for carbon steel, and 0.065 in. for stainless steel, is added to material thickness to arrive at the final plate thickness in design.

Galvanic or Bimetal Corrosion

When two dissimilar metals are immersed in a corrosive solution and a conductive connection is made between them, an electric current flows between them. As a result, the less corrosion-resistant metal, called the anode, less noble metal, or active metal, corrodes. The cathode or nobler metal does not generally corrode. For example, if a heat exchanger has copper tubes connected to a steel tube sheet, the steel will corrode due to galvanic corrosion. To avoid this, in this example, the tube sheet material should be bronze. The following table shows the galvanic series of some materials.

In Table 17.1, the higher the nobility number, the *less* noble is the material. Thus in a yacht with a Monel hull (nobility 10) and Steel rivet (Nobility 19), the rivet will fail by galvanic corrosion in seawater. Materials in the same nobility group are compatible with each other against galvanic corrosion. This table may be used to make choices when two dissimilar materials are used to fabricate equipment.

If carbon steel jacket is designed for stainless steel pipe or a stainless steel vessel, make sure that a transition pad of stainless steel is first welded to the stainless steel inner pipe or vessel and then the carbon steel jacket is welded to the transition pad. This will prevent the galvanic corrosion of the carbon steel part.

The filler metal for welding should be compatible with the parts being welded. If the use of two dissimilar metals forming a galvanic couple cannot be avoided (as in the example of the jacketed vessel cited above), then the parts with the smaller surface area should be made of a nobler metal (stainless steel as chosen), and the nobler part (the stainless steel part in this example) should be painted on the jacket side to reduce the cathode area.

Equipment should be designed for good drainage to avoid stagnation of materials after cleaning and shut down. Equipment support for insulated equipment for high-temperature gaseous operations should also be insulated to avoid corrosion at the cold spot due to condensation.

Table 17.1 Galvanic series of some materials in sea water

Nobility	Graphite/Metal/Alloys
1	Platinum
2	Gold
3	Graphite
4	Titanium
5	Silver
6	Chlorimet 3 (Ni 62, Cr 18, Mo 18), Hastelloy C (Ni 62, Cr 17, Mo 15)
7	18-8 Mo Stainless Steel (passive), 18-8 Stainless steel
8	Chromium stainless steel (11–30% Cr -passive), inconel (passive)
9	Silver solder
10	Monel (Ni 70, Cu 30), Cupronickel (Cu 60-90, Ni 40-10), Bronze (Cu-Sn), Copper, Brass (Cu-Zn)
11	Chlorimet 2 (Ni 66, Mo 32, Fe 1), Hastelloy B (Ni 60, Mo 30, Fe 6, Mn 1)
12	Inconel (active), Nickel (active)
13	Tin
14	Lead
15	Lead-Tin solders
16	18-8 Mo Stainless steel (active), 18-8 stainless steel
17	High Nickel cast iron
18	Chromium Stainless steel (Cr 13%)
19	Cast iron, steel or iron
20	2024 Aluminum (Cu 4.5, Mg 1.5, Mn 0.6)
21	Cadmium
22	Commercially pure Aluminum
23	Zinc
24	Magnesium & Magnesium alloys

Crevice Corrosion

Crevice corrosion occurs when a small volume of stagnant solution is trapped between two metal surfaces, such as in a lap joint, gasket surface, crevices under bolt, and rivet head, etc. The tube-to-tube sheet joint should be strength-welded per TEMA-B or TEMA-R instead of simple rolling to prevent crevice corrosion. Lap joints should be replaced by welded butt joints or the crevices of lap joints should be closed by continuous welding on both sides. Backing rings are used to produce full-penetration circumferential welds in piping that is accessible from only one side; but this leaves room for crevice corrosion on the inaccessible side.

Three recommended methods of welding to avoid crevice corrosion are

1. TIG (Tungsten Inert Gas) welding with gas backup

    In the TIG process the arc is formed between a pointed tungsten electrode and the workpiece in an inert atmosphere of argon or helium. The small intense arc provided by the pointed electrode is ideal for high quality and precision welding. Because the electrode is not consumed during welding, the welder does not have to balance the heat input from the arc as the metal is deposited from the melting electrode. When filler metal is required, it must be added separately to the weld pool. Power source must be constant current. Electrodes: tungsten with 1 to 4 percent thoria, or additives like lanthanum oxide. Shielding gas is argon, argon plus 2 to 5 percent hydrogen or helium.

2. Non-metallic removable backup ring, which may be carbon and ceramic

3. Consumable insert rings, which may be EB insert rings (EB is for Electric Boat Co, its developer), or flat rings.

Flat-bottom tanks should not be directly supported on a surface. They should be supported on a skirt at the perimeter and I-beams on the flat portion with non-porous caulking around the I-beams.

Pitting

Pitting is an autocatalytic anodic reaction process, and is an extremely localized attack that results in a hole in the material. It is unpredictable and appears with extreme suddenness. In general, chloride and chlorine containing ions, oxidizing metal ions such as cupric, ferric, and mercuric halides are aggressive pitters. Plain carbon steel is better than stainless steel alloys (type 304, 316) to minimize the chance of pitting in pitting applications. Hastelloy F, Nionel, Durimet 20, Hastelloy C, Chlorimet 3, and Titanium have better pitting resistance, in the order shown.

Intergranular Corrosion

This is a localized attack at the grain boundaries of metals or alloys. When chromium is added to ordinary steel to increase its resistance in many environments, such resistance comes with a price. During a heat treatment process, which includes heating the steel in the range of 510°C to 788°C, chromium carbide precipitates out at the grain boundary, and the concentration of chromium is depleted in the area adjacent to the grain boundary. The intergranular corrosion takes place in this chromium-depleted area. Knife-line attack as well as weld decay is a form of intergranular corrosion resulting from chromium carbide precipitation during welding. High-temperature heat treatment (1066°C to 1121°C) after fabrication, addition of a stabilizer such as Mo (Type 316 SS), Columbium (Type 347 SS), Titanium (Type 321 SS), and lowering of carbon content to less than 0.03 percent help intergranular corrosion resistance. The letter L in stainless steel grade specification (Type 304L, 316L, etc) indicates this extra low carbon, which increases intergranular corrosion resistance.

If the rolled surface of a stabilized or properly heat-treated austenitic steel is cut transverse to the rolling direction, the end grain is exposed. This end should be covered with an austenitic stainless steel weld to prevent end-grain attack. Equipment design and layout should be properly done so that a welder can get in position to make a sound weld joint.

Selective Leaching

Selective leaching is a form of corrosion that involves removal by leaching one of the components of an alloy in an environment. For example, when brass pipe (alloy of zinc 30 and copper 70) is used in potable water service, the inside becomes dark by selective leaching of zinc. Reducing the zinc content of the brass helps, for example, as in red brass. Graphitization or graphitic corrosion is a type of selective leaching in the corrosion of cast iron where the iron from the iron-carbon matrix is slowly leached out, leaving a porous mass of graphite of weak structural strength.

Corrosion Types and Corrosion Consideration of Materials **643**

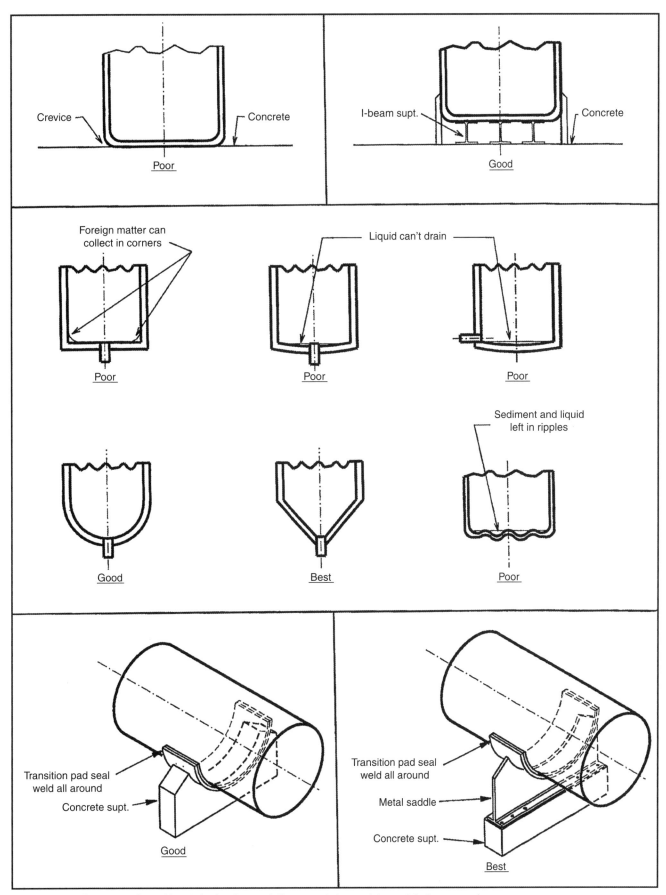

Figure 17.1 Poor, good, and best design for metallic vessels to avoid or minimize corrosion

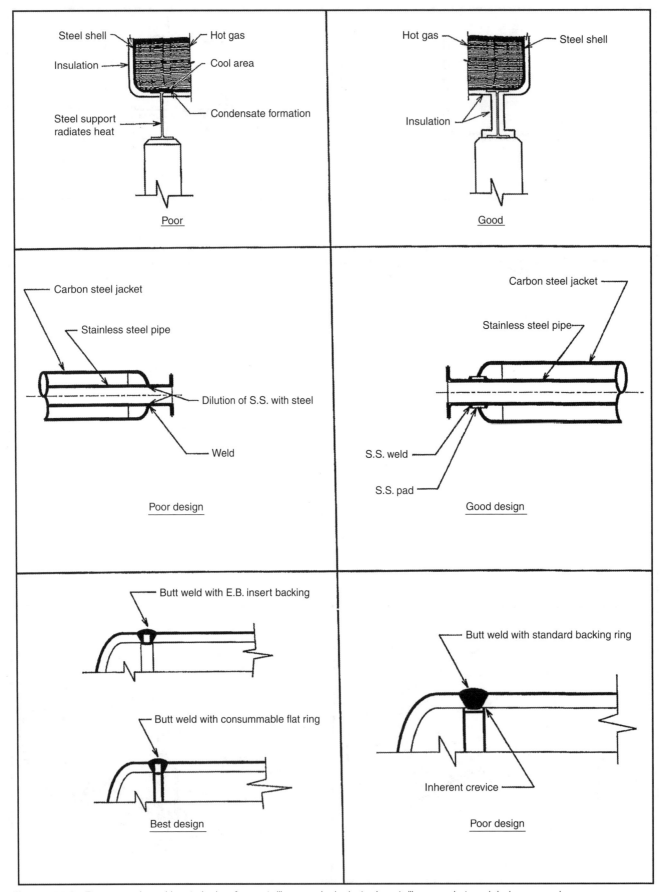

Figure 17.2 Poor, good, and best design for metallic vessels, jacketed metallic vessels to minimize corrosion

Erosion Corrosion

The relative movement between a corrosive environment and a metal causes erosion corrosion. Several factors contribute to erosion corrosion: loss of protective films on materials like stainless steels, high velocity and turbulence, impingement, velocity-induced galvanic effect, and metallurgical history of the metal or alloy. Combating erosion corrosion involves lowering the fluid velocity, installing an impingement plate, applying a coating, and cathodic protection.

Fretting is a special form of erosion corrosion that occurs at contact areas between materials under load subjected to vibration and slip such as in automotive parts and dynamic components of machinery.

Stress Corrosion

Stress corrosion cracking (SCC) is caused by the simultaneous presence of a corrosive medium and a tensile stress. Even though the surface of the material may seem to be intact, this corrosion appears in the form of fine cracks. Caustic embrittlement is a form of stress corrosion of steel. Metals may fracture due to stress corrosion fatigue, which is caused by repeated cyclic load. The mechanism of stress corrosion is not well understood. Preventative methods include heat treatment to reduce residual stress, proper choice of material (for example, inconel to replace type 304 stainless steel), application of cathodic protection, etc.

If thermal insulation is applied to austenitic stainless steel surface, make sure that the insulation does not have water-soluble chloride, or provide stainless steel band and stainless steel jacket over the entire insulation surface to prevent wetting of the insulation. This is to avoid stress corrosion cracking of the austenitic stainless steel from the water soluble chloride.

Hydrogen Damage

Hydrogen damage is a special corrosion that is caused by the diffusion of atomic hydrogen through a metal or alloys. Hydrogen damage may be divided into four types:

1. Hydrogen blistering

2. Hydrogen embrittlement

3. Decarburization

4. Hydrogen attack

The absorption of hydrogen produced by the reaction of an aqueous solution of hydrogen sulfide and iron in carbon steel in the hard weld zones causes significant loss in ductility, and the material may fail at a stress significantly below yield strength. Preventative methods include the use of coating, proper alloy, proper welding techniques, and baking of steel.

MATERIALS OF CONSTRUCTION AND THEIR BEHAVIOR

Materials of construction may be broadly divided into two groups: metals and non-metals.

Metals

The following are the most common metallic materials of constructions

1. Iron, carbon steel, and low alloy steel
2. Stainless steel
3. Nickel and high-nickel alloy steel
4. Copper, aluminum, and their alloys
5. Titanium, tantalum, zirconium, and columbium

Iron, Carbon Steel, and Low Alloy Steel

Carbon steel is an alloy of carbon, cementite (Fe_3C), and iron, carrying only minor percentages of silicon, manganese, and aluminum intentionally added to affect deoxidation and control grain size. Low alloy steels have one or more alloying elements to improve mechanical and corrosion-resistant properties over those of carbon steels.

These metals are widely used materials of construction in the petrochemical industries.

Iron and carbon steel are widely used as structural supports. Corrosion resistance depends on oxide surface film. Free graphite in cast iron is electrochemically noble, and therefore promotes corrosion in acid and salt solution. Carbon steel may be used to handle water, organic solvents and similar chemicals, and sulfuric acid at 90 percent by weight or stronger aqueous solution. Stress-relieved plain carbon steel may be used to handle caustic soda up to 75 percent and 212°F. Similar behavior is expected from plain and low alloy cast iron. Low alloy steels containing Ni show resistance to acids, alkaline, and neutral environments. Low alloy steel containing Si and Cu improves resistance to HCl.

Stainless Steels

This is a class of steels characterized by its resistance to corrosion. These steels are individually tailored to meet specific needs. Their resistance to corrosion is due to the formation of a passive oxide film at the surface, and therefore, their application includes oxidizing environments. Notable alloying elements are chromium and nickel. Stainless steels may be subdivided into four groups based on their metallurgical structures. (See Table 17.2)

1. Martensitic stainless steel
2. Ferritic stainless steel
3. Austenitic stainless steel
4. Precipitation hardening stainless steel

Nickel and High-Nickel Alloys

Commercial nickel and its alloys with elements like Cu, Cr, Mo, and Fe provide corrosion resistance to a wide range of environments. All nickel alloys have nickel as the major component, and C, Mn, Fe (except Carpenter) and Si as minor components. Besides these common elements, some nickel alloys have Cr, S, Cu, Mo, Ti, Co, Al, Ta, W, and P. Table 17.3 shows their major applications.

Table 17. 2 Characteristics summary of stainless steels

Group	AISI Type and Commercial Spec.	Major Alloying Components	Application	Remark
Martensitic	405, 410, 416, 420, 431, 440	Cr (12–17%) C (0.02–1%)	Atmospheric application, fresh water, organic exposure	Hardened metals
Ferritic	430, 430Ti	Cr, C	Architectural, automotive, high temperature	Soft metal
Austenitic	301, 302, 302B, 304, 304L, 308, 309, 309S, 310, 316, 316L, 321, 347, 348	Cr (18%), Ni(8%), Stabilizers: Ti(321), Ta(347), Cb(348), Mo(316, 347)	Petrochemical process industries	'L' indicates extra-low carbon.
Precipitation hardening	17-4PH, 15-5PH, PH13-8Mo, PH15-7Mo	Cr, Ni, Mo	Petrochemical process industries	

Copper, Aluminum, and their Alloys

Commercial copper and its alloys containing zinc, brass, and bronze, find many industrial applications. The zinc content in brass (15%–37%) is more than that in bronze (10%–12.5%).

Other alloys of copper are cupronickel, aluminum brass, aluminum bronze, etc. Copper and its alloys may be used in breweries and food industries, and with the exception of brass, in alkaline solutions at room temperature, and in neutral salt solutions. Copper can handle low-pressure steam and condensate, dry ammonia, and be used in refrigeration.

Cupronickel is most resistant in moist ammonia. Aluminum covers a broad family of metals ranging from high-purity aluminum to alloys containing as high

Table 17.3 Major application of nickel and high-nickel alloys

Alloy	Nickel, wt%	Application
Nickel 200	99.5	Caustics, halogens, Hydrogen halides, non-oxidizing salts, foods.
Nickel 201	99.5	As above.
Monel alloy 400	66.5	As above plus sulfuric acid, HF, brine, sea & brackish water. Resistant to SCC.
Monel alloy K500	66.5	As above plus sulfuric acid, HF, brine, sea & brackish water. Resistant to SCC.
Hastelloy B	61.0	Non-oxidizing acid & superior resistance to HCl.
Hastelloy C-276,	57	Pitting & oxidation resistance to aqueous corrodents. C-276 has benefits over C-22 in highly
Alloy C-22,	57	reducing medium. Alloy 59 has highest pitting resistance, and is used in HCl, H_2SO_4, HF,
Alloy 59	59	Salts, low pH, Cl^-, F^-.
Hastelloy D	82	Outstanding resistance to H_2SO_4.
Hastelloy G	44	Resistance to pitting, intergranular corrosion, SCC, oxidizing & reducing environment—versatile alloy.
Inconel alloy 600	76	Resistant to oxidation, SCC, NaOH, halogens.
Inconel alloy 625	61	Pitting & oxidation resistance to aqueous corrodents.
Inconel alloy 800	32.5	Resistant to oxidation, SCC, NaOH, halogens.
Inconel alloy 825	42	Resistance to pitting, intergranular corrosion, SCC, oxidizing & reducing environment—versatile alloy.
Carpenter 20Cb3 [symbol Cb is now replaced by Nb]	34	Resistance to pitting, intergranular corrosion, SCC, oxidizing & reducing environment—versatile alloy.
Alloy B-4	65	Pesticides production steps containing HCl.

Table 17.4 Application of aluminum alloys

Aluminum Alloy Group	Application
1XXX	The high purity members of this group are used for hydrogen peroxide and fuming nitric acid.
3XXX	Alloy 3003 is used against pitting attack.
4XXX	Alloys 4043 and 4343 are used for welding filler wire.
5XXX	Used for cryogenic equipment, ammonium nitrate, jet fuel.
6XXX	Alloy 6063 has resistance to atmospheric corrosion and is used for windows, doors, etc.

as 20 percent of other metals. The alloying elements include copper, magnesium, manganese, silicon, zinc, chromium, lead, and nickel.

Aluminum develops Al_2O_3 film on exposure to air. This film offers resistance to corrosion especially when the film reforms following an attack by a corrosive material. The corrosion resistance of aluminum and its alloys depends on the environment, and cannot be generalized. In general, aluminum alloys offer good resistance in aqueous mediums with a pH range of 4.5 to 8.5. Some peculiar behaviors of aluminum are (1) it is attacked by sodium hydroxide, but not by ammonium hydroxide and (2) it is attacked by HCl solution (pH 3), but not by fuming nitric acid.

The aluminum alloys for the process industries are divided into the following groups: 1XXX, 3XXX, 4XXX, 5XXX, and 6XXX. (Table 17.4)

Titanium, Zirconium, Tantalum, and Columbium (Niobium)

In spite of the very high cost and difficult fabrication techniques, these metals are finding commercial use in extraordinarily demanding applications.

Titanium is used for alloys containing such elements as aluminum, vanadium, manganese, chromium, tin, molybdenum, iron, niobium, and tantalum. Its resistance is due to its oxide film.

Zirconium is made in two grades: (1) Nuclear reactor grade, which is hafnium free and (2) commercial grade, which contains about 2.5 percent hafnium for nonnuclear applications. (Table 17.5)

Table 17.5 Application of Titanium, Zirconium, and Columbium (Niobium)

Metal/Alloy	Application	Disadvantage
Titanium	Environment containing inorganic chlorides such as brine, seawater, wet chlorine gas, chlorites, hypochlorites, chlorine dioxide, and nitric acid containing at least 2% water.	Pyrophoric. Do not use in liquid oxygen or solutions with high oxygen partial pressure. Do not use for hot concentrated $AlCl_3$, $CaCl_2$, $MgCl_2$, and $ZnCl_2$, dry chlorine, mineral acids except nitric acid.
Zirconium	HCl at boiling temperature and concentration. Above 300 °F, may show hydrogen embrittlement. Nitric acid at all concentrations up to boiling. Sulfuric acid up to 70% at room temperature. Chromic, phosphoric acids, and organic acids. Inorganic salt and alkaline solutions.	Impurities like ferric and sulfate ions promote corrosion and pitting. Do not use ferric and cupric chloride solutions, hydrofluoric acid, aqua regia, and wet chlorine
Tantalum and columbium (niobium)	Completely resistant to HCl in all concentrations, and high temperature and pressure. Use platinum in galvanic contact with tantalum when used in HCl and sulfuric acid at high temperature to prevent hydrogen embrittlement. May be used in nitric, phosphoric acids, liquid metals, and not highly alkaline salt solutions.	Fabrication difficulties

Tantalum and columbium coexist in their minerals. Columbium is an older name for niobium, which is shown in the periodic table. High thermal conductivity coupled with the corrosion resistance of these metals finds their application in heat exchangers.

Non-metals

The following are the most common non-metallic materials of constructions:

1. Fiberglass-Reinforced Plastics (FRP)
2. Fluorocarbons and Thermoplastics, Elastomers
3. Glass and glass-lined steels
4. Carbon and graphite

Fiber-Reinforced Plastics (FRP)

FRP is made from a resin along with a catalyst/curing system and reinforced with glass mat, or cloth. FRP is widely used as a material of construction in chemical process industries. Its corrosion resistance depends on the resin used for its manufacture. Some common resins and their applications are shown in Table 17.6.

No FRP piping or equipment should be used in flammable areas. Beware of fire hazards before selecting FRP as a material of construction.

Thermoplastics, Fluoropolymers, and Elastomers

Thermoplastics are linear polymers with little or no cross-linking. They soften when heated and regain their original shape on cooling. Unlike metals, they cannot be selected on the basis of corrosion rate. They either work or don't work in a specific environment.

Fluoropolymers are a class of paraffinic thermoplastic polymers wherein all or some of the hydrogen has been replaced by fluorine. Elastomers are homogeneous, chemical compounds made by mixing uncured natural or synthetic rubber with curing agents, accelerators, and fillers followed by vulcanization. To use this as a lining, the material undergoes a calendaring operation. Physical properties like softening point, diffusion, osmosis, and swelling should be considered before selecting a material of this category. When organic packing material is used in a distillation or absorption column, the scenario of packing agglomeration should be considered in a fire scenario. If the relief temperature is above the softening temperature of the packing, a relief device below the packed section is required to prevent overpressure of the equipment.

Table 17.6 Application of FRP Resins

Resin Type	Application
Epoxy	Water, sour crude, salt water, strong alkali, strong oxidizing solutions, jet fuels, aliphatic. Maximum temperature: 350°F.
Furan	Strong acids, concentrated alkalies, chlorinated solvents. Not applicable to oxidizing agent, nitric acid, chromic acid, concentrated sulfuric acid, chlorine, and hypochlorites. Maximum temperature: 350°F.
Phenolic	Inert to most common chemicals. Maximum temperature: 350°F.
Polyester & Vinyl ester	Non-oxidizing acids, corrosive salts, aliphatic solvents, aromatic compounds, and chlorinated solvents. Maximum temperature: 275°F.

Table 17.7 Temperature limitations of plastics

Material	Softening Temperature, °F	Embrittlement Temperature, °F
PTFE	621	−450
FEP	530	−100
PFA	590	
ETFE	520	−150
ECTFE	460	−105
PVDF	280–320	−80
PCTFE	412	−423

A variety of thermoplastics and fluoropolymers are available. They are acetyls, ABS (Acrylonitrile-Butadiene-Styrene), acrylics, cellulosoics, chlorinated polyether, PTFE (Polytetrafluoroethylene), ETFE (Ethylene tetrafluoroethylene), FEP (Fluorinated Ethylene Propylene), PCTFE (Chlorotrifluoroethylene), PVF (Polyvinylidene Fluoride), PFA (Perfluoroalkoxy), ECTFE (Polyethylene-chlorotrifluoroethylene), PVDF (Polyvinylidene fluoride), Polyamides, Polycarbonate, Polyimides, Polyolefins, Polyphenylene oxide, Polyphenylene sulfide, Polystyrene, polysulfoene, and Polyvinyl chloride.

Elastomers include natural rubber, neoprene, Butyl, SBR (styrene butadiene), EPDM (Ethylene-propylene-diene monomer), hypalon, and Nitrile. Elastomers are widely used as lining in steel equipment to offer a sturdy design. Temperature limitations of some plastic materials are shown in Table 17.7.

No plastic piping or equipment should be used in flammable areas. Beware of fire hazards (auto-ignition temperature) before selecting plastics as a material of construction. Use either low-hazard plastics or proper control measures to minimize fire hazard.

Glass and Glass-Lined Carbon Steel

Glass is an amorphous, hard, brittle substance made by fusing inorganic oxides followed by cooling. The basic inorganic oxide is silica; other oxides are the oxides of aluminum, boron, calcium, lead, magnesium, potassium, and sodium. Of the six basic types of glasses, fused silica glass, 96 percent silica glass, borosilicate glass, and aluminoborosilicate glass are most often used in the petrochemical, pharmaceutical, and food industries.

As a material of construction, glass is surpassed only by precious metals and some fluorocarbon plastics in chemical resistance. Glass is resistant to organic solvents and water in general. It is resistant to most organic and inorganic chemicals except hydrofluoric acid at any concentration and hot concentrated phosphoric acid. Alkaline solution above pH 14 at 150°F and above pH 13 at 180°F should not be used even in alkali-resistant glass. When an alkali is added in glass equipment, it should be directed to the eye of the agitator impeller for quick dispersion and neutralization. All acid salt solutions below pH 11, except fluorides, can be handled by glass.

The limitations of glass in process equipment are its brittleness and low allowable working pressure. No glass piping or equipment should be used in flammable areas. Beware of fire hazards before selecting glass as a material of construction. When hot glass is sprinkled with water, it shatters. Because of this, glass equipment should not be used in process areas where flammables are handled. If unavoidable,

Table 17.8 Minimum charging temperature of a cold product into the hot glassed side of a glass-lined vessel to avoid thermal shock

Temperature of the Glassed Side of Vessel Wall, °F	Glasteel 3300™		Glasteel 3900™	
	Minimum Charge temp. °F	Maximum (ΔT, °F)	Minimum Charge temp. °F	Maximum (ΔT, °F)
250	20	230	−10	260
300	99	201	73	227
350	178	172	155	195
400	250	150	230	170
450	318	132	300	150

this equipment should have fire-sensing valves at all connections, which will isolate them from other equipment in the event of a fire.

These limitations are overcome by using glass as a lining to fabricate process vessels, heat exchangers, and piping. Typically, jackets of glass-lined vessels are made of plain carbon steel. When acids spill on the outside of a glass-lined vessel or acid cleaning is done in the jacket of a jacketed vessel, atomic hydrogen forms due to the reaction of acid with steel. This atomic hydrogen diffuses through the metal and forms molecular hydrogen at the steel-glass interface, and finally breaks the lining. Jackets of glass-lined vessels should be cleaned using commercial alkaline descalers, such as sodium hypochlorite solution, followed by water wash.

The glass lining may be repaired by cements and tantalum plugs. Detectors may be used to identify lining failure.

The glass lining offers thermal resistance to heat transfer. It also has limitations with regard to thermal shock. Although the shock tolerance depends on the technology, one manufacturer[2] of glass-lined steel vessel offers the following guidance. This should be followed, especially when jacketed vessels use hot and cold fluids in dual services. (See Tables 17.8 and 17.9.)

Some important physical properties of glass lining:

Lining thickness: 0.044 to 0.088 in

Specific gravity: 2.5–2.7

Specific heat: 0.2 Btu/lb · °F

Thermal conductivity: 6.9 Btu-in/h · ft^2 · °F

Surface roughness: 0.0000006 ft

Table 17.9 Maximum charging temperature of a hot fluid in the jacket of a cold glass-lined vessel to avoid thermal shock

Temperature of the Glassed Side of Vessel Wall, °F	Glasteel 3300™		Glasteel 3900™	
	Maximum Charge Temp. °F	Maximum (ΔT, °F)	Maximum Charge Temp. °F	Maximum (ΔT, °F)
250	480	230	510	260
300	501	201	527	227
350	522	172	545	195
400	550	150	570	170
450	582	132	600	150

Graphite

Carbon and graphite are not the same even though the terms are used synonymously. Commercial carbon is prepared by mixing a filler substance, usually calcined petroleum coke, with a binder, usually coal tar pitch. The resultant mixture is molded or extruded in shape, and baked at 1500°F to 2500°F to form commercial carbon. When the commercial carbon is further subjected to heat treatment at 4500°F to 5400°F, the amorphous carbon is crystallized to form graphite.

Graphite as such cannot be used to make process equipment for chemical industries because of its porosity. Impervious graphite is made by impregnating porous graphite with synthetic resins such as phenolic, furan, or epoxy. The graphite is about 90 percent of finished product impregnated with resin.

Impregnated graphite is used to make heat exchangers (both shell-and-tube, and block), ejectors, piping, valves, and rupture disks etc.

Impervious graphite may be used for

- Mineral acids such as hydrochloric acid, and phosphoric acid (all concentrations), sulfuric acid (up to 90%), nitric acid (up to 20%), and hydrofluoric acid (up to 60%).

- Alkali and alkali salts (use furan or epoxy impregnated graphite), most organic acids and salts.

Impervious graphite should not be used for

- Strong oxidizers such as oleum, chromic acid, and aqua regia.

Some physical properties of impervious graphite:

Typical compressive strength: 9000 psi

Typical flexural strength: 4000 psi

Typical thermal conductivity: 1020 Btu-in./h · ft^2 · °F

Typical thermal expansion: 25.4×10^{-7} in./in · °F

Maximum operating temperature: 340°F (above this temperature, the resin decomposes)

Maximum temperature difference against thermal shock: 300°F.

LOW-TEMPERATURE APPLICATION OF METALS

The tensile strength, the yield strength, and the endurance limit of metals improve as the temperature goes down, but these improvements are not recognized by the ASME (American Society of Mechanical Engineers) code, Section VIII, which sets the criteria of acceptability when the metal is used below −20°F. To meet this criteria a test called the *Charpy V-notch impact test* is done to determine a material's resistance to brittle fracture at sub-zero temperature as specified in UG-84 of ASME Section VIII.

The impact resistance is measured in foot-pounds or inches in lateral expansion in low temperature. Guidelines for material selection for low temperature are given in Table 17.10.

Table 17.10 Low temperature application guideline

Metal	Remarks
Carbon and low alloy steel	Charpy V-notch impact test is required below −20°F. Carbon steels SA-7, SA-36, SA-113, and SA-283 shall not be used for vessels for service below −20°F.
Austenitic stainless steel	In general, the impact test is not required down to −325°F with post weld heat treatment, with the exception of 304, 304L, and 347, which can be used down to −425°F without impact test.
Wrought aluminum alloys	Good to −425°F. Below this temperature, impact test is required.
Copper, nickel, and cast aluminum alloys	Good to −325°F. Below this temperature, impact test is required.
Titanium (unalloyed)	Good to −75°F. Below this temperature, impact test is required.

MAXIMUM ALLOWABLE STRESS IN TENSION VERSUS TEMPERATURE

The maximum allowable stress of metals falls with rise in temperature. ASME Section VIII tabulates these values for various metals of different grades. These values corresponding to the design temperatures are used to calculate the thickness of the materials for the components of an equipment with a given design pressure and dimensions, and material of construction.

COST CONSIDERATIONS

While it is desirable to have the best material of construction, business consideration with regard to the economic life of the plant requires consideration of cost. Table 17.11 shows the material cost of most common metals. Some corrosion-resistant special alloys have a higher fabrication cost. Once the equipment is

Table 17.11 Material cost of common metals [Year 2005]. [ksi = 1000 lbs/in.^{2}]

Material	Tensile Strength	Yield Strength	U.S. Dollars Price/ Lb	Density, Lb/in.^{3}
Carbon Steel	60 ksi	30 ksi	$0.62	0.289
Stainless Steel 304/304L	75/70 ksi	30/25 ksi	$1.33	0.296
Stainless Steel 316/316L	75/70 ksi	30/25 ksi	$1.99	0.296
Alloy 20	80 ksi	35 ksi	$7.22	0.293
Hastelloy B3	129 ksi	59 ksi	$21.96	0.333
Hastelloy C-22	100 ksi	45 ksi	$13.32	0.314
Hastelloy C-276	100 ksi	41 ksi	$13.69	0.321
Hastelloy C-2000	100 ksi	41 ksi	$15.29	0.307
Hastelloy G-30	85 ksi	35 ksi	$14.34	0.297
Alloy 59	100 ksi	45 ksi	$14.63	0.311
Alloy 2205	90 ksi	65 ksi	$3.35	0.287
Alloy 22	100 ksi	45 ksi	$13.87	0.314
Alloy 276	100 ksi	41 ksi	$12.96	0.321
Alloy 686	100 ksi	45 ksi	$13.70	0.315
Alloy 2507	116 ksi	80 ksi	$5.75	0.287
Alloy AL-6XN	108 ksi	53 ksi	$7.16	0.291
Titanium Grade 2	50 ksi	40 ksi	$9.95	0.163
Titanium Grade 7	50 ksi	40 ksi	$21.95	0.163
Zirconium 702	55 ksi	30 ksi	$20.56	0.235
Zirconium 705	80 ksi	55 ksi	Call fabricator	0.240
Niobium	25 ksi	12 ksi	$103.00	0.309
Tantalum, 2.5%w	42 ksi	29.7 ksi	$155.00	0.601

designed and its weight is known, a fabrication cost may be estimated by multiplying the material cost.

RECOMMENDED REFERENCES

Agarwal, "Defy Corrosion Resistance with Recent Nickel Alloys," *Chemical Engineering Progress,* January, 1999.

Pfaudler Technical Data, Bulletin 1097A, the Pfaudler Company, Rochester, NY.

Process Industries Corrosion, National Association of Corrosion Engineers, Houston, TX.

CHAPTER 18

Equipment Design

OUTLINE

EQUIPMENT SPECIFICATION SHEETS 656
Operating Pressure and Temperature ■ Design Pressure and Design Temperature ■ Other Information in Equipment Specification Sheet

PRESSURE VESSELS 658
Maximum Allowable Working Pressure ■ Calculation of Thickness of Cylindrical Shells Based on Dead Weight and Wind Load

OPERATING VOLUME AND SURGE VOLUME 660

PIPING DESIGN 661
Pipe Rating ■ Pipe Schedule Numbers ■ Pipe Material Thickness ■ Nomenclature for Variables Used in the Preceding Section

PROCESS DESIGN 665
Pumping System Design ■ Temperature Rise Due To Skin Friction Under Adiabatic Condition ■ Maximum Temperature Rise Due to Motor Horsepower ■ Maximum Rate of Temperature Rise if the Centrifugal Pump is Dead-Headed ■ Minimum Flow of Centrifugal Pumps

SIZING AN AGITATOR 667
Primary Pumping Capacity, gpm ■ Apparent Superficial Velocity, ft/minute ■ Chemscale or Agitation Intensity Number, N_I ■ Tip Speed, ft/minute ■ Torque ■ Determination of Motor HP from Measured Current, Voltage, etc. ■ Types of Impeller ■ Some Considerations in Agitator Design ■ Some Steps for Preliminary Design of an Agitator System ■ Nomenclature for the Preceding Section of the Process Design

MASS TRANSFER COLUMNS 674

EQUIPMENT TESTING AND ANALYSIS 680
Non-Destructive Testing

DESIGN OPTIMIZATION 682
Degrees of Freedom for Optimization ■ One Variable Problem ■ Two or More Variables to Optimize

PROCESS FLOW SHEET DEVELOPMENT 683

OPERATING MANUAL 684
Heat Exchanger Design Tips to Avoid Trouble

RECOMMENDED REFERENCES 694

Storage vessels, fluid handling equipment, and piping are major components of a process plant. This chapter deals with the guidelines for developing design parameters of process equipment and piping. It also offers simplified methods of calculating the material thickness of process equipment for a specified design temperature, design pressure, dimensions, type of vessel, material of construction, and corrosion allowance.

Design guidelines for pumping systems, agitators, mass transfer columns, equipment testing, optimization, development of flowsheet, operating manual, and troubleshooting are also added.

EQUIPMENT SPECIFICATION SHEETS

It is the responsibility of a process engineer to prepare the process specification data sheets for the fabricators. The process equipment generally includes storage tanks, process vessels, reactors, distillation columns, pumps, compressors, heat exchangers, etc., and the associated piping. The most important pieces of information necessary for the design of equipment are described below.

Operating Pressure and Temperature

In most cases, this information is available from the process flow diagram. Include minimum, normal, and maximum temperature. Consider the effect of extreme atmospheric conditions for operating temperature.

Design Pressure and Design Temperature

The design pressure, taken at the highest connection of the pressure vessel (typically topmost flanged connection), is added to the hydrostatic head, and the resulting pressure is used to calculate the theoretical thickness of a plate. For atmospheric or low-pressure storage tanks, the selection of design pressure has implications for the setting of breathing devices or safety devices if such devices are contemplated or required by local regulation. Breathing devices are desirable to control loss of volatile substances due to change in ambient temperature. Generally breathing devices overpressure by 100 percent when fully open; therefore, if the breathing losses are to be minimized, the maximum and minimum temperatures due to process and atmospheric temperature changes should be considered. In addition, if nitrogen blanketing is used, an operating margin must be added to the maximum nitrogen pressure in the tank. As an example, suppose nitrogen blanket pressure is maintained at ±5 inches water column. Allow additional 5 inches water column operating margin, and set the breather vent at 10 inches water column. Because the vent may overpressure 100 percent when fully open, select the design pressure of the tank at 20 inches of water column and design vacuum at −20 inches of water column. In general, consideration should be given to the local barometric pressure, the dependence of vapor pressure on temperature, and the variation of vapor pressure as a

Table 18.1 Variation of liquid surface temperature due to atmospheric temperature change

Geographical Area	Maximum Liquid Temperature, °F	Maximum Vapor Temperature, °F	Minimum Vapor Temperature, °F
Gulf coast, Atlantic Seaboard, Northern Middle-West	100	140	85
Mid-Continental & South-West	115	155	100
West Coast & locations tempered by Pacific Ocean	80	120	65

result of atmospheric temperature change. The penalty of arbitrarily selecting design pressure of atmospheric storage tank for volatile liquid is the continuous breathing loss caused by the low settings of the breathing device. Table 18.1[1] gives the variation of liquid surface temperature as a result of atmospheric temperature change in the U.S.

Table 18.1 gives a general guideline only. The inside liquid temperature depends on the size and shape of the tank, product outage, paint color and condition, daily exposure to solar heat, and product temperature. An estimate of breather vent set pressure may be made by the following equation:

$$P_s = (P_a - P_{min})\frac{T_2}{T_1} + P_{max} - P_a \qquad (18.1)$$

$$P_a = 14.696\left[1 - \frac{0.0065\Delta Z}{288}\right]^{5.26433} \qquad (18.2)$$

Table 18.2 may be used as a guideline for selecting the design pressure and applicable design code.

Table 18.2 Selection of design pressure and design code

Operating Pressure (OP)	Design Pressure (DP)	Design Code	Remark
Atmospheric	±10 inch water column or twice the breather vent set pressure or set vacuum, whichever is greater	API 650	Consider Equation 18.1
10 in. wc to 15 psig	OP × 2	API 12D, 12F, 620	If DP exceeds 15 psig, use ASME VIII for metals & ASME X for FRP.
>15 psig to 160 psig	OP × 1.5	ASME VIII	Requires ASME stamp
>160 psig to 500 psig	OP × 1.3	ASME VIII	Requires ASME stamp. The stamped MAWP* should be 30% higher than operating pressure.
>500 psig	OP × 1.2	ASME VIII	Requires ASME stamp. The stamped MAWP should be 30% higher than operating pressure.

*MAWP = Maximum allowable working pressure

If the equipment is a reactor, the safety device should be (but is not necessarily required to be) set at half the MAWP to allow relief at lower exothermicity and allow higher overpressure, and the set pressure of the device should be at least 40 percent higher than the maximum operating pressure in gage. Therefore, the above multiplier of operating pressure may require adjustment. Reactive materials should never be stored in atmospheric tanks.

The design metal temperature of vessels shall be the maximum operating temperature plus a safety margin of at least 50°F. For materials whose allowable stress does not change for additional increase in design temperature, the design temperature should be increased further to take advantage of this property. For example, for carbon steel, there is only a small variation of allowable stress up to 650°F. Therefore, if the calculated design temperature is 200°F, specify 650°F as the design temperature. The implied advantage of this higher temperature comes when the material relieves under a fire condition.

Other Information in Equipment Specification Sheet

Besides operating pressure, temperature, design conditions, material of constructions, and varieties of other information that are specific to the unit operation are required. The required data are specific to the application.

PRESSURE VESSELS

The design of a pressure vessel requires the data of operating temperature and operating pressure. A safety margin is added to these operating conditions to arrive at the design temperature and design pressure. The operating temperature and operating pressure are available from process flow diagrams and are generally determined from expected variation of the parameters with due considerations of the atmospheric conditions. To arrive at the design values, allowances should be made for the alarm and interlock limits and the operating margin for the relief device. As an example, if the maximum operating pressure is 15 psig, allow 5 psig for alarm limit, and 5 psig for interlock limit. Therefore, interlock pressure = 25 psig. If a conventional relief valve is used to protect the equipment, the set pressure of the valve = 25/0.7 = 35.7 psig or rounded to 36 psig. This pressure (36 psig) should be specified as the design pressure of the equipment. When the fabricator back-calculates the maximum allowable working pressure (MAWP), which is almost invariably higher than the design pressure, the relief device may be set at the MAWP, if so desired. Therefore, unless required otherwise, unnecessary safety margin on design pressure should be avoided.

The design and construction of pressure vessels are governed by the codes of the American Society of Mechanical Engineers (ASME) and the standards of the American Petroleum Institute (API). Three sections of ASME that governs the pressure vessels are

(1) ASME SECTION I for fired pressure vessels such as boilers.

(2) ASME SECTION VIII Division 1 and 2 for unfired metallic pressure vessels. SECTION VIII Division 1 is used to design most of the pressure vessels required by the chemical industries. It employs relatively approximate formulas to specify a vessel whose design pressure (internal or external) is greater

than 15 psig but does not exceed 3000 psig with an internal diameter exceeding 6 inches. Division 2 uses more complex and rigorous mathematical analysis and is used for applications involving severe service (such as toxic chemicals), cyclic operations, design pressure exceeding 3000 psig, and so on.

(3) ASME SECTION X is used for fiber-reinforced thermosetting plastic pressure vessels for internal design pressure exceeding 15 psig but not exceeding 3000 psig, temperature range of −65°F to 250°F for non-lethal service. External pressure must not exceed 15 psig. For example, for a jacketed vessel, external design pressure may easily exceed 15 psig; therefore, coded plastic vessels are not generally jacketed.

API standards (12D, 12F, 620, 650, 650F) cover design of pressure vessels with design pressure not exceeding 15 psig.

The following sections elaborate on aspects of internal pressure per ASME SECTION VIII Division 1.

Maximum Allowable Working Pressure

Maximum allowable working pressure (MAWP) is the least of the pressure ratings of the various components of a pressure vessel. The pressure ratings of the components are back-calculated by using the code formulae from known plate thickness after allowances are made for tolerances, corrosion, thinning due to fabrication, and other loadings. The MAWP at the hot and corroded condition is valid only when the vessel is hydro-tested at 1.3 times the MAWP at new and cold condition (because of higher thickness) and stamped along with the design temperature on a plate attached to the vessel. Table 18.3 summarizes the formulas for calculating plate thickness and maximum allowable working pressure.

Calculation of Thickness of Cylindrical Shells Based on Internal Pressure, Dead Weight, and Wind Load

The thickness calculated for internal pressure as shown in Table 18.3 should be checked for the required thickness due to wind pressure.[5] (For Equations 18.3 to 18.7, see Table 18.3 next page.)

(a) Thickness for windward side (side facing the wind)

$$t = \frac{2P_w h^2}{\pi D' SE} - \frac{W}{\pi D_m SE} + \frac{PD_m}{4SE} \quad (18.8)$$

(b) Thickness for the leeward side (side away from the direction of the wind)

$$t = \frac{2P_w h^2}{\pi D' SE} + \frac{W}{\pi D_m SE} - \frac{PD_m}{4SE} \quad (18.9)$$

(c) Thickness to withstand buckling

$$t = \frac{2P_w h^2}{\pi D' S_B E} + \frac{W}{\pi D_m S_B E} \quad (18.10)$$

$$S_B = \frac{2(10^6) t_c}{D_o} \quad (18.11)$$

Table 18.3 Plate thickness and maximum allowable working pressure

	Wall Thickness inch	Maximum Allowable Working Pressure, psig	
Cylindrical shell	$t = \dfrac{PR}{SE - 0.6P}$	$P = \dfrac{SEt}{R + 0.6t}$	(18.3)
Sphere and hemispherical head	$t = \dfrac{PR}{2SE - 0.2P}$	$P = \dfrac{2SEt}{R + 0.2t}$	(18.4)
2:1 Ellipsoidal head	$t = \dfrac{PR}{SE - 0.1P}$	$P = \dfrac{SEt}{R + 0.1t}$	(18.5)
Conical head	$t = \dfrac{PR}{\cos\alpha(SE - 0.6P)}$	$P = \dfrac{SEt(\cos\alpha)}{R + 0.6t(\cos\alpha)}$	(18.6)
ASME flanged & dished head (torispherical) $(L/r = 16\ 2/3)$[1]	$t = \dfrac{0.885PL}{SE - 0.1P}$	$P = \dfrac{SEt}{0.885L + 0.1t}$	(18.7)

Notes:
1. For $L/r < 16\ 2/3$, see ASME Section VIII Division 1.
2. For external pressure, see ASME Section VIII Division 1.
3. To the material thickness calculated above, corrosion allowance and forming allowance are added, and the nearest commercial plate thickness is used for fabrication.

All vessels, irrespective of design pressure, should have a vacuum rating. Calculation of material thickness for vacuum rating of a pressure vessel per ASME VIII requires trial and error and is not presented here. However, please note that the inside shell of a jacketed vessel must stand an external pressure that equals the jacket internal design pressure plus the local barometric pressure when the inside shell of a pressure vessel with a jacket is designed for full vacuum.

The empty weight of a vessel may be approximately determined by the following equation:

$$W_e = 136(D_{mf})[HT + 0.8(D_{mf})](t_p + x) \qquad (18.12)$$

where
$x = 0.08$ for vessels with usual nozzles, a manway, and no internals
$ = 0.18$ for vessels like distillation colums with internals and more manways.
The last factor has been increased by the author from 0.15 to 0.18.

This formula applies for steel and low-alloy steel; a density correction is to be applied for other materials.

OPERATING VOLUME AND SURGE VOLUME

The operating volume of a vessel must include the largest inventory plus a safety margin. For a continuous operating process vessel, the surge volume is considered for estimating the operating volume. To this operating volume a safety margin is added to find the gross volume. Such a safety margin should include an empty volume required for vapor liquid disengagement during emergency relief under a fire scenario. For low-pressure tanks (up to 100 psig), recommended length/diameter is 3:1 to 4:1, and the corresponding value for higher pressure tanks is 6:1. Table 18.4 is a guideline to estimate the surge volume of a process vessel.

Table 18.4 Residence time for process vessel

Process Condition	(1)NLL to Empty (2)NLL to TL Residence Time	HLL to LLL Residence Time
Column liquid overhead product pumped directly into next process	(1) 10 minutes for liquid product & 5 minutes for reflux	4 minutes for liquid product, 2 minutes for reflux
Column liquid overhead to tank	(1) 5 minutes for liquid product or reflux	2 minutes for liquid product or reflux
Column overhead accumulator with only gas product	(1) 5 minutes for reflux	2 minutes for reflux
Bottoms of a side stream stripper to tank	(2) 5 minutes for product	2 minutes for product
Bottoms of a side stream stripper directly to subsequent processing	(2) 10 minutes for product	4 minutes for product
Bottoms of a column with thermosyphon reboiler to a tank	(2) 5 minutes for product	2 minutes for product
Bottoms of a column with thermosyphon reboiler to subsequent processing	(2) 5 minutes for product	4 minutes for product
Bottoms of a column with kettle-type reboiler to subsequent processing/tank	(1) 2 minutes for product	1 minute for product

Notes for Table 18.4:
In the second column (1) indicates residence time for normal liquid level to empty, (2) indicates residence time for normal liquid level to lower tangent line of equipment.
1. NLL = normal liquid level, HLL = high liquid level, LLL = low liquid level, TL = tangent line.
2. LLL should be at least 6 inches above the lower TL.
3. Minimum liquid level alarm point should be at least 12 inches above lower TL.
4. Normal liquid level in the bottom sump of a distillation column should be at least 3 ft above the bottom tangent line.
5. For the vertical knockout vessel with side-entering feed, set high liquid level 6 inches below the bottom of the feed nozzle, and minimum 14 inches above the bottom tangent line. Minimum height from the center of the feed nozzle to the upper tangent line of the vessel should be 3ft or 0.75 (diameter of vessel), whichever is greater. For horizontal knockout drum, the distance between feed inlet and vapor outlet should be 1.5 (diameter of vessel) as minimum.

PIPING DESIGN

The piping system for petrochemical plants is generally designed per piping code, ANSI B31.3.

The determination of optimum[3] pipe wall thickness requires consideration of the following factors:

- Line size as determined by hydraulic calculation.

- Phase of fluid carried by the line (liquid, vapor, sold, or a combination of phases).

- Normal operating conditions.

- Upset conditions, maximum duration of upsets, total hours of upsets per year.

- Properties of fluid to be used to test the design integrity of piping.

- Process fluid characteristics that require special preparation or treatment of the pipeline.

The code (ANSI B31.3) stipulates that the chosen design pressure and design temperature shall not be less than the most severe condition of **coincident** internal or external pressure and temperature. For example, if the most severe condition for a pipe line in one operation is 50 psig and 400°F, and in another operation, 100 psig and 150°F, then choosing 100 psig and 400°F as the most severe condition is inappropriate to design the pipe, because the selected pressure and temperature are not coincident. The code also addresses normal operation and abnormal operation, and further qualifies the design pressure by stating that the maximum difference in pressure between the inside and outside of any piping component or between any two chambers of a combination unit shall be considered, including the unintentional loss of external and internal pressure.

For example, with steam in the jacket side of a jacketed pipe, the fluid in the pipe may leak out, thereby subjecting the pipe to full external jacket pressure.

If the line is traced, electrically or by a fluid, the heating effect must be considered in establishing the design temperature.

If the process upset conditions are of short duration, the code stipulates some allowances depending on the frequency and duration of such upsets. A good example of an upset of short duration is the opening of a relief valve. Such an incidence gets notice quickly, and is likely to be corrected within 10 hours. Table 18.5 shows the duration of the upset conditions and the allowances of the code.

Table 18.5 Allowance of upset conditions in a petrochemical piping system (ANSI B31.3)

Duration per Incident in Hour not to Exceed	Total Duration in Hours per Year not to Exceed	% By which the Pressure Rating or Allowable Stress at the Upset Operating Temperature of Piping may be Exceeded
10	100	33
50	500	20

Pipe Rating

Pipes are rated according to the rating of the associated flange. For process piping, flanges are rated per ANSI B16.5. The most common primary service pressure ratings of pipe flanges in lbs are 150, 300, 400, 600, 900, 1500, and 2500. However, the flanges are generally the weakest link in a piping, which can stand much higher pressure than the nominal rating of the system. The primary service pressure rating of a flange also camouflages its strength. Table 18.6 illustrates the maximum non-shock pressure that flanges of the most common material can stand.

Table 18.6 Flange primary service pressure rating and variation of maximum non-shock service pressure rating with temperature for carbon steel, low alloy, and S.S. steel

Primary service pressure rating, lb		150	300	400	600	900	1500	2500
Hydrostatic shell test pressure, psig		425	1100	1450	2175	3250	5400	9000
Material	Service Temperature, °F	Maximum, Non-Shock Service Pressure Rating, psig						
Carbon steel,	−20 to 100	275	720	960	1440	2160	3600	6000
carbon Moly,	250	225	690	920	1380	2070	3450	5750
chrome Moly, and	450	165	650	870	1305	1955	3255	5430
Stainless steel	650	120	515	690	1030	1550	2580	4300

Table 18.7 shows some natural limits for temperature. Use this as a guideline in process design.

Table 18.7 Natural limits for temperature[7]

Temperature, °C	Application Remark
−177	Limit of liquid nitrogen as a coolant
−20	Practical lower limit of typical carbon steel. Below this temperature, special killed grade steel or stainless steel is required.
30	Lower limit for water as a coolant
60	Limit of polyvinyl chloride and polyethylene as piping and tank material
80	Heat exchangers using untreated water may severely foul with mineral deposit when turned down above this temperature.
100	Limit of polypropylene as piping and tank material. Close to practical limit for treated cooling water exchangers.
135	Approximate upper service limit of fiberglass reinforced plastics (FRP) and polyvinylidene Fluoride (PVDF) piping and vessel component.
150	Upper service limit of for Tefzel, poly(ethylene-co-tetrafluoroethylene)ETFE. Practical upper limit for polytetrafluoroethylene (PTFE or Teflon) and perfluoroalkoxy (PFA) in un-entrapped service. (See note 1 below.)
175	Upper service temperature for viton fluoroelastomer
232	Upper service limit PFA and PTFE, even when totally entrapped.(Note 2.)
315	Practical upper service limit for polyetheretherketone (PEEK), Upper service limit of Kalrez fluoroelastomer. Upper service limit for titanium in pressure retaining service. Maximum recommended service temperature of heat transfer oils.
450	Maximum service temperature for unprotected expanded graphite exposed to air. Above this temperature, only metallic and ceramic sealants may be used. Intergranular corrosion range for stainless steels.

Note 1: An example of un-entrapped service is a ring gasket placed between ordinary raised face flanges.
Note 2: An example of an entrapped service is a ring gasket or o-ring in a groove in a flange. Here the gasket is basically trapped in a metal pocket.

Pipe Schedule Numbers

The schedule numbers indicate the relative pipe thickness. The schedule number is related to internal design pressure and the allowable stress for the material of construction at a given design temperature. The higher the schedule number, the stronger the pipe.

$$\text{Schedule number} = \left(\frac{P}{S}\right)(1000) \quad (18.13)$$

where
P = internal design pressure, psig
S = Allowable stress of pipe material at the design temperature, psi

Although the schedule number varies from 5 to 160, the most common schedule numbers for process piping are 10, 40, and 80. For nominal pipe sizes up to and including 12-in., the outside diameter is greater than the nominal pipe size; for nominal pipe sizes bigger than 12-in., the outside diameter is the same as the nominal pipe size.

Pipe Material Thickness

For a metallic pipe the conservative method for calculating the pipe thickness for petrochemical application is as follows:

$$t = \frac{PD_o}{2SE} \tag{18.14a}$$

$$t = \frac{Pd}{2(SE - P)} \tag{18.14b}$$

$$t_m = t + c \tag{18.14c}$$

Depending on whether outside diameter or inside diameter is available, one of the above equations is used.

Nomenclature for Variables Used in the Preceding Section

c	Sum of inches of the mechanical allowances (thread depth or groove depth), corrosion and erosion allowance. For machined surfaces or grooves, where the tolerance is not specified, the tolerance shall be 0.02 inch in addition to the specified depth of cut.
d	Inside diameter of piping, in.
D_o	Outside diameter of piping or vessel, in.
D'	Outside diameter of shell including insulation, in.
D_m	Mean diameter of shell, in.
D_{mf}	Mean diameter of shell, ft
E	Joint efficiency, decimal, varies from 0.45 to 1.0
h	distance from top of vessel to point in consideration, ft
HT	Straight length, tangent-to-tangent, of a vessel, ft
ID	Inside diameter, ft
L	Inside crown radius of dish for ASME flanged and dished head, generally taken as equal to inside diameter of the vessel, inch
OD	Outside diameter of vessel, ft
P	Design pressure or maximum allowable working pressure, psig
P_a	Local atmospheric pressure, psia
P_s	Set pressure of safety device, psig
P_{max}	Vapor pressure of liquid at maximum liquid surface temperature, psia
P_{min}	Vapor pressure of liquid at minimum liquid surface temperature, psia
P_w	Wind pressure, psf
R	Inside radius of vessel, inch
r	Knuckle radius of a flanged and dished head, inch
S	Allowable stress of material of construction corresponding to the design temperature, psi
S_B	Allowable stress for buckling, psi

t	Pressure design wall thickness, inch, exclusive of any allowance; corrosion allowance, if applicable, and other appropriate allowances are added to this thickness, and the nearest commercially available thickness is chosen.
t_c	Corroded plate thickness, inch
t_m	Minimum thickness after adding allowances, inch
t_p	Commercial plate thickness selected for fabrication, inch
T_1	Minimum vapor space temperature, °R
T_2	Maximum vapor space temperature, °R
W	Weight of vessel, full of fluid, lb
W_{ei}	Empty vessel weight with insulation, lb
ΔZ	Elevation of the location above sea level, m (0 < Z < 11,000)
α	one-half of the included apex angle (cone angle) of a conical head

PROCESS DESIGN

Pumping System Design

The pump hydraulics and affinity laws have been covered in Chapter 4. Some additional features are included here.

Temperature Rise Due to Skin Friction Under Adiabatic Condition

Assuming skin friction is completely converted to heat, the temperature rise, ΔT, °F[K], may be given by

$$\Delta T = \frac{\Delta P}{\rho \eta C_P} \quad (18.15)$$

where
  ΔP = skin frictional pressure drop, $lb_f/ft^2[N/m^2]$,
  ρ = the density, $lb/ft^3[kg/m^3]$
  η = 778 ft·lb_f/Btu[1N·m/J],
  C_p = the heat capacity, Btu/lb·°F[J/kg·K]

Maximum Temperature Rise Due to Motor Horsepower

The equation for maximum temperature rise due to pumping, *assuming all power input goes to heat the fluid*, ΔT, °F,

$$\Delta T = \frac{TDH}{778 C_P \varepsilon_{overall}} \quad (18.16)$$

The Jacobs formula[15] for calculating temperature rise is

$$\Delta T = \frac{TDH}{778 C_p}\left(\frac{1}{\varepsilon} - 1\right) \quad (18.17)$$

where

TDH is total dynamic head, ft·lb$_f$/lb, C_p is the heat capacity, Btu/lb·°F
$\varepsilon_{overall}$ = pump efficiency × motor efficiency, decimal
ε = pump efficiency, decimal.

Maximum Rate of Temperature Rise if the Centrifugal Pump is Dead-Headed

$$\frac{dT}{d\theta} = \frac{5.1(bhp_{shut\text{-}off})}{VsC_p} \qquad (18.18)$$

$\frac{dT}{d\theta}$ = rate of temperature rise, °F/min
$bhp_{shut\text{-}off}$ = brake horsepower, HP
V = holding capacity of the pump-casing plus the piping between block valves, gallons
C_p = specific heat of the fluid being pumped, Btu/lb·°F
s = specific gravity of liquid, dimensionless.

Minimum Flow of Centrifugal Pumps

When a running pump is put on a standby mode, such as in a batch operation, a **minimum recirculation flow**, Q_{min}, is to be drawn from the pump back to the suction vessel, normally through a restriction orifice, for thermal protection of the pump.

$$Q_{min} = \frac{5.1(bhp_{shut\text{-}off})}{s[C_p \Delta T_{safe} + 0.001285(TDH_{shut\text{-}off})]} \qquad (18.19)$$

where
ΔT_{safe} = (Saturation temperature corresponding to pressure: $[P_v + (0.433s)NPSH]$, °F) −Operating temperature, °F
P_v = vapor pressure of the fluid at the pumping temperature, psia
s = specific gravity of fluid, dimensionless
TDH = total dynamic head, ft·lb$_f$/lb
C_p = heat capacity of liquid, Btu/lb·°F
$NPSH$ = net positive suction head available for the pump, ft of fluid.
Q_{min} = minimum recirculation flow of pump to prevent cavitation, gpm.

This minimum flow should be added to the maximum demand to estimate the design capacity of a batch pump, which is expected to run on a standby mode on a minimum flow through a recirculation line. The orifice at the recirculation line is to be sized for the minimum flow. The minimum flow through the bypass should also be cooled unless the vessel receiving the recirculation has enough liquid to act as a heat sink for the duration of the recirculation.

Another minimum flow, called the **mechanical minimum flow**, should be maintained to avoid excessive vibration and shaft deflection due to unbalanced radial loads. This flow is supplied by the manufacturer of the pump. The final minimum flow should be greater of minimum recirculation flow or mechanical minimum flow.

SIZING AN AGITATOR

(For explanation of symbols used, please see page 673.)

Primary Pumping Capacity, gpm

The agitator behaves like a pump. The **primary pumping capacity**, Q, is given by

$$Q = 4.33(10^{-3})N_Q N d^3 \tag{18.20}$$

$$N_Q = \frac{Q}{4.33(10^{-3})Nd^3} = \frac{Q}{(7.48)ND^3} \tag{18.21}$$

Apparent Superficial Velocity, ft/minute

The **apparent superficial velocity, or bulk velocity**, v_b, due to pumping effect of the impeller is given by

$$v_b = \frac{Q}{7.48A} \tag{18.22}$$

v_b, expressed in ft/minute, varies from 6(mild) to 60(violent). The bulk velocity is used as a measure of agitation intensity. This equation assumes that the flow is uniformly distributed over the entire cross section of the tank. In reality it does not happen so exactly, especially for radial flow impellers where the flow hits the wall and divides into two portions: one portion going up; the other, down.

Chemscale or Agitation Intensity Number, N_I

(See Table 18.9)
It is an arbitrary scale of agitation (1 = mild, 10 = violent).

$$N_I = \frac{v_b}{6} \tag{18.23}$$

From the preceding equations it can be shown that

$$N_I \propto ND \tag{18.24}$$

Tip Speed, ft/minute

It is the linear velocity of the tip of the impeller blade, expressed as

$$TS = \pi DN = 0.2618 dN \tag{18.25}$$

There is an arbitrary scale of agitation on the basis of tip speed: Low ($TS = 500 - 650$), Medium ($TS = 650 - 800$), High ($800 - 1100$). This scale is cited by Holland and Chapman.

For geometrically similar systems with a constant ratio of impeller diameter to tank diameter, a constant clearance between the impeller and the tank bottom, the maintenance of constant tip speed is a method of scaling an agitator.

Turnover time is the time to circulate the entire contents of the vessel once. The faster the contents of a tank are turned over, the more violent the agitation. It is also used as a measure of agitation intensity. The equation is

$$\theta_{turnover} = \frac{V}{Q}, \text{ min} \qquad (18.26)$$

Mixing time,[12] **Blend time**, θ, is the time to mix materials to satisfy a specified fluctuation of concentration of a key ingredient (for example, $\pm 0.1\%$).

For conventional axial flow impellers and flat blade turbines (for Reynolds number $>1 \times 10^5$),

$$\frac{N}{K}\frac{D^2}{T} = N_p = 0.85 \text{ to } 0.9 \qquad (18.27)$$

$$a = 2e^{-K\theta}$$

With $N_p = 0.9$:

$$\theta = \frac{N_p \left(\ln\frac{2}{a}\right)\left(\frac{T}{D}\right)^2}{N}$$

With $a = 0.001$, $N_p = 0.9$:

$$\theta = \frac{6.84\left(\frac{T}{D}\right)^2}{N}$$

Reynolds Number, N_{Re}, is given as

$$N_{Re} = (\text{Re in Fig 18.4}) = \frac{10.754 sNd^2}{\mu} = \frac{\rho ND^2}{\mu(0.04033)} = \frac{N_{rps}D_I^2 \rho_{si}}{\eta} \qquad (18.28)$$

The turbulent region starts somewhere between $N_{Re} \approx 10^3$ and $N_{Re} \approx 10^5$.

Power Number, N_P, is

$$N_p = \frac{1.523(10^{13})P}{sN^3 d^5} = \frac{P_I}{sN_{rps}^3 D_I^5} \qquad (18.29)$$

Generally when gravitational effects are not a factor, the power number is a function of Reynolds number for a specific impeller type.

Torque

The equation for torque, τ, is

$$\tau = \frac{63025 P}{N} \qquad (18.30)$$

$$P = \frac{\tau N}{63025}$$

$$\frac{P}{V} = \frac{\tau}{V}\left(\frac{N}{63025}\right)$$

When torque per unit volume is constant, power per unit volume is proportional to RPM for geometrically similar agitator system.

The constant torque per unit volume with equal agitation intensity is another way of scaling an agitator for a geometrically similar system. For a geometrically similar system, the impeller size increases in proportion to the tank diameter. At equal agitation intensity, as shown before, the product ND is constant. Therefore, N varies as $1/D$ for the same agitation intensity. Consequently power P varies as $1/D$. Thus if the diameter is increased five-fold, keeping geometric similarity, rpm will be decreased by one-fifth. Therefore, P/V will be one-fifth for the larger vessel when constant torque per unit volume is maintained. *This proves that scale up for equal agitation intensity should not be done using constant power per unit volume in a geometrically similar system.*

Determination of Motor HP from Measured Current, Voltage, etc.

The motor load (3 phase power) is given as

$$HP = \frac{1.73 \,(\text{Voltage})(\text{ampere})(\text{motor efficiency, fraction})(\text{power factor})}{746} \quad (18.31)$$

Types of Impeller

Axial Flow Impeller
This type of impeller produces flow along the axis of a vertical tank, i.e., parallel to the impeller shaft. Examples are marine propeller and pitched blade or axial flow turbines.

Radial Flow Impeller
This type of impeller produces flow along the radius of a vertical tank, i.e., perpendicular to the impeller shaft. An example is a flat blade turbine.

There are impellers that provide a combination of axial and radial flows.

Some Considerations in Agitator Design

Tank Geometry, Impeller Placement, and rpm.

- For blending and solids suspension, liquid-depth-to-tank-diameter ratio is 0.6 to 0.7 for optimum power.

- A single impeller usually covers liquid height of 0.5D to 2D where D is the impeller diameter. If liquid height is greater than 1.5 times the impeller diameter, then multiple impellers should be used. For turbine impellers, use the Weber formula to determine the number of turbine impellers.

$$\text{Number of turbines} = \frac{\text{liquid height}(\text{specific gravity of liquid})}{\text{Tank diameter}}$$

The height and diameter must be in the same units in the preceding equation.

- Constant speed agitator: Standard speeds of propeller units are 1750, 1150, 420, 350. Standard speeds of turbine units with HP > 3 are 280, 230, 190, 155, 125, 100, 84, 68, 56, 45, 37, 30, 25, 20, and 16.5.

Variable speed agitators use variable speed motor, eddy current magnetic couplings, belt-drive with variable spacing sheaves (Reeves drive), fluid couplings, and hydraulic motors.

- Standard four baffles are generally installed when agitators are centrally located in a vessel containing low viscosity fluid to prevent vortex and inefficient mixing. Baffle width varies from one-twelfth to one-tenth the diameter of the tank. The clearance between the baffle and the tank wall is typically equal to the width of the baffle. For marine-propeller and turbines used for blending liquids with viscosities higher than 20,000 cP, baffles are not required. When $N_{Re} < 300$ or agitators are mounted off-center, baffles are not required.

Some Steps for Preliminary Design of an Agitator System

Step 1: From a known operating volume of the vessel, calculate diameter and height of the vessel, unless they are already fixed. Try to keep diameter and liquid height equal in dimension.

Step 2: Determine the required process response, and pick an agitator type from Figure 18.1 and Table 18.8. Also decide on the ratio of impeller diameter to vessel diameter. If the system requirement is such that you need a very slow agitation with the movement of entire mass such as in crystallization and very high intensity agitation such as in pH control from the same agitator, choose Ekato Intermig or equivalent.

Step 3: Choose a Chemscale, N_I, from the required process response (See Table 18.9), determine the apparent superficial velocity, v_b, and then pumping capacity.

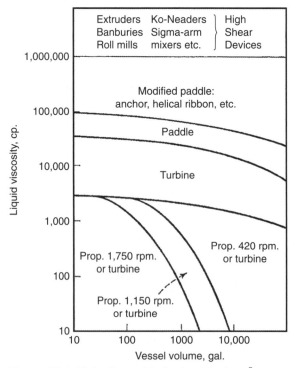

Figure 18.1 Selection guide for impeller type[9]

Table 18.8 Impeller selection guide for process[11] response. α = pitch angle of blade, δ = baffle width + clearance of baffle from the wall

Abbreviation of impeller types	Photo	Preferred geometrical dimensions and configuration	Conditions of installation	Primary flow direction	Diameter ratio d_2/d_1 [-]	Tip speed u [m/s]	Flow range	Power number [-]	Viscosity range η [Pa s]	Mixing tasks
Marine propeller		d_2/d_1 = 0.33; h_3/d_2 = 0.3; α = 25°; h_1/d_1 = 1; δ/d_1 = 0.1	propeller central with 3 baffles, off centre, pitched and side entry without baffles, also used in draft tubes	axial	0.1–0.5	2–15	turbulent	Ne = 0.35 (3-bladed); Ne = 0.85 (5-bladed); α = 24°	< 10	Blending; Suspension; Dispersion of solids; Dispersion liquid/liquid; Dispersion liquid/solid; Circulation/pumping (Arrangement in draft tube)
Pitched blade turbine		d_2/d_1 = 0.33; h_2/d_2 = 0.125; h_3/d_1 = 0.3; h_1/d_1 = 1; δ/d_1 = 0.1	central 2–4 baffles or off centre without baffles	axial-radial	0.2–0.5	3–10	turbulent/ transition range	Ne = 1.5 (6-bladed) turbulent; α = 45°	< 10	Blending; Dispersion liquid/liquid; Dispersion liquid/solid; Suspension
Flat blade disc turbine		d_2/d_1 = 0.33; h_2/d_2 = 0.2; b/d_2 = 0.25; h_3/d_1 = 0.3; h_1/d_1 = 1; δ/d_1 = 0.1	central 2–4 baffles	radial	0.2–0.5	3–7	turbulent	Ne = 4.6 (6-bladed)	< 10	Dispersion liquid–liquid; Gassing (High shear rates)
Multi-stage impulse countercurrent impeller MIG®		d_2/d_1 = 0.7; h_3/d_1 = 0.16; h_6/d_1 = 0.28; h_1/d_1 = 1; δ/d_1 = 0.1	central 2–4 baffles (d_2/d_1 < 0.7) or without baffles (d_2/d_1 > 0.7)	axial-radial	0.95	2–10	turbulent/ transition range (d_2/d_1 < 0.7) laminar (d_2/d_1 > 0.7)	Ne = 0.55 (3-stages) turbulent (d_2/d_1 < 0.7); Ne · Re = 100 laminar (d_2/d_1 > 0.7)	< 50	Homogenization; Suspension; Gassing; Heat Transfer (Gentle handling)
Interference multi-stage impulse countercurrent impeller INTERMIG		d_2/d_1 = 0.7; h_3/d_1 = 0.22; h_6/d_1 = 0.5; h_1/d_1 = 1; δ/d_1 = 0.1	central 2–4 baffles (d_2/d_1 < 0.7) or without baffles (d_2/d_1 > 0.7)	axial-radial	0.5–0.95	1–9	turbulent/ transition range (d_2/d_1 < 0.7) laminar (d_2/d_1 > 0.7)	Ne = 0.65 (2-stages) turbulent (d_2/d_1 < 0.7); Ne · Re = 110 laminar (d_2/d_1 > 0.7)	< 40	Homogenization; Suspension; Gassing; Heat Transfer (Gentle handling)
EKATO MIZER-disc® EM		d_2/d_1 = 0.33; h_3/d_1 = 0.3; h_1/d_1 = 1; δ/d_1 = 0.1	central or off centre with or without baffles	radial	0.2–0.5	8–20	turbulent/ transition range	Ne = 0.2	< 10	Dispersion liquid/liquid; Dispersion liquid/solid (High shear rates)
Helical ribbon		d_2/d_1 = 0.90; b/d_2 = 0.1; s/d_2 = 0.5; h_3/d_2 = 1; h_1/d_1 = 1	central without baffles 1 or 2 ribbons with inner screw	axial forced circulation	0.9–0.95	< 2	only laminar	Ne · Re = 440 (s/d_2 = 0.5); Ne · Re = 270 (s/d_2 = 1) in each case 2 ribbons	> 50	Heat transfer; Blending of high viscous liquids

Step 4: From the chosen ratio of impeller diameter to vessel diameter, determine the impeller diameter.

Step 5: The impeller RPM and hence the Reynolds number are unknown. Therefore, guess a pumping number from known viscosity of the system, calculate Reynolds and pumping number, and check the guessed number using Figure 18.3 or a similar figure for the chosen impeller type available from the manufacturer. Repeat until you reach an agreement with the guessed number.

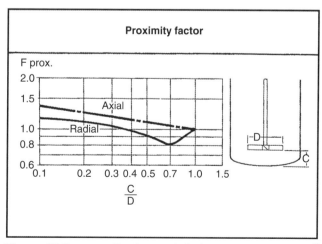

Figure 18.2 correction for proximity in power consumption[10]

Table 18.9 Selection of scale of agitation

	Required Process Response	Suggested Chemscale of Agitation N_I	Bulk Velocity, v_b, ft/min
1.0	Blending and solids suspension		
1.1	Blending fluids with specific gravity difference < 0.1 [No reaction]	1	6
1.2	Blending fluids with specific gravity difference < 0.6 [No reaction]	3	18
1.3	Blending fluids with specific gravity difference < 1 [reactor]	6–8	36–48
1.4	Blending fluids with viscosity of one component < 100 times the viscosity of the other	2	12
1.5	Blending fluids with viscosity of one component < 10,000 times the viscosity of the other [Non-reactive]	5	30
1.6	Blending fluids with viscosity of one component < 100,000 times the viscosity of the other [Critical reactors]	7–9	42–54
1.7a	Suspend solids to slurry uniformity to 98% of fluid batch height	(a) 10	(a) 60
1.7b	Suspend solids to slurry uniformity to 95% of fluid batch height	(b) 8	(b) 48
1.7c	Off bottom suspension of solids	(c) 5	(c) 30
2.0	Gas dispersion		
2.1	Coarse dispersion, not mass transfer limited	2	12
2.2	Moderate degree of gas dispersion and recirculation of dispersed bubbles	5	30
2.3	Rapid mass transfer and maximum interfacial area required.	10	60

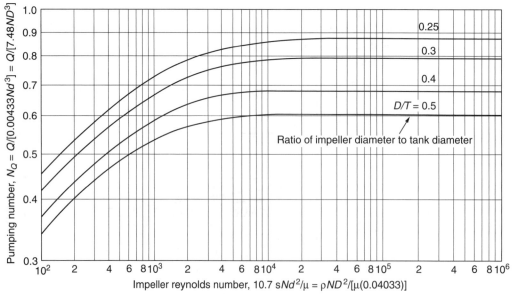

Figure 18.3 Pumping number vs impeller Reynolds number, pitched blade turbine[8]

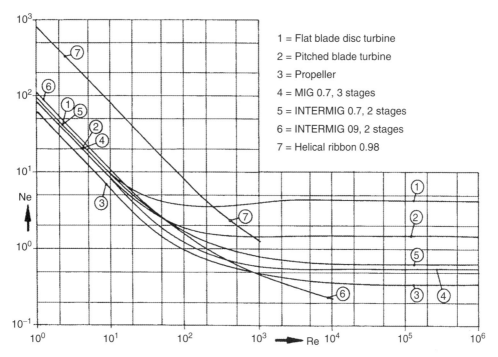

Figure 18.4 Power number Ne (= N_p in the text) vs impeller Reynolds number Re (= N_{Re} in the text) of various impeller types[11]

Step 6: Fine-tune RPM so that it is close to the suggested standard speed under the heading "**Some consideration in agitator design.**" From this suggestion and the calculated impeller Reynolds number, decide on the number and design of baffles.

Step 7: From Figure 18.4 read power number and calculate HP from the power number equation.

Step 8: Correct HP for the proximity factor from Figure 18.2, and for baffles, if necessary.

Nomenclature for the Preceding Section of the Process Design

a = amplitude of concentration variation, decimal
A = vessel cross section, ft^2
d = impeller diameter, inch
D = impeller diameter, ft
D_I = impeller diameter, m
h = liquid height, ft
K = amplitude decay constant, min^{-1}
N = rpm (revolution per minute) of impeller
N_{rps} = revolution per second of impeller
N_I = Chemscale
N_Q = pumping number (**dimensionless**). It depends on the type of impeller and impeller Reynolds number, the impeller diameter/tank diameter ratio, and is given by the manufacturer of the agitator.
N_P = power number, dimensionless, dependent on impeller design and Reynolds number.

P = agitator power, HP
P_I = agitator power, kW
Q = primary pumping capacity, gpm
s = specific gravity of fluid, dimensionless
t = tank diameter, inch
T = tank diameter, ft
TS = tip speed, ft/minute
v_b = apparent superficial velocity, ft/min
V = working capacity of the vessel, gallon
θ = mixing time, minute when N is rpm; second, when Ni in rps
$\theta_{turnover}$ = turnover time, minute
τ = torque, inch-lb
μ = viscosity, cP
ρ = density of fluid, lb/ft^3
ρ_{si} = density of fluid, kg/m^3
η = dynamic viscosity, Pa·s

MASS TRANSFER COLUMNS

Mass transfer columns are broadly divided into packed columns and plate columns. Sizing of plate columns involves selection of tray spacing, and sizing of the diameter and overall height. Plate spacing should be sufficient for the separation of the entrained liquid from the vapor before it reaches the next higher plate. Plate spacing is a function of many variables. It generally varies from 12 inches to 48 inches. Flooding limits for bubble-cap and perforated plates are plotted in Figure 18.5, while entrainment factor and overall plate efficiency for distillation

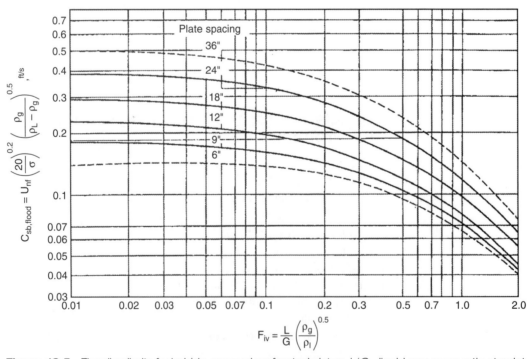

Figure 18.5 Flooding limits for bubble-cap and perforated plates. L/G = liquid-gas mass ratio at point of consideration. To convert feet per second to meters per second, multiply by 0.3048; to convert inches (symbol ") to meters, multiply by 0.0254. (From Fair, J. R., *Petro./Chem. Eng. 33* (10)45, (1969). With permission.)

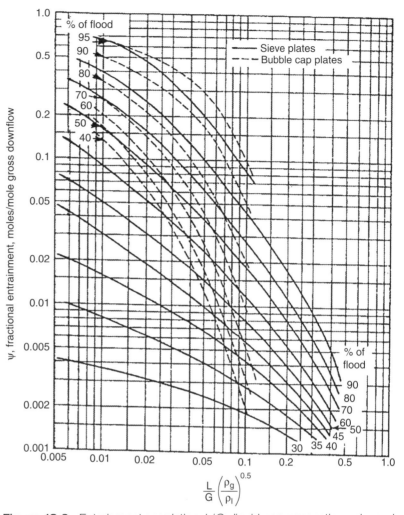

Figure 18.6 Entrainment correlation. L/G = liquid-gas mass ratio; and ρ_l and ρ_g = liquid and gas densities. (From Fair, J. R., *Petro./Chem. Eng.*, *33* (10) 45, (1961). With permission.)

are shown in Figure 18.6 and Figure 18.7. For a specified tray spacing, and known vapor and liquid flow rates, known physical properties such as liquid density, vapor density, and liquid surface tension, the flooding velocity can be calculated. Design velocity ranging from 45 to 80 percent of the flooding velocity is used to estimate the net column area from a known vapor load. Before selecting the design velocity, check Figure 18.6 so that entrainment is less than 0.1 lb of liquid per pound of liquid flow. Depending on the chosen tray type, an area required for the downcomer is to be added to get the total cross-section area of the column. Liquid loading in gpm and a recommended liquid velocity (gpm/ft^2) are used to estimate the downcomer area.

Determination of actual number of plates (N_a) requires the knowledge of overall plate efficiency (E_{oc}) and the number of theoretical plates (N_t).

$$E_{oc} = \frac{N_t}{N_a} \tag{18.32}$$

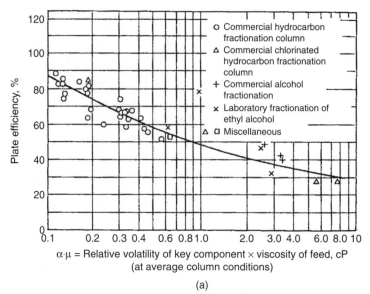

Figure 18.7 O'Conell correlation for overall efficiency E_{oc} for distillation. To convert centipoises to pascal-seconds, multiply by 10^{-3}. (O'Conell, Trans. Am. Inst. Chem. Eng., 42, 741 (1946). Reproduced with permission.)

The overall plate efficiency can be determined by empirical methods such as O'Connell correlation, Figure 18.7, Backowski correlation (Equation 18.29) in Perry[6] or by the theoretical predictive method of AIChE as shown in Perry[6] (page 18–15).

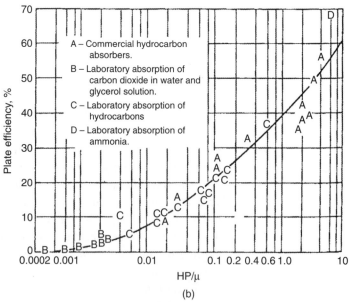

Figure 18.8 O'Conell correlation for overall column efficiency E_{oc} for absorption. To convert HP/μ in pound-moles per cubic foot–centipoises to kilogram-moles per cubic meter–pascal-second, multiply by 1.60×10^4. (O'Conell, Trans. Am. Inst. Chem. Eng., 42, 741 (1946), Reproduced by permission.)

Generally the following information is necessary for sizing a distillation column fitted with trays.

Table 18.10A Information table necessary for sizing a tray distillation column

Input Data	Name	Remark
L	Liquid flow rate, lb/h	Obtained from calculation of theoretical stages, see chapter 10
G	Vapor flow rate, lb/h	Obtained from calculation of theoretical stages, see chapter 10
ρ_l	Liquid density, lb/ft^3	Obtained from calculation of theoretical stages, see chapter 10
ρ_g	Vapor density, lb/ft^3	Obtained from calculation of theoretical stages, see chapter 10
σ	Surface tension of liquid, Dyne/cm	
α	Relative volatility of key component	
μ	Viscosity of liquid	
Ψ	Allowable fractional entrainment, lb/lb down flow (liquid flow)	
SF	System factor	See Table 18.11
FF	Flood factor	FF = 0.65 to 0.75 for diameter of column <36 inches, = 0.8 for larger diameter. Assume a trial number and verify later.
TS	Tray spacing	Assume a tray spacing for design and verify by final analysis and optimization of cost. It is available for rating an existing design.
N	Number of theoretical stages	Obtained from calculation of theoretical stages, see chapter 10
	% Flood	See FF above. Varies from 45% to 80%. Final choice requires consideration of entrainment & cost optimization.

Table 18.10B Choice of trays vs packing[17]

Factors Favoring Tray	Factors Favoring Packing
High liquid feed rate (occurs with high column pressure)	Operation under vacuum
Large column diameter (packing poses maldistribution challenges)	Lower pressure drop
Complex column (e.g. multiple draw-offs)	Small (2-3ft) column diameter
Varying feed composition	Corrosive system (wider choice of material of constructions)
Easier scale-up	Less prone to foaming
Lower overall weight	Lower liquid holdup
Entrained solids accommodate better	More amenable to batch operation

Table 18.10C Height equivalent of a theoretical plate *(HETP)* for organics[17] and maximum packing bed height

$$HETP(inch) = \begin{cases} \left[\dfrac{1100}{a_p}\right]F & \text{for modern metal packing } (\geq 1 \text{ inch}) \\[6pt] \left[\dfrac{1200}{a_p}+4\right]F & \text{for Y-type corrugated structured packing} \end{cases}$$

Where a_p is the packing surface area per unit volume, ft^2/ft^3

System (σ = Surface Tension)	F = Surface Tension Correction Factor
Organics, Surface tension, $\sigma < 25$ Dynes/cm	1
Amines, Glycols $\sigma \cong 40$ Dynes/cm	1.5
Aqueuses $\sigma \cong 60$ Dynes/cm	2

Maximum packing bed height without distributor

$$= \begin{cases} (2.5-3) \text{ Bed diameter for Raschig ring} \\ (5-8) \text{ Bed diameter for ceramic saddles} \\ (5-10) \text{ Bed diameter for slotted rings and plastic saddles} \end{cases}$$

Table 18.10D Tray characteristics[17]

Variable	Sieve Trays	Valve Trays	Bubble-Cap Trays	Dual-Flow Trays
Capacity	High	High-very high	Moderately high	Very high
Efficiency	High	High	Moderately high	Lower than other types
Turndown	~2 to 1. Not generally suitable for variable load operation	About 4–5 to 1. Some designs claim 10 to 1 or more.	Excellent, better than valve trays. Especially suitable at very low liquid rates	Low, even lower than sieve trays. Not suitable for variable load operation
Entrainment	Moderate	Moderate	High (about 3 times higher than that of sieve trays)	Low-moderate
Pressure drop	Moderate	Moderate. Older designs somewhat higher. Newer designs same as sieve trays	High	Low-moderate
Cost	Low	About 20% higher than sieve trays	High. About 2–3 times that of sieve trays	low
Maintenance	Low	Low–moderate	Relatively high	low
Fouling tendency	Low	Low-moderate	High: tendency to collect solids	Extremely low: suitable where fouling is extensive and for slurries.
Effects of corrosion	Low	Low-moderate	High	Very low
Availability of design information	Widely available	Proprietory, but information is readily available	Widely available	Available to some degree
Main applications	Most columns if turndown is not critical	1. most column 2. Service where turndown is a critical factor	1. very low flow conditions 2. If leakage must be minimized	1. Revamps where efficiency and turndown can be sacrificed 2. Highly fouling and corrosive applications.
other				Instability in large diameter (>8ft)

A step-by-step procedure is presented below for calculating the preliminary diameter of a bubble-cap tray column.

Step 1: Calculate parameter Flv and select a tray spacing.

$$Flv = \frac{L}{G}\sqrt{\frac{\rho_G}{\rho_L}} \tag{18.33}$$

Step 2: From calculated value of Flv and the selected tray spacing, read the value of $C_{sb,\text{flood}}$ (ft/s) from Figure 18.5.

Step 3: Calculate the maximum vapor velocity through the net flow area of the tray.

$$U_{nf} = C_{sb,\text{flood}} \left(\frac{\sigma}{20}\right)^{0.2} \left[\frac{\rho_l - \rho_g}{\rho_g}\right]^{0.5}, \text{ft/s} \tag{18.34}$$

Step 4: By consulting Figure 18.6 and flood factor (FF), calculate design vapor velocity, U_{nd}, ft/s.

$$U_{nd} = U_{nf}(FF) \tag{18.35}$$

Step 5: Calculate the net vapor flow area of the tray.

$$A_{\text{net,vapor}} = \frac{G}{3600 \rho_g U_{nd}}, \text{ft}^2 \tag{18.36}$$

Step 6: Choose System Factor from Table 18.11 and one or all of the following equations to calculate downcomer mass velocity in V_{dc} (gpm/ft^2).

$$V_{dc} = \begin{cases} 250(SF) \\ 41\sqrt{\rho_l - \rho_g}\,(SF) \\ 7.5\sqrt{TS(\rho_l - \rho_g)}\,(SF) \end{cases} \tag{18.37}$$

From the calculation shown above, select design downcomer mass velocity, usually the lowest value, $V_{dc,\text{design}}$.

Table 18.11 System factor for calculating downcomer area

Service	System Factor, SF
Non-foaming	1.0
Fluorine system: BF3, Freon	0.9
Moderate Foaming: oil absorber, amine or glycol regenerator	0.85
Heavy foaming: amine and glycol absorber	0.73
Severe foaming: MEK	0.6
Stable foam system: caustic regenerator	0.3

Step 7: Calculate the volumetric liquid flow rate and hence downcomer area, $A_{\text{downcomer}}$.

$$L_{\text{gpm}} = \frac{L}{(60)(8.34)\frac{\rho_l}{62.4}}, \text{ gpm} \qquad (18.38)$$

$$A_{\text{downcomer}} = \frac{L_{\text{gpm}}}{V_{\text{dc,design}}}, \text{ ft}^2 \qquad (18.38\text{a})$$

Step 8: Calculate the total area and hence the column diameter, *D*.

$$A_{\text{total}} = A_{\text{downcomer}} + A_{\text{net,vaor}} \qquad (18.39)$$

$$D = \sqrt{\left(\frac{4}{\pi}\right)A_{\text{total}}}, \text{ ft} \qquad (18.40)$$

Step 9: Finally the downcomer back-up should be verified.

EQUIPMENT TESTING AND ANALYSIS

Equipment testing may be broadly classified into two groups: mechanical testing and performance testing. Performance testing is equipment-specific, and it is generally done at the supplier's site. For example, the separation efficiency of a distillation column can be tested with a specific feed followed by collecting and analyzing the distillate and bottoms samples. For the mechanical integrity testing of pressure vessels and piping, ASME Section VIII and ANSI B31.3 are used for chemical process equipment and piping system respectively.

The inspection and testing of pressure vessels are covered in section UG-90 of ASME VIII, Division 1. A pressure vessel must be a hydro-tested to 1.3 times the MAWP (when testing fluid is a non-hazardous liquid used below its boiling point or a combustible liquid having a flash point less than 100°F). When a liquid is not acceptable as a testing medium, the equipment may be subjected to pneumatic test of 1.1 times MAWP. The metal temperature during test must be at least 30°F above the minimum design metal temperature, but not greater than 120°F. During a hydro-test, a thermal relief valve set at $1^{1}/_{3}$ times the test pressure is recommended. The following information is required on the nameplate of a pressure vessel:

1. The U symbol for all pressure vessels, [UM symbol for vessels that do not exceed 5 cu ft and 250 psi, or $1^{1}/_{2}$ cu. ft and 600 psi]

2. Certified by [name of person]

3. Maximum Allowable Working Pressure, psi____at ___°F

4. Maximum Allowable External Working Pressure, psi____at____°F

5. Minimum Design Metal Temperature___°F at _____psi

6. Manufacturer's serial number_____

7. Year built_____

8. Type of construction (W for arc or gas welded, P for pressure welded, B for brazed, RES for resistance welded)

9. Special service code (L for lethal, UB for unfired steam boiler, DF for direct firing) Radiography code (RT1, RT2, RT3, or RT4)

10. Heat treatment code (HT, PHT)

The preceding paragraphs outline the testing of the mechanical integrity of a fabricated pressure vessel. When the equipment, fabricated per ASME VIII, or API is installed for operation, it is also subject to inspection per *Pressure Vessel Inspection Code: Maintenance Inspection, Rating, Repair, and Alteration, API 510*. The inspector must be certified. Pressure relief devices are inspected and maintained in accordance with API RP 576. The inspection frequency of equipment and relief is maximum 60 months, which may be extended to 120 months after consideration of corrosion rate, remaining life, internal and external conditions, service, and consistency of operating conditions.

The remaining life of a vessel shall be calculated from the following formula:

$$\text{Remaining life (years)} = \frac{t_{actual} - t_{minimum}}{\text{corrosion rate}} \quad (18.41)$$

$$\text{Corrosion rate} = \frac{t_{previous} - t_{actual}}{\text{years between } t_{previous} \text{ and } t_{actual}}$$

where
t_{actual} = the thickness, in inches, recorded at the time of inspection for a given location or component
$t_{minimum}$ = minimum allowable thickness, in inches, for a given location or component.
$t_{previous}$ = the thickness, in inches, at the same location as t_{actual} measured during a previous inspection

Non-Destructive Testing

Besides hydro testing, which was presented in the preceding paragraphs, there are six common non-destructive testing methods to detect flaws in weld, crack, etc.

1. Radiographic examination by either x-ray or gamma radiation is used to check the quality of welds.

2. Ultrasonic techniques use vibrations with frequencies between 0.5 and 20 MHz to measure thickness of a plate.

3. Magnetic-particle examination uses magnetic flux to detect surface cracks in a plate or weld.

4. Liquid-penetrant examination involves wetting the surface using a liquid containing a dyestuff, which will penetrate the crack, followed by wiping of the surface and an application of coating material. The liquid from the crack comes out to interact with the coating, thereby showing the defect.

5. The eddy-current techniques involving the measurement of electrical impedance of an alternating-current coil.

6. The electrical resistance method.

No known method of non-destructive testing can guarantee an absence of flaw.

Destructive testing includes guided bend test, tensile strength test, and hardness test of test coupons. As the name signifies, the base metal can be damaged beyond use.

DESIGN OPTIMIZATION

The best equipment design is the least expensive and safest design that meets the specification and performs the duty. Because all businesses are motivated by profit, economic evaluation plays an important role in process design. There are many ways to reach the same destination. The best way is the least expensive way in terms of time, money, and safety.

Degrees of Freedom for Optimization

Degrees of freedom for optimization (F) refers to the number of variables that are not fixed or specified in a design and therefore are free to be adjusted for optimum design.

$$F = M - N \tag{18.42}$$

where
M = number of variables, such as equipment type, pressure, temperature, etc.
N = number of independent sources of information, such as heat and material balances

F has two components,

$$F = F_1 + F_2 \tag{18.43}$$

where F_1 = environmental degrees of freedom. These are variables whose values are fixed by the environment of the process, such as cooling water supply temperature or the inlet and outlet temperatures of the process side of a heat exchanger, etc. This is a freedom lost by the optimizer.

F_2 = economic degrees of freedom. These are variables that the designer can adjust to maximize the profitability, such as the type of an exchanger or the cooling water flow rate.

One Variable Problem

There are many situations in chemical engineering in which one variable is to be optimized. Optimum insulation thickness to conserve energy and optimum reflux ratio of a distillation column are examples. Several thicknesses of insulation may be tried to see the thickness, which gives the lowest annual cost, including the depreciated cost of insulation and the cost of lost heat. This can be done graphically. Sometimes an equation may be set up. The optimum value may be obtained analytically by taking a derivative with respect to the variable to be optimized, equating the derivative to zero, and then solving for the variable. The optimum value is then checked by taking a second derivative at the given point and examining the sign of the second derivative. The second derivative at the point of interest must be > zero so that the value of the variable at the point is the minimum.

Two or More Variables to Optimize

The solution method becomes more complex, but can be done using a similar approach.

PROCESS FLOW SHEET DEVELOPMENT

Process flow sheet is a generic term, which includes

1. *Block Flow Diagram (BFD):* a diagram showing an assembly of overall unit operations in a manufacturing or processing plant. Generally, unit operations are shown in rectangular blocks without equipment details. An overall material flow showing the destination, connectivity between interdependent units, and source of all materials and an accounting balance of inlet and outlet quantities are part of the block flow diagram. Such diagrams are generally presented to a government agency to examine the impact on environment, and are conceived at the development stage of each individual unit operation, which leads to Process Flow Diagram.

2. *Process Flow Diagram (PFD):* A process flow diagram is an elaboration of BFD. It shows, in greater detail, each unit operation along with individual equipment. It generally includes stream numbers, which identify the components of the flow, flow rate, pressure, temperature, phases (liquid/vapor/solid), and transport properties including surface tension for separation processes, which are required to size the line and select the material of construction. Concentration of critical components in ppm or pH is also shown as stream property or as a note. Minimum, normal, and maximum values of important properties such as pressure, temperature, and flow rate are also labeled against each stream to help the designer to set alarm and interlock limits and decide on the design pressure and design temperature of the associated equipment. To arrive at these ranges, consideration should be given to process variability, start-ups and shut down, emergencies, turndown, etc. If any safety factor is included, it must be communicated to avoid piling up safety factors as the design moves to the construction phase. Liberal notes should be used in PFD to communicate the thought process of creative process engineers. Similar information is also needed for the utility streams. The PFD also includes equipment identification (Tag) number, equipment name, the required duty, material of construction, and supplementary information in the form of notes to communicate to the systems engineering group for the development of piping and instrumentation diagram and design specification of equipment. The PFD of batch or semi-batch operation is much more involved. The PFD for batch/semi-batch operation must indicate which stream is batch and which stream is continuous. Each batch stream must show the batch amount and also the rate of addition as normal, minimum, and maximum as well as the sequence of addition so that lines are properly sized. A PFD is a vital piece of document for hazard and operability study, in particular for sizing emergency relief devices, and for developing the current Good Manufacturing Practices (cGMP for Pharmaceutical and bio-tech industries), Standard Operating Procedure (SOP) of a petrochemical plant, and

Piping and Instrumentation Diagram (P&ID). A PFD also identifies the environmental emissions, and may call for further studies for abatement systems.

3. *Piping and Instrumentation Diagram (P&ID):* The P&ID, otherwise known as the mechanical flow sheet, is the source of detailed engineering, and, in conjunction with the PFD and layout, serves to develop cGMP, SOP, building blocks for automation. The P&IDs are issued in stages such as preliminary, design, construction etc., with uncertain designs marked as "HOLD". The issue letter or number from the issue box must be appropriately marked to distinguish from earlier issue whenever design changes are made.

The symbols of instrumentation of conventional electrical, pneumatic, or hydraulic control follow the Instrument Society Standard (ISA) ISA-S5.1 and those of computer control system follow ISA-S5.3.

Equipment is designated by a single or double alphabetic symbol followed by a number. For example, an agitator may be designated by A-101; and a blower, BL-101. Such designation must be followed by a description and a broad specification. Thus for an agitator,

- A-101

    Reactor R-101 agitator

    Impeller size & Type: 3 ft, Pitched Blade Turbine

    Motor HP: 50 @ 3550

    Seal: Double Mechanical

    Drive: Reeves

    M/C: AISI Type 316 SS

    PO#: s20768

A line is designated by a nominal size, process identification code, line number, and insulation code. Each line must have an identification of its source and an identification of its destination. Sometimes a slope is indicated for safety reasons. Appropriate symbols are used to indicate jacketed pipe, insulation, tracing (steam, electrical, liquid), future lines, existing line, and specification break between two different types of piping specifications. Usually, the first block valve (not a check valve) or both block valves in the case of a double block valves and a bleed valve dividing the two specifications belongs to the more stringent specification. The fail position of control valves is noted. Size, type and set conditions of relief devices, range of instruments are noted.

OPERATING MANUAL

The operating manual consists of the following subjects for start-up, shut down, and maintenance:

1. Detailed process description

2. Process chemistry, if applicable

3. Start-up conditions

4. Normal operating conditions

5. Shut-down conditions

6. Maintenance aspects

7. Emergency situations and how to deal with them.

The detailed process description must be keyed to the respective P&IDs. The descriptions should be developed as one reads across the respective P&IDs in left-to-right fashion. Instrumentation and control systems and major process valves should be described as they are encountered on the P&ID.

The outlines of the operating manual are as follows:

I. Title Page

II. Table of Contents

III. Introduction
   A. Capacity

   B. On-stream factors

   C. Brief background information

   D. Major process steps

IV. Process Chemistry
   A. Brief descriptions by reaction or process steps

   B. Indicate reactions

   C. Thermodynamics

   D. Kinetics including heat of reactions

   E. By-products

V. Detail Process Description
   A. Description following the P&IDs

   B. Descriptions including
     1. Normal operations

     2. Start-up

     3. Shutdown

     4. Emergency shutdown
       a. Loss of utilities

       b. Mechanical failure

       c. Interlocks and alarm management to minimize nuisance alarms, which can lead to lack of appropriate attention thereby leading to safety and environmental hazards, loss of quality and product.

     5. Instrumentation
       a. Operating temperature, pressure, flow, level, density, pH, etc.

       b. Control limits and total range of measurement sensor determined by the control loop set point, normal range leading to alarm set point, Proven Acceptable Range (PAR) leading to interlock set point, Total range of measurement sensor.

     6. Operating variables

     7. Unique operations

8. Clean-out

9. Process safety considerations
   a. Relief devices
   b. Toxicity
   c. Explosion hazards, explosion panels, explosion suppression
   d. Fire hazards
   e. Inert gas requirement
   f. Process hazards (runaway potentials, reactivity data, explosivity data, safety factors in design, etc.)

C. Additional considerations
   1. Process equipment selection
   2. Materials of constructions
   3. Space for future installation

VI. Safety Considerations
   A. General comments
   B. Operating conditions, precautions and/or limitations
   C. Safety equipment including personal protective equipment
   D. Chemical handling and associated hazards
   E. Relief devices, explosion panels, vent, and flare system.
   F. Opening lines and equipment
   G. Leaks and fugitive emissions
   H. Interlocks and alarm descriptions

   For fully automated systems that can be defined as a safety instrumented system, refer to ISA-S84.01-1966: Application of Safety Instrumented Systems in Process Industries.

   I. Special protective systems

VII. Effluents
   A. Solids disposal
   B. Liquid effluents
   C. Vent systems
   D. Emergency discharges

VIII. Startup and Shutdown
   A. Precautions
   B. Emergency situations
   C. Rerun lines
   D. Purging
   E. Special considerations

Table 18.12A Centrifugal pump trouble-shooting

Symptom	Possible Cause of Trouble, [See Next Table for Explanation of Cause Number]
Pump does not deliver fluid	1,2,3,4,6,11,14,16,17,22,23
Insufficient pumping capacity	2,3,4,5,6,7,8,9,10,11,14,17,20,22,23,29,30,31
Insufficient developed pressure	5,14,16,17,20,22,29,30,31
Pump loses prime after start-up	2,3,5,6,7,8,11,12,13
Pump requires excessive power	15,16,17,18,19,20,23,24,26,27,29,33,34,37
Stuffing box leaks excessively	13,24,26,32,33,34,35,36,38,39,40
Packing has short life	12,13,24,26,28,32,33,34,35,36,37,38,39,40
Pump vibrates or is noisy	2,3,4,9,10,11,21,23,24,25,26,27,28,30,35,36,41,42,43,44,45,46,47
Bearings has short life	24,26,27,28,35,36,41,42,43,44,45,46,47
Pump overheats or seizes	1,4,21,22,24,27,28,35,36,41

Table 18.12B Explanation

Cause Number	Explanation
1	Pump is not primed
2	Pump or suction pipe is not completely filled with liquid
3	Suction lift too high
4	Insufficient margin between suction pressure and vapor pressure
5	Excessive amount of air or gas in the liquid
6	Air pocket in the suction line
7	Air leaks into the suction line
8	Air leaks to the pump through stuffing box
9	Foot valve too small
10	Foot valve partially clogged
11	Inlet of suction pipe insufficiently submerged
12	Water seal pipe plugged
13	Seal cage improperly located in the stuffing box, preventing sealing fluid from entering space to form seal
14	Speed to low
15	Speed too high
16	Direction of rotation wrong
17	Total head of the system is higher than the design head of the pump
18	Total head of the system is lower than the design head of the pump
19	Specific gravity of the liquid is different from the design
20	Viscosity of the liquid is different from the design
21	Operation is very low compared with design capacity
22	Parallel operation is unsuitable for this operation
23	Foreign matter in the impeller
24	Misalignment
25	Foundation not rigid
26	Shaft bent
27	Rotating part rubbing with stationary part
28	Bearings worn

(*continued*)

Table 18.12B Explanation (Continued)

Cause Number	Explanation
29	Wearing rings worn
30	Impeller damaged
31	Casing gasket defective, permitting internal leakage
32	Shaft or shaft sleeves worn or scored at packing
33	Packing improperly installed
34	Type of packing incorrect for operating conditions
35	Shaft running off-center because of worn bearing or misalignment
36	Rotor out of balance, causing vibration
37	Gland too tight, resulting in no flow of liquid to lubricate packing
38	Cooling liquid not being provided to water-cooled stuffing box
39	Excessive clearance at bottom of stuffing box between shaft casing, causing packing to be forced into pump interior
40	Dirt or grit in sealing liquid leading to scoring of shaft or shaft sleeve
41	Excessive thrust caused by mechanical failure inside pump or failure of hydraulic balancing drive, if any
42	Excessive grease or oil in antifriction bearing housing or lack of cooling, causing excessive bearing temperature
43	Lack of lubrication
44	Improper installation of antifriction bearings (damaged or unmatched bearings as a pair)
45	Dirt in bearings
46	Rusting of bearings from water in the housing
47	Excessive cooling of water cooled bearing resulting in the condensation of atmospheric moisture in bearing housing.

Heat Exchanger Design Tips to Avoid Trouble

1. The chances of achieving perfect distribution are more favorable if the tube-side fluid flows upward all the time. Therefore avoid symmetrical conical channels with inlet and outlet nozzles on the centerline.

2. In the shell side of horizontal heat exchangers, horizontally cut baffles are preferable to vertical cuts.

3. Install start-up vents at channel ends.

4. Install shell-side vent by drilling through tube sheet of a thermosyphon reboiler heated by steam.

5. Install a pair of flanges to install an orifice plate at the inlet of a reboiler, if necessary, to cut circulation rate and promote nucleate boiling.

6. All condensers have non-condensable gases. Include some non-condensable gases in the design. Allow vents at each end of the shell side or tube side to periodically purge the non-condensable gases.

7. Specify either TEMA-B (chemical service) or TEMA-R (Refinery service) for mechanical design.

8. Impingement protection is always required at the shell inlet nozzle.

9. Aim for cooling water velocity of at lest 3 ft/s. Avoid abnormal fouling factor for cooling water service: otherwise you invite the same problem you are trying to avoid.

10. Specify no-tubes-in-the-window to minimize vibration damage. You pay a higher price for a bigger shell diameter, however.

Example 18.1

Determine the set pressure of a conservation vent and suggest the design pressure of the storage tank, which will be fitted with the conservation vent. Available data are

Maximum seasonal liquid surface temperature = 100°F

Minimum seasonal liquid surface temperature = 90°F

Maximum vapor space temperature = 140°F

Minimum vapor space temperature = 85°F

Vapor pressure at 100°F = 9.6 psia

Vapor pressure at 90°F = 8.1 psia

Elevation of the tank location from sea level = 4000 ft

Solution

This problem is an application of Equations 18.1 and 18.2.

Given:

$\Delta Z = 4000$ ft $= 4000 \times 0.3048 = 1219$ m

$T_1 = 460 + 85 = 545$ R

$T_2 = 460 + 140 = 600$ R

$P_{min} = 8.1$ psia

$P_{max} = 9.6$ psia

Assumption: ideal gas law

To find: Local barometric pressure, set pressure of breather, and design pressure of the tank.

Local barometric pressure

$$P_a = 14.696 \left[1 - \frac{0.0065(\Delta Z)}{288} \right]^{5.26433}$$

$$= 14.696 \left[1 - \frac{0.0065(1219)}{288} \right]^{5.26433}$$

$$= 12.688 \text{ psia}$$

Set pressure of breather:

$$P_s = (P_a - P_{min})\frac{T_2}{T_1} + P_{max} - P_a$$

$$= (12.688 - 8.1)\frac{600}{545} + 9.6 - 12.688$$

$$= 1.963 \text{ psig}$$

Suggested tank design pressure = 1.963 × 2 = 3.926 ≅ 4 psig

Example 18.2

The following information is available to a fabricator to advise the owner about the required plate thickness, an approximate empty weight of the equipment, and the MAWP.

Inside diameter: 7 ft 6 in.

Overall tangent-to-tangent length: 60 ft

Operating temperature: 300° F

Operating pressure: 75 psig

Design pressure: 110 psig

Design temperature: 450° F

Material of construction: Carbon steel

Corrosion allowance: 0.125 in.

Joint efficiency: 0.8

Allowable stress at the design temperature: 13,800 psi

Wind velocity: 150 miles per hour at 14.7 psia and 70° F

Type of head: 2:1 ellipsoidal

Water weight in a single head volume: 3450 lbs

Insulation thickness: 3 in.

Approximate weight of insulation: 10,400 lbs.

Find the required commercial plate thickness, an approximate empty weight, and the MAWP of the distillation column. Suggest the set pressure of a safety device.

Solution

This problem is an application of Chapter 18, Table 18.3 and Equations 18.8 through 18.12.

Given:

Design pressure, P = 110 psig

Inside radius, R = (7.5 ft × 12 in/ft)/2 = 45 in.

Allowable stress, S = 13,800 psi

Joint efficiency, $E = 0.8$

Corrosion allowance, $c = 0.125$ in.

Straight length, tangent-to-tangent = 60 ft

Distance from top of vessel to the point where wind pressure is to be considered, $h = 60$ ft.

Wind velocity, $V_{wind} = 150$ mph

Calculation

Step 1: Calculate t and t_m for the shell and ends, and pick t_p

(a) For the shell

$$t = \frac{PR}{SE - 0.6P} = \frac{(110)(45)}{(13800)(0.8) - 0.6(110)} = 0.451 \text{ in.}$$

(b) For the head

$$t = \frac{PR}{SE - 0.1P} = \frac{(110)(45)}{13800(0.8) - 0.1(110)} = 0.449 \text{ in.}$$

Therefore, the shell thickness controls

$$t_m = t + c = 0.451 + 0.125 = 0.576 \text{ in.}$$

The nearest commercial plate thickness is 5/8 inch or $t_p = 0.625$ in.

Step 2: Calculate wind pressure.

$$V_{wind} = 150 \text{ mph} = \frac{150(1760)(3)}{3600} = 220 \text{ ft/s}$$

$$\rho_{air} = \frac{P_a M}{RT} = \frac{14.7(29)}{10.731(460 + 70)} = 0.075 \text{ lbs/ft}^3$$

Wind pressure, $P_w = \left[\frac{V_{wind}^2 (\rho_{air})}{2(32.17)}\right] = \frac{(220)^2 (0.075)}{2(32.17)} = 56.385 \frac{\text{lbs}}{\text{ft}^2}$

Step 3: Calculate mean diameter, outside diameter with insulation, empty weight of vessel, and water full weight of vessel.

Mean diameter = (inside diameter + outside diameter)/2

$$D_m = \left[\frac{7.5 + 7.5 + \frac{2t_p}{12}}{2}\right](12) = \left[\frac{(7.5 + 7.5 + \frac{2(0.625)}{12})(12)}{2}\right] = 90.625 \text{ in}$$

$$D_{mf} = \frac{D_m}{12} = 7.552 \text{ ft}$$

D' = outside diameter with insulation
$= 7.5(12) + 3(2) + t_p(2) = 7.5(12) + 3(2) + 0.625(2) = 97.25$ in

From equation (18-12):

$$W_e = 136(D_{mf})[HT + 0.8D_{mf}](t_p + 0.18)$$
$$= 136(7.552)[60 + 0.8(7.552)](0.625 + 0.18) = 54600 \text{ lbs}$$
$$W_{ei} = W_e + \text{insulation weight} = 54600 + 10400 = 65000 \text{ lbs}$$

Water full weight of column = weight due to water + empty weight.

Weight of water = weight of water in cylinder + weight of water in heads

$$W = \left[\left(\frac{\pi}{4}\right)(ID)^2(HT)(62.4) + \text{mass in heads}\right] + W_{ei}$$

$$= \left[\left(\frac{\pi}{4}\right)(7.5)^2(60)(62.4) + 3450(2)\right] + 65000 = 237300 \text{ lbs}$$

Step 4: Calculate thickness required for the windward side due to wind. Use empty column weight for conservative thickness. From Equation 18.8,

$$t = \frac{2P_w h^2}{\pi D'SE} - \frac{W_{ei}}{\pi D_m SE} + \frac{PD_m}{4SE}$$

$$= \frac{2(56.385)(60)^2}{\pi(97.25)(13800)(0.8)} - \frac{65000}{\pi(90.625)(13800)(0.8)} + \frac{110(90.625)}{4(13800)(0.8)}$$

$$= 0.325 \text{ inch}$$

This is less than t_c, corroded plate thickness (0.625 – 0.125 = 0.5 in), therefore, it is not the controlling thickness.

An inspection of leeward thickness will indicate that it does not control either.

Step 5: Calculate buckling thickness. From Equation 18.11,

$$S_B = \frac{2(10^6)t_c}{D_o} = \frac{2(10^6)(0.5)}{91.25} = 10960 \text{ psi}$$

From Equation 18.10,

$$t = \frac{2P_w h^2}{\pi D'S_B E} + \frac{W}{\pi D_m S_B E}$$

$$= \frac{2(56.385)(60^2)}{\pi(97.25)(10960)(0.8)} + \frac{237300}{\pi(90.625)(10960)(0.8)} = 0.247 \text{ inch}$$

The buckling thickness does not control. Therefore, 0.625 inch plate is OK.

Step 6: Calculate MAWP of the column. Note that corroded thickness is to be used for the calculation of MAWP.

(a) Based on shell

$$P = \frac{SEt_c}{R + 0.6t_c} = \frac{13800(0.8)(0.5)}{45 + 0.6(0.5)} = 121.854 \text{ psi}$$

(b) Based on head

$$P = \frac{SEt_c}{R + 0.1t_c} = \frac{13800(0.8)(0.5)}{45 + 0.1(0.5)} = 122.53 \text{ psi}$$

The weaker component is the shell. Therefore, recommended MAWP as a rounded number is 120 psig.

If this MAWP is stamped on the vessel, the relief device may be set at 120 psig.

Example 18.3

The overhead reflux accumulator of a distillation column has a 24" outside diameter feed line from the condenser. The normal operating pressure is 200 psig at 400°F. A relief valve set at 250 psig protects the accumulator and the associated piping. A scenario is considered when the relief valve will be full open and accumulate 21 percent overpressure with the temperature remaining constant at 400°F. In addition, there will be a pressure drop of 2 psi at the inlet line to the relief valve and accumulator. The feed pipe will be subjected to this overpressure. The duration of this overpressure is not expected to be more than 10 hours per upset, and no more than 10 upsets per year. If the allowable stress, corrected for joint efficiency, is 22,900 psi for the material of construction, calculate the thickness of the feed piping and the appropriate schedule number for this specific scenario. Include a corrosion allowance of 0.125 inch. The available thickness for 24" pipe is as follows: (a) Schedule 10: 0.25", (b) Schedule 40: 0.687"

Solution

This problem is an application of Chapter 18, Table 18.5, and Equations 18.13 through 18.14.

Given:

Outside diameter of pipe, $D_o = 24$ in.

Allowable stress corrected for joint efficiency, $SE = 22,900$ psi

Corrosion allowance, $c = 0.125$ in.

Calculation

Step 1: Calculate the design pressure.

(a) Calculate the upset pressure.

The upset pressure, P_{upset} = accumulated pressure + line drop.

$$P_{upset} = \left(1 + \frac{\% \text{ accumulation}}{100}\right) P_{set} + \Delta P_{line}$$

$$= \left(1 + \frac{21}{100}\right)(250) + 2$$

$$= 304.5 \text{ psig}$$

(b) Calculate the design pressure by taking advantage of the code-allowed overpressure.

For the scenario under consideration, the system may be overpressured by 33 percent. Therefore, the design pressure, P, may be calculated as follows:

$$P = \frac{P_{upset}}{1.33} = \frac{304.5}{1.33} = 228.95 \text{ psig}$$

(*Note:* If this pressure were lower than the operating pressure, the latter would be used for calculating thickness in the next step.)

Step 2: Calculate the pressure design wall thickness of pipe, t, and minimum thickness after adding allowance.

$$t = \frac{PD_o}{2SE} = \frac{228.95(24)}{2(22900)} = 0.12 \text{ in}$$

$$t_m = t + c = 0.12 + 0.125 = 0.245 \text{ in}$$

Step 3: Calculate the schedule number.

$$\text{Schedule number} = \frac{P}{S}(1000) = \frac{228.95(1000)}{22900} = 9.997$$

By comparing with thickness/schedule number data, select schedule 10 pipe.

RECOMMENDED REFERENCES

1. Varec Vapor Control Product Catalog, VAREC USA.
2. Darby, *Chemical Engineering Fluid Mechanics,* 2nd ed., Marcel Dekker, Inc., NY, p. 91, 2001.
3. Das, *Allowing for Process Upset Conditions in Piping System Design,* Plant Engineering, March 2, 1978.
4. *Chemical Plant and Petroleum Refinery Piping,* ANSI B31.3, The American Society of Mechanical Engineers, NY.
5. Raseand Barrow, *Project Engineering of Process Plants,* John Wiley & Sons, Inc., NY, 1968.
6. Perry, *Chemical Engineers' Handbook,* 6th ed., section 18, McGraw-Hill Book Company, NY, 1984.
7. Martin and Dietrich, *Fast-track Your Pilot Plant Project,* CEP, January 2005.
8. Morton, "How to design agitators for desired process response," *Chemical Engineering,* April 26, 1976.
9. Penny, "Guide to Trouble-Free Mixer," *Chemical Engineering,* June 1970.
10. EMI Incorporated, Bedford, NH.
11. *Research and Development in Mixing Technology,* EKATO, Ramsey, NJ.
12. Khang and Levenspiel, *Chemical Engineering Science,* 1976, vol. 31, Pergamon Press.
13. Branan, *The Fractionator Analysis Pocket Handbook,* Gulf Publishing Company, Houston, TX, 1978.
14. *Chemical Engineering,* February 23, 1981, pp. 83–85.

15. Jacobos, *How to select and specify Process Pumps,* Hydrocarbon Processing, June, 1965.
16. Karassik, Krutzsch, Fraser, and Messina, *Pump Handbook,* 2nd edition, McGraw Hill Book Company, NY. (a) temp rise: pp. 2.243 (b) Trouble shooting: pp. 12.16, 1985
17. Kister, "Facts at Your Fingertips," *Chemical Engineering,* April 2005.

CHAPTER 19

Engineering Economics

OUTLINE

TIME VALUE OF MONEY 697
Rate of Interest ■ Time-Scale Presentation ■ Simple Interest ■ Compound Interest ■ Nominal and Effective Annual Interest Rates ■ Annuity ■ Sinking-Fund Factor ■ Capital-Recovery Factor ■ Relation between Capital-Recovery Factor and Sinking-Fund Factor ■ Continuous Compounding of Interest ■ Capitalized Cost

CHEMICAL ENGINEERING PLANT COST INDEX (CEPCI) 706
Order-of-Magnitude Estimate of Equipment and Installed Costs ■ Estimation of Operating Cost of Production Facilities ■ Depreciation ■ Depreciation Calculations ■ Cost Comparison of Alternatives ■ After-Tax Comparisons ■ Linear Break-Even Analysis ■ Analytical Method of Optimization

RECOMMENDED REFERENCES 720
Additional Reading

There is an emphasis on economics in the P.E. examinations. In the final part of the examination, the candidates are required to answer some economics-related questions.

TIME VALUE OF MONEY

When money is borrowed by individuals, corporations, or public service organizations, they are required to pay compensation for the use of the borrowed money. The sum of money loaned is called *principal* and the compensation paid by the borrower or earned by the lender is termed *interest*.

Rate of Interest

This is the ratio of the amount of interest payment to the principal per unit of the interest period. It might also be termed as the rate of return or yield on the productive investment of capital. It is denoted by i and expressed as percent per interest period. Thus

$$i = \frac{\text{(total interest paid)}(100)}{\text{(principal)(number of interest periods)}}$$

Time-Scale Presentation

Economic disbursements on a time scale are represented as follows:

Figure 19.1 Economic disbursements on a time scale

Here 0 to n indicate the time periods, *usually in years* and $A_0, A_1, ..., A_n$ are disbursements. The zero on the time scale is the beginning of the year, and 1 is the end of one year (or first period) or beginning of the second year (or second period), and so on.

Simple Interest

When the interest earned on an investment is not reinvested with the original investment to form new interest-earning capital, the interest is called *simple interest*. Thus

$$I' = Pin \tag{19.1}$$
$$F = P(1 + in) \tag{19.2}$$

where
I' = interest
i = interest rate as fraction per interest period
n = number of interest periods
P = present worth
F = future worth.

Example 19.1

What will be the future worth after 14 months if a sum of $100 is invested today at a sample interest rate of 10 percent per year?

Solution

From Equation 19.2, $F = P(1 + in)$

$$P = 100 \quad i = 0.1 \quad n = \frac{14}{12} = 1.1667$$

Hence
$$F = \$100[1 + 0.1(1.1667)] = \$111.67$$

Compound Interest

When the interest earned on an investment is reinvested along with the original investment to earn interest, the interest earned is called *compound interest*. The compounding of interest is represented on a time scale as follows:

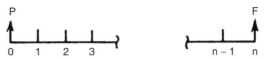

Figure 19.2 Compounding interest

Here F and P are the future and present worths, respectively. Given P, F is found by

$$F = P(1 + i)^n = P(F/P, i, n) \tag{19.3}$$

Given F, P is found by

$$P = \frac{F}{(1+i)^n} = F(P/F, i, n) \tag{19.4}$$

The terms $(F/P, i, n)$ and $(P/F, i, n)$ are the functional symbols to represent the factors $(1 + i)^n$ and $1/(1 + i)^n$, respectively. Interest tables with these and other factors are available in engineering economics texts.[1a,2a] The term $(F/P, i, n)$ is called the *single-payment future worth factor* and the term $(P/F, i, n)$ is called the single-payment present worth factor. Unless otherwise specified, the interest rate i is taken as percent per year.

Example 19.2

How many years* will it take to double an investment if the interest is 10 percent per year?

Solution

From Equation 9.3

$$F = 2P = P(F/P, 10\%, n)$$
or
$$(F/P, 10\%, n) = 2$$

From the interest table,[1a,2a] = 7.3 years approximately.

Nominal and Effective Annual Interest Rates

The nominal interest rate expresses the interest rate as percent per year. However, in many instances the compounding of interest is done more than once a year. For example, "8% compounded semiannually" means 4 percent interest is paid on the original investment at the end of the six months and it is then added to the investment to earn interest for the next six months. To convert the nominal interest rate into an effective annual interest rate (as the name signifies, the compounding is done annually in terms of an effective annual interest rate i), the following formula is used:

$$i = \left(1 + \frac{j}{m}\right)^m - 1 \tag{19.5}$$

---

*Rule of 72. This rule indicates that the number of years N over which a future value is double its present worth at the annual compound interest rate i is given by $N = 72/i$. In other words, $(1 + i)^n$ doubles about every N years. This is a rule of thumb.

where j is the nominal interest rate, fraction per period, and m is the number of the interest periods or compoundings per year. The relation between the present worth and future worth is given by

$$F = P\left(1 + \frac{j}{m}\right)^{mn} \quad (19.6)$$

where m and j are defined above and n is the number of years.

Example 19.3

A savings bank offers loans on two choices of interest:

(a) 10 percent compounded every month, and

(b) 12 percent semiannually. Which option would give a lower debt?

Solution

Effective annual interest rate, $i = (1 + j/m)^m - 1$

(a) $j = 0.1, \quad m = 12$

$$i = \left(1 + \frac{0.1}{12}\right)^{12} - 1 = 10.47 \text{ percent per year}$$

(b) $j = 0.12, \quad m = 2$

$$i = \left(1 + \frac{0.12}{2}\right)^2 - 1 = 12.36 \text{ percent per year}$$

Option (a) would give lower debt because the effective interest to be paid would be smaller.

Example 19.4

What will be the equivalent amount after 10 years, if $2000 is deposited today at an interest rate of 10 percent compounded semiannually?

Solution

There are two ways of solving this problem.

(a) Calculate the effective annual interest rate.

$$i = \left(1 + \frac{j}{m}\right)^m - 1 \quad j = 0.1 \quad m = 2$$

$$= \left(1 + \frac{0.1}{2}\right)^2 - 1 = 0.1025$$

Future worth $= F = P(1 + i)^n$
$$= \$2000(1 + 0.1025)^{10} = \$5306.60$$

(b) Equation 19.6 can be directly used.

$$j = 0.1 \quad m = 2 \quad n = 10$$

$$F = P\left(1 + \frac{j}{m}\right)^{mn}$$

$$= \$2000\left(1 + \frac{0.1}{2}\right)^{2(10)} = \$5306.60$$

Annuity

An annuity is a series of uniform end-of-period payments, such that each period is equal and the first payment is made after the first interest period. On the time scale, it looks like the following diagram (A = uniform series of payment).

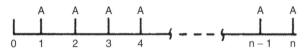

Figure 19.3 Uniform series of payments on an annuity scale

Various formulas may be developed for the annuity payment. Future worth of a uniform series payment is

$$F = A\frac{(1+i)^n - 1}{i} = A(F/A, i, n) \tag{19.7}$$

Annuity for a future worth is

$$A = F\frac{i}{(1+i)^n - 1} = F(A/F, i, n) \tag{19.7a}$$

Present worth of an annuity is

$$P = A\frac{(1+i)^n - 1}{i(1+i)^n} = A(P/A, i, n) \tag{19.7b}$$

Example 19.5

If you start saving in a bank at a rate of $1000 per year, at the end of each year from now, how much will you accumulate at the end of 10 years if the interest rate is 10 percent per year?

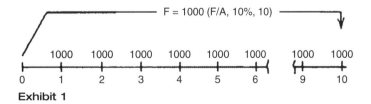

Exhibit 1

Solution

$$F = 1000(F/A, 10\%, 10) = 1000(15.937)^* = \$15,937$$

---

*In solving problems, use of the interest tables[1a,2a,3] is made. They are not reproduced in this book. The calculations may also be done with the help of a handheld calculator and interest factors.

Example 19.6

If you start saving in a bank at the rate of $1000 per year with the first savings deposit made today, how much will you accumulate at the end of 10 years if you make a total of 10 equal payments (including the first one) and the interest rate is 10 percent per year?

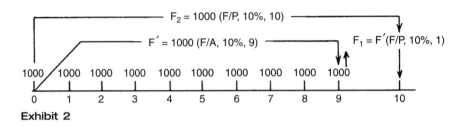

Exhibit 2

Solution

Note that the annuity formula applies to the end-of-period payment only. Hence, the first payment is not to be included in the annuity calculation. Also, a total of 10 payments, including the first one, is made, so the payment ends at the end of nine years. The future worth of the annuity made for nine years will carry the interest for one more year to calculate the future worth at the end of 10 years.

$$1000(F/A, i, n)(F/P, i, n) + 1000(F/P, i, n)$$
$$= 1000(F/A, 10\%, 9)(F/P, 10\%, 1) + 1000(F/P, 10\%, 10)$$
$$= 1000(13.579)(1.1) + 1000(2.594)$$
$$= \$17{,}530.90$$

Example 19.7

How much would I accumulate in a bank at the instant of the last payment if I made 10 annual payments of $1000 each with the first deposit made today at an interest rate of 10 percent per year?

Solution

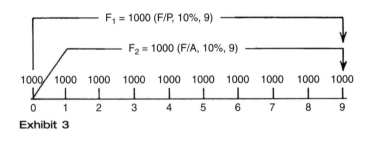

Exhibit 3

$$F = 1000(F/A, i, n) + 1000(F/P, i, n)$$
$$= 1000(F/A, 10\%, 9) + 1000(F/P, 10\%, 9)$$
$$= 1000(13.579) + 1000(2.358) = \$15{,}937$$

Sinking-Fund Factor

A sinking fund is an interest-earning fund established to accumulate a desired amount for withdrawal in the future through deposition of a uniform series of payments.

The sinking-fund factor is needed to calculate the number of the end-of-period uniform payments necessary to accumulate the desired sinking fund.

$$A = F \frac{i}{(1+i)^n - 1} = F(A/F, i, n) \tag{19.8}$$

The expression $i/[(1 + i)^n - 1] = (A/F, i, n)$ is called the sinking-fund factor.

Example 19.8

A man wants to deposit in a savings bank a certain amount of money annually by the end of the current year and for a total of 30 years. How much should he deposit annually so that he will have $30,000 by the end of 30 years if the interest rate is 10 percent?

Solution

$$A(F/A, 10\%, 30) = 30,000$$

From the interest tables,

$$(F/A, 10\%, 30) = 164.494$$

Therefore $\quad A(164.494) = 30,000$

from which $\quad A = \$182.38$

Capital-Recovery Factor

This factor is used to calculate the amount of the end-of-period payment necessary to accumulate a given present value when the number of interest periods and the interest rate are known. Substitution of the value of F (Equation 19.3) in Equation 19.8 gives

$$A = P \frac{i(1+i)^n}{(1+i)^n - 1} = P(A/P, i, n) \tag{19.9}$$

The expression $i(1 + i)^n/[(1 + i)^n - 1]$ is the capital-recovery factor and is abbreviated by $(A/P, i, n)$. The capital-recovery factor is used to calculate the capital-recovery depreciation. It is given by

$$\text{Capital-recovery factor} = (A/P, i, n) = \frac{\text{annual return}}{\text{capital invested}} \tag{19.9a}$$

Example 19.9

A savings bank is offering house mortgage loans at $8\frac{1}{2}$ percent interest compounded monthly. What will be the monthly payment for a loan of $43,000 for a 30-year mortgage if the first installment is due one month after the date of signing the deed? What will be the effective annual interest rate?

Solution

$$\text{Effective annual interest rate} = \left(1 + \frac{j}{m}\right)^n - 1$$

$$= \left(1 + \frac{0.085}{12}\right)^{12} - 1 = 8.84 \text{ percent year}$$

Number of interest periods = 30(12) = 360

$$i = \text{monthly interest rate} = \frac{0.085}{12} = 0.007083$$

$$43{,}000 = A(P/A,\ i = 0.7083\%,\ n = 360)$$

$$43{,}000 = A\frac{(1+i)^n - 1}{i(1+i)^n}$$

$$= A\frac{(1.007083)^{360} - 1}{0.007083(1.007083)^{360}}$$

$$= A(130.05844)$$

Therefore $A = \$330.62$ per month

Relation between Capital-Recovery Factor and Sinking-Fund Factor

$$\text{Capital-recovery factor} = \frac{i}{(1+i)^n - 1} + i$$

$$= \text{sinking-fund factor} + \text{interest rate} \qquad (19.9b)$$

From Equations 19.8 and 19.9, two other relations follow:

$$F = A\frac{(1+i)^n - 1}{i} = A(F/A,\ i,\ n) \qquad (19.10)$$

and

$$P = A\frac{(1+i)^n - 1}{i(1+i)^n} = A(P/A,\ i,\ n) \qquad (19.11)$$

$[(1 + i)^n - 1]/i$ and $[(1 + i)^n - 1]/i(1 + i)^n$ are called the uniform series compound-amount factor and uniform series present-worth factor, respectively.

Continuous Compounding of Interest

The compound amount formula is given by Equation 19.6 as

$$F = P\left(1 + \frac{j}{m}\right)^{mn}$$

If m (number of the interest periods per year) is increased without limit, m becomes very large and approaches infinity, and j/m becomes very small and approaches zero. This condition is called *continuous compounding* and is calculated by

$$F = P\left(1 + \frac{j}{m}\right)^{mn}_{m \to \infty} = Pe^{jn} \qquad (19.12)$$

This is the continuous-compounding single-payment formula. The continuous-compounding single-payment present-worth formula is

$$P = Fe^{-jn} \qquad (19.13)$$

Tables of the continuous-compound amount factors (e^{jn} and e^{-jn}, respectively) are available.[1c,2b] The effective continuous compounding rate becomes

$$i = \lim_{m \to \infty} \left(1 + \frac{j}{m}\right)^m - 1 = e^j - 1 \qquad (19.14)$$

Capitalized Cost

The sum of the first cost and present worth of perpetual disbursements is called the *capitalized cost*. It is used to study the long life assets such as the *railways, dams, tunnels,* and *similar* structures, which provide extended service.

The capitalized cost can be expressed by the following relation:

$$\text{Capitalized cost} = P + \frac{A}{i} + \frac{(P-L)(A/F, i, n)}{i} \qquad (19.15)$$

where

P = first cost or initial investment
A = difference between annual receipts and disbursement or simple disbursement when there is no annual receipt
i = interest rate, fraction per year
L = salvage value.

The last term of the expression (Equation 19.15) of the capitalized cost vanishes when the service life is very large (50 years or more).

Example 19.10

The initial cost of a dam is $25,000,000. The annual maintenance cost is $200,000. If the interest rate is 10 percent per year, determine the capitalized cost.

Solution

The service life of a dam is large and $(A/F, i, n) = 0$.

$$\text{Capitalized cost} = P + \frac{A}{i}$$

$$= \$25,000,000 + \frac{\$200,000}{0.1} = \$27,000,000$$

Example 19.11

How much is to be deposited now as a fixed deposit in a bank so that you would receive an annual year-end payment of $10,000 forever? Assume interest rate is 8 percent.

Solution

$$P = A(P/A, i, n=\infty) = A\frac{(1+i)^n - 1}{i(1+i)^n} \underset{n=\infty}{=} A\frac{1 - 1/(1+i)^n}{i}$$

$$= \frac{A}{i} = \frac{10,000}{0.08} = \$125,000$$

CHEMICAL ENGINEERING PLANT COST INDEX (CEPCI)

For the 40 years since its introduction, the CEPCI has been used by professionals in chemical process industries to make budget estimate (±20 to ±30%) of plant construction costs, given a base line cost. This index consists of a composite index and sub-indexes.

The structure of CEPCI-composite is as follows:

A. Equipment Index

- Heat exchangers and tanks
- Process machinery
- Pipe, valves and fittings
- Process instruments
- Pumps and compressors
- Electrical equipment
- Structural supports and miscellaneous

B. Construction Labor Index

C. Building Index

D. Engineering and Supervision Index

The Marshall & Swift Equipment Cost Index publishes trends of average cost in process industries, followed by similar trends in specific industries.

Another specialized index, called the Vatavuk Air Pollution Control Cost Index (VAPCCI), is also published to reflect the trend of equipment cost of air-pollution control systems. The indices are published in each issue of *Chemical Engineering*, published by Access Intelligence, NY.

The indices offer a baseline year with a base cost of 100. Thus the CEPCI base year is 1957–59 with the baseline cost of 100. The reported CCPCI index for 2003 is 402. As an example, if a plant cost $1 million in 1957–59, the same plant would cost (402/100) × 1 = $4.02 million (±20 to ±30%) in 2003. A word of caution: The CEPCI should not be used to estimate price escalation with a baseline figure older than five years, as shown in the cited example, to avoid large error.

The CEPCI Composite annual index for the years 1994 to 2003 is as follows:

Table 19.1 Base year (1957–59): base year index: 100

Year	1994	1995	1996	1997	1998	1999	2000	2001	2002	2003
Index	368.1	381.1	381.7	386.5	389.5	390.6	394.1	394.3	395.6	402.0

Order-of-Magnitude Estimate of Equipment and Installed Costs

The cost of equipment may be expressed in the form

$$C = K(\text{Size})^n \tag{19.15a}$$

where
  C = equipment cost
  K and n are constants related to the equipment type.

The approximate installed cost of the equipment may be obtained by multiplying the cost of the equipment by a factor (Lang factor, LF) as shown in Table 19.2. Thus the installed cost of a carbon steel heat exchanger purchased at $20,000 would be $80,000, since the LF is 4. The value of K and n can be estimated from the costs of two sizes. Given the cost of one size, the cost of another size may be estimated by using an appropriate value of n from Table 19.2.

Estimation of Operating Cost of Production Facilities

The following tabulation may be used as a guideline:

A. Direct cost

- Raw material cost
- Operating labor cost
- Direct supervisory and clerical cost
- Utilities cost

Table 19.2 Factored cost exponents and installation cost factors (Lang Factor)

Equipment	Size Basis	Value of Exponent, n	Lang Factor, LF
Pump	HP	0.52	4.0
Heat exchanger	Area	0.62	3.3(ss), 4.0(cs)
Air cooler	Area	0.75	2.2(ss), 2.3(cs)
Column	Diameter	0.65	4.0
Crystallizer	Tons/day dry solid	0.55	2.4
Plate & frame filter	Area	0.58	2.7
Rotary vacuum filter	Effective area	0.63	2.4
Rotary kiln dryer	Area	0.8	3.0
Blower & fan	Cfm	0.68	2.8
Bowl centrifuge	HP	0.73	2.4
Boiler	Lbs/h	0.7	1.8
Ball mills	Tons/h	0.7	1.8
Pulverizer	Tons/h	0.39	2.5
Atm. storage tank	Gallon	0.6	4.0
Pressure vessel	Gallon	0.6	2(ss, gl), 4(cs)
Agitator	HP	0.56	As above
Cooling tower	Gpm	0.6	1.8
Refrigeration (mech.)	Tons	0.55	1.4
Compressor	HP	0.8	3.1

Legend: ss = stainless steel, cs = carbon steel, gl = glass lined.

- Maintenance and repair cost
- Operating supplies

B. Fixed cost
- Depreciation cost
- Local taxes
- Insurance cost

C. Plant Overhead cost

D. General expenses
- Administration
- Distribution and selling
- Research and development
- Financing (interest)

Depreciation[1b,2b]

Depreciation means a decrease in the value of an asset. In the economic analysis, it may refer to the market value or value to the owner. It may also mean the systematic allocation of the cost of an asset over its useful life. Two major types of depreciation used in economics are described next.

Depreciation for Tax Computation

This is an annual fractional loss of fixed investment allowed by the government to compute the taxable income. This is beyond the control of the owner.

$$\text{Taxable income} = R - dI_F \quad (19.16)$$
$$\text{Income tax} = (R - dI_F)t \quad (19.17)$$

where
R = gross income (when cost of production excludes depreciation), $/year
d = depreciation, fraction $/$ · year
I_F = fixed investment, $
t = tax rate $/$

Depreciation for Recovery of Fixed Investment

This is an annual fraction of the fixed investment set aside by an owner or corporation to ensure that the original investment will be completely recovered in a given number of years. This depreciation and the tax are deducted from the gross income to give the net profit. The capital-recovery depreciation charge is eI_F.

$$P_N = R - eI_F - (R - dI_F)t \quad (19.18)$$

where
P_N = net profit, $/year
R = gross income when the cost of production excludes depreciation, $/year
e = depreciation for recovery of investment, $/$ · year
d = depreciation allowed by government, $/$ · year
I_F = fixed investment, $
t = tax rate, $/$ earned (for economic evaluation, use $t = 0.5$)

Depreciation Calculations

The computation of depreciation involves a consideration of (1) the cost of the asset, (2) its useful life, and (3) the salvage value of the asset at the end of its useful life.

Salvage Value

Salvage value or future worth of an asset to be realized at its disposal has to be estimated and may be

1. positive if the used item can be resold
2. zero if the used item is disposed of at no cost or if the salvage value is only a small portion of the cost of the asset
3. negative if the resale value of the asset is less than the cost of its disposal

Two other definitions are important.

Book Value. The book value B is given by

$$B = \text{purchase cost} - \text{depreciation charge}$$

Sunk Cost. Past costs generally do not affect the present or future and are therefore disregarded. These costs are called *sunk costs*. A sunk cost is given by

$$\text{Sunk cost } S = \text{book value} - \text{actual value}$$

Some of the more important methods of depreciation are discussed next.

Straight-Line Method

In this method, constant depreciation charge is made every year. Thus

$$\text{Annual depreciation} = \frac{P-L}{n} \tag{19.19}$$

where
 P = initial investment or cost
 L = salvage value
 n = economic life of the asset

$$\text{Depreciation reserve at end of } r \text{ years} = \frac{r}{n}(P-L) \tag{19.20}$$

$$\text{Book value at the end of } r \text{ years} = P - \frac{r}{n}(P-L) \tag{19.21}$$

where r is the number of years the asset is in use since its purchase.

Sum-of-the-Digits Depreciation

This method provides a larger depreciation in the early years of the life of an asset than in the later years. The sum of the digits depreciation (for any year r) is

$$\frac{\text{Digit representing remaining years of life}}{\text{Sum of the digits for entire life}}(P-L)$$

$$= \frac{n-r+1}{(1+2+3+\cdots+n)}(P-L) \tag{19.22}$$

$$= \frac{2(n-r+1)}{n(n+1)}(P-L) \tag{19.22a}$$

Book value at the end of r years is given by

$$B = (P-L)\frac{n-r}{n}\frac{n-r+1}{n+1} + L$$
$$= P - \frac{(P-L)r}{n(n+1)}[2n-(r-1)] \quad (19.23)$$

and the depreciation reserve at the end of r years is given by

$$\frac{2[r+(r+1)+(r+2)+\cdots+n](P-L)}{n(n+1)} = \left[1 - \frac{(n-r)(n-r+1)}{n(n+1)}\right](P-L) \quad (19.24)$$

Declining-Balance Depreciation

This method also allows one to have an accelerated depreciation rate in the early part of an asset's life.

$$\text{Rate of depreciation} = 1 - \left(\frac{L}{P}\right)^{1/n} \quad (19.25)$$

which requires L to be positive to be realistic. Hence, the declining-balance method cannot be applied when the salvage value of an asset is zero.

$$\text{Depreciation charge in } r\text{th year} = P\left(\frac{L}{P}\right)^{(r-1)/n}\left[1-\left(\frac{L}{P}\right)^{1/n}\right] \quad (19.25a)$$

$$\text{Book value at the end of the } r\text{th year} = P\left(\frac{L}{P}\right)^{r/n} \quad (19.25b)$$

$$\text{Depreciation reserve at end of } r\text{th year} = P\left[1-\left(\frac{L}{P}\right)^{r/n}\right] \quad (19.25c)$$

Double-Declining-Balance Method

Another declining-balance method allowed by the income tax code is to use a depreciation rate that is two times the straight-line depreciation rate. (Rates of 1.25 or 1.5 times that of the straight-line method are allowed under certain circumstances.) The double-declining-balance depreciation is

$$\frac{200 \text{ percent}}{n} \quad (19.26)$$

The double-declining-balance depreciation charge in the rth year is

$$\frac{2P}{n}\left(1-\frac{2}{n}\right)^{r-1} \quad (19.26a)$$

The book value at the end of the rth year is

$$P\left(1-\frac{2}{n}\right)^r \quad (19.26b)$$

and the depreciation reserve at the end of the rth year is

$$P\left[1-\left(1-\frac{2}{n}\right)^r\right] \qquad (19.26c)$$

When the double-declining-balance method is used and the salvage value at the end of the economic life of the asset is expected to be less than the book value predicted by the double-declining-balance method, a switch is made to the straight-line depreciation to the advantage of the taxpayer. If r_s is the year at the end of which the switch is made, the depreciation by straight-line method for the rest of the economic lifespan is calculated by the following formula:

$$\text{Annual depreciation} = \frac{P\left(1-\frac{2}{n}\right)^{r_s} - L}{n - r_s} \qquad (19.27)$$

Sinking-Fund Depreciation

The sinking fund assumes a fund in which periodic uniform deposits are made annually at some assumed interest rate. The future worth of the annuity equals the cost of the asset minus the salvage value by the end of the depreciation life of the asset.

$$\text{Annual depreciation} = (P - L)(A/F, i, n) \qquad (19.28)$$

The book value at the end of the rth year is

$$P - (P - L)(A/F, i, n)(F/A, i, r) \qquad (19.29)$$

Depreciation reserve at the end of the rth year is

$$(P - L)(A/F, i, n)(F/A, i, r) \qquad (19.30)$$

Capital-Recovery Depreciation

Annual capital-recovery depreciation is given by

$$(P - L)(A/P, i, n) + Li \qquad (19.31)$$

Note that the capital-recovery depreciation should be used to compare the time value of the alternatives. The formula for the capital-recovery depreciation is derived as follows:

$$P(A/P, i, n) - L(A/F, i, n) = P(A/P, i, n) - L[(A/P, i, n) - i]$$
$$= (P - L)(A/P, i, n) + Li$$

Annual capital-recovery depreciation is the annual sinking-fund depreciation plus the interest on initial investment, or

$$(P - L)(A/P, i, n) + Li = (P - L)(A/F, i, n) + iP \qquad (19.32)$$

Example 19.12

The installed cost of a refrigeration unit is $250,000. It has a salvage value of $20,000 after an economic life of 10 years. Using the straight depreciation method, calculate the annual depreciation, book value, and depreciation reserve at the end of 5 years.

Solution

Annual depreciation equals

$$\frac{P-L}{n} = \frac{250,000 - 20,000}{10} = \$23,000/\text{year}$$

The depreciation reserve at the end of 5 years is

$$r\left(\frac{P-L}{n}\right) = 5(23,000) = \$115,000$$

and the book value is

$$250,000 - 115,000 = \$135,000$$

Example 19.13

Compare the present worth for the depreciation using the straight-line, double-declining-balance with a switch to the straight-line at the end of the fourth year and the sum-of-the-digit depreciations for an asset valued at $14,000 at zero time with $2000 salvage value after 6 years and with an interest rate of 12 percent per year.

Solution

Straight-line depreciation is calculated as follows. The annual depreciation is

$$\frac{P-L}{n} = \frac{14,000 - 2000}{6} = \$2000/\text{year}$$

and present worth is

$$2000(P/A, 12\%, 6) = 2000(4.111) = \$8222$$

Double-declining-balance depreciation is

Annual depreciation: $\dfrac{2P}{n}\left(1-\dfrac{2}{n}\right)^{r-1}$ during rth year

First year depreciation: $\dfrac{2(14,000)}{6}\left(1-\dfrac{2}{6}\right)^{(1-1)} = \$4666.67$

Second year depreciation: $\dfrac{2(14,000)}{6}\left(1-\dfrac{2}{6}\right)^{(2-1)} = \$3111.11$

Third year depreciation: $\dfrac{2(14,000)}{6}\left(1-\dfrac{2}{6}\right)^{(3-1)} = \$2074.08$

Fourth year depreciation: $\dfrac{2(14,000)}{6}\left(1-\dfrac{2}{6}\right)^{(4-1)} = \$1382.72$

Depreciation for fifth and sixth years:

$$\frac{P\left(1-\frac{2}{n}\right)^{r_s} - L}{n - r_s} = \frac{14{,}000\left(1-\frac{2}{6}\right)^4 - 2000}{6 - 4} = \$382.72/\text{year}$$

Present worth:

  4666.67(P/F, 12%, 1) + 3111.11(P/F, 12%, 2)
  + 2074.08(P/F, 12%, 3) + 1382.72(P/F, 12%, 4)
  + 382.72(P/F, 12%, 5) + 382.72(P/F, 12%, 6) = $9413

Sum-of-digit depreciation follows.

Annual depreciation: $\dfrac{2(n-r+1)(P-L)}{n(n+1)} = \dfrac{2(6-r+1)(14{,}000 - 2000)}{6(7)}$

$$= (7 - r)(571.43)$$

Annual depreciations calculated are given in Table 19.3.

Cost Comparison of Alternatives[4,5,6]

Cost comparison of different alternatives is made in terms of the equivalent money values based on a selected interest rate and at a selected point in time.

Annual Cost Method

In this method, the initial investment, if any, is spread over the economic life by multiplication with the capital-recovery factor (A/P, i, n) or by other specified depreciation methods. The annual cost thus obtained is added to the other annual costs to obtain the total annual cost. When the alternatives involve different lifespans, the comparisons are to be made through the lowest common multiple of the life spans of the alternatives. For example, if one alternative has a 4-year lifespan and another has a 10-year lifespan, the comparison must be made for 20 years. However, in certain cases it is permissible to compare the alternatives with different lives on the basis of their own service lives, e.g., when the alternatives are compared by assuming that each asset will be replaced at the end of its useful life by an identical asset. Also, the annual costs based on the different lives of the assets are used when the least common multiple of the lives of the alternatives is unusually high (e.g., alternatives with 7- and 11-year lives have a least common multiple of 77 years) and exceeds a reasonable analysis period* for an economic study.

Table 19.3 Annual depreciations

R	1	2	3	4	5	6
Depreciation	3428.57	2857.14	2285.71	1714.29	1142.86	571.43
(P/F, 12%, r)	0.8929	0.7972	0.7118	0.6355	0.5674	0.5066
Present worth =	3428.57(0.8929) + 2857.14(0.7972) + 2285.71(0.7118)					
	+ 1714.29(0.6355) + 1142.86(0.5674) + 571.43(0.5066) = $8993					

---

*Industries with stable technologies generally use 10 to 20 years, while government agencies use 50 years or more.

Calculation of Annual Costs

Annual cost is the sum of (1) the depreciation charge for the year, (2) the annual interest charge, and (3) annual operating cost. Thus

$$\text{Annual cost} = (P - L)(A/P, i, n) + Li + \text{OC}$$

where OC signifies the annual operating costs such as maintenance. The preceding relation is based on the capital-recovery depreciation, which takes into account the compounding of the interest. An approximate annual cost of the capital recovery (CR) is based on a straight-line depreciation and an average interest and is given by

$$\text{Approximate CR} = \frac{P-L}{n} + (P-L)\frac{i}{2}\left(\frac{n+1}{n}\right) + Li$$

and approximate annual cost is given by

$$\frac{P-L}{n} + (P-L)\frac{i}{2}\left(\frac{n+1}{n}\right) + Li + \text{OC}$$

Present-Worth (PW) Method

In this method, all the uniform annual costs are converted to the present worth by multiplication with the series-payment present-worth factor $(P/A, i, n)$. Any non-uniform annual cost may be converted to its present worth by multiplication with the single-payment present-worth factor $(P/F, i, n)$. For the alternatives with different lifespans, the comparisons are to be made through the least common multiple of the lifespans of the alternative assets. However, the present-worth comparisons can also be made by (1) the study-period method, which requires the selection of a time period (assuming the best possible replacement), and (2) the service-period method. This is used when the operational requirement of an asset is limited.

Capitalized-Cost Method of Comparison of Alternatives

This method assumes a perpetual service life of an alternative or perpetual service through an infinite series of replacements of an alternative having a finite service life. The capitalized cost equals initial investment plus present worth of all perpetual costs. Thus

$$\text{Capitalized cost} = P + \frac{A}{i} + \frac{(P-L)(A/F, i, n)}{i} \tag{19.33}$$

where
- P = initial investment and/or present worth of any future investment
- A = perpetual annual cost or perpetual equivalent annual cost of a future one-time cost
- L = salvage value.

Note: When n is large (>50 years), the last term may be neglected.

Rate-of-Return Method

Rate-of-return (RR) method, also known as *internal rate of return* (IRR) and *return on investment* (ROI), is the most important method of comparing alternatives.

Rate of return provides a percentage figure of the relative yield from different ways of using the capital. The basis of the rate-of-return calculations are the present worth and annual worth. The rate of return is calculated by equating to zero the annual or present worths of cash flows involving the receipts and disbursements and by determining the interest rate that would establish the equality. Although the use of either the present worth or annual worth as the basis of calculation is correct, the rate of return is generally formulated in terms of the present worths. The calculation of i usually involves a trial-and-error procedure. If no value of i from the interest tables satisfies the equality, the correct value is determined by a linear interpolation. The relation to be established is to find i so that

$$PW(\text{receipts}) - PW(\text{disbursements}) = 0 \quad (19.34)$$

Example 19.14

A company is considering installation of two devices as follows:

	Device A	Device B	
Initial cost	$2000	$1800	
Maintenance, year · $	50	60	
Salvage value	0	0	
Annual savings, $	600	600	1st year but declining by $30 per year in subsequent years
Useful life, years	6	6	

By the method of rate of return, which device should be purchased?

Solution

$$PW(\text{benefits}) - PW(\text{costs}) = 0. \ i \text{ is to be found}$$

Device A.

PW of costs: $\qquad 2000 + 50(P/A, i, 6)$

PW of benefits: $\qquad 600(P/A, i, 6)$

Therefore $\qquad 2000 + 50(P/A, i, 6) = 600\,((P/A, i, 6)$

or $\qquad 2000 - 550(P/A, i, 6) = 0$

Note: When computing the RR by trial and error, a question often arises as to the value of i with which to begin the trials. To narrow down the calculations, divide the average annuity A by P to get A/P. For the given number of years n, find for the average A/P calculated as above the nearest interest rate from the interest tables and use this for the first trial.

In this example,

$$\text{Average } A = 600 - 50 = \$550$$

$$A/P = \frac{550}{2000} = 0.275 \quad n = 6$$

For $n = 6$ and $A/P = 0.275$, i is between 15% and 18%.

Let

$i = 15\%$	LHS = $2000 - 550((P/A, 15\%, 6) = 2000 - 550(3.784)$
	$= -81.2$
$i = 16\%$	LHS = $2000 - 550(P/A, 16\%, 6) = -26.75$
$i = 18\%$	LHS = $2000 - 550(P/A, 18\%, 6) = 2000 - 550(3.498)$
	$= 76.1$

By interpolation

$$i = 16 + 2\frac{0-(-26.75)}{76.1-(-26.75)} = 16.5 \text{ percent}$$

Device B.
Note that in this case, the savings per year involve a uniform gradient series.

PW of costs: $1800 + 60\,(P/A, i, 6)$

PW of benefits: $600(P/A, i, 6) - 30(P/G, i, 6)$

Thus $1800 + 50(P/A, i, 6) - 600\,(P/A, i, 6) + 30(P/G, i, 6) = 0$

or $1800 - 550(P/A, i, 6) + 30(P/G, i, 6) = 0$

$i = 20\%$	LHS = $1800 - 550(P/A, 20\%, 6) + 30(P/G, 20\%, 6)$
	$= 1800 - 550(3.326) + 30(6.581) = 168.13$
$i = 15\%$	LHS = $1800 - 550(P/A, 15\%, 6) + 30(P/G, 15\%, 6)$
	$= 1800 - 550(3.784) + 30(7.937) = -43.09$

By interpolation percent $i = 15 + 5\dfrac{0+43.09}{168.13+43.09} = 16.02$ percent.

The rate of return of device *A* is more.

    Maximizing the rate of return is not, however, the proper criterion to choose an alternative. For the proper application of RR (i.e., ROI) method, an incremental analysis that examines the differences in the alternatives is required. This is discussed in the next section.

Incremental Analysis Through Rate of Return by Compound Interest Method

By this method, the rate of return is the interest rate i that balances the present worth of all the annual profits with the invested capital.

$$(A/P, i, n) = \frac{\text{annual profit (return)}}{\text{invested capital}} \tag{19.35}$$

When the alternatives are compared by the rate-of-return method, the ratio of the incremental annual profit to the incremental investment is computed to determine the rate of return.

Benefit-to-Cost (B/C) Ratio Method

The *B/C* ratio of a scheme is calculated by

$$B/C \text{ ratio} = \frac{\text{annual savings or benefits}}{\text{equivalent annual cost}} \tag{19.36}$$

where

> equivalent annual cost = (initial investment)(A/P, i, n)
> + other annual cost (such as maintenance, etc.)

To justify a project, the *B/C* ratio must be greater than unity.

After-Tax Comparisons

After-tax comparisons can be made by either the annual cost or by the present-worth and the rate-of-return methods. The following definitions will be useful in following the after-tax methods.

Gross Profit

Generally, the cost of production does not include depreciation. Under this accounting procedure, gross profit is calculated by

$$R = S - C \tag{19.37}$$

where
  R = gross profit, \$/year
  S = sales revenue, \$/year
  C = cost of production excluding depreciation \$/year.

Net Profit. P_N is related to gross profit by

$$P_N = R - (R - dI_F)t - ei_F \tag{19.37a}$$

where
  I_F = fixed investment, \$
  t = tax rate, \$/\$
  e = capital-recovery depreciation, \$/\$ · year
  d = government allowed depreciation

when $\quad d = e \quad P_N = (R - dI_F)(1 - t) \tag{19.37b}$

Sometimes the cost of production includes the depreciation for the sake of simplicity. Under this accounting procedure, the gross profit, which is different from the gross profit defined in Equation 19.37, is given by

$$R' = S - C' \tag{19.37c}$$

where C' is the cost of production including depreciation. Obviously

$$C' = C + dI_F \tag{19.37d}$$
so $\qquad R' = S - C\, dI_F = R - dI_F \tag{19.37e}$

when $d = e$, from Equations 19.37b and 19.37e.

$$\text{Net profit } P_N = (R - dI_F)(1 - t) = R'(1 - t) \tag{19.37f}$$

One method of the rate-of-return analysis was given earlier. A simplified method commonly used in the process design is given below.

Return on original investment is given by

$$\text{ROI} = \frac{\text{net profit in a year}}{\text{total investment}}$$

$$= \frac{(R-dI_F)t - eI_F}{I_F + I_w} \quad \text{where } I_w \text{ is working capital}$$

$$= \frac{(R-dI_F)(1-t)}{I_F + I_w} \quad \text{when } e = d$$

$$= \frac{R'(1-t)}{I_F + I_w} \tag{19.38}$$

Generally, the cost reduction alternatives affect the cost of production and not the sales revenue, and no working capital is considered. Thus

$$\text{ROI} = \frac{(R-dI_F)(1-t)}{I_F} = \frac{[S-(C+dI_F)](1-t)}{I_F} \tag{19.39}$$

The challenge of a higher investment to cut down the cost of production of a defending scheme requiring lower investment is evaluated by estimating the rate of return on the incremental investment as follows

$$\frac{[(C+dI_F)_1 - (C+dI_F)_2](1-t)}{(I_F)_2 - (I_F)_1} \tag{19.40}$$

where subscripts 1 and 2 refer to alternatives 1 and 2, respectively. The required minimum rates of return depend on the company practice.

Payout Time Method or Payback Method

The payout time method is used in two ways:

1. Taxes and depreciation are ignored. This is the most commonly used form. The payout time T is given by

$$T = \frac{\text{fixed investment}}{\text{gross annual income}} = \frac{I_F}{R} \tag{19.41}$$

2. Taxes and depreciation are accounted for. Then T is

$$T = \frac{I_F}{R - (R-dI_F)t} = \frac{I_F}{R'(1-t) + dI_F} \tag{19.42}$$

In general, the payout time method should not be used to compare alternatives having different economic life spans.

Cash Flows in Annual Disbursement Only

Methods of economic analysis in cases when the cash flows are in annual disbursement only are illustrated by the following example.

Linear Break-Even Analysis[1b]

The break-even capacity of a plant is the capacity at which the annual revenue is equal to the annual cost of production. In other words, at break-even capacity, the gross profit is zero. Let

Q_B = break-even capacity units/time
S = revenue, $/time
R = gross profit, $/time
s = selling price, $/unit
C = total cost of production, $/time
c = variable cost, $/unit
F = fixed cost, $/time
Q = number of units produced per unit time

Then the following equations can be derived:

$$S = sQ \qquad (19.43)$$

$$C = Q_c + F \qquad (19.44)$$

$$R = S - C = Q(s - c) - F \qquad (19.45)$$

$$Q_B = \frac{F}{s-c} \quad \text{when } Q = Q_B, \quad R = 0$$

$s - c$ = contribution

$$\frac{Q - Q_B}{Q_B} = \text{margin of profit} = \frac{R}{F} \qquad (19.46)$$

Analytical Method of Optimization

This method is applied when the dependent variable in the objective function can be expressed as a function of one or more variables. Example 19.15 illustrates this method.

Example 19.15

A flat-roofed and flat-bottomed cylindrical storage tank is to be designed to store a liquid of density ρ lb/ft³ and capacity V ft³. The tank is open to the atmosphere. The cost of the roof is $$C_T$/ft² and the cost of the bottom is $$C_B$/ft². The vertical surface cost is $$kt$/ft² where k is constant and t is the wall thickness of the tank in inches. The wall thickness is calculated by

$$t = \frac{6PD}{SE}$$

where
P = internal pressure, psig
D = diameter of tank, ft
S = allowable stress, psi
E = joint efficiency.

Find the dimension of the tank for the lowest cost.

Solution

Let H be the height of tank, ft, and C_T the total cost.

$$C_T = \pi D H k t + (C_B + C_T)\left(\frac{1}{4}\pi D^2\right) \tag{19.47}$$

$$t = \frac{6PD}{SE} \tag{19.48}$$

$$P = \frac{H(\text{ft})\rho(\text{lb/ft}^3)}{144\,\text{in}^2/\text{ft}^2} = \frac{H\rho}{144}\,\text{psig} \tag{19.49}$$

Also $\quad \frac{1}{4}\pi D^2 H = V \tag{19.50}$

From Equations 19.47 through 19.50 one derives

$$C_T \frac{\pi D V k(0.04167) D V \rho}{\left(\frac{1}{4}\pi D^2\right)\left(\frac{1}{4}\pi D^2\right)(SE)} + (C_B + C_T)\frac{1}{4}\pi D^2 = \frac{0.2122 V^2 k\rho}{D^2(SE)}$$
$$+ (0.7854)(C_B + C_T)D^2$$

Employing $dC_T/dD = 0$ for minimum C_T gives

$$-\frac{0.2122(2)V^2 k\rho}{D^3(SE)} + 0.7854(2)(C_B + C_T)D = 0$$

from which

$$D = 0.721\left[\frac{V^2 k\rho}{(C_B + C_T)SE}\right]^{1/4} \tag{19.51}$$

Note: $d^2C_T/dD^2 > 0$; hence this diameter gives minimum C_T. Substituting D from Equation 19.51 in Equation 19.50, H can be determined.

In practice, the calculated diameter as obtained from Equation 19.51 is substituted in Equation 19.48 to determine the economic thickness. If the calculated thickness is less than a certain minimum allowable thickness (usually $\frac{1}{4}$-in. for steel), the minimum allowable thickness is taken and dimensions recalculated.

RECOMMENDED REFERENCES

1. Riggs, *Engineering Economics,* McGraw-Hill, NY, 1977: (a) pp. 594–609; (b) pp. 46–47; (c) p. 160.
2. Newman, *Engineering Economics Analysis,* Engineering Press, San Jose, 1976: (a) pp. 430–459: (b) p. 460.
3. Perry (ed.), *Chemical Engineers Handbook,* 5th ed., McGraw-Hill, NY, 1973, pp. 1–33 to 1–34.
4. Happel. and Jordan, *Chemical Process Economics,* 2d ed., Marcel Dekker, NY, 1975, pp. 56–62.

5. Grant and Ireson, *Principles of Engineering Economy,* 5th ed., Ronald Press, NY, 1970, pp. 201–227.
6. Peters and Timmerhaus, *Plant Design and Economics for Chemical Engineers,* McGraw-Hill, NY, 1980, pp. 302–345.

Additional Reading

7. Kurtz, *Engineering Economics for Professional Engineering Examination,* 2nd ed., McGraw-Hill, NY, 1975.
8. Uhl and Hawkins, *Technical Economics for Chemical Engineers,* A.I.Ch.E., NY, 1971.

CHAPTER 20

Plant Safety and Environmental Consideration

OUTLINE

TOXICOLOGY AND INDUSTRIAL HYGIENE 724
Threshold Limit Values (TLV) and Exposure Limits ■ Exposure to More Than One Toxicant ■ Mixture of Liquids ■ Estimation of Dilution Air

FIRE AND EXPLOSION ISSUES 727
Fire Tetrahedron ■ Dust Explosion Pentagon ■ Terminology ■ Classification of Liquids for Thermal Hazards Analysis ■ Computation of Flash Point ■ Computation of Flame Temperature ■ Calculating Flammability Limits ■ Stoichiometry of Combustion Reaction and Estimation of Flammability Limits ■ Overpressure Protection of Storage Vessel and Process Equipment ■ Emergency Relief Load Calculation and Sizing Safety Device: Conditions Considered in Emergency Relief ■ Relief Load Calculation Due to External Fire ■ Calculation of Heat Flow from Fire, or Fire Heat, (Q_{fire}) ■ Environmental Factor for Insulation ■ Sizing of Relief or Safety Valve ■ Sizing of Rupture Disc ■ Blocked Condition ■ Heat Exchanger Tube Rupture ■ Venting Capacities of Atmospheric Tanks ■ Thermal Venting ■ Deflagration Venting

HAZARD AND OPERABILITY STUDIES (HAZOP) 758
Terminology ■ Emergency Ingress and Egress

ENVIRONMENTAL CONSIDERATIONS 761

FEDERAL POLLUTION PREVENTION ACT OF 1990 762
Definitions ■ MSDS ■ Mutagen ■ pH ■ Total Organic Carbon (TOC) ■ Volatile Organic Compounds (VOC)

OZONE, FRIEND AND FOE 765
Beneficial Ozone and Its Depletion ■ Harmful Ozone and Its Creation

ABATEMENT OF AIR POLLUTION 767
VOC Control Technologies ■ Non-VOC Control Technologies

ABATEMENT OF WATER POLLUTION   768

TREATMENT OF CONTAMINATED SOIL   769

PERSONAL PROTECTIVE EQUIPMENT (PPE)   769

NOISE MANAGEMENT   770
Noise Power Level ■ Noise Pressure Level ■ OSHA Permissible Noise Exposure (OSHA 1910.95) ■ Total OSHA-Permissible Noise Dose Computation in 8-hour Working Day (OSHA 1910.95 Appendix A) ■ Other Salient Points of OSHA 1910.95 with Regard to Hearing Protection ■ Types of Noise Control in Operation and Design

SOLID WASTE MANAGEMENT   773
Hazardous Waste Incineration ■ Nuclear/Radioactive Waste ■ Flare Design ■ Environmental Permit Applications

RECOMMENDED REFERENCES   781

The purpose of this chapter is to offer a guideline, not a rigid procedure, in the safety and environmental aspects of process engineering. The user must exercise his or her judgment as to the appropriateness of a proposed method and deviate if a comparatively safe design is achieved as a result of such deviation. Under no circumstances, however, shall safety be compromised with cost. Theoretical derivations of the formulas have been avoided to get to the point as quickly as possible.

TOXICOLOGY AND INDUSTRIAL HYGIENE

The degree to which a substance is poisonous is termed its toxicity. A toxicant is any chemical substance that, when inhaled, ingested, absorbed through the skin or applied to, injected into, or developed within the body, in relatively small amounts, may cause, by its chemical reaction, damage to the body structure or disturbance to human function or even death. Most chemicals enter the body through eyes, respiratory tract, digestive tract, and skin. Two levels of toxicity are defined: (1) acute "short term" exposure that causes initiation poisoning, (2) chronic-"long period" exposure that causes anemia, leukemia, and death.

Threshold Limit Values (TLV) and Exposure Limits

There are five types of TLVs as shown below. Of these, the first three have been recognized by the American Conference of Governmental Industrial Hygienists (ACGIH).

Toxic threshold limit values (TLVs) are the greatest concentrations of a toxicant in air that can be tolerated for a given length of time without any adverse effect. Definitions of commonly used limit values are given below.

The different types of TLV's and exposure limits are as follows:

- **TLV-TWA**
  The time weighted average of a toxicant for a normal 8-hour workday or 40-hour workweek to which a worker may be exposed without adverse effect. Excursions above the limit are allowed if compensated by excursions below the limit.

Toxicology and Industrial Hygiene

- **TLV-STEL**

  The short-term exposure limit is the maximum concentration to which a worker may be exposed continuously for up to 15 minutes without adverse effect. Such effects include intolerable irritation, chronic tissue change, narcosis that increases accident proneness, impairment of self-rescue, and material reduction of worker efficiency. No more than 4 excursions are permitted per day with at least 60 minutes between exposure periods, provided that the daily TLV-TWA is not exceeded.

- **TLV-C**

  The ceiling limit is the concentration, which should not be exceeded at any time.

- **PEL**

  Permissible exposure limit is the maximum exposure allowed by OSHA 29CFR1910.1000 as the eight-hour time-weighted average. OSHA also has an excursion limit.

- **IDLH**

  Immediately dangerous to life or health. It is the maximum concentration from which one could escape within 30 minutes without escape-impairing symptoms or any irreversible health effects.

TLVs are expressed either in parts per million by volume (ppm) or milligram per cubic meter (mg/m^3).

$$1\,\text{ppm} = \frac{PM}{0.08205T}\,\text{mg/m}^3 = 0.0409M \,@\, 25°C \text{ and 1 atm}$$

$$\text{Required value in ppm} = \frac{[\text{value in mg/m}^3](0.8205T)}{PM}$$

$$\cong [\text{value in mg/m}^3]\frac{24.45}{M} \text{ at } 25°C \text{ and 1 atm}$$

$$\text{Required value in mg/m}^3 = \frac{[\text{value in ppm}](PM)}{0.8205T}$$

$$\cong [\text{value in ppm}](0.0409M) \text{ at } 25°C \text{ and 1 atm}$$

where
- P = Pressure in atm,
- T = temperature in K,
- M = molecular mass

Example 20.1

The TLV-TWA of ammonia is 25 ppm. What is the corresponding value in mg/m^3 at 1 atm, and 25°C?

Solution

$$\text{Value in mg/m}^3 = \frac{1 \times 17 \times 25}{0.08205 \times 298} = 17.4$$

Time-weighted Average (TWA)

Time-weighted average (*TWA*) of concentration is calculated by the following formula:

$$TWA = \frac{C_1 t_1 + C_2 t_2 + \cdots + C_n t_n}{t_1 + t_2 + \cdots + t_n} \tag{20.1}$$

where
C_i = Concentration
t_i = Time of Exposure

Example 20.2

Estimate the time-weighted concentration for 8 hours if a worker is exposed to concentrations of 100 ppm for 1 hr, 90 ppm for 2 hrs, and 80 ppm for 5 hrs.

Solution

$$TWA = \frac{100 \times 1 + 90 \times 2 + 80 \times 5}{1 + 2 + 5} = 85$$

Exposure to More Than One Toxicant

The effect of more than one toxicant with different *TLV-TWAs* may be calculated from

$$TLV\text{-}TWA = \sum_{i=1}^{n} \frac{C_i}{(TLV\text{-}TWA)_i} = \frac{C_1}{(TLV\text{-}TWA)_1} + \frac{C_2}{(TLV\text{-}TWA)_2} + \cdots \tag{20.2}$$

If this sum exceeds 1, then the workers are overexposed.

Example 20.3

A processing room contains 20 ppm Heptane (*TLV-TWA* = 400 ppm), 50 ppm methyl alcohol, (*TLV-TWA* = 200 ppm), 40 ppm methyl chloride (*TLV-TWA* = 50 ppm). Determine whether the combined effect of these toxicants exceeds the exposure limit.

Solution

$$\sum_{i=1}^{n} \frac{C_i}{(TLV\text{-}TWA)_i}$$
$$= \frac{20}{400} + \frac{50}{200} + \frac{40}{50}$$
$$= 1.1 > 1$$

Hence exposure limit is exceeded.

Mixture of Liquids

To determine *TLVs* in a vapor space of a mixture of liquids of known vapor pressure and composition, the equations are

$$y_i = \frac{P_i^0 x_i}{\Sigma P_i^0 x_i} \qquad (20.3)$$

$$TLV_{\text{mixture}} = \frac{1}{\Sigma \frac{y_i}{TLV_i}}$$

$$TLV_i \text{ mixture} = y_i (TLV)_{\text{mixture}}$$

where

TLV_i = threshold limit value (ppm) of pure component i (when present by itself)
P_i^0 = pure component vapor pressure of component i at the mixture temperature
TLV_{mixture} = threshold limit value of the vapor mixture of toxicants
TLV_i mixture = TLV of component i in the mixture
x_i = mole fraction of component i in the mixture of liquid
y_i = mol fraction of component i in the vapor phase.

Estimation of Dilution Air

When volatile substances are emitted in working spaces such as a laboratory, fresh air is ventilated into the working space to maintain average concentration of the volatile matter below TLV-TWA. The quantity of such dilution air in CFM (cubic feet/minute) may be estimated by

$$CFM = \frac{(3.87 \times 10^8)W}{M \times TLV \times K} \qquad (20.4)$$

where

W = mass of volatile matter evaporated/time, lb/min
M = molecular mass of the volatile matter
TLV = threshold limit value of the volatile substance, ppm
K = non-ideal mixing factor
  = 1 for perfect mixing
  = 0.1 to 0.5 for most real situations

FIRE AND EXPLOSION ISSUES

Fire and explosion prevention are key aspects of process safety.

Fire Tetrahedron

A fire is caused by the presence of sufficient quantity of each of the following three sources plus the fourth criteria.

1. Fuel, such as gasoline, hydrogen, wood, etc.,

2. Oxidizer, such as oxygen, chlorine, ammonium nitrite, etc.,

3. Ignition source, such as heat, mechanical impact, mechanical or electrical sparks or electrostatic discharge, etc.

4. Chemical chain reaction necessary to propagate fire.

Dust Explosion Pentagon

In addition to the first three sources mentioned above, two additional factors contribute to dust explosion.

1. Confinement: This can be safeguarded by explosion venting.

2. Dispersion, such as distribution of dusts in air.

Terminology

Autoignition Temperature (AIT). The temperature above which a flammable substance ignites itself by drawing sufficient amount of oxidizer and ignition from surroundings is called autoignition temperature.

Flash Point (FP). The lowest temperature at which a liquid gives off sufficient vapor to form an ignitable mixture with air near the surface of the liquid is called the flash point. At flash point, the combustion is brief.

Fire Point. The lowest temperature at which a liquid gives off sufficient vapor to form an ignitable mixture with air, and sustains combustion once ignited, is called the fire point.

Upper Flammability Limit (UFL). The highest concentration of a flammable fluid in air, above which the air-fuel mixture will not burn, is called the upper flammability limit. The fuel concentration, by definition, is expressed as follows.:

$$UFL = \frac{\text{moles fuel}}{\text{moles fuel + moles air}} \quad (20.5)$$

Lower Flammability Limit (LFL). The lowest concentration of a flammable fluid in air below which the air-fuel mixture will not burn is called the lower flammable limit. The concentration of fuel is expressed as UFL.

Lower Explosive Limit (LEL) and Upper Explosive Limit (UEL) are used interchangeably with Lower Flammability Limit (LFL) and Upper Flammability Limit (UFL).

Explosion. The bursting or rupture of an enclosure or a container due to the development of internal pressure from a deflagration is called explosion.

Deflagration. Propagation of combustion zone front at a velocity that is less than the speed of sound in the unreacted medium is called deflagration.

Detonation. Propagation of combustion zone front at a velocity that is greater than the speed of sound in the unreacted medium is called detonation.

Combustible Dust. Any finely divided material, 420 microns or less in diameter (i.e., passing through US # 40 sieve), that presents a fire or explosion hazard when dispersed and ignited in air or other gaseous oxidizer is called combustible dust.

Minimum Oxygen Concentration (MOC). The minimum percent of oxygen in air plus fuel required to propagate flame is called the MOC. This is also used to signify minimum oxidant concentration, because substances other than oxygen can cause ignition. MOC generally varies from 8 percent (vapor) to 10 percent (dust). Control points range from 4 percent (vapor) to 6 percent (dust). By definition, MOC is expressed by

$$MOC = \frac{\text{moles oxygen}}{\text{moles oxygen} + \text{moles fuel}} \quad (20.6)$$

It follows from the definition of MOC and LFL that

$$MOC = LFL \left(\frac{\text{stoichiometric moles of oxygen}}{1 \text{ mole of fuel}} \right) \quad (20.7)$$

The stoichiometric number of moles of oxygen needed to completely burn 1 mole of fuel to carbon dioxide and water can be determined from the stoichiometry of combustion reaction (parameter z in the following pages).

Minimum Ignition Energy (MIE). The minimum energy in milli Joules required to start combustion is called MIE.

Classification of Liquids for Thermal Hazards Analysis

A *flammable liquid* is one having a closed cup flash point below 100 °F (37.8 °C) and a vapor pressure not exceeding 40 psia at 100 °F. It is also known as Class I liquid.

Flammable liquids are subdivided into three classes.

	Class IA	Class IB	Class IC
Flash point °F:	< 73	< 73	≥73 < 100
Boiling point °F:	< 100	≥100	< 100
Venting device	CV & FA	CV or FA	CV or FA
for above-ground tanks	Note 1	Note 1,2,3	Note 1,2,3

Notes:

(1) CV = Conservation vent, FA = Flame arrestor
(2) Tanks of 3000 bbl capacity or less containing crude petroleum in crude producing areas, and outside above-ground atmospheric tanks under 23.8 bbl capacity shall be permitted to have an open vent.
(3) CV or FA may be omitted where conditions such as plugging, crystallization, polymerization freezing, etc., may cause obstruction resulting in the damage of the tank. Consideration should be given to heating the devices, liquid seal, or inerting.

A *combustible liquid* is one having a closed cup flash point at or above 100 °F. Combustible liquids are subdivided into three classes.

	Class II	Class IIIA	Class IIIB
Flash point °F	>100 <140	≥140 <200	≥ 200

Above-ground tanks larger than 285 bbl capacity storing Class IIIB liquids and not within the diked area or the drainage path of Class I or Class II liquids do not require emergency relief venting for fire exposure per NFPA-30 code, but may require consideration of other standards such as API-520.

Computation of Flash Point

To compute the flash point of a solution containing a single flammable or combustible liquid dissolved in inerts, do the following:

1. Find or experimentally determine the flash point of the flammable liquid in its pure state.
2. Find the vapor pressure of the flammable liquid at the flash point. Call it p_i.
3. Compute the pure component vapor pressure P_i^0 by

$$P_i^0 = p_i/x_i \qquad (20.7a)$$

   where x_i is the concentration in mole fraction of the flammable liquid in solution.
4. Determine the temperature at which the vapor pressure of the flammable liquid equals P_i^0. This temperature is the flash point of the solution.

Computation of Flame Temperature

Prevention of fires and explosion requires the knowledge of flammability zone and flame temperature. The temperature reached when a fuel is *completely* burned in air or oxygen without gain or loss of heat is called the theoretical flame temperature (TFT). The maximum adiabatic flame temperature is obtained when the fuel is *completely* burned with the theoretical amount of pure oxygen. The maximum adiabatic flame temperature, when the same fuel is burned with the theoretical amount of air under identical condition, is lower than the flame temperature obtained in pure oxygen because of the dilution effect of the inert gases such as nitrogen. Adiabatic flame temperature of actual combustion processes is lower due primarily to the excess air used to ensure complete combustion.

The Calculated Adiabatic Flame Temperature (CAFT) is determined through an adiabatic heat balance of the enthalpy of products with the enthalpy of reactants plus the heat of reaction at the standard condition at a fixed pressure.

Thus

$$\Delta H_{products} = \int_{298}^{T_f} [\Sigma n_i C_{ppi}] dT = \Delta H_{reactants} - \Delta H_{reaction}^0 \qquad (20.7b)$$

or,

$$\int_{298}^{T_f} [\Sigma n_{pi} C_{ppi}] dT = \int_{298}^{T_1} [\Sigma n_{ri} C_{pri}] dT - \Delta H_{reaction}^0 \qquad (20.7c)$$

where

T_f = Calculated adiabatic flame temperature (CAFT), K
T_1 = Initial temperature of reactant, K. If this is 298 K, $\Delta H_{reactants}$ is zero.
n_{pi} = mole of product component i
n_{ri} = mole of reactant component i
C_{ppi} = molar specific heat of product component, i = polynomial function of temperature
C_{pri} = molar specific heat of reactant component, i = polynomial function of temperature
$\Delta H_{reaction}^0$ = heat of reaction at 298 K. This is known.

The integration of the left hand side of the equation results in a polynomial function of temperature, which can be solved by trial-and-error. (See Chapter 3, Example 3.12.)

The theoretical calculation of CAFT is very complex because of the unknown nature of the equilibrium products. Usually the technique of the minimization of overall Gibbs free energy is used to determine the equilibrium composition. Software like SuperChems™, Chemcad™, and EQS4WIN™ use this technique.

The CAFT can be used to determine the flammability of a mixture. The determination of flame temperature and flame height from a pool fire of a liquid fuel is a different matter. It requires the determination of pool diameter, wind velocity, burning rate, which require the data of extinction coefficient, latent heat of vaporization, specific heat, boiling point, and initial temperature.[20]

Calculating Flammability Limits

The flammability limits of a mixture of combustible gases may be reliably predicted by the Le Chatlier formulas.

$$LFL_{mixture} = \frac{1}{\frac{y_1}{LFL_1} + \frac{y_2}{LFL_2} + \cdots + \frac{y_n}{LFL_n}} \quad (20.8a)$$

$$UFL_{mixture} = \frac{1}{\frac{y_1}{UFL_1} + \frac{y_2}{UFL_2} + \cdots + \frac{y_n}{UFL_n}} \quad (20.8b)$$

where y = mole fraction of components in the mixture on a combustible basis or inert-free basis. This is calculated by dividing the % volume of the combustible component in the mixture by the % volume of all the combustible components. Subscripts denote the components. If the LFL_i and UFL_i in the denominator are entered as %, then the answer is also obtained as %. If the % volume of the total combustibles in the mixture including the inerts falls in the flammability range of the mixture (i.e., range of $LFL_{mixture}$ to $UFL_{mixture}$), then the mixture is flammable.

The dependence of flammability limit on temperature is expressed by the correlation of Zabetakis, Lambiris & Scott.

$$\text{Flammability Limit at } T°C = \text{Flammability Limit at } 25°C$$
$$[1 \pm 0.75(T - 25)/\Delta H_c] \quad (20.9)$$

where the + sign before 0.75 is to be used for UFL, and − sign for LFL. ΔH_c is net heat of combustion in kcal/mol.

Zabetakis also developed the correlation of flammability limit and pressure.

$$UFL_p = UFL + 20.6(\log P + 1) \quad (20.10)$$

where
P = pressure in megapascals absolute = MPa, gauge + 0.101
UFL = upper flammability limit at 1 atm

Lower flammability limit has little sensitivity to pressure.

Stoichiometry of Combustion Reaction and Estimation of Flammability Limits

The stoichiometry of the combustion reaction of hydrocarbons containing C, H, and O may be represented by

$$C_mH_xO_y + zO_2 = mCO_2 + (x/2)H_2O$$

By stoichiometry,

$$z = m + x/4 - y/2 = \text{moles of oxygen per mole of fuel}$$

For hydrocarbon vapors, Jones formulas may be used to estimate flammability limits.

$$LFL = 0.55\, C_{st}$$
$$UFL = 3.5\, C_{st}$$

where LFL and UFL are in % by volume and C_{st} is the stoichiometric concentration of fuel in the combustion reaction, also in % by volume, and is defined by

$$C_{st} = \frac{(\text{moles fuel})100}{\text{moles air} + \text{moles fuel}}$$

$$= \frac{100}{\frac{\text{moles air}}{\text{moles fuel}} + 1}$$

$$= \frac{100}{\frac{\text{moles air}}{\text{moles oxygen}} \times \frac{\text{moles oxygen}}{\text{moles fuel}} + 1}$$

$$= \frac{100}{\frac{z}{0.21} + 1} \tag{20.11}$$

By combining above equations, it can be shown that:

$$LFL = \frac{55}{4.76m + 1.19x - 2.38y + 1}$$
$$UFL = \frac{350}{4.76m + 1.19x - 2.38y + 1} \tag{20.12}$$

Inerting

Inerting is the process of lowering the oxygen concentration in a combustible mixture below MOC by adding an inert gas. Three common methods of inerting are sweep-through purging, pressure purging and vacuum purging.

Sweep-through Purging

The inert gas is continuously added to a vessel at one end and withdrawn at another end, both the addition and withdrawal being at atmospheric pressure.

$$Q_v t = V \ln \frac{(C_1 - C_o)}{(C_2 - C_o)} \tag{20.13}$$

Q_v = Volumetric flow rate of inert
t = Time of purging
V = Vessel volume
C_1 = Initial concentration of oxygen
C_2 = Final concentration of oxygen
C_o = Concentration of oxygen in the inert, usually zero

Pressure Purging

The vessel is pressurized with inert gas and then vented, usually to atmospheric pressure. The process is repeated until the oxygen concentration goes below MOC.

Vacuum Purging

The content of a vessel is withdrawn by a vacuum system until a predetermined vacuum is reached. It is then filled up with the inert gas to the initial pressure. The process is repeated until the oxygen concentration goes below MOC.

The formulas for estimating purging cycles and amount of inert gas for pressure and vacuum purging are*

$$n = \frac{\ln \frac{y_0}{y_n}}{\ln \frac{P_H}{P_L}} \qquad (20.14)$$

$$\text{and} \quad M_i = \frac{n(P_H - P_L)MV}{RT} \qquad (20.15)$$

where
- n = Number of purging cycles
- y_o = Initial concentration of oxygen
- y_n = Concentration of oxygen after n purge cycles
- P_H = High pressure of the cycle, psia
- P_L = Low pressure of the cycle, psia
- M_i = Mass of inert gas required, lb
- M = Molecular mass of inert
- V = Volume of the vessel to be purged, cuft
- R = 10.73, (psia • ft³/lb mol-°R)
- T = Operating temperature, °R

Example 20.4

A 3000-gallon reactor initially filled with ambient air is to handle dusty solids. It is to be inerted using nitrogen at 95°F. Calculate the mass of nitrogen required if (a) pressure purging is used with 90 psig nitrogen, (b) vacuum purging is used. The vacuum system can produce a vacuum of 25 mm Hg absolute. (c) Sweep-through purging is used. Assume a safe final concentration of oxygen in the reactor.

Solution

Because the reactor handles dusty materials, assume final concentration of oxygen as 4 percent.

y_2 = 4 percent
y_o = 21 percent, since initially filled with ambient air

(a) Pressure purging

$$n = \frac{\ln(21/4)}{\ln(104.7/14.7)} = \frac{1.6582}{1.9632} = 0.84$$

---

*The units of y_o and y_n are the same.

Say one purge.

$$M_i = 1(104.7-14.7)(28)(3000)/(7.48)(10.72)(460+95)$$
$$= 169.88 \text{ lbs}$$

(b) Vacuum purging

$$P_H = 14.7 \text{ psia}$$
$$P_L = 14.7 \times 25/760 = 0.4836 \text{ psia}$$
$$n = \frac{\ln(21/4)}{\ln(14.7/0.4836)} = \frac{1.6582}{3.4143} = 0.49$$
$$M_i = 1(14.7-0.4836)(28)(3000)/(7.48)(10.72)(555)$$
$$= 26.83 \text{ lbs}$$

(c) Sweep-through purging

$$\text{Volume of inert} = V \ln(C_1/C_2)$$
$$= (3000/7.48)\ln(21/4)$$
$$= 665.1 \text{ cuft}$$
$$\text{Mass of inert} = 665.1 \times 14.7 \times 28/(10.72)(555)$$
$$= 46.0 \text{ lbs}$$

Overpressure Protection of Storage Vessel and Process Equipment

General Requirements

Although proper design, adequate control and instrumentation may safeguard a plant from predictable upset conditions, it is required by various codes and government acts to protect all equipment from overpressure due to unpredictable but probable upset conditions. This is usually done by installing a rupture disc or a relief valve or their combination on the equipment to be protected.

Why do we need emergency relief system hardware?

1. It is required by law, as noted in the following regulations:

   A. OSHA 1910.119—process safety management (PSM) of highly hazardous chemicals. Initial completion date of PHA was May 26, 1997. PHA is to be updated every 5 years.

   B. OSHA 1910.106—hardware design involving flammable and combustible liquid.

   C. EPA Clean Air Act Amendment (1990) Title III—control of hazardous air pollutants (HAP) through maximum achievable control technology (MACT).

2. It is required by insurance carrier to comply with NFPA-30 for Flammable & Combustible Materials that mandates consideration of fire.

3. Safety to personnel
4. To minimize business interruption
   - Loss of product
   - Loss of equipment
   - Environmental clean-up cost
   - Costly litigation

Emergency Relief Load Calculation

The sizing of the rupture disc or relief valve requires the estimation of the emergency relief load. Such relief loads may be handled by one or multiple valves. The set pressure and accumulation limits of pressure relief valves depend on the contingency and number of valves installed to handle the relief load as shown below. Accumulation is pressure rise above MAWP.

	Single-Valve		Multiple-Valve	
	Set Pressure %	Accumulation %	Set Pressure %	Accumulation %
Nonfire only				
First valve	100	110	100	116
Additional valve(s)	—	—	105	116
Fire Only				
First valve	100	121	100	121
Additional valve(s)	—	—	105	121
Supplemental* valve	—	—	110	121

All values shown above are percentages of maximum allowable working pressure (MAWP).
*Supplemental valves are used only in addition to valves sized for fire contingency.

Emergency Relief Load Calculation and Sizing Safety Device: Conditions Considered in Emergency Relief

The most common conditions considered in emergency relief load calculation are shown in Table 20.1. Internal explosion and dust explosion should also be considered. These are, however, beyond the scope of this chapter.

Water Hammer

Water hammer is a physical manifestation of trapped hydraulic shock waves when the flow through a liquid filled system is suddenly stopped. When a fluid running with a velocity of 1 ft/s is stopped suddenly in a liquid filled system, the pressure rises approximately 54 psi above the normal operating pressure. Thus when a fluid running with a velocity of 2 ft/s is stopped suddenly in a liquid filled system at a normal operating pressure of 50 psig, the pressure rises approximately 54 psi/(ft/s) above the normal operating pressure thereby increasing the peak pressure to $54 \times 2 + 50$ or 158 psi for a short duration.

Table 20.1 Relief load calculation basis for various scenarios

Item No.	Condition	Liquid Relief Load	Vapor Relief Load
1.	Fire heat input		See relief load calculation due to fire
2.	Failure of utility		
	a. Cooling water		Total incoming vapor and that generated therein.
	b. Fan-failure in air-cooled exchanger		80–100% of the vapor in.
	c. Electricity	Estimate by system study.	Estimate by system study
3.	Reflux pump failure		Total vapor to condenser
4.	Heat exchanger tube rupture. Consider this condition if the design pressure of the high pressure side is greater than 1.3 times the design pressure of the low pressure side.	Estimate flow using flow coefficient = 0.7 and flow area = 2x cross-section area of one tube. If this exceeds the normal flow, the latter should be used.	Same as liquid load. Consider critical flow.
5.	Overfilling of storage or surge vessel	Maximum liquid pump-in rate	
6.	Instrument failure such as failure of control valve.	Estimate by system study. Usually determined by maximum flow of control valve.	Same remark as under "Liquid Relief Load".
7.	Operator mistake such as blocking outlet of a positive displacement machine.	Estimate by system study. Usually the normal flow from the machine.	Estimate by system study. Usually the flow from the machine.
8.	Chemical reaction.		Vapor generation due to normal and uncontrolled condition. Consider two-phase flow.
9.	Trapped fluid in a heat exchanger.	Relief in gpm (thermal relief) $= \dfrac{\beta Q}{500 G C}$	
	where	β = Coefficient of thermal expansion of blocked fluid per °F = 0.0001 for water. Q = Maximum heat exchanger duty, Btu/h. G = Sp. Gr. of blocked fluid. C = Sp. Heat of blocked fluid Btu/lb-°F	
10.	Water hammer	Equipment may not be protected by a relief device. Use slow closing valve, or pulsation dampener	
11.	Steam hammer		May not be protected by relief device. Use slow-closing valve, or a diffusing element.

Relief devices, in particular relief valves, should not be used to control water hammer. The disk responds faster than a relief valve, which has a slower response due to spring inertia, and the disk will immediately break under water hammer conditions if the burst pressure is reached even for a split second. An otherwise well-controlled system may exhibit water hammer during start up due to ill-defined start up procedure. The best way to treat water hammer is to have a well-written start up procedure and to eliminate it by system design: (a) Starting the centrifugal pump dead-headed with control on manual by closing a modulating valve installed **before** the relief device, making sure the system can stand the dead-headed pressure, and then slowly opening the valve for the set point followed by switching the control to auto mode, (b) From a running pump, modulate the flow from a running velocity to near zero velocity and then close the valve. There are other options such as pulsation dampener, use of silent check valves at the pump discharge, hydraulic pump check valve with air cylinder, etc.

To monitor the possibility of a nuisance break of a rupture disk, a burst disk indicator with appropriate automatic action is advisable.

Steam Hammer

Steam hammer is a physical manifestation of oscillating pressure surge when the flow through a compressible fluid system is suddenly stopped. A relief device should not be used to protect equipment from steam hammer. It should be eliminated by operating procedure and system design.

Relief Load Calculation Due to External Fire

Relief load calculation due to external fire requires the estimation of the heat transfer area. Table 20.2 shows the formulae used in computing exposed areas.

To calculate the heat transfer area for fire condition, the total area calculated in Table 20.2 may be multiplied by a factor shown in Table 20.3 depending on the agency that governs the operation of the equipment. For regulatory materials (flammable material) in storage tanks, OSHA 1910.106 is mandatory. When the regulated material goes to process equipment, heat transfer area may be calculated by any of the following methods shown in Table 20.3 unless the insurance company dictates otherwise. Similarly, if the storage tank does not contain any regulated material but is exposable to fire, any of the methods shown in Table 20.3 may be used.

Table 20.2 Formulae for wetted area calculation

Equipment or Equipment Part	Total Area (A_T, Ft²)
1. Sphere shell	$A_T = \pi D^2$
2. Cylindrical shell	$A_T = \pi DL$
3. Ends	
(a) Flat	$A_T = (\pi/4) D^2$
(b) Concial	$A_T = \left(\pi \dfrac{D}{2}\right)\sqrt{(D^2/4) + h^2}$
(c) 2:1 Ellipsoidal	$A_T^* = 1.19 D^2$
(d) ASME flanged and dished	$A_T^* = 0.918 D^2$

* Excludes area of straight flange.

Except for the sphere, calculate the total area by adding the areas of the components. Formulas for ends are for one end only.

    D = Outside diameter, ft.

    L = Tangent to tangent length, ft.

This length should be adjusted per item 4 of Table 20.3 and item 3 of Table 20.5 for vertical tanks.

    h = Cone height, ft.

Table 20.3 NFPA-30 wetted area calculation table

Type of Vessel	Wetted Area, A
1. Sphere or spheroid	55% of total exposed area. [The code does not specify any elevation height limitation of equipment from grade.]
2. Horizontal tank	75% of total exposed area. [The code does not specify any elevation height limitation of equipment from grade.]
3. Rectangular tank	100% of the exposed shell and bottom area, but **excluding the top surface area**. Include bottom area if the equipment is above ground. [The code does not specify any elevation height limitation of equipment from grade.]
4. Vertical tank	The exposed shell area within first 30 ft above grade or surface where fire can be sustained. [The code implies by analogy with rectangular tank, but not explicitly, that the top surface area be excluded even if the top surface is within 30 ft from grade. The code also implies to include the bottom area, if the bottom is off the grade. However, in the author's opinion, if two-phase flow is anticipated or the material is unstable, top area should be included if the top head is within flame height, which may be greater than 30 ft.]

Note: When the surface is in contact with the grade, the surface is not considered exposed to fire. The "wetted" area does not have to be physically wet to meet the definition of the NFPA-30 code.

Calculation of Heat Flow from Fire, or Fire Heat, (Q_{fire})

Fire heat, (Q_{fire}), may be empirically determined using the appropriate code. One such code (NFPA-30) offers the following equations:[12]

$$Q_{fire} = \begin{cases} 20,000(F)(A) & \text{for } 20 < A \leq 200 \\ 199,300(F)(A)^{0.566} & \text{for } 200 < A \leq 1000 \\ 963,400(F)(A)^{0.338} & \text{for } 1000 < A \leq 2800 \\ 21,000(F)(A)^{0.82} & \text{for } A > 2800 \text{ and } MAWP > 1 \text{ psig} \\ 14.09 \times 10^6 (F) & \text{for } A > 2800 \text{ and } MAWP < 1 \text{ psig} \end{cases} \quad (20.16)$$

Use Table 20.3 to calculate wetted area, A, for above ground tanks when using Equation 20.16.

Use Table 20.4, when applicable, to select the environmental factor, F, for using in Equation 20.16.

The fire heat equation according to API-521 is different.[3]

$$Q_{fire} = \begin{cases} 21000(F)(A^{0.82}), \text{ see note 1} \\ 34500(F)(A^{0.82}), \text{ see note 2} \end{cases} \quad (20.17)$$

Note1: Applies when adequate drainage and firefighting equipment exist.
Note2: Applies when adequate drainage and firefighting equipment do not exist.

The fire heat equation according to API-2000 is similar to NFPA-30, and is as follows:[5]

$$Q_{fire} = \begin{cases} 20,000(F)(A) & \text{for } A < 200 \text{ for design pressure} \leq 15 \text{ psig} \\ 199,300(F)(A)^{0.566} & \text{for } 200 \leq A < 1000 \text{ for design pressure} \leq 15 \text{ psig} \\ 963,400(F)(A)^{0.338} & \text{for } 1000 \leq A < 2800 \text{ for design pressure} \leq 15 \text{ psig} \\ 21,000(F)(A)^{0.82} & \text{for } A \geq 2800 \text{ and } 15 \text{ psig} \geq \text{Design pressure} > 1 \text{ psig} \\ 14.09 \times 10^6 (F) & \text{for } A > 2800 \text{ and Design pressure} \leq 1 \text{ psig} \end{cases} \quad (20.18)$$

Use Table 20.5 to calculate wetted area, A, for above-ground tanks according to API-521 and API-2000, and Table 20.6 for selecting the environmental factor.

Table 20.4 NFPA-30 vent reduction factor for equipment

Environmental Condition	Factor for Materials with Heat of Combustion or Burning Rate Greater than Ethanol's	Factor for Ethanol or Materials with Heat of Combustion and Burning Rate Equal to or Less Than Ethanol's and Miscible with Water (note 4)	Factor for Materials with Heat of Combustion and Burning Rate Equal to or Less than Ethanol's but Immiscible with Water (note 4)
Bare vessel	1.0	0.5	1.0
Drainage, A > 200 ft^2	0.5	0.25	0.25
Drainage, A < 200 ft^2	1.0	0.5	1.0
Water spray (note 1)	0.3	0.15	0.3
Water spray (note 2)	0.3	0.15	0.15
Insulation (note 3)	0.3	0.15	0.3
Insulation & water spray (notes 2 & 3)	0.15	0.15 [no further reduction]	0.15 [no further reduction]

Notes:
(1) Water spray system must meet the requirement of NFPA-15. **Tanks are provided with drainage that meets the requirement of paragraph 4.3.2.3.1** [specifies drainage slope and capacity of remote impounding area]. The code implies that water spray system may be manually activated for this reduction factor.
(2) Tanks that are protected with **automatically actuated** water spray system that meets the requirement of NFPA-15. [NFPA-30 states that drainage is not required.]
(3) Insulation shall remain in place under fire exposure conditions, shall withstand dislodgement when subjected to hose stream impingement during fire exposure [the dislodgement criteria does not apply where use of solid hose streams is not contemplated or would not be practical], and the insulation shall maintain a maximum conductance (conductivity/thickness) of insulation of 4 Btu/hr · ft^2 · °F at a mean temperature of 1000°F when the outer insulation jacket or cover is at a temperature of 1660°F.
(4) No potential exposure from liquids with heat of combustion and burning rate more than ethanol's. No drainage is required for this factor to apply. Ethanol credit has been considered.
(5) Only one of the factors is applicable; no factor shall be less than 0.15.
(6) For hexane or lighter hydrocarbons, multiply the NFPA reduction factor by 3, until NFPA-30 addresses the issue of hexane fire.[17]

Environmental Factor for Insulation

It should be remembered that NFPA 30 does not give any credit for the *thickness* of a fire-proof insulation even if the conductance of the insulation is lowered below the maximum allowable value of 4 BTU/h · ft^2 · °F recommended by the

Table 20.5 API wetted area calculation table

	API-2000	API-521
1. Sphere or spheroid	55% of the total surface area or the surface area to a height of 30 ft above grade, whichever is greater	Take area for liquid inventory up to the maximum horizontal diameter or up to the height of 25 ft; whichever is greater.
2. Horizontal tank	75% of the total surface area or the surface area to a height of 30 ft above grade, whichever is greater	For liquid full vessels, take all surface area up to the height of 25 ft. For surge drums, knock out drums, process vessels, working storage, take surface area for normal operating level up to the height of 25 ft.
3. Vertical tank	Total surface of the vertical shell to a height of 30 ft plus the bottom surface area if the tank is supported above grade	For liquid full vessels, take all surface area up to the height of 25 ft. For surge drums, knock out drums, process vessels, storage vessels, take surface area for normal operating level up to the height of 25 ft. Include bottom surface area if the tank is supported above grade.
4. Fractionating column	Not specified but may be taken as vertical tank	Surface area for the normal level in bottom plus liquid holdup from all trays dumped to the normal level in the column bottom; total wetted surface up to 25 ft.

Table 20.6 API environmental factor table

Design/Configuration	Insulation Conductance (Btu/h · ft² · °F)	Insulation Thickness, (in.)	API-2000 Environmental Factor, F (May Use Equation 20.18a)	API-521 Environmental Factor, F (May Use Equation 20.18a)
Bare metal tank	—	0	1.0	1.0
Insulated tank	4.0	1	0.3	0.3
Insulated tank	2.0	2	0.15	0.15
Insulated tank	1.0	4	0.075	0.075
Insulated tank	0.67	6	0.05	0.05
Insulated tank	0.5	8	0.0375	0.0376
Insulated tank	0.4	10	0.03	0.03
Insulated tank	0.33	12	0.025	0.026
Water spray	—	—	1.0	1.0
Depressurizing and emptying facilities	—	—	1.0	1.0
Earth-covered storage above grade	—	—	0.03	0.03
Below-grade storage	—	—	0	0.0
Impounded away from tank	—	—	0.5	Not specified
Concrete tank or fireproofing	—	—	Note 1 below	Not specified

Note: (1) Use F-factor for an equivalent conductance value of insulation.
(2) If the process material is entirely gaseous, the total exposed area (instead of wetted area) and an environmental factor of 0.3 should be used for all equations (16, 17, & 18).

The definition of thermal conductance follows from Fourier's Law (Chapter 5, Equation 5.1):

$$Q = -kA(dT/dx) = -(dT)/[dx/(kA)]$$

When this is applied to conduction through insulation, the insulation resistance per unit area, by electrical analogy, is dx/k, and insulation conductance $= k/dx$, where dx = insulation thickness, in., k = thermal conductivity, Btu · in/h · ft²F. Hence the unit of conductance is Btu/h · ft²F. NFPA 30, by prescribing an upper limit of conductance, has put a lower limit of insulation thickness. API RP 521 allows consideration of insulation thickness, and prescribes a limitation on the thermal conductivity, a fundamental property of a **material**.

If approved by your industrial insurer, a credit due to higher thickness of insulation may be obtained by using the formula

$$F_{\text{APIins}} = \left[\frac{T_f - T_{\text{relief}}}{21000}\right]\left[\frac{k}{x_i}\right] \qquad (20.18a)$$

In API RP 521, the temperature, T_f, is taken as 1660 °F, which is considered an equilibrium temperature of the jacket holding the insulation in place. You can choose any realistic value of T_f by a flame temperature calculation. Experts[20] warn against using any calculated value of $F_{\text{APIins}} < 0.026$. In view of the uncertainties

Table 20.7 Sizing equation for gas, vapor or steam

Critical Flow,* Gas or Vapor	Subcritical Flow, Gas or Vapor

$A = \dfrac{W}{CKP_1 K_b} \sqrt{\dfrac{TZ}{M}}$
 	
A = Relief Area, inch^2
W = Flow, lb/h
T = Absolute temperature of inlet vapor, °R, at relief pressure
Z = Compressibility factor
M = Molecular mass of gas of vapor
 	
$C = 520 \sqrt{n \left[\dfrac{2}{n+1} \right]^{\frac{n+1}{n-1}}}$
 	
$n = \dfrac{MC_P}{MC_P - 1.99}$
 	
Cp = Specific heat, Btu/lb-°F
(If n cannot be estimated assume $C = 315$)
 	
K = Coefficient of discharge = 0.975, if unknown
P_1 = Set pressure (PSIG) × (1 + accumulation) + barometric pressure
Accumulation
   = 0.10 for unfired emergency relief
   = 0.21 for fire condition
   = 0.33 for piping
K_b = Back pressure correction factor (see Figure 20.2)
	$A = \dfrac{W}{735 FK} \sqrt{\dfrac{TZ}{MP_1 (P_1 - P_2)}}$
	
	$F = \sqrt{\left[\dfrac{n}{n-1}\right] r^{\frac{2}{n}} \left[\dfrac{1 - r^{(n-1)/n}}{1 - r}\right]}$
	
	P_2 = Back pressure, psia.
	$r = P_2/P_1$
	
	Other terms are explained under critical flow formula of gas or vapor:
	
	Note:
	Balanced-bellows relief valves that operate in the subcritical region may be sized using the critical flow formulas.
	Saturated Steam, Critical Flow
	
	$A = \dfrac{W}{515.P_1 KK_n}$
	
	$K_n = 1$ when P_1 ð 1515 psia
	Other terms are explained under critical flow formula of gas or vapor.
	
	For superheated steam, a correction factor is required; required area is higher than the corresponding area required for saturated steam.

in the stability of the insulation under prolonged fire duration, one expert[21] suggests the lower limit of $F_{APIins} = 0.075$.

Sizing of Relief or Safety Valve

Once the relief load is determined, use Table 20.7 and Table 20.8 to calculate the relief area for single-phase flows. Then use Table 20.9 to select the relief valve.

For calculation of relief load due to thermal expansion of gas caused by external fire, see reference 3.

When a relief valve is sized for viscous liquid service, it should first be sized as it was for nonviscous-type application so that a preliminary required discharge area, A, can be obtained. From manufacturers' standard orifice sizes, the next larger orifice size should be used in determining the Reynolds number, R, from the following relationship:

$$R = \dfrac{Q(2800G)}{\mu \sqrt{A}}$$

---

*See page 194 for a definition of critical flow.

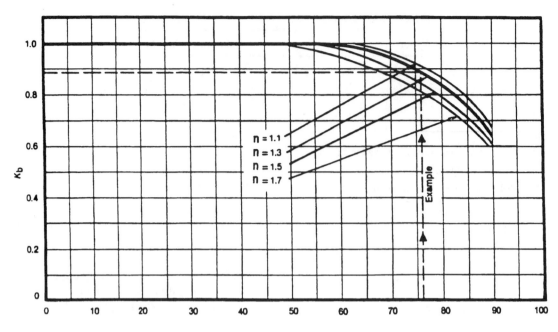

Percent of absolute back pressure $= \dfrac{P_B}{P_S + P_o} \times 100$

P_B = back pressure, in pounds per square inch absolute.
P_S = set pressure, in pounds per square inch absolute.
P_o = overpressure, in pounds per square inch absolute.

Set pressure (MAWP) = 100 pounds per square inch gauge.
Overpressure = 10 pounds per square inch gauge.
Superimposed back pressure (constant) = 70 pounds per square inch gauge.
Spring set = 30 pounds per square inch.
Built-up back pressure = 10 pounds per square inch.

Percent absolute back pressure $= \dfrac{(70 + 10 + 14.7)}{(100 + 100 + 14.7)} \times 100 = 76$

K_b (follow dotted line) = 0.89

Note: This chart is typical and suitable for use only when the make of the valve or the actual critical flow pressure point for the vapor or gas is unknown; otherwise, the valve manufacturer should be consulted for specific data. This correction factor should be used only in the sizing of conventional (non-balanced) pressure relief valves that have their spring setting adjusted to compensate for the superimposed back pressure. It should not be used to size balanced-type valves.

Figure 20.1 Constant back pressure sizing factor, K_b, for conventional safety relief valves (vapors and gases only). Reproduced courtesy of the American Petroleum Institute, Washington, D.C.

where
Q = flow rate at the flowing temperature, in U.S. gallons per minute.
G = specific gravity of the liquid at the flowing temperature referred to water
= 1.00 at 70°F.
μ = absolute viscosity at the flowing temperature, in centipoises.

When a rupture disc is used in series with the relief valve, the combination capacity is to be reduced by 10 percent per ASME code. In other words, the selected orifice area from Table 20.8 multiplied by 0.9 should be greater than or equal to the calculated area.

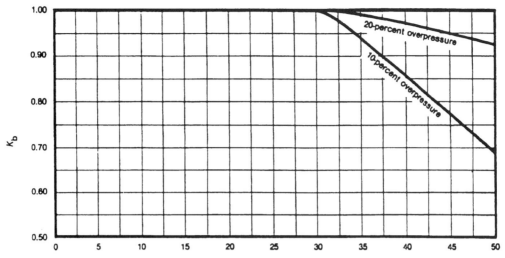

Percent of gauge back pressure = $(P_B/P_S) \times 100$

P_B = back pressure, in pounds per square inch gauge.
P_S = set pressure, in pounds per square inch gauge.

Note: The curves above represent a compromise of the values recommended by a number of relief valve manufacturers and may be used when the make of the valve or the actual critical flow pressure point for the vapor or gas in unknown. When the make is known, the manufacturer should be consulted for the correction factor. These curves are for set pressures of 50 pounds per square inch gauge and above. They are limited to back pressure below critical flow pressure for a given set pressure. For subcritical flow back pressures below 50 pounds per square inch gauge, the manufacturer must be consulted for values of K_b.

Figure 20.2 Back pressure sizing factor, K_b, for balanced-bellows pressure relief valves (vapors and gases). Reproduced courtesy of the American Petroleum Institute, Washington, D.C.

Table 20.8 Sizing equation for liquid

1. When capacity certification is required per ASME Section VIII, Div. I.:

$$A = \frac{\text{GPM}}{24.7 K_w K_v} \sqrt{\frac{G}{P_1 - P_b}}$$

2. When capacity certification is not required:

$$A = \frac{\text{GPM}}{23.56 K_p K_w K_v} \sqrt{\frac{G}{P_1 - P_b}}$$

A = Relief area, inch^2

GPM = Flow in gallon/min

G = Sp. gr.

K_p = Capacity correction factor for overpressure (see Figure 20.3)

K_w = Capacity correction factor for back pressure (see Figure 20.4)
 = 1 for conventional valve

K_v = Viscosity correction factor = 1 for nonviscous fluid (see Figure 20.5)

P_1 = Set pressure (psig) × (1 + accumulation) + barometric pressure (psia)

Accumulation

 = 0.25 when certification is not required, otherwise 10 percent

 = 0.33 for piping

P_b = Back pressure, psia

Table 20.9

Valve	Orifice Area IN.²	Valve	Orifice Area IN.²
1 D 2	0.110	3 L 4	2.853
1 E 2	0.196	4 M 6	3.60
1½ F 2	0.307	4 N 6	4.34
1½ G 2½	0.503	4 P 6	6.38
1½ H 3	0.785	6 Q 8	11.05
2 J 3	1.287	6 R 8	16.0
3 K 4	1.838	8 T 10	26.0

(The first number of the valve size indicates the nominal size in inches of the inlet, and the number following the letter code indicates the nominal size of the outlet of the relief valve. It is not unusual to have an outlet pressure rating lower than the inlet pressure rating of a relief valve.)

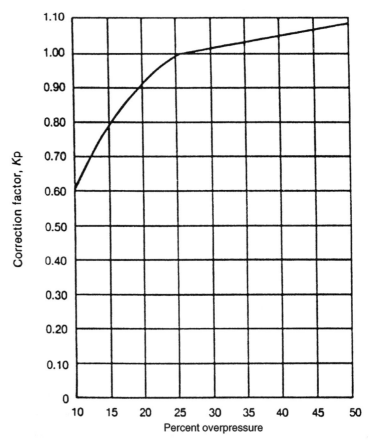

Note: The curve above shows that up to and including 25-percent overpressure, capacity is affected by the change in lift, the change in the orifice discharge coefficient, and the change in overpressure. Above 25 percent, capacity is affected only by the change in overpressure. Valves operating at low overpressures tend to chatter; therefore, overpressures of less than 10 percent should be avoided.

Figure 20.3 Capacity correction factors due to overpressure for relief and safety relief valves in liquid service. Reproduced courtesy of the American Petroleum Institute, Washington, D.C.

Fire and Explosion Issues 745

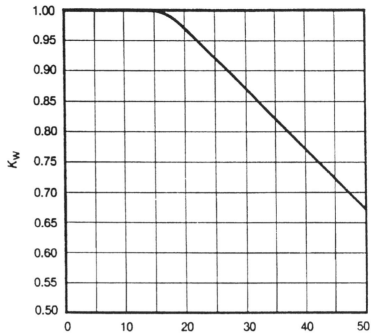

Percent of gauge back pressure = $(P_B/P_S) \times 100$
K_w = correction factor due to back pressure.
P_B = back pressure, in pounds per square inch gauge.
P_S = set pressure, in pounds per square inch gauge.
Note: The curve above represents values recommended by various manufacturers. This curve may be used when the manufacturer is not known. Otherwise, the manufacturer should be consulted for the applicable correction factor.

Figure 20.4 Variable or constant back-pressure sizing factor, K_w, for 25 percent overpressure on balanced bellows safety relief valves (liquid only). Reproduced courtesy of the American Petroleum Institute, Washington, D.C.

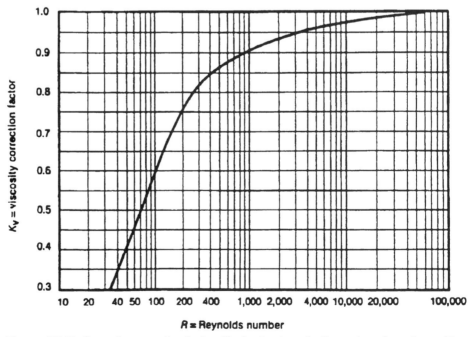

Figure 20.5 Capacity correction factor, K_V, due to viscosity. Reproduced courtesy of the American Petroleum Institute, Washington, D.C.

Pressure drop through the inlet line of the relief device should not exceed 3 percent of the differential set pressure. The outlet lines should be sized to limit the outlet pressure drop as follows: 10 percent of the differential set pressure for the conventional safety relief valve, 30 percent of the differential set pressure for balanced valves, and 50 percent of the differential set pressure for the pilot operated valves with balanced pilots. The inlet pressure drop can be minimized by using restricted lift valve; the problem can also be solved by using pilot operated valves. The differential set pressure is the difference between the set pressure and constant back pressure of a conventional valve. For balanced valves or for pilot-operated valves with balanced pilots, the differential set pressure is the same as the set pressure.

Sizing of Rupture Disc

The diameter of the rupture disc for the critical flow condition is determined by

$$d = \left(\frac{W}{146P}\right)^{\frac{1}{2}} \left(\frac{T}{M}\right)^{\frac{1}{4}} \quad (20.19)$$

P = bursting pressure in psia. This pressure does not include any accumulation.
d = diameter of rupture disc, inch.

Alternatively, the flow area required for critical flow through a rupture disc may be computed by multiplying the area obtained from the critical flow equations in Table 20.6 with 1.573. This factor is obtained by replacing the discharge coefficient (0.975) with (0.62) i.e. 0.975/0.62 = 1.573. This area is further corrected for accumulation by multiplying with the ratio of flowing pressure/set pressure, both pressures being in absolute scale. See Example 20.9. The size of the rupture disc for liquid service may be estimated using Equation 20.2 to Table 20.7 with $K_p = K_w = 1$ and P_1 as the burst pressure. No accumulation is used. A 20 percent factor of safety should be added to the calculated rupture disc area to account for the structural support (such as vacuum support, etc.). This safety factor also applies when the disc is used in combination with a relief valve.

In 1998, the ASME code has changed its rule on rupture disks. The rupture disk now can be treated as a pipe fitting with a certified K_R value as defined by Equation 4.49 in Chapter 4. The default value of K_R is 2.4.

Example 20.5

A vertical cylindrical storage tank 15′ diam. × 40′ high (tangent to tangent) with flanged-and-dished heads is used to store a flammable liquid of vapor molecular weight 90. Size the relief valve and rupture disc to protect the tank from overpressure due to external fire. Consider OSHA 1910.106 and API-520/521. Assume adequate firefighting efforts and drainage facilities exist.

Data Available

Vessel diameter:	15'0" O.D.
Vessel height:	40'0" (T-T), flanged and dished ends
Normal liquid level:	35' from bottom tangent line and the bottom tangent line is 7' above grade.

Insulation: 1" (Fire proof), thermal conductivity = 4 Btu-inch/hr · ft · ^{2}°F.
Relief valve set pressure: 130 psig, atmospheric discharge.

Table 20.10

Flowing pressure, psia	152	165	172
Flow temperature °R	450	500	550
K	312	313	315
Z	1.08	1.08	1.09
Latent heat Btu/lb.	114	112	110

Solution: [See Exhibit 1 for estimation of L]

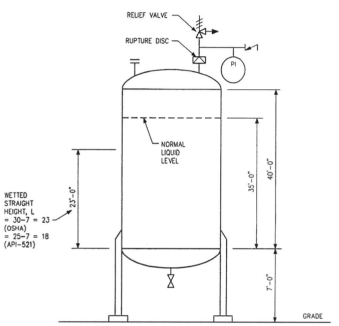

Exhibit 1 Diagram for Example 20.5

OSHA	API-520/521
Exposed area:	**Exposed area:**
$A_h = \pi DL + 0.918\ D^2$ (from Table 20.2)	$A_h = \pi(15)(18) + 0.918\ (15)^2$
$= (\pi)(15)(23) + 0.918(15)^2$	$= 1{,}054.35\ \text{ft}^2.$
$= 1{,}289.85\ \text{ft}^2$	$F = 0.075 \times 4/1 = 0.3$
$Q = 963{,}400\ F A_h^{0.338}$ (from Table 20.4)	$Q = 21{,}000\ F A_h^{0.82}$
$= (963{,}400)(0.3)(1{,}289.85)^{0.338}$	$= 21{,}000 \times 0.3 \times (1{,}054.32)^{0.82}$
($F = 0.3$ from Table 20.5)	
$= 3.253 \times 10^6$ Btu/h	$= 1.898 \times 10^6$ Btu/h

Flowing Pressure

$P_1 = 130\ (1 + 0.21) + 14.7 = 172$ psia

Corresponding Flowing Temperature

= 550°R and latent heat = 110 Btu/lb

OSHA	API-520/521
$W = \dfrac{Q}{L} = \dfrac{3253 \times 10^6}{110} = 29{,}573 \, \text{lb/hr}$	$W = \dfrac{1.898 \times 10^6}{110} = 17{,}250 \, \text{lb/hr}$
$A = \dfrac{W}{KCP_1 K_b} \sqrt{\dfrac{TZ}{M}}$	$A = \dfrac{17{,}250}{315 \times 0.975 \times 172} \sqrt{\dfrac{550 \times 1.09}{90}}$
$= \dfrac{29{,}573}{315 \times 0.975 \times 172 \times 1} \sqrt{\dfrac{550 \times 1.09}{90}}$	$= 0.843$
$= 1.445 \, \text{in.}^2$	

Because the valve is used in series with the rupture disc, a 10 percent allowance must be made. The selection would be 3K4(OSHA) or 2J3 (API-520/521) (See Table 20.8). The rupture disc diameter (OSHA) is given by

$$d = \left[\dfrac{W}{146P}\right]^{1/2} \left[\dfrac{T}{M}\right]^{1/4} = \left[\dfrac{29{,}573}{146 \times 144.7}\right]^{1/2} \left[\dfrac{550}{90}\right]^{1/4} = 1.86 \, \text{inch}$$

Flow area of the rupture disc can also be obtained from the flow area of the relief valve after corrected for the discharge coefficient and pressure.

$$A = (1.445 \times 1.573) \times (172/144.7) = 2.702 \, \text{in.}^2$$
$$\text{Diameter of the rupture disc} = [(4/\pi) \times 2.702]^{1/2} = 1.85 \, \text{in.}$$

Select 3" rupture disc in series with 3K4 relief valve or 2" rupture disc in series with 2J3 relief valve because the rupture disc in series with a relief valve should not be smaller than the inlet size of the relief valve, and 10 percent allowance must be made for the combination capacity. If the equipment is located in an area not regulated by OSHA, a smaller safety device may be used by adopting API. [Consideration of two-phase flow may require higher relief area than what is calculated above.]

Example 20.6

A pump feeds 88 GPM of a fluid at 518 psig through a shell-and-tube heat exchanger as shown in Exhibit 2. The process flows through the tube side and the coolant flows through the shell side. The tube side is designed for 650 psig at 250°F, and the shell side is designed for 100 psig at 350°F. The tubes have inside diameter of 0.62 inch. The heat duty of the exchanger is 240,000 Btu/hr. The coefficient of thermal expansion of water is 0.0001/°F. Size the relief valve for all applicable conditions and choose the final size. Assume appropriate overpressure of the relief valve. Capacity certification of the relief valve is not required.

Solution

Choose set pressure of relief valve at 100 psig. The relief valve will be located on the shell side.

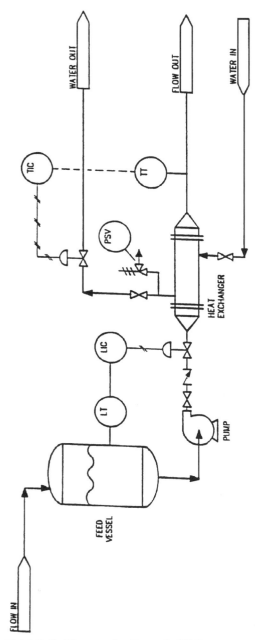

Exhibit 2 Diagram for Example 20.6

Blocked Condition

Because cooling water line has two block valves at the inlet and outlet lines, these valves may be kept closed by mistake while hot process fluid runs through the tube side.

$$\text{\underline{\textbf{Relief Load}}} = \frac{\beta Q}{500GC}$$
$$= \frac{(0.0001)(240,000)}{(500)(1)(1)}$$
$$= 0.048 \text{ GPM}$$

$$\textbf{Relief Area A} = \frac{gpm\sqrt{G}}{23.56 K_p K_w K_v \sqrt{(P_1 - P_b)}} \text{ inch}^2$$

Gpm = relief flow gallons/min = 0.048
G = Sp. gr. ≈ 1
K_p = Capacity correction due to overpressure
 = 1 for 25% overpressure
K_w = Capacity correction factor for back pressure. Assume discharge of relief valve goes to a short distance so that back pressure is small. K_w ≈ 1
K_v = Viscosity correction factor = 1
P_1 = Relief pressure = 10 psig × 1.25
P_b = Back pressure = 0 psig

$$A = \frac{0.048}{23.56 \times 1\sqrt{(125-0)}} = 0.0002 \text{ sq. inch}$$

Heat Exchanger Tube Rupture

P_1 = Design pressure of the low-pressure side = 100 psig
P_2 = Design pressure of the high-pressure side = 650 psig
1.3 P_1 = 130 psig

Because $P_2 > 1.3 P_1$, the condition of tube rupture is to be considered.

$$\text{Flow through one opening of the tube} = 236 d_0^2 C_0 \sqrt{\frac{(\Delta P)}{\rho}}$$

where
 d_0 = 0.62 inch
 C_0 = 0.7
 ΔP = (518 − 1.25 × 100) = (518 − 125) = 393 psi
 ρ = 62.4 lb/cu ft

Flow through one opening = $236(0.62)(0.7)\sqrt{393/62.4}$ = 159.4 GPM
 Because the calculated flow is greater than the normal flow of 88 GPM, the latter is to be used.

$$A = \frac{GPM\sqrt{G}}{23.56 K_p K_w K_v \sqrt{P_1 - P_b}}$$

$K_p = 1$, $K_w = K_v = 1$, $P_1 - P_b = 100 \times 1.25 - 0 = 125$

$$A = \frac{88}{23.56 \times \sqrt{125}} = 0.33 \text{ in.}^2$$

Recommended size = 1G2

Two-Phase Flow Consideration

Two-phase flow has been addressed by Design Institute of Emergency Relief System (DIERS) for sizing relief devices. We consider DIERS technology because it is required by law for all PSM-covered processes.

1. Professional engineers are required to follow RAGAGEP [Recognized And Generally Accepted Good Engineering Practice]. DIERS is one of the recognized and generally accepted good engineering practices.

2. OSHA 1910.119 Appendix C states

   "....the American Institute of Chemical Engineers has published technical reports on topics such as two-phase flow for venting devices. This type of technically recognized report would constitute good engineering practice."

   AIChE also sponsors SuperChems™, a dynamic simulation software for single and two phase flows, reactive and non-reactive systems.

3. EPA 40 CFR part 68 [General Guidance For Risk Management Programs]: Chapter 6 & 7 refer to the Center of Chemical Process Safety (CCPS), which published DIERS technology.
   The latest API RP 521 has now adopted the simplified Omega method for two-phase technology and also suggests alternative methods.

   Direct scale-up of experimental data often results in large relief area, in particular, for gassy reactive systems emphasizing the need of dynamic simulation tools for chemical process safety. Large tanks with non-foamy and non-reactive systems may be sized for all-vapor flow.

Sizing Relief Device for Two-Phase Flow During Runaway Reaction

The sizing of relief device for two-phase flow during runaway reaction requires three steps.

1. Estimation of the Rate of Heat Release per Unit Mass, q

$$q = \frac{1}{2} C_v \left[\left(\frac{dT}{dt} \right)_s + \left(\frac{dT}{dt} \right)_m \right] \quad (20.20)$$

where
q = rate of heat release per unit mass, J/kg · s
C_v = liquid heat capacity per unit mass at constant volume, J · kg · K
$(\frac{dT}{dt})_s$ = rate of temperature rise at the set pressure, K/s
$(\frac{dT}{dt})_m$ = rate of temperature rise at the relieving pressure, K/s

(*Note:* The last two parameters are determined experimentally using the vent sizing package.)

2. Estimation of the Required Mass Flux, G_T

$$G_T = \frac{0.9 \Psi \Delta H_v}{V_{fg}} \sqrt{\frac{g_c}{C_p T_s}} \quad (20.21)$$

where

G_T = mass flux, kg/m² · s
ψ = dimensionless factor dependent on length/diameter ratio of the pipe to relieve the pressure through the relief device. This is shown in Figure 20.8.
ΔH_v = latent heat of vaporization, J/kg
g_c = Newton's law conversion factor, 1 kg · m/s² · N
$v_{fg} = V_g - V_f$, m³/kg
v_g = vapor specific volume, m³/kg
v_f = liquid specific volume, m³/kg
C_p = specific heat capacity of liquid, J/kg · K (Note: 1 J = N · m)
T_s = temperature corresponding to set pressure, K

3. Computation of the Required Vent Area by the Leung Equation

$$A = \frac{m_0 q}{G_T \left[\sqrt{\frac{V}{m_0} \times \frac{\Delta H_v}{V_{fg}}} + \sqrt{C_v \Delta T} \right]^2} \quad (20.22)$$

A = vent area, m²
m_0 = mass within protected vessel before relief, kg
V = volume of the vessel, m³
ΔT = temperature corresponding to relieving pressure − temperature corresponding to set pressure, K

Example 20.7

A 3650-gallon stirred tank herbicide reactor undergoes run-away reaction on loss of cooling medium. The reactor is to be protected by a rupture disc with a set pressure of 100 psig and a relief pressure of 121 psig. The equivalent length-to-diameter ratio of the discharge pipe is 175. Determine the diameter of the rupture disc. The following information is available.

Data

Volume (V): 3,650 gallon = 13.82 m³
Reaction mass (m_o):   9,951 kg
Set temperature (T_s):   421.7 K

Data from VSP

Maximum temperature (T_m): 430.34 K
$(dT/dt)_s$:   0.876 K/s
$(dT/dt)_m$:   1.893 K/s

Physical Property Data

	100 psig	121 psig
V_f (liquid) m³/kg	0.001	0.0012
V_g (vapor) m³/kg	0.07454	0.0654
C_p (liquid) kJ/kg K	1.8634	1.9152
ΔH_v kJ/kg	392.82	380.80

Solution

Step 1. Estimate rate of heat release per unit mass.

$$q = (1/2) \, C_v \, [(dT/dt)_s + (dT/dt)_m]$$

Assuming $C_v = C_p$
$$q = (1/2)(1.8634)[0.876 + 1.893]$$
$$= 2.5798 \text{ kJ/kg} \cdot \text{s}$$
$$= 2579.8 \text{ J/kg} \cdot \text{s}$$

Step 2. Estimate the required mass flux.

$$G_T = 0.9 \Psi \frac{\Delta H_v}{V_{fg}} \sqrt{\frac{g_c}{C_p T_s}}$$

Assume: L/D of discharge pipe = 175, $\psi = 0.7$ from Figure 20.6.

$$G_T = 0.9(0.7)\frac{392820}{(0.07454 - 0.001)}\sqrt{\frac{1}{1863.4 \times 421.7}}$$
$$= 3796.257 \text{ kg/m}^2 \cdot \text{s}$$

Step 3. Estimate vent area and hence relief diameter.

$$A = \frac{m_0 q}{G_T \left[\sqrt{\frac{V}{m_0} \times \frac{\Delta H_v}{V_{fg}}} + \sqrt{C_v \Delta T}\right]^2}$$

$V_{fg} = V_g - V_f = 0.07454 - 0.001 = 0.07354 \text{ m}^3/\text{kg}$
$\Delta T = T_m - T_s = 430.34 - 421.7 = 8.64 \text{K}$

$$A = \frac{(9951)(2579.8)}{3{,}796.257 \left[\sqrt{\frac{12.82}{9.951} \times \frac{392820}{0.07354}} + \sqrt{1863.4 \times 8.64}\right]^2}$$
$$= 0.149 \text{ m}^2$$

Allowing 20 percent,

$$d = (4 \times 0.149 \times 1.2/3.14)^{0.5} = 0.4773 \text{m}$$

Venting Capacities of Atmospheric Tanks

In addition to the emergency relief requirements, the flammable liquid storage tanks need to be equipped with a vent to allow the tank to breathe during filling and emptying operations. Another purpose of venting the tank is to normalize internal tank pressures caused by expansion and contraction of the liquid in the tank due to temperature changes (thermal changes). API-RP2000 provides guidelines for calculating the venting capacities to be provided in case of both thermal and liquid movement situations. For this purpose, the flammable liquids are classified into two groups: (1) liquids with flash points below 100°F and (2) liquids with flash points above 100°F. API guideline provides the requirements of venting

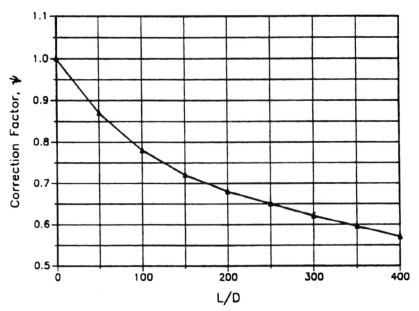

Figure 20.6 Correction factor, ψ, for two-phase flashing flow corresponding to equivalent length to diameter ratio of pipe. Crowl, Daniel A., Louvar, Joseph F., *Chemical Process Safety: Fundamentals with Applications*, 2nd edition, © 2002, p. 397. Reprinted by permission of Pearson Education, Inc., Upper Saddle River, NJ.

for normal breathing, which includes both thermal venting as well as venting required during emptying or filling of the vessel.

Thermal Venting

The venting capacities required during breathing-in and breathing-out due to temperature changes (thermal venting) are given in Table 20.11.

Venting Requirements Due to Filling and Emptying

The venting capacities required in each case are as follows:

Liquids with flash points below 100°F.

1. Out-Breathing (Pressure) Capacity—1 ft^3/h of air for each 3.5 gallons per hour of tank filling rate.

2. In-Breathing (Vacuum) Capacity—1 ft^3/h of air for each 7.5 gallons per hour of tank emptying rate.

Liquids with flash points above 100°F.

1. Out-Breathing (Pressure) Capacity—1 ft^3/h of air for each 7 gallons per hour of the tank filling rate.

2. In-Breathing (Vacuum) Capacity—1 ft^3/h for each 7.5 gallons per hour of the emptying rate.

Normal breathing rate is the sum of the thermal venting rate as given in Table 20.11 plus that calculated due to either filling or emptying the tank.

Table 20.11 Thermal venting capacities (ft³/h of air) as a function of tank capacity

Tank Capacity		Vacuum	Pressure	
Gallons	42 gal. Barrels	All Stocks	Flash Point Below 100° F	Flash Point Above 100° F
2500	60	75	75	—
5000	119	125	125	60
10,000	238	250	250	137
15,000	357	350	350	200
20,000	476	480	480	270
25,000	595	600	600	350
30,000	714	730	730	425
35,000	833	850	850	500
42,000	1000	1000	1000	600
84,000	2000	2000	2000	1200
126,000	3000	3000	3000	1800
168,000	4000	4000	4000	2400
210,000	5000	5000	5000	3000
420,000	10,000	10,000	10,000	6000
630,000	15,000	15,000	15,000	9000
840,000	'20,000	20,000	20,000	12,000
1,050,000	25,000	24,000	24,000	15,000

*Do not interpolate figures. Take larger size.

Example 20.8

Calculate the normal venting capacity for a 65,000-gallon vertical cone-roofed tank holding naphthalene (flash point 175° F) with 150 gpm filling rate and 400 gpm pump-out rate.

Solution

Pressure Venting

$$\text{Required venting for normal breathing} = \frac{150 \times 60}{7} + 1200 = 2486 \text{ ft}^3/\text{h}*$$

Vacuum Venting

$$\text{Requirements for normal breathing} = \frac{400 \times 60}{7.5} + 2000 = 5200 \text{ ft}^3/\text{h}*$$

When evaporation losses are not a factor, vents without pressure and vacuum relief valves are used. These vents are often referred to as non-conservation or free venting type vents. Pipe sizes should be selected to provide required pressure and vacuum relief under operating and emergency conditions. If it is necessary or desirable to reduce evaporation losses of the stored liquid, vents with pressure

---

*Note that the thermal venting value is not obtained by interpolation from Table 20.11, but the value for the next larger size tank is used.

and vacuum relief valves are provided. These are called conservation vents. Flame arresters should be provided on storage tanks under the following conditions:

1. On tanks containing liquids with flash points below 110° F.

2. Tanks containing liquids with flash points above 110° F if exposed to combustible occupancy, construction, or other tanks that contain liquids having flash points below 110° F.

3. Tanks where the bulk of the contents can be heated to the flash point under normal operating conditions.

Deflagration Venting

A deflagration vent is an opening in an equipment or enclosure through which pressure is relieved to prevent mechanical failure of the protected equipment. With no vent, the maximum pressure developed during a deflagration may be 6 to 10 times the initial absolute pressure. The need for deflagration venting may be eliminated[11a] by the application of prevention techniques presented in NFPA-69, Standard on Explosion Prevention Systems. The standard for deflagration vent sizing is NFPA-68, which defines two types of enclosure:

1. Low Strength Enclosure: Enclosures that are capable of withstanding reduced pressure, P_{red}, of not more than 1.5 psig (0.1 barg). Typically low strength enclosures include industrial buildings, dust collectors, etc.

2. High Strength Enclosure: Enclosures that are capable of withstanding reduced pressure, P_{red}, of more than 1.5 psig (0.1 barg). Typically high strength enclosures include vessels, silos, etc.

Venting Deflagrations for Gas Mixtures and Mists
Venting Equation for Low Strength Enclosure

$$A_v = \frac{C(A_s)}{P_{red}^{1/2}} \tag{20.23}$$

where
A_v = Area of vent, ft², [m²]
C = Venting equation constant (See Table 20.12), psig$^{1/2}$, bar$^{1/2}$
A_s = Internal surface area of enclosure of buildings, ft², [m²]
P_{red} = Reduced pressure
    = Maximum pressure developed in a vented enclosure in a vented deflagration test, psig, [barg]. P_{red} not to exceed enclosure strength, P_{es}, [0.1 barg or 1.5 psig]. P_{red} must exceed P_{stat} by 0.02 bar (0.35 psi).

Table 20.12 Venting equation constant for Equation 20.23

Fuel	C in SI or Metric, Bar$^{1/2}$	C in AES, psig$^{1/2}$
Anhydrous ammonia	0.013	0.05
Methane	0.037	0.14
Gases with fundamental velocity less than 1.3 times that of propane	0.045	0.17

Note: This formula applies when there is no vent duct, simply the deflagration panel is used. When a duct is used, please consult NFPA-68[11b].

Venting Equation of Gas or Mist Deflagrations in High Strength Enclosure

$$A_v = \left[(0.127 \log_{10} K_G - 0.0567) P_{red}^{-0.582} + 0.175 P_{red}^{-0.572} (P_{stat} - 0.1) \right] V^{2/3} \quad (20.24)$$

where

A_v = vent area, m²
K_G = Deflagration index of a gas cloud, bar · m/s, ≤ 550 bar · m/s
P_{red} = The maximum pressure developed during a vented enclosure during a deflagration test, bar, ≤ 2 bar and at least 0.05 bar > P_{stat}
P_{stat} = Static activation pressure, bat, ≤ 0.5 bar, pressure that activates (breaks) the vent enclosure (panel) when the pressure is increased slowly, at a rate < 0.15 psi/min
V = Enclosure volume, m³, ≤ 1000 m³

Other Constraints

1. The length/diameter of equipment is 2 or less.

2. Initial pressure is atmospheric.

3. No vent duct.

When the above constraints are violated, consult NFPA-68.

Venting of Deflagrations of Dusts and Hybrid Mixtures

The following equation is applicable for all enclosures, low and high strengths, for dusts and hybrid mixtures (containing moisture):

$$A_v = \left[8.535(10^{-5})(1+1.75 P_{stat}) K_{st} V^{0.75} \right] \sqrt{\frac{(1-\Pi)}{\Pi}} \quad (20.25)$$

where

A_v = vented area, m²
P_{stat} = Static burst pressure of vent, bar
K_{st} = Deflagration index, bar · m/s
V = Hazard containment volume, m³
$\Pi = \frac{P_{red}}{P_{max}}$
P_{red} = reduced pressure after deflagration (bar)
P_{max} = maximum pressure of deflagration (bar)

Constraints of the Preceding Equation

1. The length/diameter of vessel < 2

2. Initial pressure before ignition is 1 bar +/− 0.2 bar.

3. 5 bar ≤ P_{max} ≤ 12 bar

4. 10 bar · m/s ≤ K_{st} ≤ 800 bar · m/s

5. 0.1 m³ ≤ V ≤ 10,000 m³

6. No vent duct

HAZARD AND OPERABILITY STUDIES (HAZOP)

After several large disasters in industries worldwide, OSHA promulgated Process Safety Management (PSM) program (29CFR 1910.119) and EPA promulgated 40 CFR Part 68, the Risk Management Program (RMP). The word "Process" in PSM connotes using, storing, manufacturing, handling, or moving of highly hazardous chemicals as separate activities or in combination. A key provision in PSM is Process Hazard Analysis (PHA). It is a careful and critical review of what could go wrong in a plant.

The Hazard and Operability Studies (HazOp) involves investigating how a plant may run into a hazardous situation through deviations from the intended operating conditions. Generally, a group of people knowledgeable in process engineering, process chemistry, process control, process safety, process operation, mechanical integrity of equipment, and maintenance interact to identify hazardous situations. Generally no attempt is made in the HazOp study to solve a newly discovered problem unless a solution is apparent, and only problems are documented with an action list with duties assigned to knowledgeable person(s) to solve the problems at target dates. It is the responsibility of the facilitator of the HazOp meeting to generate the action list, but it is the responsibility of the project engineer to ensure that the action list is followed through.

The necessary documents generally consist of, but are not limited to,

1. Piping and Instrumentation Diagram (P&ID)
2. Process Flow Diagram (PFD), when applicable
3. Layout
4. Material and Safety Data Sheets (MSDS) of all chemicals
5. Standard Operating Procedures (SOP)
6. Reactivity Data, when applicable
7. Corrosivity data, when applicable
8. Compatibility matrix of chemicals, when applicable
9. Explosive Data, when applicable
10. Instrument sequence control, when applicable
11. Logic diagram, when applicable
12. Fabrication drawings, when applicable.

Terminology

Study Node. These are locations in a P&ID, which are considered for HazOp study. For example, the line with all instrumentation from a feed tank supplying an acid to a reactor may be considered a node.

Intentions. The intention of a node defines how the node is expected to operate in the absence of deviation. This may require such documents as P&ID, standard operating procedure, PFD, etc.

Deviations. These are departures from normal operation, which are addressed by the guide words as shown in Table 20.12.

Table 20.13 Guide word table

Guide Word	Meaning	Example
No or Not	Negation of design intent	No flow, no agitation, no inert gas
Less	Quantitative decrease	less flow, lower pressure due to pump out, less catalyst charge
More	Quantitative increase	More than batch amount, more pressure
As well as	All of the intentions achieved, but some additional activities occur	Amount of a dose as well as concentration of the active ingredient of the dose increases
Part of	Only part of the intention is achieved	One component of a mixture supplied by an outside company is missing.
Reverse	Logical opposite of intent	Reverse flow, reverse order of addition
Other than	Complete substitution. No part of the intention achieved.	Wrong component

Causes. These are reasons, which lead to the deviations.

Consequences. These are the results of the deviations.

Guide Words. These are words used to identify potential problems in the intention for which the node has been designed.

Consider a continuous reaction process in Figure 20.7. Reactant A is continuously added with reactant B to the reactor to produce product. If too little of A is added, or it is contaminated with high concentration of iron, potential runaway reaction may take place; if too much of A is added, the result is safe but off-quality product is formed. HazOp study of node 1 is illustrated.

The action list is extracted from the studies and assignments with deadlines given to the appropriate personnel. The HazOp analyses offer qualitative results.

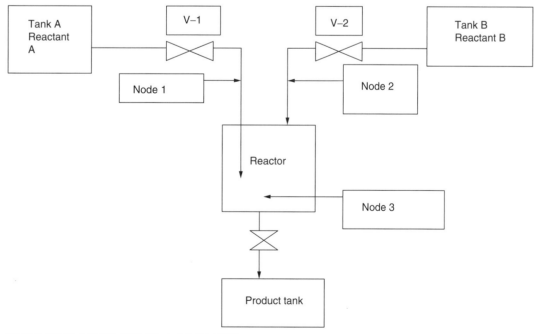

Figure 20.7 HazOp analysis example P&ID

Table 20.14 Node analysis for HazOp: Node 1: process parameter: flow

Guide word	Deviation	Consequence	Causes	Suggested Action
No	No flow	Non-reacted B, possible runaway reaction if flow stops suddenly.	(1) Valve fails closed (2) Exhausted Tank A (3) Line plugs	(1) Automate closure of valve 2 on closure of valve 1 (2) Level control & low level alarm on Tank A
Less	Less flow	Partially reacted B, possible runaway reaction.	(1) Partially closed valve (2) Partial plugging of line	(3) Flow measurements of A and B and ratio control. (4) Valve % opening indicator (5) Size emergency relief device for runaway reaction.
More	More flow	Off-quality product, no process hazard	—	—
Part of	Flow is OK, but concentration of A is lower	Excess B, potential runaway reaction.	Quality control check fails.	(6) Install in-line analyzer to check concentration
Reverse	Flow from reactor to Tank A	Potential runaway reaction in Tank A and reactor	Pressure in reactor increases due to runaway reaction	(7) Install pressure transmitter with high pressure alarm and interlock, and double on/off valves in series on node 1 to shut off supply of A on high pressure in reactor.
Other than	Wrong chemical in Tank A	Unpredictable runaway reaction.	Wrong shipment and/or wrong hook-up	(8) Organizational measure to certify the composition of chemical upon arrival, and proper hook-up to the designated receiving tank.
As well as	Flow normal, some contaminant in tank	Unpredictable runaway reaction	Contaminated shipment	See (6) and (8)

Besides HazOp studies, there are other methods of analyses:

1. What-If Analysis: This method is not very structured, but is very popular among industries. It includes such questions as (a) What if the valve gets locked in the wrong position, (b) What if the operator opens the wrong valve, etc. Results are qualitative.

2. Failure Modes, Effects, and Criticality Analysis (FMECA): The method is used to identify equipment/system failure modes and their effect on the system or plant in a tabular method. Results are qualitative.

3. Fault Tree Analysis (FTA): It is used to identify combinations of equipment failures and human error that can result in an accident event through a deductive technique. Results are qualitative.

4. Event Tree Analysis: A qualitative analysis to identify the sequences of events, following an initiating event that results in accident.

Finally, hazard quantification or hazard analysis (HAZAN) is done through quantitative estimation of expected frequencies, their consequences, and the cost of consequences

Emergency Ingress and Egress

OSHA Standards- 29 CFR 1910.146 addresses the procedure of ingress into permit-required confined spaces covered in details. It defines the acceptable entry conditions, the duties of an attendant, the authorized entrant, requirement of blanking or blinding a pipeline, ventilation, inerting, etc.

OSHA Standards- 29 CFR 1926.34 addresses the means of egress as follows:

(a) "General." In every building or structure exits shall be so arranged and maintained as to provide free and unobstructed egress from all parts of the building or structure at all times when it is occupied. No lock or fastening to prevent free escape from the inside of any building shall be installed except in mental, penal, or corrective institutions where supervisory personnel is continually on duty and effective provisions are made to remove occupants in case of fire or other emergency.

(b) "Exit marking." Exits shall be marked by a readily visible sign. Access to exits shall be marked by readily visible signs in all cases where the exit or way to reach it is not immediately visible to the occupants.

(c) "Maintenance and workmanship." Means of egress shall be continually maintained free of all obstructions or impediments to full instant use in the case of fire or other emergency.

ENVIRONMENTAL CONSIDERATIONS

Pollution is a by-product of civilization, because it is thermodynamically impossible to eliminate all wastes from the processes involved in the transformation of matter and energy required to sustain a civilized society. In the name of civilization and progress, however, humans have a tendency to rationalize defecation in their own nests.

Philosophically, pollution is misplacement of a substance in a concentration or amount that is directly or indirectly hazardous to health or life in any form. Technically, only when such misplacement violates the local or federal laws, is it called pollution. However, misplacement of substances without violating laws does not necessarily mean a healthy atmosphere.

Pollution prevention should not be left as an "end-of-pipe" problem. It should be addressed at the stage of conceptual design. Careful consideration should be given to the waste treatment cost, which is affected by the choice of solvents, starting raw materials, operating parameters and their mode of control, and work practices, including detection, monitoring, reduction, and reporting. Most pollution can be prevented, or at least minimized, when the best available control technology is considered in the beginning of the conceptual design with support from management guided by enlightened self-interest.

FEDERAL POLLUTION PREVENTION ACT OF 1990

The central points of the Federal Pollution Prevention Act of 1990 can be summarized as follows:

1. Pollution should be prevented or reduced at the source wherever possible.
2. Pollution that cannot be prevented should be recycled in an environmentally safe manner whenever feasible.
3. Pollution that cannot be prevented or recycled should be treated in an environmentally safe manner whenever feasible.
4. Disposal or other release into the environment should be considered only as a last resort and should be conducted in an environmentally safe manner.

The Act defines the following categories of waste:

1. Municipal solid wastes such as paper, glass/metal, food, plastics, etc. Disposal involves landfill, incineration, recycling, etc.
2. Industrial hazardous wastes. Disposal involves recycle, landfill, incineration, land treatment, reuse as fuels, underground injection, waste piles, waste treatment and controlled discharge to natural water, and air.

Definitions

Following are brief definitions of some key terminology used in pollution prevention.

Acceptable Daily Intake (Dose). It is the number of milligrams of a chemical per kilogram of body weight that, taken daily during an entire lifetime, appears to be without risk on the basis of all known facts at the time.

Acid Rain. All rainfalls are slightly acidic because of natural carbon dioxide. However, acid rain refers to the formation of acids due to the reaction of acidic oxides, such as oxides of sulfur and oxides of nitrogen with rainwater.

Activated Carbon. Carbon obtained from vegetable and animal sources and roasted in a furnace to develop high surface area (1000 m^2/gm) is called activated carbon.

Activated Sludge. The activated sludge is formed in the biological treatment of wastewater which is than mixed with biological culture (bacteria) in an aeration basin. Because of the growth of the bacteria and protozoa, part of the active sludge (about 90%) is incinerated, or disposed of as landfill and balance recycled.

Biochemical Oxygen Demand (BOD). It is the number of milligrams of dissolved oxygen consumed by one liter of waste for a specified incubation period in days at 20°C. The incubation period is shown as a suffix. For example, BOD_5 denotes BOD with a 5-day incubation period.

Carcinogens. Any substance that induces cancer to man or animal is called a carcinogen. A material is considered a carcinogen if (1) it is certified as a carcinogen or potential carcinogen by IARC (International Agency for Research on Cancer), (2) it is listed as a carcinogen or potential carcinogen in the *Annual Report on Carcinogen*, (3) it is regulated by OSHA as a carcinogen, or (4) a positive study has been published. Some industrially known carcinogens are asbestos, aldrin, benzene, beryllium, cadmium, carbon tetrachloride, chloroform, p-dichlorobenzene, dieldrin, DDT, formaldehyde, hexamethylenediamine, selenium, tetraethyl lead, toluene-24-diamine, trichloroethylene, and vinyl chloride.

Chemical Oxygen Demand (COD). It is the number of milligrams of oxygen which one liter of waste will absorb from a hot acidic solution of potassium dichromate. The higher the values of BOD and COD, the more dangerous the waste is to aquatic life, because the waste depletes oxygen from the water required for sustaining life. A substance with low BOD may mean that substance is not biodegradable.

Fugitive Emission. Fugitive emissions are unintentional releases from equipment such as pumps, valves, flanges, open-ended lines, filling losses, evaporation from sumps or ponds, and so on. The Environmental Protection Agency has five methods to estimate fugitive emission.

Greenhouse Effect. Combustion of fossil fuel (coal, oil, natural gas) generates carbon dioxide (about 20 billion tons/yr), which forms a layer in the earth's atmosphere. This layer traps heat radiating from the earth's surface, thereby leading to the global warming effect. Potential effects may be melting of polar ice leading to flooding in the coastal areas. Other contributors to the Greenhouse Effect are CFCs, methane, and ozone. The role of chemical engineers here is to develop alternative fuels, refrigerants, and propellants.

Hazardous Air Pollutant (HAP). Pollutants are hazardous if they are carcinogens or if exposure to them may cause serious health problems. The Clean Air Act Amendment of 1990 includes 189 HAPs.[10]

Lethal Concentration and Dose. A concentration or dose that causes death is called the lethal concentration or dose. Lethal concentration, LC_{50}, is the concentration of the substance in mg per liter that causes death within 96 hrs to 50 percent of the test group of the most sensitive important species in the locality under consideration. Lethal dose, LD_{50}, is the dose of the substance in mg per kg of body weight of a specific animal (such as a rat) that causes death to 50 percent of the sample of the animals under test. A substance with LD_{50} greater than 7 g/kg is not considered harmful.[9]

Life Cycle Analysis. Life cycle analysis of a product considers all impacts on the environment starting from the procurement of the raw material, to the generation, use, and disposal of the product.

MSDS

The Material Safety Data Sheet (MSDS) is to be provided by all facilities handling a hazardous substance. This is required by the Emergency Planning and Right to Know Act of 1986 in the USA. The data sheet contains the following information.

1. Product identification: Name, chemical formula, etc.
2. Hazardous ingredients.
3. Physical data such as appearance, color, molecular weight, boiling point, melting point, etc.
4. Fire and explosion data such as flash point, flammable limits, etc.
5. Health effects data.
6. Emergency and first aid procedure
7. Employee protection recommendation.
8. Reactivity data such as stability, polymerizability, incompatibility, and hazardous decomposition products.
9. Spill or leak procedure.
10. Special precautions and storage data.
11. Shipping data.
12. Documentation stating reason of issue, data of approval, etc.

Mutagen

Anything, such as a chemical or radiation, that changes the chromosomes of a cell so that the chromosomes of the daughter cells will be changed after cell division, is called a mutagen. A mutagen may be carcinogen or teratogen (an agent that causes birth defects: dioxin, for example).

pH

The definition of pH: It is the negative logarithm to the base 10 of hydronium ion activity.

$$pH = -\log(a_{H_3O^+}) = -\log_{10}(\gamma C_{H_3O^+}) \cong -\log_{10} C_{H_3O^+} \cong -\log_{10} C_{H^+} \quad (20.26a)$$

where
    a = activity
    γ = activity coefficient, usually 1 for dilute solution
    C = concentration in mols per liter
    H_3O^+ = hydronium ion
    H^+ = hydrogen ion

The pH for water-based solutions is normally expressed in a scale of 0 (1 normal acid) to 14 (1 normal base). This scale is arbitrary, and therefore, pH could be negative (stronger than 1 normal acid) and also greater than 14 (stronger than 1 normal base). A pH of 7 indicates a neutral solution. Because of the base 10 in logarithm scale, for example, a solution of pH at 6 is 10 times as acidic compared with a solution of pH at 7. The pH of pure water is 7 only at 25°C, and it decreases as the temperature increases because of increased dissociation. The pH of a solution is affected by the nature of the solvent, and even by neutral salts. This is because the activity coefficient is dependent on the ionic strengths of all ions, not just hydronium ions. Thus,

$$\log_{10} \gamma_{H_3O^+} = \frac{-0.5 I^{0.5}}{1 + 3 I^{0.5}} \tag{20.26b}$$

where
  I = ionic strength = $1/2 \Sigma (C_i Z_i^2)$, where the subscript denotes the ionic species, and Z denotes the ionic charge.

The acceptable pH levels of various waters[7] are as follows:

- Irrigation (4.5–9),
- Municipal secondary treatment waste, and fresh water aquatic life and wildlife (6–9),
- Public supply and recreational water (5–9), marine water (6.5–8.5).

Total Organic Carbon (TOC)

It is number of milligrams of soluble organic carbon per liter of waste. Before the analysis, the inorganic carbon materials are removed by acidification and sparging, and insoluble solids are filtered. The test involves analysis of carbon dioxide generated by burning the sample.

Volatile Organic Compounds (VOC)

Any compound of carbon, *excluding* carbon monoxide, carbon dioxide, carbonic acids, metallic carbides or carbonates, and ammonium carbonate, which participates in atmospheric photochemical reactions is termed a VOC.

OZONE, FRIEND AND FOE

Ozone is a gas consisting of three atoms of oxygen. Depending on where it exists, it can be beneficial or harmful.

Beneficial Ozone and Its Depletion

Ozone that is created naturally at a distance of 15 to 20 miles above the surface of the earth protects life on the earth by absorbing a major portion of cancer-causing ultraviolet radiation and, in so doing, releases heat in the stratosphere. CFCs (chlorofluorocarbons used as refrigerant, fire extinguisher, and propellant) escape undecomposed to the stratosphere, where, bombarded with ultraviolet radiation, they release halogens that attack ozone, breaking it down to oxygen. Nitrogen oxides released by supersonic jets also destroy ozone. Thus depletion of friendly ozone takes place, allowing more of the ultraviolet radiation to reach the surface of the earth.

The chemistry of stratospheric ozone involves a cyclical process of photochemical reaction of formation and decomposition of ozone:

(a) Formation of ozone:

$$O_2(g) + hv \rightarrow 2\ O(g)$$
$$O(g) + O_2(g) + M(g) \rightarrow O_3(g) + M^*(g) + \text{heat}$$

(b) Decomposition of ozone:

$$O_3(g) + hv \rightarrow O_2(g) + O(g)$$
$$O(g) + O(g) + M(g) \rightarrow O_2(g) + M^*(g) + \text{heat}$$

The chemistry of ozone depletion depends on the type of pollutant. With a CFC it is as follows:

(a) Photolysis or light-induced rupture of the carbon-chlorine bond and generation of atomic chlorine:

$$CF_xCl_{4-x}(g) + hv \rightarrow CF_xCl_{3-x}(g) + Cl(g)$$

(b) Reaction of atomic chlorine with ozone leading to the depletion of the latter through regeneration of atomic chlorine:

$$Cl(g) + O_3(g) \rightarrow ClO(g) + O_2(g)$$
$$ClO(g) + O(g) \rightarrow Cl(g) + O_2(g)$$

In the preceding equations, hv represents photon energy (h = Planck's constant, v = frequency of radiation), M = N_2, O_2, or H_2O, M^* = molecule M with excess energy.

Harmful Ozone and Its Creation

Although beneficial at the stratospheric level, ozone is dangerous at ground level. It is extremely reactive and toxic to breathe. It causes bronchitis and is extremely dangerous to asthma sufferers. Ozone is the key component of *smog*. Ozone is created by the oxides of nitrogen that are produced by automobiles, planes, chemical plants, burners, and so on. Oxides of nitrogen decompose by the bombardment of infiltrated ultraviolet radiation to nascent oxygen, which reacts with molecular oxygen to produce ozone. Ozone is a suicidal byproduct of civilization.

The chemistry of the formation of oxides of nitrogen (NO x) and ozone in engines and furnaces are as follows:

$$N_2(g) + O_2(g) = NO(g) \quad \text{[Inside internal combustion engine or furnace]}$$
$$2NO(g) + O_2(g) = NO_2(g) \text{ [Inside IC engine, furnace and outside]}$$
$$NO_2(g) + hv \rightarrow NO(g) + O(g) \quad \text{[Outside air]}$$
$$O(g) + O_2(g) + M(g) \rightarrow O_3(g) + M^*(g)$$

ABATEMENT OF AIR POLLUTION

Let us look briefly at methods of abating both VOC and non-VOC air pollution.

VOC Control Technologies

- *Thermal Oxidation (Incineration).* By this technology, the VOC-laden air is burned at 1300°F – to 1800°F using supplementary fuel and air, if required, to produce water and carbon dioxide.

- *Catalytic Oxidation (Incineration).* This method is the same as thermal oxidation except a catalyst is used to lower the temperature of the reaction to 700°F – 900°F, consequently reducing the fuel consumption.

- *Adsorption.* This technology uses a medium such as activated carbon or zeolite to adsorb VOC by weak intermolecular forces. The adsorbed VOC may be recovered by steam stripping or vacuum desorption and condensation.

- *Absorption.* By this method, a liquid solvent, such as water, high boiling hydrocarbons, caustic solutions (for acidic VOC), or acid, may be used in a venturi, packed/tray/spray columns to capture VOC.

- *Condensation.* This technique uses low-temperature coolants, such as brine, cryogenic fluid, or chilled water, in a surface condenser. Cryogenic fluids, like liquid nitrogen and carbon dioxide, can be directly injected into the VOC stream to effect condensation (direct contact condensation).

- *Waste Heat Recovery Boiler.* In this process, the VOC-laden air is burned in boilers to generate steam.

- *Flare.* A flare, which is generally used to control pollution during process upsets, may also be used to burn VOC-laden air.

- *Bio-Degradation.* This process uses soil or compost beds containing cultured microorganisms to convert VOCs into harmless components. The VOC-laden air must be dust free and humid.

- *Membrane Separation.* Semipermeable membranes are used to separate VOC from air due to selective permeability of the gases.

- *Ultraviolet Oxidation Technology.* By this technology, the VOC is converted to carbon dioxide and water by oxygen-based oxidants, including ozone, peroxides, and radicals such as OH and O, in the presence of ultraviolet light.

Non-VOC Control Technologies

Non-VOC air pollutants include carbon monoxide, carbon dioxide, halogens, halogen acids, nitrogen oxides, ammonia, and sulfur compounds. Many of the treatment technologies outlined for VOCs also apply for pollution control of non-VOCs.

Control of oxides of sulfur includes the following methods:

- Lime/limestone scrubbing

- Alkaline-fly-ash scrubbing

- Sodium carbonate/bicarbonate/hydroxide scrubbing

- Magnesium oxide scrubbing
- Sodium sulfite regenerative process (Wellman-Lord Process)
- Citrate process, which uses aqueous solution sodium sulfite, bisulfite, sulfate, thiosulfate, and polythionate buffered by citric acid
- Dimethyl aniline/xylidine process, which uses the compounds as a solvent to absorb sulfur dioxide
- Claus process, which uses hydrogen sulfide to react with sulfur dioxide to produce elemental sulfur.

ABATEMENT OF WATER POLLUTION

Water pollutants are classified as follows:

1. Floating pollutants, such as oil
2. Suspended pollutants, such as organic or inorganic solids
3. Dissolved pollutants, such as acids, or organic and inorganic substances.

Treatment processes for water pollution include the following:

1. Pretreatment, such as air floatation, pH adjustment, coagulation, precipitation, and flocculation
2. Clarification, such as sedimentation and removal of suspended solids
3. Filtration, such as cartridge filter, or drum, and plate-and-frame filter
4. High gradient magnetic separation (HGMS) for separation of magnetic, weakly magnetic, or non-magnetic particles
5. Aerated lagoons used for treating small quantities (<1 million gallons/day) of wastewater with BOD (150–300 mg/L)
6. Biological treatment of organic wastewater using microorganisms that convert the organics into carbon dioxide, water, and methane gas. The most common types of biological processes are activated sludge (aerobic digestion, converts organic matter to carbon dioxide and water), trickling filter, fixed activated sludge treatment (a hybrid of activated sludge and trickling bed), anaerobic digestion (converts organic matter to carbon dioxide and methane), rotating biological contactor, and nitrification. Nitrification is a biological treatment involving ammonium waste, which is converted to nitrates by two types of bacteria: nitrosomonas, which oxidizes ammonium to nitrite, and nitrobacter, which oxidizes nitrites to nitrate. Air in the aerobic digestion may be replaced by pure oxygen (UNOX process).
7. Carbon adsorption
8. Ion exchange
9. Reverse osmosis, electrodialysis, and ultrafiltration
10. Incineration
11. Stripping, as applied to ground water

12. Liquid-liquid extraction
13. Chlorination
14. Ozone treatment for organic and inorganic waste
15. Freeze concentration, evaporation, and crystallization
16. Oxyphotolysis, a combination of ozone treatment and ultraviolet light used for toxic organics, such as malathion
17. Carbon-catalyzed hydrogen peroxide treatment to remove cyanide.

TREATMENT OF CONTAMINATED SOIL

There are two main methods of treating contaminated soil:

1. Bioremediation: treatment of contaminated soil with microorganisms
2. Incineration.

PERSONAL PROTECTIVE EQUIPMENT (PPE)

The purpose of personal protective equipment is to protect a person from safety and health hazards, which cannot be engineered out of the working environment. The PPE is designed to protect vital body parts such as head, face, ears, eyes, hands, and feet.

The OSHA governances of PPE are identified in Table 20.15.

Besides the preceding list, hearing protection using earplugs, canal caps, or earmuffs are required against occupational noise exposure per OSHA 1910.95. Some salient features of this section of the OSHA governance are as follows.

Table 20.15 Personnel protective equipment governance

Area	OSHA Governance #	Brief Application
General requirement	1910.132	Hazard assessment and equipment selection, defective & damaged equipment, training, when & what PPE are necessary, how to properly don, doff, adjust, and wear PPE; care, maintenance, and disposal of PPE.
Eye and face protection	1910.133	Specifies general requirement of safety glasses, face-shields, and acceptability of clip-on protectors.
Respiratory protection	1910.134	Addresses responsible practice, employer responsibility, minimal acceptable program, storage, inspection, cleaning, repair, and selection according to the environment.
Head protection	1910.135	Addresses general requirement, & criteria of protective helmets. Helmets are required to withstand an impact of 8 lbs dropped from 5 ft height. There are three types of helmets: Class A with electric shock resistance of 2,200 volts, Class B (20,000 volts), Class C (resistant to falling objects, not electrical shock resistant).
Foot protection	1910.136	Addresses general requirement and protection from falling or rolling objects, objects piercing sole, and conformance to ANSI Z41-1991.
Electrical Protective Equipment	1910.137	Addresses design requirement of insulating blanket, matting, covers, line hose, gloves, sleeves, their electrical requirement, manufacture and marking.
Hand protection	1910.138	Specifies general requirement to prevent skin absorption, cuts, laceration, abrasion, puncture, burn, and selection of protective equipment.

NOISE MANAGEMENT

Noise Power Level

The noise power level in absolute decibels (dBA) is given by

$$Lw = 10 \log_{10}(W/W_0) \qquad (20.27)$$

where
- Lw = Sound power in dBA
- W = Intensity of sound wave, Watt/m^2
- W_0 = Threshold of human hearing, 10^{-12} Watt/m^2

If $Lw_1, Lw_2, \ldots$ are the noise power levels of machine 1, 2, etc., the combined noise power level of the machines is given by

$$Lw(\text{combined}) = 10 \log_{10}(10^{Lw1/10} + 10^{Lw2/10} + \cdots) \qquad (20.28)$$

Example 20.9

In a workshop that has a noise level of 60 decibels, a new machine with a noise level of 70 decibels is moved in. What will be the combined noise level?

Solution

$$\begin{aligned}
Lw(\text{combined}) &= 10 \log_{10}(10^{60/10} + 10^{70/10}) \\
&= 10 \log_{10}(10^6 + 10^7) \\
&= 10 \log_{10}[10^6(1 + 10)] \\
&= 10 \log_{10}(10^6) + 10 \log_{10}(11) \\
&= 60 + 10.41 \\
&= 70.41 \text{ dBA}
\end{aligned}$$

Noise Pressure Level

The noise pressure level (Lp) in absolute decibels (dBA) is given by:

$$Lp = 20 \log_{10}(P/P_0) \qquad (20.29a)$$

where
- P = root-mean-square value of sound pressure in psi
- P_0 = 0.0002 microbar = 3×10^{-9} psi

Estimation of Noise Pressure Level From Known Data

To estimate the noise pressure level (Lp) at a distance greater than that at which it was measured, use the following formula:

$$L_{p[\text{At Desired Distance}]} = L_{p[\text{At Known Distance}]} - 20\log_{10}\left[\frac{\text{Desired Distance}}{\text{Known Distance}}\right] \qquad (20.29b)$$

Example 20.10

Estimate the noise pressure level at a distance of 100 ft. when the noise pressure level at 3 ft. from the source is 120 dB.

Solution

$$\begin{aligned}
Lp(100 \text{ ft}) &= Lp(3 \text{ ft}) - 20\log_{10}(100/3) \\
&= 120 - 20\log_{10}(100/3) \\
&= 89.5 \text{ dB}
\end{aligned}$$

OSHA Permissible Noise Exposure (OSHA 1910.95)

The exposure time allowed by OSHA in an environment of noise pressure level L is given by the following formula:

$$T = 8/[2^a] \qquad (20.30)$$

where
$a = 0.2(L-90)$
T = Permitted continuous exposure time, hr
L = Noise pressure level in dBA

Total OSHA-Permissible Noise Dose Computation in 8-hour Working Day (OSHA 1910.95 Appendix A)

Constant Noise

When the noise level is constant over the entire work-shift, the noise dose is given by

$$D = 100\left[\frac{C}{T}\right] \qquad (20.31)$$

where
C = Total length of work day
T = OSHA-permitted reference duration corresponding to sound level L as given by equation (20-30)

Variable Noise

When the work-shift noise exposure is composed of two or more periods of noise at different levels, the total noise over the working day is given by

$$D = 100\left(\sum_{i=1}^{i=n}\frac{C_i}{T_i}\right) = 100\left(\frac{C_1}{T_1} + \frac{C_2}{T_2} + \cdots + \frac{C_n}{T_n}\right) \qquad (20.32a)$$

where
C_i = Total time of exposure of event i at a specific noise level, L
T_i = OSHA-permitted reference duration for the noise level, L at event i as allowed by equation
D = Noise dose as a per cent. Dose $D > 100$ implies OSHA violation

Eight-hour Time-Weighted Average of Variable Noise

The eight-hour time-weighted average sound level may be computed from the dose expressed in percent as follows:

$$TWA_{sound} = 16.61\left[\log_{10}\frac{D}{100}\right] + 90 \qquad (20.32b)$$

Other Salient Points of OSHA 1910.95 with Regard to Hearing Protection

1. A change in hearing shift [Standard Threshold Shift, STS] relative to the base line of an average 10 dB or more at 2000, 3000, and 4000 Hz in either ear of an employee, after allowance is made for hearing loss due to age, is recordable.

2. Employers shall make hearing protectors available to all employees exposed to an 8-hour time-weighted average of 85 dB or greater at no cost to employee.

Example 20.11

Estimate the continuous exposure time allowed by OSHA in an environment with noise pressure level of 105 dB.

Solution

$$a = 0.2(L - 90)$$
$$= 0.2(105 - 90) = 3$$
$$T = 8/2^3 = 1 \text{ hr}$$

Example 20.12

A worker is exposed to the following noise levels in an 8-hour working period: 50 dBA for 3 hours, 85 dBA for 4 hours, and 95 dBA for 1 hour. Determine whether the OSHA exposure limit of the worker is exceeded and compute the 8 hour time-weighted sound exposure.

Solution

Noise Level, L	Permitted Exposure Time, HR, T_i*	Total Exposure Time, HR C_i	Time Fraction Exposed C_i/T_i
50	2048	3	0.001
85	8	4	0.500
95	4	1	0.250
Total			0.750

* $T = 8/2^a$ where $a = 0.2(L-90)$

Because the summation of the time fraction exposed to various noise levels is less than 1 or 75 percent (<100%), OSHA exposure limit is not exceeded.

The eight-hour time-weighted average of sound is given by

$$TWA_{sound} = 16.61\left[\log_{10}\frac{D}{100}\right] + 90 = 16.61\left[\log_{10}\frac{75}{100}\right] + 90 = 87.9 \text{ dBA}$$

Types of Noise Control in Operation and Design

Devices such as insulation barriers, enclosures, silencers, reduced velocity, quieter valve, non-protruding gaskets, smoothing of internal welds, and increased thickness of pipe are used to reduce noise.

SOLID WASTE MANAGEMENT

The solid waste management in the U.S. is regulated by

1. The Solid Waste Disposal Act of 1965
2. The Hazardous and Solid Waste Act (HSWA) of 1984.

A hazardous substance is a material that may, if released to the surroundings, cause damage to the public health and/or the environment.

An extremely hazardous substance is a material that can, by a single exposure, cause an irreversible damage to public health and/or the environment.

Hazardous Waste Incineration

The Destruction and Removal Efficiency (DRE) of the Principal Organic Hazardous Constituents (POHC) is the mass percent of POHC removed from the waste, which should be ≥ 99 percent.

Particulates removal efficiency is determined by the maximum allowable concentration of 0.08 grains per dry standard cubic foot (gr/dscf) of the stack gas corrected to 7 percent dry volume (or approximately 50% excess air.)

In summary, the restrictions are

$$E_{POHC} \geq 99.99\%$$
$$E_{HCl} = \begin{cases} \text{Emission rate} \leq 4 \text{ lb/h} \\ \text{Scrubber Absorption Efficiency} \geq 99\% \end{cases} \quad (20.33)$$
$$C_{particulate} < 0.08 \text{ gr/dscf}$$

where
E_{POHC} = Percent removal efficiency of POHC
E_{HCl} = HCl removal or absorption efficiency
$C_{particulate}$ = Particulate concentration in emission gases

Six Elements of Waste Disposal

1. Waste generation
2. Waste handling and separation
3. Collection
4. Separation, processing, and transformation
5. Transfer and transport
6. Disposal

Four Elements of Waste Management

1. Source reduction
2. Recycling
3. Waste transformation
4. Ultimate disposal

Table 20.16 Data table for flare sizing

Variable	Name
W	Flammable vapor flow rate, lb/h
M	Average molecular weight of flammable vapor
t	Flowing temperature, °F
Z	Compressibility factor, dimensionless
$\Delta H_{c,m}$	Heat of combustion, Btu/lb of flammables
k	Specific heat ratio, dimensionless
P	Flowing pressure at the tip, psia
F	Fraction of heat radiated
K	Allowable radiation intensity at reference distance, R, Btu/h · ft²
R_R	Reference distance from the center of the flare stack, ft, at which radiation intensity is measured,
U_I	Wind velocity, ft/s
τ	Fraction of heat intensity transmitted

Nuclear/Radioactive Waste

The nuclear and radioactive wastes are wastes that constantly emit harmful radiation due to the presence of unstable isotopes. They are present in many forms such as high- and low-level wastes, transuranic waste, etc. The radioactive wastes are generated by nuclear power plants, nuclear defense, medical radiotherapy, mining, and industrial processes.

The ultimate disposal has been defined by Environmental Protection Agency (EPA) to include four processes.

1. Landfilling: area filling or trench filling
2. Land farming: disposing of waste in upper layer of the soil
3. Ocean dumping: direct dumping or in containers into the ocean
4. Deep well injection: deep injection into soil to avoid contamination of drinking water

Flare Design

An outline of the steps of flare stack design is given below.

Step 1. Collect necessary information per Table 20.16.

Step 2. Select exit velocity (flare tip velocity, U_j).

The flame[14,15,16,17] at the exit of the flare, when the heating value of the flared gas is greater than 1000 Btu/standard ft³, can be stable if the exit velocity of the flared gas is below 400 ft/s. Recommended flare tip velocity, U_j = 90 percent of 400 or 360 ft/s as the maximum value.

If the heat of combustion of the gas is given in Btu/lb, check the value in Btu/standard cubic foot from the given molecular weight to see if it is >1000 Btu/standard cubic ft. Thus

$$\rho_{standard} = \frac{P_{std}M}{RT_{std}} = \frac{14.7M}{10.731(520)} \quad (20.34)$$

$$\Delta H_{c,v} = \Delta H_{c,m}(\rho_{standard})$$

where

$\Delta H_{c,v}$ = Heat of combustion of gas, Btu/Standard cubic foot
$\rho_{standard}$ = Density of gas at standard condition, lb/Standard ft^3
$\Delta H_{c,m}$ = Heat of combustion of gas, lower heating value, Btu/lb
M = Molecular weight of gas, lb/lbmol

Step 3. Calculate speed of sound in the flared gas and from selected flare tip velocity, calculate the Mach number of the flared gas. The speed of sound in the flared gas, u_s, is given by

$$u_s = \sqrt{(kg_c RT)} \tag{20.35}$$

where

k = (specific heat at constant pressure/specific heat at constant voleume) of gas
g_c = 32.17 lb · ft(lb$_f$ · s^2), [1 (kg · m/s^2)/N]
R = 1545/M ft · lb$_f$/(lb · °R), [8314/M (N · m/kg · K)]
T = Temperature, °R, [K]
u_s = velocity of sound in the gas, ft/s, [m/s]
M = Molecular weight of gas = lb/lbmol = kg/kmol

The Mach number N_{Ma} of gas may now be calculated by

$$N_{Ma} = \frac{U_j}{u_s} \tag{20.36}$$

Step 4. From the calculated Mach number, known vapor flow rate, pressure, temperature, compressibility factor, specific heat ratio, and molecular weight, now calculate flare diameter, D:

$$D = 4.1225(10^{-3})\left(\frac{W}{PN_{Ma}}\right)^{0.5}\left[\frac{Z(t+460)}{kM}\right]^{0.25} \tag{20.37}$$

where

D = Flare tip inside diameter, ft
W = Flare gas flow rate, lb/h
P = Flowing pressure at the flare tip, psia
Z = Compressibility factor, dimensionless
t = Flowing temperature at the flare tip, °F

Step 5. Calculate flame length.

$$\begin{aligned}Q &= W(\Delta H_{c,m}) \\ L &= 0.01288Q^{0.44}\end{aligned} \tag{20.38}$$

where

Q = heat release rate, Btu/h
W = vapor flow rate, lb/h
$\Delta H_{c,m}$ = heat of combustion, lower heating value, Btu/lb
L = flame length, ft

Step 6. Calculate the flame distortion parameters Δx and Δy.

6(a): Calculate wind-to-flare-tip velocity ratio from given wind velocity and flare tip exit velocity, U_j, obtained in step 2 in consistent units:

$$\text{Wind-to-flare-tip velocity ratio} = \frac{U_\infty}{U_j} \qquad (20.39)$$

where
U_∞ = wind velocity, ft/s
U_j = flare tip velocity, ft/s

6(b): Read the x-parameter and y-parameter from Figure 20.8.

$$(x\text{-parameter})_{\text{graph value}} \equiv \sum \frac{\Delta x}{L} \qquad (20.40a)$$

$$(y\text{-parameter})_{\text{graph value}} \equiv \sum \frac{\Delta y}{L} \qquad (20.40b)$$

6(c): Calculate x and y by multiplying the graph value with L:

$$\Delta x = (x\text{-parameter})_{\text{graph value}} (L)$$
$$\Delta y = (y\text{-parameter})_{\text{graph value}} (L) \qquad (20.41)$$

where
L = flame length, ft, calculated in step 5

Step 7. Calculate the required flare (stack) height.

7(a): Calculate the diagonal parameter of the effective flare height

$$D_{\text{diagonal}} = \sqrt{\frac{\tau F Q}{4\pi K_R}} \qquad (20.42)$$

where
D_{diagonal} = diagonal length of effective flame height, ft, from the mid-point of flame to an object placed at a distance R ft from the base center line of stack.
τ = fraction of heat transmitted = 1
F = fraction of heat radiated = 0.3
Q = heat release rate, Btu/h
K_R = allowable radiation intensity at an object placed at a distance R from the centerline of stack, Btu/h · ft^2
R_R = distance from the centerline of the stack base, at the end of which, the allowable radiation intensity has been specified, ft

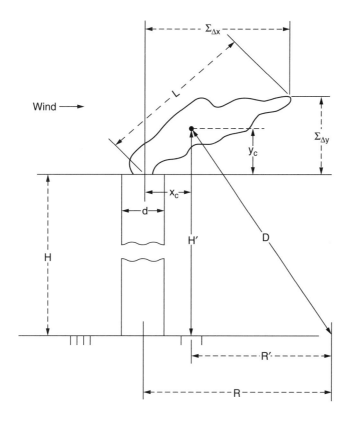

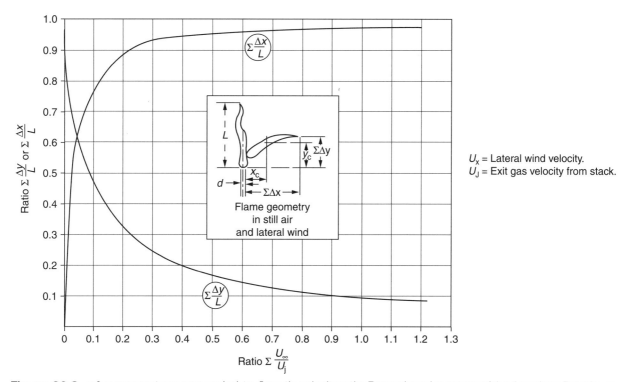

U_x = Lateral wind velocity.
U_J = Exit gas velocity from stack.

Figure 20.8 x & y-parameters versus wind-to-flare-tip velocity ratio. Reproduced courtesy of the American Petroleum Institute.

7(b) Calculate parameters R, H, and finally flare (stack) height.

$$R = R - \frac{\Delta x}{2} \tag{20.43a}$$

$$H = \sqrt{\left(D_{\text{diagonal}}^2 - R^2\right)} \tag{20.43b}$$

$$H = H - \frac{\Delta y}{2} \tag{20.43c}$$

where
H = stack height, ft

Step 8. Check the ground level concentration of vapor under flame-out condition, *i.e.*, if the flame goes out. The equations are available in metric units. Therefore, convert some calculated parameters in metric units, and finally calculate the ground level concentrations.

$$U_{j,m} = U_j(0.3048) = \text{flare tip velocity, m/s} \tag{20.44}$$

$$D_m = D(0.3048) = \text{flare tip inside diameter, m} \tag{20.44a}$$

$$U_{\infty,m} = U_\infty(0.3048) = \text{wind velocity, m/s} \tag{20.44b}$$

$$H_m = H(0.3048) = \text{flare (stack) height, m} \tag{20.44c}$$

$$Q_m = \frac{W(453.6)}{3600} = \text{vapor flow rate through flare, gm/s} \tag{20.44d}$$

$$\Delta H = \text{plume rise} = 3\frac{D_m U_{j,m}}{U_{\infty,m}}, \text{ m} \tag{20.44e}$$

$$H_{\text{eff}} = H_m + \Delta H = \text{effective stack height, m} \tag{20.44f}$$

$$C_{\max} = \frac{0.23Q_m(1000)}{U_{\infty,m} H_{\text{eff}}^2} \tag{20.45a}$$

$$ppm = \frac{C_{\max}(24.45)}{M} \tag{20.45b}$$

where
$C_{\max}$ = maximum ground level concentration under flame-out condition, mg/m^3
ppm = maximum ground level concentration under flame-out condition, ppm(v) at 25 °C and 1 atmosphere

Step 9. Calculate steam requirement for smokeless[17] operation:

$$Q_s = 1.03(10^{-3})\frac{WM}{60\rho_{\text{standard}}} \tag{20.46a}$$

where

Q_s = steam requirement, lb/min

An alternative equation:[3]

$$W_{steam} = W\left(0.68 - \frac{10.8}{M}\right) \qquad (20.46b)$$

where

W_{steam} = steam requirement, lb/h

Step 10. Take a coffee break!

Environmental Permit Applications

Title 40 (Environmental Protection Agency, EPA) of the Code of Federal Regulations (CFR) contains environmental regulations pertaining to air quality, wastewater, solid waste, and other environmental programs. Regulations for air

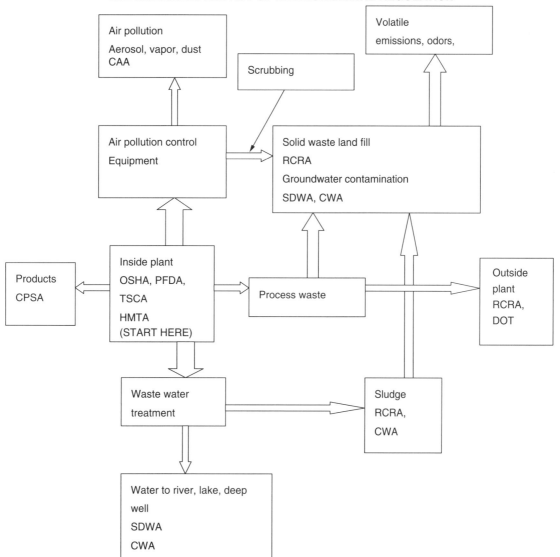

Figure 20.9 Flow chart for determining multimedia impact of environmental regulation

Table 20.17 Explanation of acronyms in the flow chart of multimedia impact of environmental regulation

Act	Explanation	Regulator	Remark
OSHA	Occupational Safety and Health Act	OSHA	Occupational Safety and Health Administration regulates workplace safety.
TSCA	Toxic Substance Control Act	EPA	Environmental Protection Agency sets and enforces regulations on discharges to environment: air, water, and land. Regulates use of pesticides, herbicides, insecticides, and toxic substances.
RCRA	Resource Conservation and Recovery Act	EPA	RECRA promotes recycling and reuse of hazardous waste, and controls their disposal.
CWA	Clean Water Act	EPA	Titles III and IV of CWA affect process industry & set concentration limits on specific chemicals. CWA regulates the permit system, NPDES (National Pollutant Discharge Elimination System), which includes municipal and industrial plants.
SDWA	Safe Drinking Water Act	EPA	Specifies maximum concentration limits on chemicals in drinking water.
CAA	Clean Air Act	EPA	Controls emissions from all sources into air: automobiles, power plants, manufacturing facilities etc. Specifies limits of emission on SO_2, particulates, hydrocarbons, NOx, oxidant, CO, lead.
CPSA	Consumer Product Safety Act	CPSA	
HMTA	Hazardous Material Transportation Act	DOT	Department of Transportation.
PFDA	Pure Food and Drug Act	FDA	Food and Drug Administration regulates the use of substances in food, drug, and cosmetic applications.

programs can be found in Title 40, Subchapter C, Parts 50–99 (40 CFR 50–40 CFR 99). 40 CFR 63 contains the National Emission Standards for Hazardous Air Pollutants (NESHAP).

The Code of Federal Regulations (CFR) for some major areas is broken down as follows:

Subchapter D, Part 100–Part 149 (Water Programs)

Subchapter E, Part 150–Part 189 (Pesticide Programs)

Subchapter F, Part 190–Part 197 (Radiation Programs)

Subchapter G, Part 201–Part 211 (Noise Abatement Programs)

Subchapter H, Part 220–Part 238 (Ocean Dumping)

Subchapter I, Part 239–Part 282 (Solid Wastes)

It should be noted that states, counties, and/or cities could have environmental regulations that are more stringent than the federal regulations. One would need to be sure that the local regulations are being complied with as well.

New facilities may require five to ten years from conception to start up. Considerable time is spent in answering environmental considerations. To avoid possible litigation and associated expenses in equipment, planning, and design, it is imperative to write an environmental impact statement (EIS) at the conceptual stage of the project. This generally requires the knowledge of the land to be used, the type of the plant, the anticipated discharges to air, water, and land, the land use pattern, geology and socioeconomic climate. The local and federal agencies as well as the environmental community review the EIS. It is also important to meet with local political and environmental groups on an informal basis to gain their support and understanding. There are examples of corporate failures to establish a plant, which had complete approval of governmental agencies, but did not gain local public support.

RECOMMENDED REFERENCES

1. Crowl and Louvar, *Chemical Process Safety: Fundamentals with Applications,* Prentice Hall, NJ, 1990.
2. *Sizing, Selection, and Installation of Pressure-Relieving Devices in Refineries, Part I*—Sizing and Selection. API Recommended Practice 520. January 2000, American Petroleum Institute, Washington, D.C.
3. *Guide for Pressure-Relieving and Depressuring Systems.* API Recommended Practice 521, January 1997, American Petroleum Institute Washington, D.C.
4. Crozier, "Sizing Relief Valves for Fire Emergencies," *Chemical Engineering,* October 28, 1985.
5. *Venting Atmospheric and Low-Pressure Storage tanks,* (Non-Refrigerated and Refrigerated), API-Standard 2000, (1998), American Petroleum Institute, Washington D.C.
6. Leung, "Simplified Vent Sizing Equations for Emergency Relief Requirements in Reactors and Storage Vessels," *AIChE Journal,* Vol 32, No 10, p1622, 1986.
7. Azad, *Industrial Waste Water Management Handbook,* McGraw-Hill Book Company, NY, 1976.
8. Mukhopadhyay and Moretti, *Current and Potential Future Industrial Practices for Reducing and Controlling Volatile Organic Compounds,* American Institute of Chemical Engineers, NY, 1993.
9. Porteous, *Dictionary of Environmental Science and Technology,* John Wiley & Sons, West Sussex, England, 1992.
10. Lipton and Lynch, *Handbook of Health Hazard Control in the Chemical Process Industry,* pp 106–112, John Wiley & Sons, Inc., NY, 1994.
11. NFPA-68, National Fire Protection Association, Quincy, MA, 2002. (a) para 6.2.7.4, (b) para 6.2.3 (c) NFPA-69,
12. NFPA-30, National Fire Protection Association, Quincy, MA, 2002.
13. Code of Federal Regulations 29.
14. USEPA: Control Technologies for Hazardous Air Pollutants, EPA/625/6-91/014.
15. Code of Federal Regulations, 40CFR 60.18: General Control device requirements
16. Code of Federal Regulations, 40CFR 264.1033: Closed-vent systems and control devices.
17. Kumar, "Design and Operate Flares Safely," *Chemical Engineering,* December 1998.
18. Guidelines for Hazard Evaluation Procedure, AIChE (CCPS).
19. Reynolds, Jeris, and Theodore, *Handbook of Chemical and Environmental Engineering Calculations,* Wiley InterScience, NY, 2005.

20. Das, *Pressure Relief: In Case of Fire, Break Assumptions.* New York: Chemical Engineering, (February 2004). [Note some errata: (1) R and K in σ of Nomenclature table should be raised to the power 4, (2) In equation 15, ε should be replaced by e, (3) In equations (17) and (18), both d and g should be raised to the power 0.5. The article also addresses calculation of flame temperature from pool fire.]
21. Hauser, PROSAF, Inc. *Personal communication.*

INDEX

% conversion, 35
−79n, 103, 110

A

absolute pressure, 9
absolute viscosity, 15
absorption, 433–457. *See also* packed towers; packings.
  column diameter, 441–442
  equation of material balance, 433–434
  equilibrium relationship, 433–444
  examples, 444–457
  height of packed column, 436–441
  $K_G a$, estimation of, from the $K_G a$ of a known system, 440
  mass transfer coefficients, 439
    estimation of, 440
  material balance, 433–444
  N_{OG}, determination of, 436–438, 439, 440–441
  N_{OL}, determination of, 438–439, 439
  N_P, determination of, 439
  operating line, 433–444
  packed absorption towers, 433–444
    packing height, 436–441
  packing size, 441, 443, 444
  packings, properties of, 443
  problem-solving, 433
  solvent flow rate, 435–436
  wetting rate, 442, 444
acceptable daily intake (dose), 762
accounting principle, 29
accumulation, 54
acentric factor, 48
acid rain, 762
activated alumina, 494
activated carbon, 493, 494, 762
activated sludge, 763
activation energies, 580
activity coefficients, 355–357
adiabatic compression, 197
adiabatic expansion, 124
adiabatic flame temperature, 99–101
adiabatic plug-flow tubular reactor, 602
adiabatic process, 113
adiabatic saturation temperature, 516
adiabatically conducted reaction, 99
adsorbents, 493. *See also* adsorption.
adsorbent/solution ratio, 498
adsorption, 493–510
  adsorbents, 493–494
    adsorbent/solution ratio, 498

bed regeneration, 503–504
BET model, 496
  degree of, 496
equilibrium relationships, 495
fixed bed, 501–510
Freundlich isotherm equation, 497, 498
isotherms, 495, 496
liquids, from, 497–510
McCabe-Thiele diagram, 500
monolayer, 495–496
multistage countercurrent operation, 500
multistage crosscurrent operation, 498–499
purification process, as, 493
regeneration of bed, 503–504
separation process, as, 493
single-stage, 497, 498
stagewise operations, 497–500
types of, 495
two stage cross current, 498–499
adsorption isotherms, 495, 496
AES system, 4, 6
agitator design, 667–674
  apparent superficial velocity, 667
  chemscale/agitator intensity number, 667
  dimensionless numbers, 668
  factors to consider, 669–670
  impeller, 669–671
  motor HP, 669
  nomenclature, 673–674
  preliminary design, 670–674
  primary pumping capacity, 667
  scale of agitation, 672
  tip speed, 667–668
  torque, 668–669
agitator intensity number, 667
AIChE method, 422
air-cooled heat exchanger, 264–267
air horsepower (fan), 199
air pollution statement, 767–768
AIT, 728
all-graphite plate exchangers, 270
allowable vapor velocity, 426
allowable velocity factor, 426, 431
Altshul Fanning friction factor, 150
aluminum/aluminum alloys, 647–648, 653
Amagat's law, 44
American Engineering System, AES 4, 6
amu, 7
analogy correlations, 338

analytical method of optimization, 719
Angstrom, 12
Aniline, 239
annual cost, 714
annual cost method, 713–714
annuity, 701
annuity for future worth, 701
annulus pipe, 251
anode, 640
Antoine equation, 39
API environmental factor table, 740
API wetted area calculation table, 739
apparent reflux ratio, 388
apparent superficial velocity, 667
approximate annual cost of the capital recovery (CR), 714
Arrhenius equation, 544–545, 548
Aspen, 411
asymmetric membrane, 312
atmospheric pressure, 9, 10
atmospheric temperature change, 657
atomic mass unit (amu), 7
atomic weight, 7
atto-, 5
austenitic stainless steel, 646, 647, 653
autocatalytic reactions, 560
autoignition temperature (AIT), 728
available NPSH, 171–172, 177, 181–182
average molecular weight, 26, 28
Avogadro's number, 7
axial flow impeller, 669
azeotropic distillation, 411
azeotropic mixtures, 349, 355, 409–412
azeotropic point, 347

B

B/C ratio method, 716–717
back-mix reactor, 570, 574, 576
Backowski correlation, 676
Backowski equation, 422
backward-feed evaporator, 281
baffle-cut segmental baffles, 252
Baker's flow regime map, 202
bang-bang controller, 621
bar, 9
barometric pressure, 656, 689
basic units, 2
basis, 31
batch heating and cooling, 268–269
batch reactor, 566, 567–568, 580–581, 591
batch recycle reactor, 582, 584
batch RO/UF concentration, 312

784   Index

batch-type filters, 291
Beattie-Bridgeman equation of state, 49
bed filter, 291–292
Benedict-Rubin-Webb equation of state, 49
beneficial ozone, 765
benefit-to-cost (B/C) ratio method, 716–717
Bernoulli equation, 135–137
BET model, 496
ß, 123, 128, 129
BFD, 683
bhp, 174, 182
bimetal corrosion, 640, 641
bimetallic fin, 266
bimolecular second-order reactions, 556–557, 562
binary azeotropic mixtures, 409–412
binary system, 346
binodal curve, 463
bio-degradation, 767
biochemical oxygen demand (BOD), 763
blackbody radiation, 214, 219
bleed-off stream, 61
blend time, 668
block diagram, 616–617
block flow diagram (BFD), 683
blocked condition, 749–750
BOD, 763
boiling, 234
boiling point, 39
boiling point diagrams, 347–349
boiling-point elevation, 280
book value, 709
bound water, 525
Boyle's law, 41
brake horsepower (bhp), 174, 182
break-even analysis, 719
breakpoint concentration, 502
breathing devices, 656
British engineering system, 3, 6
British thermal unit, 11
Brown-Martin correlation, 414
bubble-cap trays, 678–680
bubble point, 38, 358, 360
bubble pressure, 358
bubble rise velocity, 152
bubble temperature, 358
Buckingham's π method, 20–21
buckling thickness, 692
building insulation, 210, 212
bulk velocity, 667
bypass stream, 61

C

CAFT, 730–731
calculated adiabatic flame temperature (CAFT), 730–731
Calderwood equation, 68

caloric fluid temperature, 226
calorific value, 67
capacity, 188
capacity parameter (ordinate), 430–432
capital cost, 705
capital-cost method, 714
capital-recovery depreciation, 711
capital-recovery depreciation charge, 708
capital-recovery factor, 703, 704
carbon steel, 646, 653
carcinogens, 763
Carlson-Colburn equation, 356
Carnot cycle, 121
Carnot cycle COP, 271
Carnot principle, 104
Carnot refrigerator, 123, 129
Carnot reversible idealized heat engine, 104
carrier method, 308
cascade control, 616
cash flows in annual disbursement only, 718
catalyst, 591
catalytic reactor, 591
catalytic oxidation, 767
catalyzed reactions, 559
cathode, 640
cavitation, 172, 173
cavitation parameter (σ), 173
Celsius scale, 7, 8
centimeter-gram-second (cgs) system, 2, 6
centipoise, 15
centistoke, 15
centrifugal fan, 200
centrifugal pump, 666
centrifugal pump calculations, 174–175
centrifugal pump trouble-shooting, 687–688
CEPCI, 706
CEPCI base year, 706
CFCs, 763, 765
cgs system, 2, 6
characteristic equation, 632, 633
Charle's law, 41
Charpy V-notch impact test, 652
Chemcad, 731
chemical engineering plant cost index (CEPCI), 706
chemical equation, 31–32
chemical kinetics, 536. *See also* chemical reaction engineering.
chemical oxygen demand (COD), 763
chemical potential, 353–354
chemical reaction engineering, 535–612
   activation energies, 580
   Arrhenius equation, 544–545, 548
   autocatalytic reactions, 560
   back-mix reactor, 570, 574, 576

   batch reactor, 566, 567–568, 580–581, 591
   batch recycle reactor, 584
   bimolecular second-order reactions, 556–557
   catalytic reactor, 591
   chemical equilibria, 540–542
   classification of reactions, 536–537
   collision theory, 548, 549
   composite walls, 215
   CSTR. *See* stirred-tank reactor.
   effect of temperature on the rate of reaction, 544–545
   elementary/non-elementary reaction, 539
   empirical rate equations, 557–561
   FBR. *See* fluidized bed reactor (FBR).
   first-order reaction, 551, 556
   first-order reversible reactions, 561–562
   flow reactor, 566–567
   functions of chemical kinetics, 536
   gas-solid reactions, 591–593
   homogeneous catalyzed reactions, 559
   irreversible reactions, 550–561
   law of mass action, 539
   mass and energy balances, 567–579, 603
   method of differentiation, 550–551
   method of halftimes, 552
   method of initial rates, 552
   method of integration, 551–552
   method of k calculation, 553
   method of reference curves, 552
   molecularity, 539
   optimal temperature progression, 597–611. *See also* optimal temperature progression.
   order of reaction, 539
   parallel reactions, 558–559, 578–579
   PBR. *See* packed bed reactor (PBR).
   plug-flow reactor. *See* plug-flow reactor.
   product distribution and temperature, 579–593
   questions to ask, 536
   rate constant, 538
   rate equation, 537–538
   rate of chemical reaction, 537
   reactions at constant volume/temperature, 538
   recycle reactor, 579, 582–584
   reversible reactions, 561–563
   second-order reaction, 551–552, 556–557
   second-order reversible reactions, 562–563
   series reactions, 560–561, 578, 580
   steady state nonisothermal reactor design, 596–597
   stoichiometric tables, 593–597

temperature, effect of, on rate of reaction, 544–545, 544–549, 596
transition-state theory, 548
unimolecular first-order reactions, 556
zero-order reaction, 551, 558
chemical reactions, 31–32, 63–64
chemisorption, 495
Chen's equation, 87
chloroform-water-acetic acid, 461
choked flow, 195
Chu, 12
Churchill equation, 150–151
Clapeyron equation, 87
Clausius-Clapeyron equation, 87
Clausius integral, 105
Clean Air Act, 780
Clean Water Act, 780
closed loop control, 615–616, 634–637
closed system, 28, 29
coal, 67–76
cocurrent gas absorption tower, 434
COD, 763
coefficient of performance (COP), 271
coefficient of performance (ß), 123, 128, 129
Colburn J_H factor, 227–228
cold reflux, 388–391
Collins, J. J., 508
collision theory, 548, 549
columbium, 648, 649
combustible dust, 728
combustible liquid, 729
combustion of fuels, 67–76. *See also* fuels and combustion.
commercially available adsorbents, 493
complete combustion, 67
composite cylindrical resistance, 217, 218
compound amount formula, 704
compound interest, 698–699
compressibility factor chart, 46–47, 50–52
compressibility factor z, 46
compressible fluid, 193. *See also* flow of compressible fluids.
compressible fluid flow rate, 195
compression equipment, 197
concentrate, 313
concentration, 26
condensation, 235–237, 389, 767
film temperature for, 237
inside tubes, 236
horizontal, 236
vertical, 237
outside tubes, 235
vertical tubes, 235
inclined tubes, 235
conduction, 210–213
conjugate line, 463, 464
conjugate phases, 464–465
conjugate solutions, 460
consecutive reactions, 537, 578

consistent units, 16
constant molal overflow, 375–376
constant molal underflow, 379
constant molal vaporization, 379
constant noise, 771
constant pressure, 113
constant-pressure filtration, 295, 297
constant-rate filtration, 296, 298
constant-rate period drying rate, 527, 529–533
Consumer Product Safety Act, 780
contaminated soil, 769
continuous stirred-tank reactor. *See* stirred-tank reactor.
continuous back-mix reactor, 574
continuous compounding, 704–705
continuous countercurrent adsorption, 500
continuous countercurrent multistage extraction system, 470
continuous distillation with reflux, 376–379
continuous filters, 291
continuous flash distillation, 370–372
continuous-flow recycle reactor, 582
continuous RO/UF process, 312
continuous rotary disk filter, 292
continuous rotary vacuum filters, 292
control systems. *See* process control.
control valve, 187–191
sizing of, 189
controlled variables, 615
convection, 213, 233–234
convective mass transfer, 318, 333–336
convergent-divergent nozzle, 195
conversion, 35
conversion factors, 14–15
tables of, 12, 14–15
conversion of units, 12–16
conversion factors, 12, 14–15
force, 12
length, 12
mass, 12
temperature scales, 8, 12
thermal units, 12
time, 12
conversion per pass, 64
cooling towers, 520–424
COP, 271
copper and its alloys, 647, 653
Coriolis/Gyroscopic flowmeter, 142
correction factor, 250, 260
corrosion and materials of construction, 639–654
crevice corrosion, 641–642
design of metallic vessels, 643–644
erosion corrosion, 645
galvanic or bimetal corrosion, 640, 641
hydrogen damage, 645
intergranular corrosion, 642

low-temperature application of metals, 652, 653
material cost/cost considerations, 653
maximum allowable stress, 653
metals, 646–649
non-metals, 649–652
pitting, 642
selective leaching, 642
stress corrosion, 645
uniform attack, 640
cost comparison of investment projects, 713–720
after-tax comparisons, 717–718
analytical method of optimization, 719
annual cost method, 713–714
B/C ratio method, 716–717
break-even analysis, 719
capital-cost method, 714
cash flows in annual disbursement only, 718
incremental analysis, 716
payout time method, 718
present worth (PW) method, 714
rate-of-return method, 714–715
ROI, 718
costs. *See* engineering economics.
countercurrent gas absorption tower, 434
countercurrent leaching battery, 479–480
Crane formulae, 195
Crawford-Wilke flooding correlation (packed extraction towers), 474
crevice corrosion, 641–642
critical nucleate boiling heat flux, 234
critical pressure ratio, 194
critical radius, 218–219
critical velocity, 194–195
critically damped, 630
cross flow (tray tower), 428
cross-flow exchangers, 229
crude oil, 67, 68
crystallization, 263–264
crystallizer problem, 56
CSTR. *See* stirred-tank reactor.
cupronickel, 647
Cv, 157
cycle, 121
cyclical frequency, 631

D

Dalton's law, 43
Dalton's law of partial pressures, 350
Damkohler number, 571
damping coefficient, 628, 630, 631
Danckwerts surface renewal theory, 337
Darby 3-K constants, 157, 158
dead-headed, 666
decay ratio, 630
declining-balance depreciation, 710
deflagration, 728
deflagration venting, 756–757

degree of completion, 36
degrees of freedom, 54, 347
degrees of freedom for optimization, 682
dehumidification, 511
density, 24–25
depreciation, 708–713
    capital-recovery, 711
    declining-balance, 710
    double declining-balance, 710–711
    examples, 711–713
    recovery of fixed investment, 708
    salvage value, 709
    sinking-fund, 711
    straight-line method, 709
    sum-of-the-digits, 709–710
    tax computations, 708
derived units, 2
design of equipment. *See* equipment design.
design of metallic vessels, 643–644
design optimization, 682–683
design pressure, 656–658
design temperature, 656–658
destructive testing, 682
detonation, 728
deviation variables, 624
deviations, 758
dew point, 38, 358, 361–362
dew-point temperature, 513
dew pressure, 358
dew temperature, 358
diagram
    BFD, 683
    block, 616–617
    boiling point, 347–349
    enthalpy-composition, 365–367
    equilateral triangle, 461–463
    equilibrium, 347–349
    HS, 114
    Mollier, 114
    PFD, 683–684
    pH, 86, 114
    P&ID, 684
    Ponchon-Savarit (extraction), 465
    Ponchon-Savarit (leaching), 478–479, 481
    right triangle (LL extraction), 463, 464
    right triangle (leaching), 477–478
    selectivity, 465
    TS, 114
    $T\text{-}xy$ *See* diagram, boiling point.
    $x\text{-}y$. *See* diagram, equilibrium
diameter (bubble-cap tray column), 679–680
DIERS, 751
differential condensation, 389
differential distillation, 372–374
differential equation method, 21–22
differential equations, 624, 625, 634
differential set pressure, 746

differential U-tube manometer, 134
differentiation, 550–551
diffusion. *See* Molecular diffusion.
diffusion coefficients, 325
    determination of, 325–327
dilution air, 727
dimensional analysis, 18–22
    Buckingham's 6 method, 20–21
    differential equation method, 21–22
    dynamic similarity, 22
    Rayleigh's algebraic method, 18–20
dimensional formula, 16
dimensional friction factor, 252
dimensionally homogeneous equations, 16–17
dimensionless equations, 18
dimensionless numbers, 338–341
dimensions, 1–2
Dirac delta function, 625
discharge static head, 170, 181
displacement desorption, 504
distillation, 369–372
    azeotropic distillation, 411
    cold reflux, 388–391
    constant molal overflow, 375–376
    continuous flash distillation, 370–372
    defined, 370
    differential distillation, 372–374
    extractive distillation, 410–411
    feed plate location, 384–388, 414
    Fenske equation, 395
    fractional distillation, 370
    graphical stages at low concentrations, 395–409
    HETP, 423–424
    Lewis method, 379
    McCabe-Thiele method, 380–384, 392–393
    minimum reflux ratio, 387–388, 413–414
    multicomponent distillation, 411–412, 413–415
    multifeed column, 389–391
    open steam, 393–394
    optimum reflux ratio, 394–395
    packed towers, 419–426, 428–430
        random packings, efficiecy of, 424
        sizing of, 426
        structured packings, efficiency of, 424
    plate-to-plate calculations, 379–384
    Poncho-Savarit graphical method, 380
    rectification, 376, 379
    saturated steam at tower pressure, 393–394
    separation of binary azeotropic mixtures, 409–412
    side-stream withdrawal, 391–392
    simple steam distillation, 374–375
    single-still distillation, 370
    Smoker's equation, 395

    Sorel's method, 379
    stripping-column distillation, 392–393
    superheated steam, 394
    theoretical plates, 378–379, 382, 383, 407–408, 414
    total reflux, 389, 413
    tray distillation column, 677
    tray efficiencies, 421–423
    tray towers, 426–428
distillation column, 57
    overall energy balance, 386
distillation methods, 370–379
distribution coefficient, 460
distribution velocities, 148
Dittus-Boelter equation, 227
Dobbins' combination film-surface-renewal theory, 337
double-declining-balance depreciation, 710–711
double interpolation, 39, 86
double L-footed fin, 266
double pipe exchanger, 243, 251
downcomer, 427, 428, 679
downspout, 427
draft, 9
draft gauge, 134
DRE, 773
droplet diameter, 151–152
drying, 525–533
    constant period, 527, 529–533
    curve, 525, 526
    defined, 525
    equilibria, 525
    equipment, 525
    falling rate period, 527–528
    rate of, 527–528
    rate-of-drying curve, 526–527
    time of, 525, 528–529
drying curve, 525, 526
dual-flow trays, 678
Duckler's homogeneous method, 204, 206–207
Duhem's rule, 347
Duhring lines, 287
Duhring's rule, 280
Dulong's formula, 69
dumped packings, 419, 424
dust explosion pentagon, 728
dynamic similarity, 22
dynamics, 615
dyne, 2, 6

E

E_o, 421–422
economic degrees of freedom, 682
economics. *See* engineering economics.
ED, 314
effective diffusivity, 332
effective interest rate, 699
effectiveness - NTU method, 260–262

Index   **787**

EIS, 780–781
elastomers, 650
electrical current, 210
electrodialysis (ED), 314
elementary reaction, 539
embedded fin, 266
emergency relief load calculation, 735–737
emergency relief system hardware, 734–735
emissivity, 220
empirical rate equations, 557–561
endothermic reaction, 92, 537
energy
　activation, 580
　definitions of, 78
　heat, 78–79
　internal, 80–81
　kinetic, 78
　potential, 78
　thermal, 10
　units of measure, 11
energy balance and thermodynamics, 77–129
　adiabatic flame temperature, 99–101
　Carnot principle, 104
　chemical reactions, 92–129
　energy, 78–79. *See also* energy
　energy balance (closed system), 82
　energy balance (heat of reaction), 96
　energy balance (open system), 82–83
　enthalpy. *See* enthalpy *(H)*.
　enthalpy changes during phase transitions, 86–92
　enthalpy of a solution, 101–103
　enthalpy values, 85–86
　entropy. *See* entropy *(S)*.
　entropy balance, 105–106
　first law of thermodynamics, 81–82
　fugacity/fugacity coefficient, 117–121
　Gibbs free energy, 106, 110
　heat capacity, 84–85
　heat of combustion, 94
　heat of condensation, 87
　heat of formation, 93
　heat of fusion, 89
　heat of mixing, 101
　heat of reaction, 93–94
　heat of vaporization, 87–89
　Helmholtz function, 106
　integral heat of solution, 101–103
　internal energy, 80–81
　isothermal effect of pressure, 119
　laws of thermochemistry, 92
　polytropic compression, 114
　power cycles, 121–123
　refrigeration, 123–129
　residual properties, 111
　reversible process, 103
　second law of thermodynamics, 104–106

　sensible heat, 83–85
　solution thermodynamics, 116–117
　thermodynamic properties, 78
　thermodynamic properties of matter, 110
　thermodynamic property relations of ideal gas, 112–114
　third law of thermodynamics, 114–116
　work, 79
engineering economics, 697–721
　analytical method of optimization, 719
　annuity, 701
　capital cost, 705
　capital-recovery factor, 703, 704
　CEPCI, 706
　compound interest, 698–699
　continuous compounding, 704–705
　cost comparison of alternatives, 713–720. *See also* cost comparison of alternatives.
　depreciation, 708–713. *See also* depreciation.
　effective interest rate, 699
　estimation of operating cost, 707–708
　interest tables, 701n
　Lang factor, 707
　linear break-even analysis, 719
　nominal interest rate, 699
　order-of-magnitude estimate of equipment, 707
　present worth/future worth, 699–700
　rate of interest, 697–698
　rule of 72, 699n
　simple interest, 698
　sinking-fund factor, 702–703, 704
　time value of money, 697–706
　VAPCCI, 706
enthalpy *(H)*, 81
　cooling towers, 520, 524
　enthalpy-composition diagram, 365–367
　free expansion, 124
　graphical representation, 86, 114
　heat of chemical reaction, 93
　heat of mixing, 101
　humid, 514–515
　law of Hess, 93
　P-H diagram, 86, 114
　phase transitions, 86–89
　property relations, 110
　of a solution, 101
enthalpy-composition diagrams, 365–367
　in distillation, 365
　in evaporation, 282
enthalpy-entropy *(HS)* diagram, 114
enthalpy values, 85–86
entrainer, 411
entrainer gas, 263
entrainment correlation, 675
entrapped service, 663
entropy *(S)*

　defined, 105
　ideal gas, 114
　physical representation, 114
　property relations, 110
　second law of thermodynamics, 105–106
　turbine expansion, 124
entropy balance, 105–106
environmental considerations, 761–762
　air pollution statement, 767–768
　contaminated soil, 769
　definitions, 762–764
　EIS, 780–781
　Federal Pollution Prevention Act, 762
　hearing protection, 770–772
　legislation, 762, 779, 780
　MSDS, 764
　mutagen, 764
　noise/noise abatement, 770–772, 780
　overview (flow chart), 779
　ozone, 765–766
　permit application, 779–781
　pH, 764–765
　PPE, 769–772
　smog, 766
　solid waste management, 773–779
　TOC, 765
　water pollution statement, 768–769
environmental degrees of freedom, 682
environmental impact statement (EIS), 780–781
environmental permit applications, 780–782
epoxy, 649
EQS4WIN, 731
equation of state
　Beattie-Bridgeman, 49
　Benedict-Rubin-Webb, 49
　ideal gas, 40
　Peng-Robinson, 47
　Redlich-Kwong, 47
　Soave-Redlich-Kwong, 47
　van der Waal, 47, 49
　virial, 48
equilateral triangular diagram, 461–463
equilibrium condensation, 389
equilibrium diagrams, 347–349
equilibrium distillation, 370–372
equilibrium flash calculations, 358–361
equilibrium moisture, 525
equilibrium solubility relationships, 433–444
equilibrium tie line, 463
equilibrium vaporization ratios, 357
equimolar counterdiffusion, 320, 321, 324, 334, 343
equipment design, 655–695
　agitator design, 667–674. *See also* agitator design.
　centrifugal pump trouble-shooting, 687–688

design pressure/temperature, 656–658
equipment designation (name), 684
equipment specification sheet, 656–658
examples, 689–694
heat exchanger design tips, 688–689
line identification, 684
mass transfer columns, 674–680
operating manual, 684–686
operating volume, 660
optimizing the design, 682–683
packed columns, 678–679
piping design, 661–666. *See also* piping design.
pressure vessel, 658–660. *See also* pressure vessel.
process design, 665–666
process flow sheets, 683–684
pumping system design, 665–666
surge volume, 660
testing the equipment, 680–682
tray columns, 678–680
equipment designation (name), 684
equipment specification sheet, 656–658
equipment testing and analysis, 680–682
equivalent diameter, 153, 167
 for heat transfer, 228
equivalent length (Le), 154–156, 163
Erbar-Maddox correlation, 414, 415
erg, 12
erosion corrosion, 645
ethanol-water azeotrope, 349
Euler number, 22
eutectic composition, 264
eutectic temperature, 264
evaporation, 279–290
 applications, 279
 corrosion, 280
 defined, 279
 evaporator capacity, 281
 factors to consider, 279–280
 heat-transfer coefficients, 281–282
 multiple-effect evaporator, 280–281, 285–290
  forward feed, 281
  backward feed, 281
  parallel feed, 281
 single-effect evaporator, 280, 282–285
 steam economy, 282
 types of evaporators, 280
evaporation problem, 55
evaporator capacity, 281
event tree analysis, 761
excess air/oxygen, 70, 71–75
excess reactant, 35
exit marking, 761
exothermic reaction, 92
exothermic reactions, 537
experimental diffusivity data, 325

explosion, 728. *See also* fire and explosion issues.
exposure limits, 724–727
extended surface heat exchanger, 264–267
extensive properties, 29, 78
extinction-ignition curve, 606–607
extinction temperature, 607
extract phase, 460, 464–465
extract solvent, 460
extractive distillation, 410–411
extraction equipment, 472–476
extraction system. *See* Leaching; Liquid-liquid extraction.
extremely hazardous substance, 773
extruded fin, 266

F

Fahrenheit scale, 7, 8
Fair's correlation, 427
fan, 199
fan static pressure, 199
Fanning equation, 149
Fanning friction factor, 149–150, 163
 chart, 149
fault tree analysis (FTA), 761
FBR. *See* fluidized bed reactor (FBR).
Federal Pollution Prevention Act, 762
feed, 313
feed-forward control, 616
feed plate location, 384–388, 414
feed temperature extinction-ignition curve, 607
feedback control, 615–616, 634–637
Fenske equation, 395, 413
ferritic stainless steel, 646, 647
fiberglass-reinforced plastics (FRP), 649
Fick's law, 318–319
film coefficients, 226–232
 for fluids in pipes and tubes, 126
film model, 336–337
filtration, 291–306
 batch-type filters, 291
 constant-pressure, 295, 297
 constant-rate, 296, 298
 continuous filters, 291
 filter aids, 292
 filter media, 292
 filter-medium resistance, 301
 Kozeny-Carman relation, 296–302
 porosity of cake, 302–304
 problems to be solved, 292
 Ruth's equation, 293–296
 specific resistance, 301
 types of filters, 291–292
final-value theorem, 618
financial considerations. *See* engineering economics.
fire and explosion issues, 725–757
 blocked condition, 749–750
 CAFT, 730–731

combustible liquid, 729
deflagration venting, 756–757
DIERS, 751
dust explosion pentagon, 728
emergency relief load calculation, 735–737
emergency relief system hardware, 734–735
fire heat (Q_{fire}), 738
fire tetrahedron, 727–728
flame temperature, 730–731
flammability limit, 731–734
flammable liquid, 729
flash point, 730
heat exchanger tube rupture, 750–753
insulation, 739–741
inverting, 732–733
Leung equation, 752
mass flux (G_T), 751–753
NFPA-30 vent reduction factor, 739
overpressure protection, 733–734
RAGAGEP, 751
relief/safety valve, 741–746
rupture disc, 746–749
steam hammer, 737
stoichiometry of combustion reaction, 731–734
terminology, 728–729
TFT, 730
thermal venting, 754–756
vent area, 752
venting capacities of atmospheric tanks, 753–754
water hammer, 735–737
wetted area calculation, 737–739
fire heat (Q_{fire}), 738
fire point, 728
fire-proof insulation, 212, 739–741
fire tetrahedron, 727–728
first law of thermodynamics, 81–82
first-order control system, 625–628
first-order reaction, 551, 556
first-order reversible reactions, 561–562
fixed bed adsorption, 501–510
fixed bed reactor. *See* packed bed reactor (PBR).
flame temperature, 99–101, 730–731
flammability limit, 731–734
flammable liquid, 729
flange primary service pressure rating, 662
flare, 767
flare sizing, 774
flare stack design, 774–779
flash, 358
flash calculations, 358–361
flash distillation, 370–372
flash point, 728, 730
flash-vaporization problem, 359
flood factor, 426
flooding, 426, 473–474

flooding correlation (packed extraction towers), 474
flooding limits, 674
flooding pressure drop, 432
flooding velocity, 428, 675
flow across banks of tubes, 229
flow coefficient (Cv), 157
flow of compressible fluids, 193–201
   compression equipment, 197
   critical pressure ratio, 194
   critical velocity, 194–195
   fan, 199
   fan static pressure, 199
   Mach number, 193–194
   multistage compression, 197, 200
   power requirements, 198–199
   pressure drop, 195–196
   ratio of specific heats, 197
flow of fluids in pipes, 147–168. *See also* pipes-flow of fluids.
flow parameter (abscissa), 430–432
flow rate, 26
flow reactor, 566–567
   CSTR. *See* stirred-tank reactor.
   mass and energy balances, 568–576
   plug-flow reactor. *See* plug-flow reactor.
   space time, 568
   space velocity, 568
flow regimes, 201
flow system, 28
flowmeter, 142–143
flue gas, 69, 75, 76
fluid measurements, 138–143
fluid mechanics, 131–208
   Bernoulli equation, 135–137
   compressible fluids, 193–201. *See also* flow of compressible fluids.
   control valve, 187–191
   Duckler's homogeneous method, 204, 206–207
   flow of fluids in pipes, 147–168. *See also* pipes-flow of fluids.
   fluid measurements, 138–143
   fluid statics, 132–137
   Lockhort-Martinelli method, 203, 205–206
   manometers, 133–135
   mass flowmeter, 142–143
   orificemeter, 139, 141–142
   parallel and branched systems, 192–193
   permanent pressure loss. *See* pressure loss/drop.
   Pitot tube, 138–140
   pump calculations, 168–187. *See also* pump calculations.
   stagnation pressure, 138
   static pressure, 138
   surface tension, 151–152
   two-phase flow considerations, 201–207
   venturimeter, 139, 140–141
   viscosity, 132
   vortex shedding flowmeter, 143
fluid statics, 132–137
fluidized bed reactor (FBR), 567
   heat transfer, 599
   product distribution/temperature, 592
fluidized CSTR, 592
fluoropolymers, 649–650
flux of diffusing component A, 322, 323, 336
FMECA, 761
foaming, 441n
foot-pound force (ft · lb_f), 11
foot-pound-second (fps) system, 3–4, 6
force, 4, 12
forced circulation reboilers, 224, 262
forced convection, 213
forcing functions, 624–625
forward-feed evaporator, 281
fouling/fouling factor, 224–225
Fourier's law, 210
fps system, 3–4, 6
fractional conversion, 35
fractional distillation, 370
fractional/product loss, 314
free convection, 233–234
free expansion, 124, 126–127
free moisture, 525
frequency response, 625
fretting, 645
Freundlich isotherm, 497, 505
Freundlich isotherm equation, 497, 498
friction factor equations, 154
frictional losses, 149–151, 154
Froude number, 22
FRP, 649
FTA, 761
fuels and combustion, 67–76
   analysis of fuels, 68–69
   excess air/oxygen, 70, 71–75
   flue gas, 69, 75, 76
   gaseous fuels, 67, 69
   heating value of fuels, 67–68, 69
   higher heating value HHV, 67
   liquid fuels, 67, 68–69
   lower heating value LLV, 67
   Orsat analysis, 69, 72
   solid fuels, 67, 68
   stack gas, 69
   theoretical air/oxygen, 70
fugacity/fugacity coefficient, 117–121, 354–355
fugitive emission, 763
fundamental point properties, 78
fundamental units, 2
furan, 649
future worth
   annuity, 701
   present worth, and, 700
   single payment, 699

G

G, 110
$G(T)$, 605, 609–661
G-type fin, 266
gain-time constant form, 634
galvanic or bimetal corrosion, 640, 641
galvanic series of materials, 641
gas
   entrainer, 263
   heat capacity, 84
   ideal. *See* ideal gas.
   real, 45–54
gas density/specific gravity, 25
gas enthalpy transfer units, 524
gas-solid reactions, 591–593
gaseous fuels, 67, 69
g_c, 4, 5
general accounting principle, 29
generalized compressibility factor chart, 46–47, 50–52
generalized overall mass transfer coefficients, 343
Generalized property departure charts, 111, 129
Gibbs free energy, 106, 110
Gilliland correlation, 414
glass/glass-lined carbon steel, 650–651
glass heat exchanger, 271
global warming, 763
graphical presentation/methods. *See also* diagram.
   enthalpy data, 86
   McCabe-Thiele method, 380–384
   N_{OG}, 440–441
   plate-to-plate calculations, 380
   Poncho-Savarit method, 380
   thermodynamic property data, 114
graphite, 652
graphite heat exchanger, 270
Grashof number, 338
gray bodies, 214
greenhouse effect, 763
grooved fin, 266
gross heating value, 67
gross profit, 717
guide words, 759, 760
Guldberg and Waage law, 539

H

H_{OG}, 439
H_{OL}, 439
Hagen-Poiseuille equation, 16, 149, 294
halftimes, 552
HAP, 763
harmful ozone, 766
HAZAN, 761
hazard and operability studies (HazOp), 758–761

hazardous air pollutant (HAP), 763
Hazardous and Solid Waste Act (HSWA), 773
Hazardous Material Transportation Act, 780
hazardous substance, 773
hazardous waste incineration, 773
HazOp, 758–761
head-capacity curve, 174
hearing protection, 770–772
heat, 78–79, 104
heat capacity, 84–85
heat exchanger
    air-cooled, 264–267
    cross-flow, 269
    design tips, 688–689
    double pipe, 251
    extended surface, 264–267
    glass, 271
    graphite, 270
    nonmetallic, 269–271
    pressure drop, 251–260
    shell-and-tube, 249–252
    Teflon, 269
    tube rupture, 750–753
heat exchanger tube rupture, 750–753
heat generated term [G(T)], 605, 609–661
heat loss from insulated pipe to air, 217–218
heat of chemical reaction, 93
heat of combustion, 94
heat of condensation, 87
heat of formation, 93, 95
heat of fusion, 89
heat of mixing, 101
heat of reaction, 93–94, 95
heat of vaporization, 87–89
heat pump, 271
heat pump COP, 271
heat pump cycle, 271
heat removal or addition, 598–599, 605, 609–611
heat removed term [R(T)], 605, 611
heat transfer, 209–278
    batch heating and cooling, 268–269
    boiling, 234
    caloric fluid temperature, 226
    Colburn J_H factor, 227–228
    composite cylindrical resistance, 217, 218
    composite flat walls, 211–212, 215
    condensation, 235–237
    conduction, 210–213
    convection, 213, 233–234
    critical radius/maximum heat loss, 218–219
    crystallization, 263–264
    Dittus-Boelter equation, 227
    effectiveness-NTU method, 260–262
    extended surface heat exchanger, 264–267
    film coefficient of, 226–232
    flow over tube banks, 229
    fluids flowing inside/outside of pipes, 223–226
    fouling/fouling factor, 224–225
    free convection of fluids outside horizontal pipes, 233–234
    heat exchanger. *See* seat exchanger.
    heat loss from insulated pipe to air, 217–218
    heat pump, 271
    hollow sphere, 219
    insulation, 210, 212
    Kirchhoff's law, 220
    LMTD, 225–226, 249–251
    logarithmic temperature differences, 225–226, 249–251
    low, hollow cylinder, 219
    nonmetallic heat exchangers, 269–271
    overall heat transfer coefficient, 223–224
    pipe wall, 217, 218
    pressure drop in exchanger, 251–260
    radiation, 214–215, 219, 220–223
    reboilers, 262–263
    resistance thermometer, 213
    shell-and-tube exchanger, 249–252
    Sieder-Tate equation, 226, 227
    simultaneous convection-radiation, 213
    sublimation, 263
heat transfer rate, 210
heating value of fuels, 67–68, 69
heavy components, 411
height equivalent to theoretical stage (HETS), 472
Helmholtz function, 106
Henry's law, 119, 353
heterogeneous azeotrope, 349, 409
heterogeneous minimum boiling azeotrope, 410
HETP, 423–424, 678
HETS, 472
HHV, 67
Higbie penetration model, 337
high strength enclosure, 756
higher heating value, 67. *See also* HHV.
homogeneous azeotropes, 409
homogeneous catalyzed reactions, 559
homogeneous reaction, 537
Hooper 2-K factors, 157–158, 159
horsepower (hp), 11
Hougen-Watson generalized property departure charts, 111, 129
Hougen z charts, 54, 76
HQ curve, 174
HS diagram, 114
humid enthalpy, 514–515
humid heat, 514
humid volume, 514
humidification, 511. *See also* psychrometry and humidification.
humidity chart, 516–519
humidity ratio, 512
hydraulic horsepower, 182
hydraulic radius, 153
hydro testing, 680–681
hydrodynamic entry length, 151
hydrogen damage, 645

I

ideal gas, 35–40
    defined, 35
    entropy changes, 114
    equation of state, 40
    mixtures of gases, 43–45
    thermodymamic property relations, 112–114
ideal gas law, 40–43
ideal gas law constant, 41
ideal gas mixtures, 43–45
ideal mixtures, 350–351
ideal refrigeration cycle, 123
ideal stage, 476
ideal tubular reactor. *See* plug-flow reactor.
IDLH, 725
ignition-extinction curve, 606–607
ignition temperature, 607
immiscible liquids, 349–350
impact pressure, 138
impeller, 669–671
impeller diameter, 176
impeller Reynolds number, 672
impervious graphite, 652
impregnated graphite, 652
impulse function, 625
incineration, 767, 773
inclined manometer, 134, 135
incomplete reactions, 34
incremental analysis (rate of return), 716
independent reactions, 578
industrial hygiene, 724–727
industrial process analysis, 23
inert purge gas stripping cycle, 504
initial rates, 552
initial-value theorem, 618
inlet pressure drop, 746
in-line tube arrangement, 229
insulation, 210, 212
integral heat of solution, 101–103
integration, 551–552
intensive properties, 29, 78
intention (node), 758
interest, 697
interest tables, 701n
intergranular corrosion, 642
internal energy, 80–81
internal rate of return (IRR), 714
interphase mass transfer, 341–343
intrinsic energy, 80–81
inverting, 732–733

iron, 646
IRR, 714
irreversible reactions, 550–561
isenthalpic, 124
isentropic, 124, 138
isobaric process, 113
isobutanol-water system, 349
isochoric process, 113
isothermal compression, 197
isothermal effect of pressure, 119
isothermal process, 113

J

J factor methods, 338
J_H factor, 227–228
Jacobs formula, 665
Jones formulas, 732
junctions of streams, 59

K

K, 154, 155–156
K values, 357, 360, 361
$K_G a$, 440
K_v, 426, 431
Kay's method, 53
KE, 78
Kelvin scale, 7, 8
Kern relationship for film temperature, 237
kettle reboilers, 224
key components, 379, 411–412
kinematic viscosity, 15
kinetic energy, 78
kinetic energy correction factor α, 136
Kirchhoff's law, 220
Kirkbride correlation, 414
Kister-Gill plot, 429
Knudsen diffusion, 332
Kozeny-Carman relation, 296–302
Kutateladze correlation, 234

L

L-footed fin, 266
laminar flow, 148
Lang factor, 707
Langmuir equation, 504, 505
Langmuir-Hinshelwood adsorption model, 591
Langmuir isotherm, 505
Langmuir monolayer adsorption model, 495–496
Laplace transform, 617, 618, 625
latent heat, 86
law of conservation of energy, 81–82
law of corresponding states, 45
law of definite proportion, 32
law of Hess, 93
law of mass action, 539
law of multiple proportion, 32

law of reciprocal proportion, 32
laws of thermochemistry, 92
LC_{50}, 763
LD_{50}, 763
Le, 154–156, 163
Le Chatelier formulas, 731
leaching, 476–491
   calculation, 476
   calculation of equilibrium stages, 479–480
   equilibrium relationship, 476
   ideal stage, 476
   mass transfer, 490–491
   material balance equations, 480–486
   overflow/underflow, 476
   overview, 476
   Ponchon-Savarit diagram, 478–479, 489
   rate of leaching when dissolving solid, 490–491
   right triangular diagram, 477–478
   traingular diagram (material balance), 486
leaf filter, 292
Lee-Kessler generalized correlation tables, 111
Lee-Kessler z table, 52, 76
LEL, 728
length, 12
Lennard-Jones function, 326
lethal concentration (LC_{50}), 763
lethal dose (LD_{50}), 763
Leung equation, 752
Levenspiel-Kunii bubbling bed model, 592
Lewis method, 379
Lewis number, 516
Lewis-Randall rule, 118, 119
LFL, 728
LHV, 67
life cycle analysis, 764
light components, 411–412
Lightfoot's surface-stretch theory, 337
limiting reactant, 34
line identification, 684
line sizing guideline, 166–167
linear break-even analysis, 719
liquid
   adsorption, 497–510
   combustible, 729
   flammable, 729
   molecular diffusion, 323–328
   sensible heat changes, 85
   vapor pressure, 37–40
liquid fuels, 67, 68–69
liquid-liquid extraction, 459–476
   calculations, 466–472
   capacity parameter, 430
   conjugate line, 464
   conjugate phases, 464–465
   equilateral triangular diagram, 461–463
   extract phase, 464–465

   extraction equipment, 472–476
   flow parameter, 430
   liquid-liquid equilibrium, 460
   multistage countercurrent system, 470
   multistage extraction, 467
   overview, 459–460
   packed towers, 472–476
   Ponchon-Savarit diagram, 465
   raffinate phase, 464–465
   rectangular coordinates, 463, 464
   right triangle coordinates, 463, 464
   selectivity of solvent, 465
   single-stage equilibrium extraction, 466, 467
   stagewise extraction, 467
   terminology, 460
   ternary systems, 460–463
   triangular coordinates, 461–463
liquid manometers, 133–135
liquid phase adsorption, 497–510
liquid-phase mass transfer coefficients, 344
liquid relief load, 736
liquid-to-heating proportionality factor, 234
liquid-to-heating surface combination factor, 234
LL extraction. *See* liquid-liquid extraction.
LMTD, 225–226, 249–251
load disturbances, 615
loading point, 428
local barometric pressure, 656, 689
local convective mass transfer coefficient (flat plate), 338
local efficiency, 423
Lockett equation (O'Connell plot), 422
Lockhort-Martinelli method, 203, 205–206
Lockhart-Martinelli parameter, 203, 206
logarithmic mean temperature difference (LMTD), 225–226, 249–251
low alloy steels, 646, 653
low strength enclosure, 756
low-temperature application of metals, 652, 653
lower explosive limit (LEL), 728
lower flammability limit (LFL), 728
lower heating value (LHV), 67

M

Mach number, 193–194, 195
maintenance and workmanship, 761
manipulated variables, 615
manometers, 133–135
Margules equations, 356–357
Marshall & Swift Equipment Cost Index, 706
martensitic stainless steel, 646, 647
mass balance, 28–37
mass balance equation, 29–31
mass concentration, 26

mass flow rate, 26
mass flowmeter, 142–143
mass flux (G_T), 751–753
mass fraction, 26
mass transfer, 318
mass transfer columns, 674–680
mass transfer fundamentals, 317–367
  convective mass transfer, 333–336
  dimensionless numbers, 338–341
  enthalpy-composition diagrams, 365–367
  equimolar counterdiffusion, 320, 321, 324, 334
  flux of diffusing component A, 322, 323, 336
  interphase mass transfer, 341–343
  leaching, 490–491
  mass transfer in packed beds, 343–346
  molecular diffusion. *See* molecular diffusion.
  mass transfer coefficients, 335
  overall mass transfer coefficients, 342–343
  phase equilibria, 346, 353–355
  turbulent mass transfer, 336–338
  units of mass-transfer coefficients, 335, 345
  vapor-liquid equilibria, 346–365. *See also* vapor-liquid equilibria.
material balances, 23–76
  bypass stream, 61
  chemical equation, 31–32
  chemical reactions, 63–64
  concentration, 26
  density, 24–25
  distillation column, 57
  evaporation problem, 55
  flow rate, 26
  fuels and combustion, 67–76. *See also* fuels and combustion.
  gas density/specific gravity, 25
  general accounting principle, 29
  ideal gas, 40–45. *See also* ideal gas.
  mass balance, 28–37
  mass balance equation, 29–31
  mole (mol) fraction, 26
  multiple subsystems, 59–60
  phase behavior, 37–40
  problem-solving technique, 33, 54
  purge stream, 61–62
  real gases, 45–54
  recycle, 60–61
  specific gravity, 24–25
  stoichiometric problems, 33
  stoichiometry, 32
  tie component, 58
  weight fraction, 26
mass balance problems (with chemical reactions), 63–64

mass balance problems (without chemical reactions), 54–58
material safety data sheet (MSDS), 764
materials of construction. *See* corrosion and materials of construction.
mathematical modeling, 622
MAWP, 659
maximum allowable velocity, 431
maximum allowable working pressure (MAWP), 659
maximum boiling point systems, 347
maximum non-shock service pressure rating, 662
McAdams relationship for film temperature, 237
McCabe-Thiele construction in adsorption, 500
McCabe-Thiele method, 380–384, 392–393
McMillen, E. L., 299
mean diameter, 691
mean mass transfer coefficient (flat plate), 338
means of egress, 761
mechanical integrity testing, 680–681
mechanical minimum flow, 666
membrane, 307
membrane separation, 307–315
  application of membrane technology, 308
  carrier method, 308
  electrodialysis (ED), 314
  nomenclature, 315
  percent recovery, 313–314
  pervaporation, 314–315
  pore method, 308
  reverse osmosis (RO), 308–314
  ultrafiltration (UF), 314
metallic materials of constructions, 648–649
metallic vessels, 643–644
meter, 12
method of differentiation, 550–551
method of halftimes, 552
method of initial rates, 552
method of integration, 551–552
method of k calculation, 553
method of reference curves, 552
metric system, 3, 6
metric ton, 12
minimum free flow area, 229
minimum ignition energy (MIE), 729
minimum oxygen concentration (MOC), 728–729
minimum recirculation flow, 666
minimum reflux ratio, 387–388, 401, 403, 406, 413–414
minimum wetting rate
  absorption, 442, 444
  pipes–flow of fluids, 152

minus sign (–), 79n, 103, 110
mist-eliminating devices, 152
mixed-flow reactor. *See* stirred-tank reactor.
mixing time, 668
mks system, 3, 6
molal absolute humidity, 512
molal heat capacities, 94
molality, 26, 27, 310
molar concentration, 26
molar units, 6–7
molarity, 26, 310
mole, 7
mole (mol) fraction, 26
molecular diffusion, 318–332
  diffusion through polymers, 330–331
  effective diffusivity, 332
  equimolar counterdiffusion, 320, 321, 324
  experimental diffusivity data, 325
  Fick's law, 318–319
  flux of diffusing component A, 322, 323
  gases A and B plus convection, of, 320–321
  Knudsen diffusion, 332
  gases, in porous solids, 332
  liquids, in porous solids, 323, 328
  polymers, 330
  radial diffusion, 329
  solids, in, 329–332
  steady-state diffusion of A into stagnant non-diffusing B, 321–322
  steady state equimolar counter diffusion, 320
molecular mass, 7
molecular sieve zeolites, 494
molecular weight, 7
molecularity, 539
Mollier diagram, 114, 197
momentum transfer, 318
money matters. *See* engineering economics.
monolayer adsorption model, 495–496
Moody-Darcy chart, 154
Moody friction factor, 154
motor horsepower, 665, 669
MSDS, 764
multicomponent distillation, 411–412, 413–415
multicomponent system, 346
multifeed column, 389–391
multimedia impact of environmental regulation, 779, 780
multi-pass correction factor, 260
multipass shell and tube heat exchangers, 250
multiple-effect evaporator, 280–281, 285–290
multiple reactions, 578
multistage compression, 197, 200

multistage countercurrent adsorption, 500
multistage countercurrent extraction system, 480
multistage countercurrent liquid-liquid extraction system, 470
multistage countercurrent system, 470
multistage crosscurrent adsorption, 498–500
multistage extraction, 467
Murphree efficiency, 422–423
mutagen, 764

N

K_Ga, 440
N_{OG}, 436–438, 439, 440–441
N_{OL}, 438–439, 439
N_P, 439
nameplate (pressure vessel), 680–681
natural circulation reboilers, 262
natural convection, 213
natural frequency, 631
natural gas, 67
natural petroleums, 68
negative feedback, 616, 635–637
NESHAP, 780
net heating value, 67
net positive suction head ($NPSH_A$), 171–172
net profit, 717
Newton (N), 5, 6, 12
Newtonian fluid, 132
Newton's law dimensional proportionality factor, 3
Newton's law equation, 2
Newton's law proportionality factor, 132, 169, 234
Newton's second law of angular motion, 142
Newton's second law of motion, 5
NFPA-30 vent reduction factor, 739
NFPA-30 wetted area calculation table, 738
n-heptane-methyl cyclohexane-water, 461
nickel/high-nickel alloys, 646, 647
niobium, 653
nitrogen blanketing, 656
nobility number, 640, 641
node, 192
node analysis, 192
node analysis (HazOp), 760
noise/noise abatement, 770–772, 780
nominal interest rate, 699
non-destructive testing, 681–682
nonelementary reaction, 539
nongray enclosures, 221
nonideal systems, 355
nonisothermal reactors, 596–597, 599–601
nonmetallic heat exchangers, 269–271

non-metallic materials of constructions, 649–652
normal boiling point, 39
normality, 26, 310
nozzle
  concentric circular, 143
  converging-diverging, 195
  subsonic flow, 143
NPDES, 780
$NPSH_A$, 171–172, 177, 181–182
$NPSH_C$, 174
NRTL equation, 357
nuclear/radioactive waste, 774
nucleate boiling heat flux, 234
number of degrees of freedom, 54
Nusselt number, 231, 232

O

Occupational Safety and Health Act (OSHA), 780
ocean dumping, 780
O'Connell correlation, 676
O'Connell plot, 409, 422
offset, 621
once-through reboilers, 262
one variable optimization problem, 682–683
on-off controller, 621
OP, 657
open loop control, 615
open manometer, 133
open steam, 393–394
open system, 28
operating cost, 707–708
operating manual, 684–686
operating pressure (OP), 657
operating volume, 660
optimal temperature progression, 597–611
  adiabatic plug-flow tubular reactor, 602
  defined, 597
  heat addition or removal, 598–599
  heat generated term [G(T)], 605, 609–661
  heat removed term [R(T)], 605, 611
  ignition-extinction curve, 606–607
  multiple steady states, 604–611
  nonisothermal continuous-flow reactors, 602
  runaway reactions, 608–609
  steady-state plug-flow reactor with heat transfer, 602
  tubular reactors (PFR/PBR), 599
optimizing equipment design, 682–683
optimum reflux ratio, 394–395, 406
order-of-magnitude estimate of equipment, 707
order of reaction, 539
orifice, 139, 141
orifice coefficient, 161
orificemeter, 139, 141–142

Orsat analysis, 69, 72
OSHA, 780
osmosis, 308
overall heat transfer coefficient, 223–224
overall mass transfer coefficients, 342–343
overall order of reaction, 539
overall tray efficiency, 421–422
overdamped, 630
overflow, 476
overpressure protection, 733–734
overshoot, 630
ozone depletion, 766

P

packed-bed CSTR, 592
packed bed reactor (PBR), 567
  heat transfer, 599
  product distribution/temperature, 590–591, 592
packed towers
  absorption, 433–444
  design, 674–680
  distillation, 419–426, 428–430
  extraction, 472–476
  tray columns, compared, 677
packing height, determination of, 436–441, 678
packing size, 441, 443, 444
parallel and branched systems, 192–193
parallel-feed multiple-effect evaporators, 281
parallel reactions, 558–559, 578–579
partial combustion, 67
partial condensation, 389
partial pressure, 37, 323
partial vaporization, 358
particulates removal, 773
parts per million (ppm), 310
payback method, 718
payout time method, 718
PBR. See packed bed reactor (PBR).
PE, 78
Peclet number, 339
PEL, 725
Peng-Robinson equation of state, 47
per pass conversion, 64
% conversion, 35
percent absolute humidity, 512
percent absolute molar humidity, 512
percent recovery, 313–314
percent relative humidity, 513
percent yield, 36
performance curve, 174
performance testing, 680
permeability, 330
permeate, 313
permeate recovery, 313
personal protective equipment (PPE), 769–772
pervaporation, 314–315

Pesticide programs, 780
petrochemical piping system, 661–666
petroleum fuels, 68
PFD, 683–684
pH, 764–765
PH diagram, 86, 114
PHA, 758
phase, 37
phase behavior, 37–40
phase equilibria, 346, 353–355
phase rule, 347
phase transitions, 86
phenolic, 649
physical adsorption, 495
physical quantity, 1
P&ID, 684
π groups, 20
π theorem, 20–21
piezometer ring, 138
pipe expansion, 160
pipe material thickness, 664
pipe reducer, 160
pipe schedule numbers, 663–664
pipes–flow of fluids, 147–168
    bubble rise velocity, 152
    Darby 3-K constants, 157, 158
    droplet diameter, 151–152
    equivalent diameter, 153, 167
    equivalent length (Le), 154–156, 163
    Fanning friction factor chart, 149
    Fanning friction factors, 149–150, 163
    flow coefficient (Cv), 157
    frictional losses, 149–151, 154
    Hooper 2-K factors, 157–158, 159
    hydrodynamic entry length, 151
    laminar flow, 148
    minimum wetting rate, 152
    pressure drop, 160–161
    resistance coefficient (K), 154, 155–156
    Reynolds number, 147–148
    Sauter mean diameter, 151
    turbulent flow, 148
    velocity distribution, 148
piping and instrumentation diagram
    (P&ID), 684
piping design, 661–666
    example, 693–694
    factors to consider, 661
    material thickness, 664
    nomenclature, 664–665
    piping rating, 662–663
    pumping system design, 665–666
    schedule numbers, 663–664
    temperature limits, 663
piping rating, 662–663
piston flow reactor. *See* plug-flow reactor.
Pitot, Henry, 138
Pitot tube, 138–140, 146–147
pitting, 642
Pitzer acentric factor, 47, 48
plait point, 463

plant safety, 723–782
    dilution air, 727
    emergency ingress/egress, 761
    environmental considerations. *See*
        environmental considerations.
    exposure limits, 724–727
    fire and explosion issues. *See* fire and
        explosion issues.
    HazOp, 758–761
    PPE, 769–772
    TLVs, 724–727
    toxicology/industrial hygiene, 724–727
plate and frame filters, 292
plate columns, 674
plate efficiency factor, 379
plate exchangers, 225
plate spacing, 674
plate thickness, 659–660, 690–693
plate-to-plate calculations, 379–384
plug-flow reactor, 567
    diagram, 569
    mass and energy balances, 571–572, 575–576
    optimal temperature progression, 602
    product distribution/temperature, 580–581
    recycle reactor, 582
    volume, 576
pneumatic controller, 622
POHC, 773
point efficiency, 423
point functions, 81
point properties, 78
point selectivity, 36
pole-zero form of transfer function, 634
pollution, 761–762. *See also*
    environmental considerations.
polyester & vinyl ester, 649
Polymath, 603
polytropic compression, 114, 197
Poncho-Savarit graphical method, 380
pore method, 308
porous medium, 291
porous solids, 331–332
positive feedback, 616
potential energy, 78
poundal, 4, 12
power, unit of measure, 11
power cycles, 121–123
Power number, 668, 673
Poynting factor, 354
PPE, 769–772
ppm, 26, 310
practical reflux ratio, 406
Prandtl number, 227, 231, 235, 516
precipitation hardening stainless steel, 646, 647
present worth
    annuity, 701
    future worth, and, 700

    single payment, 699
    uniform series, 704
present worth method (cost comparison), 714
pressure
    absolute, 9
    atmospheric, 9, 10
    barometric, 656, 689
    constant, 113
    defined, 9
    design, 656–658
    differential set, 746
    fan static, 199
    flammability limit, 731
    isothermal effect, 119
    MAWP, 659
    operating, 657
    partial, 37
    pseudocritical, 46
    reduced, 46
    stagnation, 138
    static, 138
    triple point, 263
    units of measure, 9–10
    vapor, 37–40
pressure-enthalpy (*PH*) diagram, 86, 114
pressure head, 169
pressure loss/drop, 143–147
    compressible fluid, 195–196
    concentric circular nozzle, 143
    condenser, 253
    double pipe exchanger, 251
        annulus, 251
        inner pipe, 251
        shell side, 252
        tube side, 251
    flooding, 426, 432
    heat exchanger, 251–260
    inlet line (relief device), 746
    packed absorption towers, 441, 442
    packed distillation towers, 429–430
    pipeline, 160–161
    shell-and-tube exchanger, 251
    subsonic flow nozzle, 143
    two-phase pressure drop, 201
    venturi, 143
pressure purging, 733
pressure swing cycle, 504
pressure units, 9–10
pressure venting, 755
pressure vessel, 658–660
    ASME codes, 658–659
    hydro testing, 680
    MAWP, 659
    mechanical intensity testing, 680
    nameplate, 680–681
    plate thickness, 659–660, 690–693
    residence time, 661
Pressure Vessel Inspection Code:
    Maintenance Inspection, Rating,
    Repair, and Alteration, 681

primary pumping capacity, 667
principal, 697
principal organic hazardous constituents (POHC), 773
principle of two resistances, 341
process, 411
process control, 613–614
　advanced topics, 637
　block diagram, 616–617
　closed loop control, 615–616, 634–637
　definitions, 615–616
　development of process models, 623–625
　differential equations, 624, 625, 634
　feedback control, 615–616, 634–637
　feed-forward control, 616
　final-value theorem, 618
　first-order systems, 625–628
　forcing functions, 624–625
　initial-value theorem, 618
　Laplace transform, 617, 618, 625
　linear/nonlinear models, 624
　load disturbances, 615
　mathematical modeling, 622
　on-off controller, 621
　open loop control, 615
　overall system transfer functions, 619–620
　pneumatic controller, 622
　pole-zero form of transfer function, 634
　proportional and floating controllers, 621, 622
　proportional band, 622
　regulatory operation, 616
　Routh's array, 635, 636
　second-order systems, 628–634
　servo operation mode, 616
　stability, 634–637
　transfer functions (Laplace transform), 618
　under/overdamped, 629–630
　variables, 615, 624
　zeros of transfer function, 634
process design, 665–666
process equipment design. *See* equipment design.
process flow diagram (PFD), 683–684
process flow sheets, 683–684
process hazard analysis (PHA), 758
process piping. *See* piping design.
process plant design calculations, 59
process safety management (PSM), 758
process specification data sheets, 656–658
Properties of Gases and Liquids (Reid et al.), 45
proportional and floating controllers, 621, 622
proportional band, 622
proximate analysis, 68
pseudobinary method, 411

pseudocritical pressure, 46
pseudocritical temperature, 46PSM, 758
psychrometric chart, 516–519
psychrometric ratio, 515
psychrometry and humidifacation, 511–520
　adiabatic saturation temperature, 516
　definitions, 512–515
　psychrometric/humid chart, 516–519
　wet-bulb temperature, 515–516
pulse function, 624–625
pump affinity laws, 175–177
pump calculations, 168–187
　available NPSH, 171–172, 177, 181–182
　brake horsepower (bhp), 174, 182
　capacity, 168
　cavitation, 172, 173
　centrifugal pump calculations, 174–175
　examples, 177–187
　impeller diameter, 176
　$NPSH_A$, 171–172, 177, 181–182
　performance curve, 174
　pressure head, 169
　pump affinity laws, 175–177
　pump-head curve, 183
　shut-off head, 170
　specific speed, 173
　speed change (rpm), 176
　static discharge head, 170, 181
　static head, 168–169
　static suction head, 169, 181
　static suction lift, 169
　system head, 174
　total discharge head (TDH), 170, 182
　total suction head, 169
　velocity head, 169, 181
pump-head curve, 183
pumping system design, 665–666
Pure Food and Drug Act, 780
purge gas stripping, 504
purge stream, 61–62
PW method (cost comparison), 714

R

R(T), 605, 611
R-factor, 210, 211
radial diffusion, 329
radial flow (tray tower), 428
radial flow impeller, 669
radian frequency, 630
radiation, 214–215, 219, 220–223
radiation coefficient, 214
radiation programs, 780
radiation thermometer, 214–215
radioactive waste, 774
raffinate phase, 460, 464–465
raffinate solvent, 460
RAGAGEP, 751
Ramp function, 625

random packings, 419, 424, 429, 430, 473
rangeability, 188
Rankine scale, 7, 8
Raoult's law, 350, 351, 353
rate coefficient, 538
rate constant, 538
rate equation, 537–538
rate limiting step, 591
rate of drying, 527–528
rate-of-drying curve, 526–527
rate of interest, 697–698
rate of return, 697
rate-of-return method, 714–715
rate selectivity, 578
ratio of specific heats, 197
Rayleigh equation, 372
Rayleigh number, 231
Rayleigh's algebraic method, 18–20
reactants, 34
reaction engineering. *See* chemical reaction engineering.
real bodies, 214
real gases, 45–54
reboilers, 224, 262–263, 370
RECRA, 780
rectangular pulse function, 625
rectification, 376, 379
rectification section, 379, 390
rectification tray tower, 377
recycle, 60–61
recycle reactor, 579, 582–584
Redlich-Kwong equation of state, 47
reduced pressure, 46
reduced temperature, 46
reduced volume, 46
reference curves, 552
reflux, 379
reflux ratio 379
　cold reflux, 388–391
　minimum reflux ratio, 387–388, 413–414
　optimum reflux ratio, 394–395
　total reflux, 389, 413
refrigeration, 123–129
regeneration of bed, 503–504
relative enthalpy, 514–515
relative humidity, 513
relative volatility, 352–353
relief load calculation (external fire), 737
relief/safety valve, 741–746
remaining life (vessel), 681
residual Gibbs free energy, 111
residual properties, 111
residual volume, 111
resins, 494
resistance coefficient (K), 154, 155–156
resistance thermometer, 213
Resource Conservation and Recovery Act (RECRA), 780
restricted lift valve, 746

return on investment (ROI), 714, 718
reverse flow (tray tower), 428
reverse osmosis (RO), 308–314
reversed Carnot cycle, 123
reversible Carnot cycle, 104
reversible cyclic process, 121
reversible process, 103
reversible reactions, 561–563
Reynolds number, 147–148, 338, 668, 741
R-factor, 210, 211
Riedel's equation, 89
right triangle diagram
    leaching, 477–478
    LL extraction, 463, 464
rise time, 630
risk management program (RMP), 758
RMP, 758
RO, 308–314
RO/UF process, 312
ROI, 714, 718
rotary filter, 292
roughness factor, 156, 164, 178
round-edged orifice, 139
Routh-Hurwitz criterion, 634
Routh's array, 635, 636
Routh's test, 634
rule of 72, 699n
runaway reactions, 608–609
rupture disc, 746–749
Ruth's equation, 293–296

S

Safe Drinking Water Act, 780
safety. *See* plant safety.
safety valve, 741–746
salvage value, 709
saturated steam at tower pressure, 393–394
saturation, 38
saturation humidity, 512
saturation temperature, 38
Sauter mean diameter, 151
SCC, 645
schedule number (see pipe schedule number), 663
Schmidt number, 338, 341, 516
second law of thermodynamics, 104–106
second-order control system, 628–634
second-order reaction, 551–552, 556–557
second-order reversible recations, 562–563
segmental downcomer, 427, 428
selective leaching, 642
selectivity, 36
selectivity diagram, 465
selectivity of solvent, 465
sensible heat, 83–85
sensible heat transfer, 83
separation of binary azeotropic mixtures, 409–412
separation processes, 318

series reactions, 560–561, 578, 580
sharp-edged orifice, 139
shear stress, 132
shell and trays, 426–427
shell-and-tube exchangers, 249–252
Sherwood number, 338–341
shortcut methods (multicomponent distillation), 413–415
shut-off head, 170
SI units, 5, 6
    prefixes and symbols, 5
side-stream withdrawal, 391–392
Sieder-Tate equation, 226, 227
sieve trays, 678
silica-alumina catalyst, 591
silica gel, 494
silicon carbide, 271
simple interest, 698
simple manometer, 134, 135
simple steam distillation, 374–375
simplified Omega method, 751
Simpson's rule, 372, 426, 575, 576, 604
simultaneous convection-radiation, 213
simultaneous reaction, 537, 578
single-effect evaporator, 280, 282–285
single-K flow resistance coefficient, 157
single-payment future worth factor, 699
single-payment present worth factor, 699
single-stage adsorption, 497, 498
single-stage equilibrium extraction, 466, 467
single-still distillation, 370
sinking fund, 702
sinking-fund depreciation, 711
sinking-fund factor, 702–703, 704
sinusoidal forcing function, 625
sizing control valves, 189
skin friction, 665
slug flow, 201
smog, 766
Smoker's equation, 395
Soave-Redlich-Kwong equation of state, 47
soda ash liquor, 164
solid
    leaching, 476–491
    molecular diffusion, 329–332
    porous, 331–332
    vapor pressure, 40
solid fuels, 67, 68
solid-liquid extraction systems. *See* leaching.
solid waste management, 773–779
solubility, 264
solute rejection, 313
solution thermodynamics, 116–117
solvent extraction. *See* leaching; liquid-liquid extraction.
solvent flow rate, 435–436
sonic velocity, 195

Sorel's method, 379
space time, 568
space velocity, 568
specific gravity, 24–25
specific heat ratios, 197
specific humidity, 512
specific reaction rate (SRR), 538
specific speed, 173
specific volume, 25
specification sheet, 656–658
split flow (tray tower), 428
square expansion, 160
square reduction, 160
SRR, 538
stability, 634–637
stack gas, 69
stagewise extraction, 467
staggered tube arrangement, 229
stagnation pressure, 138
stainless steels, 646, 647
standard atmosphere, 9
standard heat of combustion, 94
standard heat of formation, 93
standard heat of reaction, 93–94, 95
    where to find values, 85–86
standard integral heat of solution, 101
standard states, 119
Stanton number, 339
state functions, 81
static discharge head, 170, 181
static head, 132, 168–169
static pressure, 138
static suction head, 169, 181
static suction lift, 169
steady state, 54
steady-state condition, 29
steady state equimolar counter diffusion, 320
steady-state mixed-flow reactor, 569. *See also* stirred-tank reactor.
steady state nonisothermal reactor design, 596–597
steady-state plug-flow reactor, 569, 571. *See also* plug-flow reactor.
steam distillation, 374–375
steam hammer, 737
steam stripping, 504
Stefan-Boltzmann law, 214
Stefan-Maxwell equation, 327
step function, 624
stirred-tank reactor, 567
    design equations, 569–571
    fluidized CSTR, 592
    heat transfer, 599–601, 604–611
    multiple reactors in series, 576–578
    parallel reactions, 578–579
    product distribution/temperature, 581
stoichiometric coefficient, 32
stoichiometric problems, 33
stoichiometric ratios, 32

stoichiometric tables, 593–597
stoichiometry, 32
stoichiometry of combustion reaction, 731–734
straight-line method of depreciation, 709
straight weir, 428
streamline flow, 227–228
stress corrosion cracking (SCC), 645
Strigle plot, 429
stripping, 379
stripping-column distillation, 392–393
stripping section, 379, 382
structured packings, 419, 424, 429, 430
study node, 758
sublimation, 40, 263
subsonic flow, 195
subsystem, 59
suction specific speed, 173
suction static head, 169, 181
suction velocity head, 169, 181
sum-of-the-digits depreciation, 709–710
sunk cost, 709
SuperChems, 731
supercritical fluids, 46
superheat, 38
superheated steam, 394
superheated vapor, 38
supersonic flow, 195
surface combination factor, 234
surface renewal theory of Danckwerts, 337
surface tension, 151–152
surge volume, 660
surroundings, 28, 29
sweep-through purging, 732
synthetic polymers, 494
system, 28–29
    definition of, 28, 78
system acetone-chloroform x-y diagram, 349
system ethyl acetate-ethanol x-y diagram, 349
system head, 174
system properties, 29, 78
system water-ibutanol, 349
systems of units, 2–6
    American engineering system, 4, 6
    British engineering system, 3, 6
    cgs system, 2, 6
    conversions. *See* conversion of units.
    fps system, 3–4, 6
    metric system, 3, 6
    overview, 6
    SI units, 5, 6

T

T-xy diagram, 347–349
tables of enthalpy values, 85–86
tank design pressure, 690
tantalum, 648, 649
tapered expansion, 160

tapered reduction, 160
Taylor expansion, 624
TDH, 170, 182
Teflon heat exchanger, 269
temperature
    adiabatic saturation, 516
    AIT, 728
    CAFT, 730–731
    constant, 113
    design, 656–658
    dew-point, 361–362, 513
    glass, 651
    low temperature application guide, 652, 653
    optimal temperature progression. *See* optimal temperature progression.
    piping, 663
    plastics, 650
    pseudocritical, 46
    rate of reaction, and, 544–545, 596
    reduced, 46
    scales, 7–9, 12
    TFT, 730
    triple point, 263
    web-bulb, 515
temperature efficiency of exchanger, 250, 260
temperature-entropy *(TS)* diagrams, 114
temperature scales, 7–8
temperature swing cycle, 503–504
temperature units, 7–9, 12
ternary liquid systems, 460–463
testing the equipment, 680–682
TFT, 730
theoretical air/oxygen, 70
theoretical flame temperature (TFT), 730
theoretical plates, 378–379, 382, 383, 407–408, 414
thermal conduction, 210–213
thermal conductivity, 210, 270
thermal convection, 213, 233–234
thermal energy, 10
thermal oxidation, 767
thermal radiation, 214–215, 219, 220–223
thermal resistance, 210
thermal units, 10–11, 12
thermal venting, 754–756
thermochemical Btu, 11
thermochemical calorie, 10
thermochemistry of chemical reactions, 92–93
thermodynamic ideal efficiency, 122
thermodynamic properties, 78
thermodynamics. *See* energy balance and thermodynamics.
thermoplastics, 649–650
thermosyphon reboiler, 262
thick orifice, 160
thin sharp orifice, 160
third law of thermodynamics, 114–116
3-K constants (Darby), 157, 158

threshold limit value (TLV), 724–727
tie component, 58
tie line, 463, 464
tie substance, 58
TIG welding, 641
time constant, 625
time of drying, 525, 528–529
time value of money, 697–706
tip speed, 667–668
titanium, 648, 653
TLV, 724–727
TLV-C, 725
TLV-STEL, 725
TLV-TWA, 724
TOC, 765
ton of refrigeration, 123
Toor-Marchello's modification of surface renewal theory, 337
torque, 668–669
total discharge head (TDH), 170, 182
total head, 170, 182
total heating value, 67
total organic carbon (TOC), 765
total reflux, 389, 413
total suction head, 169
Toxic Substance Control Act, 780
toxicant, 724
toxicity, 724
toxicology, 724–727
toxicology/industrial hygiene, 724–727
transition-state theory, 548
trapezoidal rule, 373, 575, 576
tray columns, 426–428, 674–680
tray distillation column, 677
tray efficiencies, 421–423
    Murphree, 422
    overall, E_o, 421
    point, 423
tray spacings, 427
tray towers, 426–428, 674–680
triple effect forced-circulation evaporator, 286–287
triple point, 263
triple point pressure, 263
triple point temperature, 263
Trouton's empirical approximation, 378
Trouton's rule, 89
truncated equations, 48
TS diagrams, 114
tube wall temperature, 237
tubular packed bed reactor. *See* packed bed reactor (PBR).
tubular-recycle reactor, 582
turbine expansion, 124–125
turbogrid plate, 377
turbulent boundary layer theory, 337–338
turbulent flow, 148
turbulent mass transfer, 336–338
turnover time, 668
two column scheme of fractination, 409, 410

two-constant equation, 39
two film theory, 338
two-fluid U-tube manometer, 134, 135
two-phase flow considerations, 201–207
two-phase pressure drop, 201
two-position control, 621
two-stage crosscurrent adsorption, 499
two variable optimization problem, 683
2-K factors (Hooper), 157–158, 159

U

U-tube manometers, 134, 135
UEL, 728
UF, 314
UFL, 728
ultimate analysis, 68
ultrafiltration (UF), 314
ultraviolet oxidation technology, 767
un-entrapped service, 663
unbound water, 525
uncontrolled variables, 615
underdamped, 629
underflow, 476
Underwood method, 413–414, 418
UNIFAC method, 357
uniform attack, 640
uniform series compound-amount factor, 704
uniform series present-worth factor, 704
unimolecular first-order reactions, 556
UNIQUAC equation, 357
unit operations, 318
units and dimensions, 1–22
    absolute viscosity, 15
    consistent units, 16
    conversion of units. *See* conversion of units.
    dimensional analysis, 18–22. *See also* dimensional analysis.
    dimensional formula, 16
    dimensionally homogeneous equations, 16–17
    dimensionless equations, 18
    dimensions, 1–2
    kinematic viscosity, 15
    molar units, 6–7
    pressure units, 9–10
    systems of units, 2–6. *See also* systems of units.
    temperature units, 7–9, 12
    thermal units, 10–11, 12
    work, energy and power units, 11
unmixed flow reactor. *See* plug-flow reactor.
unsteady state, 54
U.O.P. characterization factor, 68
upper explosive limit (UEL), 728
upper flammability limit (UFL), 728
U.S. ton, 12
USCS system, 4

V

V-notch weir, 428
vacuum, 9
vacuum-gauge reading, 169n
vacuum purging, 733
vacuum venting, 755
valve trays, 678
van der Waal adsorption, 495
van der Waal equation of state, 47, 49
van Laar constants, 355–356
van't Hoff equation, 310, 541
van't Hoff's coefficient, 309
VAPCCI, 706
vapor-compression refrigeration, 124–127
vapor flow rate, 431
vapor-liquid equilibria, 346–365
    activity coefficients, 355–357
    binary system, 346
    boiling point diagrams, 347–349
    bubble point, 358, 360
    chemical potential, 353–354
    dew point, 358, 361–362
    Duhem's rule, 347
    equilibrium diagrams, 347–349
    equilibrium vaporization ratios, 357
    flash calculations, 358–361
    fugacity/fugacity coefficient, 354–355
    Henry's law, 353
    ideal mixtures, 350–351
    immiscible liquids, 349–350
    Margules equations, 356–357
    multicomponent system, 346
    nonideal systems, 355
    phase equilibria, 353–355
    phase rule, 347
    Raoult's law, 350, 351, 353
    relative volatility, 352–353
    van Laar equations, 355–356
    VLE calculations, 357
        modified Raoult's law, 357
    Wilson equation, 356–357
vapor-liquid equilibrium diagram, heptane-ethyl benzene system, 373
vapor pressure, 37–40, 350
vapor pressure curve (water), 38
vapor relief load, 736
vaporization-induced fouling, 224
variable noise, 771
variables, 615, 624
virial equation of state, 48
Vatavuk Air Pollution Control Cost Index (VAPCCI), 706
velocity head, 169, 181
vena contracta, 142
vent area, 752
venting, 753–757
venting capacities of atmospheric tanks, 753–754
venting deflagrations, 756–757
venturimeter, 139, 140–141, 143–145
virial coefficients, 48
viscosity, 15, 132
VLE calculations, 357
VOC, 765
VOC control technologies, 767
volatile organic compound (VOC), 765
volatile organic compound (VOC) control technologies, 767
volume
    constant, 113
    humid, 514
    operating, 660
    reduced, 46
    residual, 111
    specific, 25
    surge, 660
volume concentration, 26
volumetric displacement of compressor, 128
volumetric flow rate, 26
vortex shedding flowmeter, 143

W

waste disposal, 773
waste heat recovery boiler, 767
waste management, 773
water hammer, 735–737
water pollution statement, 768–769
Watson's empirical equation, 87
Webber, H. A., 299
weight fraction, 26
weir, 428
wet-bulb temperature, 515–516
wetted area calculation, 737–739
wetting rate, 152
    absorption, 442, 444
    pipes – flow of fluids, 152
what-if analysis, 761
Wheatstone bridge, 213
Wilke-Chang correlation, 327, 328
Wilson equation, 356–357
Wilson plot, 246, 247
wind pressure, 691
Winn's method, 413
work
    defined, 79
    heat, compared, 104
    units of measure, 11

X

x-y diagram, 347–349

Y

yield, 36

Z

z charts, 46–47, 50–52, 54
zeolites, 494
zero-order reaction, 551, 558
zeros of transfer function, 634
zirconium, 648

ABOUT THE AUTHORS

Dilip K. Das, MS, PE, is a principal engineer in Bayer CropScience's Kansas City engineering department. He is currently in charge of emergency relief system design, and previously held process engineering positions at Ciba-Geigy, Rhone-Poulenc, and Stauffer Chemicals. He holds a BSc (Honors) in chemistry from Pakistan, a BChE (Honors) in chemical engineering from Jadavpur University, Calcutta, where he received gold medals for excellence, an MS in chemical engineering from the University of Washington, and he attended MIT for training in artificial intelligence in chemical engineering. In addition to coauthoring three review books for the FE and PE exams in chemical engineering, he is the author of several technical articles and contributing author of a book on runaway reactions, pressure relief design, and effluent handling. He holds a U.S. patent. Mr. Das is a registered engineer in the states of New York, New Jersey, Louisiana, and Missouri. He is a past chair of AIChE's Kansas City chapter, and is currently the chairman of SuperChems Technical Steering Committee under the Design Institute of Emergency Relief Systems (DIERS). He has also been published at poetry.com.

Rajaram K. Prabhudesai, PhD, PE, is currently a consulting chemical engineer. He worked as process supervisor for Badger Engineers where his responsibilities were process and plant design. Previously he was senior process engineer with Stauffer Chemical Company, where he was responsible for process design, systems engineering, process economics, and plant start-up. He also worked for AMF as principal research engineer and Coca-Cola Company as principal chemical engineer. He has written numerous papers for professional journals and contributed to Perry's *Chemical Engineers' Handbook, 5th ed.*, and Schweitzer's *Handbook of Separation Techniques for Chemical Engineers*. He has written four review books for the FE and PE exams in chemical engineering, three as co-author with Dilip Das. Dr. Prabhudesai received his MS from the University of Bombay and his PhD in chemical engineering from the University of Oklahoma.